Vermiculture Technology

Earthworms, Organic Wastes, and Environmental Management

Vermiculture Technology

Earthworms, Organic Wastes, and Environmental Management

Edited by
Clive A. Edwards
Norman Q. Arancon
Rhonda Sherman

CRC Press
Taylor & Francis Group
Boca Raton London New York

CRC Press is an imprint of the
Taylor & Francis Group, an **informa** business

CRC Press
Taylor & Francis Group
6000 Broken Sound Parkway NW, Suite 300
Boca Raton, FL 33487-2742

First issued in paperback 2021

ISBN 13: 978-1-03-223712-1 (pbk)
ISBN 13: 978-1-4398-0987-7 (hbk)

Publisher's Note
The publisher has gone to great lengths to ensure the quality of this reprint but points out that some imperfections in the original copies may be apparent.

Library of Congress Cataloging-in-Publication Data

Vermiculture technology : earthworms, organic wastes, and environmental management / [edited by] Clive A. Edwards, Norman Q. Arancon, Rhonda L. Sherman.
p. cm.
Includes bibliographical references and index.
ISBN 978-1-4398-0987-7 (hardback)
1. Earthworm culture. 2. Earthworms. I. Edwards, C. A. (Clive Arthur), 1925- II. Arancon, Norman Q. III. Sherman, Rhonda L.

SF597.E3V47 2010
631.8'75--dc22 2010037025

Visit the Taylor & Francis Web site at
http://www.taylorandfrancis.com

and the CRC Press Web site at
http://www.crcpress.com

Contents

Preface..xi
Acknowledgments.. xiii
Editors..xv
Contributors xix

Chapter 1
Introduction, History, and Potential of Vermicomposting Technology..................... 1

Clive A. Edwards

Chapter 2
Relationships between Composting and Vermicomposting 11

Jorge Dominguez and Clive A. Edwards

Chapter 3
Biology and Ecology of Earthworm Species Used for Vermicomposting..............27

Jorge Dominguez and Clive A. Edwards

Chapter 4
Discovery and Development of New Species for Vermiculture 41

Samuel W. James and Afrânio Guimaraẽs

Chapter 5
The Microbiology of Vermicomposting 53

Jorge Dominguez

Chapter 6
Small-Scale School and Domestic Vermicomposting Systems..............................67

Rhonda L. Sherman and Mary Appelhof

Chapter 7
Low-Technology Vermicomposting Systems...79

Clive A. Edwards

Chapter 8
Medium- and High-Technology Vermicomposting Systems 91

Clive A. Edwards

Chapter 9
The Potential of Vermicomposts as Plant Growth Media for Greenhouse Crop Production 103

Norman Arancon, Clive A. Edwards, Katie A. Webster, and John C. Buckerfield

Chapter 10
The Use of Vermicomposts as Soil Amendments for Production of Field Crops 129

Norman Arancon and Clive A. Edwards

Chapter 11
The Production of Vermicompost Aqueous Solutions or Teas 153

Cindy E. Salter and Clive A. Edwards

Chapter 12
The Suppression of Plant Pathogens by Vermicomposts 165

Allison L. H. Jack

Chapter 13
Use of Aqueous Extracts from Vermicomposts or Teas in Suppression of Plant Pathogens 183

Clive A. Edwards, Ahmed M. Askar, Marcus A. Vasko-Bennett, and Norman Arancon

Chapter 14
Suppression of Arthropod Pests and Plant Parasitic Nematodes by Vermicomposts and Aqueous Extracts from Vermicomposts 209

Clive A. Edwards, Ahmed M. Askar, Marcus A. Vasko-Bennett, and Norman Arancon

Chapter 15
The Use and Effects of Aqueous Extracts from Vermicomposts or Teas on Plant Growth and Yields 235

Clive A. Edwards, Ahmed M. Askar, Marcus A. Vasko-Bennett, and Norman Arancon

Chapter 16
Human Pathogen Reduction during Vermicomposting 249

Clive A. Edwards and Scott Subler

Chapter 17
Heavy Metals, Earthworms, and Vermicomposts 263

Andrew John Morgan

Chapter 18
Quality Criteria for Vermicomposts 287

Clive A. Edwards, Scott Subler, and Norman Arancon

Chapter 19
The Commercial Potential and Economics of Vermicomposting 303

Jim Jensen, Bentley Christie, and Clive A. Edwards

Chapter 20
The Production of Earthworm Protein for Animal Feed from Organic Wastes 323

Clive A. Edwards and Alan Niederer

Chapter 21
The Use of Vermiculture for Land Improvement 335

Kevin R. Butt

Chapter 22
The Potential of Earthworms Produced from Organic Wastes in Production of Pharmaceuticals 349

Mariappan Balamurugan, Kasi Parthasarathi, Lalpet S. Ranganathan, and Edwin L. Cooper

Chapter 23
The Status of Vermicomposting in North America: A Rapidly Developing Technology 369

Rhonda L. Sherman and Peter Bogdanov

Chapter 24
Vermicomposting for Businesses and Institutions 391

Rhonda L. Sherman

Chapter 25
New Developments and Insights on Vermicomposting in Spain 409

Jorge Dominguez and Manuel Aira

Chapter 26
Vermiculture and Vermicomposting in the United Kingdom 423

Kevin R. Butt and Brian Williams

Chapter 27
Vermiculture in Australia and New Zealand: From Earthworm Production to Commercial Vermicomposting 437

Katie A. Webster and John C. Buckerfield

Chapter 28
Origins and Spread of Vermicomposting in India: Focus on Sustainable Agriculture 453

Radha Kale

Chapter 29
Vermiculture in the Philippines 469

Rafael D. Guerrero III

Chapter 30
The Status of Vermicomposting in Indonesia 481

Bintoro Gunadi

Chapter 31
Vermicomposting Projects in Hong Kong 497

David John Ellery and Tse Chi Kai

Chapter 32
Vermicomposting Research and Activities in Mexico 507

Eduardo Aranda-Delgado, Isabelle Barois, Martin de los Santos Bailón, and Benito Hernández-Castellanos

Chapter 33
The Scope of Vermiculture in Cuba 533

Martha Reines Alvarez and Carlos Rodriguez Aragonés

Chapter 34
Commercial Applications of Vermiculture in China.......... 545

Sun Zhenjun

Chapter 35
Progress in Vermicomposting in Belarus, Russia, and Ukraine 565

Svetlana Maksimova

Author Index 579
Subject Index.......... 587

Preface

This book is dedicated to Mary Appelhof, who was responsible for several initiatives in the early development of vermiculture. In 1979, she organized a milestone workshop in Kalamazoo, Michigan, which addressed the breakdown of organic wastes, especially sewage biosolids, by earthworms. With funding support from the U.S. National Science Foundation, she brought together more than 40 international scientists, entrepreneurs, and commercial earthworm breeders who were interested in the potential of using earthworms to break down organic wastes, focusing on sewage biosolids. The proceedings were published in 1981 under the title *Workshop on the Role of Earthworms in the Stabilization of Organic Residues* (Volume I, Proceedings; Volume II, Bibliography), by Beech Leaf Press, Kalamazoo, Michigan.

These two publications set the scene for an expansion of further research on the breakdown of sewage wastes by earthworms at the State University of New York (SUNY) in Syracuse, New York, and was the basis of a new research program on the breakdown of agricultural organic wastes by earthworms, initiated at the Rothamsted Experimental Station, Harpenden, United Kingdom, under the leadership of Dr. Clive Edwards, supported financially by the British Agricultural Research Council. This eventually led to a major commercialization of vermiculture in the United Kingdom in the 1980s and set a basis for developments in vermiculture in many other countries.

The publications have also promoted an increase in commercial vermicomposting activity in the United States and other parts of the world. Mary then developed a personal program promoting small domestic vermicomposting systems for widespread use in homes and schools. She published a book entitled *Worms Eat My Garbage* in 1982, and by 1997 had sold more than 100,000 copies of this book through her mail-order business, Flowerfield Enterprises. She then revised it for a second edition, which sold a further 30,000 copies and also supplemented it with other paperback books aimed at interesting schoolchildren in recycling food wastes with earthworms.

In 2000, she collaborated with Professor Clive Edwards of The Ohio State University in organizing an International Symposium and Training Workshop in Kalamazoo, Michigan, under the apt title of Vermillenium, in recognition of the turn of the century. For various reasons, the proceedings were not published as planned after the symposium, and unfortunately Mary died suddenly in 2005. However, the international presentations at Vermillenium were so useful that Professor Edwards, with the support of Professor Rhonda L. Sherman of North Carolina University and Assistant Professor Norman Q. Arancon of the University of Hawaii, took manuscripts that had been written by presenters at Vermillenium, completely updated them, and recruited many new chapter authors, with the aim of compiling them into the first comprehensive review of all aspects of the innovative science of vermiculture technology.

This was the origin of the present book, which has been designed as a comprehensive reference volume covering all aspects of the new vermiculture technology. We hope that this book will provide seeds that will expand vermiculture into a major technology across the world and that will promote vermicompost-based crop production, resulting in many environmental benefits.

Clive A. Edwards
The Ohio State University

Rhonda L. Sherman
North Carolina State University

Norman Q. Arancon
University of Hawaii at Hilo

Acknowledgments

Many thanks are due to Clive Edwards' secretaries, Renata Harrison and Cisne Edwards, for all the hard work they put into getting the manuscript for *Vermiculture Technology* into a suitable form for publication. Dr. Jorge Dominguez, University of Vigo, Spain, provided advice and suggestions regarding the finalization of the structure and contents of the book. Clive Edwards' assistants, Marcus Vasko-Bennett and Ahmed Askar, assisted in the production of the manuscripts in various ways, particularly in preparation of figures and checking bibliographies, as well as being technicians in the research described in Chapters 9 and 10, 13, 14, and 15. We would like to acknowledge John C. Buckerfield of Australia, who is now deceased, for his invaluable contributions to vermiculture, including Chapter 27 in the book.

Editors

Professor Clive A. Edwards was born in Worcester, England, and holds a BSc (Hons), MSc and an honorary DSc from Bristol University, England, and MS and PhD degrees from the University of Wisconsin, United States. He worked for the British Ministry of Agriculture for four years and was a senior principal scientific officer at the Rothamsted Experimental Station, Harpenden, Herts, United Kingdom, from 1960 to 1985. He was a visiting professor at Purdue University, United States from 1966 to 1968. He was appointed as chair of the Department of Entomology at The Ohio State University in 1985 and is currently professor of entomology and environmental sciences at The Ohio State University.

Dr. Edwards has published extensively on soil ecology and ecotoxicology, environmental issues, and sustainable agriculture, and he is currently recognized as a world authority on earthworms. His book *Ecology and Biology of Earthworms*, now in its third edition, is the first comprehensive book on earthworms since Charles Darwin's *The Production of Vegetable Mould through the Action of Earthworms*, which was published in 1881. In 1996, Professor Edwards's book *Earthworm Ecology* won a Presidential Citation from the U.S. Soil and Water Conservation Society.

A prolific writer, he has published more than 480 scientific papers in addition to writing, editing, or coediting 31 books. He has been a fellow of the Royal Entomological Society of London for more than 50 years. In 1972 he was elected to fellowship of the U.K. Institute of Biology and in 1984 to the fellowship of the U.K. Royal Society of Arts, from which he received a prestigious ERAS Pollution Abatement Award.

In 1990 Professor Edwards coauthored a successful proposal for an eminent scholar position in soil ecology at The Ohio State University. In 1996 he received a Presidential Citation from the Indian Soil Ecology Society. In 1996 the Office of Academic Affairs presented him with the OSU International Outstanding Faculty Award, and he received an OSU Distinguished Scholar Award in 1998 and an OSU Distinguished Lecturer Award in 2000. In 2001 he received a Professional Career Achievement Award from the U.S. Soil Ecology Society. He received an OSU Senior Distinguished Faculty Research Award in 2002 and a Gold Medal from the British Crop Protection Council, for services to plant protection, in 2002.

Dr. Norman Q. Arancon is currently an assistant professor of horticulture at the University of Hawaii at Hilo. His interests include sustainable agriculture and horticulture, agroecology, soil ecology, hydroponics, tropical horticulture, organic farming, and sustainable development. He obtained MS and PhD degrees in environmental science from The Ohio State University, as a Fulbright scholar and a graduate research associate from 1997 to 2001. He was awarded a postgraduate diploma in agricultural studies from the University of Queensland, Brisbane, Australia, after a Travel Award from the Rotary International Foundation in 1993. He holds a bachelor's degree in agriculture with a major in crop science from Xavier University-Ateneo de Cagayan, Cagayan de Oro City, Philippines, as a Xavier Science Foundation scholar.

He worked for Xavier University as a lecturer–instructor from 1989 to 1997. From 1994 to 1997, he was appointed as research and documentation coordinator for the Sustainable Agriculture Center of the College of Agriculture, Xavier University, and later became officer-in-charge of the center in 1997.

Dr. Arancon's research at The Ohio State University included substitution of vermicomposts into commercial plant-growth media to grow horticultural crops in the greenhouse. Together with Dr. Clive Edwards and his students at the Soil Ecology Laboratory, he pioneered research on the use of vermicompost applications in the field using other commercially important vegetables, such as tomatoes, cucumbers, and peppers; ornamentals, such as marigolds and petunias; and small fruits, such as strawberries, raspberries, and grapes; and investigated the effects of vermicomposts on chemical and biochemical changes in soils. He was co-principal investigator on a USDA grant entitled "Effects and Modes of Action of Vermicomposts on Field Horticultural Crops." His work has included research on the use of vermicomposts and vermicompost "teas" on crop production and pest and disease suppression. He became a research associate professor at The Ohio State University in 2007. While at the University of Hawaii in Hilo, he continues to collaborate closely with Dr. Edwards and the Soil Ecology Laboratory at The Ohio State University and is currently designing and building a Vermiculture Center program for the Pacific Islands.

Professor Rhonda L. Sherman is a faculty member of the Department of Biological and Agricultural Engineering at North Carolina State University (Raleigh, North Carolina, United States). She provides education about and technical assistance with solid waste management throughout the United States and has assisted people in 60 countries. She has conducted numerous training courses and workshops and has created a variety of publications on solid waste management and sustainability issues for extension educators, farmers, recycling coordinators, teachers, students, commercial and institutional managers, earthworm growers, consultants, and the general public throughout North Carolina and nationally. Her key topics include vermicomposting, composting, source reduction, reuse, and recycling. Since 2001, she has offered annually the nation's only conference on large-scale vermicomposting, helping more than 600 people to start up or expand earthworm farms. Much of her work has addressed commercial and industrial waste reduction, with specific emphasis on the vermiculture, composting, construction, and lodging industries. Sherman created a curriculum and trainers' course and manual entitled *Commercial and Industrial Solid Waste Reduction Training for Community Solid Waste Managers*, which has been used to train solid waste managers in several states to develop new programs. Construction workers are learning to reduce and recycle waste created on job sites through the use of videos that Prof. Sherman produced in English and Spanish. Used-oil-recycling training sessions and support materials have resulted in the implementation of effective education programs and the reuse of thousands of gallons of used motor oil. As a result of training workshops and demonstrations, hundreds of small-scale vermicomposting systems have been set up in classrooms and homes.

She has educated thousands of school students throughout North Carolina about vermicomposting and composting, and published a fifth-grade school enrichment curriculum on vermicomposting with seven accompanying slide sets. A cofounder of the National Backyard Composting Program, Sherman was also a contributing editor to *BioCycle: Journal of Composting and Organics Recycling.*

Contributors

Dr. Manuel Aira
Departmento de Ecoloxía e Bioloxía Animal
University of Vigo
Vigo, Spain

Dr. Martha Reines Alvarez
Department of Animal and Human Biology
Havana University
Havana, Cuba

Ms. Mary Appelhof (Deceased)
Flowerfield Enterprises
Kalamazoo, Michigan

Dr. Carlos Rodriguez Aragonés
Department of Animal and Human Biology
Havana University
Havana, Cuba

Dr. Norman Arancon
College of Agriculture, Forestry and Natural Resource Management
University of Hawaii
Hilo, Hawaii

Dr. Eduardo Aranda-Delgado
Terranova Lombricultores
Coatepec, Veracruz, Mexico

Mr. Ahmed M. Askar
Soil Ecology Laboratory
Department of Entomology
The Ohio State University
Columbus, Ohio

Dr. Martin de los Santos Bailón
Departmento de Biología de Suelos
Instituto de Ecología, A.C.
Xalapa, Veracruz, Mexico

Dr. Mariappan Balamurugan
Department of Zoology
Annamalai University
Annamalai Nager, India

Dr. Isabelle Barois
Departmento de Biología de Suelos
Instituto de Ecología, A.C.
Xalapa, Veracruz, Mexico

Mr. Peter Bogdanov
VermiCo
Portland, Oregon

Mr. John C. Buckerfield (Deceased)
EcoResearch
Adelaide, Australia

Dr. Kevin R. Butt
Department of Environmental Management
University of Central Lancashire
Preston, United Kingdom

Mr. Bentley Christie
Green Waste Solutions
Elmira, Ontario, Canada

Prof. Edwin L. Cooper
Department of Neurobiology
University of California at Los Angeles
Los Angeles, California

Dr. Jorge Dominguez
Departmento de Ecoloxía e Bioloxía Animal
University of Vigo
Vigo, Spain

Prof. Clive A. Edwards
Soil Ecology Laboratory
Department of Entomology
The Ohio State University
Columbus, Ohio

Dr. David John Ellery
Sunburst Technology
Hong Kong, People's Republic of China

Dr. Rafael D. Guerrero III
Aquatic Biosystems
Bay, Laguna, Philippines

Mr. Afrânio Guimaraẽs
Minhobox
Juiz de Fora, Minas Gerais, Brazil

Dr. Bintoro Gunadi
Kwantlen Polytechnic University
Vancouver, British Columbia, Canada

Dr. Benito Hernández-Castellanos
Departmento de Biología de Suelos
Instituto de Ecología, A.C.
Xalapa, Veracruz, Mexico

Ms. Allison L. H. Jack
Department of Plant Pathology and Plant-Microbe Biology
Cornell University
Ithaca, New York

Dr. Samuel W. James
Biodiversity Institute
University of Kansas
Lawrence, Kansas

Mr. Jim Jensen
Environmental Credit Corporation
State College, Pennsylvania

Mr. Tse Chi Kai
Sunburst Biotechnology
Hong Kong, People's Republic of China

Dr. Radha Kale
Department of Zoology
University of Agriculture Sciences
Bangalore, India

Dr. Svetlana Maksimova
Institute of Zoology
National Academy of Sciences of Belarus
Minsk, Belarus

Prof. Andrew John Morgan
Cardiff School of Biosciences
Cardiff University
Wales, United Kingdom

Mr. Alan Niederer
Cham
Kanton Zug, Switzerland

Dr. Kasi Parthasarathi
Department of Zoology
Annamalai University
Tamilnadu, India

Dr. Lalpet S. Ranganathan
Shri V. L. Balakrishna Janakiammal College of Arts and Science
Kovaipudur, India

Ms. Cindy E. Salter
Organics Matter
Eugene, Oregon

Dr. Rhonda L. Sherman
Biological and Agricultural Engineering
North Carolina State University
Raleigh, North Carolina

Dr. Scott Subler
Environmental Credit Corporation
State College, Pennsylvania

Mr. Marcus A. Vasko-Bennett
Soil Ecology Laboratory
Department of Entomology
The Ohio State University
Columbus, Ohio

Ms. Katie A. Webster
EcoResearch
Stirling, South Australia

Mr. Brian Williams
ORM Professional Products
Brecon, Wales, United Kingdom

Dr. Sun Zhenjun
College of Resources and
Environmental Sciences
China Agricultural University
Beijing, People's Republic of China

CHAPTER 1

Introduction, History, and Potential of Vermicomposting Technology

Clive A. Edwards

CONTENTS

I Introduction ... 1
II Basic Principles of Vermicomposting and Vermiculture ... 2
III Establishment of Vermiculture as a Commercial Technology ... 4
IV Commercial Adoption of Vermiculture Technology ... 6
  A United States ... 6
  B Canada ... 7
  C United Kingdom ... 7
  D Russia ... 7
  E Hong Kong ... 7
  F India ... 7
V Economics of Vermicomposting Technologies ... 8
References ... 9

I INTRODUCTION

Large quantities of organic wastes are produced from agricultural production and farming systems, including animal manures, sewage biosolids, food and restaurant wastes, and industrial organic wastes. These have the potential of increasing global soil and water pollution, because they are currently disposed of by land-spreading, incineration, or into landfills. As much as 50%–60% of the total wastes that are disposed into landfills are organic wastes. If these were turned into materials useful in agriculture and horticulture, there would be great savings in primary plant nutrients and metabolic energy.

As costs of organic-waste disposal have escalated progressively and the environmental regulations on their disposal have become increasingly restrictive, there is much greater interest in exploiting organic wastes as fertilizers, soil conditioners, and amendments or as energy sources, such as for methane production. The increasing amount of organic wastes internationally brings pressures into the search for newer and environmentally acceptable methods of turning them into value-added materials.

One of the main ways of utilizing such wastes has been to use them in large-scale systems of commercial thermophilic composting in the United States, Europe, and other parts of the world. However, most of these commercial operations aim at disposing of the wastes rather than converting them into materials that can be useful in land recovery and soil improvement. The resultant composts, which can be relatively expensive to produce, do not sell for high prices in most parts of the world, and most of the commercial returns are from savings in landfill costs.

II BASIC PRINCIPLES OF VERMICOMPOSTING AND VERMICULTURE

An attractive alternative to thermophilic composting is to use certain epigeic species of earthworms to break down organic wastes and turn them into vermicomposts that can be used in horticulture and agriculture for soil improvement and as bedding-plant media. Earthworms are an important link in the food chain of many invertebrate and vertebrate animals (Macdonald 1983). Not surprisingly, early humans used earthworms both as food and as baits in hunting and fishing (Bouché 1987). The use of earthworms from dung heaps to feed domestic animals was documented in the eighteenth century (Milocco 1782), and the use of earthworms in fishing was discussed by Izaak Walton (1653). Charles Darwin (1881) first focused public attention on the great importance of earthworms in the breakdown of organic matter through the publication of his landmark book.

It was not until the twentieth century in California, and other parts of the United States, that the commercial production of earthworms for use as fishing bait began. These extremely varied commercial enterprises used the most cost-effective materials available to grow earthworms; these were usually animal wastes. A much more profitable approach to vermiculture was suggested by Oliver (1937) and Barrett (1942), who both considered that some epigeic species of earthworms, as well as the organic wastes on which they had fed, could be used by farmers and growers to improve agricultural soils and crop production. The vermiculture industry expanded rapidly, and quite soon producers were selling earthworms and vermicomposts through a range of outlets irrespective of the ways in which they were being used after they had been purchased. Unfortunately for the developing technology, some entrepreneurs introduced pyramid-selling techniques, where their profits depended on attracting people to grow earthworms, which were then sold to other potential new earthworm growers at considerable profits to the entrepreneurs. Inevitably, the claims made by many of these entrepreneurs were often wildly exaggerated, and they soon earned the disapproval of the scientific community because they tended to discredit vermiculture as a serious technology (Bouché 1987; see Chapter 23).

Fortunately for the future of vermiculture technology, more basic studies and research on the potential of earthworms for processing organic wastes began in Germany (Graff 1974) and continued in the United States in the late 1970s with Hartenstein and his associates at the State University of New York, who used them to break down sewage biosolids using research funding from the U.S. National Science Foundation (Hartenstein and Mitchell 1977). Subsequently, agricultural research in the United Kingdom that was funded by the U.K. Agricultural Research Council (Edwards et al. 1985; Edwards 1988) at Rothamsted Experimental Station, Harpenden, Herts, began an extensive research program into developing economic methods of using earthworms to break down animal and vegetable wastes into useful vermicomposts, and also to produce earthworm protein that could be used in fish farming and for domestic animal feeds (see Chapters 7, 8, 9, 10, and 20).

Many of the organic wastes produced by agriculture, farms, and modern industrial technology cause odor problems or can result in the pollution of groundwater. Early research in the Rothamsted program suggested that these problems could be alleviated by the use of earthworms, particularly *Eisenia fetida* (Savigny) and other epigeic species such as *Eisenia andrei* (Bouché), *Dendrobaena veneta* (Rosa), *Eudrilus eugeniae* (Kinberg), and *Perionyx excavatus* (Perrier), to accelerate the aerobic decomposition of organic wastes, thereby minimizing odors and pollution and producing a considerable potential profit from the sale of earthworms and also of the vermicompost for use as a plant growth medium or soil amendment (Franz 1978; Edwards et al. 1985; Edwards 1988). A wide range of human, animal, and industrial organic wastes have been used as parent materials for vermicomposting, and the resulting vermicomposts have been marketed extensively in the United States, Canada, Mexico, Italy, the Philippines, India, Australia, and elsewhere.

Edwards (1983, 2004) and Edwards and Neuhauser (1988) reported the possibilities of converting both animal and vegetable wastes into value-added materials that could be incorporated as useful amendments into agricultural soils, used as plant growth media, or used as a component of commercial horticultural bedding-plant media. For vermicomposts produced in this way to be a commercial success and attract high prices, large quantities of homogeneous organic waste materials need to be available, and there should be strict controls of the nutrient status, pH and structure of the final standardized vermicompost product must be maintained at a high level (see Chapter 16).

Several epigeic species of earthworms have been investigated for their potential to stabilize organic wastes and produce vermicomposts. Research has shown that many organic wastes can supply the large populations of microorganisms that are necessary for the growth and reproduction of species of earthworms of the genera *Eisenia*, *Eudrilus*, *Dendrobaena*, *Perionyx*, and *Pheretima* (Hartenstein, Neuhauser, and Kaplan 1979; Kaplan et al. 1980; Neuhauser et al. 1980; Edwards 2004; see Chapters 3 and 4).

The early research in the United States focused on using earthworms to stabilize sewage biosolids. However, it soon became obvious that sewage treatment plants, which often receive sewage intake from both domestic and industrial sources, and the toxic contaminants they might contain, could have effects on earthworm growth, reproduction, and survival (Hartenstein, Neuhauser, and Collier 1980), and the limitations of using contaminated materials like biosolids as feedstocks in vermiculture were assessed. Appelhof and Worden (1981) summarized the extensive research into the breakdown of sewage biosolids by earthworms and the overall commercial potential of vermiculture up to that date.

Research into vermiculture began in the United Kingdom at the Rothamsted Experimental Station, Harpenden, Herts, by Edwards and his extensive team of ecological colleagues, as well as engineers and economists, from the National Institute of Agricultural Engineering, Silsoe, Beds. This research soon established that the use of earthworms as feeds for fish, chicken, and pigs was feasible but had economic limitations, and the vermicomposts were excellent in promoting plant germination, growth, flowering, and yields. On this basis they developed a range of excellent engineering technologies for processing organic wastes into vermicomposts using earthworms (see Chapter 8).

The earthworm species used in vermiculture and vermicompost production all normally inhabit surface litter in the field (epigeic species), and in vermicomposting systems, given optimum conditions, these species restrict their activities to the top 10–15 cm (4–6 in) of an earthworm bed or system. The frequent addition of thin layers of organic wastes encourages a continuous vertical movement of the earthworms upward toward the newly applied layers of waste, leaving relatively earthworm-free vermicompost, or casts underneath. This behavior pattern was used as the basis of the development of automated continuous-flow vermicompost reactor systems (Phillips 1988), where the rates of regular addition of fresh organic wastes to the surface was equaled by the rates of removal of earthworm-worked wastes or vermicomposts from the base of the reactor. Flow rates could be adjusted, to enable an optimal earthworm biomass per unit area, in a state of equilibrium, to be reached (see Chapter 8).

Methods developed to separate earthworms from vermicomposts are numerous, and their designs usually relate to the scale of operations involved. The simplest mechanical devices to separate wastes rely on trommels with reciprocating or rotating sieves that retain the earthworms but allow smaller particles of vermicompost to pass through the meshes; others employ comblike structures through which the earthworm-worked material passes, selecting earthworms, which are hooked on the combs and drop onto a moving belt (see Chapter 8). The efficiency of such mechanical separation methods is decreased if the vermicompost is too wet because sieves and combs can become blocked if the moisture content of the vermicompost exceeds 80%.

III ESTABLISHMENT OF VERMICULTURE AS A COMMERCIAL TECHNOLOGY

The U.K. research from 1981 to 1985, which involved about 15 full-time government scientists and about 35 collaborators at experiment stations, universities, and commercial organizations, set a firm basis for vermiculture technology. This was taken up and developed by a new U.K. company, British Earthworm Technology (BET), which marketed vermicompost under the trade name of BETAGRO and received partial financial support from the U.K. Government. The company operated in the United Kingdom from 1982 to 1987. The U.K. research at Rothamsted was awarded a prestigious U.K. Pollution Abatement Technology Award by the Confederation of British Industry, U.K. Department of the Environment, and the U.K. Royal Society of Arts in 1983.

In 1984, BET organized a large symposium at Queen's College, Cambridge, under the guidance of Edwards and Neuhauser (United States). The title of the symposium was *The Use of Earthworms in Waste and Environmental Management*, and the proceedings were published under the same name (Edwards and Neuhauser 1988) by a Dutch publishing company, SPB Academic Publishing.

In 1985, Edwards was awarded a grant of £600,000 by the European Economic Community to expand vermiculture technology into other European countries. However, the grant was not taken up and used for this purpose because in 1985 Edwards moved to The Ohio State University in the United States to become a Professor and Chair of the Entomology Department. In this Department he set up the Soil Ecology Laboratory in 1990 and continued to focus on many aspects of earthworm research, including the further development of all aspects of vermicomposting technology. He received a number of large research grants from the U.S. National Science Foundation and the U.S. Department of Agriculture and other agencies. The research that was supported included further development and testing of the automated continuous-flow vermicomposting reactor system; assessments of the benefits of vermicomposts, including the identification of mechanisms that influence the growth, flowering, and yields of crops; the suppression of plant pathogens, plant parasitic nematodes, and arthropod pests by vermicomposts; and finally confirmation that most of the beneficial aspects of vermicomposts could be replicated by the use of aqueous extracts from vermicomposts, or "teas," which were much easier to use. These results were published in scientific papers.

Vermiculture technology was developed further by a research group from the University of Vigo in Spain led by Jorge Dominguez, who spent 2 years in Edwards's laboratory at The Ohio State University. This Spanish group is still a leading research team in vermiculture. Another research group, based at the Instituto de Ecología, Mexico, has been led by Isabelle Barois and Eduardo Aranda-Delgado (see Chapter 32). Pioneering vermiculture research has been done in the Philippines since 1983, led by Rafael Guerrero (see Chapter 29), and this group is still a leader in developing the technology in the tropics. Guerrero edited a book, *Vermitechnologies in Developing Countries* (Guerrero and Guerrero del Castillo 2006). The group is particularly interested in the use of earthworms as commercial fish food.

Another vermiculture research team has been at the University of Agricultural Sciences in Bangalore, India, led by Radha Kale, who has developed a range of vermiculture technologies adapted for tropical agriculture; much of her work is summarized in a book, *Vermicompost: Crown Jewel of Organic Farming* (Kale 2006; see Chapter 28). Another major vermiculture technology research group is led by

Sun Zhenjun at the China Agricultural University in Beijing, China. His group has worked on a wide range of vermiculture topics, including the use of earthworms as animal-feed protein and the potential of products from earthworms for use in the pharmaceutical industry. He has written a book, *Vermiculture and Vermiprotein* (Zhenjun 2003; see Chapter 34).

IV COMMERCIAL ADOPTION OF VERMICULTURE TECHNOLOGY

During the development of vermiculture over the last 30 years, many small and large commercial developments have been set up. Those in the United States are reviewed in Chapter 6 by Sherman. However, I will mention some of the U.S. commercial operations as well as various larger international organizations operating currently (see Chapters 23 and 24).

A United States

1. **Oregon Soil Corporation**, Oregon City, Oregon, Managing Director: Dan Holcombe
   This company has been collecting food wastes from a range of supermarkets in the Portland, Oregon, area to produce vermicomposts in automated vermicomposting reactor systems since 1991. The vermicomposts are packaged and marketed through centers in the same supermarket chain for home and garden use.
2. **R.T. Solutions LLC**, Geneseo, New York, Managing Director: Tom Herlihy
   This company, formed in 2004, operates a large, specifically designed and built facility in New York State, close to cattle farms and major highways. It processes large quantities of cattle wastes into vermicomposts in a range of automated continuous-flow vermicomposting reactor systems. The vermicomposts are marketed by the company as plant growth amendments particularly for the grape and grass industries.
3. **New Horizon Organics**, Springfield, Illinois, Managing Director: Chad Hurley
   This company, which has been operating for more than 6 years, produces vermicomposts from food wastes and cattle wastes in indoor facilities. It markets the vermicomposts as soil amendments for crop production and also as materials for the bioremediation of soils polluted by organic contaminants, such as those produced by the oil industry. It has supervised a number of bioremediation projects.
4. **Sansai International, Inc.**, Cleveland, Ohio, Managing Director: Jamie Melvin
   This company, formed in 2005, owns an 8 ha ex-automobile parts factory in Cleveland that produces and markets vermicomposts and composts, produced from food wastes from local organic markets, restaurants, and supermarkets, and also from paper wastes. Currently it uses wedge systems, but its production is expanding.
5. **Sonoma Valley Earthworm Farm**, Sonoma Valley, California, Managing Director: Jack Chambers
   This company, founded in 2000, produces vermicomposts from cattle wastes for use by the Sonoma and Napa Valley grape industry. They compost the wastes thermophilically before putting them into four continuous-flow vermicompost systems, and markets both vermicomposts and vermicompost teas.

6. **Pacific Garden Company**, Ferndale, Washington, and Millheim, Pennsylvania, President: Dr. Scott Subler
   This company was founded in 2001 by Dr. Subler, who spent 8 years working with Professor Edwards at The Ohio State University. It uses continuous-flow reactors of the type designed in the United Kingdom.

B Canada

1. **Forterra Environmental Corporation**, Toronto, Canada, Managing Director: Gary Gould
   This company, which was founded 3 years ago, produces vermicomposts in a 25,000-square-foot indoor facility and markets them for soil enrichment of golf courses, lawns, vineyards, and greenhouse soils and general soil-fertility improvements.

C United Kingdom

1. **O.R.M. Waste Management**, Brecon, Wales, Managing Director: Brian Williams
   This is one of the largest waste management companies in the United Kingdom, and its major activity is producing vermicomposts from cattle manures, in a network of centers, using *Dendrobaena veneta*, and marketing them all over the United Kingdom. The company won a Gold Medal at the Chelsea Flower Show U.K. in 2006 for the quality of their vermicompost production and use.

D Russia

1. **Green-Pik Ecoproducts Corporation**, Vladimir, Russia Managing Director: Igor Titov
   This company, formed in 2002, is producing extremely large quantities of vermicomposts and vermicompost "teas." They do the vermicomposting in very large ex-cattle-rearing buildings, each about 16,000 m^2. They use the wedge vermicomposting system designed by Edwards. They organized international conferences in 2004, 2007, and 2010 that were attended by several hundred participants.

E Hong Kong

1. **Sunburst BioTechnologies, Ltd.**, Hong Kong, Managing Director: David Ellery
   This company has been composting and vermicomposting large amounts of urban organic wastes in Australia for more than 10 years, using automated continuous-flow automated reactor systems, which they improved from the original design. They are now handling food wastes from a range of restaurants and food sources, as well as horse manure, in Hong Kong (see Chapter 31).

F India

1. **RalliGold**, Bangalore, Hyderabad, Chennai, Managing Director: Rallis India
   This operation consists of five 50 ha sites producing 50 kg bags of vermicomposts under cover from urban wastes (termed *biofertilizers*) which are marketed all over India using an innovative technology. The five sites produce 20,000 metric tons of vermicompost per annum.

V ECONOMICS OF VERMICOMPOSTING TECHNOLOGIES

The development of vermicompost technology as a solid–organic-waste management tool has had two main thrusts:

- On-site vermicomposting by individuals, businesses, farms, and institutions
- Centralized, commercial vermicomposting of source-separated resources

These operations have to take into account economic issues of scale and costs of transport of vermicomposts and raw materials.

An important issue in the United States involves the location of vermicomposting sites, and the need for permits from state agencies such as environmental protection agencies, particularly in urban areas. There are relatively few restrictions on vermicomposting operations in agricultural sites but quite extensive restrictions in urban and suburban locations. In some cases individuals and organizations who have vermicomposting operations have sidestepped permit requirements by setting up agricultural earthworm farms where the earthworms are considered to be livestock and the vermicompost produced is a by-product of this operation.

The use of organic wastes to grow earthworms is an extensive farm and garden cottage industry all over the United States and other parts of the world such as India and the Philippines. Many of these small-scale producers market the earthworm castings they produce for growing plants. Most smaller vermicomposting operations in the United States are based primarily on outdoor windrow systems, which have many economic and environmental drawbacks (see Chapter 8). They are ground-based and require large areas of land, with considerable potential for groundwater pollution with nutrients and other contaminants because they are watered regularly and usually have no protection against leaching from the earthworm beds. The vermicomposting process is slow, often taking up to 12 months to complete. The harvesting of the vermicompost is laborious and time-consuming because the earthworms in the vermicomposts have to be separated, usually by a trommel screening process, before the vermicompost is marketed. Although the initial capital outlay, other than land, is low, labor costs are high at all stages of windrow vermicomposting operations. As an economic alternative, the wedge system (Chapter 7) has been used by a number of commercial organizations. Although it involves an innovative but relatively inexpensive technology and requires less equipment, it overcomes many of the labor, economic, and environmental drawbacks associated with windrows. In particular, it uses less land, there is no leaching of nutrients into groundwater, and there is no need to separate earthworms from the vermicompost. The processing time is also shorter (3–4 months).

The automated continuous-flow vermicomposting reactor systems designed by Edwards and his colleagues (Edwards 1995, 1988, 2004), have totally different environmental, operating, and economic characteristics and requirements (see Chapter 8). The equipment has to be under cover to ensure controlled environmental conditions are maintained, and the organic-waste compartment is raised above the floor and is maintained at 80%–90% moisture content and 20°C to 35°C. Hence, there is no leaching of nutrients into the groundwater. The retention or processing time of wastes in the reactor is 30–60 days depending on the kind of waste. The economics of such automated reactors are totally different from those of other systems because they involve a need for high-initial capital inputs ($35,000 to $50,000) to build reactors, which can process up to 3 tons of waste per day, as well as various ancillary loading and transport equipment including moving, belts, macerators, and loaders for operation. However, their labor and running costs are extremely low, and reactors reach equilibrium and can be run trouble free for a number of years. One technician can run up to 12 such reactors. The capital expenditure can usually be recovered in 1–2 years (see Chapter 19). A number of smaller and relatively expensive systems based on this system have been marketed in the United States but are much less attractive economically because of their high-initial costs and low productivity.

The market for vermicomposts is extremely variable. The smallest commercial returns have been as low as $35 per ton, but a more common return is $100–$300 per ton. Returns of $500–$600 per ton are not uncommon for high-quality well-standardized and attractively packaged vermicomposts, and one company even received $1,000 per ton, but this included expensive packaging materials. There is a high demand for high-quality vermicomposts in Japan, who have been reported as willing to pay up to $300 per ton excluding the costs of transport to Japan.

REFERENCES

Appelhof, M., and Worden, D.D. 1981. *Workshop on the Role of Earthworms in the Stabilization of Organic Residues*, 9–12 April 1980. 2 vols. Beech Leaf Press, Kalamazoo, Michigan.

Barrett, T.J. 1942. *Harnessing the Earthworm*. Faber and Faber, London.

Bouché, M.B. 1987. Emergence and development of vermiculture and vermicomposting: From a hobby to an industry, from marketing to a biotechnology, from irrational to credible practices. In *On Earthworms*, ed. A.M.B. Pagliai and P. Omodeo, 519–531. Unione Zoologica Italiana (UZI), 2, Mucchi Editore, Modena, Italy.

Darwin, C.R. 1881. *The Formation of Vegetable Mould through the Action of Worms, with Observations on Their Habits*. Murray, London.

Edwards, C.A. (ed.). 2004. *Earthworm Ecology*. 2nd Ed., CRC Press, Baton Rouge, LA.

Edwards, C.A. 1995b. A historical overview of vermicomposting. *Biocycle*, June, 56–58.

Edwards, C.A. 1988. Breakdown of animal, vegetable and industrial organic wastes by earthworms. In *Earthworms in Waste and Environmental Management*, ed. C.A. Edwards and E.F. Neuhauser, 21–31. SPB Academic Publishing, the Hague, the Netherlands.

Edwards, C.A. 1983. Earthworms, organic waste and food. *Span* 26, no. 3, 106–108.

Edwards, C.A. 1982. The use of earthworms in waste disposal and protein production. *Report, Rothamsted Experimental Station for 1981*, 103–106.

Edwards, C.A., Burrows, I., Fletcher, K.E., and Jones, B.A. 1985. The use of earthworms for composting farm wastes. In *Composting of Agricultural and Other Wastes*, ed. J.K.R. Gasser, 229–242. Elsevier, Amsterdam, the Netherlands.

Edwards, C.A., and Neuhauser, E.F. 1988. *Earthworms in Waste and Environmental Management.* SPB Academic Publishing, the Hague, the Netherlands.

Franz, M.F. 1978. The economic future of the earthworm in recycling. *Compost Sci.—Land Util.* 19, 23–26.

Graff, O. 1974. Gewinnung von Biomasse aus Abfallstoffen durch Kultur des Kompostregenwurms *Eisenia foetida* (Savigny 1826). *Landbauforschung Volkenrode* 24, 137–142.

Guerrero, R., and Guerrero del Castillo (eds.). 2006. Proceedings of the International Symposium Workshop on Vermitechnology for Developing Countries, Philippines Fisheries Association, Incorporated, Los Banos, Laguna, the Philippines.

Hartenstein, R., and Mitchell, M.J. 1977. Interim report—utilisation of soil invertebrates in stabilization, decontamination and detoxification of residual sludges for treatment of wastewater. Report NSF-RANN. Grant Number ENV76-17225.

Hartenstein, R., Neuhauser, E.F., and Collier, J. 1980. Accumulation of heavy metals in the earthworm *Eisenia foetida. J. Environ. Qual.* 9, 23–26.

Hartenstein, R., Neuhauser, E.F., and Kaplan, D.L. 1979. The reproductive potential of the earthworm *Eisenia foetida. Oecologia* 43, 329–340.

Kale, R.D. 2006. *Vermicompost: Crown Jewel of Organic Farming*. Jayanthi Printery, Bangalore, India.

Kaplan, D.L., Neuhauser, E.F., Hartenstein, R., and Malecki, M.R. 1980. Physicochemical requirements in the environment of the earthworm *Eisenia foetida. Soil Biol. Biochem.* 12, 347–352.

Macdonald, D.W. 1983. Predation on earthworms by terrestrial vertebrates. In *Earthworm Ecology: From Darwin to Vermiculture*, ed. J.E. Satchell, 393–414. Chapman and Hall, London.

Milocco, A. 1782. Della maniera di nuticare e di far propagare le galline. In: Il gentiluomo coltivatore. *Venezia*, 10, 3–16.

Neuhauser, E.F., Kaplan, D.L., Malecki, M.R., and Hartenstein, R. 1980. Materials supporting weight gain by the earthworm *E. foetida* in waste conversion systems. *Agric. Wastes* 2, 43–60.

Oliver, G.S. 1937. *Our Friend the Earthworm*. Gardener's Book Club, Volume 8, California.

Phillips, V.R. 1988. Engineering problems in the breakdown of animal wastes by earthworms. In *Earthworms in Waste and Environmental Management*, ed. C.A. Edwards and E.F. Neuhauser, 111–118. SPB Academic Publishing, the Hague, the Netherlands.

Walton, I. 1653. *The Compleat Angler*. Facsimile of first edition, 1928, A. and C. Black, London.

Zhenjun, S. 2003. *Vermiculture and Vermiprotein.* China Agricultural Press, Beijing, China.

CHAPTER 2

Relationships between Composting and Vermicomposting

Jorge Dominguez and Clive A. Edwards

CONTENTS

I Introduction 12
II Composting and Vermicomposting Processes 13
III Open Composting Systems 15
    A Windrow Composting 15
    B Forced Aerated Static Piles 16
IV In-Vessel Composting 16
    A Continuous Vertical Reactors 16
    B Horizontal Reactors 17
V Open Vermicomposting Systems 17
    A Low-Cost Floor Beds 17
    B Gantry-Fed Beds 17
    C Raised Gantry-Fed Beds 18
    D Dorset Wedge-Style Beds 18
VI In-Container Vermicomposting 19
    A Bins 19
    B Batch Reactors 19
VII Organic Waste Processing Systems 19
VIII Criteria of Compost and Vermicompost Maturity 20
IX Analytical Methods to Evaluate Compost Stability 21
    A Carbon:Nitrogen Ratio 21
    B Humic Substances 21
    C Absence of Plant Inhibitors 21
    D Absence of Human Pathogens 22
    E Other Criteria 22

X Advantages of Thermophilic Composting 22
XI Advantages of Vermicomposting 23
Acknowledgment 24
References 24

I INTRODUCTION

Many environmental problems of current concern are due to the high production and local accumulations of organic wastes that are too great for the basic degradation processes inherent in nature. Until recent years the utilization of organic wastes and manures in agricultural soils was not problematic because the production was small enough to be used in limited quantities. With adequate application rates, organic wastes are a valuable resource as a soil fertilizer, providing a high content of macro- and micronutrients for crop growth, and they represent a low cost alternative to inorganic fertilizers. Environmental problems arise when the local production and accumulation of manures through intensive strategies is too great, resulting in difficulties in finding sufficient land areas for disposing of the enormous amounts of organic wastes produced. Indiscriminate spreading of large quantities of such organic wastes can damage soil fertility, can cause water pollution and odors, and may present a health risk. The potentially adverse effects of such indiscriminate applications include overfertilization, particularly with N, P, and K; ammonia toxicity for the soil biota; accumulation and concentration of heavy metals in the soil surface and soil biota; gradual increases in soil alkalinity; salt accumulation in dry conditions; establishment of anaerobiosis and anoxic decomposition pathways; input and dispersal of human pathogens; and ground water pollution.

These important environmental problems can be avoided if organic wastes are treated appropriately before their disposal or use. In this way, aerobic biodegradation can be involved to produce either a high-quality final product or simply to reduce environmental problems through a rapid processing of the waste without increased costs. Through the end of the aerobic degradation process the oxygen demand is low, the organic materials are converted to more stable products, carbon dioxide, and water are released, and heat is evolved. Under field conditions, the degradation process takes place slowly at the soil surface, without reaching high temperatures and mainly under aerobic conditions. This natural process of breakdown can be accelerated by heaping the material into windrows to avoid heat losses and thus allowing temperature increases (thermophilic composting) or by using specific species of earthworms as agents for turning, fragmentation, and aeration (vermicomposting).

The humified composts and vermicomposts rapidly attain equilibrium with the soil ecosystem without causing some of the major disruptions commonly associated with raw organic wastes. These products are valuable in agriculture as nutrient sources and in soil improvement. Currently, the science of thermophilic composting is well known and applied widely to organic-waste management.

The cultivation of earthworms in organic wastes has been termed *vermiculture*, and *vermicomposting*, the managed processing of organic wastes by earthworms to produce vermicompost, has progressed considerably in recent years. Vermicomposting has been shown to be successful for processing sewage sludge and solids from wastewater (Domínguez et al. 2000; Clark et al. 2007; Pramanik et al. 2007; Suthar 2007), paper industry waste (Elvira et al. 1996, 1998; Kaushik and Garg 2003; Gajalakshmi and Abbasi 2004), urban residues, and food and animal waste (Edwards et al. 1985; Edwards 1988; Domínguez and Edwards 1997; Atiyeh et al. 2000; Triphati and Bhardwaj 2004; Garg et al. 2006; Aira et al. 2006a, 2006b; Suthar 2007; Lazcano et al. 2008), as well as horticultural residues from plants (Gajalakshmi et al. 2005; Gupta et al. 2007; Pramanik et al. 2007; Suthar 2007) and food industry waste (Edwards 1983; Butt 1993; Nogales et al. 1999a, 1999b, 2005).

II COMPOSTING AND VERMICOMPOSTING PROCESSES

Composting and vermicomposting are two of the best-known processes for the biological stabilization of solid organic wastes. Composting involves the accelerated degradation of organic matter by microorganisms under controlled conditions, during which the organic material undergoes a characteristic thermophilic stage 45°C–65°C (113°F– 149°F) that allows sanitization of the waste by the elimination of pathogenic microorganisms. Two phases can be distinguished in composting: (a) the thermophilic stage, where decomposition takes place more intensively and which therefore constitutes the active phase of composting; (b) a maturing stage, which is marked by decreases in the temperature to the mesophilic range and where the remaining organic compounds are degraded at a slower rate. The duration of the active phase depends on the characteristics of the waste (amounts of easily decomposable substances) and on the management of the controlling parameters (aeration and watering). The extent of the maturation phase is also variable, and it is marked normally by the disappearance of phytotoxic compounds. Thermophilic composting is well established on the industrial scale for solid organic-waste treatment, although the loss of nitrogen through volatilization of NH_3 during the thermophilic stage of the process is one of the major drawbacks of the process. Through composting, the heterogeneous fresh organic material is transformed into a homogeneous and well-stabilized humuslike product (Gootas 1956; Golueke 1972; Poincelot 1975; Haug 1979; De Bertoldi et al. 1983; Zucconi and de Bertoldi 1987).

Vermicomposting involves the bio oxidation and stabilization of organic material by the joint action of earthworms and microorganisms. Although it is the microorganisms that biochemically degrade the organic matter, earthworms are the crucial drivers of the process, as they aerate, condition, and fragment the substrate, thereby drastically increasing the microbial activity. Earthworms act as mechanical blenders and by comminuting the organic matter they modify its physical and chemical status, by gradually reducing the C:N ratio and increasing the surface area exposed to microorganisms—thus making it much more favorable for microbial activity and further decomposition (Domínguez et al. 1997). Therefore two phases can also be distinguished here, (a) an active phase where the earthworms process the waste, modifying its physical state and microbial composition (Lores et al. 2006), and (b) a maturation-like phase marked by the displacement of the earthworms toward fresher layers of undigested waste, where the microorganisms take over in the decomposition of the waste. As in composting, the duration of the active phase is not fixed and will depend on the species and population density of earthworms and their ability to ingest the waste (ingestion rate). Vermicomposting is not yet adapted fully to the larger industrial scale (Domínguez et al. 1997), and since the temperature is always in the mesophilic range, pathogen removal is not completely ensured, although some studies have provided good evidence of suppression of human pathogens (Monroy et al. 2008, 2009; see Chapter 16). In some cases, organic residues require pretreatment before being vermicomposted as they may contain substances that are toxic to earthworms, such as acidic compounds (Nair et al. 2006), NH_3, and salts.

A combination of composting and vermicomposting has recently been considered as a way of achieving stabilized substrates (Tognetti et al. 2007). Composting enables sanitization of the waste and elimination of toxic compounds, and the subsequent vermicomposting reduces particle size and increases nutrient availability; in addition, inoculation of the material resulting from the thermophilic phase of composting with earthworms reduces the expense and duration of the treatment process (Ndegwa and Thompson 2001).

Both composting and vermicomposting are aerobic biodegradation processes of organic wastes that involve complex interactions between the organic waste, microorganisms, moisture, and oxygen contents. The waste material normally contains indigenous mixed populations and communities of microorganisms. When the moisture content and oxygen concentration are brought to a suitable level, microbial activity increases. In addition to oxygen and water, microorganisms require a source of carbon, macronutrients such as N, P, and K, micronutrients, and certain amounts of trace elements for their normal growth and reproduction. These requirements are provided by the organic waste materials. By using the organic matter as a food source the microorganisms reproduce rapidly and release carbon dioxide, water, some organic products, and energy. Some of this energy produced is consumed during the metabolism processes with the remainder released as heat.

Thermophilic composting is a biological process in which the active agents are microorganisms; consequently, the successful outcome of the composting process depends on the presence of the appropriate microbial population—preferably indigenous—and on the provision for suitable conditions for microbial activity (Zucconi and de Bertoldi 1987). Several aspects related to nutrition and microbial succession characterize the composting process. The primary effect of the microbial succession is the establishment of a pattern in such way that one group of organisms paves the way for a succeeding group. A parallel well-known feature, syntrophy, sometimes referred also as synergy, has been recognized in microbiological terminology and refers to nutritional and metabolic interactions between two or more groups of bacteria when growing as a mixed culture. Through syntrophy or synergy, metabolic end products produced by one group of organisms may be used as nutrients by the following ones. Thus, the combined activities of two or more different types of organisms growing together may result in final products that are quantitatively or qualitatively very different from the total sum of the activities of each individual organism.

Collaborative or syntrophic decomposition of organic matter is the normal course of events in both composting and vermicomposting of organic wastes. However, in both natural decay and vermicomposting systems, a huge variety of microorganisms and soil invertebrates grow and interact, contributing to the "cycle of matter." The vermicomposting system sustains complex food webs, and different chemical forms of several nutrient elements become modified into long-lived organic compounds that are important for nutrient dynamics as well as plant growth regulators.

Vermicomposting is a mesophilic, completely aerobic decomposition process and, as such, is almost entirely the result of microbial activities. Earthworms and other soil animals do not have cellulolitic systems developed enough to digest plant material and therefore much of the animal nutrition depends on the action of microbes, either free-living or associated with their guts. Although microoganisms are responsible for the biochemical degradation of organic matter in the vermicomposting process, earthworms are important in conditioning the substrate and promoting microbial activity. They can be considered as mechanical blenders because they break down organic material, increase the surface area exposed to microbes, and move fragments and bacteria-rich excrement through the waste profile, homogenizing the organic material. The vermicomposting process per se is due to microbial activities, with the earthworms having also a strong influence, but the whole soil fauna community also plays an important role in these processes through its interactions with soil microbes and therefore must be considered in a holistic perspective.

III OPEN COMPOSTING SYSTEMS

A Windrow Composting

This consists of placing the mixtures of different raw organic materials in long, narrow piles or windrows that are turned mechanically on a regular basis to aerate them. Turning alone often does not ensure adequate and consistent oxygenation. Within an hour after turning, oxygen levels within a pile often drop drastically, and microbial activity is accordingly reduced. For this reason, the pile must be turned frequently, leading to technical and economic problems. In most modern systems there is recognition that the different phases of the process require different aeration rates. The results of this are turning programs that, for example, recommend turning every 3–4 days for the first 2–3 weeks and on a weekly basis thereafter. When a particular turning sequence does not cause local problems (e.g., odor) and the final product is "acceptable," there is no reason to modify it (Stentiford 1996). On a more rational basis, systems using temperature as a turning indicator have been developed. Briefly, these work on the basis that as soon as a certain temperature 55°C–60°C (131°F–140°F) is achieved in the pile core, the material must to be turned.

B Forced Aerated Static Piles

Forced aeration systems are intended to supply air to the composting mass using pressurized air systems (typically using low head, <150 mm, high-volume fan units) rather than turning the mass. There are basically three ways to oxygenate the piles:

1. **Bottom Suction**
   The air is drawn through the pile by the imposition of negative pressure. In this kind of ventilation height is a critical factor. With piles higher than 2.5–3.0 m (8–9 ft) it is almost impossible to obtain uniform aeration. These piles must be blanketed with an insulating layer (usually cured compost) to ensure a uniform distribution of temperature (Finstein et al. 1983).
2. **Bottom Blowing**
   Aeration is provided by blowing air through the pile (positive pressure). This method tends to cool down and dry the bottom layers of the pile, leaving the outer layers warm and moist (Finstein et al. 1983).
3. **Alternative Ventilation**
   In these systems bottom blowing aeration is alternated with bottom suction aeration. The alternative flows of air movement lead to a homogenization of temperature and moisture gradients throughout the pile (Pereira Neto et al. 1991).

IV IN-VESSEL COMPOSTING

In-vessel composting refers to a group of methods that confine the mass to be composted inside a building, container, or vessel. In-vessel systems are designed to promote rapid digestion rates by careful monitoring and control of the composting process; although these systems can produce an end product more quickly, they are more complex and relatively costly to build, operate, and maintain. There are a variety of in-vessel methods with different combinations of vessels, aeration devices, and turning mechanisms. Among these, the most widely utilized are the following.

A Continuous Vertical Reactors

Usually the materials are loaded up from the top of the reactor and discharged from the bottom. Oxygenation is provided by forcing air up from the bottom through the composting mass. These reactors can process large amounts of material (as much as 2000 m^3 (2616 yd^3)) and may be as high as 9 m (28 ft); however, the height is extremely critical, and masses higher than 3 m (10 ft) lead to serious problems in ventilation, by hypo- or hyperventilation (Haug 1993).

B Horizontal Reactors

The materials are arranged along the length of the unit, and the height never exceeds 2–3 m (6–9 ft). The principal advantage of these systems is the feasibility for controlling the process, it being possible to shorten the thermophilic stage. Because the oxygen is supplied either by turning or by aeration the mass to be composted can be oxygenated uniformly and the temperature easily controlled (Haug 1993).

V OPEN VERMICOMPOSTING SYSTEMS

A Low-Cost Floor Beds

The traditional methods of vermicomposting have been based on beds or windrows on the ground containing materials up to 45 cm (1.5 ft) deep, but such methods have numerous drawbacks. They require large areas of land for large-scale production and are relatively labor-intensive, even when machinery is used for adding materials to the beds, watering, and harvesting the products. More important, such systems process organic wastes relatively slowly (Edwards 1988). Outdoor windrows or beds with simple walls are the simplest type of process. The size of such beds is flexible, but the width should not exceed 2.4 m (8 ft), which allows the entire bed to be inspected easily, without the need to walk on the bed, and is also compatible with the sizes of many suitable covering and construction materials. The length is less important and depends on the area available. The beds can be laid on soil that is freely draining and does not suffer waterlogging. Concrete areas are ideal for earthworm-processing systems since they provide a firm surface for tractor operations. However, it is essential for precautions to be taken to prevent too much water from entering the beds and to allow excess water to drain away from the bed. Usually, such floor beds are covered with permeable materials, and the covers are removed only for watering and addition of new waste materials. It is necessary to feed prepared organic wastes on top of the bed surface in thin layers of 5–10 cm (2–4 in) every week or so, depending on the type of waste. In addition, it is necessary to retain a light and open texture in the waste (Price 1987; Phillips 1988).

B Gantry-Fed Beds

An important principle to improve the efficiency of processing of organic wastes by earthworms is to add the wastes to the beds in thin layers of 2.5–5.0 cm (1–2 in) at frequent intervals. This can be done readily by adding the wastes by means of an overhead gantry running on wheels on the top of the walls of the beds.

This gradual addition of waste minimizes the generation of heat during composting and ensures that earthworms are continually processing the fresh wastes close to the surface.

C Raised Gantry-Fed Beds

Earthworms are usually confined to the top 10–15 cm (4–6 in) of the bed. The efficiency and rate of processing the wastes can be considerably increased by placing the bed above the ground. If the bed has a mesh underneath, the earthworm-processed organic matter can be sieved by mechanical action, such as a breaker bar, and can be collected using a moving belt or a slurry scraper. If the waste is added to the top of the bed in thin layers from a mobile gantry daily and collected at the base, a continuous waste-processing system can be obtained. Such systems can be relatively sophisticated by a complete mechanization and automation of the addition and collection systems of the waste. Such automated continuous-processing reactors have been operating successfully for as long as 3–4 years (Price 1987; Phillips 1988; see Chapter 8).

D Dorset Wedge-Style Beds

The design of this type of bed is quite different from the relatively shallow flat beds described previously. Waste is added to the angled leading edge in shallow layers. Because of the large amounts of waste added, the benefit of this type of bed is that the wedge tends to be more self-heating. It is particularly suitable for cattle waste, and separated pig solids. It is especially recommended for processing organic wastes in winter, and the depth of the wedge should not exceed 1 m (3.1 ft).

The floor of a Dorset wedge has the same requirements as the outdoor beds. It should be free-draining and firm enough to withstand quite heavy machinery. The only essential wall required is the removable rear wall, which should be sound enough to enable the front-end loader to dig the bed. Side walls can help to maintain the appropriate depth of the waste. Once the majority of the leading edge of the bed appears to be earthworm-worked, a new layer of fresh waste is added. This is usually done by adding wastes at the top of the leading edge and then raking the waste down onto the whole surface. Care should be taken to avoid unnecessary compaction of the bed; with long beds, a side-discharge muck spreader can be used. The amount of waste added to the wedge should be lower during the summer than in the winter months to prevent overheating.

Because of the shape of the Dorset wedge, it is not important to specify the dimensions or time to harvest the bed, and the best indicator is when there is at least a load of approximately 20 m^3 (26.2 yd^3) that can be harvested. Most earthworms are probably in the top 15 cm (6 in) of the leading edge, and therefore this material has to be removed in order to harvest the earthworm-worked material behind, which is collected and loaded up for transport. The remaining inoculum is used to form a new wedge as before (Price 1987; Phillips 1988; see Chapter 7).

VI IN-CONTAINER VERMICOMPOSTING

A Bins

Other systems refer to the use of bins or large containers, often stacked in racks. Although these and other small-scale systems are widely used, they have drawbacks when applied on a larger scale. They require considerable machinery for handling and lifting, and also there are several problems related to addition of water and additional layers of new material and to drainage. Edwards (1988) discussed batch vermicomposting in stacked boxes or containers and suggested that it is too labor-intensive, since the batches have to be moved in order to add more wastes or water.

B Batch Reactors

Much more promising techniques have used containers provided with legs. These allow adding the feedstock at the top from modified spreaders or mobile gantries and collecting the vermicompost mechanically at the bottom after being sieved through the mesh floors using breaker bars. Such methods were developed and tested at the National Institute for Agricultural Engineering (Silsoe, England) and are currently being used at several places in the United States and elsewhere, ranging from relatively low-technology systems using manual loading and collection to completely automated systems using hydraulically driven continuous-flow reactors. Such reactors process 1 m (3.1 ft) deep layers of suitable organic wastes in 30–60 days (Edwards and Bohlen 1996). Although these systems require more capital outlay, the cost of the reactor can usually be returned in 1–3 years (see Chapter 19), and they can be operated on a large scale with minimal labor requirements. An automated reactor processing 1000 tons of waste per year can be built for \$35,000–\$50,000, and the lower-technology systems cost much less. Detailed economic studies at Silsoe have shown that such reactors have a much greater economic potential to produce a high-quality plant growth media in a shorter time and more efficiently than windrows or ground beds. An even larger-scale system called the Sovadec was developed by Marcel Bouché in southern France and was based on similar principles, involving separation and sorting of the wastes followed by composting, vermicomposting, and sieving. This system can convert 27% of a total urban waste stream into a valuable vermicompost.

VII ORGANIC WASTE PROCESSING SYSTEMS

A summary of the recommended values for the important factors in composting and vermicomposting processes are given in Tables 2.1 and 2.2. The main task is to translate these factors into economic and reliable composting systems. The complexity of the composting and vermicomposting equipment and the problems of how to approach the optimum values of these factors may vary considerably from simple heap processes to large-scale mechanized reactors.

Table 2.1 Recommended Values of Process Factors

Composting	
Process Factor	**Values**
C:N ratio of wastes	25:1 to 30:1
Initial particle size	10–15 mm (0.4–0.6 in) for agitated systems and forced aeration
Moisture content	55%–60% (higher value possible when using bulking agents such as straw or wood chips)
Oxygenation	0.6–1.8 m^3.day.kg (0.78–2.4 yd^3.day.2.2 lb) volatile solids or maintain oxygen level at 10%–18%. Feedback temperature or oxygen control of air blower possible in forced aeration systems.
Temperature	55°C–60°C (131–140°F)
Agitation	No agitation to periodic turning in simple systems. Short bursts of vigorous agitation in mechanized systems.
Windrow size	Any length; 1.5 m high × 2.5 m wide (4.9 × 8.2 ft) wide for natural aeration heaps. Heap size can be increased for forced aeration.
Reactor size	Height is extremely critical, and masses higher than 3 m (3.1 yd) can lead to serious problems in ventilation.
Human pathogens	Killed after the thermophilic stage of the process
Time taken	The self-heating and the thermophilic phase (about a week) are followed by several months of "curing" at mesophilic temperatures

Table 2.2 Characteristics of Vermicomposting

Process Factor	Values
C:N ratio of wastes	25:1 to 30:1
Initial particle size	10–20 mm (0.4–0.8 in) (higher values slow down the process)
Moisture content	80%–85% (limits 60%–90%)
Oxygen	Earthworms maintain aerobic conditions
Temperature	15°C–25°C (limits 4°C–30°C) (59–77°F) (limits 39–86°F)
pH	>5 and <9
Ammonia content of wastes	Low: <0.5 $mg \cdot g^{-1}$
Salt content of wastes	Low: <0.5%
Windrow size	Any length and width 50 cm high (higher values slow down the process or can even stop it long)
Reactor size	40 m long × 2.4 m wide (128 ft long × 8 ft wide) × lm deep. Wastes should be added in thin layers 5–10 cm (2–4 in.)
Human pathogens	Killed after 70 days of vermicomposting
Time taken	From 4 to 12 months in the windrows to 30–60 days in the continuous reactor systems

VIII CRITERIA OF COMPOST AND VERMICOMPOST MATURITY

Both composting and vermicomposting transform fresh organic wastes into useful products that are rich in available nutrients for plant growth, poor in readily biodegradable carbon, almost depleted in phytoinhibitory substances, and relatively free of plant and human pathogens. Subjectively, a mature compost should be dark brown or black, with a granular, spongy, or fibrous texture, and smell like mold or soil. A mature vermicompost should also be dark black, usually finely divided peat-like material with excellent structure, porosity, aeration, and drainage properties and high moisture-holding capacity (see Chapter 18).

IX ANALYTICAL METHODS TO EVALUATE COMPOST STABILITY

A Carbon:Nitrogen Ratio

The C:N ratio is one of the most widely utilized parameters to follow the development of material undergoing a composting or vermicomposting process, and it varies remarkably depending on the feedstocks and by itself can hardly give reliable indications of compost maturity. The C:N ratio of a mature compost or vermicompost should ideally be around 10, but this is hardly ever achieved due to the presence of recalcitrant organic compounds, or materials that decompose poorly. The most critical aspect related to the C:N ratio is whether further decomposition of the mature products in the soil will result in the release of mineral N or cause competition with plants for the N in the soil solution. C:N ratios ranging between 35.6–50.8 cm (14–20 in) in mature composts and vermicomposts are therefore acceptable as long as their further decomposition is slow and does not use up additional N from the soil.

B Humic Substances

The quantity and quality of compost and vermicompost humic substances have often been studied in the last 10 years (Roletto et al. 1985; Sequi et al. 1986; Saviozzi et al. 1988), suggesting possible parameters and evaluation indexes for compost maturity. During the maturation process, humic substances evolve qualitatively with an increasing predominance of humic acids over fulvic acids; the ratio between these two is considered an important index of compost maturity; and it is considered that it should be more than 1 in a mature compost.

C Absence of Plant Inhibitors

Perhaps the best indicator of a compost's or vermicompost's maturity is the absence of bio-inhibitory aliphatic acids and phenolics, which can be detected by chromatography or by using seed germination tests. In the latter test small seeds (*Lepidium sativum*) are placed in a Petri dish on filter paper that has been soaked in a water extract of the underlying final material, and they should germinate to the same percentage level as those placed on paper soaked in distilled water. Other methods have been devised: Zucconi, Pera, Forte, and De Bertoldi (1981) and Zucconi, Pera, Forte, Monaco, et al. (1981) evaluated phytotoxicity as a function of *L. sativum* seed germination and root growth; Wong (1985) used *Brassica parachinensis* seeds; and Kuboi and Fujii (1984) tested 20 plant species in liquid shaking culture.

D Absence of Human Pathogens

It has been proposed that a compost and vermicompost should be considered hygienic if 100 g (3.5 oz) of sample do not contain *Salmonella*, human viruses, infective parasitic helminthic eggs, and no more than 5×10^4 fecal coliforms and 5×10^5 fecal streptococci (see Chapter 16).

E Other Criteria

Organic nitrogen mineralization is a useful parameter to determine readily biodegradable nitrogen compounds, and therefore it can be correlated inversely with compost stability. The N mineralization assay evaluates organic matter stability as a function of the existing equilibrium between organic and mineral N during a 10-day incubation period. Respiration rates are related directly to the speed of microbial metabolism; therefore, they are related inversely to compost maturity. Consequently, in the early degradation stages, corresponding to a fast microbial activity due to the most readily fermentable organic fractions, respiration is very high. Later on, after a decrease in the biological activity, respiration decreases strongly until it reaches low values that remain constant in a stable final product.

X ADVANTAGES OF THERMOPHILIC COMPOSTING

The composting process typically reduces the pH of the end products, and the compost acts like a buffer in the soil. In-vessel composting allows collection of ammonia, which can be recycled as fertilizer to avoid atmospheric pollution. Compost can also provide a high content of available plant nutrients and improves soil physical properties such as water-holding capacity, cation-exchange capacity, soil aeration and permeability, and water infiltration, which all significantly contribute to reduce soil erosion and losses of nutrients by surface runoff. Composting produces a noticeable rapid volume and weight reduction as a consequence of organic matter mineralization, and water losses by evaporation; in addition, nutrient content per volume and unit of weight increases. Odor problems and proliferation of flies and rats can be minimized with effective management. Composting produces antagonists to plant pathogens and also weed seeds, and pathogenic organisms can be eliminated by the high temperatures occurring during the process. The composting process increases the amounts of humified compounds, and, although the total amount of heavy metals increases (as a consequence of the carbon losses by mineralization during the process), the amounts of bio-available heavy metals tend to decrease due to the formation of stable complexes with these polymerized substances (see Chapter 17).

Organic wastes recycled as composts also reduce the exploitation of limited resources such as inorganic fertilizers and peat and decrease the costs of disposal of the organic waste. Spreading compost on agricultural land can lead to a more uniform distribution of the nutrients, and unlike addition of raw organic wastes, no phytotoxic effects on seedlings and plant roots occur. For horticultural purposes, composts represent an economic alternative to peat and mosses. The duration of the composting process can vary depending on the characteristics of the final product to be obtained. In some cases simply removing pollution from the initial waste to obtain a "fresh compost" could be enough although the final compost may not be of agricultural value, and this process could then be completed in less than a week. When the objective is to obtain a high-quality compost it is necessary to continue the process until the maturation or curing of the compost is completed (the self-heating and the thermophilic phase, lasting about a week, are followed by several months of "curing" at mesophilic temperatures). Maturation is a passive stage that does not require any treatment and can take months.

XI ADVANTAGES OF VERMICOMPOSTING

Earthworms can break down organic matter very rapidly, resulting in stable, nontoxic vermicomposts with a better structure, microbial content, and available nutrient content than composts. These have a potentially high economic value as soil conditioners or media for plant growth. Although the best final products and the shortest residence times are obtained by high-technology systems, the low-technology ones can be easily adapted and managed in small farms or livestock operations. Vermicompost is a finely divided, peatlike material with a low C:N ratio, excellent structure, porosity, aeration, drainage, and moisture-holding capacity, and it supplies a suitable mineral balance, improves plant nutrient availability, and could act as complex-nutrient-source granules.

Similarly to composting processes, vermicomposting reduces waste bulk density, and recent research also showed that it greatly reduces populations of pathogenic microorganisms (see Chapter 16). It is generally accepted that the thermophilic stage during the composting process eliminates human pathogens, but it has been shown that human pathogens are also eliminated during vermicomposting, probably by means of an antagonism mechanism.

As an aerobic process, both composting and vermicomposting lead to N mineralization, but the presence of earthworms in vermicomposting increases and accelerates the N mineralization rate. Moreover, the humification rates that take place during the maturation stage are higher and faster during vermicomposting, resulting in a greater decrease of bio-available heavy metals. There is circumstantial evidence that the final product may contain hormone-like compounds or plant growth regulators that could accelerate plant growth and crop yields (see Chapter 9).

The main purpose when applying composting and vermicomposting technologies to organic-waste management has been to decrease landfill disposal and to obtain value-added products that can be suitable for commercialization. For this reason, many of the other applications of these processes have been disregarded and poorly studied. Therefore, we consider these two processes of utmost importance for stabilizing organic wastes, and at the same time solving or at least minimizing those environmental problems that could arise from their disposal. In many cases there is no need to complete the procedure, but, depending on the composition and characteristics of the initial waste to be treated, the processes could be extended, and then the end products would be of a much higher quality.

The thermophilic composting process seems suitable for the rapid treatment of large amounts of organic wastes, in order to eliminate contamination problems more quickly than the traditional low-technology vermicomposting systems. However, the newer vermicomposting continuous reactor systems seem to be equally applicable to rapid large-scale organic-waste processing. Traditional batch and bed vermicomposting systems may be an alternative, inexpensive way to avoid environmental problems and at the same time obtain a valuable organic fertilizer. Vermicomposting, whether low or high technology, may have an important role in organic-waste management, and it is possible to suggest that vermicomposting and composting are not necessarily mutually exclusive and could be used in sequence to take advantage of the unique and valuable features of each.

ACKNOWLEDGMENT

This research was partially supported by a Xunta de Galicia grant (07MRU023383PR) and the Spanish Ministerio de Ciencia e Innovaciòn (CTM2009-08477).

REFERENCES

De Bertoldi, M., Vallini, G., and Pera, A. 1983. The biology of composting: A review. *Waste Manage. Res.* 1, 157–176.

Edwards, C. A. 1988. Breakdown of animal, vegetable and industrial organic wastes by earthworms. In *Earthworms in Waste and Environmental Management*, ed. C. A. Edwards and E. F. Neuhauser, 21–31. SPB, the Hague, the Netherlands.

Edwards, C. A., and Bohlen P.J. 1996. *Biology and Ecology of Earthworms*. 3rd ed. Chapman and Hall, London.

Finstein, M. S., Miller, F. C., Strom, P. F., Mac Gregor, S. T., and Psarianos, K. M. 1983. Composting ecosystem management for waste treatment. *Biotechnology* 1, 347–353.

Golueke, C. G. 1972. *Composting: A Study of the Process and Its Principles*. Rodale Press, Emmaus, PA.

Gootas, H. B. 1956. *Fundamentals of Composting Process*. University of California, Berkeley, CA.

Haug, R. T. 1979. Engineering principles of sludge composting. *J. Water Pollut. Control Fed.* 51, 2189–2206.

Haug, R. T. 1993. *The Practical Handbook of Compost Engineering*. Lewis, Boca Raton, FL.

Kuboi, T., and Fujii, K. 1984. A new method for seedling assay of phytotoxic substances: Liquid shaking culture. *Soil Sci. Plant Nutr.* 30, 209–218.

Pereira Neto, J. T., Nobrega, C. C., and Stentiford, E. I. 1991. A hybrid mode of aeration for aerated static pile composting systems. In the Proceedings of 7th International Conference on Solid Waste Management and Secondary Materials, Philadelphia, PA, December 1991.

Phillips, V. R. 1988. Engineering problems in the breakdown of animal waste by earthworms. In *Earthworms in Waste and Environmental Management*, ed. C. A. Edwards andE. F. Neuhauser. SPB, the Hague, the Netherlands, 444–448.

Poincelot, R. P. 1975. *The Biochemistry and Methodology of Composting*. Bulletin 754, Connecticut Agricultural Experiment Station, New Haven.

Price, J. S. 1987. Development of a vermicomposting system. *Proc. 4th Int. CIEC Sym. Agric. Waste Manage. Environ. Prot.* 1, 294–300.

Roletto, E., Barberis, R., Consiglio, M., and Iodice, R. 1985. Chemical parameters for evaluating compost maturity. *BioCycle* 26, 46–47.

Saviozzi, A., Riffaldi, R., and Levi-Minzi, R. 1988. Maturity evaluation of organic wastes. *BioCycle* 29, 54–56.

Sequi, P., De Nobili, M., Leita, L., and Cercignani, G. 1986. A new index of humification. *Agrochimica* 30, 175–178.

Stentiford, E. I. 1996. Composting control: Principles and practice. In *The Science of Composting*, ed. M. de Bertoldi, P. Sequi, B. Lemmes, and T. Papi, 49–59. Chapman and Hall, Glasgow.

Wong, M. H. 1985. Phytotoxicity of refuse compost during the process of maturation. *Environ. Pollut.* 37, 159–174.

Zucconi, F., and De Bertoldi, M. 1987. Composting specifications for the production and characterization of compost from municipal solid waste. In *Compost: Production, Quality and Use*, ed. M. De Bertoldi, M. P. Ferranti, P. L´Hermite, and F. Zucconi, 30–51. Elsevier Applied Science, London.

Zucconi, F., Pera, A., Forte, M., and De Bertoldi, M. 1981. Evaluating toxicity of inmature compost. *BioCycle*, 22, 54–57.

Zucconi, F., Pera, A., Forte, M., Monaco, A., and De Bertoldi, M. 1981. Biological evaluation of compost maturity. *BioCycle*, 22, 27–29.

CHAPTER 3

Biology and Ecology of Earthworm Species Used for Vermicomposting

Jorge Dominguez and Clive A. Edwards

CONTENTS

I Introduction ... 28
II Earthworm Species Suitable for Vermicomposting ... 29
A Temperate Species ... 29
1 *Eisenia fetida* (Savigny 1826) and *Eisenia andrei* (Bouché 1972) ... 29
2 *Dendrodrilus rubidus* (Savigny 1826) ... 32
3 *Dendrobaena veneta* (Rosa 1886) ... 32
4 *Lumbricus rubellus* (Hoffmeister 1843) ... 33
5 *Drawida nepalensis* (Michaelsen 1907) ... 33
B Tropical Species ... 34
1 *Eudrilus eugeniae* (Kinberg 1867) ... 34
2 *Perionyx excavatus* (Perrier 1872) ... 34
3 *Polypheretima elongata* (Perrier 1872) ... 35
III Influence of Environmental Factors on Survival and Growth of Earthworms ... 35
A Temperature ... 36
B Moisture Content ... 36
C pH ... 36
D Aeration ... 37
E Ammonia and Salts ... 37
IV Predators, Parasites, and Pathogens of Earthworms ... 38
Acknowledgment ... 38
References ... 38

I INTRODUCTION

Earthworms are macroscopic clitellate oligochaete annelids that live in soil. They are segmented worms, bilaterally symmetrical, with an external gland (clitellum) for producing the egg case (cocoon), a sensory lobe in front of the mouth (prostomium), and an anus at the end of the animal body, with a small number of bristles (setae) on each segment. They are hermaphrodite animals, and reproduction normally occurs through copulation and cross-fertilization, following which each of the mated individuals produces cocoons containing 1–20 fertilized ova. The resistant cocoons, which are tiny and roughly lemon-shaped, with shape differing between species, are usually deposited near the soil surface, except in dry weather when they are laid at deeper layers. Cocoons hatch after an incubation period that varies according to the earthworm species and environmental conditions. Hatchling earthworms, unpigmented and only a few millimeters in length on emerging from the cocoons, gain their adult pigmentation within a few days. Assuming favorable conditions, they reach sexual maturity within several weeks after emergence. Mature individuals of most vermicomposting species can be distinguished easily by the presence of the clitellum, the pale- or dark-colored swollen band located behind the genital pores. The clitellum secretes the fibrous cocoon, and the clitellar gland cells produce a nutritive albuminous fluid that fills the cocoon. Earthworms display indeterminate growth and can continue to grow in size after completing their sexual development although they do not add segments.

According to Reynolds and Wetzel (2004), there are more than 8300 species in the Oligochaeta, of which about half are terrestrial earthworms. The most common earthworms in Europe, North America, western Asia, and many other parts of the world belong to the family Lumbricidae, whereas in West Africa, many of the common earthworms belong to the family Eudrilidae. In South Africa there is the Microchaetidae, in Australia and other parts of eastern Asia, the Megascolecidae, and the family Glossoscolecidae predominate in Central and South America.

Different species of earthworms have different life histories, occupy different ecological niches, and have been classified, on the basis of their feeding and burrowing strategies, into three ecological categories: epigeic, anecic, and endogeic (Bouché 1977). Endogeic (soil feeders) and anecic species (burrowers) live in the soil and consume a mixture of soil and organic matter, and thus excrete organomineral feces. Epigeic species of earthworms are litter dwellers and litter transformers; they live in organic soil horizons, in or near the surface litter, and feed primarily on coarse particulate organic matter. They ingest large amounts of undecomposed litter and excrete holorganic fecal pellets. These species are small in body size and uniformly pigmented with high metabolic and reproductive rates, which represent adaptations to the highly variable environmental conditions at the soil surface. In tropical regions, epigeic earthworms can also be found in the axils of Bromeliaceae plants.

II EARTHWORM SPECIES SUITABLE FOR VERMICOMPOSTING

Epigeic species of earthworms, with their natural ability to colonize organic wastes; high rates of consumption, digestion, and assimilation of organic matter; tolerance to a wide range of environmental factors; short life cycles; high-reproductive rates; and endurance and tolerance of handling, show good potential for vermicomposting. Few earthworm species display all these characteristics, and in fact only five have been used extensively in vermicomposting *Eisenia andrei* (Savigny), *Eisenia fetida* (Bouché), *Dendrobaena veneta* (Savigny), and, to a lesser extent, *Perionyx excavatus* (Perrier), and *Eudrilus eugeniae* (Kinberg). Characteristics and life history aspects of eight common species of earthworms are summarized in Table 3.1.

A Temperate Species

1 Eisenia fetida *(Savigny 1826) and* Eisenia andrei *(Bouché 1972)*

These lumbricid earthworm species are those most commonly used in vermicomposting and vermiculture mainly because they are ubiquitous with a worldwide distribution and colonize organic substrates naturally, their life cycles are short, they have a wide temperature- and moisture-tolerance range, and they are resilient earthworms that can be readily handled.

E. fetida corresponds to the striped or banded morph, with the area around the intersegmental groove having no pigmentation and appearing pale or yellow; hence, its common names of "brandling" or "tiger" earthworm; whereas *E. andrei*, the common "red" worm, corresponds to the uniformly reddish morph. Aside from the differences in pigmentation, the two species are morphologically similar and their overall requirements the same. Their reproductive performances and life cycles do not differ significantly, although growth rates and cocoon production are slightly higher in *E. andrei*. The problem of their taxonomic status remained unresolved for a long time, and moreover in much of the current literature both species are termed indiscriminately as *E. fetida* or *E. foetida*, the latter an illegal or inaccurate emendation of the former name, and it is often not clear which of the two species is being referred to. We have confirmed that they are two different biological species, reproductively isolated, and that they are also two different phylogenetic species. The reproductive isolation was confirmed after studying the offspring viability from inter- and intraspecific crosses of both species (Domínguez et al. 2005). Additionally, fully resolved and well-supported phylogenetic trees based on mitochondrial (COI) and nuclear DNA sequences (28S) confirmed that they are different phylogenetic species (Pérez-Losada et al. 2005). This evidence implies important considerations; thus in vermiculture or vermicomposting *E. andrei* is more often recommended since its growth and reproduction rates are higher. Moreover, the two species are syntopic, commonly living in mixed colonies in dung and vermicompost heaps, and therefore hybridization is possible. The existence of postcopula but not precopula isolation in sympatric populations clearly affects the population dynamics

Table 3.1a Comparison of Some Aspects of the Biology of the Vermicomposting Species

	Eisenia fetida	*Eisenia andrei*	*Dendrobaena rubida*	*Dendrobaena veneta*
Color	Brown and buff bands	Red	Reddish purple	Reddish and purple bands
Size of adult earthworms	4–8 mm × 50–100 mm (0.016–0.03 in × 1.9 × 3.0 in)	4–8 mm × 50–100 mm (0.016–0.03 in × 1.9 × 3.0 in)	3–4 mm × 35–60 mm (0.11–0.16 in × 1.3–2.4 in)	5–7 mm × 50–80 mm (0.2–0.27 in × 2–3.15 in)
Mean weight of adults	0.55 g (0.01 oz)	0.55 g (0.01 oz)	0.25 g (0.008 oz)	0.92 g (0.032 oz)
Time to maturity (days)	28–30	21–28	54	65
Number of cocoons day^{-1}	0.35–0.5	0.35–0.5	0.20	0.28
Mean size of cocoons	4.85 mm × 2.82 mm (0.03 × 0.11 in)	4.8 mm × 2.82 mm (0.03 × 0.11 in)	3.19 mm × 1.97 mm (0.12 × 0.07 in)	3.14 mm × 1.93 mm (0.12 × 0.075 in)
Incubation time (days)	18–26	18–26	15–40	42.1
Hatching viability (%)	73–80	72	85	20
Number of worms cocoon^{-1}	2.5–3.8	2.5–3.8	1.67	1.10
Self-fertilization	+	+	+	—
Life cycle (days)	45–51	45–51	75	100–150
Limits and optimal Tª	25°C (0°C–35°C) (77°F)(32°F–95°F)	25°C (0°C–35°C) (77°F)(32°F–95°F)	—	25°C (15°C–25°C) (77°F)(59°F–77°F)
Limits and optimal moisture	80%–85% (70%–90%)	80%–85% (70%–90%)	—	75% (65%–85%)

Table 3.1b Comparison of Some Aspects of the Biology of the Vermicomposting Species

	Drawida nepalensis	*Eudrilus eugeniae*	*Perionyx excavatus*	*Lumbricus rubellus*
Color	—	Reddish brown	Reddish brown	Reddish brown
Size of adult earthworms	—	5–7 mm × 80–190 mm	4–5 mm × 45–70 mm	4 mm × 70–150 mm
Mean weight of adults	0.82 g (0.02 oz)	2.7–3.5 g (0.09–0.12 oz)	0.5–0.6 g (0.01–0.02 oz)	0.80 g (0.02 oz)
	Drawida nepalensis	***Eudrilus eugeniae***	***Perionyx excavatus***	***Lumbricus rubellus***
Time to maturity (days)	34–42	40–49	28–42	74–91
Number of cocoons day^{-1}	0.15	0.42–0.51	1.1–1.4	0.07–0.25 mm (0.003–0.01 in)
Mean size of cocoons	—	—	—	3.50 mm × 2.46 mm (0.13 × 0.10 in)
Incubation time (days)	24	12–16	18	35–40
Hatching viability (%)	75–88	75–84	90	60–70
Number of worms cocoon^{-1}	1.93	2–2.7	1–1.1	1
Self-fertilization	+	—	—	—
Life cycle (days)	100–120	50–70	40–50	120–170
Limits and optimal temperature	—	25°C (16°C–30°C) 77°F(61°F–86°F)	25°C–37°C (77°F–99°F)	—
Limits and optimal moisture	—	80% (70%–85%)	—	—

by reducing the fitness of the individuals. For this reason, in applied aspects it is important to keep the two species separated (Domínguez et al. 2005) although they often occur in cultures together.

The life cycle and population biology of *E. fetida* and *E. andrei* in different organic wastes have been investigated by several authors (Watanabe and Tsukamoto 1976; Hartenstein et al. 1979; Edwards 1988; Reinecke and Viljoen 1990; Domínguez et al. 1997; Domínguez and Edwards 1997; Domínguez et al. 2000; Monroy et al. 2006). The optimum temperature for growth of both species is 25°C(68°F), and although they can tolerate a wide range of moisture conditions, the optimum moisture content for these species is 85%. In optimum conditions the length of their life cycles (from newly-laid cocoon through clitellate adult earthworm) ranges from 45 to 51 days. The time for hatchlings to reach sexual maturity varies from 21 to 30 days. Copulation in these species, which takes place beneath the soil or waste surface, has been mentioned by various authors since 1845 and has been observed more often than in any other megadrile species. Cocoon laying starts 48 hours after copulation, and the rate of cocoon production is 0.35–0.5 day^{-1}. The hatching viability is 72%–82%, and the incubation period ranges from 18 to 26 days. The number of young earthworms hatching from viable cocoons varies from 2.5 to 3.8 depending on the temperature. In controlled conditions, the average life span is 594 days at 18°C(64.4°F) and 589 days at 28°C(82.4°F) with a maximum life expectancy between 4.5 and 5 years, although under natural conditions it may be considerably shorter.

2 Dendrodrilus rubidus (*Savigny 1826*)

Dendrodrilus rubidus is a holarctic earthworm species belonging to the family Lumbricidae with a relatively cosmopolitan distribution. It is an epigeic earthworm with a clear preference for highly organic soils, and it has also been found in organic substrates such as rotting wood and straw, pine litter, compost, peat, and nearby sewage tanks and manure. Although some specific aspects of their biology have been investigated (Gates 1972; Sims and Gerard 1985; Bengtsson et al. 1986; Cluzeau and Fayolle 1989; Elvira et al. 1996), it is not an earthworm usually used in vermicomposting or vermiculture. The mean time for hatchlings of *D. rubidus* to reach sexual maturity is 51 days, and the mean rate of cocoon production is 0.2–0.4 cocoons earthworm^{-1} day^{-1} (Bengtsson et al. 1986; Cluzeau and Fayolle 1989; Elvira et al. 1996). Hatching success for cocoons is 85%, with a mean incubation time of 22 days and an average of 1.7 hatchlings (between 1 and 3) emerging from each cocoon (Elvira et al. 1996). *Dendrodrilus rubidus* can complete its life cycle within 75 days, and according to Cluzeau and Fayolle (1989), one of the factors that contribute to the high reproduction rate of this species is that its reproduction may be biparental, amphimitic, or uniparental, either by parthenogenesis or by self-fertilization.

3 Dendrobaena veneta (*Rosa 1886*)

This species is a large earthworm that can also survive in soil with potential for use in vermiculture; although it is not very prolific, it grows very rapidly (Edwards

1988; Viljoen et al. 1991). A number of commercial vermicomposting companies use this species, and it may have particular potential for protein-production systems and for breeding for use in field soil improvement.

Dendrobaena veneta (also sometimes called *Eisenia hortensis*) is a robust earthworm that can tolerate much wider moisture ranges than many other species and has a preference for mild temperatures 15°C–25°C (59°F–68°F). Its life cycle can be completed in 100–150 days, and 65 days is the average time to reach sexual maturity. Mean cocoon production has been reported as 0.28 day^{-1}, but the hatching viability seemed low (20%), and the mean period of cocoon incubation period is 42 days. The mean number of earthworms hatching from each viable cocoon is about 1.1 (Lofs-Holmin 1986; Viljoen et al. 1991, 1992; Muyima et al. 1994).

4 Lumbricus rubellus (*Hoffmeister 1843*)

Lumbricus rubellus is usually found in moist soils, particularly those to which animal manures or sewage solids have been applied (Cotton and Curry 1980). In surveys of commercial earthworm farms in the United States, Europe, and Australia, earthworms sold under the name *L. rubellus* were usually always *E. fetida* or *E. andrei*.

There are little data about its moisture and temperature requirements and preferences, although it is known that it clearly prefers moist conditions and can survive cold temperatures well. The optimal temperature for growth is 18°C (64.4°F), and suboptimal temperatures are less harmful than supraoptimal ones. *L. rubellus* has a relatively long life cycle (120–170 days) with a slow growth rate and a long maturation time (74–91 days). The mean cocoon production rate varies from 0.07 to 0.25 cocoons earthworm^{-1} day^{-1}, and hatching viability is 60%–70%. After an incubation period of 35–40 days, one single earthworm emerges from each cocoon (Cluzeau and Fayolle 1989; Elvira et al. 1996). The low maturation and reproductive rate indicate that it is not ideal for use in vermicomposting, although its size and vigor could make it of potential interest as fish bait or for land-improvement purposes.

5 Drawida nepalensis (*Michaelsen 1907*)

This is a temperate earthworm species not widely utilized in vermicomposting although it has affinity for organic matter and shows some suitable characteristics for vermiculture. Although this species has a lower growth rate and produces fewer cocoons and hatchlings per cocoon than most of the other vermicomposting species, it has a relatively short life cycle (100–120 days) and can reproduce, like *E. fetida*, without mating. The mean time to sexual maturity is 34–42 days, and the cocoon production is 0.15 cocoons worm^{-1} day^{-1}. Its hatching viability is 75%–88%, and the mean incubation time is 23.6 days. The mean number of young earthworms hatching from viable cocoons is 1.93 (Kaushal and Bisht 1992, 1995).

B Tropical Species

1 Eudrilus eugeniae (*Kinberg 1867*)

This earthworm species is native to Africa, but it has been bred extensively in the United States, Canada, and elsewhere for the fish bait market where it is commonly called the "African night crawler." It is a large earthworm that grows extremely rapidly, is reasonably prolific, and under optimum conditions can be considered as ideal for production of animal feed protein. Its main disadvantages are its narrow temperature tolerance and sensitivity to handling. *E. eugeniae* has high-reproduction rates (Bano and Kale 1988; Edwards 1988) and is capable of decomposing large quantities of organic wastes quickly and incorporating them into the topsoil (Neuhauser et al. 1979, 1988; Edwards 1988).

It shows preference for high temperatures, with maximum biomass production occurring at 25°C–30°C (77°F–86°F), while the growth rates were very low at 15°C (59°F) (Loehr et al., 1985; Viljoen and Reinecke 1992; Domínguez et al. 2001). Its use in outdoor vermiculture may therefore be limited to tropical and subtropical regions, unless winter temperatures are controlled. It can tolerate moisture contents between 70% and 85%, the optimum being 80%–82%. Domínguez et al. (2001) reported that individuals continued to increase in weight with virtually no mortality for 22 weeks. Reinecke et al. (1992) reported continuous growth and maximum weight up to 21 weeks at 25°C (77°F). The life cycle of *E. eugeniae* ranges from 50 to 70 days, and its life span can be 1–3 years. Sexual maturity is attained within 40–49 days, and a week after this period the individuals start to lay cocoons, between 0.42 and 0.51 cocoons day^{-1} (Viljoen and Reinecke 1989; Reinecke et al. 1992; Reinecke and Viljoen 1993; Domínguez et al. 2001). This is, together with *E. fetida* and *E. andrei*, a more rapid rate of development than for any other species of earthworm that has been reported to date and makes a very fast rate of population multiplication possible. The cocoon incubation period ranges from 12 to 16 days, and hatching success from 75% to 84%, with the mean number of earthworms per cocoon between 2 and 2.7 (Viljoen and Reinecke 1989; Reinecke et al. 1992; Reinecke and Viljoen 1993; Domínguez et al. 2001).

2 Perionyx excavatus (*Perrier 1872*)

Perionyx excavatus is an earthworm commonly found over a large area of tropical South Asia (Stephenson 1930; Gates 1972) although it has also been transported to Europe and North America. This is an epigeic species that lives solely in organic wastes, and high-moisture contents and adequate amounts of suitable organic material are required for populations to become fully established and to process organic wastes efficiently.

This tropical earthworm is extremely prolific, and it is almost as easy to handle as *E. fetida* and very easy to harvest. Its main drawback is its inability to withstand low-temperature conditions, but for tropical conditions it seems an ideal species. Although it has a shorter maturation and incubation time than *E. eugeniae*, its fecundity is higher. It is a very common species in Asia and is used in vermiculture in India, the Philippines, and Australia. The life cycle and the potential of this species for breaking down organic wastes have been documented by various authors under controlled conditions (Kale et al. 1982; Reinecke and Hallatt 1989; Hallatt et al. 1990; Reinecke et al. 1992; Hallat et al. 1992; Edwards et al. 1998). *P. excavatus* does not grow much at low temperatures although it can survive them 4°C (39.2°F), but it is less susceptible to high temperatures over 30°C (86°F) than *E. eugeniae*. Even in tropical areas, *P. excavatus* does not grow during low-winter temperatures but can survive the high-summer temperatures, whereas *E. eugeniae* has a much narrower tolerance range for temperature and cannot survive either the extreme low winter or the high summer temperatures. The life cycle of *P. excavatus* takes 40–50 days. Sexual maturity is attained within 20–28 days, and the mean cocoon production is 2.8 cocoons earthworm^{-1} day^{-1}, the mean incubation time of cocoons at 25°C (77°F) is 18 days, the hatching success is high (85%–90%), and usually only one hatchling emerges from each cocoon (Kale et al. 1982; Reinecke and Hallatt 1989; Hallatt et al. 1990, 1992; Reinecke et al. 1992; Edwards et al. 1998).

3 Polypheretima elongata (*Perrier 1872*)

This tropical earthworm species has been tested for the treatment of solid organic wastes, including municipal and slaughterhouse wastes; human, poultry, and dairy manures; and mushroom compost in India. A project in India using this species claimed to have a commercially viable facility for the "vermistabilization" of 8 tons (7.8 t) of solid wastes day^{-1}; they developed a "vermifilter" (packed with vermicompost and live earthworms) that produces reusable water from sewage sludge, manure slurries, and organic wastewater from food processing. *P. elongata* appears to be restricted to tropical regions and may not survive severe winters. There seem to be no detailed data about its life cycle available in the literature.

III INFLUENCE OF ENVIRONMENTAL FACTORS ON SURVIVAL AND GROWTH OF EARTHWORMS

Cocoon production, rates of development, and growth of earthworms are all critically affected by environmental conditions. Epigeic earthworms are relatively tolerant to the environmental conditions of the organic wastes, so quite simple low management windrow or bed systems have been used extensively to process these wastes. However, it has been demonstrated clearly that these earthworms have well-defined limits of tolerance to environmental parameters, such as moisture and temperature, and that the wastes are processed much more efficiently within a relatively narrow range of favorable chemical and environmental conditions. If these limits are greatly exceeded, earthworms may move to more suitable zones in the waste, leave it, or die, so that these wastes are processed very slowly.

A Temperature

Earthworms have fairly complex responses to changes in temperature. Neuhauser et al. (1988) studied the potential of several earthworm species to grow in sewage sludge, and they concluded that all these species have a range of preferred temperatures for growth, ranging between 15°C (59°F) and 25°C (77°F). In their studies, cocoon production was more restricted by temperature than growth, and most of the cocoons were laid at 25°C (77°F). Edwards (1988) studied the life cycle and optimal conditions for survival and growth of *E. fetida*, *D. veneta*, *E. eugeniae*, and *P. excavatus*. Each of these four species differed considerably in terms of response and tolerance to different temperatures. The optimum temperature for *E. fetida* was 25°C (77°F), and its temperature tolerance was between 0°C (32°F) and 35°C (95°F). *Dendrobaena veneta* had a rather low-temperature optimum and rather less tolerance to extreme temperatures. The optimum temperatures for *E. eugeniae* and *P. excavatus* were around 25°C (77°F), but they died at temperatures below 9°C (48.2°F) and above 30°C (86°F). Optimal temperatures for cocoon production were much lower than those more suitable for growth for these species.

Temperatures below 10°C (50°F) generally result in reduced or little feeding activity; and below 4°C (39.2°F), cocoon production and development of young earthworms ceases completely. In extreme temperature conditions earthworms tend to hibernate and migrate to deeper layers of the windrow for protection. Earthworms can also acclimate to temperature in autumn and survive the winter, but they cannot survive long periods under freezing conditions unless they are in protective cells. The unfavorable effect of high temperatures (above 30°C (86°F)) on most species of earthworms is not entirely a direct effect because these warm temperatures also promote chemical and microbial activities in the substrate, and the increased microbial activity tends to consume the available oxygen, with negative effects on the survival of earthworms.

B Moisture Content

There are strong relationships between the moisture content of organic wastes and the growth rate of earthworms. In vermicomposting systems, the optimum range of moisture contents for most species has been reported to be between 50% and 90%. *Eisenia fetida* and *E. andrei* can survive in moisture ranges between 50% and 90%, but they grow more rapidly between 80% and 90% in organic wastes (Edwards 1988; Domínguez and Edwards 1997).

C PH

Most epigeic earthworms are relatively tolerant to pH and can tolerate pH levels of 5–9, but when given a choice in the pH gradient, they move toward the more acid material, with a pH preference of 5.0.

D Aeration

Earthworms lack specialized respiratory organs, and oxygen and carbon dioxide diffuse through their body wall. Thus, earthworms are very sensitive to anaerobic conditions. *E. fetida* have been reported to migrate in high numbers from a water-saturated substrate in which oxygen has been depleted, or in which carbon dioxide or hydrogen sulfide has accumulated.

E Ammonia and Salts

Earthworms are very sensitive to ammonia and cannot survive in organic wastes containing high levels of this cation (e.g., fresh poultry litter). They also die in wastes with large quantities of inorganic salts. Both ammonia and inorganic salts have very sharp cutoff points between toxic and nontoxic (<1 mg· g^{-1}(0.016.1b^{-1}) of ammonia and <0.5% salts) (Edwards 1988). However, organic wastes containing high levels of ammonia can become acceptable after the removal of ammonia by a period of precomposting or by leaching with water.

Outside the limits of these environmental parameters, both earthworm activity and the rates of processing of the organic wastes decrease dramatically; for maximum vermicomposting efficiency, wastes should be preconditioned to make them suitable for earthworms. The optimal conditions for breeding *E. fetida* and *E. andrei* are summarized in Table 3.2. These characteristics do not differ too much from those suitable for other species (Edwards 1988).

In addition of environmental conditions, earthworm population density affects rates of earthworm growth and reproduction. Even when the physical–chemical characteristics are ideal for vermicomposting, problems can develop due to overcrowding. Reinecke and Viljoen (1990) in studies with *E. fetida* reared in cow manure and Domínguez and Edwards (1997) studying the growth and reproduction of *E. andrei* in pig manure found that, when grown at different population densities, the earthworms in the crowded containers grew more slowly and with a lower final bodyweight, although the total weight of earthworm biomass produced per unit of waste was greater. Maturation rate was also affected by the stocking rate; thus, earthworms of the same age developed the clitellum at different times in the different stocking rates.

Table 3.2 Optimal Conditions for Breeding *E. fetida* and *E. andrei* in Organic Wastes

Condition	Requirements
Temperature	15°C–20°C (limits 4°C–30°C) (59–68°F) (limits 39–86°F)
Moisture content	80%–90% (limits 60%–90%)
Oxygen	Aerobicity
Ammonia content of the waste	Low: < 1 mg·g^{-1} (0.016 oz.1b^{-1})
Salt content	Low: < 0.5%
pH	5–9

IV PREDATORS, PARASITES, AND PATHOGENS OF EARTHWORMS

Earthworms can be the target for a wide range of vertebrate predators and are attacked by several parasites and pathogens. Earthworms are an important component of the diet of many vertebrate predators. They are preyed on by many species of birds and mammals. Centipedes, ants, carabid and staphylinid beetles, and their larvae also prey on earthworms.

Earthworms have many internal parasites, including Protozoa, Platyhelminthes, Rotatoria, nematodes (Nematoda), and fly larvae. Bacteria, such as *Spirochaeta* sp. and *Bacillus* sp., and fungal pathogens have been reported to parasitize earthworms, although little is known of their effects on their hosts. The most common protozoan parasites of earthworms belong to Gregarina. These protozoa have been found in many different parts of the body of earthworms, including the alimentary tract, coelom, blood system, testes, spermathecae, seminal vesicles, and even inside the cocoons. A number of ciliate protozoa and platyhelminth worms also infest the bodies of earthworms although few cause them any serious harm. There are several instances of platyhelminth worms being found in the bodies of earthworms. Many nematodes occur in the tissues of earthworms; few seem to cause serious damage, and often the earthworm is merely acting as an intermediate host for them. Some nematodes are carried passively by earthworms. Others develop within the earthworm, and some are true parasites, with the earthworm being the sole host. Some mites feed on and destroy the developing earthworm cocoon.

ACKNOWLEDGMENT

This research was partially supported by a Xunta de Galicia grant (07MRU023383PR) and the Spanish Ministerio de Ciencia e Innovaciòn (CTM2009-08477).

REFERENCES

Bano, K., and Kale, R. D. 1988. Reproductive potential and existence of endogenous rhythm in reproduction of earthworm *Eudrilus eugeniae*. *Proc. Zool. Soc.* (Calcutta) 38, 9–14.

Bengtsson, G., Gunnarsson, T., and Rundgren, S. 1986. Effects of metal pollution on the earthworm *Dendrobaena rubida* (Sav.) in acidified soils. *Water Air Soil Pollut.* 28, 361–383.

Bouché, M. B. 1977. Strategies lombriciennes. In *Soil Organisms as Components of Ecosystems* (eds. U. Lohm and T. Persson). *Biol. Bull.* 25, 122–132.

Cluzeau, D., and Fayolle, L. 1989. Croissance et fécondité compareés de *Dendrobaena rubida tenuis* (Eisen, 1874), *Eisenia andrei* (Bouché, 1972) et *Lumbricus rubellus rubellus* (Hoffmeister, 1843) (Oligochaeta, Lumbricidae) en élevage contrôlé. *Rev. Ecol. Biol. Sol.* 26, 111–121.

Cotton, D. C. F., and Curry, J. P. 1980. The response of earthworm populations (Oligochaeta, Lumbricidae) to high applications of pig slurry. *Pedobiologia* 19, 425–438.

Domínguez, J., Briones, M. J. I., and Mato, S. 1997. Effect of diet on growth and reproduction of *Eisenia andrei* (Oligochaeta, Lumbricidae). *Pedobiologia* 41, 566–576.

Domínguez, J., and Edwards, C. A. 1997. Effects of stocking rate and moisture content on the growth and maturation of *Eisenia andrei* (Oligochaeta) in pig manure. *Soil Biol. Biochem.* 29, 743–746.

Domínguez, J., Edwards, C. A., and Ashby, J. 2001. The biology and ecology of *Eudrilus eugeniae* (Kinberg) (Oligochaeta) bred in cattle wastes. *Pedobiologia* 45, 341–353.

Domínguez, J., Ferreiro, A., and Velando, A. 2005. Are *Eisenia fetida* (Savigny, 1826) and *Eisenia andrei* Bouché, 1972 (Oligochaeta, Lumbricidae) different biological species? *Pedobiologia* 49, 81–87.

Edwards, C. A. 1988. Breakdown of animal, vegetable and industrial organic wastes by earthworms. In *Earthworms in Waste and Environmental Management*, ed. C. A. Edwards and E. F. Neuhauser, 21–31. SPB, the Hague, the Netherlands.

Edwards, C. A., Domínguez, J., and Neuhauser, E. F. 1998. Growth and reproduction of *Perionyx excavatus* (Perr.) (Megascolecidae) as factors in organic waste management. *Biol. Fertil. Soils* 27, 155–161.

Elvira, C., Domínguez, J., and Mato, S. 1996. The growth and reproduction of *Lumbricus rubellus* and *Dendrobaena rubida* in cow manure. Mixed cultures with *Eisenia andrei*. *Appl. Soil Ecol.* 5, 97–103.

Gates, G. E. 1972. Burmese earthworms. An introduction to the systematics and biology of megadrile oligochaetes with special reference to Southeast Asia. *Trans. Am. Philos. Soc. N. S.* 62, 1–326.

Hallatt, L., Reinecke, A. J., and Viljoen, S. A. 1990. The life cycle of the oriental compost worm *Perionyx excavatus* (Oligochaeta). *S. Afr. J. Zool.* 25, 41–45.

Hallatt, L., Reinecke, A. J., and Viljoen, S. A. 1992. Moisture requirements in the life cycle of *Perionyx excavatus* (Oligochaeta). *Soil Biol. Biochem.* 24, 1333–1340.

Hartenstein, R., Neuhauser, E. F., and Kaplan, D. L. 1979. Reproductive potential of the earthworm *Eisenia foetida*. *Oecologia* 43, 329–340.

Kale, R. D., Bano, K., and Krishnamoorthy, R. W. 1982. Potential of *Perionyx excavatus* for utilizing organic wastes. *Pedobiologia* 23, 419–426.

Kaushal, B. R., and Bisht, S. P. S. 1992. Growth and cocoon production of *Drawida nepalensis* (Oligochaeta). *Biol. Fertil. Soils* 14, 205–212.

Kaushal, B. R., and Bisht, S. P. S. 1995. Growth and cocoon production by the earthworm *Drawida nepalensis* (Oligochaeta: Moniligastridae) in oak and pine litter. *Pedobiol.* 39, 417–422.

Loehr, R. C., Neuhauser, E. F., and Malecki, R. 1985. Factors affecting the vermi-stabilization process. Temperature, moisture content and polyculture. *Water Res. Technol.* 19, 1311–1317.

Lofs-Holmin, A. 1986. Processing of municipal sludges through earthworms (*Dendrobaena veneta*). *Swedish J, Agric. Res.* 16, 67–71.

Monroy, F., Aira, M., Domínguez, J., and Velando, A. 2006. Seasonal population dynamics of *Eisenia fetida* (Savigny, 1826) (Oligochaeta, Lumbricidae) in the field. *C.R. Biol.* 329, 912–915.

Muyima, N. Y. O., Reinecke, A. J., and Viljoen-Reinecke, S. A. 1994. Moisture requirements of *Dendrobaena veneta* (Oligochaeta), a candidate for vermicomposting. *Soil Biol. Biochem.* 26, 973–976.

Neuhauser, E. F., Kaplan, D. L., and Hartenstein, R. 1979. Life history of the earthworm *Eudrilus eugeniae*. *Rev. Ecol. Biol. Sol.* 16, 525–34.

Neuhauser, E. F., Loehr, R. C., and Malecki, M. R. 1988. The potential of earthworms for managing sewage sludge. In *Earthworms in Waste and Environmental Management*, ed. C. A. Edwards and E. F. Neuhauser, 9–20. SPB Academic Publishing, the Hague, the Netherlands.

Pérez-Losada, M., Eiroa, J., Mato, S., and Domínguez, J. 2005. Phylogenetic species delimitation of the earthworms *Eisenia fetida* (Savigny, 1826) and *Eisenia andrei* Bouché, 1972 (Oligochaeta, Lumbricidae) based on mitochondrial and nuclear DNA genes. *Pedobiologia* 49, 317–324.

Reinecke, A. J., and Hallatt, L. 1989. Growth and cocoon production of *Perionyx excavatus* (Oligochaeta). *Biol. Fertil. Soils* 8, 303–306.

Reinecke, A. J., and Viljoen, S. A. 1990. The influence of worm density on growth and cocoon production of the compost worm *Eisenia fetida* (Oligochaeta). *Rev. Ecol. Biol. Sol* 27 (2), 221–230.

Reinecke, A. J., Viljoen, S. A., and Saayman, R. J. 1992. The suitability of *Eudrilus eugeniae*, *Perionyx excavatus* and *Eisenia fetida* (Oligochaeta) for vermi-composting in Southern Africa in terms of their temperature requirements. *Soil Biol. Biochem.* 24, 1295–1307.

Reynolds, J. W., and Wetzel, M. J. 2004. *Nomenclatura Oligochaetologica. Supplementum Quartum. A catalogue of names, descriptions and type specimens of the Oligochaeta.* Illinois Natural History Survey Special Publication.

Stephenson, J. 1930. *The Oligochaeta.* Oxford University Press, Oxford, U.K.

Viljoen, S. A., and Reinecke, A. R. 1989. Life-cycle of the African nightcrawler, *Eudrilus eugeniae* (Oligochaeta). *S. Afr. J. Zool.* 24, 27–32.

Viljoen, S. A., and Reinecke, A. R. 1992. The temperature requirements of the epigeic earthworm species *Eudrilus eugeniae* (Oligochaeta). A laboratory study. *Soil Biol. Biochem.* 24, 1345–1350.

Viljoen, S. A., Reinecke, A. J., and Hartman, L. 1991. Life-cycle of the European compost worm *Dendrobaena veneta* (Oligochaeta). *S. Afr. J. Zool.* 26, 43–48.

Viljoen, S. A., Reinecke, A. J., and Hartman, L. 1992. The temperature requirements of the epigeic earthworm species *Dendrobaena veneta* (Oligochaeta)- a laboratory study. *Soil Biol. Biochem.* 24, 1341–1344.

Watanabe, H., and Tsukamoto, J. 1976. Seasonal change in size, class and stage structure of lumbricid *Eisenia foetida* population in a field compost and its practical application as the decomposer of organic waste matter. *Rev. Ecol. Biol. Sol* 13, 141–146.

CHAPTER 4

Discovery and Development of New Species for Vermiculture

Samuel W. James and Afrânio Guimaraẽs

CONTENTS

I Introduction 41
II Why Diversify in Earthworm Species? 42
III How New Earthworm Species for Vermicomposting or Culture Have Been Identified 44
IV Different Species of Earthworms Suitable for Vermiculture 45
A *Dichogaster annae* (Horst 1893) 46
B *Amynthas gracilis* (Kinberg 1867) 48
C *Metaphire schmardae* (Horst 1883) 50
Acknowledgment 51
References 51

I INTRODUCTION

Throughout the world, vermiculture and vermicomposting are growing in importance for various economic and ecological purposes, details of which are provided elsewhere in this volume. Though the applications are diverse, the collection of earthworm species involved is not. Most vermiculture operations use one of four species, *Eisenia fetida* (Savigny 1826), *Eisenia andrei* (Bouché 1972), *Perionyx excavatus* (Perrier 1872), or *Eudrilus eugeniae* (Kinberg 1867). These all belong to the ecological category of species of epigeic earthworms (Bouché 1977) inhabiting purely organic matter microhabitats in nature. A few other species are cultured on smaller scales and/or in a more restricted geographical extent, such as *Lampito mauritii* (Kinberg 1867), *Dendrobaena veneta* (Rosa 1886), or *Dendrobaena hortensis* (Michaelsen 1890).

In spite of this very small number of species, many people involved in rearing earthworms do not have a clear idea which earthworm species exist in their culture beds. This taxonomic ignorance makes the vermiculture industry susceptible to market manipulation by similarly ignorant salesmen or, even worse, by persons clever enough to exploit the unschooled customer. This situation is entirely avoidable by simple methods of earthworm identification and perhaps by the creation of a certification system to ensure the identity and purity of cultures. Although certification may seem unnecessary at present, if more species are added to the brief list already mentioned, the diversity might overtax the ability of nontaxonomists to identify their earthworms, or the earthworms that they wish to purchase, correctly.

If the industry moves toward using a greater diversity of cultured earthworm species, serving a greater diversity of end uses and markets, correct identification will become more important. An example of this is the medicinal leeches of Europe, in which patents were based on incorrectly identified specimens (Siddall et al. 2007). This was discussed with respect to earthworms by James (2009). DNA bar coding (Hebert et al. 2003; http://www.barcoding.si.edu/DNABarCoding.htm) of earthworms is now known to work quite well (Chang et al. 2008; King et al. 2008; Rougerie et al. 2009), which is to say that the intraspecific divergence of the cytochrome oxidase I bar code region is considerably less than the interspecific divergence (0%–6% and 8%–20%, respectively). Most of the commonly cultured earthworm species already have DNA bar code database entries, so it is feasible to identify an unknown species by molecular methods. As new species are used by the industry, they should also be added to the DNA bar code database for the protection of organizations introducing a new cultured species and for the protection of people who would purchase these earthworms or products derived from them.

Diversification of the group of cultured earthworms currently used may seem unnecessary from some perspectives but a highly laudable goal to others. In this Chapter we present the case for the latter, while acknowledging the validity of arguments for the former. We also present some interesting newly-cultured species, how the proper conditions for culture were discovered, and why those species were chosen for culturing.

II WHY DIVERSIFY IN EARTHWORM SPECIES?

We offer the following motivations for finding new earthworm species for vermiculture: the reduction of risks associated with monoculture, broadening of the substrates usable in vermicomposting, and culture of earthworms usable for purposes other than the traditional ones of vermicompost production and fish bait. In the last aspect we can envision such uses as feeds for pets or aquarium fishes, a use already in effect in Brazil; the use of earthworms as model organisms or as test organisms in environmental monitoring; and the production of pharmaceutical compounds or preparations.

Monocultures carry the risk that a pest or disease will spread rapidly to attack the monocultured species, and a narrow genetic base will not provide resistant genotypes able to weather the crisis. Examples from plant production abound. It is even clear that monocultures facilitate the evolution of new pests and diseases (Strong 1979; Rey et al. 1981) and that the greater the extent of cultivation of a particular crop, the greater the number of new problems that can arise. This occurs where novel plant types are introduced to areas with no history of cultivation of that crop and there are no relatives of that crop from which indigenous pests can jump (Strong 1979). Although in such cases, there is usually a period of freedom from problems, eventually some pest finds the new resource. Vermiculture appears to be in the preproblem state, regardless of earthworm species or regional location of culture beds. This could be because few diseases affect earthworms generally, though we know rather little about this aspect save for a few parasites, or because the extent of the monoculture(s) is still quite small, or because the cultures are physically isolated from one another, unlike the cultivation of an important crop. The latter two factors, if truly related to the present lack of problems, will change if vermiculture becomes as significant a technology as it should be. The considerable experience of this Chapter's second author in vermiculture indicates that pest and disease problems are almost always due to management errors by the human operators of vermicultures. However, for the time being, there is no definitive evidence of monoculture-related problems to motivate a push for diversification. While we cannot predict that problems will arise, we can at least offer our suggestions for what do to in order to prevent or cure the problems, should they arise.

Given that we know little about the generalist or specialist nature of parasites and diseases affecting cultured earthworms, we should attempt to find the greatest possible diversity of culturable species, looking at diversity on a genus or family scale. To find additional potential vermicomposting species, one could explore many species of *Eisenia*, for example. Many *Eisenia* species inhabit the same natural microhabitat (under bark of fallen trees, other organic accumulations of litter) as *E. fetida* and *E. andrei*, so it is reasonable to hypothesize that they will be culturable under similar artificial conditions.

Cryptic species, or underdescribed species, are known within many Lumbricidae (Jaenike 1982; King et al. 2008; Rougerie et al. 2009), including *E. fetida* (Perez-Losada et al. 2005). We should be aware that it is possible that other so-called earthworm species may turn out to consist of two or more species, using a general lineage species concept (de Queiroz 2007) or the biological species concept. In such cases, the actual diversification gained may amount to zero such as the discovery by Perez-Losada et al. (2005) of a new cryptic lineage within *E. fetida* (not *E. andrei*, but a third, unnamed species), which simply identified a lineage within cultured populations. This third species could be advantageous to the industry if it has different cultural requirements or resistance to diseases than its named relatives, but this is not yet known. Additionally, we have detected some minor morphological variation within *Perionyx* populations cultured in various locations, suggesting that not everything assumed to be *P. excavatus* actually is this species.

In contrast, if one wanted to find another species susceptible to the same problems as the known species, the cryptic species or other congeners would be the choice. It is probably better to look in other genera rich in epigeic species, such as

Dendrobaena and *Eisenia* among the European Lumbricidae (Csuzdi and Zicsi 2003), plus the North American genus *Bimastos*, arboreal dichogastrine species (Acanthodrilidae) of the northern Neotropics and equatorial Africa, arboreal and histisol-dwelling Brazilian Ocnerodrilidae (James and Brown 2006), tree-trunk and histisol-dwelling *Kynotus* from Madagascar, at least 50 species of *Perionyx* (Blakemore 2006) and other megascolecid earthworms of comparable habitats in the humid tropical parts of South and Southeast Asia, and various other epigeic species from other places and in other families. Some examples from Australia are *Anisochaeta buckerfieldi* (Blakemore 1997), *Anisochaeta tenax* (Fletcher 1886), and *Anisochaeta dorsalis* (Fletcher 1887; Blakemore 2006).

Besides using diversification to avoid the perils of large monocultures, there is also diversification for exploiting new opportunities. Different feedstocks could be used from various organic waste streams, if suitable earthworms could be found to consume them. The bait market could adapt to regional preferences of sport fishing, and pharmaceutically useful worms could be cultured. All of these possibilities have been discussed in James (2009) and need not be repeated in detail here. In most imaginable cases there will always be vermicompost as a by-product of growing earthworms, so the question comes down to finding earthworms suitable for the other uses and developing appropriate culture practices.

III HOW NEW EARTHWORM SPECIES FOR VERMICOMPOSTING OR CULTURE HAVE BEEN IDENTIFIED

One effective method of discovering earthworms suitable for vermicomposting, and potentially for the diversification of compostable substrates, is to explore compost heaps, manure piles, and other organic refuse concentrations in places where earthworms can have access to them. Mindy Jaffe (Waikiki Worms) has done this on the Hawaiian island of Oahu (in litt.). Her results have not yet provided any breakthroughs, but she has found several species, among them *Gordiodrilus elegans* (Beddard 1892) and *Dichogaster* spp., living in organic matter. In most cases more than one species was found per location, raising another question: Would there be any benefit to the final product of vermicomposting to have a community of two or more earthworm species involved in the process? Further exploration of "open-access" vermicomposting seems worthwhile. All are subject to the randomness of the colonization process, which is accentuated by the fact that in nearly all cases the earthworms involved are nonnative species.

Only where the vermicompost sites are near areas still inhabited by native epigeic earthworms could one hope to discover new vermicomposters from the indigenous earthworm fauna. One way to do this would be to create vermicompost pile "traps" in natural areas, leave them for long periods, and sample them periodically for earthworms. This saves the trouble of catching sufficient numbers of earthworms and experimenting with culture techniques, and it can be done by any villager willing to haul the necessary materials some distance into a forest. Scientists can become involved in the second stage, that of working out the details of methods of culture for the earthworms that self-select for the vermicompostable materials available to the local human population. This brings us to the most difficult topic, the development of the best culture techniques. The scientific literature is relatively silent on the subject of failed attempts. These typically do not get published, so it is hard to benefit from others' mistakes, and we are perhaps doomed to repeat some of them. From what we have heard, endogeic earthworm species are typically hard to culture (see Chapter 22). On a more optimistic note, we can benefit from the success of those who have developed new culture methods for earthworms not previously raised in an organized fashion. The Brazilian enterprise Minhobox currently has 12 earthworm species under cultivation, most of them not previously cultured. This is the most diversified vermiculture enterprise of which we are aware, and points the way to the future. The second author of this chapter is the principal of this company. Here we describe the evolution of Minhobox, the process used to develop cultures of novel earthworm species, and several specific examples of culture methods.

IV DIFFERENT SPECIES OF EARTHWORMS SUITABLE FOR VERMICULTURE

Minhobox began with an unconventional attitude toward vermiculture, namely, the adoption of animal science techniques as taught in universities for ordinary breeding of livestock species. In its early days, the enterprise used the conventional technique of raising commonly cultured earthworms (*E. andrei* and *E. eugeniae*) on plots of soil covered with various types of organic residues. However, much more than transforming the organic residues from agricultural industries and animal raising into something ecological and profitable, Minhobox was motivated to refine the techniques of raising earthworms. There was also an interest in using endemic Brazilian species, in accordance with the location of the enterprise in Minas Gerais, Brazil.

After a few months of using conventional techniques, Minhobox encountered typical problems with the traditional vermiculture methods, which reduce the productivity and numbers of usable earthworms. In order to make cultures more efficient, in 1993 the second author of this chapter developed a technique of vertical vermiculture in boxes, which was named Minhobox (*minhoca*, Brazilian for "earthworm," plus box; also a pun on a type of small Brazilian roadside shop known as a *minibox*) and gave its name to the enterprise. This started to solve the problems of ordinary breeding in plots of ground, especially the otherwise-difficult process of separating the earthworms from the vermicompost. Later on, to respond to a larger demand by the market, a similar technique of vermiculture was developed in horizontally arrayed beds named Minhobed, utilizing the principal benefits of vertical methods. Besides separating the earthworms in a very practical way, both techniques do not need periodic rehydrations, protect the earthworms from predators, utilize the installation space more effectively, free the vermicompost from weeds, make handling the materials more practical, increase the profitability of the enterprise, impose rational controls on the production, and, most of all, allow the use of an uncommonly cultured species of earthworms.

The first to be introduced to these vermiculture methods were two less common earthworm species in the Brazilian market: *P. excavatus* and *E. fetida*. The search continued for alternative earthworm species that would provide new and important characteristics in the production of vermicomposts, vermicompost aqueous extracts, fish bait, earthworm flour, and live food and that did not present the restrictions and disadvantages of traditionally raised earthworms.

With a wider variety of species, including the alternative ones, one can adopt the most appropriate species in each situation according to the following criteria:

- The different climatic conditions—expanding vermiculture in various countries
- The kind of available raw material in the area or region where vermiculture will be set up
- The product of the vermiculture to be prioritized (vermicompost, fish bait, earthworm flour, or live food)
- The endemism, specifically in controlled inoculation in agricultural fields, to avoid the ecological harm of introducing an exotic species to a new site

Presently, besides the common species *E. andrei*, *E. eugeniae*, *P. excavatus*, and *E. fetida*, Minhobox raises several uncommon species that will be discussed in the following, plus *Octodrilus complanatus* and another not-yet-identified species of the Ocnerodrilidae family. Here we present informal case studies of three species in cultivation, *Dichogaster annae* (Horst 1893), *Amynthas gracilis* (Kinberg 1867), and *Metaphire schmardae* (Horst 1883).

A *Dichogaster annae* (Horst 1893)

The unpaved areas used for organic residue pretreatment at the Minhobox production facility allow access by creatures from the soil, including some species of earthworms. Among them, the detritivorous and cosmopolitan species *D. annae*, originally from the central part of the African continent and accidentally introduced into Brazil, among other places, is frequently found in the piles of animal manures undergoing precomposting. These adventitious earthworm populations were used as the breeding stock for experiments in rearing *D. annae*.

First, the starting stock of earthworms was taken to the enterprise laboratory for evaluation of their potential commercial uses and investigation of the best ways to breed them. For 3 years Minhobox studied the substrates and the most favorable temperatures for their development, the most appropriate enclosures for maintaining them, and the population density for maximizing their breeding compost production.

Adapted from *P. excavatus* vermiculture in enclosed horizontal beds, *D. annae* is cultured in boxes separated according to the stages of the life cycle. This is a significant innovation compared to ordinary vermiculture, in which all life stages are in one unit. One section is reserved for accommodating the parent colonies with population densities that maximize reproduction, equivalent to 3000 earthworms per 100 L (26.4 gal) of substrate. A short time before the completed incubation period of the cocoons resulting from the first round of copulations, the breeding-stock earthworms are directed to new stages of copulation, and the cocoons are taken to incubation sections to wait for the viable ones to hatch out. The hatchlings are then taken to development sections where they are selected according to the desired length.

This species was developed for the live food market for aquarium fishes. Commercially available live food for ornamental fish is rarely found in the market, and, if found in aquarium stores, it does not have a long shelf life. In addition, the majority of the food species are not always reared but are often captured in polluted streams, thus risking the survival of the fish in the aquariums. The most frequent commercially raised earthworm species, the red earthworms (*E. andrei* and *E. fetida*), present some disadvantages when served as live food for aquarium fishes: Besides being relatively long and thick when fully grown and too big to be swallowed by small fish such as betta, guppies, or neons, these earthworms naturally excrete a liquid with a repulsive smell and taste, similar to garlic, supposedly for protection against predators. Knowing the urgent demand for quality live food in the aquarium market, we tried to develop a breeding technique of an uncommon earthworm species that presented favorable characteristics for the nutrition of aquarium fish and can be packed for longer durations.

So, young *D. annae* was launched into the Brazilian aquarium market by Minhobox, packed in a substrate composed of sanitized cellulose polymers and supplied with a nutritious solution that keeps them hygienic, deodorized, and alive for at least 2 months in the package, with minimum care. Besides being a species that can be eaten by small fish and is equally palatable for marine and freshwater fish, the *D. annae* hatchling earthworms are naturally agitated immediately after manipulation, increasing their attractiveness to fish as live food. The people who have aquaria now have a nutritious and entertaining alternative in the stores to feed their fish: Besides exercising their mobility, the *D. annae* provides high protein in the diet, supplies natural pigment, and acts as a mineral and vitamin supplement, containing iron and vitamin B.

Being a detritivorous earthworm species and thus requiring a substrate rich in organic material, *D. annae* is an earthworm that also produces high-quality vermicompost with a peculiarity: Its excrement is composed of minute cylindrical grains. Consequently, even when this species is fully grown, its vermicompost has a smaller granulation compared to those produced by bigger species generally bred in captivity. Presuming that a smaller organic-matter granulation is related to a higher incorporation into the soil and higher absorption capacity by plant root systems, the *D. annae* vermicast, especially that originated in Minhobox enclosures populated by growing hatchlings, is characterized as an excellent organic soil amendment.

Dichogaster annae has some unusual qualities providing challenges to the vermiculture operator. It perceives changes in weather such as a coming cold front and has more intense reactions to a frontal system compared to the common species used in vermiculture. Under the typical characteristics of this weather phenomenon, such as a temperature drop followed by thunder and heavy rains, this earthworm emerges to the surface of the substrate and escapes out of control, wandering seemingly at random with no search for shelter. Escape behavior is expressed strongly during packing as live food for fish: The ones selected immediately try to crawl out of the uncovered jars. If *D. annae* is bred in areas where cold fronts are frequent, the boxes must have a cover with an efficient ventilator made of microholes and a good seal able to prevent the escape of the recently born earthworms, because the escaping behavior is common to all ages of earthworms.

The establishment of a *D. annae* culture must take into consideration the local climate, because during the colder months of the year or on days when the temperature drops below 15°C (59°F) they become less prolific and vermicompost production is reduced. However, this can be overcome with heating and other measures to ensure a stable and adequate temperature. Culture in stacked boxes is more exposed to temperature changes and therefore needs closer control of ambient temperature in the culture facility.

B *Amynthas gracilis* (Kinberg 1867)

This Asian earthworm has characteristics unlike other cultured species. It is epiendogeic and therefore does not produce vermicompost with the same content and effects on plants as the ordinarily cultured species such *Eisenia* spp., *E. eugeniae*, and *P. excavatus*. *Amynthas gracilis* has intense and variable iridescence over a greenish background color and is of a size reasonable for fish bait. It can also withstand summer and subtropical temperatures. This earthworm has the predictable behavior of thrashing around intensely when removed from its substrate, giving rise to its common names such as "ballerina," "jumper," "disco," and "crazy," for example. It moves rapidly on the ground in a snake-like manner, searching for cover and protection from predators or drying. This high level of activity makes it agitate when hooked, so that it becomes attractive to the fish. Another part of its predator-avoidance behavior is tail autotomy, or self-amputation of a portion of the tail, typically that part grasped by the predator or human handling the earthworm. However, by taking care to grasp the whole earthworm or take it by the head end, this can be avoided. Its durability on the hook is also considerable, maintaining the body whole without fragmentation even after dying and being repeatedly bitten by small fish.

If packed in a ventilated container that enables the entrance of fresh air and expelling of waste gases, and filled with the substrate in which the earthworms were raised, *A. gracilis* survives during transportation and fishing trips for at least 20 days. Simple maintenance such as periodically moistening the substrate, keeping the ventilation unobstructed, and maintaining the earthworms at moderate temperatures, will ensure and extend the survival of the packed bait earthworms. In Minhobox, the production of *A. gracilis* is performed with a method of vermiculture in plastic-enclosed horizontal beds (Minhobed) with some adaptations to conform to their preferred type of substrate, their life cycle, and their vertical migrations.

Although able to survive for a reasonable period of time in a substrate exclusively composed of organic residues, *A. gracilis* requires layers of mineral soil for better survival and reproductive efficiency. However, the process is more complex than in ordinary vermiculture. The culture of *A. gracilis* must be constituted of beds reserved for separately culturing three distinct phases of production: the reproduction by the breeding stock, the incubation of the cocoons, and the development of the hatchlings.

Considering these requirements, Minhobox raises this earthworm in beds filled with two layers of substrate separated by a screen: below, soil suitable to maintain this earthworm, and above, as sources of nourishment, a precomposted mix of animal- and vegetable-originated residues. This is the reproduction-bed content and construction. The final result of *A. gracilis* activity, in a bed designed for reproduction, is a mixture of the two layers with intermediate texture and coloration of both. Because the final priority is the multiplication of earthworms for sale, the population density that will generate the highest rate of copulation and cocoon production is the most important factor for success in this kind of vermiculture.

At the end of the reproduction phase, which coincides with the initial crop of cocoons nearing hatching, the upper layer, which is preferred by *A. gracilis* as the place to lay its cocoons must be removed, in preference to the egg-laying by *A. gracilis*, by lifting the screen and upper layer and taking it to the incubation bed. For the adults remaining in the lower layer, another fresh upper layer must be furnished, effectively resetting the conditions necessary for high rates of reproduction. It is important that this change be concluded before the cocoons hatch, not permitting hatchlings to mix with the adults, increase the population density, and thereby negatively affect reproduction. At the end of the incubation phase, which is observed by seeing the hatching of the majority of the cocoons, all the substrate that stores the recently hatched earthworms must be taken to the development beds, where they must be kept until harvesting, when they have reached the bait-worm characteristics. In the development beds, the contents are irrigated with permeable hoses. The ends of the beds are exposed to ensure ideal humidity during the longest phase of the rearing.

Being a cosmopolitan earthworm, widely distributed in the tropical, subtropical, and warmer temperate zones, *A. gracilis* can be used in controlled inoculation projects to improve soil conditions in agricultural fields. This species constantly emerges to the surface of the soil looking for food and then returns to the mineral soil, generating important burrows for water infiltration. They take their casts deeper into the soil, incorporating organic material in a faster and more efficient way than natural dislocation would. To promote earthworm proliferation in soils lacking earthworms, a culture bed near the farming place must be established, prioritized for reproduction. The difference between this purpose and the rearing of bait would consist of taking the incubation and development phases to the field, where the cocoons will be incubated and the hatchlings will grow in the field soil. Of course, if one wishes to work with larger numbers of breeding stock, then the earthworm-production process for bait should be followed until numbers are sufficient to quickly populate the area of the farm.

As a cautionary note, it is pointless to spread this cocoon and young earthworm bearing material on soil if the environmental conditions are not taken into consideration. Preferably, this earthworm species must already be tested in the particular agricultural environment, indicating its suitability to the specific conditions of the soil and cropping systems. Furthermore, it would be environmentally incorrect to introduce *A. gracilis* to a region where it does not already occur, because it is an invasive species.

Assuming these conditions are met, the incubation-bed material must be placed in small cavities near the plants, followed by deposition of organic material and frequent irrigation to ensure the development of the hatchlings and their future proliferation when they reach the reproductive stage. However, to ensure better success of the controlled earthworm inoculation, soil tillage must be limited, because *A. gracilis* is intolerant to disturbances in its habitat.

The resistance of this species to agricultural practices that utilize chemical products is relatively high. In a guava cultivation visited by the second author, in Ubá, a city in the southeastern region of Brazil, a high population of *A. gracilis* was living on the surface of the soil covered with fallen leaves and branches under the guava tree crowns. The survival of these earthworms was favored by periodic irrigations and by fertilization with vegetable residues originating from the pruning and hoeing of the corridors. Apparently the population was not harmed by fortnightly spraying of pyrethroid insecticides, which are relatively harmless to earthworms.

C *Metaphire schmardae* (Horst 1883)

Originally from East Asia, this species of earthworm is often found in the soils of the southeastern region of Brazil, where the founding specimens of Minhobox's stock were collected. They are often found with *A. gracilis* but in smaller populations. This observation motivated the enterprise to test its raising in a similar way to *A. gracilis*. As an epiendogeic species, *M. schmardae* should not be used as a vermicompost producer. However, this earthworm species presents some favorable characteristics for the fish bait industry, similar to *A. gracilis*: It lures the fish by thrashing about and by its attractive color, withstands packing in small bait containers, has a good size for fishing hooks, and does not easily fragment when bitten by smaller fishes.

Because it is very brightly colored and attractive in the adult stage, like other beautiful species of earthworms, *M. schmardae* has the potential to build the unexploited and promising market of pets and/or educational applications. Keeping them between transparent culture panels covered by removable opaque covers would, besides permitting observation of their development, reproduction, mixing of layers of soil, construction of galleries, and generation of excrement, allow the raiser to appreciate its beautiful pigmentation of bright green stripes.

According to the habitat of *M. schmardae* in the soil, Minhobox tested edaphic conditions corresponding to those of natural populations and attempted to maximize those characteristics for reproduction. In a general form, the culture conditions resemble those of the three life stages of *A. gracilis* but with some important differences. The substrate is composed of two layers in the reproduction and development boxes, separated by a screen: below, mineral soil collected where they live as in nature, and above, a thinner layer, essentially composed of organic matter. The boxes used for storing them are low and covered with a cover that is ventilated. The two layers are changed on different schedules, the top layer that nourishes the earthworms more frequently.

Irrigation is not applied to the reproduction boxes. The characteristics of the boxes and the short length of time adopted for the renovation of the substrates protect them from dehydration, even though the initial moisture content of 70% gently decreases.

After the reproduction time, the upper layer is removed for incubation; then, at hatching, it is placed in a box as the lower layer with more soil, and an upper organic-matter layer is added. Layers are replaced as necessary based on the earthworms' feeding activity. Development continues until earthworms are bait size or adult, and then they are separated and sold or used as breeding stock.

The techniques developed by Minhobox for rearing epiendogeic worms have considerable potential to expand vermiculture into new applications. Rearing of other species, such as the temperate-zone *Lumbricus rubellus* as fish bait, could be made more effective. The use of earthworms in ecotoxicology or other environmental roles, such as bioremediation, could be much more effective. The reliance on epigeic earthworm species in ecotoxicology is justified primarily because they are easy to culture, not because they are typical of ordinary soil conditions. The ability to culture soil-dwelling species effectively should improve the availability of earthworms in testing, because one would be able to conduct the tests in real soil environments.

Bioremediation using earthworms to rehabilitate damaged environments, such as reclamation of mine tailings, restoration of urban soils in depopulating cities of the northern United States, creation of populations in chemically treated agricultural land, or other applications, also requires soil-dwelling earthworms. The only way to do this at present is to gather earthworms from existing populations, a very labor-intensive task unlikely to achieve adequate economic returns. The ability to collect founding populations and to efficiently multiply them, as done by Minhobox, is a significant advance.

ACKNOWLEDGMENT

The senior author is grateful for the support of the U.S. National Science Foundation award DEB-0516439.

REFERENCES

Blakemore, R. J. 2006. *Cosmopolitan Earthworms—An Eco-Taxonomic Guide to the Peregrine Species of the World.* 2nd ed. VermEcology, Yokohama, Japan.

Bouché, M. B. 1977. Strategies Lombriciennes. *Ecol. Bull.* 25: 122–132.

Chang, C.-H., Rougerie, R., and Chen, J.-H. 2008. Identifying earthworms through DNA barcodes: Pitfalls and promise. *Pedobiologia* 52: 171–180.

Csuzdi, C., and Zicsi, A. 2003. *Earthworms of Hungary (Annelida: Oligochaeta, Lumbricidae).* Pedozoologica Hungarica No. 1. (Csuzdi, C. and Mahunka, S. eds.) Hungarian Natural History Museum, Budapest, Hungary.

de Queiroz, K. 2007. Species concepts and species delimitation. *Syst. Biol.* 56: 879–886.

Hebert, P. D. N., Cywinska, A., Ball, S. L., and deWaard, J. R. 2003. Biological identifications through DNA barcodes. *Proc. Royal Soc. London, B* 270: 313–321.

Jaenike, J. 1982. "Eisenia foetida" is two biological species. *Megadrilogica* 4: 6–8.

James, S.W. 2009. Taxonomy, biodiversity and new vermitechnology application. In (Eds) C.A. Edwards, R. Jeyaraaj., I.A. Jeyaraaj. *Proc. Indo-U.S. workshop on vermitechnology in Human Welfare 4–7 June 2007*, Kongunadu Science and Arts College, Coimbatore, India.

King, R. A., Tibble, A. L., and Symondson, W. O. C. 2008. Opening a can of worms: Unprecedented sympatric cryptic diversity within British lumbricid earthworms. *Mol. Ecol.* 17: 4684–4698.

Perez-Losada, M., Eiroa, J., Mato, S., and Dominguez, J. 2005. Phylogenetic species delimitation of the earthworms *Eisenia fetida* (Savigny, 1826) and *Eisenia andrei* Bouche, 1972 (Oligochaeta, Lumbricidae) based on mitochondrial and nuclear DNA sequences. *Pedobiologia* 49: 317–324.

Rey, J. R., McCoy, E. D., and Strong, D. R. 1981. Herbivore pests, habitat islands and the species/area relationship. *Am. Nat.* 6: 611–622.

Rougerie, R., Decaens, T., Deharveng, L., Porco, D., James, S.W., Chang, C.H., Richard, B., Popatov, M., Sudhardjonor and Herbert, P.D.N. 2009. DNA barcodes for soil animal taxonomy *Pesquisa Agropecuaria*, Brasileira, 44(8): 789–801.

Siddall, M. E., Trontelj, P., Utevsky, S. Y., Nkamany, M., and MacDonald, K. S. 2007. Diverse molecular data demonstrate that commercially available medicinal leeches are not *Hirudo medicinalis. Proc. Royal Soc. B* 274: 1481–1487.

Strong, D. R. 1979. Biogeographical dynamics of insect-host plant communities. *Annu. Rev. Entomol* 24: 89–119.

CHAPTER 5

The Microbiology of Vermicomposting

Jorge Dominguez

CONTENTS

I What is Vermicomposting? 53
II Vermicomposting Food Web 55
III The Process of Vermicomposting 55
IV Effects of Earthworms on Microbial Communities during Vermicomposting 56
A Microbial Biomass 57
B Bacterial and Fungal Growth 59
C Effects of Earthworms on the Activity of Microbial Communities 59
D Effect of Earthworms on Total Coliform Bacteria during Vermicomposting 61
E Effect of Earthworms on the Composition of Microbial Communities 63
V Conclusions 64
Acknowledgment 64
References 65

I WHAT IS VERMICOMPOSTING?

Although Darwin (1881) first drew attention to the great importance of earthworms in the decomposition of dead plants and the release of nutrients from them, it was necessary to wait more than a century until this was taken seriously as a field of scientific knowledge or even a real technology. Vermicomposting is a mesophilic bio-oxidative process in which detritivorous earthworms interact intensively with microorganisms and soil invertebrates within the decomposer community, strongly affecting decomposition processes, accelerating the stabilization of organic matter, and greatly modifying its physical and biochemical properties (Edwards and Bohlen 1996; Domínguez 2004; Edwards et al. 2004). Microorganisms produce the enzymes that cause the biochemical decomposition of organic matter, but earthworms are the crucial drivers of the process as they are involved in the indirect stimulation of microbial populations through fragmentation and ingestion of fresh organic matter, which results in a greater surface area available for microbial colonization, thus dramatically increasing microbiological activity. Earthworms also modify microbial biomass and activity through stimulation, digestion, and dispersion in the casts (Figure 5.1) and interact closely with other biological components of the vermicomposting system, thereby affecting the structure of the microflora and microfauna communities (Domínguez et al. 2003; Lores et al. 2006). Thus, the decaying organic matter in vermicomposting systems is a spatially and temporally heterogeneous matrix of organic resources with contrasting qualities that result from the different rates of degradation that occur during decomposition.

Vermicompost, the end product of vermicomposting, is a finely divided peat-like material with high porosity and water-holding capacity and a low C:N ratio; it contains many nutrients in forms that are readily taken up by plants. High rates of mineralization occur in the organic matter-rich earthworm casts, which greatly enhances the availability of inorganic nutrients, particularly ammonium and nitrates but also phosphorus, potassium, calcium, and magnesium for plants. Vermicompost also contains plant growth hormones produced by microorganisms and plant growth regulators such as humates, in the production of which microorganisms also play a role.

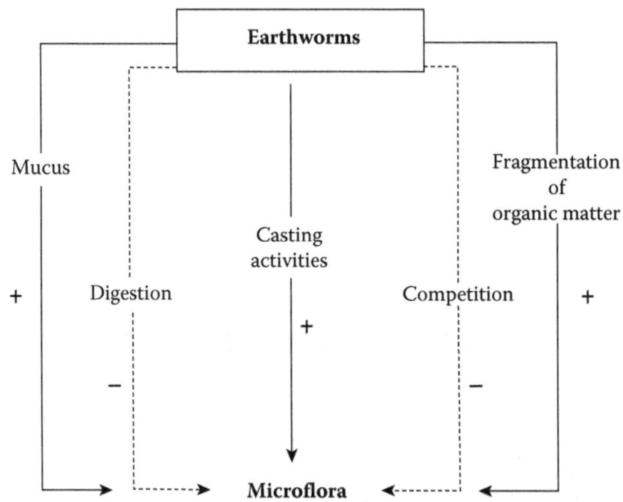

Figure 5.1 Positive (+) and negative (–) effects of earthworms on microbial biomass and activity.

II VERMICOMPOSTING FOOD WEB

Vermicomposting systems sustain a complex microbial and invertebrate food web that results in the recycling of organic matter and release of nutrients. Biotic interactions between decomposers (i.e., bacteria and fungi) and the soil fauna include competition, mutualism, predation, and facilitation, and the rapid changes that occur in both functional diversity and substrate quality are the main properties of these systems (Sampedro and Domínguez 2008). The most numerous and diverse members of this food web are microorganisms, although there are also abundant protozoa and many invertebrates of varying sizes, including nematodes, microarthropods, and large populations of earthworms (Monroy 2006; Sampedro and Domínguez 2008). These invertebrates cover a range of trophic levels—some feed primarily on microbes (bacteriovores and fungivores), on organic waste (detritivores), or on a mixture of organic matter and microbes (microbio-detritivores), whereas others feed on animals (carnivores) or across different trophic levels (omnivores); (Sampedro and Domínguez 2008).

The primary consumers of the vermicomposting food web are the microorganisms (mainly bacteria, fungi, and ciliates) that break down and mineralize organic residues. Microorganisms are the most numerically abundant and diverse members of the vermicomposting food web and include many thousands of different organisms. Secondary and higher-level consumers, that is, the soil invertebrates, including the earthworms, exist together with microbes, feeding on and dispersing them throughout the organic matter. As organic matter passes through the gizzard of the earthworms, it becomes finely ground prior to digestion. Endosymbiotic microbes produce extracellular enzymes that degrade cellulose and phenolic compounds, enhancing the degradation of ingested material; and the degraded organic matter passes out of the earthworm's body in the form of casts. As earthworms feed on decaying organic wastes, their burrowing and tunneling activities aerate the substrate and enable water, nutrients, oxygen, and microbes to move through it; their feeding activities increase the surface area of organic matter that microorganisms act on. As decomposers die, more food is added to the food web for other decomposers.

Earthworms accelerate decomposition processes during vermicomposting (Aira et al. 2006, 2007), but it is not clear from where they obtain their energy inputs (decaying organic matter, microorganisms, microfauna, or a combination of these). They may utilize different strategies ranging from nonselective substrate feeding to grazing, and they have the ability to shift between living and nonliving carbon sources (Domínguez et al. 2003).

III THE PROCESS OF VERMICOMPOSTING

The vermicomposting process includes two different phases involving the activity of earthworms, (a) an active phase during which earthworms process wastes, thereby modifying their physical state and microbial composition (Lores et al. 2006) and (b) a maturation-like phase marked by the displacement of the earthworms toward fresher layers of undigested waste, during which the microbes take over the decomposition of the earthworm-processed waste (Domínguez 2004; Lazcano et al. 2008). The duration of the active phase is not fixed, and it depends on the species and population densities of earthworms and the rates at which they ingest and process the wastes.

The effect of earthworms on the decomposition of organic waste during the vermicomposting process is, in the first instance, due to gut-associated processes (GAPs). These processes include all the modifications that the decaying organic matter and the microorganisms undergo during transit through the earthworms' intestines. These modifications include the addition of sugars and other substances, modification of the microbial diversity and activity, modification of the microfaunal populations, homogenization, and the intrinsic processes of digestion, assimilation, and production of mucus and excretory substances such as urea and ammonia, which constitute a readily assimilable pool of nutrients for microorganisms. Decomposition is also enhanced through the action of endosymbiotic microbes that reside in the earthworm gut. These microbes produce extracellular enzymes that can degrade cellulose and phenolic compounds, thereby further enhancing the degradation of ingested material. Other physical modifications of the substrate are caused by the burrowing activities of earthworms, including aeration and homogenization of the substrate, which also favor microbial activity and further decomposition (Domínguez 2004). The proximate activities of earthworms enhance the mineralization of both carbon and nitrogen in the substrate significantly, and such effects are in proportion to the earthworm population densities (Aira et al. 2008).

Upon completion of GAPs, the resultant earthworm casts undergo cast-associated processes (CAPs), which are more closely associated with the aging processes, the action of the microflora and microfauna presents in the substrate, and the physical modification of the egested materials. During these processes the effects of earthworms are mainly indirect and derived from the GAPs. It is important to note that in vermicomposting systems, earthworm casts are almost always mixed with material not ingested by the earthworms, and the final vermicompost consists of a mixture of the two different fractions. During this aging process, vermicompost reaches its optimum in terms of biological properties that promote plant growth and suppress plant diseases. Currently, there is insufficient information regarding when this optimum is achieved, how we can determine it in each case, and whether this optimum has some kind of expiration date. It is important to note that it is possible that the optimal quality may be achieved only in natural ecosystems built from a correct site-specific balance of soil, plants, microorganisms, macroorganisms including earthworms, and climate. However, it is not possible yet to determine easily when a vermicompost is optimal and thus, this can be known only after its application.

IV EFFECTS OF EARTHWORMS ON MICROBIAL COMMUNITIES DURING VERMICOMPOSTING

Microorganisms are the main agents of biochemical decomposition, whereby earthworms are involved in the indirect stimulation of microbial populations through fragmentation of organic matter, that is, by increasing the surface area available for microbes. Earthworms also modify the microbial populations through digestion, stimulation, and dispersion in casts. Therefore, it is necessary to establish the effects of earthworms on microorganisms, because whether the earthworms stimulate or depress the microbiota, or modify the structure and function of microbial communities, they can have quite different effects on the rates and form of decomposition of organic matter. To address these questions we performed an experiment in our laboratory with mesocosms, filled with cow manure either with 10 mature earthworms belonging to *E. andrei* or without earthworms ($n = 5$ each). We used cow manure as the substrate, which is known to support a dense decomposer food web (Sampedro and Domínguez 2008). The mesocosms consisted of 2 L plastic jars filled with 200 g (fresh weight) of substrate. We used the epigeic earthworm *E. andrei* (Bouché 1972), which is distributed widely and easy to manage under laboratory conditions. We allowed mature earthworms (375 ± 7 mg; mean individual fresh weight ± standard error of the mean) to shed their gut contents on moistened tissue paper for 24 h at room temperature before the experiment. We covered the jars (containing the substrate and the earthworms) with perforated lids and stored them at random in a scientific incubator (20°C and 90% humidity); after 1 month, the earthworms were removed, and vermicompost and control samples were collected and processed immediately for microbial analyses. Viable microbial biomass was determined as the sum of all identified phospholipid fatty acids (PLFAs). The structure of the microbial community was assessed by PLFA analysis, and some specific PLFAs were used as biomarkers to determine the presence and abundance of specific microbial groups. Microbial community function was determined by measuring the bacterial and fungal growth rates by the incorporation of radioactively labeled leucine into proteins and radioactively labeled acetate into the fungal-specific lipid ergosterol, respectively. The metabolic quotient, a parameter that evaluates the efficiency of microorganisms in utilizing organic C compounds, was also determined.

A Microbial Biomass

Microbial communities were characterized by PFLA profiles (Zelles 1999) and these analyses revealed that earthworm activity impacted greatly on microbial community structure and function. We found that the activity of earthworms reduced the viable microbial biomass, measured as the total content of PLFAs, after 1 month of vermicomposting (Figure 5.2a); the presence of earthworms reduced total microbial biomass by approximately four to five times relative to the control without earthworms. Certain specific PLFAs can be used as biomarkers to determine the effect of earthworms on the presence and abundance of specific microbial groups. The sum of PLFAs characteristic of gram-positive bacteria (iso/anteiso branched-chain PLFAs), gram-negative bacteria (monounsaturated and cyclopropyl PLFAs), and actinomycetes (10Me branched PLFAs) were chosen to represent the bacterial biomass; and the fungal biomarker 18:2ω6,9 was used to indicate fungal biomass (Frostegård and Bååth 1996; Zelles 1997).

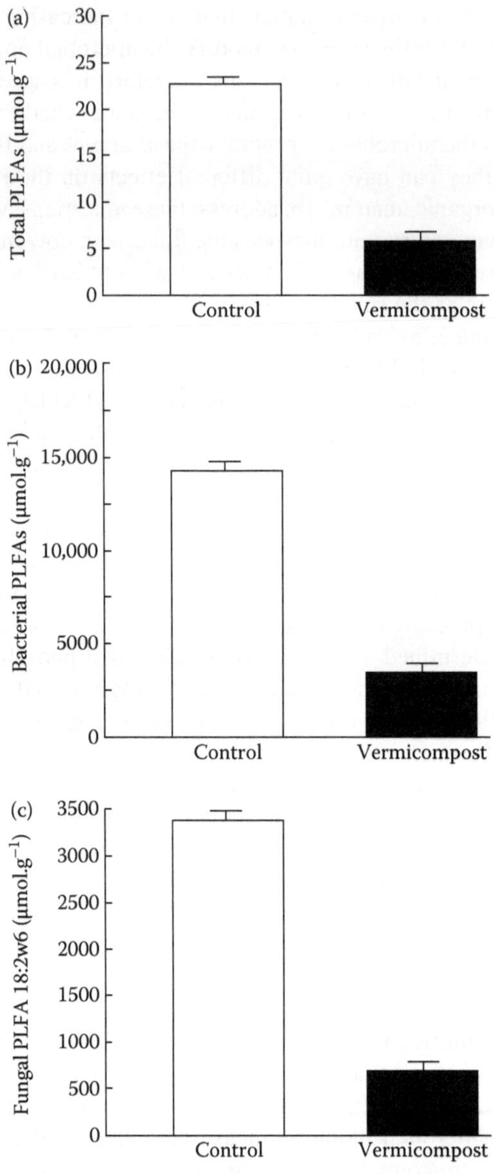

Figure 5.2 Impact of earthworms (*Eisenia andrei*) on microbial communities after vermicomposting in cow manure (for 1 month). (a) Total phospholipid fatty acids (PLFAs), a measure of microbial biomass; (b) bacterial biomass, calculated as the sum of the bacterial PLFA markers: i14:0, i15:0, a15:0, i16:0, 16:1ω5, 16:1ω7, i17:0, a17:0, 10Me18:0, 18:1ω7, cy17:0, and cy19:0; and (c) PLFA 18:2ω6,9, a measure of fungal biomass. Values are means ± SE. Control is the same treatment without earthworms.

The abundance of both bacteria and fungi was drastically reduced by the earthworms after 1 month of vermicomposting (Figure 5.2b and c). Earthworms can reduce microbial biomass directly by selectively feeding on bacteria and fungi (Schönholzer et al. 1999) or indirectly by accelerating the depletion of resources available for the microbes.

B Bacterial and Fungal Growth

In our studies bacterial growth was estimated by the use of the leucine incorporation technique (Bååth et al. 2001), and fungal growth with the acetate-in-ergosterol incorporation technique (Bååth 2001). Earthworm activity greatly decreased bacterial growth rates but did not affect fungal growth rates after 1 month of vermicomposting (Figure 5.3). Animal manures are microbe-rich environments in which bacteria constitute the largest fraction, with fungi mainly present as spores; moreover, the first stages of decomposition in these organic wastes are dominated mainly by bacteria because of the availability of water and easily decomposable substrates. Hence, the activity of earthworms is expected to affect the bacterial growth rates to a greater extent than the fungal growth rate. In addition, carbon availability is a limiting factor for earthworm growth, and it has been reported that earthworms and microorganisms may compete for carbon resources (Tiunov and Scheu 2004); thus, earthworm activity may have reduced the quantity of resources available for microbial communities, and consequently the bacterial growth rates. The fungal growth rate was expected to decrease during the maturation stage, when depletion of more recalcitrant compounds takes place.

C Effects of Earthworms on the Activity of Microbial Communities

There is extensive evidence in the literature suggesting that earthworms and other soil invertebrates grazing on microorganisms enhance microbial activity in the first instance. As a result of this activity, earthworms reduce the later availability of these resources for the microbial communities, and consequently their activity. Thus, in our experiment, the microbial activity, measured as basal respiration, decreased after 1 month of vermicomposting with the earthworm species *E. andrei* (Bouché) (Figure 5.4a).

Organic carbon taken up by the heterotrophic microbial communities is partitioned between microbial cell biomass production, metabolite excretion, and respiration. The proportion of substrate carbon retained as microbial biomass, relative to carbon respired as CO_2, depends on the efficiency of microbial growth (i.e., the efficiency with, which substrates are incorporated into biomass and by-products), as well as on the degree of protection of microbial biomass in the organic matrix and on the rate of decomposition of bacterial and fungal by-products by other microorganisms. Thus, the lower the microbial growth efficiency or the less protected the biomass, the greater the amount of carbon lost as CO_2. The metabolic quotient

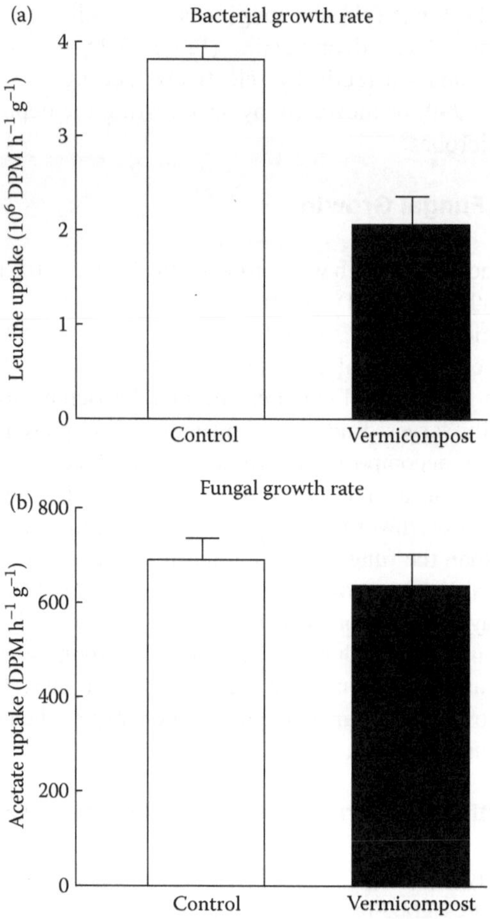

Figure 5.3 Impact of earthworms (*Eisenia andrei*) on microbial growth after vermicomposting of cow manure (for one month). (a) Bacterial growth rate estimated as incorporation of ac-in-erg. Values are means ± SE. Control is the same treatment without earthworms.

or specific activity of the microbial biomass (qCO_2; microbial respiration per unit biomass) can be used as a measure of microbial efficiency; higher values of qCO_2 indicate that microbial communities are under conditions of higher stress. Thus, less of the energy yielded by substrate metabolism can be used for biosynthetic purposes. An important portion of this energy will be expended on cell maintenance and lost as CO_2. Earthworm activity reduced the metabolic quotient after 1 month of vermicomposting (Figure 5.4b), indicating that microbial communities used the available energy more efficiently in the presence of earthworms. As a consequence, the system functioned much better, as shown by the large increases in the rate of decomposition of organic matter (Figure 5.5a) and in the rates of nitrogen mineralization (Figure 5.5b).

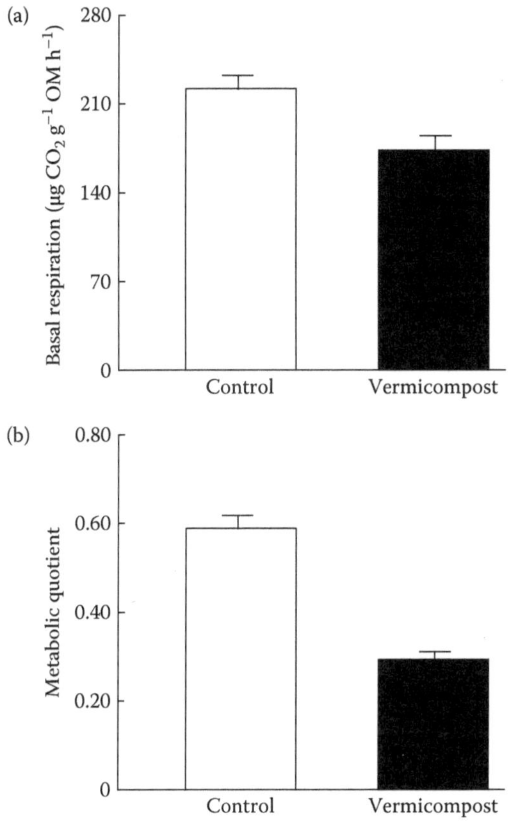

Figure 5.4 Effect of earthworms (*Eisenia andrei*) on microbial community function after vermicomposting of cow manure (for 1 month). (a) Basal respiration, a measure of microbial activity, and (b) the metabolic quotient estimated as the amount of CO_2 released from the sample per unit of biomass. Values are means ± SE. Control is the same treatment without earthworms.

D Effect of Earthworms on Total Coliform Bacteria during Vermicomposting

Earthworms also greatly reduced populations of total coliform bacteria during vermicomposting. Passage through the guts of the earthworm species *Eisenia andrei*, *Eisenia fetida*, and *Eudrilus eugeniae* reduced the populations of total coliforms by 98% relative to that in fresh pig slurry (Figure 5.6a; Monroy et al. 2008, 2009). The same drastic reduction in the populations of total coliforms was also found in another experiment after 2 weeks of vermicomposting with *E. fetida* (Monroy 2006). The reductions in total coliforms were similar to those reported by Eastman et al. (2001) for these and other human pathogens, which indicates the effectiveness of vermicomposting in reducing the levels of human pathogens during stabilization of biosolids and other organic wastes. As discussed earlier, digestion

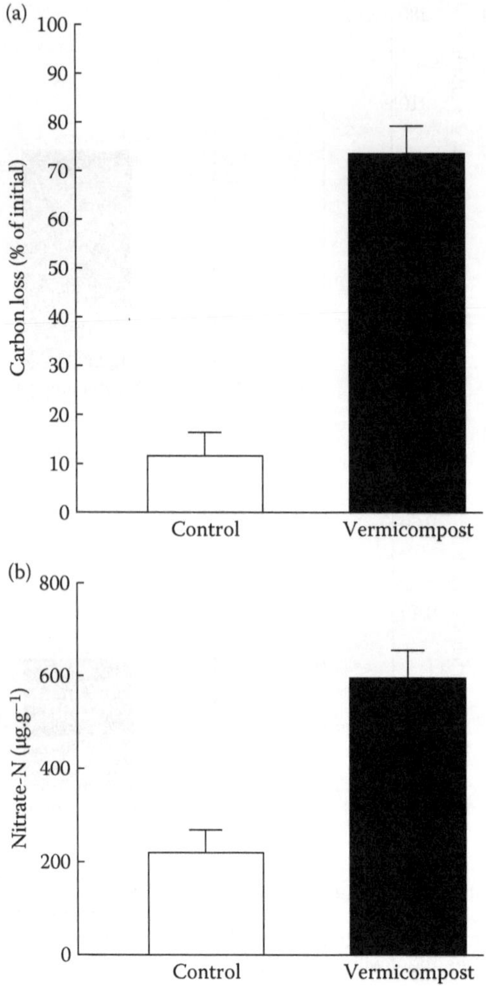

Figure 5.5 (a) Carbon loss (percentage of initial) after 1 month of vermicomposting of cow manure, as affected by the presence of the earthworm *Eisenia andrei.* (b) Effect of earthworms (*Eisenia andrei*) on the amount of Nitrate-N produced after vermicomposting of cow manure (for 1 month). Values are means ± SE. Control is the same treatment without earthworms.

of decaying substrates by earthworms decreases the availability of nutrients for microorganisms, thereby decreasing microbial numbers in the casts and altering the pattern of microbial composition (G. G. Brown 1995). There is increasing evidence that earthworms have a specific gut microflora (Karsten and Drake 1995; Horn et al. 2005), and the decrease in total coliforms also may be related to competitive interactions between coliforms and microorganisms that are specific to the earthworm gut (B. A. Brown and Mitchell 1981). Moreover, a negative effect of the passage through the earthworm gut on enterobacteria such as *Serratia marcescens*,

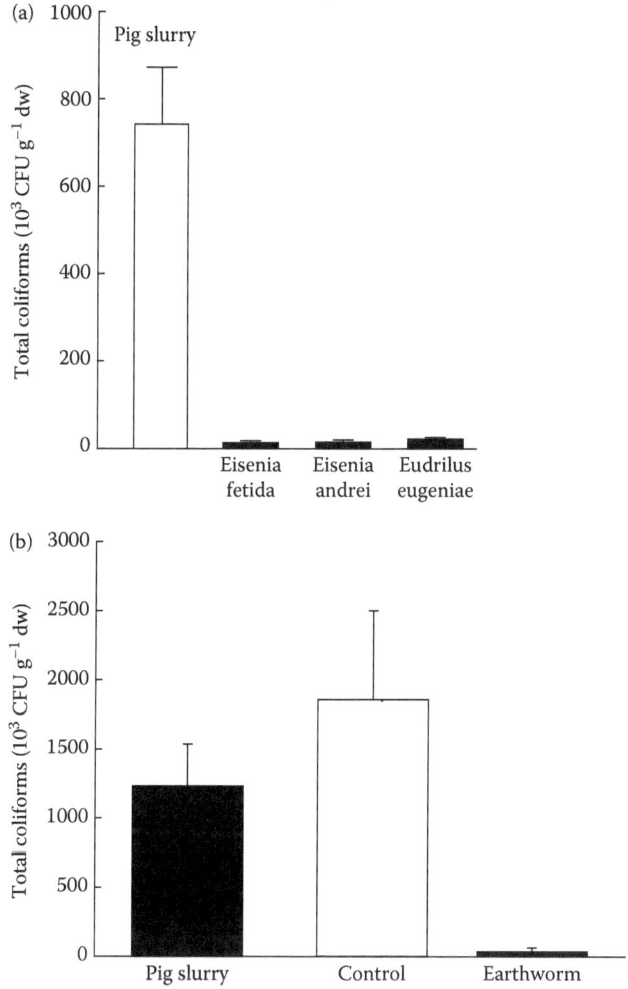

Figure 5.6 Effects of earthworms on total coliforms in pig slurry after (a) the transit through the gut of three species of epigeic earthworms and (b) 2 weeks of vermicomposting with *Eisenia fetida.* Values are means ± SE. Control is the same treatment without earthworms (dw: dry weight).

Escherichia coli, and *Salmonella enteridis* has been observed by several authors (Brown and Mitchell 1981), which suggests the occurrence of selective effects on the ingested microorganisms.

E Effect of Earthworms on the Composition of Microbial Communities

The discriminant analysis of 25 PLFAs (i14:0, 14:0, i15:0, a15:0, 15:0, i16:0, 16:1ω9, 16:1ω7, 16:1ω5, 16:0, 10Me16:0, i17:0, a17:0, cy17:0, 17:0, 10Me17:0, 18:2ω6,9,

18:1ω9, 18:1ω7, 18:0, 10Me18:0, cy19:0, 20:4ω6, 20:5ω3, 20:3ω6) clearly differentiated between the vermicomposts obtained using three different epigeic earthworm species (*E. andrei*, *E. fetida*, and *Perionyx excavatus*), irrespective of which manure type (cow, horse, or rabbit) was used in vermicomposting. This indicates that different PLFA profiles were associated with the vermicomposts, not related to the type of animal manure used but rather to the earthworm species and/or their endosymbiotic gut microflora. Moreover, the separation between vermicomposts and control substrates (manures processed without earthworms) was also very clear, indicating that earthworms play a key role in shaping the structure of the microbial community in organic wastes during the vermicomposting process. Similar results were also found with fatty acid methyl esters (FAMEs) profiles (Lores et al. 2006). From this perspective, and since different vermicomposts produced by different earthworm species and from different types of organic wastes contain an enormous and specific variety of microorganisms, it is possible to obtain specific vermicomposts for different practical applications. This may be especially important in producing plant container growth media for impoverished and/or intensively fertilized soils.

V CONCLUSIONS

Vermicomposting is a bio-oxidative process in which detritivorous earthworms interact intensively with microorganisms in decomposition processes, accelerating the stabilization of organic matter and greatly modifying its physical and biochemical properties. Earthworms are crucial drivers of the process since they are involved in the indirect stimulation of microbial populations through fragmentation and ingestion of fresh organic matter. Earthworms reduce overall microbial biomass and activity during the vermicomposting process. The activity of epigeic earthworms drastically reduces the viable microbial biomass during the vermicomposting process, and this reduction is proportionally higher for fungi than for bacteria, possibly because earthworms may use fungi for food selectively. After 1 month of vermicomposting the bacterial growth rates decrease in the substrate whereas the fungal growth rates are not affected. Microbial activity can be measured as decreases in basal respiration after vermicomposting. Earthworm activity helps microbial communities use the available energy more efficiently and plays a key role in shaping the structure of the microbial communities in organic wastes during the vermicomposting process. This evidence indicates that detritivorous earthworms directly modulate the decomposer community composition in the short term, thereby accelerating the decomposition of organic matter.

ACKNOWLEDGMENT

This research was partially supported by a Xunta de Galicia grant (07MRU023383PR) and the Spanish Ministerio de Ciencia e Innovaciòn (CTM2009-08477).

REFERENCES

Aira, M., Monroy, F., and Domínguez, J. 2006. Changes in microbial biomass and microbial activity of pig slurry after the transit through the gut of the earthworm *Eudrilus eugeniae* (Kinberg, 1867). *Biol. Fertil. Soils* 42:371–376.

Aira, M., Monroy, F., and Domínguez, J. 2007. *Eisenia fetida* (Oligochaeta: Lumbricidae) modifies the structure and physiological capabilities of microbial communities improving carbon mineralization during vermicomposting of pig manure. *Microbiol. Ecol.* 54:662–671.

Aira, M., Sampedro, L., Monroy, F., and Domínguez, J. 2008. Detritivorous earthworms directly modify the structure, thus altering the functioning of a microdecomposer food web. *Soil Biol. Biochem.* 40:2511–2516.

Bååth, E. 2001. Estimation of fungal growth rates in soil using ^{14}C-acetate incorporation into ergosterol. *Soil Biol. Biochem.* 33:2011–2018.

Bååth, E., Pettersson, M., and Söderberg, K. H. 2001. Adaptation of a rapid and economical microcentrifugation method to measure thymidine and leucine incorporation by soil bacteria. *Soil Biol. Biochem.* 33:1571–1574.

Brown, B. A., and Mitchell, M. J. 1981. Role of the earthworm, *Eisenia foetida*, in affecting survival of *Salmonella enteriditis* ser. *typhimurium*. *Pedobiologia* 22:434–438.

Brown, G. G. 1995. How do earthworms affect microfloral and faunal community diversity? *Plant Soil* 170:209–231.

Darwin, C. 1881. *The Formation of Vegetable Mould through the Action of Worms with Observations on Their Habits*. Murray, London.

Domínguez, J. 2004. State of the art and new perspectives on vermicomposting research. In: Edwards, C.A. (ed.), *Earthworm Ecology* 2nd Edition, 401–424. 2nd ed. CRC Press, Boca Raton, FL.

Domínguez, J., Parmelee, R. W., and Edwards, C. A. 2003. Interactions between *Eisenia andrei* (Oligochaeta) and nematode populations during vermicomposting. *Pedobiologia* 47:53–60.

Eastman, B. R., Kane, P. N., Edwards, C. A., Trytck, L., Gunadi, B., Stermer, A. L., and Mobley, J. R. 2001. The effectiveness of vermiculture in human pathogen reduction for USEPA biosolids stabilization. *Compost Sci. Util.* 9:38–49.

Edwards, C. A., and Bohlen, P. J. 1996. *Biology and Ecology of Earthworms*. Chapman and Hall, London.

Edwards, C.A., Domínguez, J. & Arancon, N.Q. 2004. The influence of vermicomposts on plant growth and pest incidence. In Shakir, S.H. and Mikhail, W.Z.A. (eds), *Soil Zoology for Sustainable Development in the 21st Century*, 397–420, El Cairo, Egypt.

Frostegård, Å., and Bååth, E. 1996. The use of phospholipid fatty acid analysis to estimate bacterial and fungal biomass in soil. *Biol. Fertil. Soils* 22:59–65.

Horn, M. A., Ihssen, J., Matthies, C., Schramm, A., Acker, G., and Drake, H. L. 2005. N_2O-producing bacteria isolated from the gut of the earthworm *Aporrectodea caliginosa*. *Int. J. Syst. Evol. Microbiol.* 55:1255–1265.

Karsten, G. R., and Drake, H. L. 1995. Comparative assessment of the aerobic and anaerobic microflora of earthworm guts and forest soils. *Appl. Environ. Microbiol.* 61:1039–1044.

Lazcano, C., Gómez-Brandón, M., and Domínguez, J. 2008. Comparison of the effectiveness of composting and vermicomposting for the biological stabilization of cattle manure. *Chemosphere* 72:1013–1019.

Lores, M., Gómez-Brandón, M., Pérez-Díaz, D., and Domínguez, J. 2006. Using FAME profiles for the characterization of animal wastes and vermicomposts. *Soil Biol. Biochem.* 38:2993–2996.

Monroy, F. 2006. Efecto das miñocas (clase Oligochaeta) sobre a comunidade descompoñedora durante o proceso de vermicompostaxe. PhD diss., Universidade de Vigo, Spain.

Monroy, F., Aira, M., and Domínguez, J. 2008. Changes in density of nematodes, protozoa and total coliforms after transit through the gut of four epigeic earthworms (Oligochaeta). *Appl. Soil Ecol.* 39:127–132.

Monroy, F., Aira, M., and Domínguez, J. 2009. Reduction of total coliform numbers during vermicomposting is caused by short-term direct effects of earthworms on microorganisms and depends on the dose of application of pig slurry. *Sci. Total Environ.* 407:5411–5416.

Sampedro, L., and Domínguez, J. 2008. Stable isotope natural abundances ($\delta^{13}C$ and $\delta^{15}N$) of the earthworm *Eisenia fetida* and other soil fauna living in two different vermicomposting environments. *Appl. Soil Ecol.* 38:91–99.

Schönholzer, F., Dittmar, H., and Zeyer, J. 1999. Origins and fate of fungi and bacteria in the gut of *Lumbricus terrestris* L. studied by image analysis. FEMS *Microbiol. Ecol.* 28:235–248.

Tiunov, A. V., and Scheu, S. 2004. Carbon availability controls the growth of detritivores (Lumbricidae) and their effect on nitrogen mineralization. *Oecologia* 138:83–90.

Zelles, L. 1997. Phospholipid fatty acid profiles in selected members for soil microbial communities. *Chemosphere* 35:275–294.

Zelles, L. 1999. Fatty acid patterns of phospholipids and lipopolysaccharides in the characterization of microbial communities in soil: a review. *Biol. Fertil. Soils* 29:111–129.

CHAPTER 6

Small-Scale School and Domestic Vermicomposting Systems

Rhonda L. Sherman and Mary Appelhof

CONTENTS

I Introduction 67
II History of Small-Scale Vermicomposting 68
III Advantages of Household Earthworm Bins over Backyard Composting 69
  A The Basic Small-Scale System 69
  B Suitable Vermicomposting Containers 70
IV Types of Earthworm Bedding 73
V Earthworm Systems for Classrooms and Households 74
VI Types of Food Waste Suitable for Vermicomposting 75
VII Procedures for Vermicomposting Food Wastes 75
  A Bin Setup 75
  B Burying Garbage 76
  C Signs of Reproduction 76
VIII Earthworm Castings or Vermicomposts 76
  A Sideways Separation 76
  B Light Separation 76
  C Vertical Separation 77
IX Vermicomposting in the Classroom 77
X Conclusions 77
References 77

I INTRODUCTION

Vermicomposting on a small scale is becoming increasingly popular in households and school classrooms. Small indoor vermicomposting systems can handle up to 1–3 kg (2–6 lb) of food waste in a week and provide 40–60 L (10 to 15 gal) of vermicompost per year to provide a nutrient-rich humus source for garden and house plants. Systems located outdoors can handle larger amounts of food waste, but these are influenced by extremes in weather. The primary purpose of vermicomposting food residues (uneaten food and food-preparation scraps) is to convert them to a useful product in an environmentally sound manner. Other benefits include diverting household organic waste from landfills, reducing the need for transport and centralized processing of such materials, and reducing water and electricity use by keeping food scraps from in-sink garbage disposals. Vermicomposting also reduces global warming because food residuals are the second-largest source of methane (a greenhouse gas) in landfills, and at 34% of all methane emissions, landfills are the biggest human-related source of methane in the United States (U.S. Environmental Protection Agency (EPA) 2009b). Users of household vermicomposting systems tend to be enthusiastic about their earthworm bins, partly because it makes them feel good to do something useful with the food they would otherwise throw away.

The availability of good instructional materials has made vermicomposting projects an excellent school classroom activity for children in grades ranging from kindergarten through high school. Having an earthworm bin in the classroom provides an opportunity for children to learn about organic decomposition, the soil food web, composting, and how earthworms and their castings are beneficial for soil and plant growth. Children experience waste recycling in action when they eat an apple, feed the core to earthworms, and use the earthworm castings to fertilize lettuce that they can then eat on their luncheon sandwich.

II HISTORY OF SMALL-SCALE VERMICOMPOSTING

The first publication describing a technique for using earthworms to process household food residuals was *Let an Earthworm Be Your Garbage Man* (Hopp 1954). It required a strong back and the willingness to remove several sq. feet (sq. meters) of sod, prior to edging the hole with cement blocks to create the earthworm composting area. After protecting the pile of excavated soil with straw to keep it from freezing, the homeowner added food waste to the pit daily, weekly, or monthly and covered the deposit with soil and straw. The rich food source attracted earthworms, which processed and mixed the diverse materials, turning them into a rich vermicompost.

A small guide also published in 1954 was entitled *With Tails We Win!* (Crowe and Bowen 1954). The authors built an early earthworm bin from concrete blocks in their basement that was 270 × 120 × 75 cm (9 ft × 4 ft × 2.5 ft). The bin had a sloping floor leading to a drain, and bedding consisting of two bushels (34.6L) each of cow manure and leaves. Crowe and Bowen vermicomposted large quantities of garbage in the bin without any odor, rotating daily deposits of about a gallon (3.4L) of table scraps, melon rinds, eggshells, and coffee grounds.

These publications inspired Mary Appelhof to set up a basement vermicomposting bin made of a galvanized metal stock tank with a drainage hole and limestone gravel for control of acidity. Oak planks separated gravel from the peat moss and manure bedding. Redworms (*Eisenia fetida*) thrived in this environment, fed by the manure and gallons of food wastes buried in the bin. Spring harvest of the bin yielded rich black vermicompost and a healthy crop of earthworms. A brochure describing the technique, *Basement Worm Bins Reduce Garbage and Produce Potting Soil* (Flowerfield Enterprises 1973) received national publicity, eliciting a few hundred responses sent with self-addressed stamped envelopes.

Flowerfield Enterprises received a National Science Foundation grant under the Small Business Innovation/Research Program to conduct a feasibility study of using earthworms for garbage disposal (Flowerfield Enterprises 1981). Surveying over 800 individuals with varying levels of knowledge about vermicomposting, the project team found that a third expressed a willingness to use earthworms to process food wastes and another third were willing to learn more about vermicomposting. The researchers were surprised to learn that only 25% "couldn't stand the thought" of using earthworms. Additionally, the more respondents learned about vermicomposting techniques, the more likely they were to want to try it.

To meet the need for a good manual describing how to use redworms to process household kitchen waste, Appelhof published *Worms Eat My Garbage* privately (Appelhof 1982, 1988). During the first 18 months, 4000 copies were sold, and a total of 5000 more copies were sold over the next 5 years. From 1989 to 1991, 5000 copies of the book were sold each year, then interest in vermicomposting escalated in 1992, doubling the number of copies sold. That number tripled in 1993, and in subsequent years 15,000 copies were sold annually. By 1997, Appelhof had sold 100,000 books. That year, Flower Press published an expanded and revised edition. Today, with 165,000 copies sold, and translations into Korean, French, German, and Japanese, *Worms Eat My Garbage* remains the definitive guide to household-scale vermicomposting.

III ADVANTAGES OF HOUSEHOLD EARTHWORM BINS OVER BACKYARD COMPOSTING

Household vermicomposting offers several advantages over backyard composting, including the following:

- Effective vermicomposting can take place using small volumes of wastes.
- Earthworms do the turning and aeration, so less labor is required to aerate the system.
- A contained system prevents access by pests such as raccoons, rats, and dogs.
- Vermicompost has more available nutrients and plant growth hormones for plants than traditional thermophilic compost.
- Indoor vermicomposting saves cold trips to the compost pile in winter.
- Apartment dwellers are able to vermicompost.

A The Basic Small-Scale System

A typical household-scale vermicomposting system consists of (a) a suitable container, (b) bedding, (c) earthworms (*E. fetida*) or other species, and (d) a proper environment (Appelhof 1997). Organic food waste is buried in moistened bedding in the bin. The earthworms, together with millions of microorganisms, feed on the garbage and transform it into black, odorless earthworm castings, or feces, or vermicompost. As long as the system has enough oxygen and moisture, and temperatures remain between freezing and 32°C (90°F), the bedding disappears, the organic wastes breakdown, and the earthworms multiply. Periodically, earthworms and vermicompost (a mixture of earthworm castings, partially decomposed waste, and uneaten bedding) can be harvested. Vermicompost can be used as a top dressing for house plants, as an ingredient in potting mixes, or as an immediate source of fertilizer for transplants and seed beds in gardens.

B Suitable Vermicomposting Containers

Many types of containers are suitable for vermicomposting (Sherman 1994, 1997). A bin can be constructed of untreated, nonaromatic wood, or a plastic container may be used. A wooden box works better than a plastic one for outdoor vermicomposting, because it will keep earthworms cooler in the summer and warmer in the winter. However, a wooden bin is less portable than a plastic one because it is a lot heavier. Homemade plastic earthworm bins are a lot less expensive than wood, but they tend to accumulate excess moisture. Commercial units specially designed for vermicomposting may be purchased from a number of vendors. Most single commercial units resemble a plastic storage box with a lid and holes for aeration and drainage. However, the Worm Swag vermicomposter is a unique invention from Australia (Figure 6.1). A large hoop covered with black cloth has a large green sack hanging from it; the bottom is tied with a string. The unit can be hung from overhead or placed on a stand. When harvesting, the string is untied and the vermicompost is squeezed out of a short cylinder (see photo).

Aeration is essential so homemade bins must have holes drilled around their perimeter near the top of the container. 1.3 cm (0.5 in) diameter holes may be left open or covered with window screen applied with a hot glue gun. Some people prefer to use a 7.6 cm (3 in) hole-saw to drill one hole in each end of the bin and insert 7.6 cm (3 in) louvered soffet vents (available in the gutter section of a hardware store) in the holes.

Liquid should not be allowed to accumulate in the bottom of the bin; however, since this often occurs, it is recommended that holes be drilled in the floor of the container. About six 1.3 cm (0.5 in) diameter holes should be adequate. A plastic tray may be placed under the earthworm bin to collect excess moisture. This leachate is not "earthworm tea" or "compost tea," as it is mistakenly commonly called. The leachate is liquid that has passed through undigested organic material; thus, it may contain pathogens or excess nutrients that may be harmful to plants. The leachate should be dumped on weeds or flushed down the toilet.

Plastic containers used for earthworm bins should not be clear or translucent because earthworms are sensitive to light. A plastic bin should be washed thoroughly and rinsed before bedding and earthworms are added. A tight-fitting lid is necessary to (a) help keep the earthworms in the bin, (b) protect the earthworms from light, (c) prevent moisture in the bedding from evaporating, and (d) keep predators out of the container.

Figure 6.1 Worm Swag. (Courtesy of Swag Industries Pty Ltd.)

Earthworm bins may be kept indoors or outside. People have vermicomposting systems in a variety of locations inside their homes, in addition to basements, breezeways, and garages. Earthworm bins may be kept in living rooms, kitchens, bathrooms, bedroom closets, and family rooms. Because there is no odor and earthworms work quietly, guests are often surprised to discover that the living room coffee table is also an earthworm bin.

Care must be taken in choosing an outdoor location for an earthworm bin. The bin should be placed in the shade where it is protected from the direct sun. If temperatures inside the earthworm bin rise above 35°C (95°F), the earthworms may try to leave the bin or die. It is easier to protect the bin from cold temperatures than from excessive heat. There are several ways to keep the earthworms from getting too cold. The vermicomposting system may be insulated with hay bales, rigid board insulation, or blankets. It can be heated by placing an aquarium heater in water inside a gallon jug that has been cut in half or by applying electric heat mats or tape approved for damp locations inside of the earthworm bin.

The earthworm bin size depends on the amounts of food residuals generated by the household. The general rule of thumb is to provide 0.1 m^2 (1 ft^2) of surface area for each 0.2 kg (0.5 lb) of garbage generated. A bin that is 0.3 m (2 ft) long by 0.3 m (2 ft) wide by 20 cm (8 in) deep should be adequate for two people who produce about 1.5 kg (3.5 lb) of food scraps each week. A larger bin that is 0.9 cm (3 ft) by 0.6 cm (2 ft) is suitable for four to six people who may generate about 2.7 kg (6 lb) of waste per week.

Some people choose to buy or make stacking earthworm bins. Advantages of these types of bins are that harvesting the vermicompost and managing leachates is easier. Plans for making this type of bin and other earthworm bins may be found on the Internet. Alternatively, sideways separation bins can also be made or purchased. Commercially successful stacking bins include the Can-O-Worms (Figure 6.2; also sold as Worm Condo) and the Worm Factory (Figure 6.3; and similar variations called Worm Tower, Worm Chalet, Gusanito Worm Farm, Wormtopia Worm Farm, and Verme Tower).

These continuous-flow systems come with two to four removable stacking trays and a separate lid. The base of the earthworm bin is a reservoir with a spigot for leachate collection and easy removal. To operate the bin, the first tray is set on top of the reservoir and filled halfway with moist bedding. Earthworms are placed on the bedding, and the lid is put on the bin (food is added a couple of days later). Add food and bedding to the first tray (as described in the following) until it is almost full. Then place the second tray on top of the first, fill it halfway with moist bedding, and bury food waste in it. Leave the first tray alone, and add food scraps as they disappear from the second tray. The earthworms will move up into the second tray to eat the food. Continue this procedure as you add the third and fourth trays (if you have them). The earthworms will keep moving up into whichever tray is receiving food waste. Eventually, almost all of the earthworms will migrate to the upper tray(s), and

Figure 6.2 Can-O-Worms. (Courtesy of Triformis Corporation, a distributor of vermicomposting bins, www.triformis.com)

Figure 6.3 Worm Factory. (Courtesy of The Worm Factory® by Nature's Footprint, Inc.)

the first tray will contain vermicompost that can be emptied and added to soil. The first tray can then be added to the top of the stack.

IV TYPES OF EARTHWORM BEDDING

The functions of earthworm bedding are to hold moisture, offer a suitable medium in which the earthworms can live and function, and provide a place to bury garbage. Contrary to common belief, soil is not a major component of earthworm bedding. Although earthworms need a small amount of soil, two handfuls are sufficient to provide grit for their gizzards in household-scale systems. Soil also contains bacteria, fungi, and other microorganisms that will inoculate the medium with a greater diversity of microorganisms than will be provided from the food residuals and bedding alone. Shredded newspaper, machine-shredded office paper, leaf mold, or mixtures of these materials make satisfactory beddings which cost nothing. Sufficient water must be added to the bedding to make it damp but not wringing wet. If paper crinkles, then it is too dry.

Recommended bedding for household systems is shredded newspaper because anyone can obtain it at no cost. Commercially-available bedding is coir (coconut fiber) which can be purchased in compressed blocks. This is easy to use; one simply places the block in water. The fiber quickly absorbs water and expands to make an aesthetically pleasing bedding in which earthworms live very well. Coir is a waste product from the coconut industry in tropical islands where disposal of the residues is a problem. Although environmental costs for transportation are a minor concern, use of this renewable resource makes more sense than using peat moss, which may be too acidic and difficult to obtain.

About half of the earthworm bin should be filled with moist bedding. Newspaper or office paper should be shredded first into strips up to one-inch wide the length of the paper. The paper will tear easily and neatly if torn "with the grain" in which the paper was made. For most newspapers, hold it up as if you were going to read it and tear a narrow strip downward at the edge. If the paper does not tear into a neat strip, turn the paper sideways and that should work. To moisten the bedding, fill a bucket two-thirds of the way full of water and place the bedding in the bucket for 15 minutes. Bedding needs to be immersed in water for a reasonable time period to become fully saturated. After 15 minutes of soaking, pull handfuls of bedding out of the bucket and gently squeeze most of the water out. Then fluff out the bedding as it is placed into the earthworm bin. The bedding will soon be consumed by the earthworms so be prepared to replace it when you can no longer cover the food scraps with at least two inches of bedding. Simply repeat the steps in the previous paragraph.

V EARTHWORM SYSTEMS FOR CLASSROOMS AND HOUSEHOLDS

Vermicomposting requires the use of the proper type of earthworm—a species whose niche in nature is to process large amounts of organic material. These are known as epigeic species. Redworms (*E. fetida*) are the most suitable, known variously as manure worms, red wigglers, tiger worms, and other common names. To reduce the confusion caused by local earthworm names, more people are becoming comfortable with the scientific name, *E. fetida*, and its close relative, *Eisenia andrei*. Most redworm cultures available from commercial growers are one or the other or a mixture of these species (see Chapter 3).

As with commercial vermiculture systems, household-scale vermicomposting bins should never be set up with earthworms dug from the garden. Soil-burrowing endogeic earthworm species have different ecological niches. They aerate the soil with their burrows, they mix organic material with soil, and their activity helps to create a spongy texture in soil that holds water better than compacted soil. However, soil-burrowing earthworms will not be able to process the amount of organic material placed in a vermicomposting bin nor reproduce quickly enough. People who do not obtain the proper species of earthworms may conclude that vermicomposting does not work when the problem is that they are using the wrong species of earthworm.

At least 1000 earthworms (about 0.5 kg (1 lb)) should be added to the earthworm bin. Purchase them from an earthworm grower because some bait shops may not sell *E. fetida*, and if they do, they sell about 30 earthworms in each container. Hence to obtain 1000 worms, at least 33 containers would need to be purchased!

The earthworms procured will probably have been raised in a different environment, such as another type of bedding or maybe even a dissimilar climate if the earthworms were received by mail order. Whether the earthworms arrived by airplane or automobile, they will surely have been shaken up by the journey. Therefore, when the earthworms are put into their new home, they may try to flee this foreign environment. To keep the earthworms in the earthworm bin, shine a light overhead for several days. Since earthworms are adverse to light, they will stay in the vermicomposting system. After a few days to one week, the earthworms should be accustomed to their new environment and choose to stay inside the earthworm bin, and thus the light may be removed.

VI TYPES OF FOOD WASTE SUITABLE FOR VERMICOMPOSTING

More than 25% of the food prepared by Americans is thrown away, amounting to 44 billion kg (96 billion lb) of food waste annually. The U.S. EPA estimated in 2007 that 12.5% of the waste generated by American households was food residuals (uneaten food and food-preparation scraps). Less than 3% of food waste is recovered, and the rest is discarded in landfills or incinerators (U.S. EPA, 2009a).

All kinds of organic materials can be put into a vermicomposting bin, but people have the most success with vegetable and fruit waste, tea leaves, and coffee grounds (bags and filters are okay too). Eggshells add calcium to the bin, but they should be pulverized with a rolling pin first. Bread and pizza dough should be moistened before being placed in the earthworm bin. Chopping or grinding food scraps is recommended, as smaller particles will break down faster. Do not add the following organic materials to household vermicomposting systems: meat, bones, grease, dairy products, and nut butters (the high protein content can attract flies, rats, and other pests); citrus fruit and rinds (they can make the bin acidic and attract red mites); and human and pet feces (these may contain pathogens).

People expect offensive odors when they start burying their food waste in a small earthworm bin. Most users express surprise that there is little to no odor; odor problems can be reduced by disturbing the bedding as little as possible when burying waste, by covering it with bedding, and by depositing the waste in different locations each time. Overloading the system and lack of adequate aeration are the most frequent causes of odor problems.

VII PROCEDURES FOR VERMICOMPOSTING FOOD WASTES

A Bin Setup

Once all components are assembled—an aerated container, two handfuls of soil, 0.9 kg (1 lb) of *E. fetida*, bedding, and food waste—it takes less than half an hour to set up the bin for the first time. Paper should be shredded, soaked in a bucket for 15 minutes, wrung out, and fluffed out as it is placed in the earthworm bin. Soil is then mixed in with the bedding. Earthworms are placed directly on top of the bedding with the organic medium in which they were shipped. They quickly burrow down into the bedding away from the light. Use a garden fork to draw aside enough bedding to form a hole in which to bury some food waste. Cover the garbage with 5–10 cm (2–4 in) of bedding to prevent fruit flies and odors. It is that simple.

B Burying Garbage

Most people bury garbage in bins once or twice a week—preferably sequentially in different locations in the bin each time. Whenever food scraps are buried, monitor the moisture level in the bedding to enable adjustments as needed. If the bedding crinkles because it is dry, use a water spritzer to moisten it. If the bin seems to be soggy and waterlogged, add dry shredded bedding to absorb some moisture and leave the lid off for a few hours to allow moisture to evaporate.

C Signs of Reproduction

After several weeks, earthworm cocoons will be visible in the earthworm bin. They are yellowish and about the size of a match head (2–3 mm (0.15 in) in diameter). After 2–3 months more than twice as many earthworms should be present. Overpopulation will not be a problem, but if the bin begins to seem crowded after several months, some earthworms could be donated to others wanting to start vermicomposting (neighbors, teachers, and Cooperative Extension agents) or to go fishing.

VIII EARTHWORM CASTINGS OR VERMICOMPOSTS

As vermicomposting takes place, dark and crumbly vermicompost will accumulate gradually on the bottom of the container. In 3–4 months, the vermicompost can be harvested by separating it from the earthworms and removing it from the bin. There are several methods for sorting this valuable nutrient-laden humus to use for growing plants.

A Sideways Separation

Bury food scraps on only one side of the earthworm bin for a few weeks, and most of the earthworms will migrate to that side of the bin. Then the vermicompost can be removed from the other side of the bin that has not been receiving food. Fresh bedding can be added on this side and the process repeated. After both sides are harvested, food may be added to both sides of the bin again.

B Light Separation

Empty the contents of the earthworm bin onto a plastic sheet where there is strong sunlight or artificial light. Wait 10 minutes for the earthworms to move down away from the light, and then skim off the top layer of vermicompost. Remove more vermicompost every 10 minutes. Return the earthworms to the bin in fresh bedding prepared ahead of time. Vermicompost that is harvested should be placed in a separate container with pinprick holes that allow air inside yet prevent moisture from evaporating.

C Vertical Separation

The easiest way to separate vermicompost from earthworms is to start with a stacking bin system as described earlier.

IX VERMICOMPOSTING IN THE CLASSROOM

Vermicomposting has tremendous appeal for children, many of whom seem to have a natural fascination with earthworms. An earthworm bin in the classroom not only provides a way to keep lunch and snack wastes out of the landfill but also offers many opportunities for learning. Vermicomposting fits perfectly with state requirements for students to learn about habitats and ecosystems—communities of organisms and their interaction with the environment. A classroom earthworm bin can be used to help students learn about science, math, language arts, music, history, social studies, geography, and other subjects. Vermicomposting curriculum guides for all ages are available for purchase or may be downloaded from the Internet.

Classroom space constraints often make household-scale vermicomposting bins appropriate, although the actual volume of food waste that can be fed to the vermicomposting earthworms is limited. That is why many schools are choosing to expand their vermicomposting efforts by having one or more large earthworm bins to process food waste from their cafeterias. This trend is described further in Chapter 24, "Vermicomposting for Businesses and Institutions."

X CONCLUSIONS

It is not possible to describe all of the household-scale vermicomposting systems that exist. A common denominator for developing interest in a particular area seems to be one enthusiastic individual who has a vermicomposting bin going and cannot stop talking about it. Frequently, these enthusiasts find they are talking about earthworms so much they write a book or start teaching others how to vermicompost. Others are intrigued enough that they pursue processing of larger quantities of waste using windrows or automated flow-through systems such as those discussed in other chapters. The consequence is that vermicomposting bins on a household scale seem to be growing in number as fast as the well-fed earthworms that inhabit them.

REFERENCES

Appelhof, M. 1988. "Domestic Vermicomposting Systems." In *Earthworms in Waste and Environmental Management*, ed. Clive A. Edwards and Edward F. Neuhauser, 157–161. The Hague, Netherlands: SPB Academic Publishing.

Appelhof, M. 1982. *Worms Eat My Garbage*. Kalamazoo, MI: Flower Press.

Appelhof, M. 1997. *Worms Eat My Garbage,* 2nd ed. Kalamazoo, MI: Flower Press.

Appelhof, M., Harris, B. L., and Fenton, M. F. 1993. *Worms Eat Our Garbage: Classroom Activities for a Better Environment.* Kalamazoo, MI: Flower Press.

Crowe, M., and Bowen, G. 1954. *With Tails We Win!* Eagle River, WI: Shields.

Flowerfield Enterprises. 1973. *Basement Worm Bins Reduce Garbage and Produce Potting Soil.* Kalamazoo, MI: Flowerfield Enterprises.

Flowerfield Enterprises. 1981. "Biodegradable Solid Waste Conversion into Earthworm Castings." Grant No ISP-800975 prepared for the National Science Foundation Appropriate Technology Program, Small Business Innovation Research Program.

Hopp, H. 1954. *Let an Earthworm Be Your Garbage Man.* Home, Farm, and Garden Research Associates. Eagle River, WI: Shields.

Sherman, R. 1997. *Worm Away Your Cafeteria Food Scraps.* AG-551. Raleigh, NC: Cooperative Extension Service.

Sherman, R. 1994. *Worms Can Recycle Your Garbage.* AG-473-18. Raleigh, NC: Cooperative Extension Service.

U.S. Environmental Protection Agency (EPA). 2009a. *Basic Information about Food Scraps, Wastes—Resource Conservation—Common Waste & Materials—Organic Materials.* http://www.epa.gov/osw/conserve/materials/organics/food/fd-basic.htm

U.S. Environmental Protection Agency (EPA). 2009b. *Methane Sources and Emissions.* http://epa.gov/methane/sources.html.

CHAPTER 7

Low-Technology Vermicomposting Systems

Clive A. Edwards

CONTENTS

I Introduction 79
II Windrow Vermicomposting Systems 81
III Batch Systems of Vermicomposting 82
IV Earthworm-Harvesting Equipment 83
  A Typical Rotating Trommel 83
  B Improved Comb-Type Mechanism of Separating Earthworms from Vermicompost 84
V Domestic Vermicomposting Systems 85
VI Wedge Vermicomposting Systems 86
VII Diseases and Predators of Earthworms in Vermicomposting Systems 88
  A Common Vermicompost Bed Organisms 88
  B Invertebrate Earthworm Enemies 89
  C Vertebrate Earthworm Predators 89
  D Diseases and Parasites of Earthworms 89
References 90

I INTRODUCTION

In response to the extremely large amounts of organic wastes currently produced, thermophilic composting technology has become increasingly popular for large-scale processing and disposal of a wide range of organic wastes. However, the process does not always produce high-quality products that have good potential for soil and land improvement, and most thermophilic composts do not have great economic value. Over the last 30 years, interest has increased progressively in the potential of a related process that involves the use of earthworms to break down organic wastes in a mesophilic process. In 1881, Charles Darwin first drew attention to the great importance of earthworms in the breakdown of dead plant organic materials and the release of the essential nutrients they contain in his book *The Formation of Vegetable Mould through the Action of Worms.* Many of his conclusions have been confirmed and utilized extensively during the last century (Edwards and Bohlen 1996). However, only in recent years has the considerable potential of using earthworms in systems of breaking down organic wastes to produce vermicomposts been explored in more depth to full commercial ventures (Edwards and Neuhauser 1988).

Some species of earthworms inhabit organic litter on the soil surface. They fragment organic wastes extremely rapidly and increase microbial activity in them dramatically; these are termed *epigeic* earthworms. The main difference between thermophilic composting and vermicomposting is that whereas composting is an aerobic process that can reach temperatures of 60°C–70°C (140°F–158°F), vermicomposting systems are mesophilic and must be maintained at temperatures below 35°C (95°F). Exposure of the earthworms to temperatures above this, even for relatively short periods, will kill them, and, to avoid such overheating in vermicomposting systems, very careful management of the wastes is required. Epigeic earthworms are very active and will consume organic wastes located in a relatively narrow horizontal aerobic layer of 10–15 cm (4–6 in), that is, close to the surface of a bed or container, very rapidly. The critical key to successful vermicomposting lies in adding organic wastes to the surface in successive thin layers at frequent intervals, so that any thermophilic heating that may occur does not become excessive and, if well managed, this low level of heating will maintain the activity of the earthworms at a high level of efficiency through colder periods in temperate countries, since vermicomposting works best at temperatures between 20°C and 25°C (68°F–77°F).

Almost any kinds of agricultural, urban, or industrial organic wastes can be used for vermicomposting, but some may need some form of preprocessing before use to make them acceptable to earthworms. Such preliminary treatments can involve washing, precomposting, macerating, or mixing the organic matter. Organic manures and wastes from the brewing, soft drink, processed potato, and paper industries; sewage biosolids; and yard, garden, and food wastes are particularly suitable for vermicomposting. Often, mixtures of several different wastes can be processed more readily than single types of wastes, are easier to maintain aerobically with an acceptable moisture content, and result in a better product.

The species of earthworms that are used for vermicomposting, termed epigeic species, can consume organic wastes very rapidly and fragment them into much finer particles by passing them through a grinding gizzard inside their mouth that all earthworms possess. The earthworms obtain their nourishment from the microorganisms that grow on the organic waste rather than the wastes themselves; at the same time, they promote further microbial activity in the wastes, so that the earthworm casts, or vermicomposts, that they produce are much more fragmented and very much more microbially active than the organic wastes that the earthworms consume. During this process, the important plant nutrients that the wastes contain, particularly N, P, K, and Ca, are released and converted into forms that are much more soluble and readily available to plants than those in the original waste. The retention time of the waste in the earthworm gut is short, at most a few hours, and very large quantities of organic matter are often passed through an average population of earthworms more than once.

In the traditional thermophilic composting process using windrows, organic wastes have to be turned regularly, or aerated in some way, to maintain aerobic conditions in the waste. This may often involve extensive engineering and machinery to process the organic wastes as rapidly as possible on a large scale. In the vermicomposting process, the earthworms, which can survive only under aerobic conditions, take over both the roles of turning over the waste and maintaining it in an aerobic condition, thereby lessening the need for expensive engineering to achieve these aims.

A major constraint to vermicomposting is that, in contrast to traditional composting, which is a thermophilic process that can raise temperatures in the waste to more than 65°C (149°F), vermicomposting systems must be maintained at temperatures below 35°C (95°F) to avoid killing the earthworms. The processing of organic wastes by earthworms occurs most rapidly at temperatures between 15°C and 25°C (60°F to 79°F) and at moisture contents of 70% to 90%. Outside these limits, earthworm activity and productivity and the rates of waste processing can fall off, and for maximum efficiency, the wastes should be maintained as close to these environmental limits as possible, which usually means keeping the systems indoors or under cover. The earthworms are also sensitive to certain environmental conditions in the wastes. In particular, earthworms are very sensitive to ammonia, salts, and other chemicals. For instance, they will die quickly if exposed to wastes containing more than 0.5 mg of ammonia per g^{-1} of waste and more than 0.5% salts (Edwards and Neuhauser 1988). However, salts and ammonia can be washed out of organic wastes readily or decreased by thermophilic precomposting. Contrary to common belief, earthworms do not have many serious natural enemies, diseases, or predators, and they can survive exposure to many adverse conditions; this will be discussed more later and also in other chapters.

II WINDROW VERMICOMPOSTING SYSTEMS

In the United States and Canada there is a very extensive, but relatively small-scale, cottage industry that grows earthworms for fish bait in a variety of organic wastes. These use, almost exclusively, outdoor ground beds or windrows. Such systems require large areas of land for large-scale earthworm and vermicompost production and are relatively labor-intensive, even when machinery is used for adding the organic wastes to the beds and harvesting the vermicomposts. More importantly, windrow systems process wastes relatively slowly, taking anywhere from 6 to 18 months to process a layer 45 cm deep (18 in), particularly when winters are cold. Since this is usually an outdoor process, there is evidence that a large proportion of the essential plant nutrients, which are in a relatively soluble form, are either washed out of the organic matter or can volatilize from it during this long processing period.

Table 7.1 Characteristics of Windrow Vermicomposting Systems

Benefits	• Low capital outlay
	• Easily managed
Drawbacks	• Labor-intensive
	• Needs large areas of land
	• Slow processing time
	• Considerable loss of nutrients through leaching and volatilization
	• Impossible to harvest vermicompost without earthworms

Such nutrient losses are undesirable, since they can contribute to groundwater pollution, and result in a poor, low-nutrient vermicompost product with relatively poor potential as a plant growth medium.

Windrow vermicomposting is common all over the United States and elsewhere in the world. However, it is labor- and land-intensive, is seasonal, and takes place outdoors. Moreover, some enterprises harvest only half the vermicompost produced, leaving the rest as an earthworm inoculum for the next windrow to be set up. Thus, it requires the separation of earthworms from the finished vermicompost using rotating trommels or other equipment. Windrows are ground-based and require large areas of land, and they have potential for groundwater pollution with nutrients and contaminants, since they are watered regularly and usually have little protection against leaching. Windrows should have a firm concrete base to avoid harvesting soil with the vermicompost. It helps to minimize watering by using a permeable top cover such as sacking or bamboo strips held together with twine; these prevent evaporation but allow watering. Under no circumstances should plastic or polyethylene sheeting be used to cover windrows because it turns the upper layers of the windrow anaerobic, so that the earthworms leave the windrow and lie below the cover where there is oxygen. The processing is slow, taking 4–18 months to complete. The harvesting of the vermicompost is laborious and time-consuming, since the earthworms in the waste have to be separated, usually by some form of screening process, before marketing. Although the initial capital outlay, other than land, is low, labor costs are high at all stages of the operation (Table 7.1).

III BATCH SYSTEMS OF VERMICOMPOSTING

Methods of batch vermicomposting in large or small stacked boxes or containers have been discussed by Edwards and Neuhauser (1988), Edwards (2004), and Carver and Christie (2007), who suggested that most of the batch processing methods he tested were too labor-intensive because the units had to be moved to add more wastes in thin layers to the surface of the unit. It is also difficult to access the earthworms in them or add water to them because they are usually stacked one above the other in racks to economize on space.

Table 7.2 Characteristics of Vermicompost Batch System

Benefits	• Needs relatively little space
Drawbacks	• Considerable expenditure on containers and container-moving equipment
	• Difficult to maintain optimal moisture conditions with sprays
	• Labor-intensive
	• Harvesting of vermicompost without earthworms impossible without separating them from the vermicompost in a screening or trommel system

Batch vermicomposting can be done in any size of container. There have been various attempts to develop improved batch systems and modular container systems. However, batch systems all have the same major disadvantage as windrows, that is, the need for a labor-intensive separation of earthworms from the vermicompost before it can be used or marketed. Additionally, there is a need to inoculate earthworms into each new container when a new batch system is started (Table 7.2).

IV EARTHWORM-HARVESTING EQUIPMENT

When organic wastes are broken down into vermicomposts by earthworms in windrows or batch systems, there is an essential step of extracting the earthworms from the vermicompost before it can be marketed. This is usually achieved by passing the processed organic wastes or vermicomposts through equipment with rotating screens set on a slope with a cone-shaped solid unit at the end (Figure 7.1). These systems are commonly termed trommels.

A Typical Rotating Trommel

Different-sized mesh apertures can be used in trommels to screen the vermicomposts into different particle sizes. The better models have a range of screen sizes, with the smallest screen mesh at the loading end and increasingly larger aperture meshes further down the slope. A typical unit has a smallest screen size of 2 mm (0.079 in) and a largest mesh of 0.5 cm (0.2 in). The diameter of the screens can be as large as required, but a typical unit has a diameter of about 60 cm (23.6 in), and they are usually rotated by an electric motor at about 10 revolutions per minute (rpm). The funnel-shaped metal end has a maximum diameter of about 1 m (40 in). The vermicomposts can be collected into containers along the length of the trommel. At the funnel end, the earthworms adhere to the smooth metal surface as it rotates and drop off into a collection tray from the highest part in the metal cone. It is important that the vermicompost not be too moist to avoid compacting into balls that will not pass through the mesh.

Figure 7.1 Rotating screen earthworm/waste-separating system or trommel.

B Improved Comb-Type Mechanism of Separating Earthworms from Vermicompost

Engineers at the U.K. National Institute of Agricultural Engineering, Silsoe, Beds, designed a much more efficient earthworm waste separator based on a comb system (Price and Phillips 1990; Figure 7.2). Steel combs constructed from nails 5.8 cm (2.2 in) long and 2.5 mm (0.1 in) in diameter are fitted to the inside of a cylinder 2.5 m (8.2 ft) long and 0.8 m (3 ft) in diameter. The combs are spaced at 50 mm (2 in)spacing axially and 60 mm (2.3 in) spacing circumferentially. They are set radially but with a trailing angle of 10° (in the prototype separator, masonry nails are used, hammered through undersized holes in a PVC cylinder). In a later model, the cylinder was rolled from steel sheet, and the comb teeth were made from spring steel, held in place with special steel clips.

The cylinder rests on rollers in a support frame, with jacks at one end to enable the whole machine to be tilted to an angle from the horizontal. A drive unit is provided to rotate the cylinder at rates up to 10 rpm. A 4 m (4.3 yd) long by 0.5 m (20 in) wide conveyor belt is placed coaxially through the cylinder and fitted with a simple hopper at its rear end. The general arrangement of the components is shown in Figure 7.3.

In operation, the rear of the machine is jacked up, and the cylinder is set to rotate. The conveyor belt (B) draws a thin layer of earthworm-rich vermicompost from the hopper (A), the layer being swept off (B) into the rotating cylinder (D) by the deflector (C). Earthworms become draped over the combs moving through the vermicompost and are carried upward until the combs are nearly inverted. At this point they drop off and can be collected on the conveyor belt, forward of the deflector (C), and are transported forward (E) into a receptacle (F). Earthworm-free vermicompost tumbles forward within the cylinder because of its inclination from the horizontal and emerges at the end into a receptacle (G).

Figure 7.2 The Silsoe comb-type earthworm/vermicompost separator.

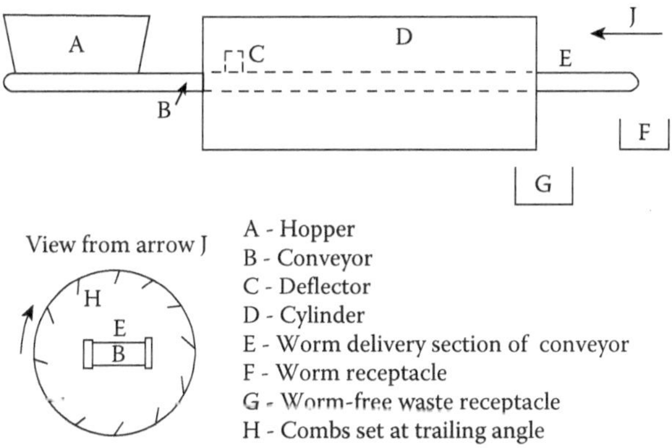

Figure 7.3 General arrangement of components of Silsoe comb-type earthworm separator.

The most suitable rotational speed for the cylinder is about 7 rpm, and the inclination of the machine about 2 degrees from horizontal. The machine can be operated continuously without any buildup of wet vermicompost. Coarse or lumpy vermicomposts are broken up successfully by the movement of the combs. The machine operates with greatest efficiency with an input of 5 L (10.57 pts) of vermicompost per minute but is still quite efficient with inputs of 15 L (31.7 pts) of vermicompost per minute.

V DOMESTIC VERMICOMPOSTING SYSTEMS

Small-scale systems of vermicomposting, for disposal of domestic household and food wastes, have been used extensively in homes, schools, and even jails (Appelhof

Figure 7.4 Eliminator domestic vermicomposting system.

1997) (see Chapter 6) (Table 7.3). They range from simple containers with perforated lids for aeration to more sophisticated commercially produced stacking systems of different sizes and complexities, usually with mesh bottoms so processed vermicompost can fall into the lower container. These include circular stacking systems such as the Can-O-Worms, the Worm Wigwam System, a rectangular stacking system called Worm Factory, and systems such as the Eliminator (Figure 7.4), which has a side-operated breaker bar and collection drawer at the base, an opening glass door for inspection of the vermicompost, a hinged lid to allow addition of wastes. These commercial systems have attracted the interest of some local urban-waste authorities, some of whom have encouraged home owners to use them, often by donating them to the users on condition that they put no food waste into the main organic-waste-disposal stream. The benefits and drawbacks of domestic vermicomposting system are summarized in Table 7.3.

VI WEDGE VERMICOMPOSTING SYSTEMS

A Dorset wedge vermicomposting system was designed by Edwards and colleagues at Rothamsted with others from the National Institute of Agricultural

Table 7.3 Domestic Vermicomposting Systems

Benefits	• Removal of significant amounts of organic wastes from landfill disposal
	• Provision of plant growth media and soil additives for home use
	• Removal of odor from food wastes
	• Useful educational tool
Drawbacks	• Need for careful management, especially moisture control
	• Possible breeding of flies

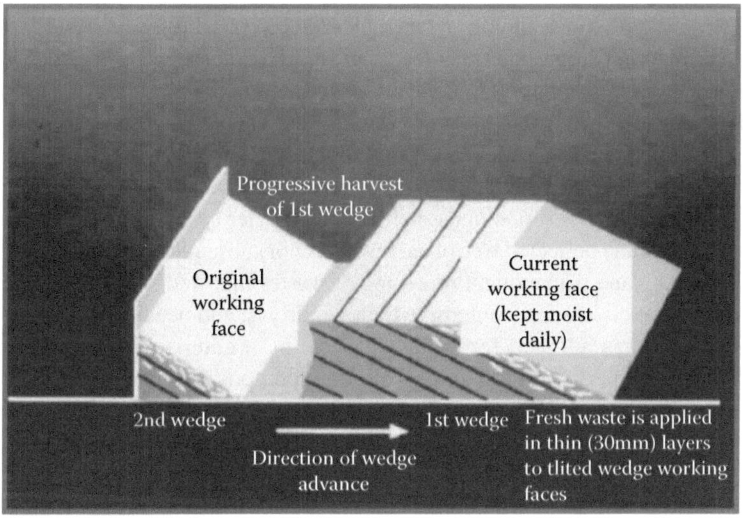

Figure 7.5 Principles of the wedge vermicomposting system.

Engineering in the United Kingdom. It is based on adding successive thin layers of 5–10 cm (2–4 in) of organic waste at a 45 degree angle from a vertical removable barrier (Figure 7.5). The wedge system can be of any width or length but should be limited in height to about 1.2–1.5 m (1.1–1.6 yd) for ease of adding organic wastes to the leading edge. It should be situated on a concrete base or on some other solid surface from which vermicompost can be collected easily. The system starts with waste at a layer at an angle of 45 degrees against a removable barrier. Partially vermicomposted organic waste containing 9 kg (19.8 lb) (wet weight) of *E. fetida* (or other species) m^{-2} to a depth of about 15 cm (6 in) is used to start the system. The surface is kept moist (80% moisture content) to a depth of 15 cm (6 in) by a fine water spray applied twice daily to the leading edge, which is covered by a layer of material such as sacking that can pass moisture. Additional layers of waste 2–3 cm (0.8–1.2 in) thick at an angle of 45 degrees are added daily, and then the pile is covered again.

Table 7.4 Characteristics of Wedge Systems

Benefits	• Low capital outlay • Needs less labor than windrows • Faster processing time • Much less leaching or volatilization of nutrients • Easy to harvest the vermicompost without earthworms
Drawbacks	• Relatively slow processing time • Need for machinery to apply wastes at frequent intervals

The earthworms move rapidly from the older layers of fully processed organic waste or vermicompost into the fresh feedstock material at the wedge surface, so that the entire earthworm population is always concentrated in the top 15 cm (6.0 in) below the leading surface of the wedge. At convenient intervals (e.g., every 1 to 2 months), the removable barrier can be taken away and replaced about 60 cm (0.6 yd) behind the leading face of the wedge, so that no earthworms are removed when the vermicompost is collected. All of the vermicompost behind this barrier can be removed with front loader machinery and collected free of earthworms, for subsequent drying to 35%–45% moisture, sieving, and packaging. Processing of wastes in a wedge system takes about 3–4 months, is much less labor-intensive than windrows, and needs much less space.

This system was adopted for large-scale use by a company called Green-Pyk in Russia that markets large quantities of vermicomposts produced in very large cattle sheds 100 m × 30 m (109 yd × 32.8 yd) using cattle wastes as feedstock. The company also markets aqueous extracts from vermicomposts, or "teas," which are produced on-site. The benefits and drawbacks of the wedge system are summarized in Table 7.4.

VII DISEASES AND PREDATORS OF EARTHWORMS IN VERMICOMPOSTING SYSTEMS

A Common Vermicompost Bed Organisms

Since vermicomposting systems use various forms of organic matter as feedstocks, as the basis of vermicompost production, other invertebrates that live on decaying organic matter may also infest the vermicomposting system. Most of these actually improve the productivity of the system, although people unfamiliar with them may suspect them of attacking the earthworms. Often this is because they are attracted to dead earthworms. The invertebrates commonly found in vermicomposting systems are the following:

- Enchytraeids or pot worms are small white worms closely related to earthworms but only about 2–5 mm (0.08–0.2 in) long. They should not be mistaken for unsegmented nematodes or eelworms.

- Detritivore Acarina or mites mostly live on decaying organic matter and cause no harm to earthworms.
- Collembola, or springtails, may be white or colored and range in length from 1–5 mm (0.04–0.2 in); some species can spring in the air. They cause no harm in beds, and they feed on and help to break down decaying organic matter.
- Sowbugs, woodlice, or isopods are grayish invertebrates about 5 mm (0.2 in) long with segmented plates on their backs and possessing many legs. When disturbed they can roll up into balls. They also live on decaying organic matter.
- Millipedes, some of which are cylindrical and some flat-backed, from 1.3–5.0 cm (0.51–2.0 in) in length, are segmented invertebrates with a pair of legs to each segment. They are usually dark-colored and feed on decaying matter, which helps the vermicomposting process.

B Invertebrate Earthworm Enemies

Several parasites and predatory mites can attack the cocoons of earthworm, but these are not very common in vermicomposting systems. A cluster fly, *Pollenia rudis*, that parasitizes earthworms is quite common in urban areas in the United States, but there have not been many reports of attacks on earthworms from vermicomposting systems. In recent years, several species of planarian flatworms, such as *Bipalium kewensis* and *Bipalium pennsylvanicum*, have been reported as specific predators of earthworms, but there are few reports that they are causing serious problems in vermicomposting systems.

Some large predatory carabid beetles can attack earthworms or their cocoons, but these are unlikely to be found in earthworm beds. Three species of carnivorous slugs or mollusks can prey on earthworms but are not common. Centipedes are not often found in vermicomposting systems. They have many segments and many legs, with one pair to each segment, and may be flat backed or thin, cylindrical, and elongated up to 5 cm (2.0 in) long. They are general predators of other invertebrates that could prey on cocoons or kill immature earthworms but do not usually attack earthworms.

C Vertebrate Earthworm Predators

Earthworms are attractive as food to many species of birds or mammals in the wild, but these are seldom a problem in indoor vermicomposting systems. Vertebrates that feed exclusively on earthworms are moles, which may occasionally invade vermicomposting systems and cause considerable damage.

D Diseases and Parasites of Earthworms

Earthworms have many internal parasites, including nematodes or eelworms, a wide range of protozoa, small tapeworms, and fly larvae. However, there is little evidence that any of these invertebrates cause any earthworm mortality. A wide range of microorganisms occur in the bodies of earthworms, but there are very few reports of any of them causing diseases in vermicomposting systems.

REFERENCES

Appelhof, M. 1997. *Worms Eat My Garbage.* 2nd ed. Flowerfield Enterprises LlC, Kalamazoo, MI.

Carver, D., and Christie, B. 2007. *The Biology and Business of Raising Earthworms: The Complete Guide to Worm Farming*. Bugdugle Ltd, Wellington, New Zealand.

Darwin, C. 1881. *The Formation of Vegetable Mould through the Actions of Worms with Observations of Their Habits*. Murray, London.

Edwards, C. A. (Ed.). 2004. *Earthworm Ecology.* 2nd ed. CRC Press, Boca Raton, FL.

Edwards, C. A., and Bohlen, P. J. 1996. *Biology and Ecology of Earthworms.* Chapman & Hall, London.

Edwards, C. A., and Neuhauser, E. F. (Eds.). 1988. Breakdown of animal, vegetable, and industrial organic wastes by earthworms. In: *Earthworms in waste and Environmental Management*, 21–31. SPB Publishing bv. the Hague, the Netherlands.

Price, J. S., and Phillips, V. R. 1990. An improved mechanical separator for removing live worms from worm-worked organic wastes. *Biol. Wastes* 33: 25–37.

CHAPTER 8

Medium- and High-Technology Vermicomposting Systems

Clive A. Edwards

CONTENTS

I Introduction 91
II Manually Operated Systems: Raised Continuous-Flow Vermicomposting Beds 93
III Fully Automated Continuous-Flow Vermicomposting Reactor Systems 94
IV Indian Design of Top-Loading Vermicomposting System 98
V Complete Urban Waste-Recycling Vermicomposting Systems 99
VI Earthworm Toilet Systems 99
VII Economics of Alternative Vermicomposting Systems 100
References 100

I INTRODUCTION

The principles behind vermicomposting are relatively simple and related to those involved in traditional thermophilic composting. Certain species of earthworms, termed *epigeic* species, can consume organic residuals and wastes very rapidly and fragment them into much finer particles by passing them through a grinding gizzard in their mouths, an organ that all earthworms possess (Darwin 1881). The earthworms derive their nourishment from the microorganisms that grow on the organic materials they consume. At the same time, they promote further microbial activity in the residual wastes so that the fecal material, or "casts," that they produce is much more fragmented and microbially active than the materials the earthworms consume. During this process, the important plant nutrients in the organic materials, particularly N, P, K, and Ca, are released from the organic matter and converted through microbial action into forms that are much more soluble and available to plants than those in the parent compounds.

The retention time of the waste in the gut of the earthworm is short. Earthworms can digest several times their own weight each day, and large quantities of organic matter are passed through an average population of earthworms in a vermicomposting system so that in a finished vermicompost all of the organic matter has passed through the earthworms (see Chapter 2). In the traditional thermophilic aerobic composting process, the organic materials in windrows have to be turned regularly or aerated in some way, usually in containers, to maintain aerobic conditions. This often may involve extensive engineering to process the residual wastes as rapidly as possible on a large scale. In vermicomposting, the earthworms, which can survive only under aerobic conditions, take over the roles of both turning and maintaining the organics in aerobic conditions, thereby lessening much of the need for expensive engineering; they have been called "ecological engineers" (Edwards and Bohlen 1996).

The major constraint is that, in contrast to traditional composting (a thermophilic process that can raise temperatures in the waste to more than 70°C (158°F)), vermicomposting is a mesophilic process; vermicomposting systems must be maintained at temperatures below 35°C (95°F). Exposure of the earthworms to temperatures above this, even for relatively short periods, will kill them. Avoiding such overheating requires careful management and addition of wastes in thin, carefully managed layers. Earthworms are active and consume organic materials in a relatively narrow layer of 15–20 cm (6–9 in) below the surface of a compost heap or bed. The key to successful vermicomposting lies in adding materials to the surface of piles or beds in thin, successive layers so that any thermophilic-composting heating that occurs does not become excessive. The heating, however, should be managed to maintain the activity of the earthworms at high levels of efficiency. Generally speaking, adding 2.5 cm (1 in) of material to a system every day or two is usually enough, depending on the types of feedstock used in vermicomposting systems.

The processing of organic materials by most epigeic species of earthworms occurs most rapidly at temperatures between 15°C and 25°C (60°F to 79°F) and at moisture contents of 70%–90%. Outside these limits, earthworm activity and productivity, and thus the rate of waste processing, fall off dramatically. For maximum efficiency, the feedstock should be maintained as close to these environmental limits as possible, hence should be under cover in a building or insulated plastic tunnel.

Almost any agricultural, urban, or industrial organic waste can be used for vermicomposting, but many may need some form of preprocessing to make them acceptable to earthworms. Such preliminary treatments can involve washing, precomposting, macerating, or mixing. Often, mixtures of several different materials can be processed more readily than individual wastes, and mixtures are usually easier to maintain aerobically and can result in a better, if less standardized, product. This is primarily due to the wet and sloppy characteristics of some waste materials, which require the addition of a bulking agent for better handling. Organic cattle, pig, and horse wastes, residuals from the paper industry, food wastes, sewage biosolids, and urban organic wastes are particularly suitable for vermicomposting.

The traditional smaller-scale methods of vermiculture described and reviewed in Chapters 7 and 24 are mainly based on beds or windrows on the ground containing materials to a depth of about 50 cm (18 in), but such methods have numerous drawbacks. They require large areas of land for large-scale production and are relatively labor-intensive, even when appropriate machinery is used for adding materials to the beds. More important, such systems process organics relatively slowly, taking anywhere from 6 to 18 months to completely break them down (see Chapter 7). There is good evidence that a large proportion of the essential plant nutrients, which are in a relatively soluble form, are washed out during such a period. Moreover, a significant proportion of nutrients can volatilize during such a long processing period. Such nutrient losses are undesirable, particularly in relation to groundwater pollution, and result in a poor-quality product.

II MANUALLY OPERATED SYSTEMS: RAISED CONTINUOUS-FLOW VERMICOMPOSTING BEDS

It is essential for successful vermicomposting that organic wastes are added to beds or other systems in thin layers 2–3 cm (0.8–1.2 in) frequently to prevent thermophilic composting from developing (Edwards and Neuhauser 1988; Edwards 2004; Edwards and Arancon 2004). This can be done in beds about 1 m (3 ft) deep and 4–5 m (12–15 ft) wide with floor drainage. However, such systems have to be harvested when vermicomposting is complete using front-loading or other appropriate mechanical equipment. They will still require a stage of separation of earthworms from the processed organic waste using trommels or comb systems as described in Chapter 7.

The key principle of all efficient earthworm-based, organic-waste-processing systems, whatever their design configuration, is to add small, 1–3 cm (1–1.5 in) layers of semisolid waste (75%–90% moisture content) to the surface of a system daily. This gradual addition of organic matter avoids any overheating developing from a thermophilic phase of composting, although with good management sufficient heat from thermophilic composting can be generated to keep the system operating through colder spells in the winter in colder regions. The system can be managed to keep temperatures in the building or tunnel lower in summer using fans and higher in winter within a range of 18°C–22°C (64–72°F). The earthworms remain active in the aerobic upper 10–15 cm (4–6 in) of waste, taking it through their guts, breaking it down into fine fragments and particles by a grinding gizzard, mixing it, and keeping it aerobic. As additional layers of waste are added at daily intervals, earthworms gradually move up vertically through the reactor so that they always remain in the top aerobic 10–15 cm (4–6 in) of the organic waste.

All of the most efficient vermicomposting techniques have used containers raised on legs above the ground. These allow waste feedstock to be added at the top from mobile gantries and be collected mechanically at the bottom through mesh floors using breaker bars to drop it to the floor. Such methods were developed and tested extensively by engineers at the National Institute for Agricultural Engineering,

Silsoe, in collaboration with soil ecologists from Rothamsted Experimental Station in England. The methods they designed ranged from relatively low-technology systems using manual loading and collection to completely automated systems. The fully processed vermicompost can be collected from the bottom of a system by mechanically releasing the lower levels of vermicompost through a mesh grid, by use of a breaker bar that travels along the length of the reactor, so that it can be collected with no loss of earthworms, which remain in the upper layers of the chamber.

III FULLY AUTOMATED CONTINUOUS-FLOW VERMICOMPOSTING REACTOR SYSTEMS

The full-scale continuous-flow reactor systems designed by Edwards at Rothamsted and colleagues from the National Institute of Agricultural Engineering (Edwards 2004) are 40 m (128 ft) long by 2.4 m (8 ft) wide, built in 2.4 m (8 ft) by 2.4 m (8 ft) modules (Figure 8.1), on a metal frame, with plywood or plastic sides sealed against moisture. The body of the reactor is 1 m (3.2 ft) deep, standing on legs 1 m (3.1 ft) high (Figure 8.2; Edwards 1998), and it must be maintained in a building under cover to control environmental processing conditions. It has a hydraulically winch-driven, gear-operated feed hopper (Figure 8.3) that runs on the rails at the top of the sides of the reactor and can automatically add a thin layer 1–3 cm (1–1.2 in) of waste to the surface of the reactor every 1–2 days in one or more horizontal transit passes. The base of the upper part of the reactor has a metal mesh floor with a 5 cm × 10 cm (2–4 in) mesh aperture. Situated immediately above the perforated floor is an electrically winched breaker bar (Figures 8.4 through 8.6), which can be pulled along the full length of the reactor in either direction. As this bar moves along the length of the reactor, it disturbs the lowest 2.5–5.0 cm (1–2 in) of processed waste to release this layer through the perforated base to the ground. This processed

Figure 8.1 Basic reactor module (2.44 m × 2.44 m (8 ft × 8 ft)).

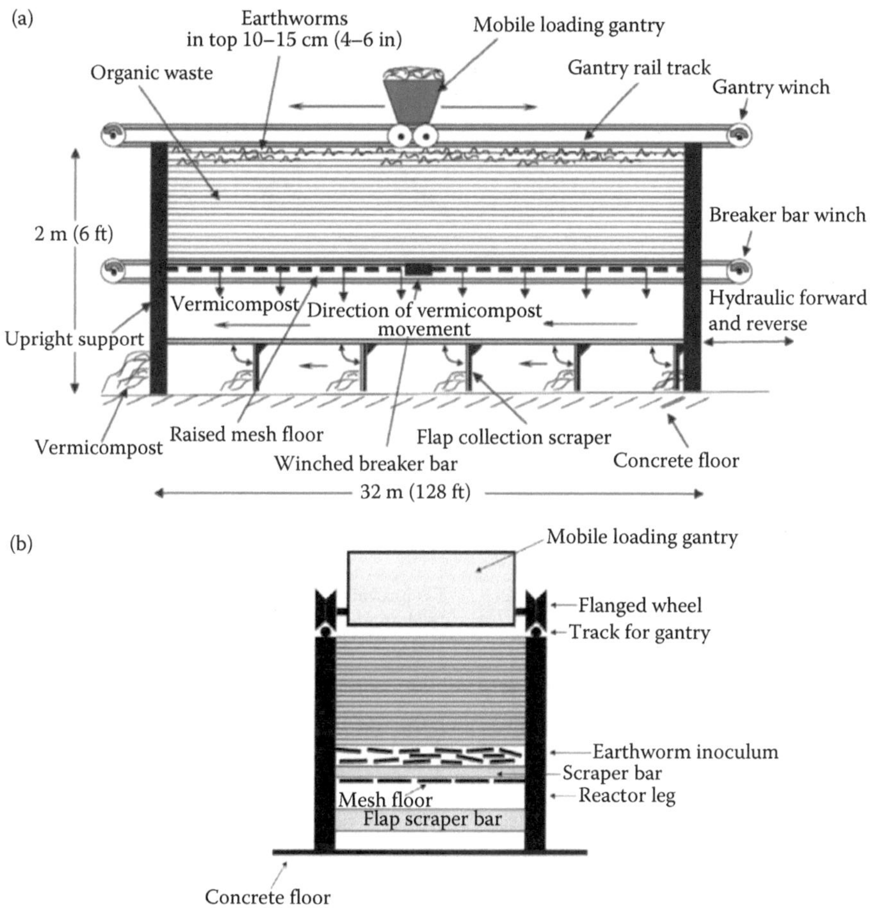

Figure 8.2 (a) Freestanding reactor (side view) and (b) Freestanding reactor (end view).

waste can then be brought to one end of the reactor daily by a hydraulically driven, reversible scraper dairy-barn-cattle-waste flap-bar system (Figure 8.5) or moving belt. The reactor functions best with an equilibrium loading of about 9 kg.m^{-2} (2 lb.ft^{-2})of earthworms (*Eisenia fetida*). A single full-scale reactor of this size, which has a 30–60-day retention time for the waste to pass from the top to bottom, can process more than 1000 tons (984 t) of animal, plant, human, or food waste or sewage biosolids per annum or up to 3 tons (3 t) per day depending on the type of waste. If multiple reactors are used, there are considerable economic savings in construction and operation.

Such reactors have operated successfully in the United Kingdom, United States, Hong Kong, and Australia for periods of up to 12 years (Edwards 1998). The economics of reactor systems is discussed in Chapter 19. The earthworm populations in such reactors tend to reach an equilibrium biomass of about 9 kg.m^{-2} (2 lb. ft^{2}),

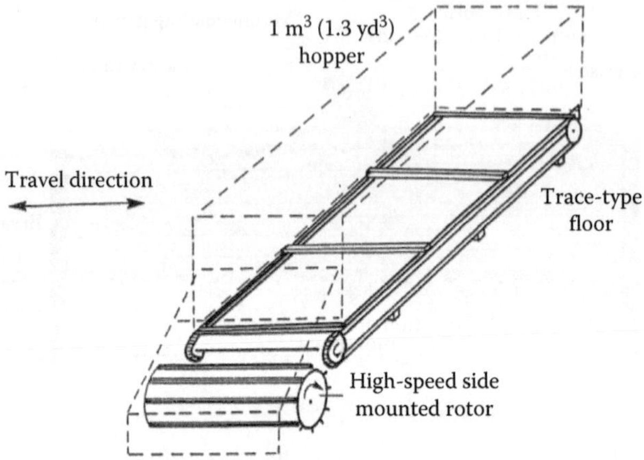

Figure 8.3 Gantry feed hopper with full-width rotor.

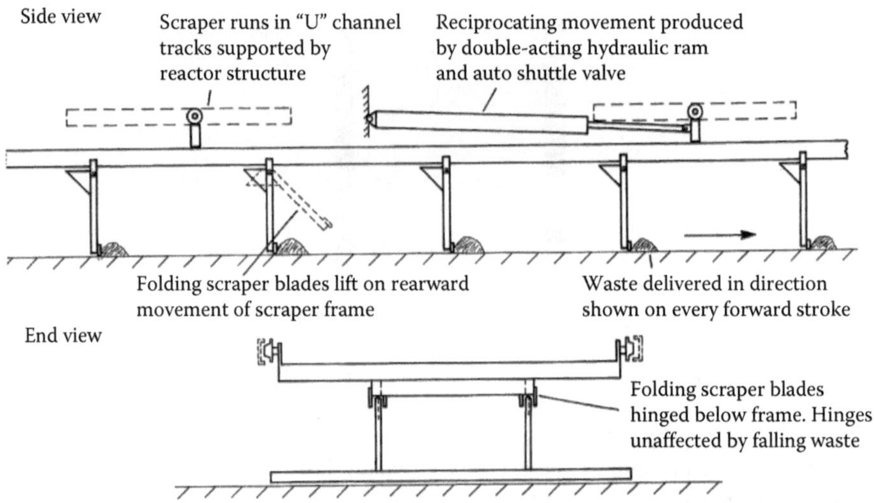

Figure 8.4 Reciprocating scraper vermicompost-collection system (not to scale).

which may be attained as a mixed-age population of earthworms. These reactors can process fully the whole 1 m (3 ft) depth of suitable organic wastes that they contain in about 30 to 60 days, depending on the kind of waste being processed (Edwards 1995, 1998). Economic studies have shown that such reactors have a much greater economic potential to produce high-grade plant growth media, with little loss of material through leaching or volatilization, very quickly and much more efficiently than either windrows, wedge systems, or ground beds.

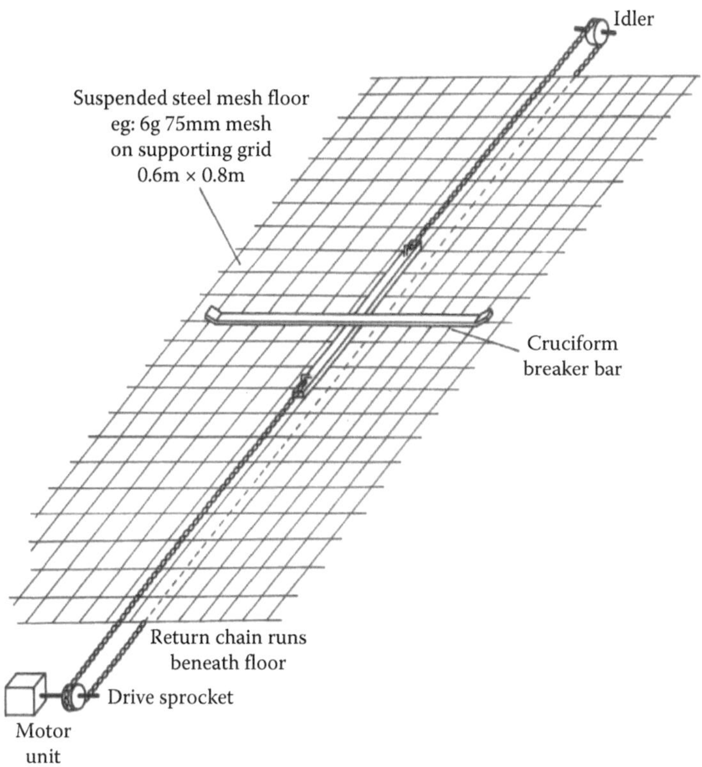

Figure 8.5 Layout of breaker-bar discharge floor with one possible drive system.

Figure 8.6 Breaker-bar discharger and winch unit.

IV INDIAN DESIGN OF TOP-LOADING VERMICOMPOSTING SYSTEM

An interesting system of vermicomposting has been designed by the Rallis Fertilizer Company in India. The system is designed to process urban wastes, including food wastes, into vermicomposts in a continuous-flow pattern using an innovative concept (see Chapters 8 and 28). The processing units consist of two concrete walls 1 m (3 ft) deep on a concrete base, and the units are 3 m (9.2 ft) wide. Down the center of each unit, there is a line of upright pipes about 1 m (3.1 ft) high with multiple perforations, which are linked to an air-compression system running the length of each bed (Figure 8.7).

The systems use the tropical earthworm *Eudrilus eugeniae* to process food and urban organic wastes. The waste is added daily in 2.5–5.0 cm (1–2 in) layers, and the upper surface is kept moist by regular sprays. Each system is under cover so that moisture and temperature in the beds can be controlled. Organic wastes continue to be added above the heights of the bed walls until the waste is about 60 cm (2 ft) higher than the walls. At this point no more waste is added, watering of the bed stops, and compressed air is pumped into the bed, which makes it aerobic again. Within a few days, all of the earthworms retreat into the lower part of the bed. Then the surface layer of vermicompost, flush with the walls, can be removed with front-loading equipment. It is then screened through a very large rotating trommel and marketed in 50 kg (110 lb) plastic bags.

Five groups of vermicomposting systems of this kind are being operated in India. Each site has about 40 ha (100 acres) of land. Two sites are near Bangalore, two in Chennai, and one in Hyderabad. All sites process about 6000 (6096 tons) metric tons of urban wastes per day. Total vermicompost production is 20,000 metric tons (20,320 t) per annum, and total thermophilic compost production is 250,000 metric tons (254,000 tons) per annum. The vermicompost is sold under the trade name of Ralligold and the compost as Black Gold. These are marketed by a company called Caramandal Fertilizer Co.

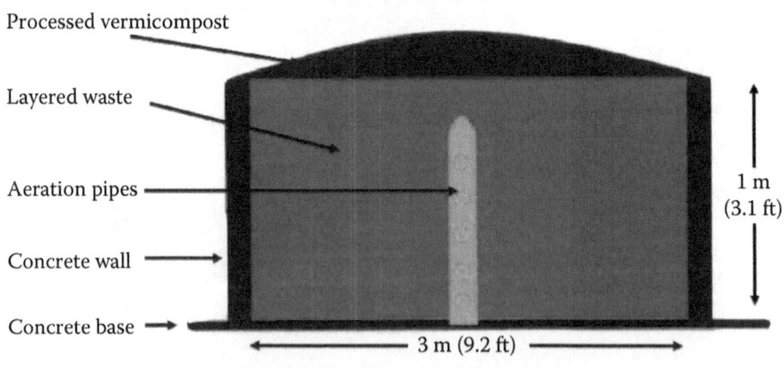

Figure 8.7 Indian top-harvesting vermicomposting system.

V COMPLETE URBAN WASTE-RECYCLING VERMICOMPOSTING SYSTEMS

A vermiculture-based urban waste-recycling system that can handle all the waste from a community was developed in Montelimar in southern France, under a system termed NATURBA with the commercial name of SOVADEC. This involves processing all the total urban waste stream from a community of 40,000 people by first passing it through a rotating selector that breaks up plastic materials and removes them by heating cables, followed by manual sorting, separation of rolling objects such as bottles, and separation of ferrous metal objects with magnets. The waste is then transported to a thermophilic composting system and kept there for 30 days, followed by vermicomposting in a very deep continuous-flow vermicompost system for about 60 days (Figure 8.8); the earthworms are then removed and the vermicompost is packaged. This system was able to turn as much as 27% of a total urban waste stream into vermicompost. This vermicompost can be marketed profitably and adds to the commercial potential of complete waste recycling very considerably.

VI EARTHWORM TOILET SYSTEMS

Various systems of self-maintaining human toilets based on earthworms, as an alternative to septic systems, have been developed in various parts of the world. The most successful is the Dowmus Composting Toilet. This system is well engineered, completely odor free, and built in Australia. It needs little maintenance over several years and has been adopted for general use in some Australian state parks (Windust 1994) (Figure 8.9).

Figure 8.8 Vermicomposting unit of the SOVADEC waste-recycling system.

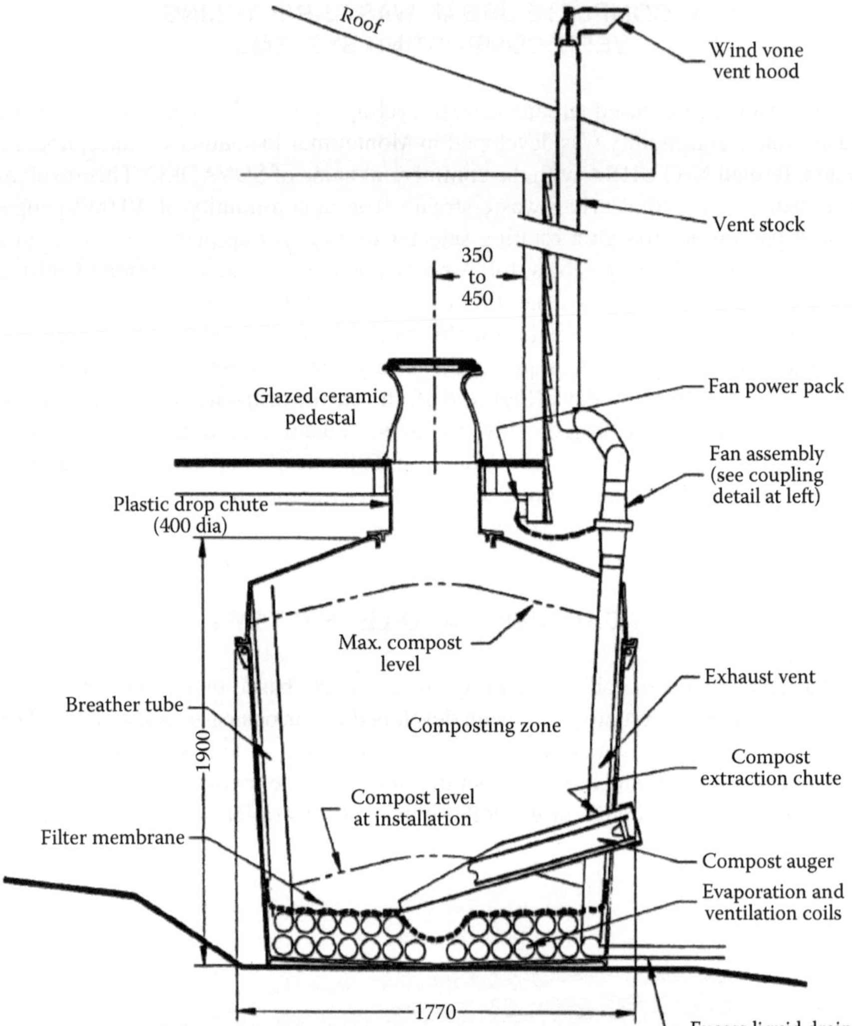

Figure 8.9 Australian Dowmus vermicomposting toilet (all units cms (2.5 in)).

VII ECONOMICS OF ALTERNATIVE VERMICOMPOSTING SYSTEMS

The use of organic wastes to grow earthworms and produce vermicomposts is an extensive farm and garden cottage industry in the United States and other parts of the world. Many of these small-scale producers market the earthworm castings they produce locally for growing plants. Most such operations in the United States are based primarily on outdoor windrow systems, which have many economic and environmental drawbacks as discussed in Chapter 7. They are ground-based and require large areas of land, and they have potential for groundwater pollution with nutrients and other contaminants, because they are watered regularly and usually have no protection against leaching from the beds. The process is slow, often taking 6–12 months to complete. The harvesting of the vermicompost is laborious and time-consuming because the earthworms in the waste have to be separated, usually by a screening or trommel process, before the vermicompost is marketed. Although the initial capital outlay, other than land, is low, its labor costs are high at all stages of operation. As a good economic alternative, the wedge system designed by Edwards and colleagues, described in Chapter 7, has been used by a number of organizations. Although it uses an innovative but relatively inexpensive technology and requires less equipment, it overcomes many of the labor, economic, and environmental drawbacks associated with windrows. In particular, it uses less land, there is no leaching into groundwater, and there is no need to separate earthworms from the vermicompost. The processing time is also shorter (3–4 months).

The automated continuous-flow reactor system designed by Edwards and his colleagues (Edwards 1995, 1998) has totally different environmental, operating, and economic characteristics. The equipment has to be maintained under cover to ensure controlled environmental conditions, and the organic-waste compartment is raised above the floor and is maintained at 70%–80% moisture content and 20°–35°C; it is important that there is no leaching of nutrients. The retention or processing time of most organic wastes in the reactor is 30–60 days. The economics of such automated reactors (discussed in Chapter 19) are totally different from those of other systems because they involve a need for high initial capital inputs for reactors ($35,000–$50,000) each which can process up to 3 tons of waste per day, as well as ancillary loading and transport equipment, including: moving belts, macerators, and loaders for operation. However, their labor and running costs are extremely low, and the earthworm populations in reactors reach equilibrium and can usually be run trouble-free without adding or removing earthworms for a number of years. The capital expenditure can usually be recovered in 1–2 years. A number of relatively expensive smaller systems based on the principles of this system have been marketed in the United States but are much less attractive economically.

REFERENCES

Darwin, C. 1881. *The Formation of Vegetable Mould through the Action of Worms, with Observation of Their Habits*. Murray, London.

Edwards, C. A. (Ed.). 2004. *Earthworm Ecology.* 2nd ed. CRC Press, Boca Raton, FL.

Edwards, C. A., and Arancon, N. Q. 2004. The use of earthworms in the breakdown of organic wastes to produce vermicomposts and animal feed protein. In *Earthworm Ecology*, ed. C. A. Edwards, 345–381. 2nd ed. CRC Press, Boca Raton, FL.

Edwards, C. A., and Bohlen, P. J. 1996. *Biology and Ecology of Earthworms* (3rd Ed.). Chapman and Hall, London.

Edwards, C. A., and Neuhauser, E. F. (Eds.). 1988. *Breakdown of Animal, Vegetable, and Industrial Organic Wastes by Earthworms* 21–23, SPB Publishing, The Hague, The Netherlands.

Edwards, C.A. 1995. Historical overview of vermicomposting. *Biocycle*, 56–58.
Edwards, C.A. 1998. Vermicomposting's potential: Small to Large-scale operations, *Resource Recycling*, *Section A*, 16–18.
Windust, A. 1994. *Worms Downunder, Downunder: For Farm, Garden, Schools, Profit, and Recycling*. Allscape, Mandurang, Victoria, Australia.

CHAPTER 9

The Potential of Vermicomposts as Plant Growth Media for Greenhouse Crop Production

Norman Arancon, Clive A. Edwards, Katie A. Webster, and John C. Buckerfield

CONTENTS

I Introduction 103
II Nature of Vermicomposts 104
  A Characteristics of Vermicomposts 104
  B Physical Properties of Vermicomposts 104
  C Chemical Properties of Vermicomposts 105
  D Biological Properties of Vermicomposts 105
III Effects of Vermicomposts on the Growth of Greenhouse Crops 106
IV Factors Influencing Increases in Growth and Yield 117
  A Plant Growth Hormones in Vermicomposts 117
  B Humates in Vermicomposts 121
References 124

I INTRODUCTION

Animal manures, food wastes, yard waste, and sewage sludges have long been recognized in agriculture as beneficial organic soil amendments for the maintenance of soil fertility to support crop growth. The new approaches to the use of organic amendments in farming have proved to be an effective ecological means of improving the physical, chemical and biological properties of soils. Organic matter is an excellent source of plant-available nutrients, and its addition to soil maintains high microbial populations and activities (Pascual et al. 1997; Zink and Allen 1998) with increased values of biomass C, basal respiration, biomass C to total organic C ratios, and metabolic quotients (qCO_2). Many crop yields have increased with the improvements in soil quality resulting from addition of organic wastes. Significant yield increases have been reported using mulches from coffee husks (Bwamiki et al. 1998), as well as increases in crop productivity using animal manures and hay residues (Johnston et al. 1995). The important roles of organic amendments to soil and their potentially positive effects on crop yields make them a valuable component of farm soil nutrient-management programs in organic and alternative agriculture and also in traditional agriculture.

A process related to composting, which can improve even further the beneficial utilization of organic wastes, is vermicomposting. Various workers have examined the suitability of vermicomposts as plant growth media and have addressed their potential commercial value in the horticultural industry. This Chapter focuses on the physical and biological characteristics of vermicomposts and their uses in the greenhouse. The Chapter also discusses the possible mechanisms by which vermicomposts can influence plant growth.

II NATURE OF VERMICOMPOSTS

A Characteristics of Vermicomposts

Vermicomposts are products derived from the accelerated biological degradation of organic wastes by interactions between earthworms and microorganisms. Earthworms consume and fragment the organic wastes into finer particles by passing them through a grinding gizzard, and they derive their nourishment from the microorganisms that grow on the organic matter. The process accelerates the rates of microbiological decomposition of the organic matter, increases microbial populations, and alters the physical and chemical properties of the material, leading to accelerated humification, during which the unstable organic matter is fully oxidized and stabilized (Albanell et al. 1988; Orozco et al. 1996).

B Physical Properties of Vermicomposts

From a morphological point of view, vermicomposts are dark and homogeneous, with a mull-like soil odor. They have greatly increased surface areas, providing more microsites for microbial decomposing organisms, and strong adsorption and retention of nutrients (Shi-wei and Fu-zhen 1991). Albanell et al. (1988) reported that vermicomposts tend to have pH values near neutrality, which may be due to the production of CO_2 and the organic acids produced during microbial metabolism.

Elvira et al. (1996) reported that humification rates were increased significantly in paper-pulp mill sludge processed by the earthworm *Eisenia andrei* (Bouché). The transformations into humic compounds during passage through the earthworm gut showed that the rates of humification of ingested organic matter were intensified during gut transit (Kretzschmar 1984).

C Chemical Properties of Vermicomposts

Edwards and Burrows (1988) and Arancon and Edwards (2004) reported that vermicomposts, especially those produced from animal waste manures, usually contained greater quantities of mineral elements than commercial plant growth media, and many of these elements were in forms that could be taken up more readily by the plants, such as nitrates, exchangeable P, and soluble K, Ca, and Mg. Orozco et al. (1996) reported that vermicomposting of coffee pulp increased the availability of nutrients such as P, C, and Mg. Werner and Cuevas (1996) reported that most vermicomposts contained adequate amounts of macronutrients, micronutrients, and trace elements of various kinds, but amounts inevitably depended on the type of the parent earthworm feedstock. Businelli et al. (1984) reported similar differences in vermicomposts' chemical compositions, based on the substrate used to produce them. In their experiments, the largest elemental values were recorded in vermicomposts from a mixture of cattle and horse manure with 38.8% organic C, 2.7% total N, and 1080 mg. kg^{-1} (0.017 oz. lb^{-1} No_3–N). Edwards (1988) reported larger amounts of mineral nutrients in vermicomposts compared to a commercial plant growth medium. The wastes they investigated were separated cattle solids, separated pig solids, cattle solids on straw, pig solids on straw, duck solids on straw, and chicken solids on shavings. These materials contained mineral contents (% dry weight) ranging from 2.2–3.0 N, 0.4–2.9 P, 1.7–2.5 K, and 1.2–9.5 Ca, whereas the commercial plant growth medium had only 1.80 N, 0.21 P, 0.48 K, and 0.94 Ca. The quantity and quality of the nutrients in vermicomposts can be explained by the accelerated mineralization of organic matter, increased microbial activity, breakdown of polysaccharides, and higher rates of humification achieved during vermicomposting (Albanell et al. 1988; Elvira et al. 1996). In investigations into the bioconversion of solid paper-pulp mill sludge by earthworms, it was reported that the total carbohydrate content decreased while the total extractable C, nonhumified fraction, and humification rates increased by the end of the experiment (Elvira et al. 1997).

D Biological Properties of Vermicomposts

The main qualities of vermicomposts that make them superior organic soil amendments are their excellent biological properties. These properties have been linked not only to nutrient cycling in the soil or plant growth media but also to their ability to supply crops with plant growth hormones (PGHs) such as auxins, gibberellins, and cytokinins and also fulvic and humic acids functioning as plant growth regulators (PGRs). Vermicomposts are rich in bacteria, actinomycetes, fungi (Edwards 1983; Tomati et al. 1987; Werner and Cuevas 1996), and cellulose-degrading bacteria (Werner and Cuevas 1996). In addition, Tomati et al. (1983) reported that earthworm castings, obtained after sludge digestion, had large microorganism populations, especially bacteria.

III EFFECTS OF VERMICOMPOSTS ON THE GROWTH OF GREENHOUSE CROPS

The utilization of vermicomposts in the horticultural industry has increased rapidly since the publication of a range of positive results from their use in greenhouse experiments. Improvements in the growth and development of greenhouse crops after addition of vermicomposts, results in much greater economic returns to the grower, due to faster germination rates, earlier flowering, larger yields, and better-quality crops. Economic returns could also come by savings from the replacement of costly peat-based plant growth media. Some of our research attempted to elucidate the mechanisms involved in these plant responses. The possible mechanisms for increased crop productivity include a combination of factors such as improved physicochemical qualities of plant growth media after substitution of vermicomposts into them, biological changes in the growth media, and the production of plant growth hormones and plant growth regulators. In greenhouse experiments with vegetables, Edwards and Burrows (1988) reported that vermicomposts improved seedling emergence over that in a commercial plant growth medium, using a wide range of test plants such as peas, lettuce, wheat, cabbages, tomatoes, and radishes. Additionally, the growth of ornamental shrubs such as *Eleagnus pungens*, *Cotoneaster conspicua*, *Pyracantha*, *Viburnum bodnantense*, *Chaemaecyparis lawsonia*, *Cupressocyparis leylandii*, and *Juniperus communis* was better in a vermicompost mixture than in a commercial plant growth medium. They also reported that some ornamental plants, such as chrysanthemums, salvias, and petunias, flowered earlier compared to those grown in commercial peat and plant growth media. Plant growth was promoted, even in response to a 5% substitution of a 50:50 mixture of pig and cattle manure vermicomposts into a commercial plant growth medium. Similar results were obtained by Atiyeh et al. (1998), who demonstrated that vermicomposts increased vegetable and ornamental seedling growth at relatively low concentrations, even when all needed nutrients were available.

In a greenhouse pot trial by Buckerfield et al. (1999), using 0%–100% mixtures of vermicompost and sand, similar trends in growth responses were reported. Although the germination of radishes decreased by less than 50% with increasing vermicompost concentrations, growth and radish harvest weights increased directly in proportion to the application rates of vermicomposts, with the yields of plants grown in 100% vermicompost being up to 10 times greater than those grown in 10% vermicompost. The best growth of tomatoes, lettuce, and peppers was reported at substitution rates of 10%, 8%, and 6%, respectively, using a duck waste vermicompost and peat mixture (Wilson and Carlile 1989). It was shown further that rates of plant nutrient uptake correlated with rates of plant growth and development.

Subler et al. (1998) reported increases in plant growth in mixtures of vermicomposts with a soilless medium, MM360, in comparison to that in two commercial growth media produced from biosolids and yard waste, using common bedding plants, tomatoes, and marigolds as test plants. In their experiments, considerable differences in chlorophyll contents were observed during the early stages of marigold growth in response to vermicompost applications. At the end of the trial, significant increases, in leaf areas and in total plant weights were reported in response to combinations of 10% vermicompost and 90% MM360, compared with those in the control 100% MM360. Significant increases in tomato seedling weights, at substitution rates of 10% and 20% MM360, were also reported. Raspberries grown in 80% MM360 substituted with 20% pig manure vermicompost, produced the greatest shoot dry weights. In an investigation by Scott (1988), using hardy nursery stocks, 20%–50% substitutions of vermicompost from cattle manure, pig manure, and duck waste into a peat:sand growth medium, with a full application of nutrients, produced better growth than a peat:sand mixture used as a control. Chan and Griffiths (1988) reported stimulating effects of pig manure vermicomposts on the growth of soybeans (*Glycine max*), particularly in terms of increased root lengths, lateral root numbers, shoot lengths, and internode lengths of seedling plants. Another rooting experiment using vermicomposts showed that the establishment of *Vanilla planifolia* cuttings was much better in vermicomposts than in other planting media used, such as coir pith and sand (Siddagangaiah et al 1996). Similar responses were observed for cloves (*Syzygium aromaticum*) and black peppers (*Piper nigrum*) sown in a 1:1 ratio of vermicompost and soil mixtures (Thankamani et al. 1996). Black pepper cuttings grown in vermicomposts were significantly taller and had more leaves than those grown in commercial potting mixtures. Plant heights, numbers of branches, and the longest taproots were on cloves grown in the vermicompost mixture. Vermicomposts produced from coir dust also increased the yields of onions, *Allium cepa* (Thanunathan et al. 1997).

The effects of vermicomposts produced from pig manure on the growth and yields of tomatoes were investigated by Atiyeh et al. (2000). Vermicomposts were substituted into a commercial greenhouse media, MM360, at the rates of 0% (control), 10%, 20%, and so on at 10% increments up to 100% (with all nutrient levels balanced). The germination rates of tomato seeds increased significantly upon substitution of 20%, 30%, and 40% vermicompost into MM360. Substitution of 10%–50% vermicompost into MM360 increased the dry weights of tomato seedlings significantly compared to those grown in the 100% MM360 controls. The largest marketable yields 5.1 kg (11.2 lb) per plant in terms of tomato quality were in response to the substitution of 20% vermicompost into MM360, and mean weights of tomato fruits in a blend of 80% MM360 with 20% vermicompost were 12.4% greater than those grown in 100% MM360. Substitution of MM360 by 10%, 20%, and 40% vermicompost decreased the proportions of tomato fruits that were nonmarketable and produced more large-size diameter > 6.4 cm (2.5 in) than small-size diameter <5.8 cm (2.3 in) fruits (Figure 9.1).

It was most interesting, in those experiments, that seedlings grown in 100% pig manure vermicompost were significantly shorter, had fewer leaves, and weighed less than those grown in the 100% MM360 controls. This confirmed earlier conclusions by Atiyeh et al. (1998) that vermicomposts did not always increase crop growth and yields more in response to substitutions of larger concentrations of vermicompost than to smaller concentrations. Atiyeh et al. (2002) documented the effects of vermicomposts produced from pig manure, using similar rates of substitution into MM360, on marigolds as a test crop in the greenhouse with all needed nutrients

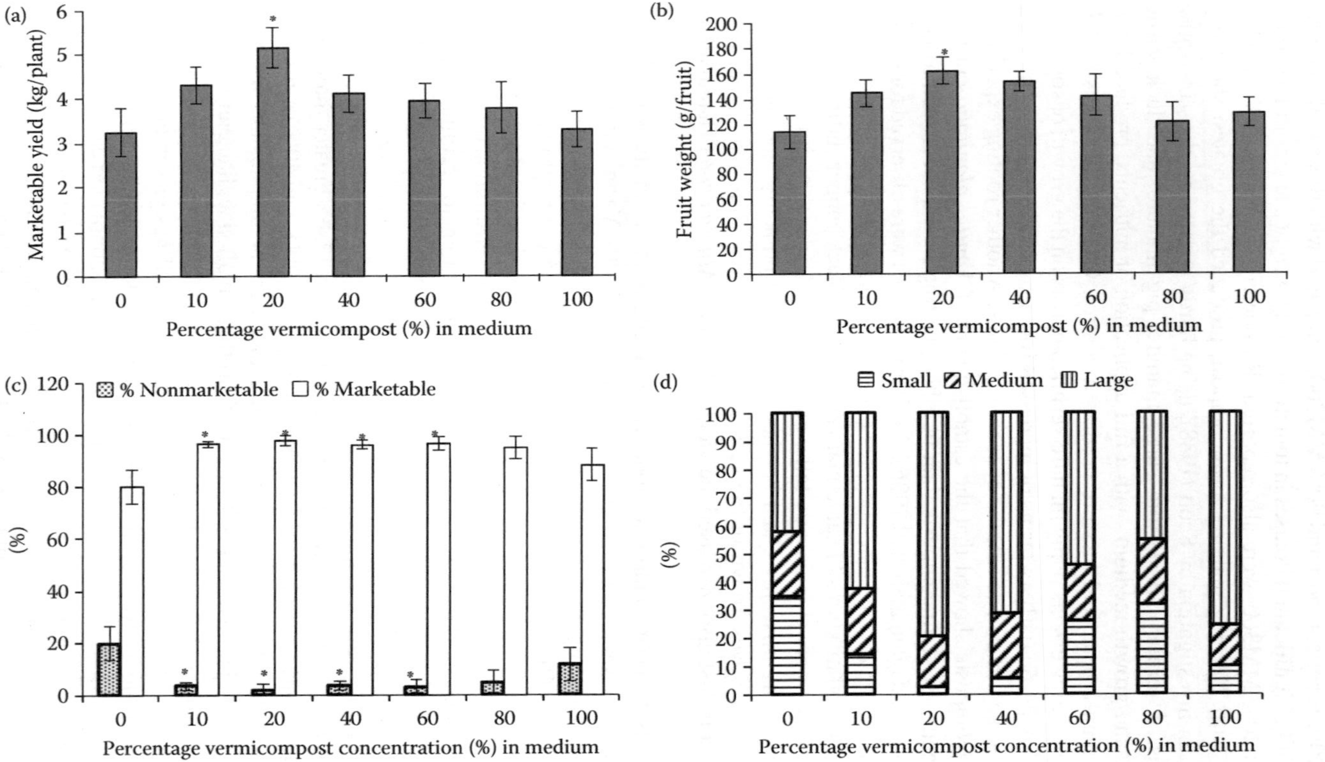

Figure 9.1 Yield and sizes of tomato fruits (mean ± SE) produced in a standard commercial potting medium (MM360) substituted with different concentrations of pig manure vermicompost. Columns followed by * are significantly different from the MM360 control (0% vermicompost) at $p \leq 0.05$. (Plants received all needed nutrients.)

provided. The greatest vegetative growth resulted from substitution of MM360 by 30% and 40% pig manure vermicompost, and the smallest growth responses were in the potting mixtures with 90% and 100% vermicompost. The most flower buds were produced by the potting mixtures containing 40% pig manure vermicompost and 60% MM360, and the fewest in the potting mixtures containing 100% vermicompost. Marigolds grown in MM360 substituted with 90% and 100% pig manure vermicompost had the fewest flowers. After substitution of 80%, 70%, 60%, 50%, 40%, 30%, 20%, and 10% MM360 with 20%, 30%, 40%, 50%, 60%, 70%, 80%, and 90% vermicompost, the marigold roots were all larger than those of plants grown in the MM360 controls (Figure 9.2).

In the Soil Ecology Laboratory at The Ohio State University, vermicomposts produced from food wastes were tested on bell peppers (*Capsicum annum*) using similar substitution rates into MM360 (Arancon et al. 2004). Peppers were grown in 30 L (7.93 gal) polystyrene pots in the greenhouse, watered regularly with tap water, and fertilized three times a week with 20-10-20 (200 ppm N). Peters Professional Plant Nutrient solution to provide all nutrients the plants needed. Peppers grown in potting mixtures containing 40% food waste vermicompost and 60% MM360 yielded 45% higher fruit weights and had 17% greater mean numbers of fruits than those grown in 100% MM360. Peppers grown in potting mixtures containing 40% food waste vermicompost and 60% MM360 produced the largest fruit compared to all the peppers planted in the other vermicompost and MM360 mixtures. A mixture of only 10% vermicompost in 90% MM360 was sufficient to increase both pepper fruit yields and fruit sizes, compared to peppers planted in 100% MM360 (Figure 9.3). The mean heights, numbers of buds, and numbers of flowers of peppers grown in potting mixtures containing 10%–80% vermicompost also increased. A trend of decreased growth and yield parameters in response to the highest rates of vermicompost substitution, similar to that reported for other crops by Atiyeh et al. (2002), was also documented for peppers.

Using similar substitution rates of vermicomposts into MM360, three types of vermicomposts (produced from cattle manure, food wastes, and paper wastes) were investigated for their effects on the growth and flowering of petunias (Arancon et al. 2008). As in similar research with other crops in the Soil Ecology Laboratory at The Ohio State University (Atiyeh et al. 2002; Arancon et al. 2004), plants in all of the treatments received all needed inorganic nutrients. The main aim of supplying all required nutrients was to rule out possible confounding effects of nutrients in the growth responses of test plants. There was a consistent tendency for the petunia seeds to germinate faster, often significantly so, when vermicomposts were in the mixtures, with seedlings often emerging even as early as 6 days after sowing in all types of vermicomposts (Figure 9.4a, b, and c).

Petunias germinated poorly in mixtures of 80%, 90%, and 100% cattle manure vermicomposts with 20%, 10%, and 0% MM360 and did not survive more than a few weeks of growth in the greenhouse. However, petunias grown in mixtures containing 10%–60% cattle manure vermicompost produced significantly greater ($p \leq 0.05$) shoot dry weights than those grown in 100% MM360 (Figure 9.5a) independent of the nutrient supply. Petunias grown in the mixture of 40% cattle manure vermicompost

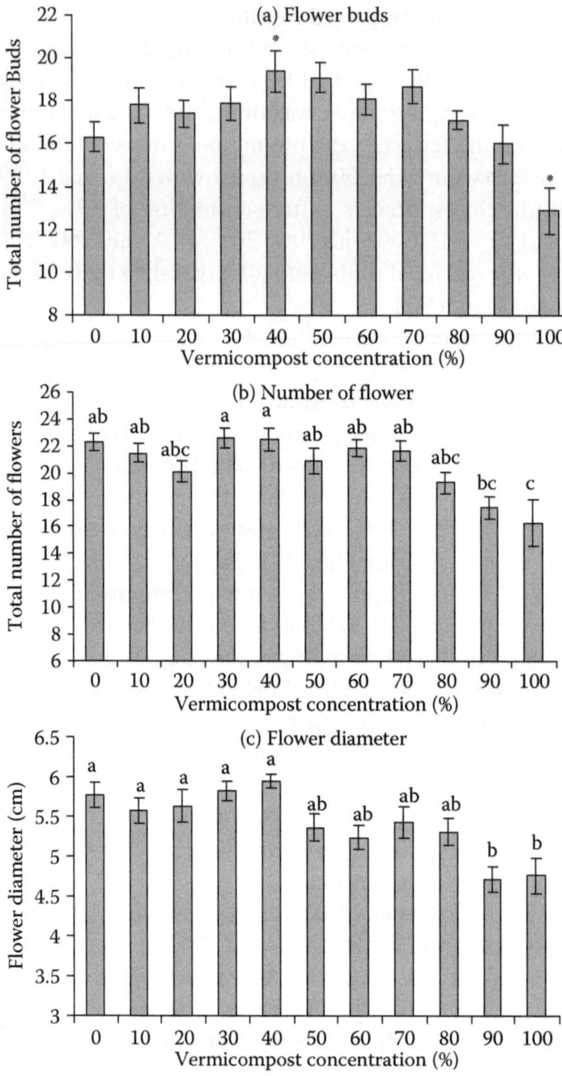

Figure 9.2 **(a)** Total number of flower buds and (b) flowers, as well as (c) flower diameter (means ± SE), produced by marigolds grown in commercial potting medium (MM360) substituted with different concentrations of pig manure vermicompost. Means marked by an asterisk (*) are significantly different from the control (0% vermicompost) at $p \leq 0.05$. Plants had all needed nutrients. Means designated by same letters do not differ significantly ($p < 0.05$).

and 60% MM360 produced much larger shoot dry weights, significantly greater than the shoot dry weights of petunias grown in MM360 with substitutions of 10%, 20%, 30%, 50%, and 0% cattle manure vermicompost, independent of the nutrient supply. Root dry weights showed a similar trend, with petunias grown in 10%–60% cattle manure vermicompost substituted into MM360 producing significantly greater dry

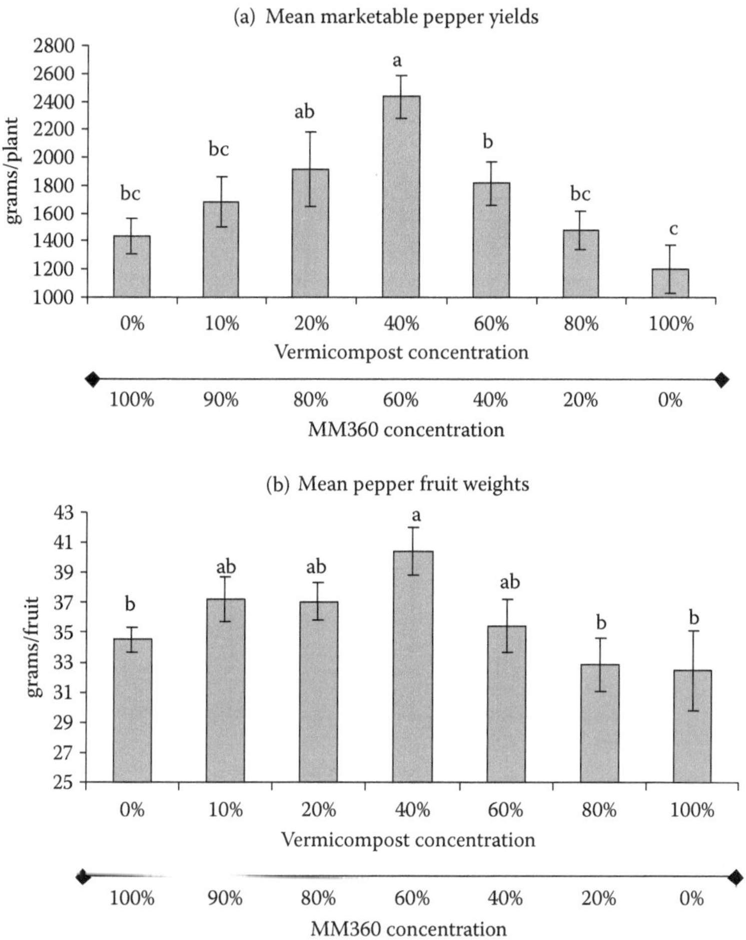

Figure 9.3 Mean yields (a) and mean fruit weights (b) of peppers produced in standard commercial medium (MM360) substituted with different concentrations of food waste vermicompost. Columns labeled with the same letter(s) do not differ significantly ($p < 0.05$). (Plants had all needed nutrients.)

root weights than those grown in MM360 only (control) (Figure 9.5b), independent of availability of nutrients.

Petunias grown in 30% and 40% substitutions of food waste vermicompost into MM360 had significantly greater root dry weights than those grown in MM360 (control) (Figure 9.6a), but the shoot dry weights did not differ significantly based on the rates of vermicompost substitution (Figure 9.6b). Petunias grown in substitutions of 10%–60% paper waste vermicompost into MM360 produced significantly greater shoot dry weights than those grown in 100% MM360 or in mixtures containing from 70% to 100% paper waste vermicompost in MM360 (Figure 9.7a). Petunias grown in substitutions of 10% to 90% paper waste vermicompost into MM360 produced

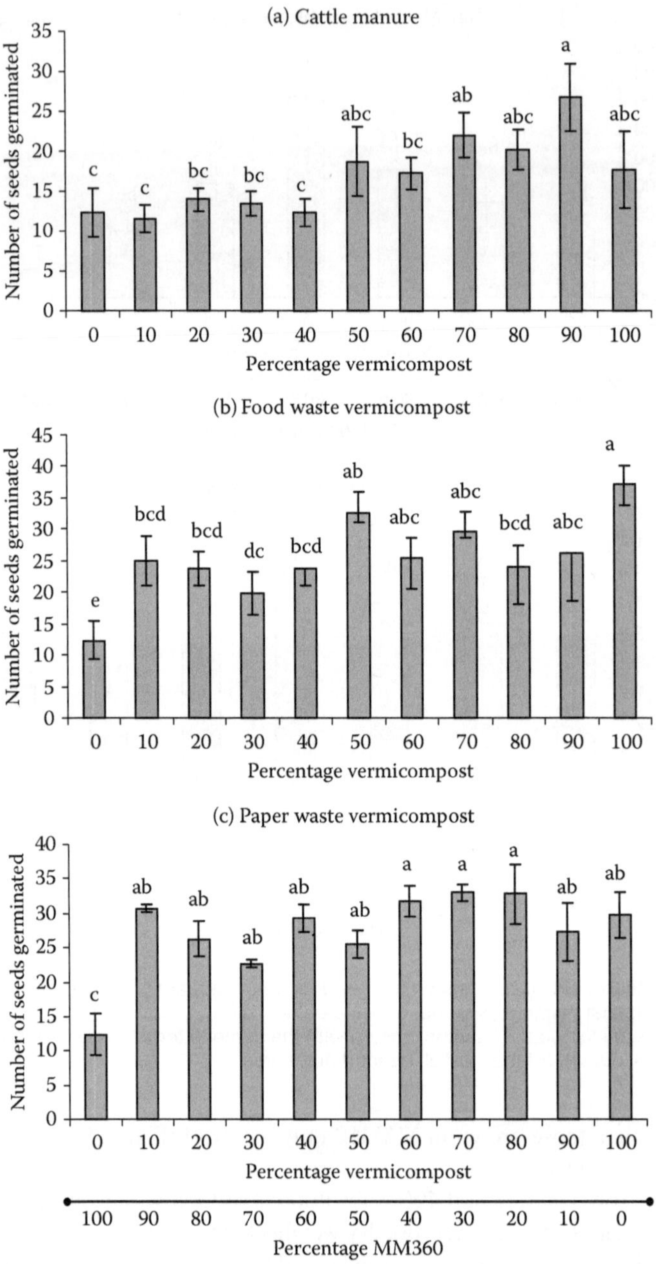

Figure 9.4 Number of petunia seeds germinating 6 days after sowing in standard commercial medium (MM360) substituted with different concentrations of (a) cattle manure vermicompost, (b) food waste vermicompost, and (c) paper waste vermicompost. Columns labeled with the same letter(s) do not differ significantly ($p < 0.05$). (Increases are independent of the nutrient supply.) (All plants received all needed nutrients.)

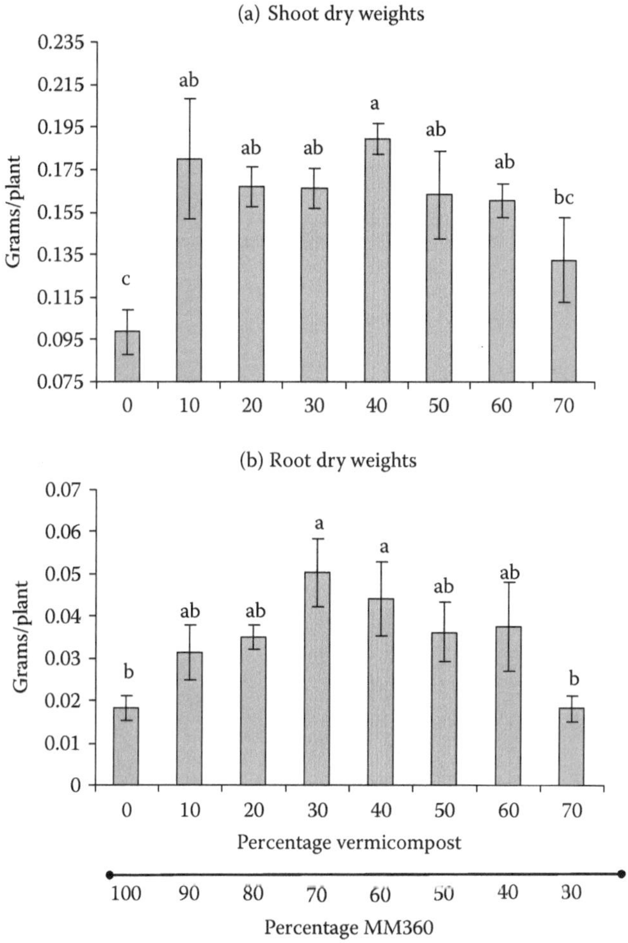

Figure 9.5 Mean petunia (a) shoot dry weights and (b) root dry weights in a commercial medium (MM360) substituted with different concentrations of cattle manure vermicompost 42 days after sowing. Columns labeled with the same letter(s) do not differ significantly ($p < 0.05$). (Increases are independent of the nutrient supply.)

significantly greater root dry weights than those grown in 100% MM360 with no vermicompost added or grown in 100% vermicompost only (Figure 9.7b).

There were significantly more flowers on petunias grown in 20%, 30%, and 40% cattle manure vermicompost substituted into 80%, 70%, and 60% MM360 (Figure 9.8a). Substitutions of 20%, 30%, and 40% food waste vermicompost into MM360 resulted in significantly more flowers than substitution of greater amounts of food waste vermicompost (Figure 9.8b). Petunias grown in a mixture of 40% paper waste vermicompost and 60% MM360 produced significantly more flowers than those grown with paper waste vermicompost substituted at other rates (Figure 9.8c). All these increases in flower production were independent of nutrient availability.

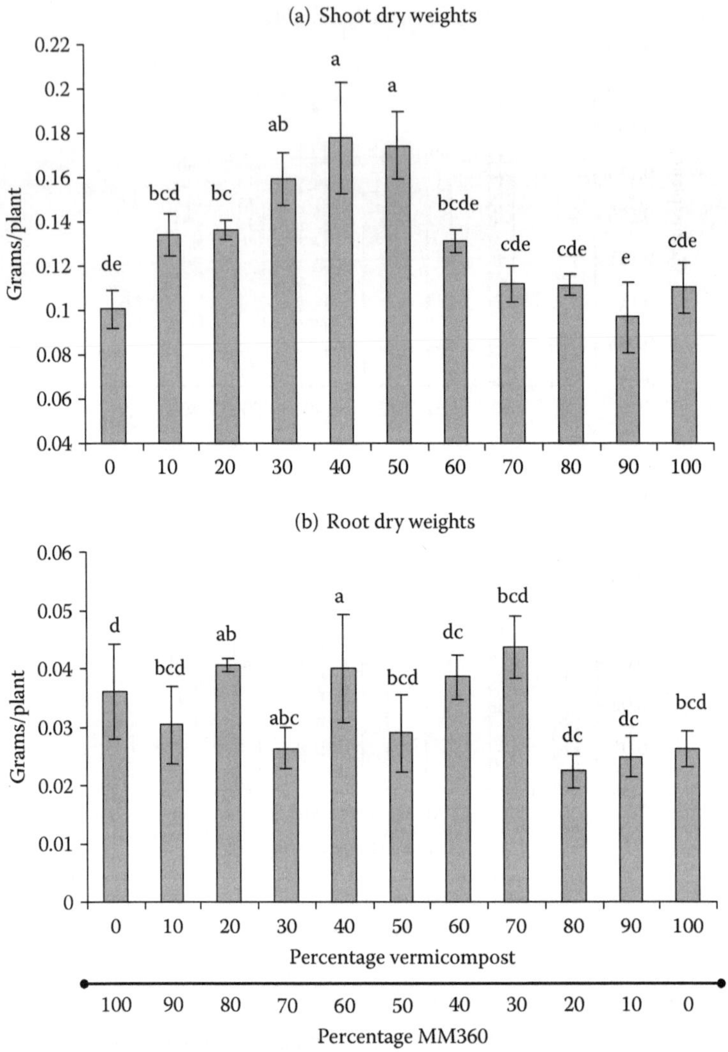

Figure 9.6 Mean petunia (a) shoot dry weights and (b) root dry weights in a commercial medium (MM360) substituted with different concentrations of food waste vermicompost 42 days after sowing. Columns labeled with the same letter(s) do not differ significantly (p <0.05). (Increases are independent of the nutrient supply.)

Similar experiments using commercially produced cattle manure vermicompost processed by the earthworm *D. veneta* (Rosa), substituted into peat-based compost growth media at rates of 10%, 20%, 40%, and 100% (v/v), were conducted on tomatoes (Roberts et al. 2007). These experiments documented fruit quality, including damage due to blossom end rot and fruit cracking, in addition to growth, yields, and vitamin C contents of the crop. Vermicomposts significantly increased tomato germination rates (by 176%) and improved the marketability of fruits at substitution

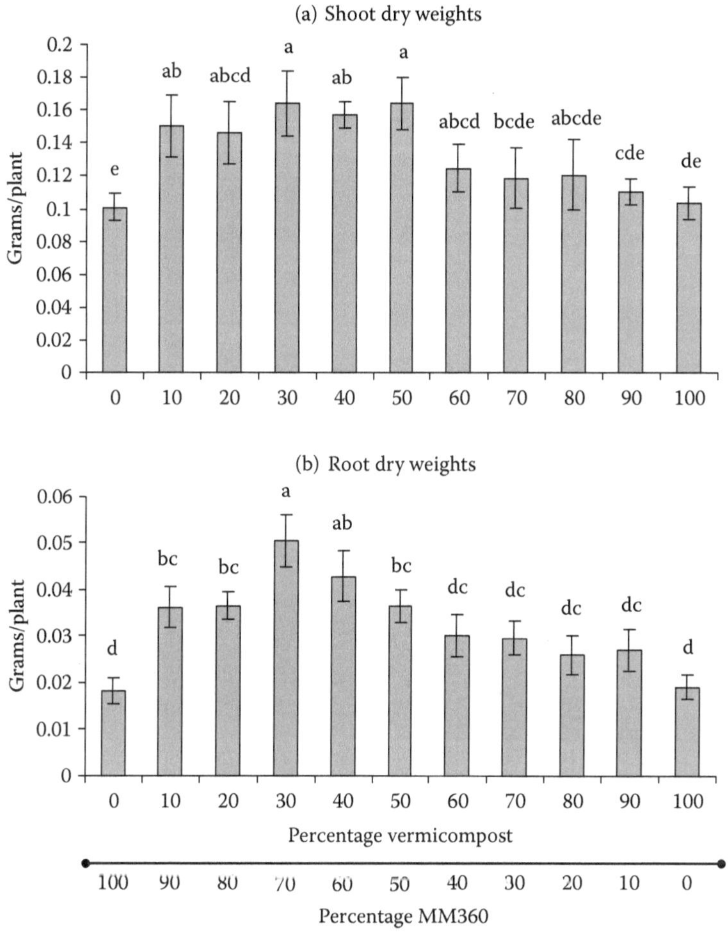

Figure 9.7 Mean petunia (a) shoot dry weights and (b) root dry weights in a commercial medium (MM360) substituted with different concentrations of paper waste vermicompost 42 days after sowing. Columns labeled with the same letter(s) do not differ significantly ($p < 0.05$). (Increases are independent of the nutrient supply.)

rates of 40% and 100%, due to a lower incidence of physiological disorders (blossom end rot and fruit cracking). Total fruit yields, marketable fruit yields, fruit numbers, individual fruit weights, and vitamin C concentrations were unaffected by the substitutions of vermicompost.

We pointed out earlier that substitution rates of less than 50% vermicompost into commercial growth media such as MM360 seemed to produce the most positive responses from test crops in the greenhouse consistently (Atiyeh et al. 2002; Arancon et al. 2004). These results were typical of those from another investigation that tested only low vermicompost substitution rates for growing greenhouse and bedding plants, including tomatoes, French marigolds, bell peppers, and cornflowers (Bachman and

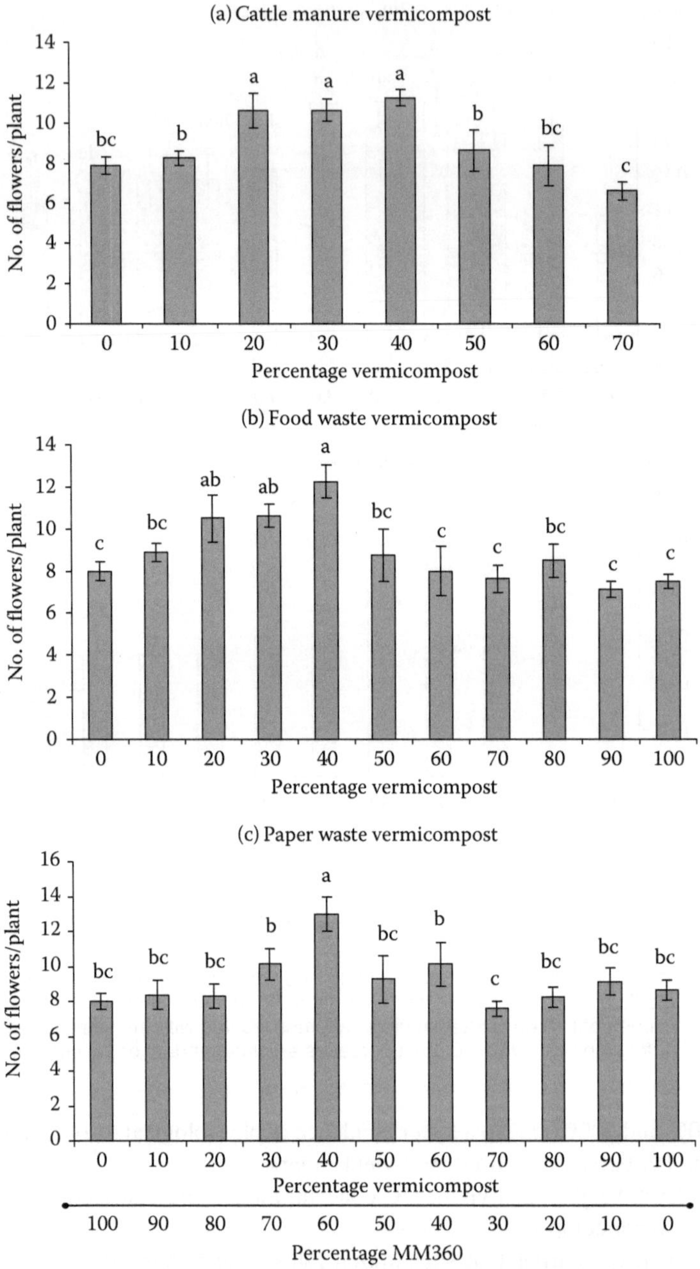

Figure 9.8 Mean numbers of flowers on petunias grown in the commercial medium (MM360) substituted with different concentrations of (a) cattle manure vermicompost, (b) food waste vermicompost, and (c) paper waste vermicompost 79 days after sowing. Columns labeled with the same letter(s) do not differ significantly ($p < 0.05$). (Increases are independent of the nutrient supply.)

Metzger 2008). These crops were used to test the effects of vermicompost produced from pig manure, which was substituted into MM360 at rates of 0%, 10%, and 20%. The incorporation of pig manure vermicompost into plant growth media, at rates up to 20%, increased shoot and root weights, leaf areas, and shoot:root ratios of both tomatoes and French marigold seedlings. When seedlings of tomatoes, French marigolds, and cornflowers were transplanted into six-cell packs, there was greater growth in media substituted with vermicompost than in the control MM360, and the greatest growth occurred when vermicompost was added to both the germination and transplant media. These effects increased when seedlings in the transplant media were irrigated with water containing inorganic fertilizers.

The impacts of applications of compost and vermicompost produced from water hyacinths was assessed in terms of growth and flowering of crossandra (Gajalakshmi and Abbasi 2002). Applications of vermicompost led to statistically significant improvements in the growth and flowering of crossandra compared to those of untreated plants. Zaller (2007) investigated the effects of vermicomposts produced from organic food and cotton wastes, applied at rates of 0%, 20%, 40%, 60%, 80%, and 100% (v/v) into a fertilized commercial peat potting substrate, on three varieties of tomato seedlings. Vermicompost applications influenced the emergence rates of and elongation of seedlings significantly. Root:shoot ratios were affected by vermicompost amendments for two varieties at the seedling stage and one field-grown tomato variety. However, morphological parameters (circumference, dry matter content, peel firmness) and chemical fruit characteristics (contents of C, N, P, K, Ca, Mg, l-ascorbic acid, glucose, fructose) were improved significantly by vermicompost amendments to the seedling substrates. In other experiments, a tomato variety was used as a test crop to investigate the effects of vermicomposts produced from sheep manure on tomato growth and yields (Gutierrez-Miceli et al. 2007). The experiment involved adding vermicompost into soil in proportions of 0:1, 1:1, 1:2, 1:3, 1:4, and 1:5 (v/v). Addition of vermicompost increased plant heights significantly but had no significant effect on the numbers of leaves or yields 85 days after transplanting. Yields of tomatoes were significantly greater when the ratio of vermicompost to soil was 1:1, 1:2, or 1:3.

IV FACTORS INFLUENCING INCREASES IN GROWTH AND YIELD

A Plant Growth Hormones in Vermicomposts

Several factors have been linked to increases in the growth and germination of plants grown in greenhouse media that were substituted with vermicomposts, independent of nutrient supply. In some experiments, when there was no additional fertilization other than the nutrients in the vermicomposts, plant growth and development were directly proportional to the substitution rates of vermicomposts into a soilless plant growth medium MM360 (Atiyeh et al 1998). This investigation pointed out that the effects of the vermicomposts on the growth and development of greenhouse crops were not related directly to the nutrient contents of vermicomposts. One-half of all the plants received full inorganic fertilization to account for any effects that could have been caused by differences in nutrition. However, there were increases in growth even at a low substitution rate of 5% vermicompost into 95% MM360 compared to plants in 100% MM360 (control). The percentage total porosity, percentage air space, pH, and ammonium concentrations of the container medium all decreased significantly after substitution of MM360 with pig manure vermicompost; whereas the bulk density, container capacity, electrical conductivity, overall microbial activity, and nitrate concentrations all increased with increasing substitutions of vermicompost. Increases in electrical conductivity due to salt content were suggested as a possible reason for the decreases in plant growth at the higher vermicompost substitution rates.

Wilson and Carlile (1989) also reported a trend for decreased growth in response to applications of high concentrations of vermicompost. Similar observations were made by Bachman and Metzger (2008), who reported that media containing 10% vermicompost promoted the greatest increases in plant growth. Work at the Soil Ecology Laboratory at The Ohio State University (Arancon et al. 2008) clearly showed that mineral nutrition could be virtually eliminated as the main factor influencing the effects of vermicompost on the growth and flowering of plants such as petunias, since plants received all the nutrients they needed. Tissue-N did not differ between treatments (Table 9.1), which supports the conclusion that the nutritional status of the plants was not responsible for the increases in growth and rates of flowering. For instance, shoot growth was significantly slower in petunias planted in MM360 substituted with 70% to 100% food waste vermicompost and paper waste vermicompost, compared to those that were grown in MM360 substituted with 40% food waste and paper waste vermicompost. In previous experiments, we concluded that some of the slower growth rates of plants at high rates of vermicompost substitution were possibly a response to higher concentrations of plant growth hormones, such as auxins, produced by microorganisms in vermicomposts. Plant growth hormones can have either positive or negative effects on plant growth (Arancon et al. 2006).

Auxins are a class of plant growth hormones that are produced endogenously in the actively growing regions of plants such as shoot tips and axillary buds. They are involved in cell division, elongation, and plant tissue development, hence they are hormones responsible for the principal growth patterns of plants termed apical dominance, phototropism, and geotropism. It has been established that exogenous applications of synthetic auxins can increase plant growth dramatically at lower application rates but could deter growth and development if applied at higher concentrations (Hopkins and Hüner 2004). Optimum growth and development of plants are often observed when exogenous auxins are applied within the range of 10^{-7}–10^{-3} μm (Taiz and Zeiger 2006). Concentrations greater than 10^{-3} μm often produced stunted plant growth. The growth pattern of plants, in response to a range of concentrations of auxins, usually resembles a bell-shaped curve, which is very similar to the bell-shaped growth patterns of plants grown in a range of concentrations of vermicomposts. Commonly, plant growth increased as vermicompost substitution rates increased up to about 40% and decreased when substitution rates were greater

Table 9.1 **The concentrations (means ± SE) of mineral nitrogen and microbial biomass in MM360, substituted with different concentrations of cow manure vermicompost, at transplanting and 79 days after transplanting, and organic nitrogen in shoot tissues 79 days after transplanting**

Percentage Cow Manure Vermicompost in MM360	Mineral-N		Microbial Biomass-N		Tissue-N
	At Transplanting	79 days after transplanting	At Transplanting (μg g^{-1})	79 days after transplanting	79 days after transplanting
Control	171 ± 2.5 de	51 ± 8 a	427 ± 34 e	329 ± 31 cd	4.8 ± 0.5
10	8 ± 21 f	33 ± 6 abc	719 ± 250 de	369 ± 63 c	4.9 ± 0.1
20	121 ± 6 e	19 ± 9 bc	771 ± 48 cde	471 ± 140 abc	4.6 ± 0.1
30	221 ± 6 cd	15 ± 7 c	1452 ± 249 a	421 ± 72 bc	4.5 ± 0.2
40	236 ± 12 c	33 ± 8 abc	1147 ± 31 abc	109 ± 153 de	4.7 ± 0.3
50	340 ± 48 bc	33 ± 2abc	1165 ± 229 ab	95 ± 17 de	4.4 ± 0.3
60	292 ± 5 b	29 ± 3 bc	1214 ± 70 ab	709 ± 120 a	4.2 ± .2
70	408 ± 16 a	30 ± 3 bc	1502 ± 78 ab	100 ± 41 de	4.6 ± 0.2
80	423 ± 3 a	36 ± 7 ab	1184 ± 4 ab	65 ± 28 e	4.6 ± 0.3
90	464 ± 21 a	33 ± 4 abc	1166 ± 39 ab	658 ± 55 ab	4.6 ± 0.2
100	459 ± 17 a	23 ± 7 bc	1056 ± 3 bcd	362 ± 47 c	4.7 ± 0.2
P	***	*	***	***	**ns**
LSD (Alpha = 0.05)	**56**	**18**	**389**	**285**	**–**

Note: Means followed by the same letters do not significantly differ ($p < 0.05$).

than 50% (Figures 9.1, 9.2, 9.3, 9.4, 9.6, 9.7, and 9.8; Atiyeh et al. 2001; Arancon et al. 2004; Arancon et al. 2008).

The likelihood of the existence of a nonnutritional plant growth factor in vermicomposts was reinforced further by the increased growth of seedlings germinated in vermicompost, being maintained in experiments where the seedlings were transplanted into six-pack trays without any added vermicompost. Grappelli et al. (1987), Tomati and Galli (1995), and Bachman and Metzger (2008) all reported increases in plant growth in response to vermicompost that were biological in nature and related them to possible increases in beneficial enzymatic activities, increased populations of beneficial microorganisms, or the presence of biologically active plant growth–regulating substances such as plant growth regulators or plant growth hormones in the vermicomposts.

Tomati et al. (1983, 1987, 1988, 1990); Grappelli et al. (1987); and Tomati and Galli (1995) tested vermicomposts produced from organic wastes by earthworms as media for growing ornamental plants and mushrooms. They concluded that the plant growth increases that occurred in all of their experiments in response to vermicomposts were much too large to be explained purely on the basis of the nutrient contents of the vermicomposts. Moreover, the growth changes observed included stimulation of rooting, dwarfing, changed times of flowering, and lengthening of internodes. They also compared the growth of petunia, begonia, and coleus in response to aqueous vermicompost extracts to that from adding plant growth hormones such as auxins, gibberellins, and cytokinins to soil. They concluded that there was excellent evidence of potential hormonal effects produced by earthworm activity, and this conclusion was supported by the high levels of cytokinins and auxins they found in the vermicomposts. Krishnamoorthy and Vajrabhiah (1986) reported, in laboratory experiments involving large earthworm populations, that seven species of earthworms could promote the production of cytokinins and auxins in organic wastes very dramatically. They also demonstrated significant positive correlations ($r = 0.97$) between earthworm populations and the levels of cytokinins and auxins present in 10 different field soils and concluded that levels of earthworm activity correlated strongly with PGR production.

Edwards and Burrows (1988) reported that the growth of 28 ornamentals and vegetables in plant growth media, produced by the processing of organic wastes by the earthworm *E. fetida,* was much greater than that in commercially available plant growth media. These increases were too great to be explained solely through influence of earthworm activity on plant nutrient quality and availability. They reported that the growth of ornamentals was influenced significantly even when the earthworm-processed organic wastes were diluted 20:1 with other suitable materials and the nutrient content balanced to that of comparable media. Moreover, the growth patterns of the plants, including leaf development, stem and root elongation, and flowering of biennial ornamental plants in the first season of growth, all indicated the likelihood of some biological factor other than nutrients, such as the production of plant growth–influencing substances like plant growth hormones. There is little evidence that earthworms can synthesize plant hormone-like metabolites independently. However, many workers have reported great increases in the numbers of bacteria, actinomycetes, and fungi in freshly deposited earthworm casts, compared with their numbers in the surrounding soil or organic matter (Edwards and Bohlen 1996). Such multiplications may be due to the great increases in numbers of microorganisms occurring during passage through the earthworm's intestine, because the food fragmented by the earthworms forms a richer substrate for greatly increased microbial activity or because fragmentation of organic matter in the earthworm's gizzard increases the available surface area for microbial activity (Edwards and Arancon 2004). The ability of microorganisms, including bacteria, fungi, yeasts, actinomycetes, and algae, to produce biologically active substances, such as PGHs, in appreciable quantities has been fully confirmed (Arshad and Frankenberger 1993; Frankenberger and Arshad 1995). The endogenous plant growth hormones that microorganisms can produce include: auxins, gibberellins, cytokinins, ethylene, and abscisic acid.

Many free-living nitrogen-fixing microorganisms are common in the earthworm gut where they are probably only in transit and are excreted with earthworms' casts when they are present in high concentrations (K. Lee 1985). Arancon et al. (2008) confirmed that such greater production of microbial biomass was well documented in potting media that were substituted with vermicomposts (Table 9.1), and this was evident in terms of effects on plant growth even at the transplant stage after 79 days of growth.

In addition, other factors may contribute to the effects of vermicomposts on plant growth. During vermicomposting, earthworms may produce metabolites that can increase plant growth by enhancing the plant's protein-synthesizing capacity. Aldag and Graff (1975), studying oats grown in soil mixtures inoculated with *E. fetida*, found marked increases in dry matter production, total protein, and protein N in the plant tissues. Atlavinyte and Vanagas (1982) reported similar results for barley and rye grown in soils with *Aporrectodea caliginosa* and *Lumbricus terrestris* present. Tomati et al. (1987) described increases in wheat yields grown in soil in the presence of casts obtained from *E. fetida*. Graff and Makeschin (1980) compared the protein content of ryegrass plants grown in the burrows of *L. terrestris* with that of plants grown in artificial burrows made with a long rod. They reported that the earthworm activity increased the dry matter and protein yields of ryegrass plants significantly. Earthworm casts also increased protein synthesis in lettuce by 24% and in radishes by 32% (Tomati et al. 1990). The improved N metabolism that they reported was suggested to be due possibly to an increase in nitrate reductase activity, which can regulate nitrate availability for plants (Tomati et al. 1990). Nitrate reductase is influenced by concentrations of nitrate in soil as well as by PGRs (Hageman 1979). Wareing (1982) and T. Scott (1984) reported that microbial-hormone-like substances could play an important role by regulating ionic absorption and inducing some enzymatic activities.

B Humates in Vermicomposts

It has been demonstrated that various plant growth-promoting compounds, particularly PGRs, are produced during vermicomposting. Atiyeh et al. (2002)

suggested that PGHs such as auxins, kinetins, and gibberellins may be relatively transient in soil because of their high solubility and rapid rates of breakdown in ultraviolet light, rendering these compounds relatively ineffective in influencing plant growth in soil. They hypothesized that PGHs could be absorbed by humates and fulvates in vermicomposts and released gradually on a time scale synchronized closely with plant growth. Atiyeh et al. (2002) showed clearly that humates could influence plant growth considerably. Vermicomposts originating from animal manure, sewage sludges or paper-mill sludges have been reported to contain large amounts of humic and fulvic substances (Albanell et al. 1988; Petrussi et al. 1988; Senesi et al. 1992; Garcia et al. 1995; Masciandaro et al. 1997; Elvira et al. 1998). Studies of the effects of humic substances on plant growth, under conditions of adequate mineral nutrition, have produced positive plant growth increases consistently (Chen and Aviad 1990). For instance, applications of humic substances to soils increased the dry matter yields of corn and oat seedlings (Y. S. Lee and Bartlett 1976; Albuzio et al. 1994); numbers and lengths of tobacco roots (Mylonas and McCants 1980); dry weights of shoots, roots, and nodules of soybean, peanut, and clover plants (Tan and Tantiwiramanond 1983); and vegetative growth of chicory plants (Valdrighi et al. 1996); it also induced shoot and root formation in tropical crops grown in tissue culture (Goenadi and Sudharama 1995).

Humates can be extracted from organic materials such as vermicomposts by an acid/alkali fractionation technique (Valdrighi et al. 1996), yielding approximately 4 gm humic acids per kilogram (0.064 $oz.lb^{-1}$) of vermicompost (Atiyeh et al. 2001). Typical growth responses that workers in the Soil Ecology Laboratory at The Ohio State University obtained after treating plants with humic substances include increased growth in response to increasing concentrations of humic substances, but with some decreases in growth at the higher concentrations of the humic materials (Atiyeh et al. 2002; Figure 9.9). This growth response pattern resembles that of the effects of a range of application rates of vermicompost to tomatoes (Atiyeh et al. 2001). In other research (Atiyeh et al. 2002), humic acids were extracted from pig manure vermicompost, using the alkali/acid fractionation procedure, and mixed with a soilless container medium (MM360) to provide a range of application rates of 0, 50, 100, 150, 200, 250, 500, 1000, 2000, and 4000 mg of humate kg^{-1} (0, 0.0008, 0.0016, 0.0024, 0.0032, 0.004, 0.008, 0.016, 0.032, 0.064) $oz.lb^{-1}$, dry weight of container medium, and tomato seedlings were grown in the mixtures.

This incorporation of a range of concentrations of vermicompost-derived humic acids into MM360 increased the rates of growth of tomato and cucumber plants significantly, in terms of plant heights, leaf areas, and shoot and root dry weights, with a bell-shaped pattern of responses similar to that produced by vermicomposts. Plant growth tended to be increased by treatments of the plants with 50–500 mg kg^{-1} (0.0008–0.008 $oz.lb^{-1}$) humic acids, but growth often decreased significantly when the concentrations of humic acids derived in the container medium exceeded 500–1000 mg kg^{-1} (0.008–0.016 $oz.lb^{-1}$) (Figure 9.9; Atiyeh et al. 2002; Arancon et al. 2006).

Humic substances have also been shown to increase the yields of corn, oats, soybeans, peanuts, clover, chicory plants, and other tropical crops (Y. Lee and Bartlett 1976; Mylonas and Muscolo 1980; Nardi et al. 1988; Muscolo, Felicim, et al. 1993;

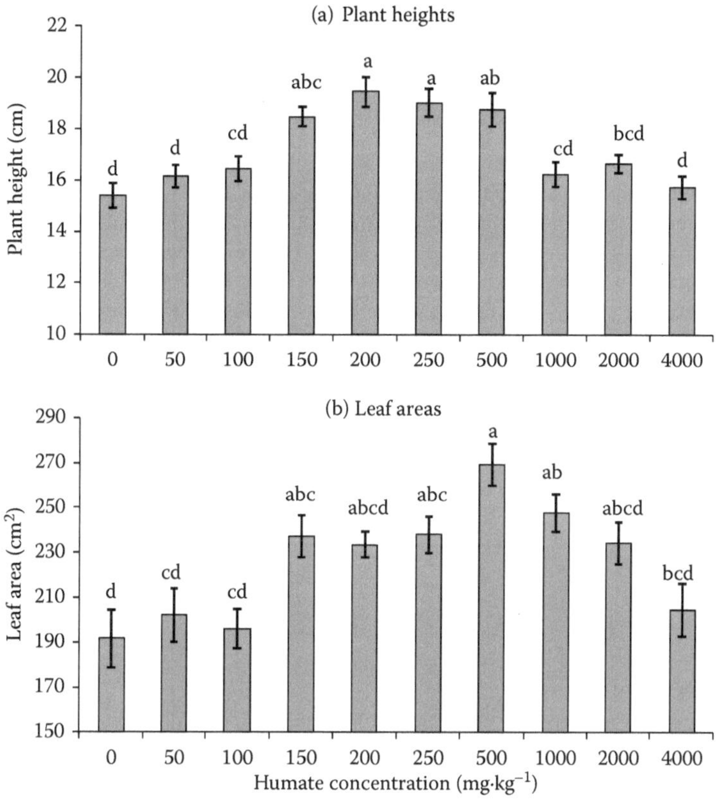

Figure 9.9 Average plant heights (a) and leaf areas (b) of tomato seedlings (mean ± standard error) grown in a standard commercial potting medium (MM360) containing different concentrations of vermicompost-derived humic acids, with all needed nutrients supplied. Columns labeled with the same letters do not differ significantly ($p < 0.05$).

Albuzio et al. 1994; Muscolo, Panuccio, et al. 1996; Valdrighi et al. 1996; Hayes and Wilson 1997; Muscolo, Bovalo, et al. 1999). The plant hormone–like activity of humic acids produced from vermicomposts has been shown to be the most probable mechanism, through plant growth hormones adsorbed onto the complex structure of humic acids (Canellas et al. 2000). The adsorption of plant growth hormones onto humates would allow these relatively transient compounds to persist in soil over the life of crops, be released slowly, and thereby have much greater effects on plant growth over a considerably longer period. In support of their hypothesis, Canellas et al. (2000) identified exchangeable auxin groups attached to humic acids that had been extracted from cattle manure vermicompost, following a detailed structural analysis. These complexes enhanced root elongation, lateral root emergence, and plasma membrane H+ ATPase activity of maize roots. Hence, there is an impressive body of evidence suggesting that this is the key mechanism explaining the dramatic effects of vermicomposts on plant germination, growth, flowering, and yield.

REFERENCES

Albanell, E., Plaixats, J., and Cabrero, T. 1988. Chemical changes during vermicomposting (*Eisenia fetida*) of sheep manure mixed with cotton industrial wastes. *Biol. Fertil. Soils* 6: 266–269.

Albuzio, A., Concheri, G., Nardi, S., and Dell'Agnola, G. 1994. Effect of humic fractions of different molecular size on the development of oat seedlings grown in varied nutritional conditions. In *Humic Substances in the Global Environment and Implications on Human Health*, eds. N. Senesi and T. M. Mianom, 199–204. Elsevier Science, Amsterdam.

Aldag, R., and Graff, O. 1975. N-Fraktionen in Regenwurmlösung und deren Ursprungsböden. *Pedobiologia* 15: 151–153.

Arancon, N. Q., and Edwards, C. A. 2004. Vermicomposts suppress plant pest and disease attacks. *BioCycle* March: 51–53.

Arancon, N. Q., Edwards, C. A., Atiyeh, R., and Metzger, J. D. 2004. Effects of vermicomposts produced from food wastes on the growth and yields of greenhouse peppers. *Biores. Technol.* 93: 139–144.

Arancon, N. Q., Edwards, C. A., Babenko, A., Cannon, C., Galvis, P., and Metzger, J. D. 2008. Influences of vermicomposts, produced by earthworms from cattle manure, food waste, and paper waste on the germination, growth and flowering of petunias in the greenhouse. *Appl. Soil. Ecol.* 39: 91–99.

Arancon, N. Q., Edwards, C. A., and Bierman, P. 2006. Influences of vermicomposts on field strawberries: Part 2. Effects on soil microbiological and chemical properties. *Biores. Technol.* 97: 831–840.

Arshad, M., and Frankenberger, W. T. 1993. Microbial production of plant growth regulators. In *Soil Microbial Ecology*, ed. F. B. Metting, Jr., 307–347. Marcel Dekker, New York.

Atiyeh, R. M., Arancon, N. Q., Edwards, C. A., and Metzger, J. D. 2002. The influence of earthworm-processed pig manure on the growth and productivity of marigolds. *Biores. Technol.* 81: 103–108.

Atiyeh, R. M., Subler S., and Edwards, C. A. 1998. Growth of tomato plants in vermicomposted hog manure. In *Proceedings of the 6th International Symposium on Earthworm Ecology Pedobiologia* 43: 724–728.

Atiyeh, R. M., Subler, S., Edwards, C. A., and Metzger, J. D. 2001. Pig manure vermicompost as a component of a horticultural bedding plant medium: Effects on physicochemical properties and plant growth. *Biores. Technol.* 78: 11–20.

Atiyeh, R. M., Subler, S., Edwards, C. A., Bachman, G., Metzger, J. D., and Shuster, W. 2000. Effects of vermicomposts and composts on plant growth in horticultural container media and soil. *Pedobiologia* 44: 579–590.

Atlavinyte, O., and Vanagas, J. 1982. The effect of earthworms on the quality of barley and rye grain. *Pedobiologia* 23: 256–262.

Bachman, G. R., and Metzger, J. D. 2008. Growth of bedding plants in commercial potting substrate amended with vermicompost. *Biores. Technol.* 99: 3155–3161.

Buckerfield, J. C., Flavel, T., Lee, K. E., and Webster, K. A. 1999. Vermicomposts in solid and liquid form as plant-growth promoter. *Pedobiologia* 43: 753–759.

Businelli, M., Perucci, P., Patumi, M., and Giusquiani, P. L. 1984. Chemical composition and enzymatic activity of some worm casts. *Plant Soil* 80: 417–422.

Bwamiki, D. P., Zake, J. Y. K., Bekunda, M. A., and Woomer, P. L. 1998. The use of coffee husks as an organic amendment to improve soil fertility in Ugandan banana production. In *Carbon and Nutrient Dynamics in Natural and Agricultural Tropical Ecosystems*, ed. L. Bergstrom and H. Kirchmann, 115–121. CAB International, Wallingford, UK.

Canellas, L. P., Olivares, F. L., Okorokova, A. L., and Facanha, A. R. 2000. Humic acids isolated from earthworm compost enhance root elongation, lateral root emergence, and plasma H+ -ATPase activity in maize roots. *Plant Physiol.* 130: 1951–1957.

Chan, P. L. S., and Griffiths, D. A. 1988. The vermicomposting of pre-treated pig manure. *Biol. Wastes* 24: 57–69.

Chen, Y., and Aviad, T. 1990. Effects of humic substances on plant growth. In *Humic Substances in Soil and Crop Sciences: Selected Readings*, eds. P. MacCarthy, C. E. Clapp, R. L. Malcolm, and P. R. Bloom, 161–186. ASA and SSSA, Madison, WI.

Edwards, C. A. 1988. Breakdown of animal, vegetable and industrial organic wastes by earthworms. In *Earthworms in Waste and Environmental Management*, ed. C. A. Edwards and E. F. Neuhauser, 21–31. SPB Academic Publishing, the Hague, the Netherlands.

Edwards, C. A. 1983. Utilization of earthworm composts as plant growth media. In *International Symposium on Agricultural and Environmental Prospects, of Earthworms*, ed. U. Tomati, and A. Grappelli, 57–62. Rome, Italy.

Edwards, C. A., and Arancon, N. Q. 2004. The use of earthworms in the breakdown of organic wastes to produce vermicomposts and animal feed protein. In *Earthworm Ecology*, ed. C. A. Edwards, 345–438. 2nd ed. CRC Press, Boca Raton, FL.

Edwards, C. A., and Bohlen, P. J. 1996. *The Biology and Ecology of Earthworms.* 3rd ed. Chapman and Hall, London.

Edwards, C. A., and Burrows, L. 1988. The potential of earthworm composts as plant growth media. In *Earthworms in Environmental and Waste Management*, ed. C. A. Edwards and E. F. Neuhauser, 211–220. SPB Academic Publishing, the Hague, the Netherlands.

Elvira, C., Goicoechea, M., Sampedro, L., Mato, S., and Nogales, R. 1996. Bioconversion of solid paper-pulp mill sludge earthworms. *Biores. Technol.* 57: 173–177.

Elvira, C., Sampedro, L., Benitezm, E., and Nogales, R. 1998. Vermicomposting of sludges from paper mill and dairy industries with *Eisenia andrei*: A pilot-scale study. *Biores. Technol.* 63: 205–211.

Elvira, C., Sampedro, L., Dominguez, J., and Mato, S. 1997. Vermicomposting of wastewater sludge from paper-pulp industry with nitrogen rich materials. *Soil Biol. Biochem.* 29: 759–762.

Frankenberger, W. T., Jr., and Arshad, M. 1995. *Phytohormones in Soil: Microbial Production and Function.* Marcel Dekker, New York.

Gajalakshmi, S., and Abbasi, S. A. 2002. Effect of the application of water hyacinth compost/vermicompost on the growth and flowering of *Crossandra undulaefolia* and on several vegetables. *Biores. Technol.* 85: 197–199.

Garcia, C., Ceccanti, B., Masciandaro, G., and Hernandez, T. 1995. Phosphatase and b-glucosidase activities in humic substances from animal wastes. *Biores. Technol.* 53: 79–87.

Goenadi, D. H. and Sudharama, I. M. 1995. Shoot initiation by humic acids of selected tropical crops grown in tissue culture. *Plant Cell Rep.* 15: 59–62.

Graff, O., and Makeschin, F. 1980. Crop yield of rye-grass influenced by the excretions of three earthworm species. *Pedobiologia* 20: 176–180.

Grappelli, A., Galli, E., and Tomati, U. 1987. Earthworm casting effect on *Agaricus bisporus* fructification. *Agrochimica* 21: 457–462.

Gutierrez-Miceli, F. A., Moguel-Zamudio, B., Abud-Achila, M., and Gutierrez-Oliva, V. F. 2007. Vermicompost as a soil supplement to improve growth, yield and fruit quality of tomato (*Lycopersicon esculentum*). *Biores. Technol.* 98: 2781–2786.

Hageman, R. H. 1979. Integration of nitrogen assimilation in relation to yield. In *Nitrogen assimilation of plants: Proceedings of 6th Long Ashton Symposium*, ed. E. J. Hevitt and C. V. Cutting. 591–612. Academic Press, London.

Hayes, M. H. B., and Wilson, W. S. (eds.). 1997. *Humic Substances, Peats and Sludges: Health and Environmental Aspects.* Royal Chemical Society, U.K.

Hopkins, W. G., and Hüner, N. P. A. 2004. The plant hormones: Control and developments. *Introduction to Plant Physiology,* 341–365. Wiley, New York.

Johnston, A. M., Janzen, H. H., and Smith, E. G. 1995. Long-term spring wheat response to summerfallow frequency and organic amendment in southern Alberta. *Can. J. Plant Sci.* 75: 347–354.

Kretzschmar, A. 1984. Vermicomposting and humification by earthworms in relation to soil fertility: research priorities for their management and utilization. In *International Conference in Waste Management*, Cambridge.

Krishnamoorthy, R. V., and Vajrabhiah, S. N. 1986. Biological activity of earthworm casts: An assessment of plant growth promoter levels in casts. *Proc. Indian Acad. Sci. (Anim. Sci.)* 95: 341–351.

Lee, K. E. 1985. *Earthworms, Their Ecology and Relationships with Soils and Land Use.* Academic Press, Sydney, Australia.

Lee, Y. S., and Bartlett, R. J. 1976. Stimulation of plant growth by humic substances. *Soil Sci. Soc. Am. J.* 40: 876–879.

Masciandaro, G., Ceccanti, B., and Garcia, C. 1997. Soil agro-ecological management: Fertirrigation and vermicompost treatments. *Biores. Technol.* 59: 199–206.

Muscolo, A., Bovalo, F., Gionfriddo, F., and Nardi, S. 1999. Earthworm humic matter produces auxin-like effects on *Daucus carota* cell growth and nitrate metabolism. *Soil Biol. Biochem.* 31: 1303–1311.

Muscolo, A., Felicim, M., Concheri, G., and Nardi, S. 1993. Effect of earthworm humic substances on esterase and peroxidase activity during growth of leaf explants of *Nicotiana plumbaginifiolia. Biol. Fertil. Soils* 15: 127–131.

Muscolo, A., Panuccio, M. R., Abenavoli, M. R., Concheri, G., and Nardi, S. 1996. Effect of molecular complexity and acidity of earthworm faeces humic fractions on glutamate dehydrogenase, glutamine synthetase, and phosphenolpyruvate carboxylase in *Daucus carota* a II cell. *Biol. Fertil. Soils* 22: 83–88.

Mylonas, V. A., and McCants, C. B. 1980. Effects of humic and fulvic acids on growth of tobacco. I. Root initiation and elongation. *Plant Soil* 54: 485–490.

Nardi, S., Arnoldi, G., and Dell'Agnola, G. 1988. Release of hormone-like activities from *Alloborophora rosea* and *Alloborophora caliginosa* feces. *J. Soil Sci.* 68: 563–657.

Orozco, S. H., Cegarra, J., Trujillo, L. M., and Roig, A. 1996. Vermicomposting of coffee pulp using the earthworm *Eisenia fetida*: Effects on C and N contents and the availability of nutrients. *Biol. Fertil. Soils* 22: 162–166.

Orozco-Castillo, C., Chalmers, K. J., Waugh, R., and Powell, W. 1994. Detection of genetic diversity and selective gene introgression in coffee using RAPD markers. *Theor. Appl. Genet.* 87: 934–940.

Pascual, J. A., Garcia, C., Hernandez, T., and Ayuso, M. 1997. Changes in the microbial activity of an arid soil amended with urban organic wastes. *Biol. Fertil. Soils* 24: 429–434.

Petrussi, F., de Nobili, M., Viotto, M., and Sequi, P. 1988. Characterization of organic matter from animal manures after digestion by earthworms. *Plant Soil* 105: 41–46.

Roberts, P., Jones, G. E., and Jones, D. L. 2007. Yield responses of wheat (*Triticum aestivum*) to vermicompost. *Compost Sci. Util.* 15: 6–15.

Scott, M. A. 1988. The use of worm-digested animal waste as a supplement to peat in loamless composts for hardy nursery stock. In *Earthworms in Environmental and Waste Managment.* Ed. C. A. Edwards and E. F. Neuhauser, SPB Academic Publishing, the Hague, the Netherlands.

Scott, T. K. 1984. Hormonal regulation of development. In *Encyclopedia of Plant Physiology*, 10. Springer, Berlin.

Senesi, N., Saiz-Jimenez, C., and Miano, T. M. 1992. Spectroscopic characterization of metal-humic acid-like complexes of earthworm-composted organic wastes. *Sci. Total Environ.* 117–118: 111–120.

Shi-wei, Z., and Fu-zhen, H. 1991. The nitrogen uptake efficiency from ^{15}N labeled chemical fertilizer in the presence of earthworm manure (cast). In *Advances in Management and Conservation of Soil Fauna*, ed. G. K. Veeresh, D. Rajagopal, and C. A. Viraktamath, 539–542. Oxford and IBH, New Delhi.

Siddagangaiah, B. A., Vadiraj, M. R., Sudharshan, M. R., and Krishnakumar, V. 1996. Standardization of rooting media for propagation of vanilla (*Vanilla planifolia* Andr.). *J. Spices Aromatic Crops* 5: 131–133.

Subler, S., Edwards, C. A., and Metzger, J. 1998. Comparing vermicomposts and composts. *BioCycle* 39: 63–66.

Taiz, L., and Zeiger, E. 2006. *Physiology* (4th ed.), Sinauer Associates, Sunderland, MA.

Tan, K. H., and Tantiwiramanond, D. 1983. Effect of humic acids on nodulation and dry matter production of soybean, peanut, and clover. *Soil Sci. Soc. Am. J.* 47: 1121–1124.

Thankamani, C. K., Sivaraman, K., and Kandiannan, K. 1996. Response of clove (*Syzygium aromaticum* (L.) Merr. & Perry) seedlings and black pepper (*Piper nigrum* L.) cuttings to propagating media under nursery conditions. *J. Spices Aromatic Crops* 5: 99–104.

Thanunathan, K., Natarajan, S., Senthilkumar, R., and Arulmurugan, K. 1997. Effect of different sources of organic amendments on growth and yield of onion in mine spoil. *Madras Agric. J.* 84: 382–384.

Tomati, U., and Galli, E. 1995. Earthworms, soil fertility and plant productivity. *Proc. Int. Coll. on Soil Zool. Acta Zool. Fenn.* 196: 11–14.

Tomati, U., Galli, E., Grappelli, A., and Dihena, G. 1990. Effect of earthworm casts on protein synthesis in radish (*Raphanus sativum)* and lettuce (*Lactuca sativa*) seedlings. *Biol. Fertil. Soils* 9: 288–289.

Tomati, U., Grappelli, A., and Galli, E. 1988. The hormone-like effect of earthworm casts on plant growth. *Biol. Fertil. Soils* 5: 288–294.

Tomati, U., Grappelli, A., and Galli, E. 1987. The presence of growth regulators in earthworm-worked wastes. In *On Earthworms. Proceedings of International Symposium on Earthworms. Selected Symposia and Monographs*, ed. A. M. Bonvicini Paglioi and P. Omodeo, 423–435. Unione Zoologica Italiana, Mucchi, Modena.

Tomati, U., Grappelli, A., and Galli, E. 1983. Fertility factors in earthworm humus. In: *Proceedings of the International Symposium on Agricultural and Environmental Prospects in Earthworm Farming*, 49–56. Publication Ministero della Ricerca Scientifica e Technologia, Rome, Italy.

Valdrighi, M. M., Pera, A., Agnolucci, M., Frassinetti, S., Lunardi, D., and Vallini, G. 1996. Effects of compost-derived humic acids on vegetable biomass production and microbial growth within a plant (*Cichorium intybus*)-soil system: A comparative study. *Agric. Ecosyst. Environ.* 58: 133–144.

Wareing, P. F. 1982. Plant growth substances 1982. In *Proceedings of the 11th International Conference on Plant Growth Substances,* 669–678. Academic Press, London.

Werner, M., and Cuevas, R. 1996. Vermiculture in Cuba. *BioCycle* 37: 61–62.

Wilson, D. P., and Carlile, W. R. 1989. Plant growth in potting media containing worm-worked duck waste. *Acta. Horticult.* 238: 205–220.

Zaller, J. G. 2007. Seed germination of the weed *Rumex obtusifolius* after on-farm conventional, biodynamic and vermicomposting of cattle manure. *Ann. Appl. Biol.* 151: 245–249.

Zink, T. A., and Allen, M. F. 1998. The effects of organic amendments on the restoration of a disturbed coastal sage scrub habitat. *Restoration Ecol.* 6: 52–58.

CHAPTER 10

The Use of Vermicomposts as Soil Amendments for Production of Field Crops

Norman Arancon and Clive A. Edwards

CONTENTS

I Introduction 129
II Effects of Vermicomposts on the Growth of Field Crops 131
III Methods of Vermicompost Applications to Field Soils and Their Effects on Plant Growth and Yields 131
IV Rates of Vermicompost Applications and Their Effects on Plant Growth and Yields 133
 A Peppers 134
 B Tomatoes 137
 C Strawberries 137
 D Grapes 137
V Effects of Vermicomposts Produced from Different Materials on Plant Growth and Yields 138
VI The Economics of the Use of Vermicomposts in Field Crops 140
VII Physicochemical and Biological Changes in Soils in Response to Vermicompost Applications 141
 A Physicochemical Changes in Soils 141
 B Biological Changes in Soil 147
VIII Conclusions 148
References 148

I INTRODUCTION

Earthworms are important soil-dwelling organisms that contribute greatly to increasing fertility of soils to support productive crop and natural plant growth. They create channels aerating soil that allow efficient gas exchange between soil and plant roots. They support symbiosis among soil-dwelling microorganisms and promote overall microbial activities. Earthworms also facilitate the mineralization of nutrients during the breakdown of organic matter, thereby supplying nutrients for plant growth. Earthworm biology, ecology, and benefits in natural and agricultural soils have been documented in numerous publications since the Darwin's 1881 book *The Formation of Vegetable Mould, through the Action of Worms.* It has been demonstrated clearly that some species of earthworms are specialized to live in decaying organic matter; these are termed *epigeic* species (Edwards and Bohlen 1996). Some earthworms are able to process sewage sludges and biosolids from wastewater; brewery wastes; processed potato wastes; waste from the paper industries; food wastes from supermarkets and restaurants; animal manures from poultry, pigs, cattle, sheep, goats, horses, and rabbits; and horticultural residues from dead plants, yard wastes, and wastes from the mushroom industry (Edwards and Neuhauser 1988; Edwards and Arancon 2004). The ability of some species of earthworms and microorganisms to process organic wastes has been developed in a technical process termed *vermicomposting.*

Vermicomposts are finely divided peatlike materials with high porosity, aeration, and drainage and good water-holding capacities (Edwards and Burrows 1988; Edwards and Arancon 2004). The chemical composition of vermicomposts is determined partly by the degree of earthworm activity and the conditions they have been exposed to but mainly by the composition of the parent wastes used. Compared to the parent materials, vermicomposts have a much lower C:N ratio, lower content of soluble salts, greater cation-exchange capacities, and increased humic acid contents due to the transformation of organic substances into stable humic compounds (Albanell et al. 1988; Edwards and Bohlen 1996). Most plants prefer a growth medium that is slightly acidic (pH 6–6.5). During vermicomposting, the pH of organic wastes decreases slightly, tending toward neutrality (Albanell et al. 1988; Buchanan et al. 1988). For instance, Mainoo et al. (2009) vermicomposted pineapple wastes using *Eudrilus eugeniae* and produced a vermicompost that had a pH between 7.2 and 9.2 from an acidic pineapple waste material of pH 4.4.

Vermicomposts contain much larger populations of bacteria (5.7×10^7), fungi (22.7×10^4), and actinomycetes (17.7×10^6) compared with those in conventional thermophilic composts (Nair et al. 1997). Vermicomposts' outstanding physicochemical and biological properties make them excellent materials to use as amendments to greenhouse container growth media, as organic fertilizers, or as soil amendments for various horticultural crops. Actinobacteria, which have potential to suppress fungal pathogens in plants, were reported to be the dominant communities in extracts from paper-sludge vermicomposts (Yasir et al. 2009). Gopal et al. (2009) reported that vermicomposted coconut leaves and cow manure had a greater range of microbial communities than the original substrate. They reported that vermicompost was conducive to multiplication of aerobic heterotrophic bacteria, actinomycetes, *Trichoderma* spp., and *Azotobacter.* These results were confirmed by Sen and Chandra (2009), where pressmud, trash, bagasse, and cattle manure were processed into thermophilic composts or vermicomposts using *E. eugeniae.* Principal component analysis (PCA) of enzymatic activities and community-level physiological profiles revealed differences in the functional response of microbial communities in composts and vermicomposts during their maturation phases. Vivas et al. (2009) reported that vermicomposting of olive-mill waste produced more dehydrogenase and other enzyme activities than the parent material or thermophilic compost from the same materials (see Chapter 5).

II EFFECTS OF VERMICOMPOSTS ON THE GROWTH OF FIELD CROPS

We demonstrated clearly, in the results discussed in Chapter 9, that substitution of solid vermicomposts into plant growth media such as MM 360 can increase the production of greenhouse crops dramatically, producing faster germination rates, seedling growth, increased leaf areas and plant heights, increased numbers of flowers, more fruits, and greater overall yields. These plant growth and yield parameters were achieved by greenhouse plants with low vermicompost-substitution rates into soilless plant growth media when all nutrients required were provided. Obvious changes in the growth media after vermicompost substitutions included changes in physical, chemical, and biological properties. Of these, biological properties such as increases in microbial biomass and the presence of plant growth regulators (PGRs) and plant growth hormones (PGHs) are the main mechanisms that could increase crop production (see Chapter 9). This Chapter reviews the uses and effects of vermicompost applications in the field, focusing on the following areas:

- Methods of vermicompost applications to field soils and their effects on plant growth and yields
- Rates of vermicompost applications in the field and their effects on plant growth and yields
- Effects of vermicomposts produced from different materials on plant growth and yields
- The economics of the use of vermicomposts on field crops
- Finally, the effects of vermicompost applications on the chemical, physical, and biological properties of field soils and the mechanisms by which vermicomposts affect plant growth and yield are discussed

III METHODS OF VERMICOMPOST APPLICATIONS TO FIELD SOILS AND THEIR EFFECTS ON PLANT GROWTH AND YIELDS

One of the earliest methods was to raise seedlings in vermicompost growth media in the greenhouse before transplanting them out into field soils, with the aim to carry over the residual effects of vermicomposts on seedlings to the later growth stages of crops. The hypothesis behind this practice was that faster germination and increased seedling growth in vermicompost-substituted growth media should produce better transplants and crop growth and yields after transplanting to the field. When cabbages were grown in compressed blocks made from pig waste vermicompost, after transplanting to the field, they were larger and more mature at harvest compared to those produced in a commercial blocking material (Edwards and Burrows 1988). This method certainly has considerable commercial applicability since the current methods of vermicompost production command a relatively high price for vermicomposts. To date, little other research has examined the residual or carryover effects of vermicomposts from season to season in field-grown crops.

A surface application of vermicomposts derived from grape marc, spread under rows of grapevines and covered with straw and paper mulches, increased the yields of a grape variety (pinot noir) by 55% (Buckerfield and Webster 1998). The increases in yields included large increases in both grape bunch-weights and bunch numbers and no losses in flavor. In an experiment at a second site, animal manure vermicompost applications within rows of grapes, under a straw mulch, increased chardonnay grape yields by up to 35%; vermicompost applications tended to have a much greater effect on yields when applied under straw mulches than when applied directly to the soil surface, probably through decreased degradation of vermicomposts by exposure to sun and air. In later experiments, Buckerfield and Webster (1998) reported that a single application of vermicompost to grapes in this way still had positive residual effects on grape growth and yields after 5 years. Venkatesh et al. (1998) reported that yields of Thompson seedless grapes were significantly greater when vermicomposts were applied directly into the rows of grapes. Seyval grapes had greater marketable yields, more fruit clusters per vine, and bigger berry sizes after applications of food waste and paper waste vermicomposts, directly within 20 cm of the row area of 10-year-old vines covered with straw (Arancon et al. unpublished data; Figure 10.1).

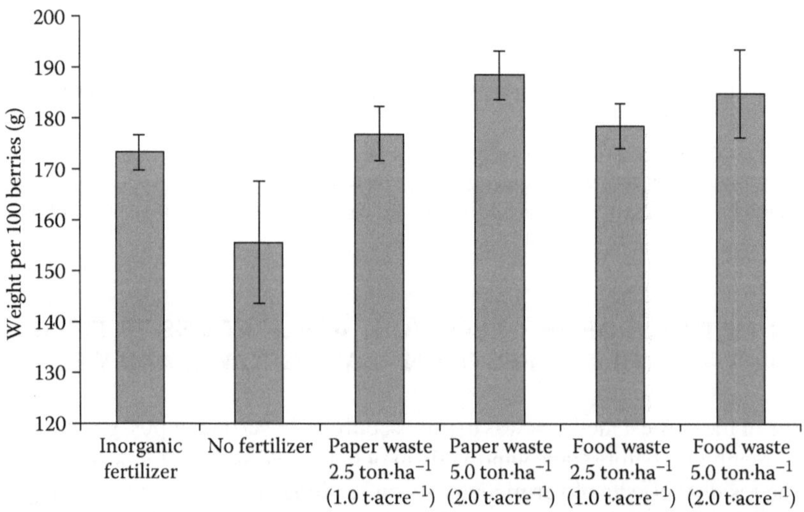

Figure 10.1 Marketable yield of Seyval grapes after application of food waste and paper waste vermicompost at 2.5 and 5.0 tons.ha^{-1} (1.0 and 2.0 t·acre^{-1}) of application.

These vermicomposts were applied to soil in which amounts of macronutrients were equalized in all plots. Other direct applications of vermicomposts into field soils produced increased growth and yield of crops such as okra (Ushakumari et al. 1999), bananas (Athani et al. 1999), rice (Vasanthi and Kumaraswamy 1999), tomatoes (Patil et al. 1998), potatoes (Mrinal et al. 1998), sunflowers (Devi et al. 1998), sugarcane (Zende et al. 1998), mulberries (Murarkar et al. 1998), China asters (Nenthra et al. 1999), peppers and tomatoes (Arancon et al. 2003), and strawberries (Arancon et al. 2004). In soils planted with peppers (Arancon et al. 2005), tomatoes (Arancon et al. 2003), and strawberries (Arancon et al. 2005) vermicomposts were incorporated into the top 15 cm of raised beds close to the rows using a rototiller and then covered with black polyethylene plastic mulch on the day of transplanting. No other vermicompost applications were made during the growth cycle of these crops except during the transplanting day of the second year.

Vermicomposts and inorganic fertilizer were applied in amounts adequate to achieve uniform NPK for all crops in all plots. For N, residual amounts of total and extractable N from the previous year's cropping were analyzed 1–2 weeks prior to the applications of inorganic fertilizer and vermicompost during the second year. Food waste, cow manure, and paper waste vermicomposts were also analyzed for total and available N content. These analyses were the basis for calculating the rates of inorganic fertilizer to be applied, to complete the balanced N, P, and K treatment for each plot. Vermicompost treatments were broadcast and mixed into the plots at least 2 weeks prior to planting. At transplant, 50% of the total inorganic fertilizer needed to complete the available-N requirement in each treatment was applied together with the vermicompost treatment. This was to allow the vermicompost to react with soil and the inorganic fertilizer. After 2 weeks, another soil analysis was made to assess the N status of the soil. From the results of this analysis, the remaining amounts of inorganic fertilizer needed to complete the available-N requirement equivalent to that in the full inorganic treatment were applied to vermicompost plots at transplanting.

IV RATES OF VERMICOMPOST APPLICATIONS AND THEIR EFFECTS ON PLANT GROWTH AND YIELDS

The rates of vermicompost applications to field crops have varied considerably across different areas and regions. Most of these applications were made in conjunction or combination with inorganic fertilization practices, and we could find few reports from purely organic farming. Most of these applications were made with the aim of determining well-balanced rates of joint application of inorganic fertilizer and vermicompost that would give the best plant growth responses, yields, and economic returns. In some experiments, the vermicompost was an addition, not just a substitution, to a current fertilization program, and the aim was to investigate whether the growth and yield increases were enough to justify such additions. The amounts of vermicompost additions into field soils that increased crop yields ranged from as low as 2 ton·ha^{-1} (0.8 t·acre^{-1}) to as high as 10 ton·ha^{-1} (4.0 t·acre^{-1}), and inorganic fertilizer applications ranged from 50%–100% of the recommended amount per crop. However, it is difficult to separate the confounding effects of inorganic fertilizers and vermicomposts on plant growth and yield, as joint sources of nutrition. To overcome this difficulty, some workers have aimed at ruling out the effects of nutrients in increasing growth and yield of test plants, by balancing the major elements between vermicomposts and inorganic fertilizers. This method involved careful laboratory analyses to quantify the elemental contents of vermicomposts used. Inorganic fertilizers were then used to supplement the remainder of the recommended rates of N, P, and K fertilization for a particular crop (Arancon et al. 2003, 2005).

Two field experiments were at The Ohio State University Research Center at Piketon, Ohio, in 1999 and 2000. The experimental site had a soil type of a DoA-Dole silt loam. Commercially produced cattle manure, food waste, and paper waste vermicomposts were applied to soils, in two sets of plots, at rates of either 10 ton·ha^{-1} (4 t·acre^{-1}) or 20 ton·ha^{-1} (8 t·acre^{-1}) in 1999 and 5 ton·ha^{-1} (2 t·acre^{-1}) or 10 ton·ha^{-1} (4 t·acre^{-1}) in 2000. A set of replicate control plots received a full recommended rate of inorganic fertilizers. A second set of vermicompost replicate plots received 20 ton·ha^{-1} (8 t·acre^{-1}) of traditional thermophilic leaf composts in 1999 and 10 ton·ha^{-1} (4 t·acre^{-1}) in 2000.

A Peppers

Raised soil beds 1.5 m (6.5 ft) wide, 5.5 m (18.3 ft) long, and 0.15 m (0.5 ft) high were constructed, and the same plots were used in the second year. Vermicomposts were applied into the row before transplanting the peppers and supplemented with amounts of inorganic fertilizer to correspond with the recommended full rate of 95–95 kg NK ha^{-1} (85–85 lb NK acre^{-1}) that was applied to the inorganic fertilizer–treated plots. Vermicomposts, composts, and inorganic fertilizers were incorporated into the top 15 cm (6 in) of the beds in the plots with a rototiller. It is important to note that vermicomposts were applied only on the raised beds close to the established rows of crops. Soil-analytical tests showed that the available P was 48 ppm (Bray P1 Method) an adequate amount for commercial pepper production of peppers, so no P fertilizer was applied in either year. The full fertilizer rates applied to the inorganic fertilizer control were 95 kg. ha^{-1} (85 lb·acre^{-1}) total N (as ammonium nitrate) and 95 kg. ha^{-1} (85 lb·acre^{-1}) total K (as potassium chloride), of which 34 kg N ha^{-1} (30 lb N acre^{-1}) (in the form of a urea-ammonium nitrate solution) was applied to all plots through ferti-irrigation 30 days after transplanting. In the second year, the full preplant rate of 95 kg. ha^{-1} (85 lb·acre^{-1}) total N was applied to all plots to correct for lower amounts of immediately available N in the organically amended plots early in the growing season of the first year.

In 1999, peppers planted in plots treated with paper waste and cow manure vermicomposts at 10 ton·ha^{-1} (4 t·acre–1) or 20 ton·ha^{-1} (8 t·acre–1) yielded significantly more than those grown in plots treated with inorganic fertilizers only (Figure 10.2), except for those in plots treated with food waste vermicomposts. Soils treated only with inorganic fertilizers produced peppers with significantly fewer marketable peppers and lower shoot dry weights than peppers grown in the vermicompost- or

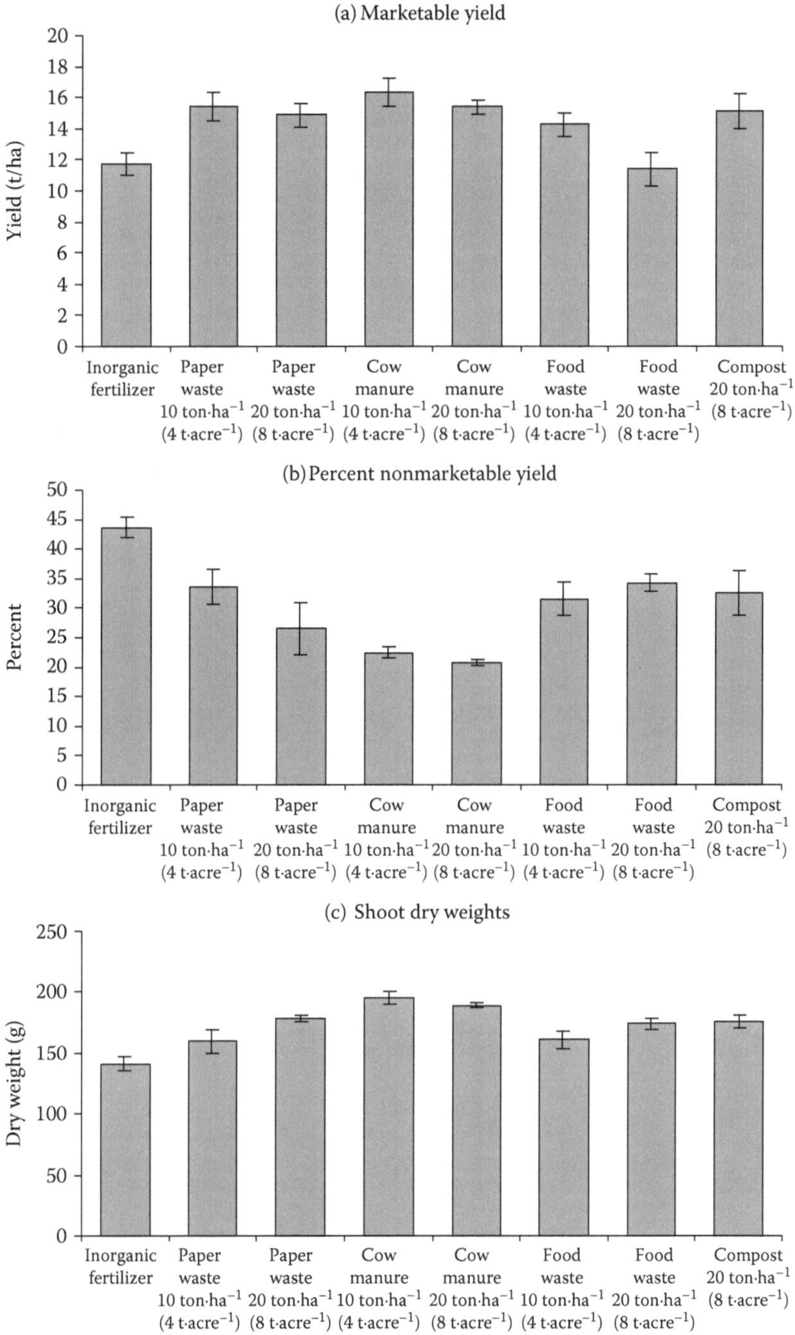

Figure 10.2 Marketable yields of peppers (a), percentage nonmarketable yields (b), and shoot dry weights (c) (means ± SE) at harvest in 1999. Standard errors are given at the top of each column.

compost-treated plots with balanced nutrients. Marketable yields in 2000 were significantly greater for peppers grown in soils that received both vermicompost treatments and inorganic fertilizer amendments than for those treated with only inorganic fertilizers (Figure 10.3). Soils treated with cattle manure vermicomposts, applied at a rate of 10 ton·ha^{-1} (4 t·acre^{-1}), produced pepper plants with significantly greater shoot

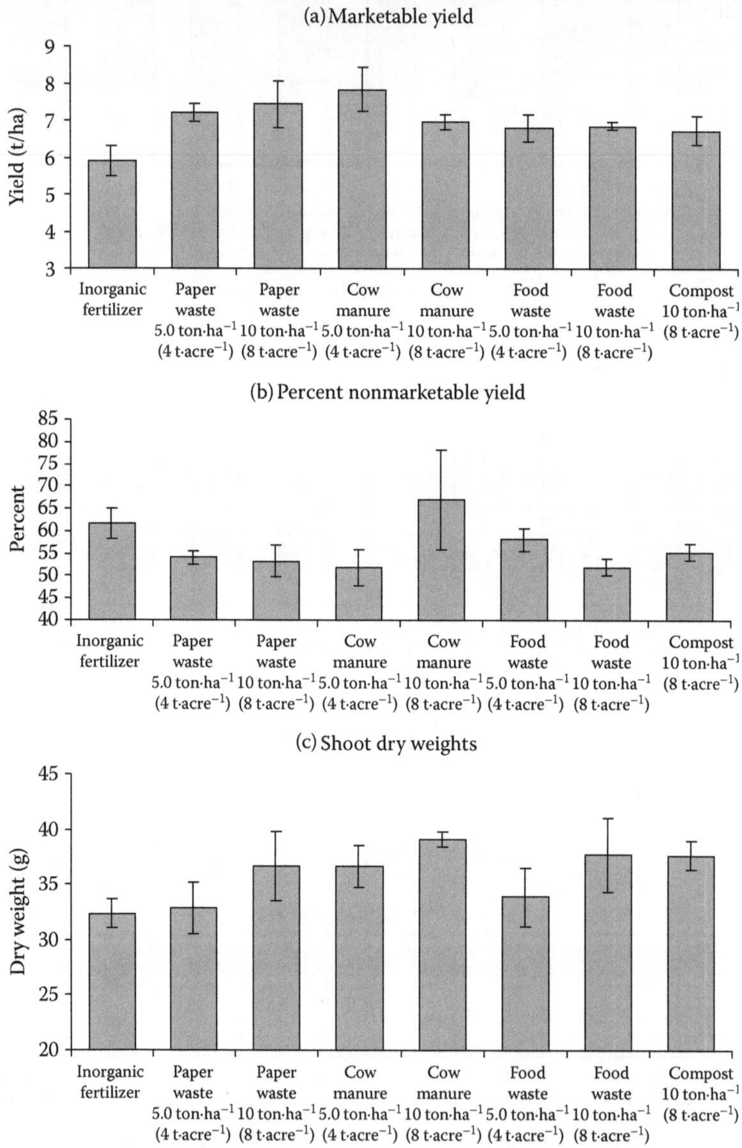

Figure 10.3 Marketable yields of peppers (a), percentage nonmarketable yields (b), and shoot dry weights (c) (means ± SE) at harvest in 2000. Standard errors are given at the top of each column.

dry weights than those of plants in soils treated with only inorganic fertilizers but similar to all other treatments (Figure 10.3).

B Tomatoes

Tomatoes were grown in the same way as peppers using 80–75–75 kg NPK ha^{-1} (71–67–67 lb) as the recommended rate of fertilization, which was used in balancing the nutrients between those plots receiving vermicomposts and those receiving inorganic fertilizers (Arancon et al. 2003). The same types and rates of vermicompost applications were used in the experiments for both the first and second year. *A priori* orthogonal contrast analyses of the differences between the yields of inorganically fertilized tomatoes and the yields of those receiving vermicomposts showed that tomato plants receiving vermicomposts produced greater marketable yields of tomatoes. Moreover, there were significant differences in marketable yields between tomato plants receiving different types of vermicomposts.

C Strawberries

Strawberries were also raised in the same way as peppers and tomatoes, using a different recommended rate of balanced fertilization, 85–155–125 kg (76–138–162 lb) NPK ha^{-1} (acre^{-1}). Commercial food waste and recycled paper vermicomposts were applied at rates of 10 ton·ha^{-1} (4 t·acre^{-1}) and 5 ton·ha^{-1} (2 t·acre^{-1}) in 2000 to strawberries in experiments at Piketon and Fremont, Ohio. The marketable strawberry yields in the vermicompost plots were consistently and significantly larger than those in plots treated with the inorganic fertilizer only although all received the same overall level of nutrient inputs. Yields increased significantly in response to supplemental vermicompost applications compared to those from strawberries that received only inorganic fertilizers (Arancon et al. 2004; Figure 10.4). Vermicompost applications increased strawberry growth and yields significantly, including increases of up to 37% in leaf areas, 37% in plant shoot biomass, 40% in numbers of flowers, 36% in numbers of plant runners, and 35% in marketable fruit weights. A similar field experiment using strawberries was reported by Singh et al. (2008). In his experiments four application rates of vermicompost 2.5, 5.0, 7.5, and 10 ton·ha^{-1} (1.0, 2.0, 3.0 and 4.0 t·acre^{-1}) were supplemented with inorganic fertilizers to balance the fertilizer requirements of strawberries. The best results were obtained in response to 7.5 ton·ha^{-1} (3 t·acre^{-1}) vermicompost applications, which produced larger marketable yields and reduced physiological disorders to the strawberries.

D Grapes

Ten-year-old Seyval grapes were grown with food waste and paper waste vermicomposts applied at a low rate of 2.5 ton·ha^{-1} (1.0 t·acre^{-1}), supplemented with inorganic fertilizers (as with peppers, tomatoes, and strawberries) and mulched with straw within the row. Grapes receiving vermicomposts produced comparable yields to those of vines receiving a full rate of inorganic fertilizers (Figure 10.1).

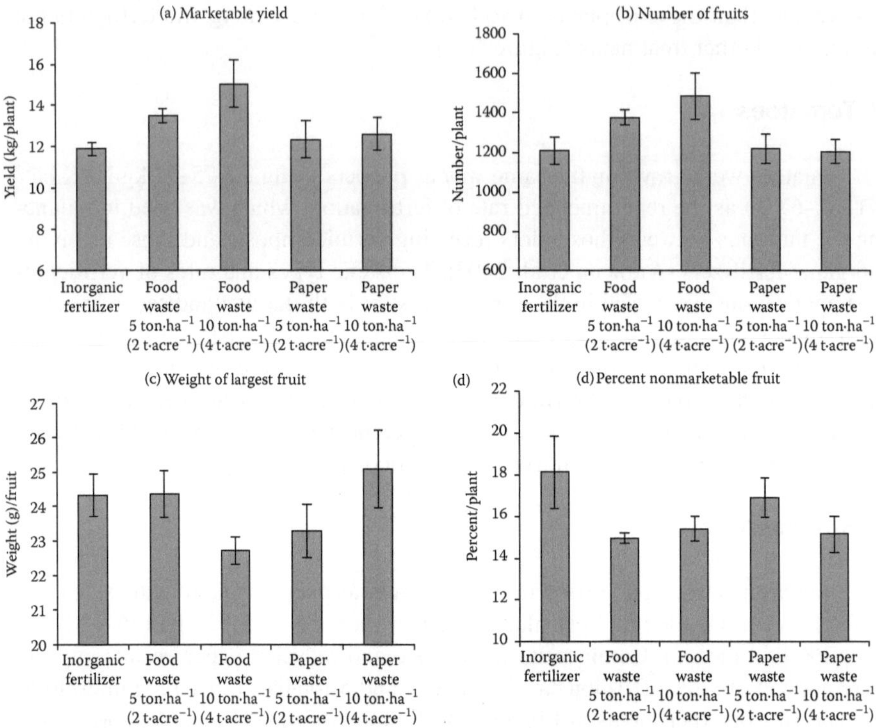

Figure 10.4 Yield and yield components (means ± SE) of strawberries in Site 1 (Piketon, Ohio). Standard errors are given at the top of each column.

V EFFECTS OF VERMICOMPOSTS PRODUCED FROM DIFFERENT MATERIALS ON PLANT GROWTH AND YIELDS

It has been well established by the Soil Ecology Laboratory at The Ohio State University that crops planted in soils that had been previously treated with vermicomposts, in combination with any fertilization program, usually produce greater yields than those that received only inorganic fertilizers. However, there were also differences in yields between crops that were treated with vermicomposts produced from different parent wastes. The main chemical, physical, and biological qualities of vermicomposts depend largely on the parent organic wastes.

Table 10.1 summarizes differences in the chemical composition of the vermicomposts. Different kinds of vermicomposts interact differently with different soils and hence may influence crop growth and yields differently. The effects of such interactions on crop yields are typical of those from the Soil Ecology Laboratory at The Ohio State University, where two or three kinds of vermicomposts were used at the same time in at least five experiments on a wide range of crops, including peppers (Arancon et al. 2005), tomatoes (Arancon et al. 2003), strawberries (Arancon et al. 2005), grapes, raspberries, and cucumbers (Arancon et al. in litt.). All of these

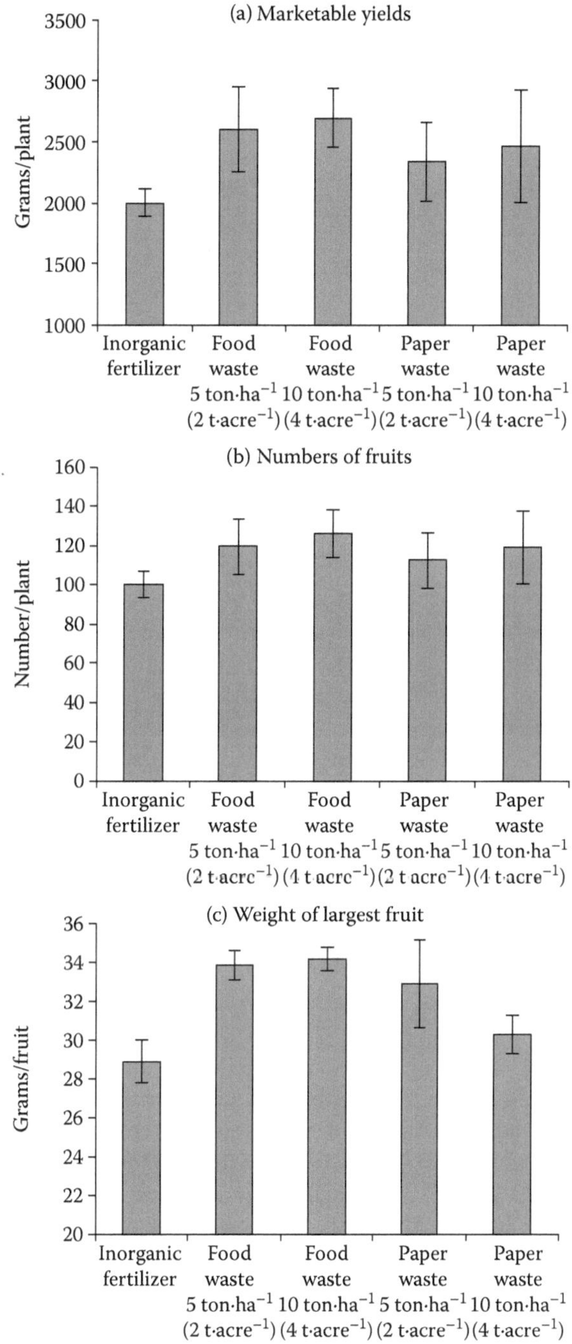

Figure 10.5 Yield components (means ± SE) of strawberries in Site 2 (Freemont, Ohio). Standard errors are given at the top of each column.

Table 10.1 Chemical Composition of Cattle, Food, and Paper Waste Vermicomposts

	N (%)	P (%)	K (%)	B (μg.g⁻¹)	Ca (μg.g⁻¹)	Fe (μg.g⁻¹)	Mg (μg.g⁻¹)	Mn (μg.g⁻¹)	Na (μg.g⁻¹)	S (μg.g⁻¹)	Zn (μg.g⁻¹)
Food waste vermicompost	1.3	2.7	9.2	23	18,614	23,264	4364	610	842	2587	279
Cattle manure vermicompost	1.9	4.7	1.4	58	23,245	3454	5802	160	3360	5524	516
Paper waste vermicompost	1.0	1.4	6.2	31	9214	17,811	7661	447	613	1929	127

experiments ran for at least two cropping seasons. In the first year of a field trial, peppers grown on soils treated with food waste vermicomposts applied at 20 ton·ha^{-1} (8 t·acre^{-1}) produced significantly lower yields than those treated with paper waste vermicompost applied at 10 ton·ha^{-1} (4 t·acre^{-1}). In the second year, peppers grown in soils treated with cattle manure vermicompost applied at rates of 5 ton·ha^{-1} (2 t·acre^{-1}) and food waste vermicompost applied at rates of 10 ton·ha^{-1} (4 t·acre^{-1}) had significantly fewer nonmarketable peppers than peppers planted in soils treated with cow manure vermicompost at 10 ton·ha^{-1} (4 t·acre^{-1}) (Figure 10.4).

In the strawberry experiments, apparent differences could also be observed in the effects of the two kinds of vermicomposts used: paper waste and food waste vermicomposts. Strawberry leaf areas, number of runners, marketable yields, and numbers of fruits were significantly greater ($p \leq 0.05$) for plants grown in plots treated with food waste vermicomposts than for plants grown in plots treated with paper waste vermicomposts (Figure 10.5).

VI THE ECONOMICS OF THE USE OF VERMICOMPOSTS IN FIELD CROPS

The commercial values of vermicomposts vary considerably based on their parent organic-waste sources, ranging through animal manures, biosolids, horticultural and food wastes, and industrial organic wastes. Vermicomposts are currently sold at prices from about $50–$500 (U.S.) a ton. Food waste vermicomposts can cost from about $300 per ton, cow manure vermicomposts about $100 per ton, and paper waste vermicomposts $50 per ton. Such costs are critical factors influencing the overall profitability of any crop production using vermicomposts as soil amendments. Clearly, based on the recommended application rates, it is only economical to use vermicomposts to treat more valuable horticulture crops such as ornamentals, vegetables, and fruit crops. It would not be economical to use them on cereals or other agricultural field crops although they are effective at quite low application rates.

Calculations of the cost and return of producing strawberries and peppers with vermicomposts can help to determine the economic practicability of using vermicomposts rather than using full inorganic fertilization. The production costs presented in Table 10.2 for strawberries were based on a normal yearly growing cycle for each crop and corresponding increases in crop yields, as a result of vermicompost applications, with the cost of vermicomposts factored in. Economic calculations showed that the application of paper waste vermicomposts to strawberries at the rate of 2.5 ton·ha^{-1} (1.0 t·acre^{-1}) (dry weight) would be more profitable than applying food waste vermicomposts at the same rate (Table 10.2), although plants treated with food waste vermicomposts produced greater yields compared to plants treated with paper waste vermicomposts. Such cost and return calculations for bell peppers also indicated that the use of cow manure and paper waste vermicomposts, based on our cost estimates, was more profitable than the use of food waste vermicomposts applied at the same rate (Table 10.3). Paper waste vermicompost applications were more profitable due to their relatively low cost, compared to those of cow manure and food waste vermicomposts, although yields of peppers grown with paper waste vermicompost amendments increased less than those grown with cow manure vermicompost additions. Therefore, correct current prices and sources of vermicomposts are an important factor to consider in deciding the most appropriate vermicompost, as well as application rates for a particular crop and season. As vermicomposting gains in popularity it seems likely that vermicompost-production costs will decrease with widespread development of higher-technology vermicompost-production systems. We still need to identify the minimum effective application rates of vermicomposts for many crops and soils.

VII PHYSICOCHEMICAL AND BIOLOGICAL CHANGES IN SOILS IN RESPONSE TO VERMICOMPOST APPLICATIONS

A Physicochemical Changes in Soils

The improvements in growth and yields of crops grown in potting media in greenhouses or in field soils that had been substituted or amended with vermicomposts could be attributed to several factors. First, vermicomposts contribute to overall improvements in the physicochemical and biological characteristics of the field soils that favor better plant growth.

Field experiments at The Ohio State University (Arancon et al. 2003) demonstrated that soils treated with mixtures of vermicomposts, which were supplemented by recommended rates of inorganic fertilizers and planted with tomatoes, usually tended to have greater amounts of total extractable N, orthophosphates, dehydrogenase enzyme activity, and microbial biomass N than soils that received equivalent amounts of inorganic fertilizers only (Arancon et al. 2005). However, in the second year of the experiments, there was even more microbial biomass N and orthophosphates in soils to which mixtures of vermicomposts and inorganic fertilizers were applied than in those that received inorganic fertilizers only (Figure 10.6). In soils

Table 10.2 Comparison of Economic Costs and Returns of Strawberry Production Using Full Inorganic Fertilization or Vermicomposts as Organic Amendments

Operation	Cost per Acre U.S.	Materials		Hand Labor		COST PER ACRE (U.S. $)			
		Type	Cost	Hours	U.S.	IF	FW 21% IN.	CW 39% IN.	PW 31% IN.
LAND PREPARATION									
Subsoil	$38.8					38.75	38.75	38.75	38.75
Disc 2x	$11.5					23	23	23	23
Land plane	$12					12	12	12	12
Borders, cross check and break borders	$17.8					17.75	17.75	17.75	17.75
Flood irrigate		water	14.6			14.56	14.56	14.56	14.56
Fertilizer double-spread	$8	227kg (500 lb.)11-52-0	63.8			71.75	880	936	155
Disc 2x	$11.5					23	23	23	23
Triplane	$11					11	11	11	11
List beds	$13.5					13.5	13.5	13.5	13.5
TOTAL LAND PREPARATION (U. S. $)						225.31	1033.6	1089.6	308.56
GROWING PERIOD									
Drip system and tape		Drip system	700	20	$155	855	855	855	855
Install plastic mulch	$55	Plastic mulch	110			165	165	165	165
Metam sodium via drip		Metam sodium	100	4	$31	131	131	131	131
Transplanting		17 M plants	850	40	$310	1160	1160	1160	1160
Fertilizer (via drip)		400 lb. N @ .35	140			140	26.25	8.82	52.5
Drip maintenance		350 lb. Phosphorus	91			91	0	0	0
Irrigate 20x		Chemicals	30			30	30	30	30
		Water 4 ac/ft	58.2	16	$124	182.24	182.24	182.24	182.24
insect control 7x & 3x drip	$9	Insecticides	280			343	343	343	343
Remove drip tape and plastic				20	$155	155	155	155	155
Disc out beds	$11.5					11.5	11.5	11.5	11.5
TOTAL GROWING PERIOD (U.S.$)						3263.74	3059	3041.6	3085.2

GROWING PERIOD AND LAND PREPARATION COSTS (U.S. $)			3489.05	4092.6	4131.1	3393.8
Land Rent (net acres)			225	225	225	225
Cash Overhead	17 % of preharvest costs and land restoration		631.39	631.39	631.39	631.39
TOTAL PREHARVEST COSTS (U.S.$)			4345.44	4948.9	4987.5	4250.2
Harvest cost						
Pick, haul, pack, cool, and sell	1300crates/acre @	4.40 per crate	5720	6921.2	7950.8	7493.2
TOTAL OF ALL COSTS (U.S.$)			10065.4	11870	12938	11743
Yield 1300 crates/acre @ 10.00/30 lb crate			13000	15730	18070	17030
NET RETURNS (U.S. $)			2934.56	3859.9	5131.7	5286.6

Note: *IF: Inorganic fertilizer; CW: Cattle waste vermicompost; PW: Paper waste vermicompost; FW: Food waste vermicompost.

Table 10.3 Comparison of Economic Costs and Returns of Pepper Production Using Full Inorganic Fertilization or Vermicomposts as Organic Amendments

	Full Inorganic	Food Waste Vermicompost	Paper Waste Vermicompost
Yield Increase	–	**21.5%**	**16%**
Operating Costs (U.S. $)			
Plants	120	120	120
Seed-Cover Crop	1	1	1
Fertilizer	76	35.72	31.92
Vermicompost		500	125
Services-Soil and Tissue Tests	60	60	60
Chemicals	171	171	171
Fuel and Machine Operating	250	250	250
Repairs	57	57	57
Straw	88	88	88
Marketing, Promotion	162	162	162
Overheads	145	145	145
Labor	365	390	390
Total Operating Costs	1495	1979.72	1600.92
Fixed Costs (U.S. $)			
Land Investment	104	104	104
Machinery Investment	125	125	125
Machinery Depreciation	169	169	169
Irrigation Eqpt. Investment	75	75	75
Irrigation Eqpt. Depreciation	35	35	35
Total Fixed Costs (U.S. $)	508	508	508
TOTAL COSTS (U.S. $)	2003	2487.72	2108.92
YIELD 3000 lbs @ $1/lb	3000	3645	3480
NET RETURNS (U.S. $)	997	1157.28	1371.08

Note: Yield increases of 21.5% and 16% resulting from the applications of food waste and paper waste vermicomposts, respectively, were factored into the calculations.

planted with strawberries, the amounts of total extractable N, microbial biomass N, and dissolved organic N were not different statistically between all treatments at the end of the growth cycle of strawberries, but there were more orthophosphates in those soils that received vermicompost and fertilizer treatments than in soils treated with only inorganic fertilizers.

The amounts of total extractable N in soils from the vermicompost-treated plots did not differ significantly from those in the inorganic fertilizer control soils. The amounts of total extractable soil N consisted mainly of nitrate-N rather than ammonium N. The marked decreases in extractable N in soils from the inorganic

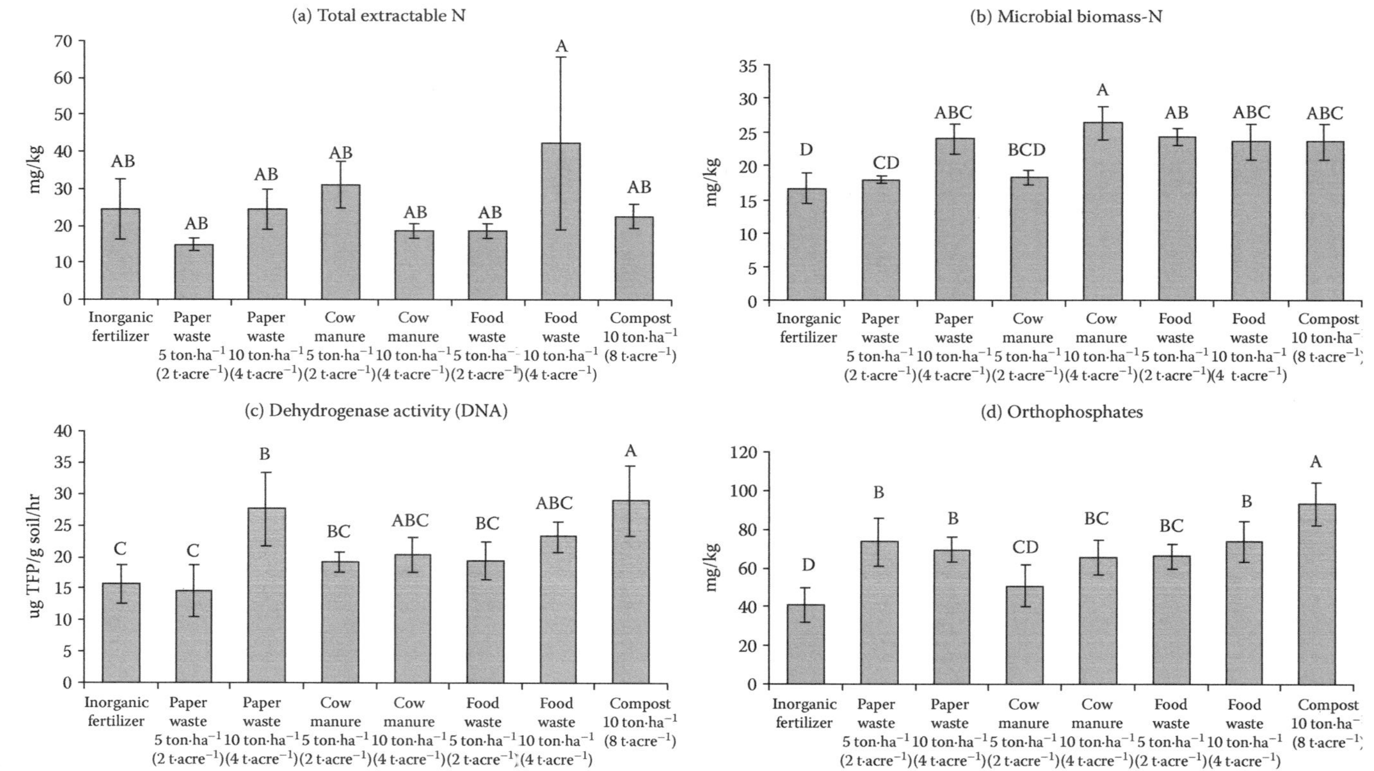

Figure 10.6 Total extractable-N, microbial biomass-N, dehydrogenase activity (DHA), and orthophosphates (means ± SE) in soils planted with peppers at harvest in 2000. Columns designated the with same letter(s) are not significantly different at $p < 0.05$. (From Arancon, N.Q., Edwards, C.A., Bierman, P., Metzger, J.D., and Lucht, C., *Pedobiol.* 49: 297–306, 2005. With permission.)

fertilizer control plots and the paper waste vermicompost–treated plots, relative to soils from the food waste vermicompost–treated plots, may have been due to the greater amounts of total C and N in the food waste vermicompost, which could have provided more N for mineralization. Hence, the food waste vermicompost could have produced more residual N in soil than the inorganic fertilizer or materials with a lower overall content of total N, such as paper waste vermicompost, did. There have been other reports of increases of N in soil after additions of vermicomposts (Nenthra et al. 1999). Amounts of soil N increased significantly after incorporating vermicomposts into soils (Kale et al. 1992; Nenthra et al. 1999; Sreenivas et al. 2000), and the amounts of P and K available also increased in other experiments (Venkatesh et al. 1998).

A similar trend was recorded for soil orthophosphates, because soils from all the vermicompost- and inorganic fertilizer–treated plots contained significantly more orthophosphates than soils from the inorganic-fertilizer-only plots at harvest, although all plots had received equal amounts of P at transplanting. This implies that the continuous inputs of orthophosphates to the soil were probably due to slow rates of release from the vermicomposts. Both types of vermicomposts contained similar amounts of total P, which may explain why there were similar amounts of soluble P in soils from the vermicompost-treated plots at harvest. A trend for amounts of dissolved organic N to decrease in soils from the vermicompost-treated plots may have been the result of mineralization of nutrients from N pools, since those water-soluble compounds, which occur in the organic fraction of vermicomposts, are degraded easily by microorganisms (Cook and Allan 1992). In our experiments, increases in the amounts of orthophosphates in soil from the vermicompost-treated plots could be explained by significant correlations between the microbial biomass N and orthophosphates, indicating that release of P was due largely to the activity of soil microorganisms. Marinari et al. (2000) reported similar increases in phosphates in soil after applications of organic amendments.

The overall fate of the different nutrients in the soil could have been influenced by several factors. Nutrient uptake by plants, together with leaching and the activity of microorganisms, may have been factors (Carlile and Wilson 1993). Hence, the differences in the residual N and P between treatments may have been due to differences in rates of plant absorption of nutrients or rates of leaching of particular nutrients, or to microbial immobilization of these nutrients. Rates of leaching of N differed among the treatments; the inorganic control soils had the fastest rates of leaching of N, because they contained less organic matter than soils from the vermicompost-treated plots. Similarly, decreased leaching of nitrates has been reported from compost-treated soils (Maynard 1989). Leaching of nutrients could also have been decelerated by microbial nutrient immobilization. Since there were many more soil microorganisms in soils treated with vermicomposts, they could sequester nutrients and use them for metabolic activities. The immobilization of nutrients in our experiments may be explained by increases in microbial biomass in the soils treated with vermicomposts; such increases in microbial biomass were greater in soils receiving the higher rates of vermicompost applications, that is, 10 ton·ha^{-1} (4.0 t·acre^{-1}).

B Biological Changes in Soil

Field experiments at The Ohio State University (Arancon et al. 2005) demonstrated that dehydrogenase enzyme activity and microbial biomass were usually greater in vermicompost- and inorganic fertilizer–treated plots than in the control plots that received equivalent amounts of inorganic fertilizers only (Figure 10.6). Microbial biomass has often been used as a factor to define soil quality in long-term experiments (Albaladejo and Diaz 1990) and may be an early and sensitive indicator of soil ecological stress or a need for soil restoration in long-term field experiments (Jenkinson and Ladd 1981; Paul 1984). Pascual et al. (1999) reported significant increases in microbial biomass C in response to sewage waste applications to agricultural soils. The additions of vermicomposts to the soil, especially applications of food waste vermicompost, increased microbial biomass N significantly. The increases in soil microbial biomass that occurred apparently did not influence the supply of nutrients to the strawberry plants by nutrient immobilization, since there was even more residual N and P in soils from the vermicompost-treated plots than in soils from the inorganic control plots, and there were positive correlations between the availability of these nutrient elements and microbial biomass.

The ecological significance of microbial activity in soils cannot be evaluated by simply assessing populations of microorganisms or their activities. We also measured dehydrogenase enzyme activity, since these enzymes are considered to be excellent indices of overall microbial activity (Nannipieri et al. 1990). In field experiments at The Ohio State University, the dehydrogenase activity increased in soils treated with vermicomposts, particularly those treated with food waste vermicompost. Levels of dehydrogenase activity correlated positively and significantly with amounts of soil microbial biomass, especially toward the later growth stages of the strawberry plants. There have been many reports that organic fertilizers can increase soil microbiological activity (Bolton et al. 1985; Fraser et al. 1988; Marinari et al. 2000) so it is not surprising that vermicomposts can also do this. Increases in soil dehydrogenase activity at the later strawberry growth stage, 110 days after transplanting in our experiments, were probably due to mineralization of N and P because of the positive correlations between dehydrogenase activity, microbial biomass, and concentrations of N and P. Dehydrogenase activity in soils followed a different pattern, because dehydrogenase activity was greater in the control soils treated with only inorganic fertilizer than in the soils treated with both vermicompost and inorganic fertilizer at transplanting. A significantly lower soil dehydrogenase activity, which occurred in vermicompost-treated plots at transplanting, may also have been due to an inhibitory effect resulting from the introduction of nonindigenous soil microorganisms from the vermicomposts to the microflora, which may have triggered competition between microorganisms.

However, this could have been overcome at the later stages of strawberry growth, especially at harvest.

Gaind and Nain (2009) reported similar increases in microbial biomass in addition to increases in the available P and N content of wheat soils that had been treated with vermicomposts at 3 ton·ha^{-1} (1.2 t·acre^{-1}). Tejada et al. (2009) reported improvement of physical (structural stability and bulk density), chemical (exchangeable Na percentage), and biological (soil microbial biomass-C, soil respiration, and soil enzymatic activity) properties of soils after application of vermicomposts to field soils at rates of 5 ton and 10 ton·ha^{-1} (1.0 and 4.0 t·acre^{-1}). Maheswarappa et al. (1999) and Jeyabal and Kuppuswamy (2001) reported increased amounts of organic C, improvements in pH, decreased soil bulk densities, improved soil porosities and water-holding capacities, and increased microbial populations and dehydrogenase activity of soils in response to vermicompost treatments.

VIII CONCLUSIONS

Clearly the use of vermicompost in field soil is much more complex, in terms of methods of application, location, and relation to the plant roots, seasonal effects, and residual effects, than its use in growing greenhouse crops. In the greenhouse, soils, amendments, growing conditions, and application methods are much more standardized. We still need to identify the lowest effective economical application rates for different crops, soils, and regions. It seems likely that aqueous vermicompost extracts, or "teas," will be much easier to use on field crops than solid vermicomposts. Teas can be applied as soil drenches at appropriate intervals in the life of the crop. For further information on the use of teas to improve crop germination, growth, and yields, see Chapter 15.

REFERENCES

Albaladejo, J., and Diaz, E. 1990. Degradación y regeneración del suelo en el litoral mediterraneo español: Experiencias en el Proyecto Lucdeme. In *Degradación y Regeneración del Suelo en Condiciones Ambientales Mediterraneas*, ed. J. Albaladejo, M. A. Stocking, and E. Diaz, 191–212. CSIC, Murcia, Spain

Albanell, E., Plaixats, J., and Cabrero, T. 1988. Chemical changes during vermicomposting (*Eisenia fetida*) of sheep manure mixed with cotton industrial wastes. *Biol. Fertil. Soils* 6: 266–269.

Arancon, N.Q., Edwards, C.A., Bierman, P., Metzger, J., Lee, S., and Welch, C. 2003. Applications of vermicomposts to tomatoes and peppers grown in the field and strawberries grown under high plastic tunnels. *Pedobiologia* 47: 731–735.

Arancon, N.Q., Edwards, C.A., Bierman, P., Metzger, J.D., and Lucht, C. 2005. Effects of vermicomposts produced from cattle manure, food waste and paper waste on the growth and yields of peppers in the field. *Pedobiologia* 49: 297–306.

Arancon, N.Q., Edwards, C.A., Bierman, P., Welch, C. and Metzger, J.D. 2004. Influences of vermicomposts on field strawberries: 1. Effects on growth and yields. *Biores. Technol.* 93: 145–153.

Athani, S.I., Hulamanai, N.C., and Shirol, A.M. 1999. Effect of vermicomposts on the maturity and yield of banana. *S. Indian Hortic.* 47: 1–7.

Bolton, H., Papendick, R.I., and Bezdicek, D.F. 1985. Soil microbial biomass and selected soil enzyme activities: Effects of fertilization and cropping practices. *Soil Biol. Biochem.* 17: 297–302.

Buchanan, M.A., Russell, G., and Block, S.D. 1988. Chemical characterization and nitrogen mineralization potentials of vermicompost derived from differing organic wastes. In *Earthworms in Waste and Environmental Management*, ed. C.A. Edwards and E.F. Neuhauser, 231–239. SPB Academic Publishing, the Hague, the Netherlands.

Buckerfield, J.C., and Webster, K.A. 1998. Worm-worked waste boosts grape yields: Prospects for vermicompost use in vineyards. *Aust. N.Z. Wine Ind. J.* 13: 73–76.

Carlile, W.R., and Wilson, D.P. 1993. Microbial activity in media containing worm-worked duck waste. *Acta Hortic.* 342.

Cook, B.D., and Allan, D.L. 1992. Dissolved organic matter in old field soils: Total amounts as a measure of available resources for soil mineralization. *Soil Biol. Biochem.* 24: 585–594.

Darwin, C., 1881. *The Formation of Vegetable Mould, through the Action of Worms* John Murray, London.

Devi, D., Agarwal, S.K., and Dayal, D. 1998. Response of sunflower (*Helianthus annuus*) to organic manures and fertilizers. *Indian J. Agron.* 4: 469–473.

Edwards, C.A., and Arancon, N.Q. 2004. Interactions among organic matter, earthworms and microorganisms in promoting plant growth. In *Functions and Management of Soil Organic Matter in Agroecosystems*, ed. C.A. Edwards, F. Magdoff, and R. Weil, 327–376. CRC Press, Boca Raton, FL.

Edwards, C.A., and Bohlen, P.J. 1996. *The Biology and Ecology of Earthworms.* 3rd ed. Chapman & Hall, London.

Edwards, C.A., and Burrows, I. 1988. The potential of earthworm composts as plant growth media. In *Earthworms in Environmental and Waste Management*, ed. C.A. Edwards and E.F. Neuhauser, 211–220. SPB Academic Publishing, The Hague, the Netherlands.

Edwards, C.A., and Neuhauser, E.F. (Eds.). 1988. *Earthworms in Waste and Environmental Management.* SPB Academic Publishing, the Hague, the Netherlands.

Fraser, D.G., Doram, J.W., Sahs, W.W., and Lesoing, G.W. 1988. Soil microbial populations and activities under conventional and organic management. *J. Environ. Qual.* 17: 585–590.

Gaind, S., and Nain, L. 2009. Chemical and biological properties of wheat soil in response to paddy straw incorporation and its biodegradation by fungal inoculants. *Biodegradation* 18: 496–503.

Gopal, M., Gupta, A., Sunil, E., and Thomas, G. 2009. Amplification of plant beneficial microbial communities during conversion of coconut leaf substrate to vermicompost by *Eudrilus* sp. *Curr. Microbiol.* 59: 15–20.

Jenkinson, D.S., and Ladd, J.N. 1981. Microbial biomass in soil: Measurement and turnover. In *Soil Biochemistry*, ed. E.A. Paul and J.N. Ladd, 415–471. Dekker, New York.

Jeyabal, A., and Kuppuswamy, G. 2001. Recycling of organic wastes for production of vermicompost and its response in rice–legume cropping system and fertility. *Eur. J. Agron.* 15: 153–170.

Kale, R.D., Mallesh, B.C., Kubra, B., and Bagyaraj, D.J. 1992. Influence of vermicompost application on the available macronutrients and selected microbial populations in a paddy field. *Soil Biol. Biochem.* 24: 1317–1320.

Maheswarappa, H.P. Nanjappa, H.V., and Hegde, M.R. 1999. Influence of organic manures on yield of arrowroot, soil physico-chemical and biological properties when grown as intercrop in coconut garden. *Ann. Agric. Res.* 20: 318–323.

Mainoo, N., Barrington, S., Whalen, J., and Sampedro, L. 2009. Pilot-scale vermicomposting of pineapple wastes with earthworms native to Accra, Ghana. *Biores. Technol.* 100: 5872–5875.

Marinari, S., Masciandaro, G., Ceccanti, B., and Grego, S. 2000. Influence of organic and mineral fertilizers on soil biological and physical properties. *Biores. Technol.* 72: 9–19.

Maynard, A. 1989. Agricultural composts as amendments reduce nitrate leaching. *Frontiers Plant Sci.* 24: 24.

Mrinal, S., Rajkhowa, D.J., and Saikia, M. 1998. Effect of planting density and vermicomposts on yield of potato raised from seedling tubers. *J. Indian Potato Assoc.* 25: 141–142.

Murarkar, S.R., Tayade, A.S., Bodhade, S.N., and Ulemale, R.B. 1998. Effect of vermicomposts on mulberry leaf yield. *J. Soils Crops* 8: 85–87.

Nair, S.K., Naseema, A., Meenakumari, S.K., Prabhakumari, P., and Peethambaran, C.K. 1997. Microflora associated with earthworms and vermicomposting. *J. Trop. Agric.* 35: 93–98.

Nannipieri, P., Grego, S., and Ceccanti, B. 1990. Ecological significance of the biological activity in soil. *Soil Biol. Biochem.* 6: 293–355.

Nenthra, N.N., Jayaprasad, K.V., and Kale, R.D. 1999. China aster (*Callistephus chinensis* (L.) *Ness*) cultivation using vermicomposts as organic amendment. *Crop Res. Hisar.* 17: 209–215.

Pascual, J.A., Hernandez, T., and Garcia, C. 1999. Lasting microbiological and biochemical effect of the addition of municipal solid waste to an arid soil. *Biol. Fertil. Soils* 30: 1–6.

Patil, M.P., Humani, N.C., Athani, S.I., and Patil, M.G. 1998. Response of new tomato genotype Megha to integrated nutrient management. *Adv. Agric. Res. India* 9: 39–42.

Paul, E.A. 1984. Dynamics of organic matter in soils. *Plant Soil* 76: 275–285.

Sen, B., and Chandra, T.S. 2009. Do earthworms affect dynamics of functional response and genetic structure of microbial community in a lab-scale composting system? *Biores. Technol.* 100: 804–811.

Singh, R.R., Sharma, S., Kumar, R.K., Gupta, R., and Patil, R.T. 2008. Vermicompost substitution influences growth, physiological disorders, fruit yield and quality of strawberry (*Fragaria x ananassa* Duch.). *Biores. Technol.* 99: 8507–8511.

Sreenivas, C., Muralidhar, S., and Rao, M.S. 2000. Vermicomposts: A viable component of IPNSS in nitrogen nutrition of ridge gourd. *Ann. Agric. Res.* 21: 108–113.

Tejada, M., Hernandez, M.T., and Garcia, C. 2009. Soil restoration using composted plant residues: Effects on soil properties. *Soil Tillage Res.* 102: 109–117.

Ushakumari, K., Prabhakumari, P., and Padmaja, P. 1999. Efficiency of vermicomposts on growth and yield of summer crop okra (*Abelmoschus esculentus* Moench). *J. Trop. Agric.* 37: 87–88.

Vasanthi, D., and Kumaraswamy, K. 1999. Efficacy of vermicomposts to improve soil fertility and rice yield. *J. Indian Soc. Soil Sci.* 47: 268–272.

Venkatesh, P.B., Patil, S., Patil, C.V., and Giraddi, R.S. 1998. Effect of in situ vermiculture and vermicomposts on availability and plant concentration of major nutrients in grapes. *Karnataka J. Agric. Sci.* 11: 117–121.

Vivas, A., Moreno, B., Garcia-Rodriguez, S., and Benitez, E. 2009. Assessing the impact of composting and vermicomposting on bacterial community size and structure, and microbial functional diversity of an olivemill waste. *Biores. Technol.* 100: 1319–1326.

Yasir, M., Aslam, Z., Kim, S.W., Lee, S.W., Jeon, C.O., and Chung, Y.R. 2009. Bacterial community composition and chitinase gene diversity of vermicompost with antifungal activity. *Biores. Technol.* 100: 4396–4403.

Zende, G.K., Ruikar, S.K., and Joshi, S.N. 1998. Effect of application of vermicomposts along with chemical fertilizers on sugar cane yield and juice quality. *Indian Sugar* 48: 357–369.

CHAPTER 11

The Production of Vermicompost Aqueous Solutions or Teas

Cindy E. Salter and Clive A. Edwards

CONTENTS

I Introduction 153
II Basic Tea-Production Concepts 154
  A Vermicompost Sources 154
  B Aeration and Agitation/Stirring 155
  C Supplemental Nutrients (Additives) 156
  D Duration of the Extraction and Brewing Process 156
  E Water Quality and Temperature 157
III Tea-Production Systems 157
  A Passive Steep 157
  B Mechanical Systems 158
IV Vermicompost Tea Benefits and Application Methods 160
V Compost Tea, Vermicompost Tea, and the National Organic Program (NOP) 161
VI Conclusions 161
  A Recommendations of the Compost Tea Task Force 161
  B Vermicomposting 163
  C Composting 163
References 163

I INTRODUCTION

Vermicompost aqueous solutions—commonly referred to as vermicompost "teas"—can be defined simply as water extracts of solid vermicomposts from which microorganisms, soluble nutrients, and plant-beneficial substances are converted into a liquid form. Vermicompost tea can be used in a wide range of horticultural and agricultural systems to elicit plant growth and pest- and disease-management responses through a variety of mechanisms. One of the unique features of vermicompost tea is that, unlike its solid counterpart, it can be applied directly to plant foliage. It is also used as a soil drench and has been shown to be effective in relatively small quantities (see Chapters 13, 14, and 15).

Vermicompost tea is a subset of a broader category of aqueous extracts known as *compost teas* or *compost extracts* (Brinton 1995). While the production and application technology and the modes of action of these extracts may be similar, *vermicompost teas* are more precisely characterized by the organic substrates, vermicomposts, from which the beneficial substances are extracted. Just as vermicomposts have proved to be a consistently prolific source of plant-beneficial compounds, increasing numbers of grower testimonials and some recent scientific studies suggest that vermicompost teas are similarly effective.

II BASIC TEA-PRODUCTION CONCEPTS

The basic concept of vermicompost tea production is simple:

$$\text{Vermicompost tea} = \text{Vermicompost} + \text{Water} + \text{Time}$$

The chemical and biological characteristics of vermicompost tea may differ with changes in inputs and other process variables. To establish a production system with a consistent outcome, it is advantageous to hold as many of these variables constant as possible. Vermicompost tea production inputs and processing variables include the following:

- Vermicompost sources
- Mechanical aeration and stirring/agitation
- Supplemental nutrients (additives)
- Duration of the extraction and brewing process
- Water quality and temperature

A Vermicompost Sources

Because vermicompost teas are basically the concentrated essence of vermicomposts, the parent source of the vermicompost is perhaps the most important variable in achieving an aqueous extract with consistent quality. Vermicompost characteristics vary according to types of feedstock and other production variables discussed in more detail in other chapters. A reliable source of vermicompost with consistent characteristics will help minimize variability in the aqueous extracts produced from it.

The ratio of vermicompost to water used in producing vermicompost tea is a variable that can be held constant. Ratios between 5% (1:20) and 20% (1:5) solid vermicompost to water have proved to be effective in laboratory and greenhouse trials (see Chapters 13, 14, and 15). Regardless of the ratio of vermicompost to water used in the production process, the final extract can be diluted further as needed for specific applications.

B Aeration and Agitation/Stirring

Many versions of the extracts derived from composts, vermicomposts, manures, and other organic materials have been used beneficially in agriculture for many years (Scheuerell and Mahaffee 2002, 2004). Some of these crude extracts were made by simply steeping the material in water for days or even weeks, usually without mechanical stirring or aeration. The practice of introducing air into the solution (and/or agitating, recirculating, or stirring the solution) during the extraction process is a fairly recent development. The rationale for introducing air into the extraction process is to encourage the proliferation and survival of aerobic microorganisms in the aqueous solution or tea, and also to decrease the culture and development of anaerobic microorganisms that may produce metabolic by-products unfavorable to plant growth in the aqueous extract. The first commercial aerated compost/vermicompost tea system, shown in Figure 11.1, was introduced in 1996. There are now a number of commercial aerated systems available on the market with a wide range of sizes and capabilities, and many custom-designed, homemade systems are in use around the world. (For more detail on aerated compost tea systems, see later in this chapter.)

Figure 11.1 First commercially available compost tea system (Growing Solutions' Microb-brewer, 12-gallon and 50-gallon models, introduced in 1997). (Courtesy of Growing Solutions Inc.)

C Supplemental Nutrients (Additives)

The advent of aerated tea brewing equipment and widespread use of teas was accompanied by a common practice of adding various additives or supplemental nutrients to the solution as microbial food sources. Although such additives are not necessary for the extraction phase of vermicompost tea production, they may affect the subsequent growth of microorganisms that are extracted into the solution. The practice of adding such additives to the process was based on the theory that an increased active microbial biomass would enhance the efficacy of the final product. Examples of supplemental materials that have been used include molasses (Duffy et al. 2004), humic acids, kelp, rock powders, fish emulsions, and a variety of other ingredients. When they are used, the quantity of supplemental nutrients used in compost/vermicompost tea production is usually small relative to the amount of compost/vermicompost used. There are commercial blends of additives offered by tea equipment manufacturers that are formulated and prescribed specifically for their equipment design and capacity.

There is considerable debate within the realm of compost/vermicompost tea enthusiasts, scientists, regulators, and other interested parties regarding the use of additives, due to the potential of carbohydrate-rich materials to promote the growth of undesirable microorganisms, especially human pathogens that could be transferred to food crops, thereby posing potential consumer health risks. The primary argument is that even a minuscule quantity of human pathogens present in a solid substrate such as compost or vermicompost could be multiplied to unsafe levels in the presence of supplemental nutrients, simple sugars such as molasses in particular. (See later section on National Organic Program (NOP) and pathogen safety.) The debate is less pertinent to nonfood crops or to food crops with long intervals (>90 days) between tea application and harvest. The scientific literature on the mechanisms of action of compost and vermicompost teas, either with or without amendments, is indeed very sparse at this time.

D Duration of the Extraction and Brewing Process

The amount of time it takes to produce vermicompost tea varies from hours to days, depending on the type of brewing systems used, objectives, and production variables. The physical extraction of soluble components and microorganisms from vermicomposts occurs fairly rapid and is less time-dependent than the actual "brewing" process. It is relevant to note that in some cases the brewing step is eliminated after a relatively brief "extraction-only" process. This production scenario is effectively exempt from regulation by the National Organic Standards Board (NOSB) per the task force recommendations (2002, 2004; see section at the end of this chapter).

A commonly-used and convenient brewing time is 24 hours, with continuous aeration and agitation, although this varies widely according to the type and size of equipment and other factors. The temperature of the water (see later in this chapter) affects microbial growth and therefore impacts the amount of time necessary to achieve the desired microbial concentration and movement of nutrients into teas. Although brew time probably does not significantly impact the concentrations of soluble components that are extracted into solution, because it is a biological process the chemical components may be subsequently influenced by biological activity. There is a point of diminishing return in the production process by which time the rates of microbial growth may decline due to limiting factors. Monitoring the chemical and biological characteristics of the solution can help to establish optimal brew times.

E Water Quality and Temperature

The physical and chemical characteristics of the water used for compost tea production can be significant variables. Temperature affects the types of organisms that will grow, as well as their rates of growth. Temperature also affects the concentrations of dissolved oxygen in water because water that is too warm has a limited capacity to contain oxygen, directly affecting the amount of oxygen that is available to microorganisms. Dissolved or suspended solids, pesticides, heavy metals, pathogens, chlorine, and other components of the water supply can all affect the growth of microorganisms to varying degrees.

III TEA-PRODUCTION SYSTEMS

Vermicompost teas can be produced using equipment configurations ranging from the most basic homemade system to elaborate commercial systems and everything in between. The majority of tea-production systems fall into one of the following types:

- Passive steep (nonmechanical)
- Mechanical systems
    - Stirring
    - Recirculating
    - Forced air
    - Hybrid
    - Extractors

A Passive Steep

Early production of teas employed a “passive steep” approach, wherein the compost or vermicompost was placed in an open-weave bag and submerged in water, then left to steep passively with periodic manual stirring over a period of days or even weeks. Although not mechanically aerated, the passive steep systems nonetheless are not necessarily anaerobic since oxygen is passed into the solution via surface air exchange when it is stirred manually.

B Mechanical Systems

The desire to accelerate the brewing process and to make it more consistent and "user friendly" led to the development of mechanically aerated tea-production systems, thereby launching the modern day compost tea industry. There are now homemade and commercially manufactured mechanical versions with a wide range of features and capacities. Homemade or farm-made units, of which there are infinite varieties, can be as effective as commercial ones but generally speaking are one of a kind. Desirable features of commercial compost/vermicompost tea systems include ease of use and maintenance, safety, energy efficiency, and quality and consistency in production from one tea batch to the next. In short, these systems help to control some of the many variables inherent in tea production and make it more accessible to a broader user base.

Mechanically aerated tea systems are designed to provide optimum conditions for the extraction and growth of aerobic microorganisms (bacteria, fungi, actinomycetes, protozoa), as well as the extraction of the nutrients and soluble organic compounds from the compost or vermicompost source. The basic components of an aerated tea system are a liquid holding tank, a mesh container to hold the compost or vermicompost, and an air delivery system. Aeration serves two purposes: It provides required oxygen for the microorganisms, and homogeneous mixing and agitation aids in the physical extraction process. The capacity of these systems varies from less than 5–1000 gallons or more. Figure 11.2 is a schematic drawing of an

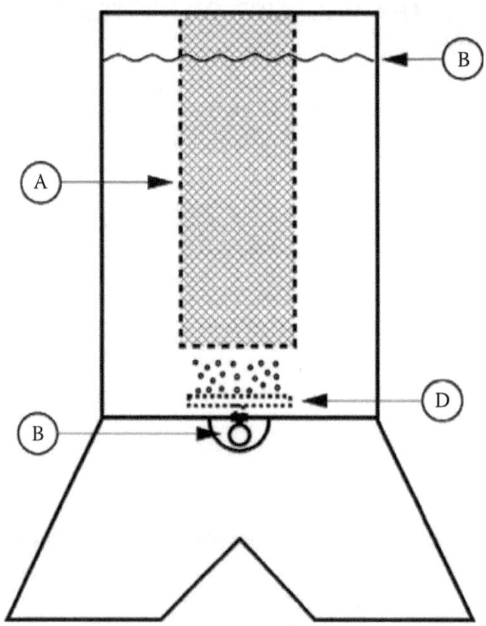

Figure 11.2 Schematic of tea brewer using diffusion disk aeration. A, mesh basket containing compost/vermicompost; B, tea discharge valve; C, water level; D, aeration disk. (Courtesy of Growing Solutions, Inc.)

aerated tea brewer that uses fine bubble diffusion technology to provide an aeration and mixing.

Tea systems can be categorized further by the mechanical means used to stir, agitate, or aerate the solution. Most systems use one or more of the following technologies:

- *Recirculation*: Recirculating systems include a mechanical pump that continually recirculates the solution through the holding tank, creating a mixing action and entraining surface air into the system by the vortex created by the recirculation action. A conical tank design is often associated with a recirculating system.
- *Stirring*: In these systems a motor powers a mechanical paddle with a stirring action that creates a vortex to agitate, mix, and entrain oxygen into the solution.
- *Forced air*: These systems are the most common of the currently available commercial tea systems. They include a variety of blower types and air-flow rates depending on tank capacity. In addition to blower motors, there are a number of different air delivery systems that actually distribute the air from the blower into the solution. Figure 11.3 shows an example of a commercially available forced-air system utilizing fine bubble diffusion technology.
- *Hybrid systems*: As the name implies, hybrid tea-production systems incorporate one or more of the preceding mechanical elements to provide oxygen or agitation to the solution.
- *Extractors*: Some systems are designed with the singular objective of extraction, and these may differ in design from other tea systems that are designed to increase microbial growth and biomass. Most tea systems can, in effect, be operated as an extractor unit by simply reducing the brewing time.

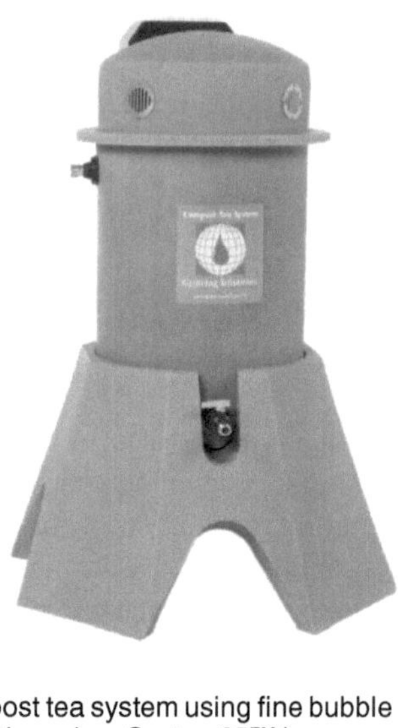

Figure 11.3 Forced-air compost tea system using fine bubble diffusion technology. (Courtesy of Growing Solutions, Inc. System25™.)

IV VERMICOMPOST TEA BENEFITS AND APPLICATION METHODS

The past few years have seen widespread innovation and maturation in vermicompost tea technologies, uses, and applications. Vermicompost teas are now being produced and used in large-scale agriculture, viticulture, orchards, horticulture, nurseries, turf greens, commercial landscaping, and home gardens. Rates, frequency, and modes of application vary according to the cropping system, pest and disease pressures, and existing conditions.

Vermicompost tea is primarily known for its ability to boost soil microbiological activity by adding millions of bacteria, fungi, actinomycetes, and protozoa along with the by-products of their metabolism. The active components in vermicompost tea include microorganisms, water-soluble fulvic acids (building blocks for humic acids) and particulate humates, plant-growth hormones produced by microorganisms, and materials that improve the availability of micronutrients (chelating effect). Soluble nutrients present in vermicompost tea nourish plants directly as well as feeding existing soil microorganisms.

While clearly a different product from vermicompost, vermicompost tea shares many of its beneficial attributes. Like vermicompost, it increases soil biological activity and provides beneficial organic compounds. Unlike vermicompost, it can be applied directly to plant foliage as well as to the soil. Direct foliar application of vermicompost tea provides nutrients that may be utilized directly by the plant while also introducing a diverse array of microorganisms that colonize leaf surfaces. An increase in scientific studies is showing this to be instrumental in the disease-suppression capabilities of vermicompost teas.

Many growers use both soil and foliar applications of vermicompost teas. Whether applied to the foliage or to the soil, existing spray or irrigation equipment can usually be adapted for teas. In some cases it is necessary to filter or strain the vermicompost tea to accommodate spray nozzles—drip emitters in particular. It is not uncommon for growers to add other nutrients or compatible amendments, if needed, as a tank mix to the tea mixture to economize on spray applications.

Growers using vermicompost teas report positive effects in plant vigor, bud break, fruit color, root volume, and pest and disease resistance (see Chapters 13, 14, and 15). In many instances, these effects have enabled them to reduce their use of fertilizers, fungicides, pesticides, and other costly inputs. Another approach used by some growers is to use vermicompost tea to accelerate the decomposition of plant residues.

Knowing how and when to apply the tea is important to the success of any production strategy. The ideal number, rate, and timing of applications, as well as the mode of application, varies with crop, soil type, weather, and disease pressures. Growers typically start with a small area of a crop to learn how to use it properly and then gradually incorporate vermicompost tea into their overall biofertility program.

V COMPOST TEA, VERMICOMPOST TEA, AND THE NATIONAL ORGANIC PROGRAM (NOP)

Certified organic growers are subject to the U.S. NOP implemented in October 2002 (NOSB 2002). The NOP is a federal law that requires all organic food products to meet the same standards and be certified under the same certification process. As a sanitation precaution, the NOP regulates the use of manure and manure-based composts in edible crops, stipulating pre-harvest intervals between applications of these materials and harvest of the edible crop. Because the implications for compost tea on this issue in the rule were unclear, the National Organic Standards Board (NOSB 2004) appointed a 13-member scientific panel—the Compost Tea Task Force in 2003–2004—and relegated to them the task of clarifying the issues and making recommendations. In a report, the Compost Tea Task Force identified why increasing numbers of organic gardeners, farmers, and vermicompost and compost tea equipment manufacturers are turning to compost and vermicompost tea as a unique approach to biofertility management. A primary reason for using compost and vermicompost tea is to transfer microbial biomass, fine particulate organic matter, and soluble chemical components of compost or vermicompost into an aqueous phase that can be applied to plant surfaces and soils in ways not possible or economically feasible with solid compost or vermicompost (NOSB 2004.)

The report concluded that compost and vermicompost teas can and do provide benefits to plants and the environment, and it offered guidelines to produce and apply compost and vermicompost tea safely. For example, the scientists recommended a number of common sense measures and quality assurance tests to avoid transferring pathogens from poor-quality materials to edible plants.

VI CONCLUSIONS

Vermicompost teas show great promise for horticulture and agriculture, especially organic growers but should be viewed like any other input and not as a silver bullet. Used correctly, vermicompost teas are a powerful tool, particularly when used in combination with an integrated fertility and pest-management program. As in any emerging industry, there is a need for scientific research to validate the experiences of practitioners and to expand on existing studies that have documented the benefits of plant health, growth, and vigor from the use of vermicompost teas.

A Recommendations of the Compost Tea Task Force

1. Potable water must be used to make compost tea and for any dilution before application.
2. Equipment used to prepare compost tea must be sanitized before use with a sanitizing agent as defined by 21 CFR 178.1010.

3. Compost tea should be made with compliant compost or vermicompost, using the NOSB Compost Task Force Guidelines set forth on April 18, 2002, for thermal compost and vermicompost, or compost as defined in section 205.203 (c) (2) of the NOP rules. For compost tea, this applies to 100% plant-feedstock materials in addition to manure feedstocks because nonmanure compost feedstocks may harbor high levels of fecal bacteria.
4. Compost teas made **without compost tea additives** can be applied without restriction.
5. Compost teas made **with compost tea additives** can be applied without restriction if the compost tea-production system (same compost batch, additives, and equipment) has been pretested to produce compost tea that meets the Environmental Protection Agency recommended recreational water-quality guidelines for a bacterial indicator of fecal contamination. These indicators and the passing criteria are *Escherichia coli* (126 CFU/100 mL) or enterococci (33 CFU/100 mL). At least two compost teas batches must be tested using accepted methodology, with the average population their indicator bacteria across compost tea batches used as the measurement of passing. Each new batch of compost would require that the system-quality assurance pretest be conducted again as indicated. After it passes again, compost tea from the system can be used without restriction.

   If compost teas made **with compost tea additives** has not been pretested for indicator bacteria, their use on food crops is restricted to the 90/120-day preharvest interval. Crops not intended for human consumption, ornamental plants, and grain crops intended for human consumption are exempt from bacterial testing and 90/120 day preharvest interval restrictions. In the view of the task force, educating producers about the potential for contamination and its impacts on public health and marketing, as well as how this recommended quality assurance testing system would avoid potential contamination, will provide compelling incentives for producers to follow the rules.
6. Compost extracts—any mixture of compost, water, additives, and adjuvants that are not held for more than 1 hour before use—may be applied without restriction.
7. Raw manure extracts or teas may be applied to the soil with a 90/120-day preharvest restriction, foliar applications are prohibited.
8. Compost leachates may be applied to the soil with a 90/120-day preharvest restriction, but foliar applications are prohibited.
9. Compost tea is not allowed for the production of edible seed sprouts.
10. The emerging national acceptance of compost tea as a biologically based crop-production tool by organic as well as conventional growers clearly indicates the need for further scientific investigation to validate the benefits and concerns of compost tea use. The Task Forceunanimously urges USDA and its agencies to strongly support additional research on the potential for crop contamination and plant disease/pest control by compost tea. There is an urgent national need to address critical data gaps, uncertainties, and variability in existing data that limited the evaluation of potential crop contamination by the current task force. Data are urgently needed to provide science-based recommendations on how compost tea-production and application practices impact potential crop contamination, while at the same time preserving the means for improving plant health and vigor. Critical issues requiring further data include compost quality, compost tea additives, temperature and duration of compost tea production, and the population dynamics of human pathogens in microbially diverse agro-ecosystems relative to preharvest intervals for application of compost tea.

B Vermicomposting

Vermicomposting is a process of earthworms digesting organic matter to transform the material into a beneficial soil amendment. Vermicompost was defined by the NOSB Compost Task Force (NOSB 2002) as follows: "Vermicompost is acceptable if (i) made from only allowed feedstock materials, except for incidental residues that will not lead to contamination, (ii) aerobicity is maintained by regular additions of thin layers of organic matter at 1–3 day intervals, (iii) moisture is maintained at 70–90%, and (iv) duration vermicomposting is at least 12 months for outdoor windrows, 4 months for indoor container systems, 4 months for angled wedge systems, or 60 days for continuous flow reactors." In addition to these recommendations for making vermicompost tea acceptable to the NOP, it would also be acceptable if the organic feedstock was precomposted for a period according to the task force guidelines prior to vermicomposting.

C Composting

Composting is a managed process in which organic materials, including animal manure and other residuals, are decomposed aerobically by microbial action. *Thermophilic* composting refers to the time-limited, self-heating process in which heat generated by microbial respiration is retained in the mass of a pile or windrow such that vulnerable pathogenic microorganisms are destroyed. Compost is defined by the NOSB Compost Task Force (NOSB 2002) as "Compost, in addition to that described in Section 205.203 (c) (2), is acceptable if (i) made only from allowed feedstock materials, except for incidental residues that will not lead to contamination, (ii) the compost undergoes an increase in temperature to at least 35°C (31°F) and remains there for a minimum of 3 days, and (iii) the compost pile is managed to ensure that all of the feedstock heats to the minimum temperature."

REFERENCES

Brinton, W.F. 1995. The control of pathogenic fungi by use of compost teas. *Biodynamics* Jan/Feb: 5–12.

Duffy, B., Sarreal, C., Subbaras, R., and Stanker, L. 2004. Effect of molasses on regrowth of *E. coli* 0157:H7 and *Salmonella* in compost "teas". *Compost Sci. Util.* 12(1), 93–96.

National Organic Standards Board (*NOSB*). 2004. Compost Tea Task Force Report. Agricultural; Marketing Service USDA. www.ams.usda.gov/nosb/meetings/compostteataskforce.pdf.

National Organic Standards Board (*NOSB*). 2002. Compost Tea Task Force Recommendations as amended by NOP. http://www.ams.usda.gov/nosb/archives/crop/recommendations.html.

Scheuerell, S., and Mahaffee, W. 2002. Compost "teas": Principles and prospects for plant disease control. *Compost Sci. Util.* 10 (4), 313–335.

Scheuerell, S., and Mahaffee, W. 2004. Compost tea as a container medium drench for suppressing seedling damping-off caused by *Pythium ultimum. Phytopathology* 9A, 1156–1163.

CHAPTER 12

The Suppression of Plant Pathogens by Vermicomposts

Allison L. H. Jack

CONTENTS

I Introduction 165
II The Role of Earthworms in Suppression of Soilborne Plant Diseases......... 167
III Suppression of Plant Pathogens with Vermicompost Amendments............. 168
IV Conclusions and Directions for Future Research 175
References.. 177

I INTRODUCTION

Soils where plant disease incidence is low, even when the pathogen is present in high numbers, or disease-suppressive soils, were first described in the late 1800s (Huber and Schneider 1982), and the biological nature of this phenomenon was documented in the 1950s (Menzies 1959). Since then, researchers have been working to understand the nature of this interaction between a host plant, a pathogen, and the surrounding microbial communities, so that agricultural soils can be managed in a way that encourages suppression of disease (Curl 1988; Whipps 2001; Mazolla 2002; Sánchez-Moreno and Ferris 2007). The original conceptual underpinnings of the modern organic agriculture movement include an emphasis on maintaining soil organic matter to support biologically-based soil and plant health (Howard 1943, 1945). Among the numerous available methods for managing soil organic matter are the practices of cover cropping and addition of organic materials from external sources, that is, compost or vermicompost amendments. In addition to increasing soil organic matter, the highly diverse microflora of composts and vermicomposts can suppress plant diseases when used as soil amendments. Disease-suppressive composts and vermicomposts have been well documented in the scientific literature (see reviews by Hoitink and Fahy 1986; Weltzien 1989; Hoitink et al. 1991; Litterick et al. 2004); however, the use of compost for disease management is plagued by inconsistency in practice. Only relatively recently have earthworms and vermicomposts been investigated for their ability to suppress plant disease. This Chapter reviews the current state of knowledge on the roles of earthworms and vermicomposts in biologically based plant disease management and points out directions for future research efforts.

Before addressing this topic, it is important to understand the basics of earthworm digestion and how earthworms interact with the soil microbiota. Earthworms affect overall microbial activity and diversity in the soil environment both directly, through feeding and secreting casts, and indirectly through various modifications of soil organic matter (Brown 1995; Binet et al. 1998; Brown et al. 2000). Earthworms derive nutrients from microorganisms associated with organic matter breakdown in the earthworm's gut. Digestion of these materials relies on a complex interplay between the enzymes and mucus secreted by the earthworm's digestive system, enzymes produced by the gut microbiota, and microbially mediated decomposition of organic matter in the surrounding soil. As ingested soil microbes travel the earthworm intestinal tract, their numbers are affected in the crop and gizzard through grinding, then they eventually increase in the hindgut due to the secretion of mucus, which provides a source of readily available carbon (Hartenstein et al. 1981; Trigo et al. 1999). This intestinal mucus aids digestion as the stimulated gut microbes secrete exoenzymes to break down the more recalcitrant organic matter in ingested soil (Drake and Horn 2007). Intestinal mucus continues to stimulate microbial activity outside the earthworm in excreted casts, a process sometimes described as the earthworm's "external rumen" (Lavelle et al. 1995). Increases in microbial activity sparked by the intestinal mucus lead to the further decomposition of soil organic matter in and around each cast. When the cast and surrounding soil are reingested, additional nutrients are made available to the earthworm through this advanced decomposition (Brown et al. 2000). In this mutualistic relationship, the soil microbiota benefit from the additional carbon sources excreted by the earthworm in the form of mucus, and the earthworm benefits from the increased decomposition of the soil organic matter is carried out by these microorganisms.

The microbial community composition changes significantly while ingested material travels through the earthworm intestinal tract (Sampedro and Whalen 2007); however, it is not totally clear whether earthworms possess symbiotic gut microbiota that are not otherwise found in the soil environment as is the case for many other soil invertebrates. Some researchers have found no evidence for gut-specific bacteria in *Lumbricus terrestris* L. using molecular profiling (Egert et al. 2004), while others have found multiple fatty acid bacterial biomarkers in *L. terrestris* gut samples that are not present in the surrounding soil (Sampedro and Whalen 2007). Culture-based studies have revealed that the gut of starved *Eisenia fetida* (Savigny) contained a fermentative bacterium, *Aeromonas hydrophila*, not found in the food source of the earthworm (Toyota and Kimura 2000). There is a well-established precedent for symbiotic associations with bacteria in lumbricid earthworms. *Eisenia fetida* has a highly specific symbiotic relationship with *Acidovorax* spp., which are recruited selectively into nephridia in early stages of embryogenesis and are passed vertically through generations via the cocoon albumin (Davidson and Stahl 2006, 2008). Although there is evidence that other bacterial species in the cocoon albumin eventually colonize the gut of the developing earthworm (Davidson and Stahl 2008), it is not yet clear if similar highly specific symbiotic relationships exist between gut microbes and earthworms. This is an important area for future research, as any applied use of vermicomposts for plant disease suppression will benefit from a more complete understanding of the role of substrate-derived and potentially symbiotic microbes in earthworm digestion, and the selective effects that earthworms can have on microbial community structure in their soil and compost environments.

II THE ROLE OF EARTHWORMS IN SUPPRESSION OF SOILBORNE PLANT DISEASES

Earthworms can modify soil microbial community structure depending on the types of organic matter present (Enami et al. 2001). Their interactions can impact plant health in agroecosystems because earthworms may shift the soil microbial community from one that allows the development of disease if a pathogen is present to one that does not. Earthworms of the genus *Aporrectodea* (Orley) have been studied for their associations with the reduction of symptoms of several soilborne plant diseases. For example, the presence of *A. rosea* (Eisen) and *A. trapezoides* (Dugés) was correlated with a reduction in symptoms of *Rhizoctonia solani* on wheat in an Australian field soil (Stephens, Davoren, Ryder, Doube, and Correll 1994) and on clover and perennial rye grass in a pot study (Stephens and Davoren 1997). These earthworm species were also associated with suppression of disease caused by *Gaeumannomyces graminis* var. *tritici* on wheat; however, this suppression did not occur at higher inoculum levels (Stephens and Davoren 1995). Similar experiments in Canada found a slight but not statistically significant ($p \leq 0.08$) suppression of *G. graminis* var. *tritici* on wheat by *A. trapezoides*, *A. tuberculata (Eisen)*, and *A. rosea* (Clapperton et al. 2001). There is a possibility that earthworm-associated disease suppression is region and/or soil specific since *Aporrectodea* spp. were correlated with suppression of *R. solani* in only one of two soils tested (Stephens, Davoren, Ryder, Doube, and Correll 1994; Stephens and Davoren 1995). Earthworm-associated disease suppression has been documented in additional pathosystems in greenhouse experiments; however, field research in this area continues to be complicated by the difficulties associated with excluding earthworms from plots (Elmer 2009; Elmer and Ferrandino 2009).

Earthworms may act as vectors for the dispersal of disease-suppressive microorganisms in soil. For example, *A. trapezoides* spread the biocontrol bacterium *Pseudomonas corrugata* (effective against *G. graminis* var. *tritici* on wheat) to a depth of 9 cm (3.54 in) after surface inoculation in pots compared to a depth of 3 cm (1.2 in) achieved in no-earthworm controls. The presence of earthworms was also correlated with an increase in colonization of wheat roots by *P. corrugata* (Stephens, Davoren, Ryder, and Doube 1993). Adding barley straw extended the earthworm-associated increase in *P. corrugata* population levels from 29 to 63 days (Stephens, Davoren, and Hawke 1995). Attempts have been made to develop a biocontrol inoculation system where pelletized manures spiked with *P. corrugata* were applied as "earthworm food" with the idea that earthworms would graze in the area of the nutrient-rich pellet and spread the bacterium through the soil (Schmidt et al. 1997). Populations of specific bacteria may persist in the earthworm gut as evidenced by their presence in casts much longer than expected given their relatively short gut-retention times. Casts of *Octolasion cyaneum* (Enterion) were found to have detectable populations of *Pseudomonas fluorescens* for up to 12 days after the earthworms were moved into a substrate initially free of the bacterium (Clegg et al. 1995).

In addition to the concept of using earthworms to distribute biocontrol agents, attempts to engineer broadly suppressive vermicomposts by altering the initial organic materials have been made. This potential applied use of earthworms is based on the knowledge that earthworm diet can affect the microbial community structure in casts (Egert et al. 2004); however, no clear pattern has emerged on the relationships between feedstock and the population levels of a wide range of "beneficial" bacteria and fungi. For example, substrates that increased the numbers of fluorescent pseudomonads in casts also decreased the numbers of *Trichoderma* spp., a genus containing many effective biocontrol agents (Gopal et al. 2009). This approach may be feasible only in specific pathosystems where the key groups of disease-suppressive microorganisms have been identified. In addition to stimulating the activities of and/or dispersing disease-suppressive microorganisms, earthworms may also directly decrease the viability of plant pathogens. Gut transit through *E. fetida* significantly reduced infectivity and damaged proteins of cowpea mosaic virus (CPMV) and tobacco mosaic virus (TMV; Amaravadi et al. 1990). Some earthworm polysaccharides have in vitro activity against common plant pathogens; however, since a whole-earthworm extract was used in these studies, it is not clear whether these compounds would ever come into direct contact with soil or gut microorganisms (Wang et al. 2007). Along with using earthworm castings as soil and potting media amendments, intentionally increasing earthworm populations or using earthworms to distribute biocontrol agents may become future biologically based disease-control strategies.

III SUPPRESSION OF PLANT PATHOGENS WITH VERMICOMPOST AMENDMENTS

Vermicomposts have not been as extensively studied as thermophilic composts for their ability to suppress plant diseases; however, a small body of literature does exist on this topic. Many of these studies consist of a small trial with a single pathogen that is published as a brief technical report or as part of a larger field experiment (Wright et al. 1999; Rivera, Wright, Lopez, and Guastella 2001; Chaoui et al. 2002; Bhadoria et al. 2003; Rivera, Wright, Lopez, and Fabrizio 2004; Asciutto et al. 2006; Singh et al. 2008). See Table 12.1 for details of these studies (see Chapter 13).

Table 12.1 Disease-Suppressive Vermicompost Results in the Literature

Reference	Feedstock	Amendment Rate/Significant Suppression +[a] = yes, –[a] = no					Substrate	Crop	Pathogen
(Kannangara et al. 2000)[b]	dairy manure separated solids	5	10	20	30	40	yellow cedar sawdust	Cucumber (*Cucumis sativa* cv. 'Corona')	*Fusarium oxysporum* f. sp. *radicis cucumerinum*
(Joshi et al. 2009)	unspecified	10 t. ha^{-1}[c] –(2005)	–(2006)				soil	French bean (*Phaseolus vulgaris* L.)	*Rhizoctonia solani*
		10 t. ha^{-1}[c] +(2005)	+(2006)				soil		*Phaeoisariopsis griseola*
(Rivera et al. 2004)	cattle manure	25 +[d]	50 +	75 +	100 +		soil	white pumpkin (*Cucurbita maxima*)	*Rhizoctonia solani*
(Asciutto et al. 2006)	unspecified	25 –[e]	50 –	75 –	100 –		unspecified potting media	bedding ornamental (*Impatiens wallerana*)	*Rhizoctonia solani*
(Rodriguez Navarro et al. 2000)	cattle manure	10 +	20 +	30 +	40 +		black earth, chicken manure, rice husks (70:20:10)	Gerbera daisy (*Gerbera jamesonii*)	*Phytophthora dreschleri and Fusarium oxysporum*
		10 + fert –	20 + fert +	30 + fert –	40 + fert –				
(Singh et al. 2008)	vegetable waste and cattle manure	2.5 t a^{-1} +	5 t a^{-1} +	7.5 t a^{-1} +	10 t a^{-1} +		soil	strawberry (*Fragaria x ananassa* cv. 'Chandler')	*Botrytis cinerea*
(Rivera et al. 2001)	unspecified	25 +	50 +	75 +	100 +		soil	eggplant (*Solanum melongena* cv 'Florida market')	*Rhizoctonia solani*
(Bhadoria et al. 2003)	unspecified	3.2 t ha^{-1} +					soil	rice (*Oryza sativa* cv. 'Pusa basmati')	*Rhizoctonia solani*

(*Continued*)

Table 12.1 Disease-Suppressive Vermicompost Results in the Literature (Continued)

Reference	Feedstock	Amendment Rate/Significant Suppression +[a] = yes, –[a] = no				Substrate	Crop	Pathogen
(Wright et al. 1999)	unspecified	25 +	50 +	75 +	100 +	soil	Autumn squash (*Cucurbita maxima*)	*Rhizoctonia solani*
(Chaoui et al. 2002)	food waste	10 +	20 +	40 +		Metro mix 360	cucumber (*Cucumis sativa*)	*Pythium ultimum*
		10 +	20 +	40 +	70 +		radish (*Raphanus sativus*)	*Rhizoctonia solani*
	food waste	5 t a^{-1} +	10 t a^{-1} +					
	paper waste	5 t a^{-1} +	10 t a^{-1} –			soil	strawberry (*Fragaria x ananassa* cv. 'Chandler')	*Verticillium* spp. (naturally infested soil)
(Szczech et al. 1993)	Cattle manure	10 +	20 +	100 +		Peat	Tomato (*Lycopersicon esculentum*)	*Phytophthora nicotianae* var. *nicotianae*
		10 –[f]	100 +[f]				Tomato (*Lycopersicon esculentum*)	*Fusarium oxysporum* f. sp. *lycopersici*
		root dip[g] +[f]					Cabbage (*Brassica oleracea* cv. 'Ditmarska')	*Plasmodiophora brassicae*
(Szczech and Smolinska 2001)	Sheep manure	50 +	100 +			Peat	Tomato (*Lycopersicon esculentum* cv. 'Remiz')	*Phytophthora nicotianae* var. *nicotianae*
	Cattle manure	50 –	100 +					
	Horse manure	50 +	100 +					
	Sewage sludge	50 +	100 +[d]					

(Szczech 1999)	Cattle manure	20	100			Peat	Tomato (*Lycopersicon esculentum* cv. 'Remiz')	*Fusarium oxysporum* f. sp. *lycopersici*
		–	+					
		15	30			7:2:1 pine bark compost, pine sawdust, and brown coal powder		
		+	+					
		30				7:3 pine bark compost brown coal powder		
		+						
(Sahni, Sarma, and K. Singh 2008)	Vegetable waste, leaf litter	10	25	50		sterile field soil	Chickpea (*Cicer arietinum* cv. 'Avrodhi')	*Sclerotium rolfsii*
		+	+	+				
			P. irregular	*P. ultimum*	*R. solani*			
(Scheuerell et al. 2005)	Cattle manure, food waste, paper,	50	+	+	–	Sunshine mix # 1 (Sun Gro) peat perlite medium	Cucumber (*Cucumis sativus* cv. 'Marketmore 76')	*Pythium irregulare*
		50	+	+	–			*Pythium ultimum*
	Dairy manure	50	–	+	–	Sphagnum peat, vermiculite	Cabbage (*Brassica oleracea* cv. 'Cheers')	*Rhizoctonia solani*
	Dairy manure, straw, eggshells	50	+	+	–			
(Villa-Briones et al. 2008)	Unspecified	7.5 t. ha^{-1} (2.99 t. + acre^{-1})	12.5 t. ha^{-1} (4.99 t. + acre^{-1})			Soil	Tomato (*Lycopersicon esculentum* Mill.)	*Nacobbus aberrans*
		+	+					

[a] + = significant suppression; – no significant suppression.
[b] In this case a thermophilic compost made from the same feedstock was suppressive toward the pathogen.
[c] Vermicompost applied in conjunction with nonaerated VC extract applied every 15 days.
[d] 25% amendment was only significantly suppressive at 15°C, not 22°C.
[e] 25% amendment had significantly higher disease than the control (nonamended).
[f] No statistics were run on these trials; however, a trend of suppression was seen with vermicompost amendments.
[g] Vermicompost used as a component of a root dip before transplanting, clay:vermicompost:water 3:8:5–6.

Overall, most of the vermicomposts suppressed disease in the pathosystems tested, which indicates the considerable future potential of this biologically based disease management practice. However, it is important to note that statistically significant increases in seedling survival during research trials may not be agronomically relevant (Asciutto et al. 2006). In addition, vermicompost amendment rates of 75%–100% (v:v) into transplant media may significantly suppress disease symptoms but may not be commercially viable for transplant production (Wright et al. 1999). Vermicomposts high water-holding capacity can decrease oxygen availability to roots, while high electrical conductivity can sometimes lead to germination problems in certain crop species when incorporated into soil at rates over 15%–20% (Kannangara et al. 2000). Even with multiple positive results, vermicompost-mediated disease suppression remains highly variable and can depend on the feedstock (Szczech and Smolinska 2001), type of potting medium (Szczech 1999), pathosystem (Szczech et al. 1993; Scheuerell et al. 2005), temperature (Rivera, Wright, Lopez, and Fabrizio 2004), presence of synthetic fertilizers (Rodríguez Navarro et al. 2000), and amendment rates (Asciutto et al. 2006). This variability presents a formidable barrier to wider use of vermicompost for disease suppression in the horticulture industry. Trials of vermicompost batches in different pathosystems may provide useful practical information for specific production systems but do not offer any explanation of why disease suppression fails to occur in some circumstances.

Results from more in-depth research reports have started to increase our understanding of the use of vermicompost; however, at this point the data available seem to pose more questions than answers. For example, Szczech and Smolinska (2001) quantified *Phytophthora nicotianae* inoculum during their disease-suppression bioassays. Each tomato seed was sown in a medium infested with roughly equivalent amounts of *P. nicotianae* inoculum amendment with vermicompost significantly reduced disease symptoms. At the end of the trials, *P. nicotianae* inoculum was quantified again and was equal to or higher than the nonamended peat control (Szczech and Smolinska 2001). In this case, it appears the vermicompost amendments suppressed disease symptoms without reducing the pathogen inoculum in the substrate. However, using dilution plating to quantify *P. nicotianae* inoculum does not reveal which life stages of the pathogen were present (i.e., zoospores, zoosporangia, oospores, hyphae, etc.), nor does it offer further insight into the mechanism of suppression. It is possible that vermicompost-derived microbes may have prevented the germination of infective propagules by modifying seed or root exudates as is the case with some biocontrol agents (Windstam and Nelson 2008), but without a more detailed investigation it is impossible to know which microbial interactions are leading to the observed suppression of disease symptoms.

Measuring inoculum densities during plant disease-suppression trials may be one way to further our understanding of the interactions between the plant host, the pathogen, and the vermicompost-associated microorganisms. However, as with all measurements, we must constantly reevaluate our assumptions about what such data actually mean. If an inoculum density is not always reduced when significant suppression of disease symptoms is observed, then what does this measurement contribute to our understanding of the interactions? For example, a similar result to that of Szczech and Smolinska was found for vermicompost-mediated suppression of *R. solani* on *Impatiens* spp.; that is, there were no significant differences in pathogen inoculum density between treatments at the end of the experiment (Asciutto et al. 2006). In contrast, Kannangara and colleagues found that while the incorporation of only 5% vermicompost into a potting medium reduced colony-forming units of *Fusarium oxysporum* f. sp. *radicis cucumerinum* by an order of magnitude, disease levels were no different than in the nonamended medium (Kannangara et al. 2000). In their study, thermophilic compost made from the same dairy manure feedstock also reduced pathogen inoculum by an order of magnitude in addition to significantly suppressing *F. oxysporum* symptoms. For future studies, it may be helpful to quantify the inocula associated with the surface of germinating seeds and/or plant roots to ascertain not only whether the pathogen is surviving in the media but also whether it is able to colonize the host and produce disease symptoms. In some cases, compost-mediated disease suppression can be explained by interactions that occur on the surface of germinating seeds within short time frames (Chen and Nelson 2008), so exploring the relationships in the immediate infection area in different pathosystems may be more fruitful than any analysis of bulk media.

Several main strategies have been used in attempting to characterize the nature of thermophilic compost-mediated disease suppression: (1) correlating overall measurements of the compost microbial community with the degree of suppression observed in a bioassay (respiration potential, ammonia production, microbial activity, etc.), (2) isolating bacteria and fungi from composts and testing their suppression of a plant pathogen in vitro, and (3) screening the compost for microorganisms known or thought to be involved in suppression and correlating this to suppression observed in a bioassay. Using overall measurements of the suppressive substrate's microbial community has been a successful strategy for specific pathogens. For example, high- microbial activity (Lumsden et al. 1987; Chen et al. 1988; Craft and Nelson 1996; Erhart et al. 1999; van Os and van Ginkel 2001) and low-organic-matter decomposition (Boehm et al. 1993; Stone et al. 2001) have correlated well with suppression of *Pythium* spp. in a wide variety of substrates. However, there is some evidence that these correlations may be pathogen specific. A survey of 30 compost samples, including four vermicomposts, reported that respiration potential, measured as CO_2 evolution over a 7-day period, could be correlated with suppression of *Pythium ultimum* and *P. irregulare* but did not correlate with suppression of *R. solani* (Scheuerell et al. 2005).

Isolation and in vitro screening of bacterial and fungal cultures have been carried out on disease-suppressive vermicomposts. Out of 200 bacterial and 100 fungal isolates screened, a higher proportion of those able to suppress the growth of *Fusarium oxysporum* occurred in the suppressive vermicompost compared to the conducive peat (Szczech 1999). However, in vitro suppression of fungal growth does not guarantee that an isolate will be able to successfully colonize the host's spermosphere or rhizosphere or will have the same suppressive activity in the presence of the host. To complicate matters further, it is not clear that a high number of individually suppressive isolates will be suppressive when combined in a single application. McKellar and Nelson (2003) reported that 40% of isolates from a suppressive compost and 87% of isolates from a conducive compost were able to suppress *P. ultimum* when used as a cottonseed treatment. When all of the isolates from each compost sample were mixed and applied together as a seed treatment, the group with the higher percentage of individually suppressive isolates was less suppressive than the group with a lower percentage of individually suppressive isolates (McKellar and Nelson 2003). Only a small number of isolates were screened in this case, but it would be worth investigating this phenomenon further. Disease-suppressive composts may provide examples of emergent properties inherent in complex microbial communities where we cannot predict the actions of the whole by totaling what we know about the actions of the parts.

Screening composts and other substrates for microorganisms known to be involved in the suppression of specific pathogens is common in the disease-suppression literature. In some cases the presence of target organisms is correlated with disease suppression, for example, *Bacillus* spp. (Postma et al. 2000) and *Pseudomonas* spp. (Lumsden et al. 1987) with *Pythium aphanidermatum* in cucumber. However in other cases groups known to be associated with suppression of *P. aphanidermatum* in cucumber, for example *Pseudomonas* spp., were present in higher numbers in disease-conducive than in disease-suppressive substrates (Postma et al. 2000). Even if the presence of the target organism or group of organisms correlates with the degree of disease suppression measured with a bioassay, there is still no confirmation of their actual involvement in suppression. Given the complexity of this phenomenon, it would not be surprising to find a unique mechanism at work in each pathosystem. Hence, reliable predictive factors of disease-suppressive potential in any substrate remain elusive.

Aside from their potential disease-suppressive characteristics, vermicomposts have been investigated for their potential use as complementary amendments and/or carriers for biocontrol agents. A wide body of knowledge exists on biocontrol organisms, as reviewed by Kloepper et al. (1999), Whipps (2001), Sturz and Christie (2003), Compant et al. (2005), and Berg (2009), and this area of research has led to many commercially available biocontrol products (Nelson 2004). Application methods are constantly being refined, and there is considerable interest in composts and vermicomposts in particular, as substrates that could not only support the growth and proliferation of biocontrol organisms but also potentially enhance disease control by providing additional degrees of suppression. For example, Sahni and colleagues documented a vermicompost capable of providing low but significant levels of suppression of *Sclerotium rolfsii* on chickpeas: a 12%–32% reduction in seedling mortality depending on the soil amendment rate (Sahni, Sarma, D. Singh, et al. 2008). The biocontrol agent *Pseudomonas syringae* strain PUR46 provided a 52% reduction in seedling mortality, while the combination of vermicompost and seed treatment with *P. syringae* provided an 88% reduction in seedling mortality. Applying a foliar spray of oxalic acid and zinc sulfate, amending soil with 25% (v:v) vermicompost, and treating seeds with *P. fluorescens* led to complete suppression of disease symptoms over a 30-day period (Sahni, Sarma, and K. Singh 2008). However, not all biocontrol agents and vermicomposts are compatible. After 300 days of storage, *Pseudomonas fluorescens* strain Pf-D populations in a nonsterilized vermicompost were two orders of magnitude lower than those in a sterilized vermicompost (Bora and Deka 2007), a result that limits its use as a carrier and/or storage medium for this biocontrol agent. However, vermicompost feedstocks can be manipulated to create specific conditions conducive to growth of a biocontrol organism. Pereira and colleagues amended cow manure vermicompost with molasses, rice husks, and sugarcane bagasse in an attempt to increase the growth rates and survival of *Bacillus subtilis* and *Trichoderma harzianum* (Pereira et al. 1998). The physiological requirements of the two organisms were so distinct, however, that no single batch of vermicompost could support the growth of both. If vermicomposts are ever to be effectively used as carriers for biocontrol agents, the compatibility of each combination of biopesticide and vermicompost type must be established for both viability during storage and effectiveness in the field.

Vermicompost amendments have been applied in conjunction with a wide variety of other disease-management strategies. However, it is not always clear what specific benefit, if any, the vermicompost provided due to how data are reported. For example, amending soil with vermicompost can increase the efficiency of solarization in the inactivation of *Sclerotium cepvorum* sclerotia (Pereira et al. 1996), but the vermicompost was never tested without solarization to determine its inherent suppressive qualities. Bora and Deka applied *P. fluorescens* as a seed treatment, root dip, and soil amendment at transplanting to control *Ralstonia solanacearum* on tomato successfully (Bora and Deka 2007). The soil application of *P. fluorescens* included a 25.1 ton ha^{-1}(10 t·$acre^{-1}$) amendment of nonsterilized vermicompost as the carrier. Without control treatments, consisting of vermicompost applied without the biocontrol agent or the biocontrol agent applied in an inert substrate, it is impossible to know whether the documented suppression was due to the vermicompost alone, the biocontrol agent alone, or a combination of the two. The inclusion of multiple control treatments should be encouraged for future integrated disease-management studies with vermicompost so that contributions of individual treatments can be documented without ambiguity.

IV CONCLUSIONS AND DIRECTIONS FOR FUTURE RESEARCH

The potential of vermicomposts to suppress disease in a variety of pathosystems has been extensively documented. With the recent rise in interest in biologically based plant disease management in both certified organic and conventional systems, expanded support for research in this field is now available, which will help build our knowledge base in this area. Carrying out trials of different materials to test for disease suppression in a variety of pathosystems can play an important role in this research effort; however, an increase in our understanding of how this suppression works is crucial to developing effective practices using these materials (Borneman and Becker 2007). Clearly, our lack of understanding of the factors that lead to the observed variability in compost-mediated disease

Table 12.2 Proposed Basic Characteristics of Vermicomposts That Should Be Reported in Future Studies

Vermicompost

- Feedstock: organic waste used in vermicomposting process, actual proportions if available
- Commercial source
- Type of vermicomposting system: continuous flow-through beds, windrows, etc.
- Species of earthworm used
- Storage conditions before use in study

Application of Vermicompost

- Specific rate of amendment
  - For soil amendments, was it incorporated or top-dressed? How was it applied?
- Substrate amended
  - Other potting media components or soil type and location

Note: Potential phytotoxicity should always be ruled out as phytotoxic amendments may suppress plant pathogens but are not relevant for use in actual production systems.

suppression continues to be a major impediment to the widespread adoption of these materials in different plant production systems (Ben-Yephet and Nelson 1999). The shorter production times for vermicompost (<70 days compared to 6 months for thermophilic compost) may provide an incentive for large-scale production to move toward indoor facilities with high levels of process control. This, combined with the use of a single feedstock year-round, may enhance the batch-to-batch consistency of these materials. The research community, vermicompost industry, and growers who are using these materials need to work together to increase our understanding of the biology of this system while also helping vermicomposters produce consistently suppressive materials that can be incorporated easily into existing plant production systems. The challenges of registering composts and vermicomposts as biopesticides will eventually need to be addressed by multiple stakeholders.

I hope that this Chapter will serve as a basis for researchers and practitioners to build a more comprehensive body of knowledge in this area. Given the highly variable nature of these materials, our field of study needs a coordinated research effort where results of studies can be compared and where everyone working in this field is aware of what others are doing. Reporting important details such as the vermicompost source, feedstock, production method, and so on is an easy place to start (Table 12.2). In addition, simple laboratory-management practices like freezing large batches of vermicompost and thawing small amounts before use can greatly reduce experiment-to-experiment variability. Vermicomposts are similar to composts but are also unique in many ways given the complex relationship between earthworms and microbial communities. A coordinated research effort could help transform vermicompost-mediated disease suppression from an interesting phenomenon into a widespread effective control practice.

REFERENCES

Amaravadi, L., M. S. Bisesi, and R. F. Bozarth. 1990. Vermial virucidal activity: Implications for management of pathogenic biological wastes on land. *Biol. Wastes* 34:349–358.

Asciutto, K., M. Rivera, E. Wright, D. Morisigue, and M. Lopez. 2006. Effect of vermicompost on the growth and health of *Impatiens wallerana*. *ΦYTON: Int. J. Exp. Bot.* 75:115–123.

Ben-Yephet, Y., and E. B. Nelson. 1999. Differential suppression of damping-off caused by *Pythium aphanidermatum*, *P. irregulare*, and *P. myriotylum* in composts at different temperatures. *Plant Dis.* 83 (4):356–360.

Berg, G. 2009. Plant–microbe interactions promoting plant growth and health: Perspectives for controlled use of microorganisms in agriculture. *Appl. Microbiol. Biotechnol.* 84 (1):11–18.

Bhadoria, P. B. S., Y. S. Prakash, S. Kar, and A. Rakshit. 2003. Relative efficacy of organic manures on rice production in lateritic soil. *Soil Use Manage.* 19 (1):80–82.

Binet, F., L. Fayolle, and M. Pussard. 1998. Significance of earthworms in stimulating soil microbial activity. *Biol. Fertil. Soils* 27 (1):79–84.

Boehm, M. J., L. V. Madden, and H. A. J. Hoitink. 1993. Effect of organic matter decomposition level on bacterial species diversity and composition in relationship to *Pythium* damping-off severity. *Appl. Environ. Microbiol.* 59 (12):4171–4179.

Bora, L. C., and S. N. Deka. 2007. Wilt disease suppression and yield enhancement in tomato (*Lycopersicon esculentum*) by application of *Pseudomonas fluorescense*-based biopesticide (Biofor-Pf) in Assam. *Indian J. Agric. Sci.* 77 (8):490–494.

Borneman, J., and J. O. Becker. 2007. Identifying microorganisms involved in specific pathogen suppression in soil. *Annu. Rev. Phytopathol.* 45:153–172.

Brown, G. G. 1995. How do earthworms affect microfloral and faunal community diversity? *Plant Soil* 170:209–231.

Brown, G. G., I. Barois, and P. Lavelle. 2000. Regulation of soil organic matter dynamics and microbial activity in the drilosphere and the role of interactions with other edaphic functional domains. *Eur. J. Soil Biol.* 36 (3–4):177–198.

Chaoui, H. I., C. A. Edwards, A. Brickner, S. S. Lee, and N. Q. Arancon. 2002. Suppression of the plant diseases, *Pythium* (damping-off), *Rhizoctonia* (root rot) and *Verticillium* (wilt) by vermicomposts. *Proceedings of Brighton Crop Protection Conference—Pests and Diseases*, 2, 8B-3, 711–716.

Chen, M.-H., and E. B. Nelson. 2008. Seed-colonizing microbes from municipal biosolids compost suppress *Pythium ultimum* damping-off on different plant species. *Phytopathology* 98 (9):1012–1018.

Chen, W., H. A. J. Hoitink, A. F. Schmitthenner, and O. H. Tuovinen. 1988. The role of microbial activity in the suppression of damping-off caused by *Pythium ultimum*. *Phytopathology* 78:314–322.

Clapperton, M. J., N. O. Lee, F. Binet, and R. L. Conner. 2001. Earthworms indirectly reduce the effects of take-all (*Gaeumannomyces graminis* var. *tritici*) on soft white spring wheat (*Triticum aestivum* cv. Fielder). *Soil Biol. Biochem.* 33 (11):1531–1538.

Clegg, C. D., J. M. Anderson, H. M. Lappin-Scott, J. D. van Elsas, and J. M. Jolly. 1995. Interactions of a genetically modified *Pseudomonas fluorescens* with the soil feeding earthworm *Octolasion cyaneum* (Lumbricidae). *Soil Biol. Biochem.* 27 (11):1423–1429.

Compant, S., B. Duffy, J. Nowak, C. Clement, and E. A. Barka. 2005. Use of plant growth-promoting bacteria for biocontrol of plant diseases: Principles, mechanisms of action, and future prospects. *Appl. Environ. Microbiol.* 71 (9):4951–4959.

Craft, C. M., and E. B. Nelson. 1996. Microbial properties of composts that suppress damping-off and root rot of creeping bentgrass caused by *Pythium graminicola*. *Appl. Environ. Microbiol.* 62 (5):1550–1557.

Curl, E. A. 1988. The role of soil microfauna in plant-disease suppression. *CRC Crit. Rev. Plant Sci.* 7 (3):175–196.

Davidson, S. K., and D. A. Stahl. 2006. Transmission of nephridial bacteria of the earthworm *Eisenia fetida*. *Appl. Environ. Microbiol.* 72 (1):769–775.

Davidson, S. K., and D. A. Stahl. 2008. Selective recruitment of bacteria during embryogenesis of an earthworm. *ISME J.* 2:510–518.

Drake, H. L., and M. A. Horn. 2007. As the worm turns: The earthworm gut as a transient habitat for soil microbial biomes. *Appl. Environ. Microbiol.* 61 (1):169–189.

Egert, M., S. Marhan, B. Wagner, S. Scheu, and M. W. Friedrich. 2004. Molecular profiling of 16S rRNA genes reveals diet-related differences of microbial communities in soil, gut, and casts of *Lumbricus terrestris* L. (Oligochaeta : Lumbricidae). *FEMS Microbiol. Ecol.* 48 (2):187–197.

Elmer, W. H. 2009. Influence of earthworm activity on soil microbes and soilborne diseases of vegetables. *Plant Dis.* 93 (2):175–179.

Elmer, W. H., and F. J. Ferrandino. 2009. Suppression of Verticillium wilt of eggplant by earthworms. *Plant Dis.* 93 (5):485–489.

Enami, Y., S. Okano, H. Yada, and Y. Nakamura. 2001. Influence of earthworm activity and rice straw application on the soil microbial community structure analyzed by PLFA pattern. *Eur. J. Soil Biol.* 37:269–272.

Erhart, E., K. Burian, W. Hartl, and K. Stich. 1999. Suppression of *Pythium ultimum* by biowaste composts in relation to compost microbial biomass, activity and content of phenolic compounds. *J. Phytopathol.-Phytopathol. Ztg.* 147 (5):299–305.

Gopal, M., A. Gupta, E. Sunil, and G. V. Thomas. 2009. Amplification of plant beneficial microbial communities during conversion of coconut leaf substrate to vermicompost by *Eudrilus* sp. *Curr. Microbiol.* 59 (1):15–20.

Hartenstein, F., E. Hartenstein, and R. Hartenstein. 1981. Gut load and transit time in the earthworm *Eisenia foetida*. *Pedobiol.* 22:5–20.

Hoitink, H. A. J., and P. C. Fahy. 1986. Basis for the control of soilborne plant pathogens with composts. *Appl. Environ. Microbiol.* 24:93–114.

Hoitink, H. A. J., Y. Inbar, and M. J. Boehm. 1991. Status of compost-amended potting mixes naturally suppressive to soilborne diseases of floricultural crops. *Plant Dis.* 75 (9):869–873.

Howard, A. 1943. *An Agricultural Testament*. New York: Oxford University Press.

Howard, A. 1945. *Farming and Gardening for Health or Disease*. London: Faber and Faber.

Huber, D. M., and R. W. Schneider. 1982. The description and occurrence of suppressive soils. In *Suppressive Soils and Plant Disease*, ed. R. W. Schneider. St. Paul, MN: American Phytopathological Society. 1–7.

Joshi, D., K. S. Hooda, J. C. Bhatt, B. L. Mina, and H. S. Gupta. 2009. Suppressive effects of composts on soil-borne and foliar diseases of French bean in the field in the western Indian Himalayas. *Crop Prot.* 28 (7):608–615.

Kannangara, T., R. S. Utkhede, J. W. Paul, and Z. K. Punja. 2000. Effects of mesophilic and thermophilic composts on suppression of *Fusarium* root and stem rot of greenhouse cucumber. *Can. J. Microbiol.* 46 (11):1021–1028.

Kloepper, J. W., R. Rodriguez-Kabana, G. W. Zehnder, J. F. Murphy, E. Sikora, and C. Fernandez. 1999. Plant root–bacterial interactions in biological control of soilborne diseases and potential extension to systemic and foliar diseases. *Australas. Plant Pathol.* 28 (1):21–26.

Lavelle, P., C. Lattaud, D. Trigo, and I. Barois. 1995. Mutualism and biodiversity in soils. *Plant Soil* 170 (1):23–33.

Litterick, A. M., L. Harrier, P. Wallace, C. A. Watson, and M. Wood. 2004. The role of uncomposted materials, composts, manures, and compost extracts in reducing pest and disease incidence and severity in sustainable temperate agricultural and horticultural crop production—A review. *Crit. Rev. Plant Sci.* 23 (6):453–479.

Lumsden, R. D., E. R. Garcia, J. A. Lewis, and G. A. Frias-T. 1987. Suppression of damping-off caused by *Pythium* spp. in soil from the indigenous Mexican Chinampa agricultural system. *Soil Biol. Biochem.* 19:501–508.

Mazolla, M. 2002. Mechanisms of natural soil suppressiveness to soil diseases. *Antonie van Leeuwenhoek* 81:557–564.

McKellar, M. E., and E. B. Nelson. 2003. Compost-induced suppression of *Pythium* damping-off is mediated by fatty-acid-metabolizing seed-colonizing microbial communities. *Appl. Environ. Microbiol.* 69 (1):452–460.

Menzies, J. D. 1959. Occurrence and transfer of a biological factor in soil that suppresses potato scab. *Phytopathology* 49:648–652.

Nelson, E. B. 2004. Biological control of oomycetes and fungal pathogens. In *Encyclopedia of Plant and Crop Science* edited by R. M. Goodman. 137–140. New York: Marcel Dekker.

Pereira, J. C. R., G. M. Chaves, L. Zambolim, K. Matsuoka, R. S. Acuna, and F. R. D. Vale. 1996. Control of *Sclerotium cepivorum* by the use of vermicompost, solarization, *Trichoderma harzianum* and *Bacillus subtilis*. *Summa Phytopathol.* 22 (3–4):228–234.

Pereira, J. C. R., G. M. Chaves, L. Zambolim, K. Matsuoka, R. S. Acuna, and F. R. D. Vale. 1998. Survival of *Trichoderma harzianum* and *Bacillus subtilis* in vermicompost. *Summa Phytopathol.* 24 (3–4): 231–238.

Postma, J., Mjeim Willemsen-de Klein, and J. D. van Elsas. 2000. Effect of the indigenous microflora on the development of root and crown rot caused by *Pythium aphanidermatum* in cucumber grown on rockwool. *Phytopathology* 90 (2):125–133.

Rivera, M. C., E. R. Wright, M. V. Lopez, and M. C. Fabrizio. 2004. Temperature and dosage dependent suppression of damping-off caused by *Rhizoctonia solani* in vermicompost amended nurseries of white pumpkin. *ΦYTON: Int. J. Exp. Bot.*, 73:131–136.

Rivera, M. C., E. R. Wright, M. V. Lopez, and G. S. Guastella. 2001. Use of vermicompost to suppress *Rhizoctonia solani* in nurseries of eggplant, 2000. *Biol. Cult. Tests Control Plant Dis.* 16 (V88).

Rodríguez Navarro, J. A., E. Zavaleta Mejía, P. Sánchez García, and H. González Rosas. 2000. The effect of vermicompost on plant nutrition, yield and incidence of root and crown rot of gerbera (*Gerbera jamesonii* H. Bolus). *Fitopatología* 35:66.

Sahni, S., B. K. Sarma, D. P. Singh, H. B. Singh, and K. P. Singh. 2008. Vermicompost enhances performance of plant growth-promoting rhizobacteria in *Cicer arietinum* rhizosphere against *Sclerotium rolfsii*. *Crop Prot.* 27 (3–5):369–376.

Sahni, S., B. K. Sarma, and K. P. Singh. 2008. Management of *Sclerotium rolfsii* with integration of non-conventional chemicals, vermicompost and *Pseudomonas syringae*. *World J. Microbiol. Biotechnol.* 24 (4):517–522.

Sampedro, L., and J. K. Whalen. 2007. Changes in the fatty acid profiles through the digestive tract of the earthworm *Lumbricus terrestris* L. *Appl. Soil Ecol.* 35 (1):226–236.

Sánchez-Moreno, S., and H. Ferris. 2007. Suppressive service of the soil food web: Effects of environmental management. *Agric. Ecosyst. Environ.* 119 (1–2):75–87.

Scheuerell, S., D. M. Sullivan, and W. Mahaffee. 2005. Suppression of seedling damping-off caused by *Pythium ultimum*, *Pythium irregulare*, and *Rhizoctonia solani* in container media amended with a diverse range of Pacific Northwest compost sources. *Phytopathology* 95:306–315.

Schmidt, O., B. M. Doube, M. H. Ryder, and K. Killham. 1997. Population dynamics of *Pseudomonas corrugata* 2140R lux8 in earthworm food and in earthworm casts. *Soil Biol. Biochem.* 29 (3–4):523–528.

Singh, R., R. R. Sharma, S. Kumar, R. K. Gupta, and R. T. Patil. 2008. Vermicompost substitution influences growth, physiological disorders, fruit yield and quality of strawberry (*Fragaria* x *ananassa* Duch.). *Biores. Technol.* 99 (17):8507–8511.

Stephens, P. M., and C. W. Davoren. 1995. Effect of the lumbricid earthworm *Aporrectodea trapezoides* on wheat grain yield in the field, in the presence or absence of *Rhizoctonia solani* and *Gaeumannomyces graminis* var. *tritici*. *Soil Biol. Biochem.* 28 (4–5):561–567.

Stephens, P. M., and C. W. Davoren. 1997. Influence of the earthworms *Aporrectodea trapezoides* and *A. rosea* on the disease severity of *Rhizoctonia solani* on subterranean clover and ryegrass. *Soil Biol. Biochem.* 29 (3–4):511–516.

Stephens, P. M., C. W. Davoren, and B. G. Hawke. 1995. Influence of barley straw and the lumbricid earthworm *Aporrectodea trapezoides* on *Rhizobium meliloti* L5-30r, *Pseudomonas corrugata* 2140r, microbial biomass and microbial activity in a red–brown earth soil. *Soil Biol. Biochem.* 27 (11):1489–1497.

Stephens, P. M., C. W. Davoren, M. H. Ryder, and B. M. Doube. 1993. Influence of the lumbricid earthworm *Aporrectodea trapezoides* on the colonization of wheat roots by *Pseudomonas corrugata* strain 2140R in soil. *Soil Biol. Biochem.* 25 (12):1719–1724.

Stephens, P. M., C. W. Davoren, M. H. Ryder, B. M. Doube, and R. L. Correll. 1994. Field evidence for reduced severity of *Rhizoctonia* bare-patch disease of wheat, due to the presence of the earthworms *Aporrectodea rosea* and *Aporrectodea trapezoides*. *Soil Biol. Biochem.* 26 (11):1495–1500.

Stone, A. G., S. J. Traina, and H. A. J. Hoitink. 2001. Particulate organic matter composition and *Pythium* damping-off of cucumber. *Soil Sci. Soc. Am. J.* 65 (3):761–770.

Sturz, A. V., and B. R. Christie. 2003. Beneficial microbial allelopathies in the root zone: The management of soil quality and plant disease with rhizobacteria. *Soil Tillage Res.* 72 (2):107–123.

Szczech, M. M. 1999. Suppressiveness of vermicompost against *Fusarium* wilt of tomato. *J. Phytopathol.* 147:155–161.

Szczech, M., W. Rondomanski, M. W. Brzeski, U. Smolinska, and J. F. Kotowski. 1993. Suppressive effect of a commercial earthworm compost on some root infecting pathogens of cabbage and tomato. *Biol. Agric. Hortic.* 10:47–52.

Szczech, M., and U. Smolinska. 2001. Comparison of suppressiveness of vermicompost produced from animal manures and sewage sludge against *Phytophthora nicotinae* Breda de Haan var. *nicotinae*. *J. Phytopathol.* 149:77–82.

Toyota, K., and M. Kimura. 2000. Microbial community indigenous to the earthworm *Eisenia foetida*. *Biol. Fertil. Soils* 31 (3–4):187–190.

Trigo, D., I. Barois, M. H. Garvin, E. Huerta, S. Irisson, and P. Lavelle. 1999. Mutualism between earthworms and soil microflora. *Pedobiologia* 43 (6):866–873.

van Os, G. J., and J. H. van Ginkel. 2001. Suppression of *Pythium* root rot in bulbous iris in relation to biomass and activity of the soil microflora. *Soil Biol. Biochem.* 33 (11):1447–1454.

Villa-Briones, A., E. Zavaleta-Mejia, M. Vargas-Hernandez, O. Gomez-Rodriguez, and S. Ramirez-Alarcon. 2008. Vermicompost incorporation for the management of *Nacobbus aberrans* in tomato (*Lycopersicon esculentum* Mill.). *Revista Chapingo Serie Ciencias Forestales y del Ambiente* 14 (3):249–255.

Wang, C., Z. Sun, Y. Liu, D. Zheng, X. Liu, and S. Li. 2007. Earthworm polysaccharide and its antibacterial function on plant-pathogen microbes *in vitro*. *Eur. J. Soil Biol.* 43:S135–S142.

Weltzien, H. C. 1989. Some effects of composted organic materials on plant health. *Agric. Ecosyst. Environ.* 27 (1–4):439–446.

Whipps, J. M. 2001. Microbial interactions and biocontrol in the rhizosphere. *J. Exp. Bot.* 52:487–511.

Windstam, S., and E. B. Nelson. 2008. Differential interference with *Pythium ultimum* sporangial activation and germination by *Enterobacter* cloacae in the corn and cucumber spermospheres. *Appl. Environ. Microbiol.* 74 (14):4285–4291.

Wright, E. R., M. C. Rivera, A. L. Cheheid, M. C. Fabrizio, and J. Mosedale. 1999. Control of *Rhizoctonia solani* in nurseries of autumn squash by the amendment with vermicompost. *Biol. Cult. Tests Control Plant Dis.* 14:183.

CHAPTER 13

Use of Aqueous Extracts from Vermicomposts or Teas in Suppression of Plant Pathogens

Clive A. Edwards, Ahmed M. Askar, Marcus A. Vasko-Bennett, and Norman Arancon

CONTENTS

I Introduction 184
A Thermophilic Composts 184
B Thermophilic Compost Teas 184
C Vermicomposts 185
D Vermicompost Teas 185
II Research at The Ohio State University on Suppression of Root Pathogens of Vegetables by Vermicompost Teas 186
A Production of Aqueous Vermicompost Extracts (Teas) 186
B Design of Root Pathogen Experiments 186
C Effects of Food Waste Vermicompost Teas on *Fusarium oxysporum* Attacking Tomatoes 187
D Effects of Food Waste Vermicompost Teas on *Phytophthora capsici* Attacking Cucumbers and Tomatoes 187
1 Cucumber Experiments with *Phytophthora capsici* 187
2 Tomato Experiments with *Phytophthora capsici* 189
E Effects of Food Waste Vermicompost Teas on *Rhizoctonia solani* Attacking Cucumbers and Tomatoes 190
1 Cucumber Experiments with *Rhizoctonia solani* 190
2 Tomato Experiments with *Rhizoctonia solani* 192

F Effects of Food Waste Vermicompost Teas on *Pythium ultimum* Attacking Cucumbers and Tomatoes 193
1 Cucumber Experiments with *Pythium ultimum* 193
2 Tomato Experiments with *Pythium ultimum* 193
III Research at The Ohio State University on Suppression of Foliar Pathogens of Vegetables by Vermicompost Aqueous Extracts (Teas) 195
A Design of Foliar Pathogen Experiment 195
B Effects of Food Waste Vermicompost Teas on *Plectosporium tabacinum* Attacking Vegetables 196
1 Cucumber Experiment with *Plectosporium tabacinum* 196
2 Tomato Experiment with *Plectosporium tabacinum* 196
C Effects of Food Waste Vermicompost Teas on *Botrytis cinerea* Damage 198
1 Cucumber Experiment with *Botrytis cinerea* 198
2 Tomato Experiment with *Botrytis cinerea* 200
D Effects of Food Waste Vermicompost Teas on *Sclerotinea rolfsii* Damage 200
1 Cucumber Experiment with *Sclerotinea rolfsii* 200
2 Tomato Experiment with *Sclerotinea rolfsii* 202
E Conclusions on Effects of Vermicompost Tea Sprays on Foliar Pathogens 202
IV Mechanisms by Which Vermicompost Teas Suppress Plant Pathogens 203
References 205

I INTRODUCTION

A Thermophilic Composts

It has been well established that plants grown with the addition of traditional thermophilically composted organic wastes tend to have lower incidence of plant disease, probably due to the development and multiplication of microorganisms antagonistic to plant pathogens (Hoitink and Kuter 1986; Chen et al. 1987). Specific diseases that have been controlled by thermophilic composts include *Fusarium* (Liping et al. 2001; Kannangara et al. 2001), *Pythium* (Chen et al. 1987), *Phytophthora* (Hoitink and Kuter 1986), and *Rhizoctonia* (Kuter et al 1983). Thermophilic composts can suppress soilborne plant diseases, particularly those caused by root rot pathogens in container systems (Hoitink et al. 1997, and in the field (Drinkwater et al. 1995; Stone et al. 2003). They can also suppress *Sclerotinia*.

B Thermophilic Compost Teas

The first systematic scientific studies on the use of thermophilic compost "teas" for plant disease control were during the 1980s at the University of Bonn, Germany, under the direction of Professor Weltzien. These studies utilized preparations made by steeping compost with 5–10 parts water from 3 days to several weeks, and a number of plant diseases were controlled successfully with such thermophilic extracts. Compost tea efficacy can be affected by the source and age of compost, production practices including the ratio of compost to water, duration of production, addition of microbial growth substances (Scheuerell and Mahaffee 2004), and application with spreader-stickers (Scheuerell 2002; see Chapter 11). The effects of different compost tea production and application practices on plant disease management have been reviewed thoroughly (Scheuerell and Mahaffee 2002). Most studies have addressed the effect of teas on foliar diseases of plants. However, several reports have indicated the potential of using compost teas for suppressing root rot pathogens.

Heat-treating a compost tea eliminated all suppression of pathogens in vitro, indicating the important role of large populations of microorganisms in plant disease suppression (Tränkner 1992). Recent research has demonstrated the utility of compost tea as a soil-less media drench for controlling damping-off of cucumbers caused by *Pythium ultimum* (Scheuerell and Mahaffee 2004). Scheuerell and Mahaffee (2006) reported the suppression of *Botrytis cinerarea* on geraniums and of *P. ultimum* on cucumbers by compost teas. To date, foliar disease suppression imparted by thermophilic composts has been attributed to a systemic induction of plant host defenses that leads to an enhanced state of resistance to the pathogen under greenhouse conditions (Zhang et al. 1996, 1998; Khan et al. 2004, Alfano et al. 2007; see Chapter 12).

C Vermicomposts

Nakamura (1996) reported suppression of *Plasmodiophora brassicae*, *Phytophthora nicotianae* (tomato late blight), and *Fusarium lycopersici* (tomato fusarium wilt) on tomatoes by vermicomposts. It seems probable that species of antagonistic microorganisms that can decrease plant disease incidence may be produced in vermicomposts during the vermicomposting process and that these microorganisms are a main mechanism in plant pathogen suppression. Huelsman and Edwards (1998) reported the suppression of diseases of cucumbers and peppers by vermicomposts. Rodríguez et al. (2000) demonstrated suppression of fungal pathogens such as *Rhizoctonia solani*, *Phytophthora drechsleri*, and *Fusarium oxysporum* by the incorporation of vermicomposts into the soil. Chaoui et al. (2002) reported the suppression of the plant pathogens *Pythium*, *Rhizoctonia*, and *Verticillium* by vermicomposts. Arancon and Edwards (2004) discussed the extent of potential suppression of plant diseases by solid vermicomposts, and Edwards et al. (2006) described the reported effects of several types of vermicomposts on plant diseases (see Chapters 12 & 14).

D Vermicompost Teas

Studies by Nakasone et al. (1999) revealed that aqueous extracts from vermicomposts (teas) suppressed the mycelial growth of *Sclerotinium cinerea*, *Sclerotinia sclerotiorum*, *Sclerotinium rolfsii*, *R. solani*, and *F. oxysporum*. Zaller (2006) reported that foliar sprays of vermicompost teas suppressed late blight (*Phytophthora*) infestations on tomatoes. Heat-treating a vermicompost tea eliminated all suppression of pathogen growth *in vitro*, indicating the important role of large populations and diversity of microorganisms produced by vermicomposts in plant disease suppression (Tränkner 1992). More recent research has demonstrated the utility of compost and vermicompost teas as soilless media drenches for controlling damping-off of cucumbers caused by *P. ultimum* (Scheuerell and Mahaffee 2004).

II RESEARCH AT THE OHIO STATE UNIVERSITY ON SUPPRESSION OF ROOT PATHOGENS OF VEGETABLES BY VERMICOMPOST TEAS

A research program to assess the suppression of root pathogens on tomatoes and cucumbers began in the Soil Ecology Laboratory at The Ohio State University in 2007, funded by the U.S. Department of Agriculture.

A Production of Aqueous Vermicompost Extracts (Teas)

The various important issues associated with the preparation and use of compost and vermicompost teas are discussed in detail in Chapter 11. In preliminary research on vermicompost teas in our laboratory, we made the important discovery that teas that were actively aerated during production had much greater effects on plant disease suppression than those that were not. Aerated teas also performed much more consistently than nonaerated teas. Hence, for all of our experiments we standardized our vermicompost tea production to use only aerated vermicompost teas (see Chapters 11 and 15).

All of our experimental treatments in greenhouse experiments consisted of a range of three concentrations of aqueous vermicompost extracts, namely, 5%, 10%, and 20%, and their effects on plant pathogens were always compared with those of a deionized water control. The vermicompost aqueous extracts were produced in commercial vermicompost-extract-brewing equipment called the Growing Solutions System 10™ Compost Brewing Equipment (Growing Solutions, Inc., Oregon), with a maximum capacity of 37.5 L (79.2 pt). A 20% aqueous vermicompost solution was prepared by placing 7.5 L (15.8 pt) of food waste vermicompost in a suspended mesh container in the brewing equipment containing 30 L (63.4 pt) of water. This was extracted for 24 h while aerating continually. The 20% aqueous extract was diluted to 10% (v:v) or 5% (v:v) for use in the experiments. The aqueous vermicompost extracts were used within 24 h of preparation to minimize any loss of microbial activity. The 5%, 10%, and 20% aqueous vermicompost extracts and a deionized water control were applied as drenches to bring the growth medium (Metro Mix 360 [MM 360]) to field capacity at sowing and at weekly intervals thereafter.

B Design of Root Pathogen Experiments

For all of the experiments in the Soil Ecology Laboratory at The Ohio State University that were designed to assess the effects of vermicompost aqueous extracts on root pathogens of tomatoes or cucumber. The plants were grown in the greenhouse in MM 360 and inoculated with plant pathogens under experimental conditions. The

Table 13.1 Root Rot Plant Pathogen Inoculum Densities that were Incorporated into the Soil-Less Medium before Planting Test Plants

Root Rot Pathogens	Plant Host	Pathogen Inoculum Density
Fusarium oxysporum	Tomato	10^8 spores in 5 mL (0.17 oz) water drenched onto media
Phytophthora capsici	Tomato and cucumber	10^8 sporangia in 5 mL (0.17 oz) water drenched onto media
Rhizoctonia solani	Cucumber and tomatoes	0.1% (v:v) *Rhizoctonia* cultured ground rice inoculum
Pythium ultimum	Cucumber and tomato	0.1% (v:v) *Pythium* potato-soil inoculum

soil pathogens, crop plants, and pathogen inoculum density used for each pathogen-plant combination are summarized in Table 13.1. All of the drench treatments were applied to four replicate 10 cm diameter pots, each sown with either eight tomato or eight cucumber seeds and thinned out to produce four plants per pot. Teas were applied as soil drenches (5%, 10%, and 20% extracts and a deionized water control), up to the field capacity of the medium, at sowing and at weekly intervals thereafter, until harvesting. Pots were arranged on greenhouse benches in a completely randomized design. The numbers of seedlings emerging were recorded. Roots were washed and rated for root pathogen damage severity using a five-point scale (0, no damage; 5, total damage). The final dry weights of aboveground tissues and roots were determined after oven-drying at 55°C (131°F) for 72 h. Leaf areas were measured on a Licor leaf area meter.

C Effects of Food Waste Vermicompost Teas on *Fusarium oxysporum* Attacking Tomatoes

The soil-less medium (MM 360) in which the plants were grown was supplied with nutrients from Peters Professional Solution three times weekly to provide all needed nutrients. Test pots were then drenched with *F. oxysporum* (10^8 spores per 5 mL (0.17 oz) water; Table 13.1) at the two-leaf stage and then were watered with a range of food waste vermicompost teas (5%, 10%, and 20%) or a deionized water control weekly. All three concentrations of vermicompost teas (5%, 10%, and 20%) suppressed the *F. oxysporum* damage to roots significantly ($p < 0.05$; Figure 13.1) compared with the deionized water control.

D Effects of Food Waste Vermicompost Teas on *Phytophthora capsici* Attacking Cucumbers and Tomatoes

1 Cucumber Experiments with Phytophthora capsici

The cucumber plants were grown in MM 360 in the same way as described for the tomatoes in the *F. oxysporum* experiment, and the *P. capsici* pathogen inocula were drenched as 10^8 sporangia in 5 mL (0.169 oz) water applied to the plant growth

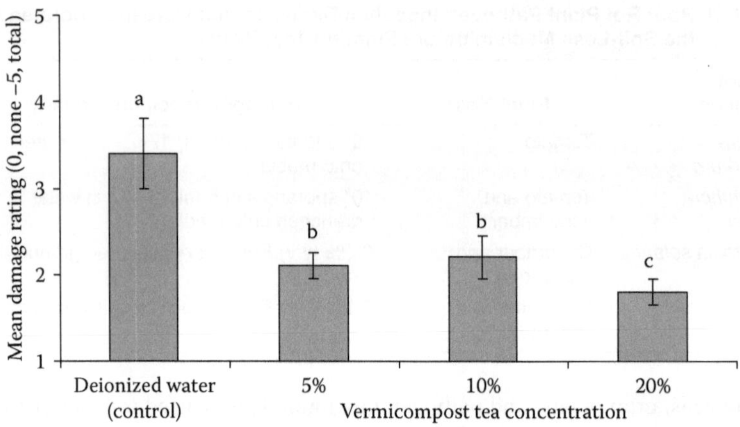

Figure 13.1 Suppression of *Fusarium oxysporum* damage on tomato roots by soil applications of vermicomposts on plants grown in MM 360 with all needed nutrients supplied. Columns represent mean damage on a scale of 1, low, to 5, total (means ± SE). Columns with different letters are significantly different ($p \leq 0.05$).

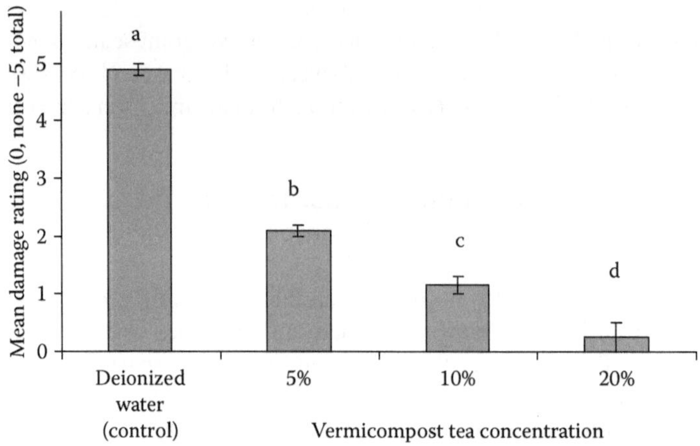

Figure 13.2 Mean root damage ratings of *Phytophthora capsici* on cucumbers (mean ± SE). Columns with different letters are significantly different ($p \leq 0.05$). Plants were grown in MM 360 with all needed nutrients supplied.

medium at the two-leaf stage. Three concentrations of vermicompost teas were drenched into the medium (5%, 10%, and 20%) to bring it to field capacity at sowing and weekly thereafter; other experimental details were similar to those of *F. oxysporum* experiments. All three concentrations of vermicompost teas suppressed damage ratings of *P. capsici* on cucumbers significantly ($p \leq 0.05$; Figure 13.2) mean plant heights of the attacked cucumbers significantly ($p \leq 0.05$; Figure 13.3) compared with the deionized water controls as a result of the pathogen suppression.

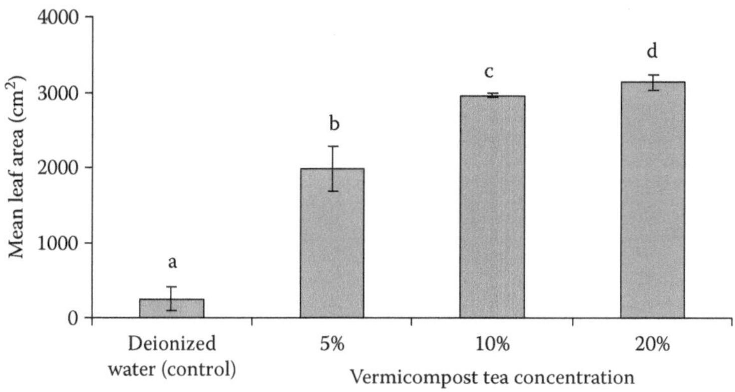

Figure 13.3 Mean leaf areas of cucumbers attacked by *Phytophthora capsici* (means ± SE). Columns with different letters are significantly different ($p \leq 0.05$). Plants were grown in MM 360 with all needed nutrients supplied.

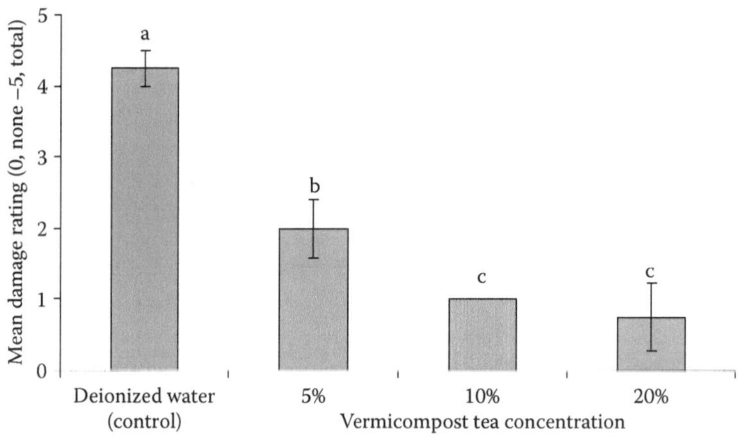

Figure 13.4 Mean damage ratings of *Phytophthora capsici* on tomatoes (means ± SE). Columns with different letters are significantly different ($p \leq 0.05$). Plants were grown in MM 360 with all needed nutrients supplied.

2 *Tomato Experiments with* Phytophthora capsici

Tomato plants were grown in the same way described in the *F. oxysporum* experiments, and the pathogen was applied as a soil drench at the same concentrations as in the cucumber experiment together with a deionized water control. The pathogen *P. capsici* caused considerable damage (nearly 90%) to tomatoes (Figure 13.4), which was significantly suppressed ($p \leq 0.05$) by all of the food waste vermicompost tea spray concentrations. For mean leaf areas (Figure 13.5) only the 20% tea application rate produced tomato plants with leaf areas significantly greater than the deionized water control ($p \leq 0.05$).

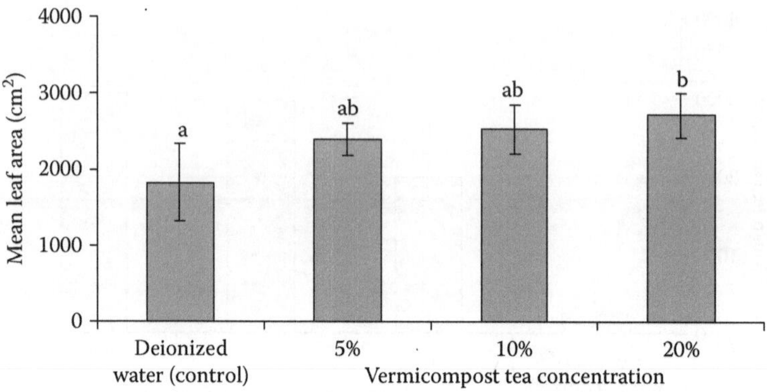

Figure 13.5 Mean leaf areas of tomatoes attacked by *Phytophthora capsici* (means ± SE). Columns with different letters are significantly different ($p \leq 0.05$). Plants were grown in MM 360 with all needed nutrients supplied.

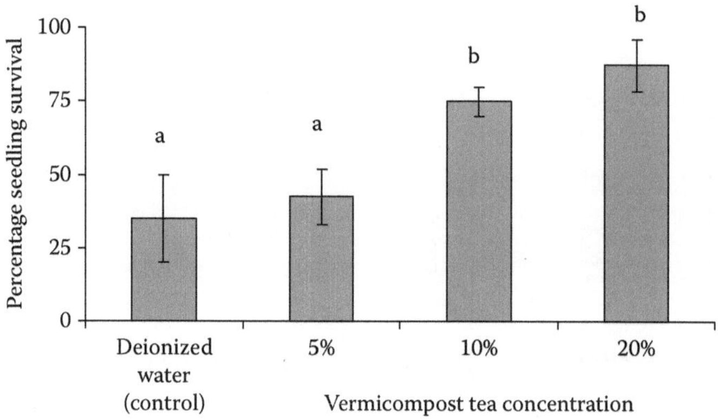

Figure 13.6 Suppression of cucumber seedling losses to *Rhizoctonia solani* by soil applications of vermicompost teas (means ± SE). Columns with different letters are significantly different ($p \leq 0.05$). Plants were grown in MM 360 with all needed nutrients supplied.

E Effects of Food Waste Vermicompost Teas on *Rhizoctonia solani* Attacking Cucumbers and Tomatoes

1 Cucumber Experiments with Rhizoctonia solani

Cucumber plants were grown in the same way as described for the *F. oxysporum* experiments. The *R. solani* inoculum was cultured in a 0.1 (v:v) ground rice medium and applied as an aqueous drench to the MM 360 growth medium. All plants were treated with 5%, 10%, or 20% vermicompost tea. The initial effects of the *Rhizoctonia* inoculations were to decrease the numbers of cucumber seedlings surviving the pathogen quite dramatically; only 35% of seedlings surviving in the deionized control water treatment (Figure 13.6) compared with nearly 90%

survival in response to the 20% tea treatment. The 10% and 20% vermicompost teas increased seedling survival after pathogen attack significantly ($p \leq 0.05$). All the concentrations of vermicompost tea soil drenches except 5% suppressed damage by *R. solani* to cucumbers significantly ($p \leq 0.05$; Figure 13.7) and increased the leaf areas ($p \leq 0.05$; Figure 13.8).

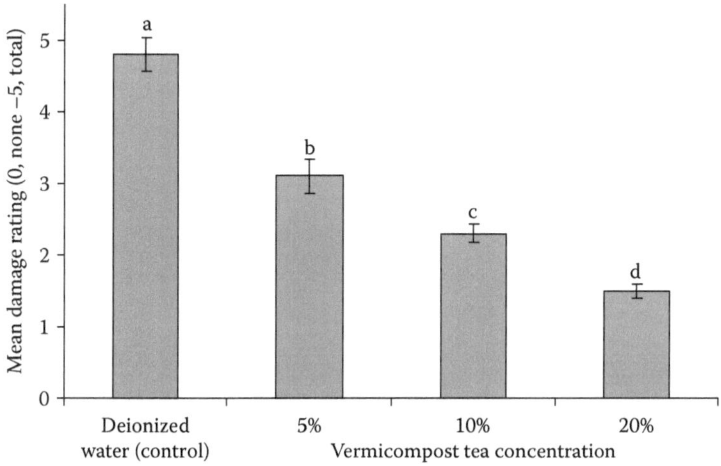

Figure 13.7 Mean damage ratings of *Rhizoctonia solani* on cucumbers by soil applications of vermicompost teas (means ± SE). Columns with different letters are significantly different ($p \leq 0.05$). Plants were grown in MM 360 with all needed nutrients supplied.

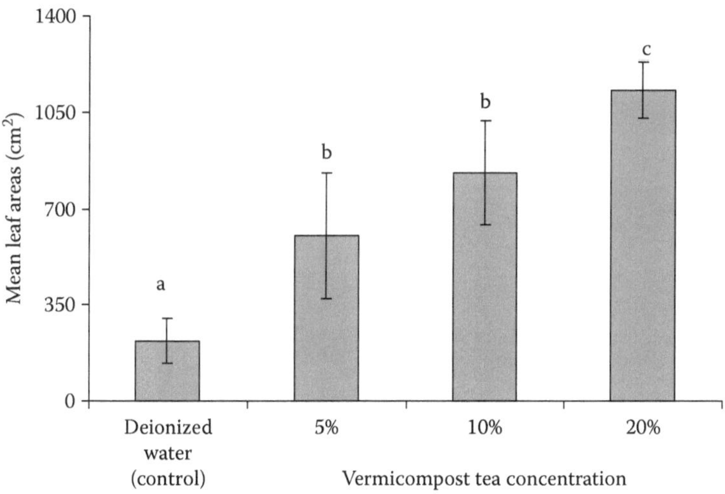

Figure 13.8 Mean leaf areas of cucumbers attacked by *Rhizoctonia solani* (means ± SE). Columns with different letters are significantly different ($p \leq 0.05$). Plants were grown in MM 360 with all nutrients supplied.

2 *Tomato Experiments with* Rhizoctonia solani

Tomato plants were grown in the same way as in the *F. oxysporum* experiment, and the *R. solani* inocula was applied in the same way and at the same concentrations as in the cucumber experiment. All the concentrations of vermicompost tea soil drenches suppressed damage by *R. solani* to tomatoes significantly ($p \leq 0.05$; Figure 13.9) and increased the tomato leaf areas ($p \leq 0.05$; Figure 13.10).

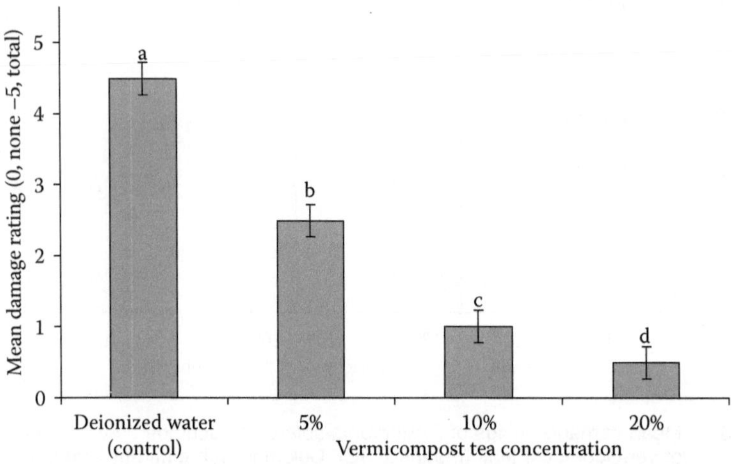

Figure 13.9 Mean damage ratings of tomatoes attacked by *Rhizoctonia solani* (means ± SE). Columns with different letters are significantly different ($p \leq 0.05$). Plants were grown in MM 360 with all nutrients supplied.

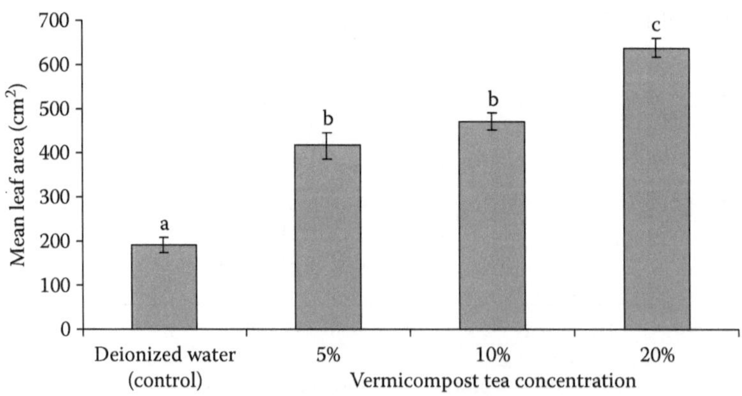

Figure 13.10 Mean leaf areas of tomatoes attacked by *Rhizoctonia solani* (means ± SE). Columns with different letters are significantly different ($p \leq 0.05$). Plants were grown in MM 360 with all nutrients supplied.

F Effects of Food Waste Vermicompost Teas on *Pythium ultimum* Attacking Cucumbers and Tomatoes

1 *Cucumber Experiments with* Pythium ultimum

The cucumber seedlings were grown as in the *F. oxysporum* experiment in MM 360 and the *P. ultimum* applied as a 0.1% (v:v) potato soil inoculum. The initial effects of *P. ultimum* on the cucumber seedlings were to decrease the percentage survival of seedlings in the deionized water control dramatically, to only 15%. The effects of the vermicompost teas in suppressing the pathogens and increasing seedling survival are summarized in Figure 13.11. The vermicompost teas at 5%, 10%, and 20% all increased the survival of the cucumber seedlings significantly ($p \leq 0.05$). In particular, the 20% vermicompost tea treatments had very significant effects in increasing seedling survival (Figure 13.11).

All three food waste vermicompost tea application rates suppressed *P. ultimum* damage to cucumbers significantly ($p \leq 0.05$; Figures 13.12 and 13.13). They also increased leaf areas significantly ($p \leq 0.05$; Figure 13.14) compared with the deionized water control. There was almost 100% damage in the deionized water control.

2 *Tomato Experiments with* Pythium ultimum

The tomato plants were grown in the same way as in the *F. oxysporum* experiment and the *P. ultimum* applied in the same way as in the cucumber experiment as a 0.1% potato soil inoculum. The mean damage ratings of *P. ultimum* on tomatoes in response to the teas are illustrated in Figure 13.15. All of the food waste vermicompost tea applications suppressed damage by *P. ultimum* significantly ($p \leq 0.05$) and also increased plant leaf area significantly ($p \leq 0.05$; Figure 13.16). The suppression

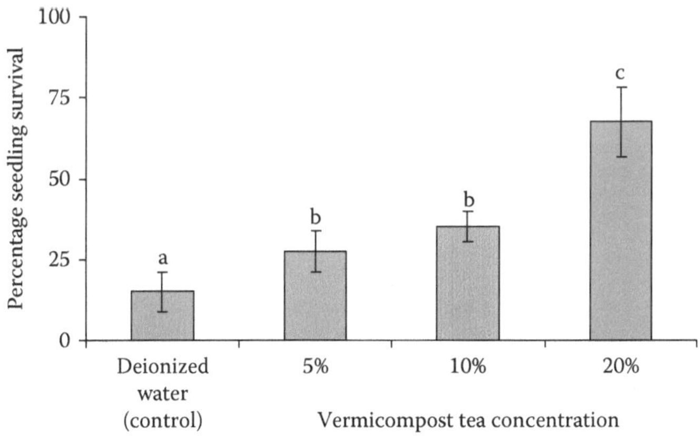

Figure 13.11 Suppression of cucumber seedling losses to *Pythium ultimum* by soil applications of vermicompost teas on cucumbers (means ± SE). Columns with different letters are significantly different ($p \leq 0.05$). Plants were grown in MM 360 with all needed nutrients supplied.

Figure 13.12 Effects of a range of vermicompost teas on *Pythium ultimum* attacking cucumbers, compared with deionized water control. (VC = vermicompost.)

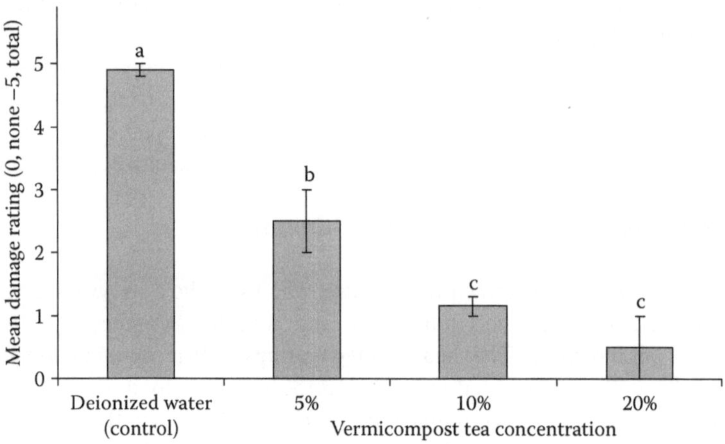

Figure 13.13 Mean damage ratings of cucumbers attacked by *Pythium ultimum* (means ± SE). Columns with different letters are significantly different ($p \leq 0.05$). Plants were grown in MM 360 with all needed nutrients supplied.

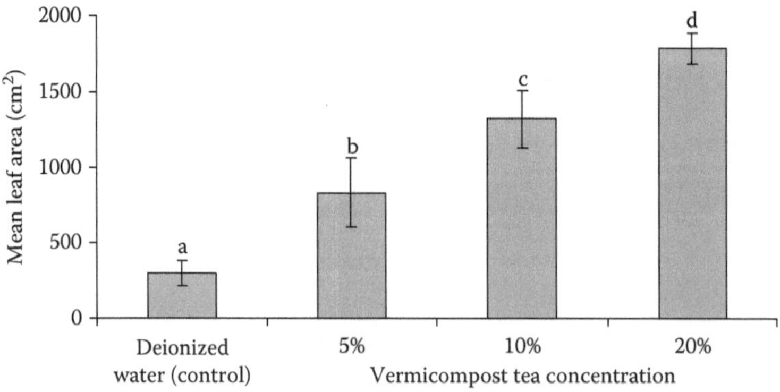

Figure 13.14 Mean leaf areas of cucumbers attacked by *Pythium ultimum* (means ± SE). Columns with different letters are significantly different ($p \leq 0.05$). Plants were grown in MM 360 with all needed nutrients supplied.

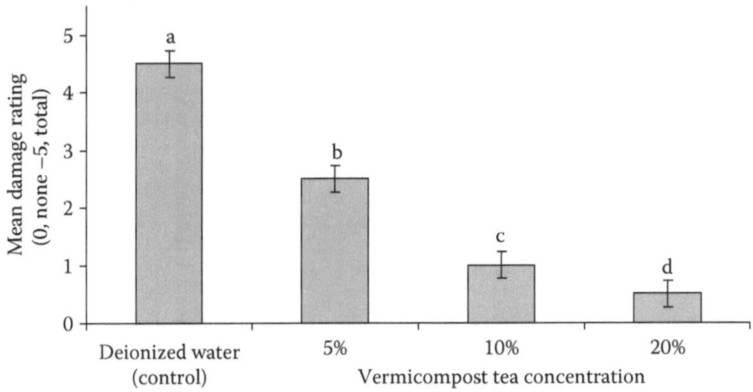

Figure 13.15 Mean damage ratings of *Pythium ultimum* on tomatoes (means ± SE). Columns with different letters are significantly different ($p \leq 0.05$). Plants were grown in MM 360 with all needed nutrients supplied.

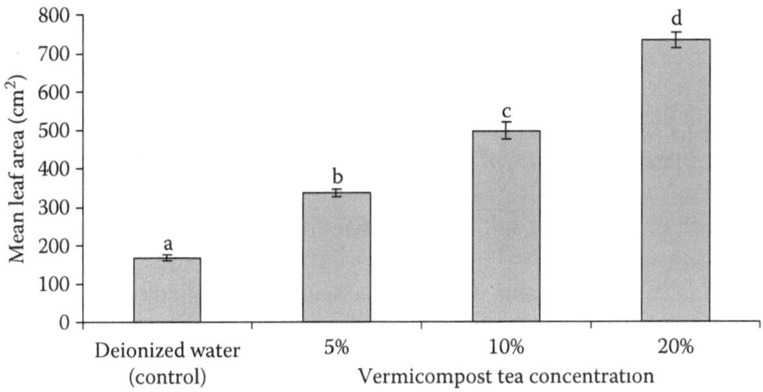

Figure 13.16 Mean leaf areas of *Pythium ultimum* on tomatoes (means ± SE). Columns with different letters are significantly different ($p \leq 0.05$). Plants were grown in MM 360 with all needed nutrients supplied.

of this pathogen by the vermicompost teas was very impressive, especially for the 10% and 20% tea treatments.

III RESEARCH AT THE OHIO STATE UNIVERSITY ON SUPPRESSION OF FOLIAR PATHOGENS OF VEGETABLES BY VERMICOMPOST AQUEOUS EXTRACTS (TEAS)

A Design of Foliar Pathogen Experiment

The foliar pathogens, crop plants, foliar pathogen inoculum densities, and age of plants that were used for each pathogen–plant combination are summarized in

Table 13.2 Foliar Plant Pathogens, Crops, and Pathogen Inoculum Densities Applied to the Foliage of Cucumbers and Tomatoes

Foliar Pathogens	Host Plant	Pathogen Inoculum Density	Plant Age at Pathogen Inoculation
Plectosporium tabacinum	Cucumber Tomato	10^4 spores.mL^{-1} water (0.16 in^{-3})	Two true leaves
Botrytis cinerea	Cucumber Tomato	10^5 spores.mL^{-1} water (0.16 in^{-3})	28 days after sowing
Scleretonia rolfsii	Cucumber Tomato	10^3 spores.mL^{-1} water (0.16 in^{-3})	14 days after sowing
Verticillium wilt	Tomato	10^{18} spores.5mL^{-1} water (0.8 in^{-3})	2 days after sowing

Table 13.2. The range of vermicompost teas and the control deionized water treatments were sprayed onto the crop foliage until run-off, 1 day before and 7 and 14 days after inoculating plants with the foliar plant pathogens. Greenhouses were maintained at 25°C (77°F) and more than 75% relative humidity to provide optimal conditions for infection by the pathogens. The numbers of lesions per leaf and percentage leaf area with disease symptoms were recorded after 14 days.

B Effects of Food Waste Vermicompost Teas on *Plectosporium tabacinum* Attacking Vegetables

1 *Cucumber Experiment with* Plectosporium tabacinum

Tomato and cucumber plants were sprayed with an inoculum of 10^4 *P. tabacinum* spores per 1 mL^{-1} (0.06 in^3) of water when the plants had two true leaves. A range of food waste vermicompost aqueous solutions (teas) were sprayed onto the foliage until run-off 1 day before and 7 and 14 days after inoculation with the pathogen. Foliar disease symptoms were rated based on the number of lesions per leaf and percentage leaf areas with disease symptoms on a scale of 0 (no symptoms) to 5 (total damage). After disease assessment the mean leaf areas were measured on a Licor leaf-area meter.

The effects of a range (5%, 10%, and 20%) of food waste vermicompost teas on the foliar pathogens were compared with those of a deionized water control. The vermicompost teas suppressed the rating of *Plectosporium* damage to cucumbers significantly at all doses ($p \leq 0.05$ level), compared with the deionized water control (Figure 13.17). The tea sprays increased mean cucumber leaf areas significantly ($p \leq 0.05$; Figure 13.18).

2 *Tomato Experiment with* Plectosporium tabacinum

The experiment was set up in the same way as in the cucumber experiment, and the *Plectosporium* inoculants, which consisted of 10^4 spores per mL deionized water,

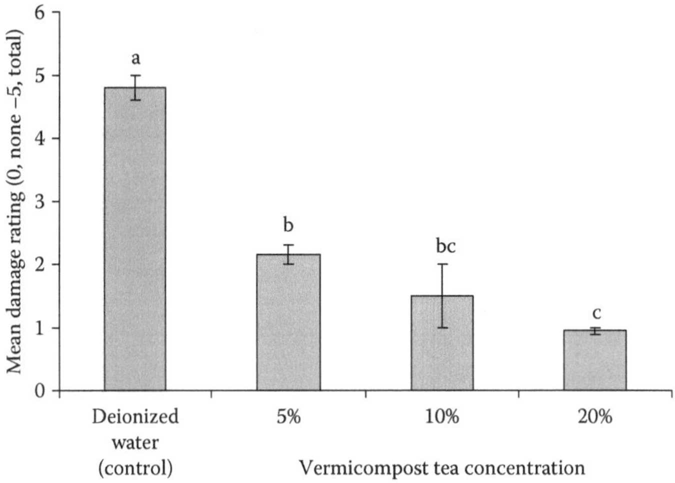

Figure 13.17 Mean damage rating of *Plectosporium tabacinum* on cucumbers (means ± SE). Columns with different letters are significantly different ($p \leq 0.05$). Plants were grown in MM 360 with all needed nutrients supplied.

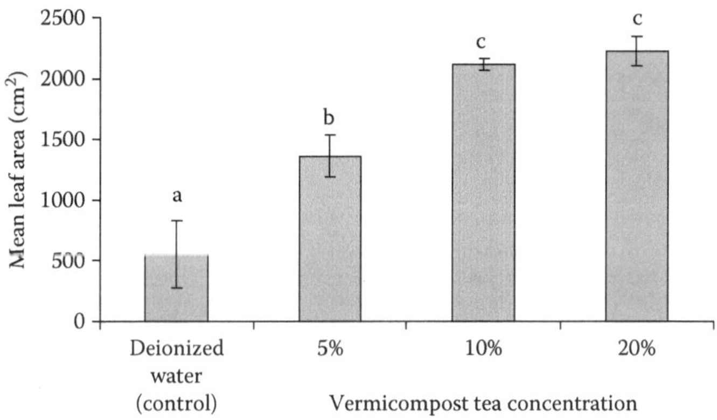

Figure 13.18 Mean leaf areas of cucumbers attacked by *Plectosporium tabacinum* (means ± SE). Columns with different letters are significantly different ($p \leq 0.05$). Plants were grown in MM 360 with all needed nutrients supplied.

were sprayed on the foliage of the plants at the two-true-leaf stage. The vermicompost teas were sprayed on the foliage until run-off 1 day before and 7 and 14 days after inoculation with the pathogen. All the concentrations of vermicompost teas decreased *P. tabacinum* damage ratings significantly ($p \leq 0.05$; Figure 13.19) and increased mean leaf areas significantly ($p \leq 0.05$; Figure 13.20) compared with the deionized water control.

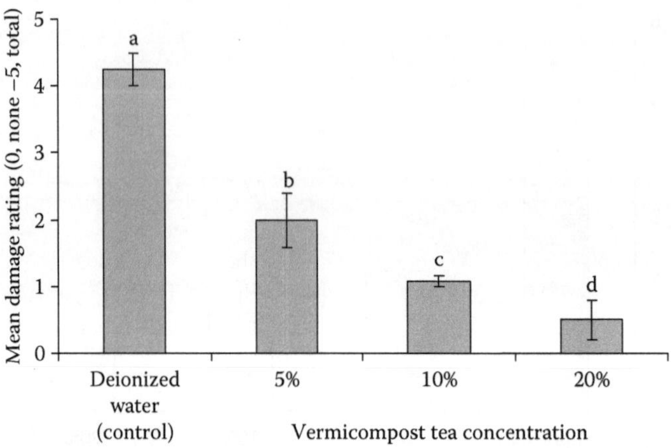

Figure 13.19 Mean damage ratings of *Plectosporium tabacinum* on tomatoes (means ± SE). Columns with different letters are significantly different ($p \leq 0.05$). Plants were grown in MM 360 with all needed nutrients supplied.

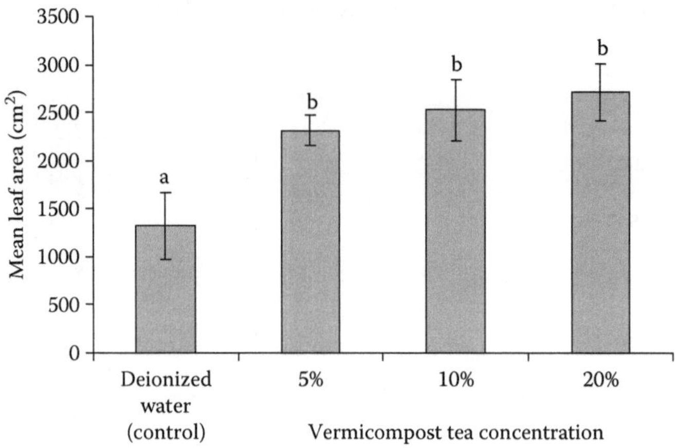

Figure 13.20 Mean leaf areas of tomatoes attacked by *Plectosporium tabacinum* (means ± SE). Columns with different letters are significantly different ($p \leq 0.05$). Plants were grown in MM 360 with all needed nutrients supplied.

C Effects of Food Waste Vermicompost Teas on *Botrytis Cinerea* Damage

1 *Cucumber Experiment with* Botrytis cinerea

Cucumber plants were inoculated with 10^5 *B. cinerea* spores per 1 mL^{-1} of water 28 days after sowing. A range of concentrations of food waste vermicompost teas were sprayed onto the foliage until run-off 1 day before and 7 and 14 days after inoculation of the plants with the pathogen. Foliar disease damage on the cucumber-plants was rated based on the number of lesions per leaf and percentage leaf areas with disease symptoms on a scale of 0 (no symptoms) to 5 (total damage). Leaf areas were measured on a Licor leaf-area meter.

The effects of a range of food waste vermicompost teas, compared to a deionized water control, on ratings of damage to cucumbers by *B. cinerea* are summarized in Figure 13.21, and the associated increases in leaf areas are summarized in Figure 13.22. All of the concentrations suppressed the *Botrytis* damage significantly ($p \leq 0.05$).

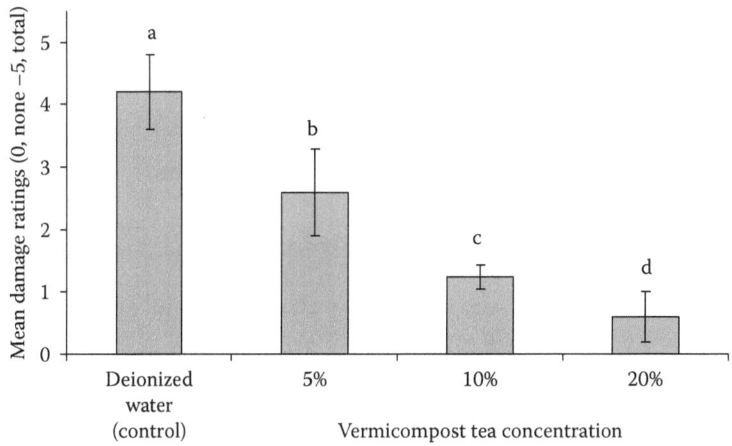

Figure 13.21 Mean damage ratings of *Botrytis cinerea* on cucumbers (mean ± SE). Columns with different letters are significantly different ($p \leq 0.05$). Plants were grown in MM 360 with all needed nutrients supplied.

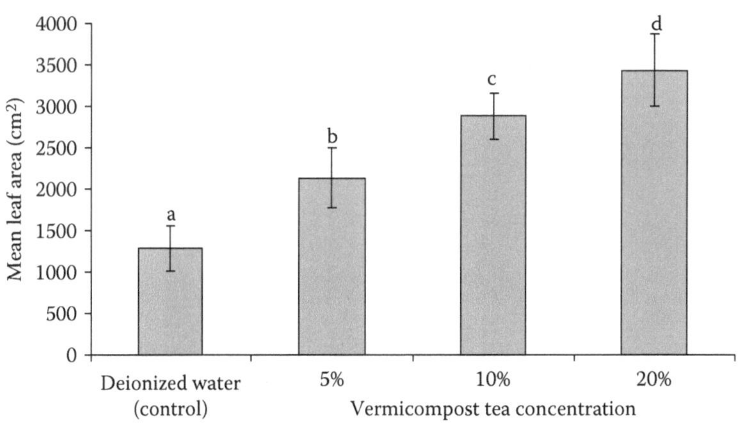

Figure 13.22 Mean leaf areas of *Botrytis cinerea* on cucumbers (mean ± SE). Columns with different letters are significantly different ($p \leq 0.05$). Plants were grown in MM 360 with all needed nutrients supplied.

2 *Tomato Experiment with* Botrytis cinerea

Tomato plants were inoculated with 10^5 *B. cinerea* spores 1 mL^{-1} (0.06 in^3) of water 28 days after sowing. A range of food waste vermicompost teas were sprayed onto the foliage until run-off 1 day before and 7 and 14 days after inoculation with the pathogen. Symptoms of foliar disease damage to the tomato plants were rated based on the number of lesions per leaf and percentage leaf areas with disease symptoms, on a scale of 0 (no symptoms) to 5 (total damage). Leaf areas were measured on a Licor leaf-area meter.

The effects of a range of food waste vermicompost teas compared to a water control on damage to tomatoes by *B. cinerea* are summarized in Figure 13.23, and the associated increases in leaf areas are summarized in Figure 13.24. All of the tea concentrations suppressed the *Botrytis* damage significantly ($p \leq 0.05$).

D Effects of Food Waste Vermicompost Teas on *Sclerotinea Rolfsii* Damage

1 *Cucumber Experiment with* Sclerotinea rolfsii

The foliage of the plants was inoculated with *S. rolfsii* (10^3 spores per 1 mL^{-1} (0.16 in^{-3}) water) and then were sprayed with, one of a range of vermicompost tea concentrations or a deionized water control, weekly. All tea treatments were applied to four replicate 10 cm diameter pots, each sown with four tomato seeds to produce four tomato plants. Pots were arranged on greenhouse benches in a completely

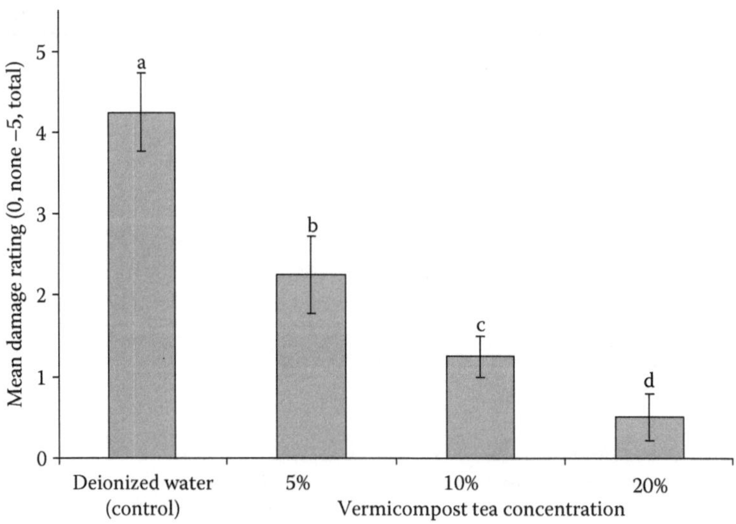

Figure 13.23 Mean damage ratings of tomatoes attacked by *Botrytis cinerea* (means ± SE). Columns with different letters are significantly different ($p \leq 0.05$). Plants were grown in MM 360 with all needed nutrients supplied.

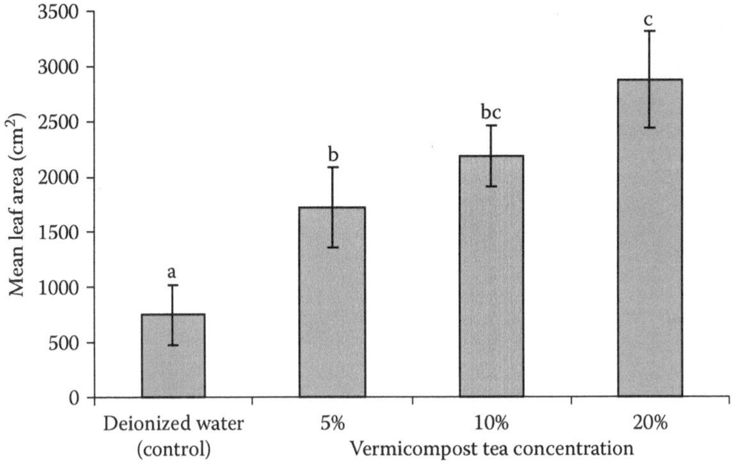

Figure 13.24 Mean leaf areas of tomatoes attacked by *Botrytis cinerea* (experiment 4; means ± SE). Columns with different letters are significantly different ($p \leq 0.05$). Plants were grown in MM 360 with all needed nutrients supplied.

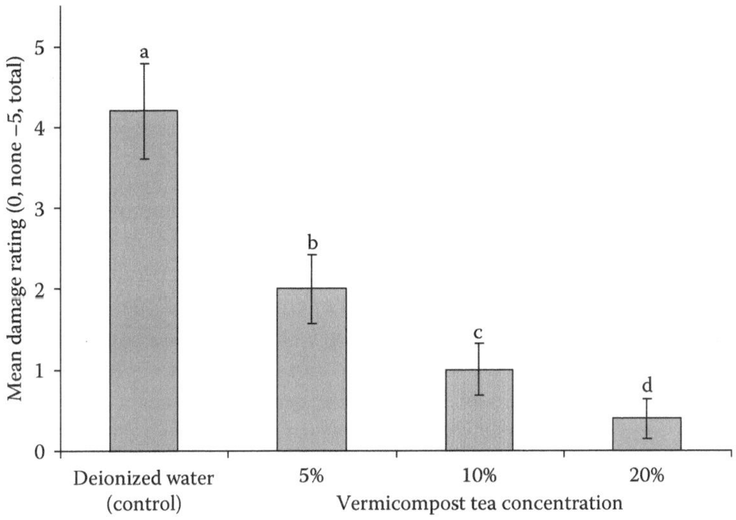

Figure 13.25 Effects of food waste vermicompost tea drenches on *Sclerotinia rolfsii* damage ratings to cucumbers (experiment 2; mean ± SE). Columns with different letters are significantly different ($p \leq 0.05$). Plants were grown in MM 360 with all needed nutrients supplied.

randomized design. The numbers of seedlings emerging were recorded, and each seedling was rated for root rot severity using a five-point scale (0, no symptoms; 5, total damage). The final dry weights of above ground and root tissues were determined. The suppression of percentage damage ratings to cucumbers by *S. rolfsii* is summarized in Figure 13.25, and increases in leaf areas in Figure 13.26.

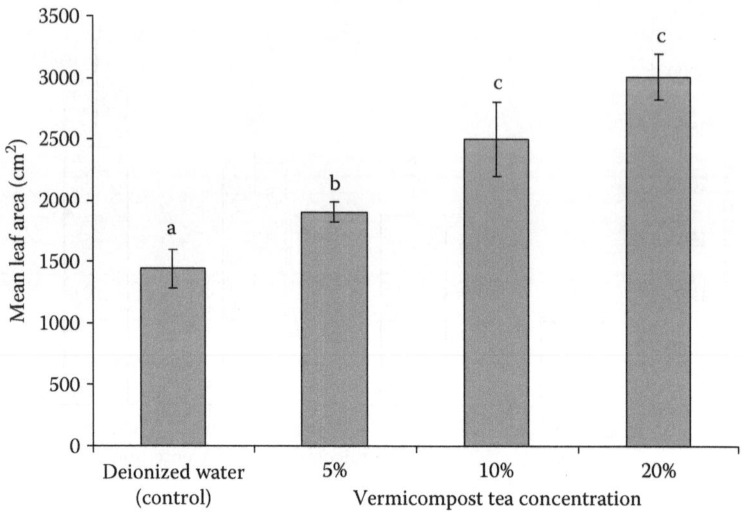

Figure 13.26 Effects of food waste vermicompost tea drenches with *Sclerotinia rolfsii* on leaf areas of cucumbers (mean ± SE). Columns with different letters are significantly different ($p \leq 0.05$). Plants were grown in MM 360 with all needed nutrients supplied.

2 *Tomato Experiment with* Sclerotinea rolfsii

The plants were inoculated with 10^3 *S. rolfsii* spores in 1 mL^{-1} (0.03 oz) water and then were sprayed with a range of vermicompost teas compared with a deionized water control weekly. All tea spray treatments were applied to four replicate 10 cm diameter pots, each sown with four tomato seeds to produce four tomato plants. Pots were arranged on greenhouse benches in a completely randomized design. The numbers of seedlings emerging were recorded, and each seedling was rated for root rot severity using a five-point scale (0, no symptoms; 5, total damage). The final dry weights of aboveground and root tissues were determined. Leaf areas were measured on a Licor area measure. The mean damage ratings of *S. rolfsii* on tomatoes were all significantly less in response to all vermicompost tea application rates ($p \leq 0.05$; Figure 13.27). For mean leaf areas the 10% and 20% vermicompost tea application increased leaf areas significantly ($p \leq 0.05$; Figure 13.28).

E Conclusions on Effects of Vermicompost Tea Sprays on Foliar Pathogens

Clearly, food waste vermicompost tea sprays were very effective in suppressing foliar plant diseases, which are very difficult to control by chemical means. This is of particular interest and value to the organic farmer or gardener who cannot use chemical pesticides. It appears that to ensure satisfactory suppression, it may be

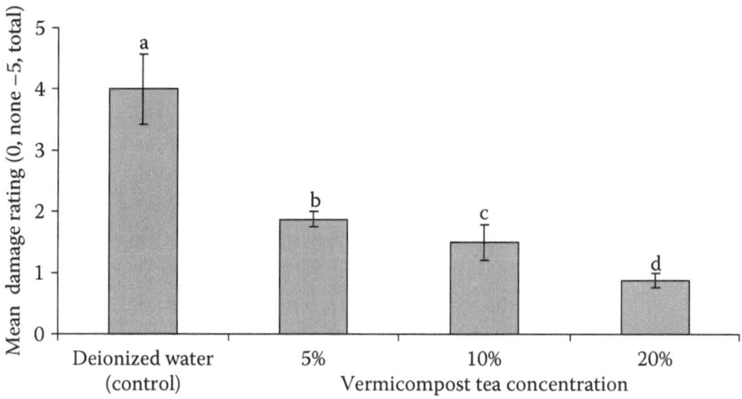

Figure 13.27 Mean damage ratings of *Sclerotinia rolfsii* on tomatoes (mean ± SE). Columns with different letters are significantly different ($p \leq 0.05$). Plants were grown in MM 360 with all needed nutrients supplied.

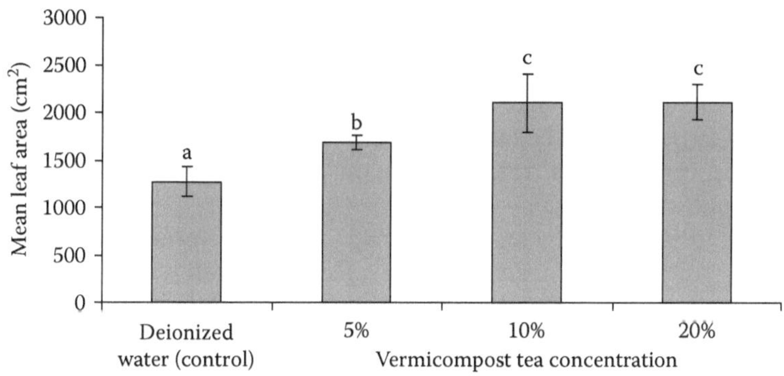

Figure 13.28 Mean leaf areas of tomato plants attacked by *Sclerotinia rolfsii* (mean ± SE). Columns with different letters are significantly different ($p \leq 0.05$). Plants were grown in MM 360 with all needed nutrients supplied.

preferable to use 10% or 20% food waste vermicompost teas applied twice, either as a drench for root pathogens or, as a foliar spray for foliar pathogens.

IV MECHANISMS BY WHICH VERMICOMPOST TEAS SUPPRESS PLANT PATHOGENS

Plant diseases are major problems in horticultural enterprises and are tremendously challenging to the horticultural industry. Even with the combination of complex management strategies, including the modification of cultural practices and use of new and resistant varieties, plant pathogens constantly adapt, resulting in new challenges for growers. One strategy that has shown great potential in plant pest and disease management is the development of ecologically based strategies aimed at increasing the growth and diversity of soil-dwelling and epiphytic microbial populations (Koike et al. 2000). A practice within this strategy is the use of soil amendments such as composts, vermicomposts, and their aqueous extracts (teas) to produce what is referred to as substrate-induced resistance. We have demonstrated that vermicomposts are high-quality soil amendments containing large microbial populations and communities. The production of teas from vermicomposts can multiply both soil-dwelling and epiphytic microbial populations that can suppress plant pathogens.

Several mechanisms of specific suppression by vermicompost and vermicompost teas have been proposed, including (a) destruction of pathogen propagules, (b) prevention of propagule germination, (c) antibiosis, (d) hyperparasitism, (e) competition for nutrients, or (f) competition for infection sites (Stone et al. 2004). In practice there is probably is a combination of multiple effects, including competition for nutrients, antibiosis, extracellular enzymes, and parasitism, acting directly on the pathogens in the bulk soil or the rhizosphere or indirectly through host-mediated induced resistance (Noble and Coventry 2005). The other type of suppression is a general suppression that cannot be linked to any individual organism. A wide range of aerobic organisms may be involved, and a range of population effects may lead to suppression of plant pathogens. However, it is difficult to determine the exact suppression mechanisms since a compost, or vermicompost, represents a "microbial community structure rather than a single species" (Boulter et al. 2002).

Effective disease antagonists that may occur in vermicomposts include *Trichoderma hamatum, Flavobacterium balustinum, Pseudomonas aeruginosa, P. fiuorecens, P. putida, P. stutzeri, Xanthomonas maltophilia, Janthinobacterium lividum, Enterobacter cloacae, E. agglomerans, Bacillus cereus, B. mycoides,* and *B. subtilis* (Hoitink and Fahy 1986; Dowling et al. 1996; O'Sullivan and O'Gara 1992; Shanahan et al. 1992). Several of these microorganisms can induce systemic resistance to plant disease (Han et al. 2000; Krause et al. 2003). Alfano et al (2007) demonstrated that systemic resistance induced in tomato by *Trichoderma hamatum* 382 was linked to the up-regulation of genes for several different stress-related proteins. A diverse array of bacteria and several fungi have been identified that can significantly suppress pathogens. Some suppressive properties include secretion of hydrolytic enzymes that attack pathogen membranes (e.g., chitinase, protease, β-1,3, glucanase), iron-chelating siderophores that limit the availability of iron for the growth of pathogens, cyanide, and antibiotics (Mazzola 2002). Vermicomposts produced through mesophilic processes have the advantage over thermophilic composts of promoting a greater diversity of microorganisms, thereby offering more pathogen-suppressive properties than solid vermicomposts. A more detailed discussion on the microbial qualities of vermicomposts can be found in Chapter 5.

REFERENCES

Alfano, G., Lewis Ivey, M.L., Cakir, C., Bos, J.I.B., Miller, S.A., Madden, L.V., Kamoun, S., and Hoitink, H.A.J. 2007. Systemic modulation of gene expression in tomato by *Trichoderma hamatum* 382. *Phytopathology* 97: 429–437.

Arancon, N.Q., and Edwards, C.A. 2004. Vermicomposts suppress plant pest and disease attacks. *Biocycle*, March: 51–53.

Boulter, J.I., Trevors, J.T., and Boland, G.J. 2002. Microbial studies of compost: Bacterial identification and their potential for turfgrass pathogen suppression. *World J. Microbiol. Biotechnol.* 18: 661–671.

Chaoui, H., Edwards, C.A., Brickner, A., Lee, S., and Arancon, N.Q. 2002. Suppression of the plant parasitic diseases: *Pythium* (damping off), *Rhizoctonia* (root rot) and *Verticillium* (wilt) by vermicompost. In *Proceedings Brighton Crop Protection Conference—Pests and Diseases*, 8B-3: 711–716.

Chen, W., Hoitink, H.A., Schmitthenner, A.F., and Touvinen, O.H. 1987. The role of microbial activity in suppression of damping off caused by *Pythium ultimum*. *Phytopathology* 78, 314–322.

Dowling, D.N., Sexton, R., Fenton, A., Delany, I., Fedi, S., McHugh, B., Callanan, M., Moënne-Loccoz, Y., and O'Gara, F. 1996. Iron regulation in plant-associated *Pseudomonas fluorescens* M114: Implications for biological control. In *Molecular Biology of Pseudomonads*, ed. Nakazawa, K., Furukawa, K., Haas, D., and Silver, S., 502–511. ASM Press, Washington, DC.

Drinkwater, L.E., Letourneau, D.K., Workneh, F., van Bruggen A.H., and Shennan, C. 1995. Fundamental differences between conventional and organic tomato agroecosystems in California. *Ecol. Appl.* 5: 1098–1112.

Edwards, C.A., Arancon, N.Q., and Greytak, S. 2006. Effects of vermicompost teas on plant growth and disease. *Biocycle* 47: 28–31.

Han, D.Y., Coplin, D.L., Bauer, W.D., and Hoitink, H.A.J. 2000. A rapid bioassay for screening rhizosphere microorganisms for their ability to induce systemic resistance. *Phytopathology* 90: 327–332.

Hoitink, H.A.J., and Fahy, P. 1986. Basis for the control of soilborne plant pathogens with composts. *Ann. Rev. Phytopathol.* 24: 93–114.

Hoitink, H.A.J., and Kuter, G.A. 1986. Effects of composts in growth media on soil-borne pathogens. In *The Role of Organic Matter in Modern Agriculture*, ed. Chen, Y., and Avnimelech, Y., 289–306. Martinus Nijhoff, Dordrecht, the Netherlands.

Hoitink, H.A.J., Stone, A.G., and Han, D.Y. 1997. Suppression of plant diseases by compost. *Hortic. Sci.* 32: 184–187.

Huelsman, M.F., and Edwards, C.A. 1998. Management of disease in cucumbers (*Cucumis sativus*) and peppers (*Capsicum annum*) by using composts as fertility inputs. In *Proceedings Brighton Pest Control Conference: Pests and Diseases*, 8D-15, 881–886.

Kannangara, T., Utkhede, R.S., Paul, J.W., and Punja, Z.K. 2001. Effect of mesophilic and thermophilic composts on suppression of *Fusarium* root and stem rot of greenhouse cucumber. *Can. J. Microbiol.* 46: 1021–1028.

Khan, J., Ooka, J.J., Miller, S.A., Madden, L.V., and Hoitink, H.A.J. 2004. Systemic resistance induced by *Trichoderma hamatum* 382 in cucumber against Phytophthora crown rot and leaf blight. *Plant Dis.* 88: 280–286.

Koike, T., Yazaki, K., Funada, R., Maruyama, R., Mori, Y., and Sasa, K. 2000. Forest health and vitality in northern Japan—a history of larch plantation. *Res. Note, Fac. Forestry, Univ. of Joensuu* 92: 49–60.

Krause, M.S., De Ceuster, T.J.J., Tiquia, S.M., Michel, F.C., Jr., Madden, L.V., and Hoitink, H.A.J. 2003. Isolation and characterization of rhizobacteria from composts that suppress the severity of bacterial leaf spot of radish. *Phytopathology* 93: 1292–130.

Kuter, G.A., Nelson, G.B., Hoitink, H.A., and Madden, L.V. 1983. Fungal population in container media amended with composted hardwood bark suppressive and conductive to Rhizoctonia damping-off. *Phytopathology* 73: 1450–1456.

Liping, M.A., Qiao, X., Gao, F., and Hao, B. 2001. Control of sweet pepper *Fusarium* wilt with compost extract and its mechanism. *Yingyong Yin Huanjing Shengwu Xnebao* 7: 84–87.

Mazzola, M. 2002. Mechanisms of natural soil suppressiveness to soil-borne diseases. *Antonie van Leeuwenhoek* 81: 557–564.

Nakamura, Y. 1996. Interactions between earthworms and microorganisms in biological control of plant root pathogens. *Farming Japan* 30: 37–43.

Nakasone, A.K., Bettiol, W., and de Souza, R.M. 1999. The effect of water extracts of organic matter on plant pathogens. *Summa Phytopathol.* 25: 330–335.

Noble, R., and Coventry, E. 2005. Suppression of soil-borne plant diseases with composts: A review. *Biocontrol Sci. Technol.* 15: 3–20.

O'Sullivan, D.J., and O'Gara, F. 1992. Traits of fluorescent *Pseudomonas* spp. involved in suppression of plant root pathogens. *Microbiol Rev.* 56: 662–676.

Rodríguez, J.A., Zavaleta, E., Sanchez, P., and Gonzalez, H. 2000. The effect of vermicomposts on plant nutrition, yield and incidence of root and crown rot of gerbera (*Gerbera jamesonii H. Bolus*). *Fitopatologia* 35: 66–79.

Scheuerell, S. 2002. *Compost teas and compost-amended container media for plant disease control*. PhD diss., Oregon State University, Corvallis, Oregon.

Scheuerell, S., and Mahaffee, W. 2006. Variability associated with suppression of gray mold (*Botrytis cinerea*) on geranium by foliar applications of nonaerated and aerated compost teas. *Am. Phytopathol. Soc.* 90: 1201–1208.

Scheuerell, S.J., and Mahaffee, W.F. 2004. Compost tea as a container medium drench for suppressing seedling damping-off caused by *Pythium ultimum*. *Phytopathology* 94: 1156–1163.

Scheuerell, S., and Mahaffee, W. 2002. Compost tea: Principles and prospects for plant disease control. *Compost Sci. Util.* 10(4): 313–338.

Shanahan, P., O'Sullivan, D.J., Simpson, P., Glennon, J.D., and O'Gara, F. 1992. Isolation of 2,4-diacetylphloroglucinol from a fluorescent pseudomonad and investigation of physiological parameters influencing its production. *Appl. Environ. Microbiol.* 58: 353–358.

Stone, A.G., Scheurell, S.J., and Darby, H. M. 2004. Suppression of soilborne diseases in field agricultural systems: Organic matter management, cover cropping and other cultural practices. In: *Soil Organic Matter in Sustainable Agriculture*, eds. Magdoff, F., and Weil, R. R., 131–177. CRC Press, Boca Raton, FL.

Stone, A.G., Vallad, G.E., Cooperband, L.R., Rotenberg, D., Darby, H.M., James, R.V., Stevenson, W.R., and Goodman, R.M. 2003. The effect of organic amendments on soilborne and foliar diseases in field grown snap bean and cucumber. *Plant Dis.* 87: 1037–1042.

Tränkner, A. 1992. Use of agricultural and municipal organic wastes to develop suppressiveness to plant pathogens. In: *Biological Control of Plant Diseases: Progress and Challenges for the Future*, eds. Tjamos, E.S., Papavizas, G.C., and Cook, R.J., 35–42. Plenum Press, New York.

Zaller, J. 2006. Foliar spraying of vermicompost extracts: Effects on fruit quality and indications of late-blight suppression of field-grown tomatoes. *Biol. Agric. Hortic.* 24: 165–180.

Zhang, W., Dick, W.A., and Hoitink, H.A.J. 1996. Compost-induced systemic acquired resistance in cucumber to *Pythium* root rot and anthracnose. *Phytopathology* 86: 1066–1077.

Zhang, W., Han, D.Y., Dick, W.A., Davis, K.R., and Hoitink, H.A.J. 1998. Compost and compost water extract-induced systemic acquired resistance in cucumber and Arabidopsis. *Phytopathology* 88: 450–455.

CHAPTER 14

Suppression of Arthropod Pests and Plant Parasitic Nematodes by Vermicomposts and Aqueous Extracts from Vermicomposts

Clive A. Edwards, Ahmed M. Askar, Marcus A. Vasko-Bennett, and Norman Arancon

CONTENTS

I Introduction 210
II Suppression of Arthropod Pests by Solid Vermicomposts 210
III Suppression of Arthropod Pests by Aqueous Solutions from Vermicomposts (Teas) 213
A Cucumber Beetles and Tobacco Hornworms 213
B Aphids, Mealybugs, and Spider Mites 217
IV Mechanisms by which Vermicomposts and Vermicompost Aqueous Solutions (Teas) Suppress Arthropod Pests 219
A Solid Vermicomposts 219
B Vermicompost Teas 220
V Effects of Vermicomposts on Soil Nematode Communities 223
VI Suppression of *Meloidogyne hapla* on Tomatoes by Aqueous Extracts from Vermicomposts (Teas) 226
VII Suppression of *Meloidogyne hapla* on Cucumbers by Aqueous Extracts from Vermicomposts (Teas) 227
VIII Mechanisms by which Vermicompost Teas May Suppress Plant Parasitic Nematode Populations 229
References 230

I INTRODUCTION

Vermicomposts are produced from organic wastes through interactions between earthworms and microorganisms, and they are utilized as plant growth media and soil amendments. They are produced by an aerobic and mesophilic process involving complex interactions between earthworms and microorganisms; this process stabilizes the organic matter, lowers the C to N ratio, and makes the nutrients that the vermicompost contains available to plants. It has been shown in the literature that adding various types of organic amendments, including manures and thermophilic composts, to soils can decrease arthropod and nematode pest incidence, lower populations of these pests, and lessen damage to plants growing in the amended soils (Chellemi 2002; Altieri and Nicholls 2003; Atkinson et al. 2004).

II SUPPRESSION OF ARTHROPOD PESTS BY SOLID VERMICOMPOSTS

The first reports of vermicomposts suppressing arthropod pests were by Arancon and Edwards (2004) and Arancon et al. (2005), who showed that solid vermicomposts suppressed cabbage white caterpillar attacks on cabbages. They also reported significant suppression of mealybug attacks (*Pseudococcus* sp.) on cucumbers and tomatoes, of two-spotted spider mite (*Tetranychus urticae*) attacks on bush beans and eggplants, and of attacks by aphids (*Myzus persicae*) on cabbages by low-application rates of food waste vermicomposts (Arancon et al. 2005, 2007).

Yardim et al. (2006) reported the effects of food waste vermicomposts on populations of adult striped cucumber beetles (*Acalymma vittatum*) and spotted cucumber beetles (*Diabotrica undecimpunctata*) on cucumbers and of tobacco hornworms (*Manduca quinquemaculata*) on tomatoes. In his paper, vermicomposts were evaluated in both greenhouse and field experiments for their effects on the pest populations as well as on the damage the pests caused. In the field, cucumber and tomato plants were grown, with two different application rates (2.5–5.0 ton.ha^{-1} (5–10 t.acre^{-1}) of food waste vermicompost compared with one recommended rate of inorganic fertilizer, in a complete randomized block design field experiment. All treatments were balanced for NPK. Field cucumber beetle populations were suppressed significantly ($p \leq 0.05$) on cucumber plants treated with food waste vermicompost, at both application rates, compared with those on plants treated with only a recommended rate of inorganic fertilizer (Figure 14.1). Damage by tobacco hornworm caterpillars to tomato fruits in the field was decreased significantly by both application rates of food waste vermicompost, compared to that on plants grown with only the recommended rate of inorganic fertilizer (Figure 14.2). In the greenhouse, cucumber and tomato plants were grown in a soil-less medium, Metro Mix 360 (MM 360), substituted with 0%, 20%, or 40% food waste vermicompost, and were exposed to standardized pest attacks by releasing known numbers of pests into nylon mesh cages. In the greenhouse, both the 20% and 40% vermicompost substitution rates decreased damage to cucumber foliage by cucumber beetles (Figure 14.3).

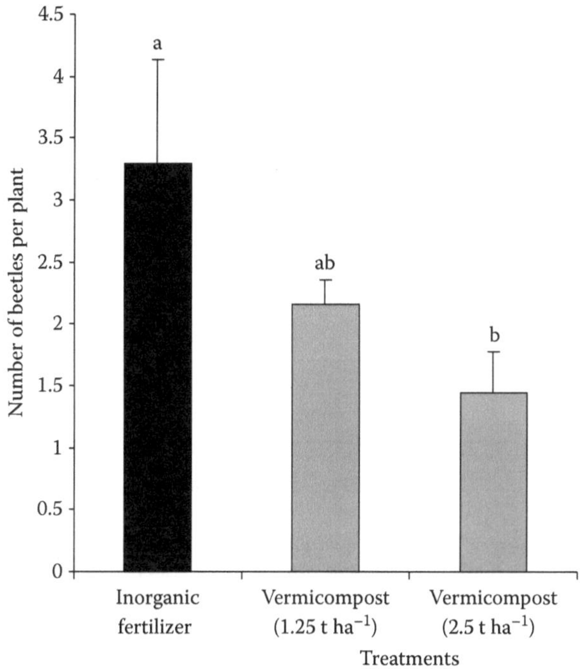

Figure 14.1 Numbers of striped cucumber beetles (*Acalymma vittatum*) per plant (mean ± SE) on cucumber plants in the field in response to food waste vermicompost or inorganic fertilizer applications. Plants were grown with all needed nutrients. Means designated by the same letter are not significantly different ($p \leq 0.05$).

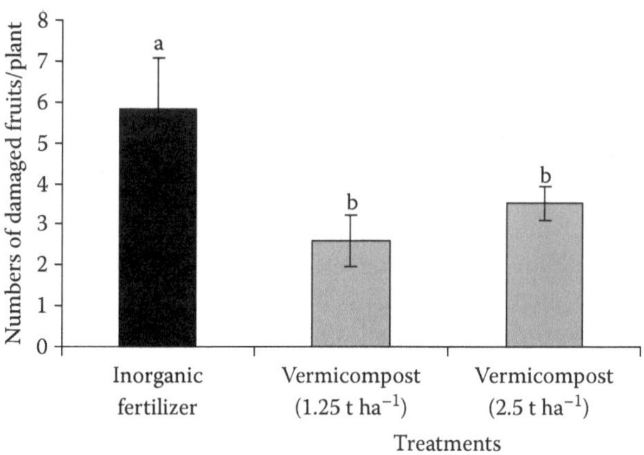

Figure 14.2 Tobacco hornworm (*Manduca quinquemaculata*) caterpillar damage (mean ± SE) to tomato fruits in the field in response to food waste vermicompost or inorganic fertilizer applications. Plants were grown with all needed nutrients. Means designated by the same letter are not significantly different ($p \leq 0.05$).

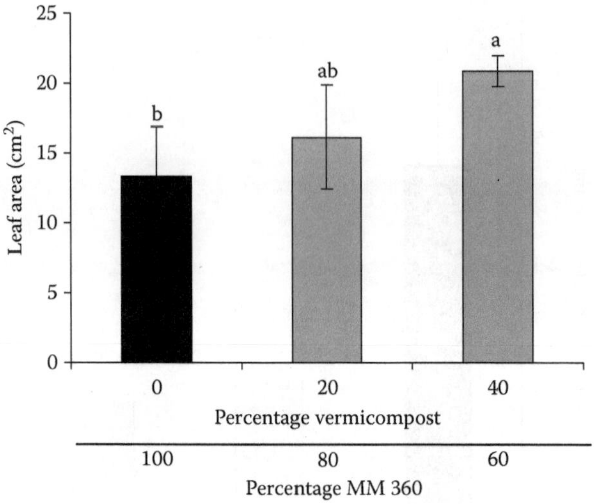

Figure 14.3 Leaf areas of cucumber seedlings (mean ± SE) after exposure to striped cucumber beetle (*Acalymma vittatum*) infestations in greenhouse cages for 2 days, in response to substitutions of vermicomposts into MM 360. Plants were grown with all needed nutrients. Means designated by the same letter are not significantly different ($p \leq 0.05$).

In other experiments at the Soil Ecology Laboratory at The Ohio State University, a vermicompost, produced commercially from food wastes, was tested for its capacity to suppress populations of, and damage to tomato and cucumber plants by, two-spotted spider mites (*Tetranychus urticae*), mealybugs (*Pseudococcus* sp.), and aphids (*Myzus persicae*), in the greenhouse (Arancon et al. 2007). A range of mixtures of food waste vermicomposts and a soil-less bedding-plant growth medium, MM 360, was tested in cages 40 × 40 × 40 cm (15.7 × 15.7 × 15.7 in) with a 0.2 mm (0.05 in) mesh aperture, into which known numbers of greenhouse-bred arthropod pests were released. The crops tested were cucumbers and tomatoes for mealybugs, bush beans and eggplants for spider mites, and cabbages for aphids. In all experiments, 10 cm (4 in) diameter pots, each containing one seedling grown in the same (MM 360) vermicompost mixture, were exposed to either 50 mealybugs, 100 spider mites, or 100 aphids in cages, with each cage treatment replicated four times per treatment. The five growth-medium mixtures tested were (i) 100% MM 360, (ii) 90% MM 360 with 10% vermicompost, (iii) 80% MM 360 with 20% vermicompost, (iv) 60% MM 360 with 40% vermicompost, and (v) 20% MM 360 with 80% vermicompost. Almost all of the mixtures containing vermicomposts suppressed the arthropod pest populations and decreased pest damage significantly ($p \leq 0.05$), compared with the MM 360 controls. Not only did the vermicomposts make the plants less attractive to the pests, but they also had considerable effects on pest reproduction over time. The effects of the vermicompost substitutions tended to be least on spider mites, intermediate on mealybugs, and greatest on aphids; however, this may relate to the motility of the pests, as well as to the suppression potential of vermicomposts. Possible

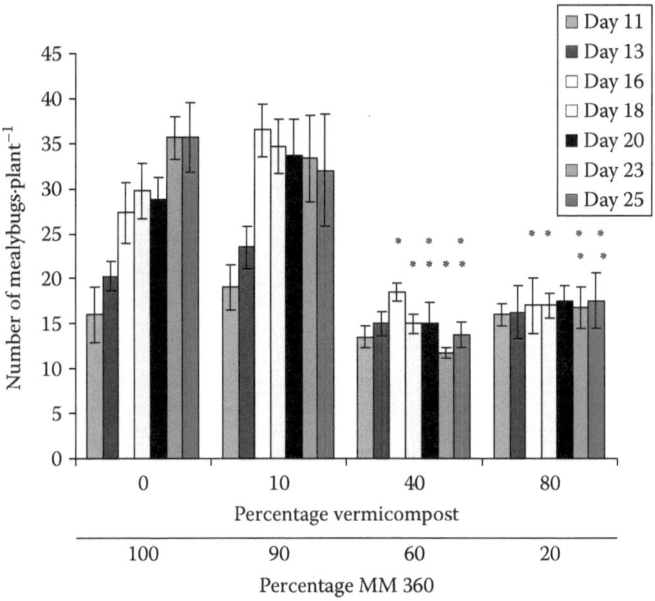

Figure 14.4 Development of mealybugs (*Pseudococcus* sp.) on tomato plants grown in the greenhouse with mixtures of food waste vermicompost and MM 360 and supplied with all needed nutrients. Populations were recorded on seven dates over 25 days after mealybug infestations. Significantly different means are designated * ($p \leq 0.05$) or ** ($p \leq 0.01$).

mechanisms for the suppression discussed include the form of N available in the leaf tissues, the effects of vermicomposts on micronutrient availability, and the possible uptake of water-soluble phenols by the plants from vermicomposts, thereby making the tissues unpalatable and affecting pest reproduction and survival.

The development of mealybug (*Pseudococcus* sp.) populations on tomatoes grown in MM 360 substituted with 0, 10, 40, and 80% solid food waste vermicompost is illustrated in Figure 14.4. Damage ratings for eggplants due to two-spotted spider mites (*T. urticae*) in response to substitutions of 20% and 40% food waste vermicompost substitutions into MM 360 are shown in Figure 14.5. The development of aphid (*M. persicae*) populations in response to substitutions of 20% or 40% food waste vermicompost into MM 360 is summarized in Figure 14.6.

III SUPPRESSION OF ARTHROPOD PESTS BY AQUEOUS SOLUTIONS FROM VERMICOMPOSTS (TEAS)

A Cucumber Beetles and Tobacco Hornworms

Other research at the Soil Ecology Laboratory at The Ohio State University has demonstrated significant suppression of plant parasitic nematodes by soil drenches

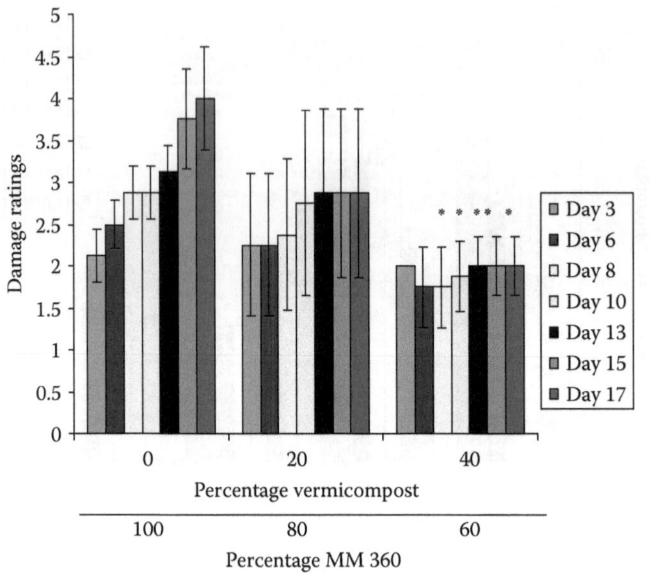

Figure 14.5 Ratings of damage by spider mites (*T.urticae*) (mean ± SE) to eggplants grown in the greenhouse in mixtures of food waste vermicompost and MM 360 and supplied with all needed nutrients. Populations were recorded on seven dates over 25 days after mealybug infestations. Means designated * are significantly different ($p \leq 0.05$).

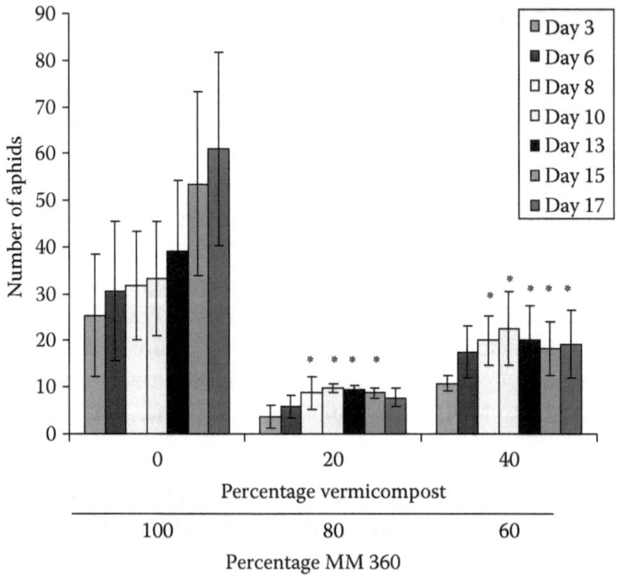

Figure 14.6 Development of aphid populations (*Myzus persicae*) on cabbages grown in the greenhouse in mixtures of food waste vermicompost and MM 360 and supplied with all needed nutrients. Measurements were made on seven dates over 12 days after infestation. Means designated * are significantly different ($p \leq 0.05$).

of aqueous extracts (teas) produced from food waste vermicomposts (Edwards et al. 2007) and of aboveground foliar pests by soil drenches of aqueous extracts (Edwards, Arancon, Vasko-Bennett, Askar, Keeney, and Little 2009; Edwards, Arancon, Vasko-Bennett, Askar, and Keeney 2009).

The greenhouse experiments that are reported in this Chapter describe the effects of aqueous extracts (teas) produced from food waste-based vermicomposts in commercial tea-brewing equipment on populations of and damage by cucumber beetles (*A. vittatum*) on cucumber plants (*Cucumis sativa*) and the caterpillars of tobacco hornworm (*M. sexta*) on tomato plants (*Lycopersicon esculentum*). The plants were grown in cages 40 × 40 × 40 cm; 2 mm mesh (15.7 × 15.7 × 15.7 in mesh) in the greenhouse. Eight cucumber beetles were released onto four cucumber plants per cage, or eight tobacco hornworms were released onto four tomato plants per cage. The soil in the pots in which the plants were growing was drenched with either deionized water (control) or 5%, 10%, or 20% aqueous vermicompost extracts weekly for 2 weeks.

The aqueous vermicompost extracts were prepared in commercial tea-brewing equipment from vermicomposts produced from supermarket food wastes. The ratio of vermicompost to water was one to four (v:v), to produce a 20% aqueous solution that could be diluted to 5% and 10% concentrations (see Chapters 11 and 13). The effects of vermicompost tea soil drenches applied at dilutions of 5%, 10%, and 20% were compared with those of deionized water, in the suppression of cucumber beetles (*A. vittatum*) attacking cucumbers and tobacco hornworms (*M. sexta*) attacking tomatoes in greenhouse-cage experiments. Tomatoes and cucumber seedlings were germinated and grown for four weeks in 25 cm diameter pots containing a soil-less growth medium (MM 360); they were thinned to four plants per pot. They were placed under mesh cages, with one pot containing four plants in each treatment cage. At germination, plants were treated with soil drenches of 5%, 10%, or 20% vermicompost extract or a deionized water control, to bring the medium to field capacity, and thereafter at weekly intervals. A complete Peter's Professional Nutrient Solution was applied weekly to all plants. In each experiment, eight cucumber beetles or eight tobacco hornworms were released onto the leaves of cucumber or tomatoes in each cage (i.e., two pests per test plant). All treatments were replicated four times per pest experiment, in a randomized complete-block design. Numbers of pests were counted and damage rated (0, none, to 5, total) on Days 1, 3, 5, 7, 9, 11, 13, and 14 after the release of pests into the cages.

The effects of the vermicompost tea soil drenches on the tobacco hornworm populations (after 1, 3, 5, 7, 9, 11, and 14 days) are summarized in Figure 14.7. The effects of the vermicompost tea soil drenches on cucumber beetles (after 1, 3, 5, 7, 9, 11, 13, and 14 days) are summarized in Figure 14.8.

All of the concentrations of vermicompost extracts tested suppressed the establishment of the two pests on the plants and the damage caused by them significantly ($p \leq 0.05$). The higher the concentration of the vermicompost aqueous extracts (teas) applied, the greater the suppression of the pests. We concluded that the most likely cause for the unpalatability of the plants to the pests was the uptake of soluble phenolic compounds into the plant tissues from the vermicompost aqueous extract (tea) soil drenches applied to MM 360. These compounds are known to make plants unattractive to pests and to affect pest reproduction and survival rates.

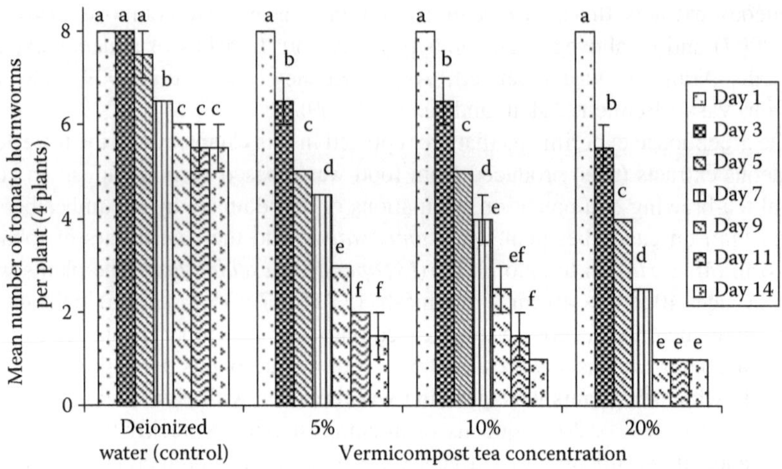

Figure 14.7 Effects of a range of soil drenches (*Manduca Sexta*) with aqueous extracts from food waste vermicomposts on numbers of tobacco hornworms attacking tomato plants in the greenhouse (means ± SE). Plants were grown in MM 360 with all needed nutrients supplied. Means followed by the same letters are not significantly different ($p \leq 0.05$).

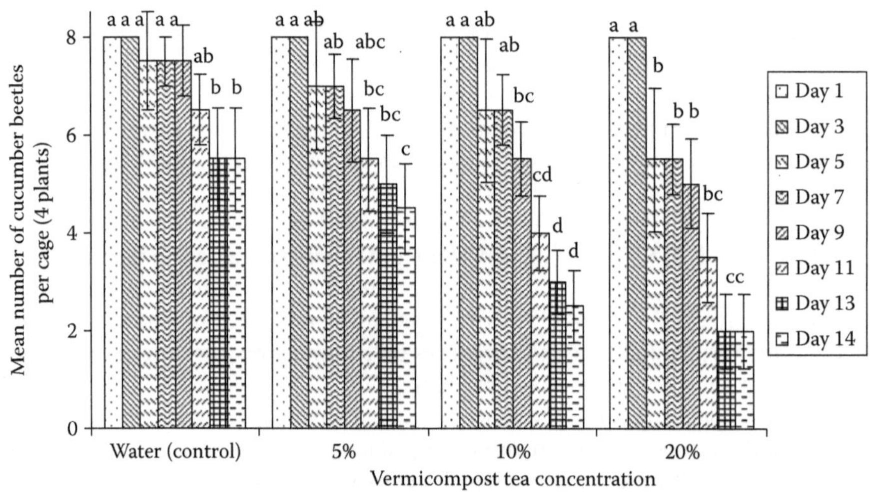

Figure 14.8 Changes in numbers of cucumber beetles (*Acalymma vittatum*) on cucumber plants treated with a range of food waste vermicompost aqueous extracts (teas) (means ± SE). Plants were grown in MM 360 with all needed nutrients supplied. Means followed by the same letter are not significantly different ($p \leq 0.05$).

B Aphids, Mealybugs, and Spider Mites

In another series of greenhouse experiments three species of arthropod pests—peach aphids (*Myzus persicae*), citrus mealybugs (*Planococcus citri*), and two-spotted spider mites (*Tetranychus urticae*)—were caged with tomato and cucumber plants grown in MM 360 to which a range of aqueous extracts (teas) from food waste vermicomposts were applied as soil drenches (Edwards, Arancon, Vasko-Bennett, Askar, Keeney, and Little 2009). The designs of the experiments were similar to those used with tobacco hornworms on tomatoes and striped cucumber beetles on cucumbers, except that 100 pests (25 per plant) were released into each cage instead of 8 pests.

Changes in aphid populations on tomatoes in response to three concentrations of food waste vermicompost teas are summarized in Figure 14.9. Changes in citrus mealybug populations on tomatoes in response to three concentrations of food waste vermicompost teas are shown in Figure 14.10. Vermicompost teas' effects on damage by two-spotted spider mites are presented in Figure 14.11.

The three arthropod pests that we tested in the second set of experiments, to assess the effects of the aqueous vermicompost extracts (teas) on arthropod pest suppression on tomatoes and cucumbers (green peach aphids, citrus mealybugs, and two-spotted spider mites), are all important pests of these crops. The availability

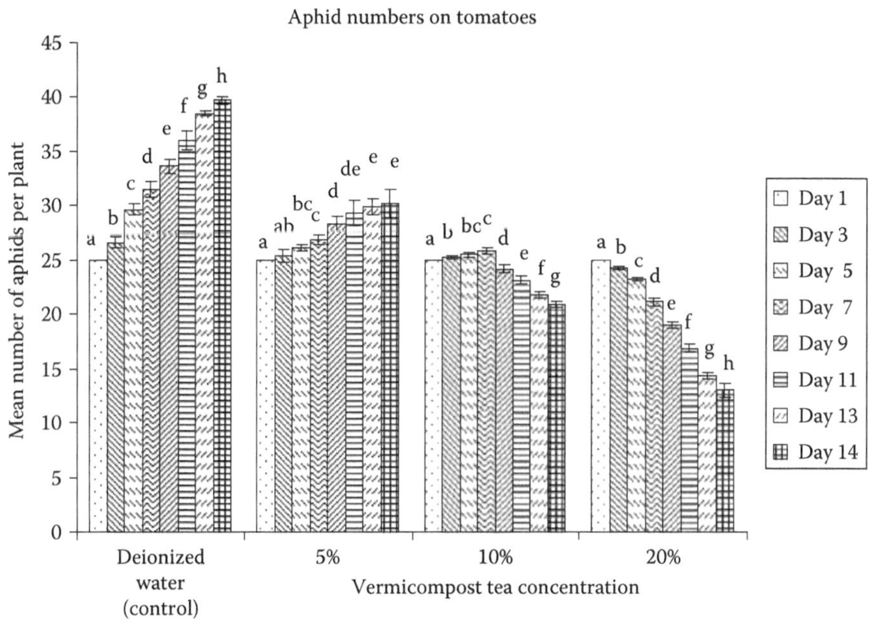

Figure 14.9 Effects of a range of aqueous vermicompost extracts or deionized water (control) on green peach aphid (*Myzus persicae*) numbers on tomatoes. All needed nutrients were supplied. Means followed by the same letters are not significantly different ($p \leq 0.05$).

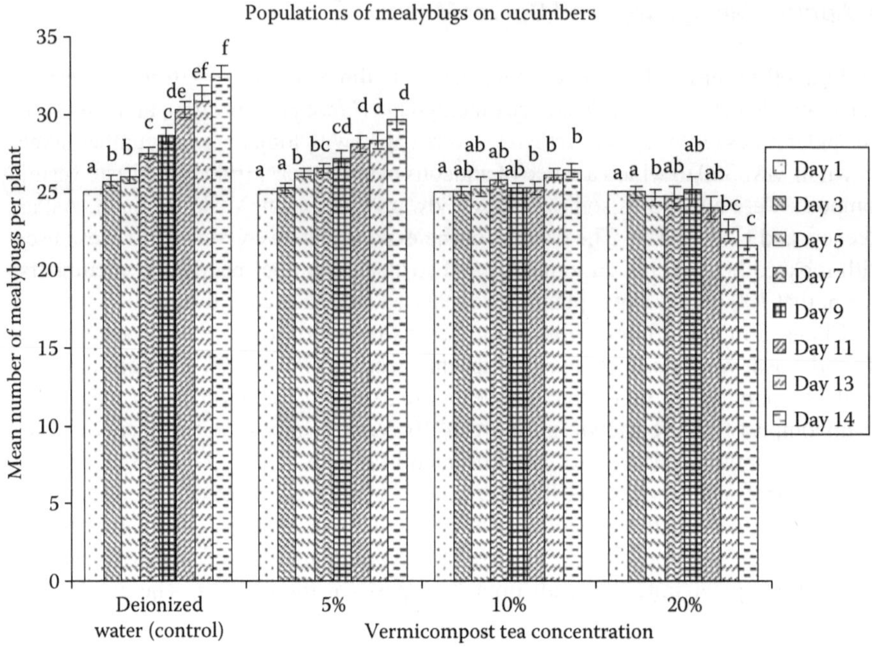

Figure 14.10 Effects of a range of concentrations of aqueous extracts from food waste vermicomposts on citrus mealybugs (*Planococcus citri*) on cucumbers. All needed nutrients were supplied (means ± SE). Means followed by the same letters are not significantly different ($p \leq 0.05$).

of such innovative organic control measures, in the form of aqueous vermicompost extracts (teas), would be particularly welcome to organic growers who are prohibited from using inorganic pesticides on their crops. Vermicomposting is a process that can be carried out at a range of scales using fairly simple to high technologies (Edwards and Arancon 2004), using a variety of relatively inexpensive equipment for vermicompost extract preparation, which is available commercially.

The overall effects of the aqueous vermicompost solutions on both numbers and damage by all three of these pests were dramatic, significant, and consistent on both crops, tomatoes and cucumbers. Clearly, weekly applications of these aqueous extracts, or teas, to tomatoes and cucumbers as soil drenches had three major effects on the arthropod pests. Firstly, since all three species of pests tested in the experimental cages had a free choice to infest any of the test plants, it seems that all application concentrations of the aqueous extracts made both tomatoes and cucumber plants much less attractive for feeding to all three species of pests. The highest application concentration of 20% aqueous extract tea had major impacts on the extent of infestation by all three pests. In a number of the experiments even the 5% extracts made the plants relatively unattractive to the pests.

Secondly, the weekly soil drenches of vermicompost extracts applied to the soilless medium in which the plants grew must also have interfered with the reproduction

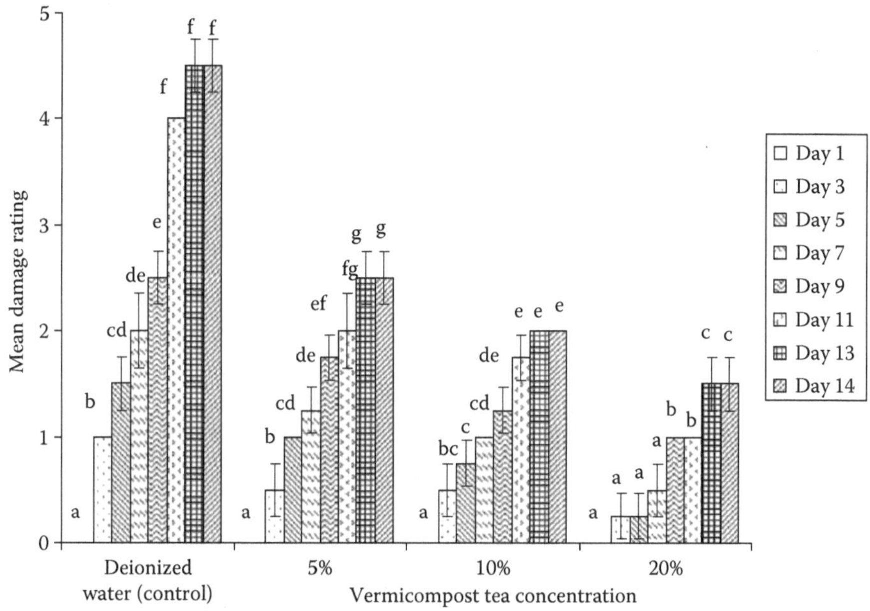

Figure 14.11 Damage ratings (0–5 total) of two-spotted spider mites (*Tetranychus urticae*) on cucumbers treated with a range of soil-applied vermicompost aqueous extracts or deionized water (control). All plants received all needed nutrients. (Means ± S.E) Means followed by the same letters are not significantly different ($p \leq 0.05$).

patterns of the green peach aphids, citrus mealybugs, and two-spotted spider mites, since the increased rates of application of an aqueous vermicompost extracts decreased the rates of reproduction of all three pests. The numbers of all three species leveled off or decreased, particularly in response to the higher-application rates of extracts. Finally, there was consistent evidence that the higher-application rates of vermicompost extracts caused the pests to either leave the plants or die, since overall numbers of the pests on the crops decreased significantly with time in response to these higher-application rates of vermicompost aqueous extracts.

IV MECHANISMS BY WHICH VERMICOMPOSTS AND VERMICOMPOST AQUEOUS SOLUTIONS (TEAS) SUPPRESS ARTHROPOD PESTS

A Solid Vermicomposts

It was discussed in Chapters 9 and 14 that the microbiological properties of vermicomposts can render soil suppressive to arthropod pests. The mostly likely mechanisms of arthropod suppression stem from improvements in the crop's ability to tolerate, if not deter, attacks from arthropods. These were demonstrated by the marked decreases of the populations of these arthropod pests in a few days after they been introduced onto plants that had been grown with solid vermicomposts. Depending on the pest species and the crop, the amounts, timing, and types of inorganic or organic fertilization used can either stimulate or suppress pest populations. Excess fertilization, especially with N, can promote succulent vegetative growth that often provides a microenvironment favorable for increased development of pests and damage.

Our results demonstrate that substitution of solid vermicomposts into MM 360 had considerable influences on the intensity of attacks by aphids, mealybugs, spider mites, cucumber beetles, and tobacco hornworms. It is unlikely that vermicomposts provided some essential nutrient elements that could have either increased the plant's resistance to pests or made the plants less palatable to pests because all plants received balanced amounts of nutrients.

Although there is an extensive literature on the use of organic wastes to reduce the incidence of insect infestations, relatively few research reports deal with the suppression of insect pest attacks by vermicomposts. Ramesh (2000) reported decreased attacks by sucking pests in response to vermicomposts. Rao et al. (2001) and Rao (2002) reported considerable decreases in populations of jassids, aphids, coccinellid beetles, and spider mites on groundnuts grown in soils amended with vermicomposts, compared to those grown in soils amended with inorganic fertilizer. Biradar et al. (1998) reported decreases in populations of *Heteropsylla cubana* after growing *Leucaena leococephala* with vermicomposts.

Variations in the nutritional status of plant tissues, with respect to nutrients other than N, could have influenced pest resistance or caused increased susceptibility to insect attacks, due to deficiencies of K, Ca, Bo, Zn, and Si (Patriquin et al. 1995). The assimilation of nutrients could have followed a different pattern between vermicompost-treated plants and those of inorganically fertilized plants, and this could have affected the production of endogenous secondary metabolites that have known roles in the suppression of attacks from arthropod pests by modifying their feeding habits and affecting their digestive systems. One of the known plant secondary metabolites that have deterrent effects on arthropod attacks are the phenolic compounds. It is well known that phenolic substances are distasteful to secondary decomposers in soil systems and inhibit the breakdown of dead plant materials by invertebrates (Heath and Edwards 1964). Simmonds (1998) reviewed the modification of insect feeding behavior by phenolics and nonprotein amino acids as well as the general inhibition of insect pest feeding.

B Vermicompost Teas

The pests that we tested in our vermicompost aqueous extract (tea) experiments to assess these extracts' effects on pest suppression were green peach aphids, citrus mealybugs, two-spotted spider mites, cucumber beetles, and tobacco hornworms, which are all important pests of these crops. The amount of suppression of all these pests by the aqueous vermicompost extracts was dramatic. Innovative organic control measures, based on aqueous vermicompost extracts (teas), would be particularly welcome to organic growers, who are prohibited from using inorganic pesticides on their crops.

The overall effects of the aqueous vermicompost extracts, or teas, on both numbers of and damage by aphids, mealybugs, spider mites, cucumber beetles, and tobacco hornworms were dramatic, significant, and consistent on both tomatoes and cucumbers. Clearly, weekly applications of these aqueous extracts to tomatoes and cucumbers as soil drenches produced three major effects. Firstly, since all three species of sucking pests in experimental cages had a free choice to infest the test plants, it seems that all application rates of the aqueous extracts made both tomato and cucumber plants much less attractive to all of the pests (Figures 14.7 through 14.11). The highest application rate of 20% aqueous extract stopped virtually all pest infestations and the 10% extract had major impacts on the extent of infestation. In some of the experiments even the 5% aqueous extracts or teas made the plants relatively unattractive to pests.

Secondly, the weekly soil drenches of vermicompost extract to the soil in which the plants grew must also have interfered with the reproduction patterns of aphids, mealybugs, and spider mites-(Figures 14.9 through 14.11), since the increases in rates of application of teas decreased the rates of reproduction of the pests, because numbers of all three species leveled off or decreased, especially in response to the higher-application rates of teas. Finally, there was consistent evidence that the higher application rates of vermicompost extracts caused the pests to either leave the plants or die, since total numbers of pests on the crops decreased significantly with time in response to the higher-application rates of teas.

This raises the question of possible mechanisms of how these vermicompost teas may influence the responses of the pests when the teas are taken up into the plants. Many reports in the literature describe organic nutrient sources as decreasing numbers of pest arthropods (Culliney and Pimentel 1986; Eigenbrode and Pimentel 1988; Patriquin et al. 1995; Morales et al. 2001; Yardim and Edwards 2003; Phelan 2004). It has also been suggested that these effects are due to the uptake into plants of phenols from organic manures (Ravi et al. 2006). However, although such mechanisms may account for the suppression of pest attacks by solid vermicomposts, they do not account for suppression by teas. Teas are applied to the soil in which the crops grow as drenches and materials; they contain materials that can pass readily from the solid vermicomposts into the teas, including soluble nutrients, free enzymes, a wide range of microorganisms, and water-soluble phenols.

Vermicomposts support much greater microbial diversity and activity than thermophilic composts, because organic wastes that have been fragmented by earthworms have a much greater surface area and can support considerably greater microbial activity. To consider the mechanisms of pest suppression by teas further, it is important to discuss which of the materials passing into the extracts could be responsible for the pest suppression, since the pests' responses must depend on uptake of these materials into the tomato and cucumber plants from the soil drenches. The suppression could not be caused by uptake of soluble nutrients since all of the plants were supplied regularly with all the nutrients that they needed from Peter's Nutrient Solution, which was applied to the experimental plants three times a week.

Possibly, some free enzymes in the aqueous extract could have influenced the pest suppression, but this is very unlikely on the scales and consistencies of suppression seen in these experiments. For instance, the enzyme chitinase has been reported to be present in some vermicomposts (Hahn 2001), and it is feasible that this could affect arthropod pest molting, but there seem to be no reports in the literature of this enzyme having any effects on pests. We reviewed possible mechanisms by which microorganisms that were taken up into the tissues of plants could influence arthropod pest feeding but could not find any relevant reports.

The other possible mechanism by which vermicomposts and similar organic materials might suppress attacks by arthropod pests on crop plants could be to change the pests' feeding responses, due to soluble phenolic substances that could be taken up from soil drenches of teas into plants. It is well known that soluble phenolic substances are distasteful to invertebrate organic-matter decomposers and can inhibit the breakdown of dead and decaying plant materials by invertebrates in soils (Edwards and Heath 1963; Heath and Edwards 1964). An endogenous phenoloxidase enzyme has been extracted from an earthworm, *Lumbricus rubellus*, that is sometimes used in vermicomposting, and this compound can bioactivate organic compounds to form phenols, such as *p*-nitrophenol (Park et al. 1996). Polychlorinated phenols and their metabolites have been reported from a range of soils containing earthworms (Knuutinen et al. 1990). Vinken et al. (2005) reported that monomeric phenols could be absorbed by humic acids in the gut of earthworms.

Phenolics have been identified as common insect antifeedants (Koul 2008). Stevenson et al. (1993) reported inhibition of the development of *Spodoptera litura* caterpillars by a phenolic compound present in wild groundnuts. Summers and Felton (1994) proposed that *Lepidoptera* larval feeding was decreased by oxidative stress caused by phenolic compounds in plants. Haukioja et al. (2002) reported that phenolic materials in plant tissues decreased the rates of consumption of tissues by a geometrid caterpillar, *Epirrita autumnata*. Phenols deterred feeding by southern armyworms, *Spodoptera eridania* (Lindroth and Peterson 1988). Kurowska et al. (1990) reviewed the effects of 46 phenols as insect repellents and feeding deterrents and concluded that many can have significant effects on foliar pest attacks. Bhonwong et al. (2009) reported that polyphenol oxidases in tomatoes could produce resistance to cotton bollworms (*Helicoverpa armigera*) and beet armyworms (*Spodoptera exigua*). They concluded that these chemicals had a clear feeding-deterrent effect on pests. Tomato phenol oxidase in tomato plants slowed down feeding and rates of growth of the common cutworm, *Spodoptera litura* F. (Mahanil et al. 2008). Plant phenolics affected the rates of development and survival of the autumn moth, *Epirrita autumnata* (Hawida et al. 2007).

Chrzanowski (2008) reported that phenolic acids in blackcurrant and sour cherry leaves retarded the reproduction and fecundity of grain aphids (*Sitobian avenae*). Eleftherianos et al. (2006) reported that changes in the levels of phenols in maize and barley plants decreased the fecundity of the cereal aphids *Rophalosiphum padi* and *S. avenae*. Galls induced by *Pemphigus populi* aphids on *Populus nigra* were suppressed by large amounts of phenols in the plant tissues (El-Akkad and Zalat 2000).

These diverse results all point to the probability that water-soluble phenols that are extracted from the vermicompost during aqueous extractions and may be taken up into plants from soil receiving drenches of vermicompost teas could be the most likely mechanisms by which vermicompost aqueous extracts can suppress pest attacks. We hypothesize that the decreases in insect pest numbers and in damage to plants grown with vermicompost teas that we recorded could be attributed to the presence of water-soluble phenolic compounds in plants grown with vermicompost tea applications that make the plants less attractive to pests and interfere with their reproduction.

Although these conclusions are based only on circumstantial evidence, the evidence makes it seem extremely likely that water-soluble phenols, passing from vermicomposts into teas and then into plants, may be the main mechanism by which vermicompost aqueous extracts influence the suppression of pest feeding, reproduction, and mortality by aphids, mealybugs, and spider mites reported in this chapter.

V EFFECTS OF VERMICOMPOSTS ON SOIL NEMATODE COMMUNITIES

There is a very extensive scientific literature demonstrating that additions of various kinds of organic matter to soils tends to decrease populations of plant parasitic nematodes (Addabdo 1995; Akhtar 2000; Akhtar and Malik 2000). Also, a number of reports indicate that traditional thermophilic composts can suppress plant parasitic nematode populations (McSorley and Gallaher 1995; Sipes et al. 1999; Miller 2001). Reneco et al. (2009) reported nematode suppression by two composts applied to soils at various doses (0%, 1%, 2.5%, 5%, and 10% w/w). There was a high-negative correlation between the numbers of nematodes and the application rate. There have also been reports in the scientific literature of vermicomposts suppressing populations of plant parasitic nematodes. Swathi et al. (1998) demonstrated that 1.0 kg m^{-3} (1.7 $lb.yd^{-3}$) of vermicompost suppressed attacks of *Meloidogyne incognita* in tobacco plants. Morra et al. (1998) reported partial control of *M. incognita* by vermicomposts in a tomato-zucchini rotation. Ribeiro et al. (1998) reported that vermicomposts decreased the numbers of galls and egg masses of *M. javanica*. A number of other reports show vermicomposts as suppressing the plant parasitic nematode *M. incognita* (Pandey 2005; Saikia et al. 2007; Masheva et al. 2009).

Research in the Soil Ecology Laboratory at The Ohio State University (Arancon et al. 2002, 2003) involved field experiments that showed that solid vermicomposts had dramatic effects on nematode populations and communities in field soils in which tomatoes (*L. esculentum*), bell peppers (*Capsicum annuum grossum*), strawberries (*Fragaria ananassa*), or grapes (*Vitis vinifera*) were grown.

Plots were replicated four times for each treatment with one of three different types of vermicompost: Vermicomposts based on commercial cattle manure, food waste, and paper waste were applied at rates of 40 $ton.ha^{-1}$ (20 $t.acre^{-1}$) to tomatoes or peppers. Food waste- and paper waste-based vermicomposts were applied at 2.5 $ton.ha^{-1}$ or 5 $ton.ha^{-1}$ (5 $t.acre^{-1}$ or 10 $t.acre^{-1}$) to strawberries and grapes. One set of control replicate plots received the full recommended rate of inorganic fertilizer for all crops. A second set received recommended rates of traditional thermophilic composts in the tomato and pepper experiments. The grape experiment had one set of replicated control plots treated with the full recommended rates of inorganic fertilizer and another with no fertilizer. All of the vermicompost- and compost-treated plots were supplemented with inorganic fertilizers to balance the initial nutrient supply, as far as possible, with that in the inorganic fertilizer treatment (except for the unfertilized control). Vermicomposts, composts, and inorganic fertilizers were surface-applied to the top 15 cm (6 in) of the beds in the tomato, pepper, grape, and strawberry plots and incorporated with a rototiller, and vermicompost treatments were surface-applied incorporated with a rototiller on and covered with a straw mulch in the grape plots.

The effects of the vermicomposts and inorganic fertilizers on field populations of plant parasitic nematodes are summarized in Figure 14.12, and those on fungivorous nematodes in Figure 14.13. In all of the vermicompost-treated soils, populations of plant parasitic nematodes were suppressed significantly for all crops ($p < 0.05$) compared to those in plots treated with inorganic fertilizers only. Populations of plant parasitic nematodes were consistently larger in plots treated with inorganic fertilizer than in those to which no inorganic fertilizer was applied and only vermicompost

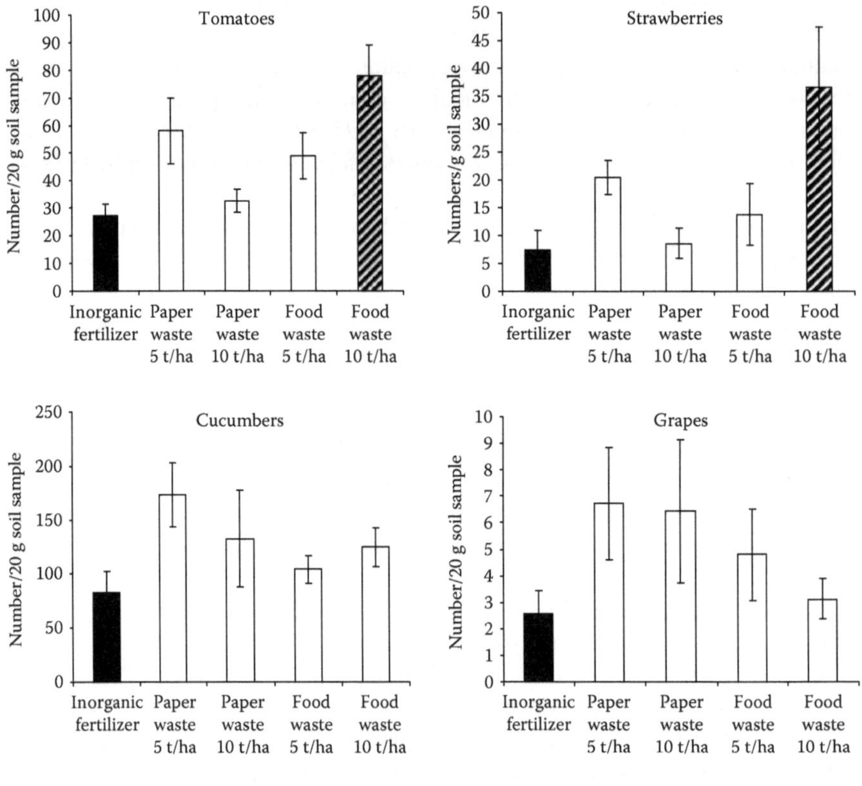

Figure 14.12 Effects of inorganic fertilizers, thermophilic composts, and vermicomposts on populations of plant parasitic nematodes (means ± SE).

was used (Figure 14.12). The populations of fungivorous nematodes were significantly larger in response to most of the vermicompost treatments, compared to those in plots treated with only inorganic fertilizer, in the tomato, cucumber, strawberry, and grape experiments (Figure 14.13). The effects increase the competitive ability of fungivorous nematode populations compared to plant parasitic nematodes.

However, the mechanisms that caused these changes in nematode populations are still only speculative. Predator–prey interactions may have contributed to the lower density of plant parasitic nematodes in vermicompost-treated plots. Vermicomposts may increase numbers of predatory or omnivorous nematodes, or arthropods such as mites, which prey selectively on plant parasitic nematodes (Bilgrami 1996). Vermicomposts can promote the growth of nematode-trapping fungi and fungi that attack nematode cysts and may thereby influence populations of plant parasitic nematodes (Kerry 1988). Additionally, some rhizobacteria colonize roots and kill plant parasitic nematodes by producing enzymes and toxins that are toxic to them (Siddiqui and Mahmood 1999), and nematodes may be killed by the release of toxic substances such as hydrogen sulfide, ammonia, and nitrites, produced during vermicomposting (Rodriguez-Kabana 1986).

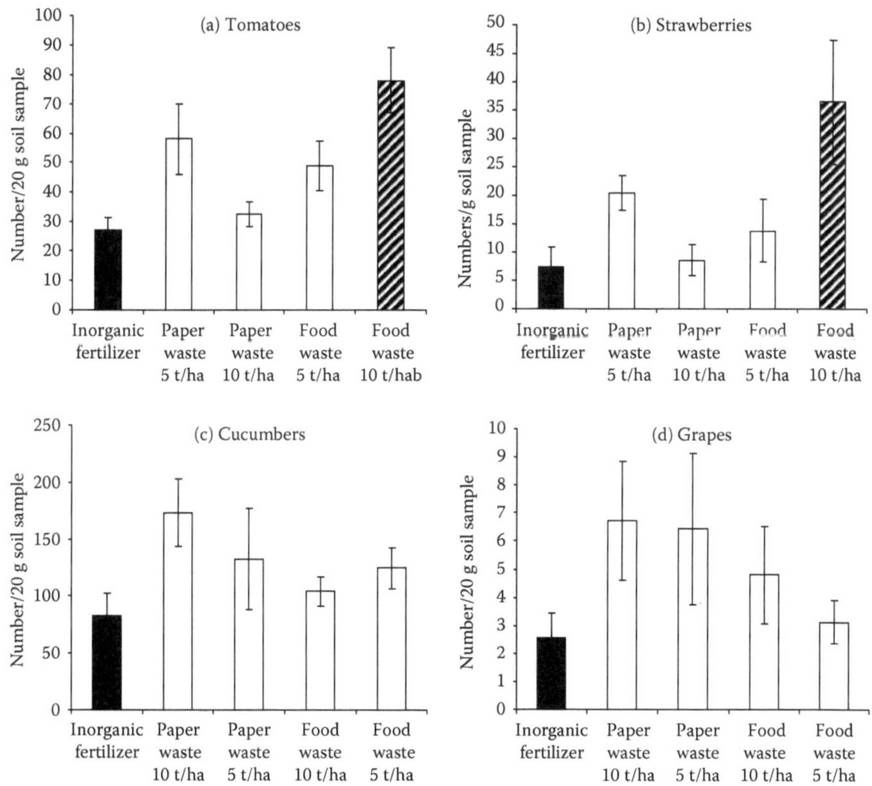

Figure 14.13 Effects of inorganic fertilizers, thermophilic composts, and vermicomposts on populations of fungivorous nematodes (means ± SE).

We found the greatest influence of vermicomposts to be on fungivorous nematode populations. Fungi serve as a major source of food for earthworms, and earthworms facilitate dispersal of fungi by excreting fungal spores in their casts (Edwards and Fletcher 1988), which may explain the greater impact of vermicomposts on fungivorous than on bacterivorous nematode populations.

VI SUPPRESSION OF *MELOIDOGYNE HAPLA* ON TOMATOES BY AQUEOUS EXTRACTS FROM VERMICOMPOSTS (TEAS)

Meliodogyne hapla is a serious nematode pest of many crops and is widely distributed globally. In greenhouse experiments, 6-week-old tomato seedlings were transplanted into 10 cm diameter pots containing a sand:loam soil (1:3 v:v) mixture treated with the experimental treatments. Drench treatments of 5%, 10%, or 20% food waste vermicompost tea or a control deionized water treatment were applied at transplanting and every 2 weeks thereafter. One week later 10,000 *M. hapla* eggs were added to each pot in suspension in deionized water. The eggs were collected from cultures maintained on tomatoes. Each treatment was replicated four times. Pots were arranged on benches in a completely randomized design. The greenhouse was maintained at 25°C (77°F). Plants were watered with deionized water and Peter's Nutrient Solution three times weekly. Thirty days after infestation, soil was removed from the pots to assess the extent of root damage and the number of root galls. The washed roots were rated for the number of root knot galls and the number of galls per unit wet weight of roots assessed.

The effects of food waste vermicomposts and 20% thermophilic compost teas on the mean dry shoots of tomato plants inoculated with *M. hapla* are summarized in Figure 14.14. All the three application rates of food waste vermicompost tea increased the shoot weights of tomato plants significantly ($p \leq 0.05$), compared with the deionized water control. The three vermicompost tea application rates all increased mean leaf areas significantly ($p \leq 0.05$; Figure 14.15) compared with the deionized water control. The growth of tomato plants inoculated with *M. hapla* and treated with a range of vermicomposts or 20% thermophilic compost are illustrated in Figure 14.16. The differences in tomato plant growth between treatments, especially in response to the vermicompost teas, were spectacular. The higher the concentration of vermicompost tea, the less was the response of the tomato plant to the nematode; this demonstrates clearly the vermicomposts' dramatic suppression of the nematode populations and damage.

The number of nematode root knot galls on tomato plants in response to the vermicompost tea applications is illustrated in Figure 14.17. The suppression of the nematode galls in response to application of the vermicompost teas was extremely dramatic and significant ($p \leq 0.05$) and led to considerable increases in tomato growth. Clearly, the effects of the food waste vermicompost teas on the suppression of *M. hapla* damage were dramatic, and so was the suppression of the number of galls on the tomato roots, as is illustrated in Figure 14.17. Since the roots of the plants grown with ionized water were so small compared with those receiving vermicompost teas,

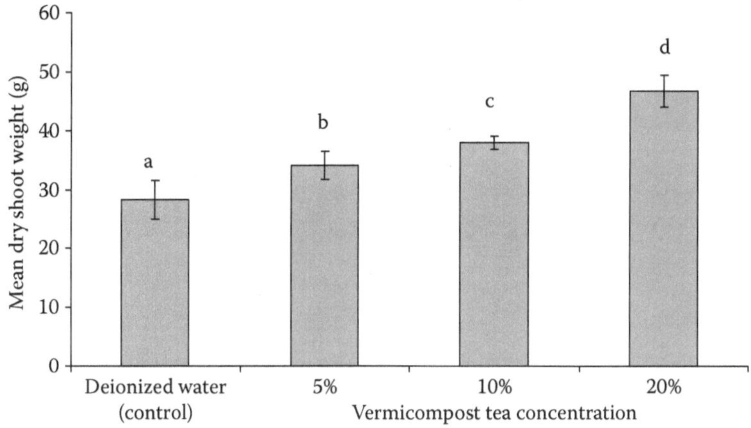

Figure 14.14 Mean dry shoot weights of tomatoes grown in MM 360 with all needed nutrients supplied, infested with *Meloidgyne hapla*, and treated with soil drenches of vermicompost tea (mean ± SE). Columns with different letters are significantly different ($p \leq 0.05$).

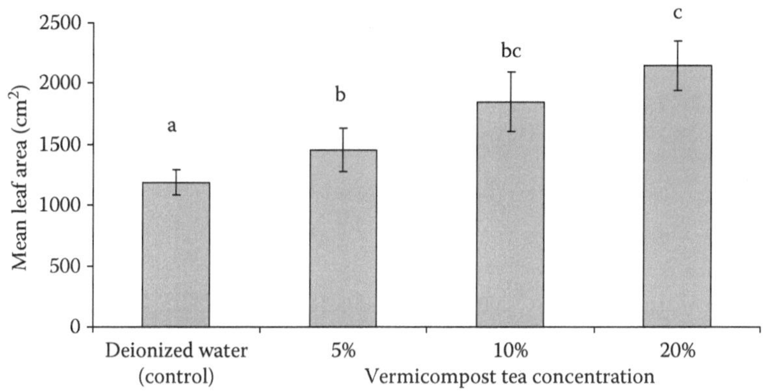

Figure 14.15 Mean leaf areas of tomato plants grown in MM 360 with all needed nutrients, infested with *Meloidgyne hapla*, and treated with soil drenches of vermicompost tea (mean ± SE). Columns with different letters are significantly different ($p \leq 0.05$).

the data were expressed as the number of galls per gram of roots (wet weight), but the size of the roots is a more relevant index of the overall damage.

VII SUPPRESSION OF *MELOIDOGYNE HAPLA* ON CUCUMBERS BY AQUEOUS EXTRACTS FROM VERMICOMPOSTS (TEAS)

In the experiment assessing the effects of food waste vermicomposts on *M. hapla* on cucumbers, the same experimental procedures were followed as in the first experiment on the effects of vermicompost aqueous extracts on *M. hapla* on tomatoes. The

Figure 14.16 A comparison of tomato plants infested with *Meloidogyne hapla* and treated with thermophilic compost teas or a range of food waste vermicompost teas. Plants were grown in MM 360 with all needed nutrients supplied. (VC-vermicompost.)

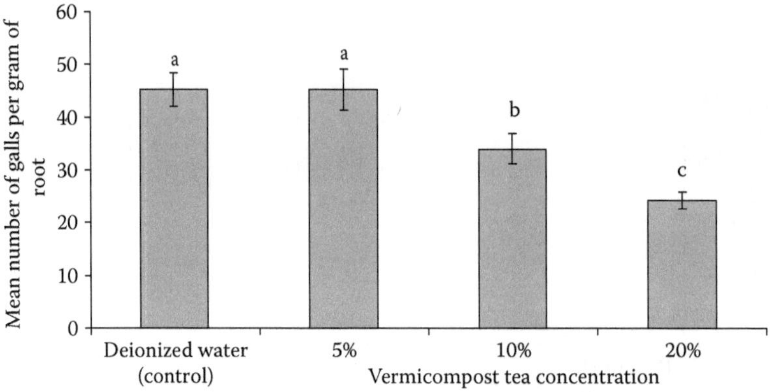

Figure 14.17 Mean numbers of *Meloidgyne hapla* galls (mean ± SE) on tomato roots infested with the nematodes and treated with soil drenches of vermicompost tea. Columns with different letters are significantly different ($p \leq 0.05$). All plants were grown in MM 360 and received all needed nutrients.

procedure for infesting the cucumber plants by adding *M. hapla* eggs was the same as used previously for the tomato experiments on this nematode species. The mean increased dry shoot weights above aboveground plants in response to vermicompost applications are shown in Figure 14.18, and the mean increased leaf areas of the treated plants are summarized in Figure 14.19.

The numbers of galls on the cucumber roots (Figure 14.20) were decreased significantly ($p \leq 0.05$) by the 10% and 20% application rates of food waste vermicompost teas compared with the water control ($p \leq 0.05$). Similarly, only the 10% and 20% tea applications increased mean leaf areas (Figure 14.19). Only the 10% and

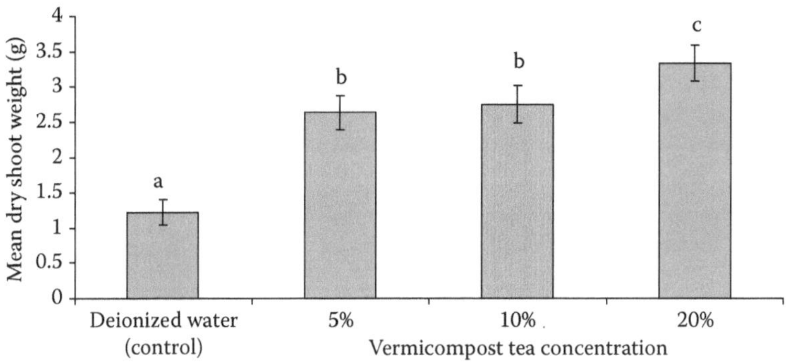

Figure 14.18 Mean dry shoot weights of cucumbers grown in MM 360 with all needed nutrients supplied, infested with *Meloidgyne hapla*, and treated with soil drenches of vermicompost tea (mean ± SE). Columns with different letters are significantly different ($p \leq 0.05$).

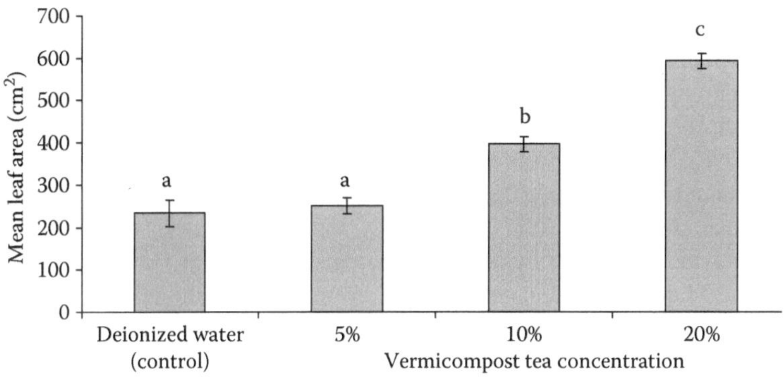

Figure 14.19 Mean leaf areas of cucumbers infested with *Meloidgyne hapla*, grown in MM 360 with all needed nutrients, and treated with soil drenches of vermicompost tea (mean ± SE). Columns with different letters are significantly different ($p \leq 0.05$).

20% food waste vermicompost tea applications increased cucumber leaf areas and fresh shoot weights (Figure 14.18) significantly ($p \leq 0.05$). However, all three applications of food waste vermicompost teas increased fresh root weights significantly compared with the water controls ($p \leq 0.05$). In terms of the heights of the cucumber plants (Figure 14.18), all three application rates (5–10, and 20%) increased the mean heights of the cucumber plants significantly (Figure 14.19).

VIII MECHANISMS BY WHICH VERMICOMPOST TEAS MAY SUPPRESS PLANT PARASITIC NEMATODE POPULATIONS

The mechanisms that caused changes in plant parasite nematode populations as a result of the applications of organic-matter amendments, including vermicomposts,

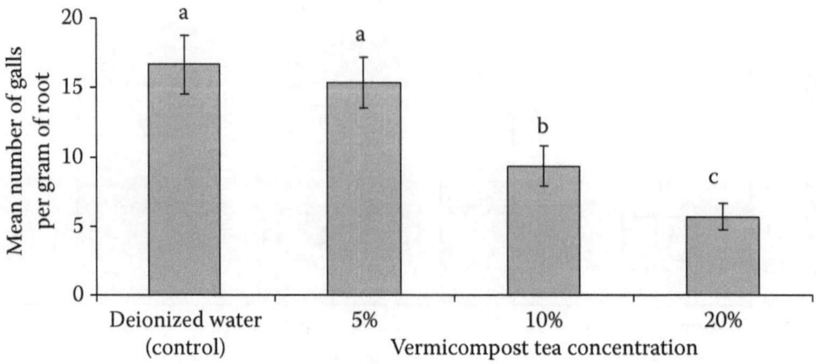

Figure 14.20 Mean numbers of *Meloidgyne hapla* galls (mean ± SE) on cucumber roots infested with the nematode and treated with soil drenches of vermicompost teas or deionized water. Plants were grown in MM 360 and received all needed nutrients.

are still only speculative. Greater predator–prey pressures may also have contributed to the lower-population density of plant parasitic nematodes in vermicompost-treated plots. Vermicomposts may increase numbers of predatory or omnivorous nematodes, or arthropods such as mites, that prey selectively on plant parasitic nematodes (Bilgrami 1996). Vermicomposts can promote the growth of nematode-trapping fungi and fungi that attack nematode cysts and may thereby influence populations of plant parasitic nematodes (Kerry 1988). Additionally, some rhizobacteria colonize roots and kill plant parasitic nematodes by producing enzymes and toxins that are toxic to them (Siddiqui and Mahmood 1988); nematodes may be killed by the release of toxic substances such as hydrogen sulfide, ammonia, and nitrites, produced during vermicomposting (Rodriguez-Kabana 1986). We found the greatest influence of vermicomposts to be on fungivorous nematode populations. Fungi serve as a major source of food for earthworms, and earthworms facilitate dispersal of fungi by excreting fungal spores in their casts (Edwards and Fletcher 1988), which may explain the greater impact of vermicomposts on fungivorous than on bacterivorous nematode populations.

REFERENCES

Addabdo, T.D. 1995. The nematicidal effect of organic amendments: A review of the literature 1982–1994. *Nematol. Mediterr.* 23: 299–305.

Akhtar, M. 2000. Approaches to biological control of nematode pests by natural products and enemies. *J. Crop Prod.* 3: 367–395.

Akhtar, M., and Malik, A. 2000. Role of organic amendments and soil organisms in the biological control of plant parasitic nematodes: A review. *Biores. Technol.* 74: 35–47.

Altieri, M.A., and Nicholls, C. 2003. Soil fertility and insect pests: Harmonizing soil and plant health in agroecosystems. *Soil Tillage Res.* 72: 203–211.

Arancon, N.Q., and Edwards, C.A. 2004. Vermicomposts suppress plant pest and disease attacks. *Biocycle* March: 51–53.

Arancon, N.Q., Edwards, C.A., Bierman, P., Metzger, J.D., and Lucht, C. 2005. Effects of vermicomposts produced from cattle manure, food waste and paper waste on the growth and yields of peppers in the field. *Pedobiologia* 49: 297–306.

Arancon, N.Q., Edwards, C.A., Oliver, T.J., and Byrne, R.J. 2007. Suppression of two-spotted spider mite (*Tetranychus urticae*) mealy bugs (*Pseudococcus spp*) and aphid (*Myzus persicae*) populations and damage by vermicomposts. *Crop Prot.* 26: 26–39.

Arancon, N., Edwards, C.A., Yardim, F., and Lee, S. 2002. Management of plant parasitic nematodes by use of vermicomposts. In *Proceedings of Brighton Crop Protection Conference- Pests and Diseases*, 8B-2: 705–710.

Arancon, N.Q., Yardim, E., Edwards, C.A., and Lee, S. 2003. The trophic diversity of nematode communities in soils treated with vermicomposts. *Pedobiologia* 47: 736–740.

Atkinson, D., Litterick, A.M., Walker, K.C., Walker, R., and Watson, C.A. 2004. Crop protection—what will shape the future picture? *Pest Manage. Sci.* 60: 105–112.

Bhonwong, A., Stout, M.J., Attajarusit, J., and Tantasawat, P. 2009. Defensive role of tomato polyphenol oxidases against collar bollworm (*Helicoverpa armigera*) and beet armyworm (*Spodoptera exigua*). *J. Chem. Ecol.* 35: 28–38.

Bilgrami, A.L. 1996. Evaluation of the predation abilities of the mite *Hypoaspis calcuttaensis*, predaceous on plant and soil nematodes. *Fundam. Appl. Nematol.* 20: 96–98.

Biradar, A.P., Sunita, N.D., Teggel, R.G., and Devarandavadgi, S.B. 1998. Effect of vermicompost on the incidence of subadult psyllid. *Ins. Environ.* 4: 55–56.

Chellemi, D.O. 2002. Nonchemical management of soilborne pests in fresh market vegetable production systems. *Phytopathology* 92: 1367–1372.

Chrzanowski, G. 2008. Influence of phenolic acids isolated from blackcurrant and sour cherry leaves on grain aphids (*Sitobion avenae* F.). *Pol. Pestycydy* 1–2: 127–133.

Culliney, T.W., and Pimentel, D. 1986. Ecological effects of organic agricultural practices on insect populations. *Agric. Ecosyst. Environ.* 15: 253–256.

Edwards, C.A., and Arancon, N.Q. 2004. The use of earthworms in the breakdown of organic wastes to produce vermicomposts and animal feed protein. In *Earthworm Ecology*, ed. C.A. Edwards, 345–438. 2nd ed. CRC Press, Boca Raton, FL.

Edwards, C.A., Arancon, N.Q., Emerson, E., and Pulliam, R. 2007. Suppressing plant parasitic nematodes and arthropod pests with vermicompost "teas". *Biocycle* 48: 38–39.

Edwards, C.A., Arancon, N.Q., Vasko-Bennett, M., Askar, A., Keeney, G., and Little, B. 2009. Suppression of green peach aphid (*Myzus persicae*) (Sulz.), citrus mealy bug (*Planococcus citri*) (Risso), and two spotted spider mite (*Tetranychus urticae*) (Koch.) attacks on tomatoes and cucumbers by aqueous extracts from vermicomposts. *Crop Prot.* 29: 80–93.

Edwards, C.A., Arancon, N.Q., Vasko-Bennett, M., Askar, A., and Keeney, G. 2009. Effect of aqueous extracts from vermicomposts on attacks by cucumber beetles (*Acalymna vittatum*) (Fabr.) on cucumbers and tobacco hornworm (*Manduca sexta*) (L.) on tomatoes. *Pedobiologia,* 53: 141–1481.

Edwards, C.A., and Fletcher, K.E. 1988. Interactions between earthworms and microorganisms in organic matter breakdown. *Agric. Ecosys. and Environ.* 20: 235–249.

Edwards, C.A., and Heath, G.W. 1963. The Role of Soil Animals in Breakdown of Leaf Material. In *Soil Organisms*, ed. J. Doeksen and J. van der Drift, 76–84. North Holland, Amsterdam.

Eigenbrode, S.D., and Pimentel, D. 1988. Effects of manure and chemical fertilizers on insect pest populations on collards. *Agric. Ecosyst. Environ.* 20: 109–125.

El-Akkad, S., and Zalat, S. 2000. *Populus* galls induced by *Pemphigus* aphids in Sinai. Egypt. *J. Biol.* 2: 15–19.

Eleftherianos, I., Vamvatsikos, P., Wards, D., and Gravanis, F. 2006. Changes in the levels of plant total phenols and free amino acids induced by two cereal aphids and effects on aphid fecundity. *J. Appl. Entomol.* 130: 15–19.

Hahn, G.E. 2001. Methods of using worm castings for insect repellency. US Patent 6475503.

Haukioja, E., Ossipov, V., and Lempa, K. 2002. The interactive effects of leaf maturation and phenolics on consumption and growth of a geanetrid moth. *Futrand. Exp. Appl.* 104: 125–136.

Hawida, S., Kapari, L., Ossipov, V., Ramtala, M.J., Ruuhola, T., and Haukioja, E. 2007. Foliar phenolics are differently associated with *Epirrita autumnata* growth and immunocompetence. *J. Chem. Ecol.* 33: 1013–1023.

Heath, G.W., and Edwards, C.A. 1964. Some methods for assessing the activity of soil animals in the breakdown of leaves. *Pedobiologia* 4: 80–87.

Kerry, B. 1988. Fungal parasites of cyst nematodes. In *Biological Interactions in Soil*, ed. C.A. Edwards, B.R. Stinner, D. Stinner, and S. Rabatin, 293–306. Elsevier, Amsterdam.

Knuutinen, J., Palm, H., Hakala, H., Haimi, J., Huhta, V., Salminen, J. 1996. Polychlorinated phenols and their metabolites in soils and earthworms of a sawmill environment Chemosphere. 20: 609–623.

Koul, O. 2008. Phytochemicals and insect control: An antifeedant approach. *Crit. Rev. Plant Sci.* 27: 1–24.

Kurowska, A., Gora, J., and Kalemba, D. 1990. Effects of plant phenols on insects. *Pol. Wiadomosci Chemiczne* 44: 399–409.

Lindroth, R.L., and Peterson, S.S. 1988. Effects of plant phenols on performance of southern armyworm larvae. *Oecologia* 75: 185–189.

Mahanil, S., Attajarusit, J., Stout, M.J., and Thipayong, P. 2008. Overexpression of tomato phenol oxidase increases resistance to the common cutworm. *Plant Sci.* 174: 456–466.

Masheva, S., Yankova, V., Tringovska, I., and Kanazirska, V. 2009. Application of some bioproducts for improvement and protection of greenhouse tomato from soil pests. *Int. Soc. Hortic. Sci.* 765–770.

McSorley, R., and Gallaher, R.N. 1995. Effect of yard waste compost on plant parasitic nematode densities in vegetable crops. *J. Nematol.* 27: 245–249.

Miller, P.M. 2001. Reducing field populations of several plant-parasitic nematodes by leaf mold composts and some other additives. *Plant Dis. Rep.* 61: 328–331.

Morra, L., Palumbo, A.D., Bilotto, M., Ovieno, P., and Picascia, S. 1998. Soil solarization: Organic Fertilization and grafting contribute to build an integrated production system in a tomato-zucchini sequence. *Colture-Protte* 27: 63–70.

Morales, H., Perfecto, I., and Ferguson, B. 2001. Traditional fertilization and its effect on corn insect populations in Guatemalan Highlands. *Agric. Ecosyst. Environ.* 84: 145–155.

Pandey, R. 2005. Management of *Meloidogyne incognita* in *Artemisia pallens* with bio-organics. *Phytoparasitica* 33: 304–308.

Park, S.R., Cho, E.J., Yu, K.H., Kim, Y.S., Suh, J.J., and Chang, C.S. 1996. Endogenous phenol oxidase from an earthworm *Lumbricus rubellus. Tongmul Hakoehi* 39: 36–46.

Patriquin, D.G., Baines, D., and Abboud, A. 1995. Diseases, pests and soil fertility. In *Soil Management in Sustainable Agriculture*, ed. H.F. Cook and H.C. Lee, 161–174. Wye College Press, Wye, UK.

Phelan, P.L. 2004. Connecting below-ground and above-ground food webs: The role of organic matter in biological buffering. In *Soil Organic Matter in Sustainable Agriculture*, ed. F. Magdoff and R.R. Well, 199–226. CRC Press, Boca Raton, FL.

Ramesh, K. 2000. *Digital Photoelasticity: Advanced Techniques and Applications.* Springer, Berlin.

Rao, K.R. 2002. Induce host plant resistance in the management of sucking pests of groundnut. *Ann. Plant Prot. Sci.* 10: 45–50.

Rao, K.R., Rao, P.A., and Rao, K.T. 2001. Influence of fertilizers and manures on the population of coccinellid beetles and spiders in groundnut ecosystem. *Ann. Plant Prot. Sci.* 9: 43–46.

Ravi, M., Dhandapani, N., Sathiah, N., and Murugan, M. 2006. Influence of organic manures and fertilizers on the incidence of sucking pests of sunflower, *Helianthus annuus* L. *Ann. Plant Prot. Sci.* 14: 41–44.

Reneco, M., Sasanellin, N., and Šalamun P. 2009. The effect of two compost soil amendments, based on municipal green and penicillin production wastes, on plant parasitic nematodes. *Helminthologia* 46: 190–197.

Ribeiro, C.F., Mizobutsi, E.H., Silva, D.G., Pereira, J.C.R., and Zambolim, L. 1998. Control of *Meloidognye javanica* on lettuce with organic amendments. *Fitopatologia Brasileira* 23: 42–44.

Rodriguez-Kabana, R. 1986. Organic and inorganic amendments to soil as nematode suppressants. *J. Nematol.* 18: 129–135.

Saikia, J., Aparajita, Borah, and Bora, B.C. 2007. Efficacy of organic amendments in combination with the nematicide for management of *Meloidogyne incognita* in Brinjal. *Indian J. Nematol.* 37: 2–8.

Siddiqui, Z.A., and Mahmood, I. 1999. Role of the bacteria in the management of plant parasitic nematodes: A review. *Biores. Technol.* 69: 167–179.

Simmonds, M.S.J. 1998. Chemoecology: The legacy left by Tony Swain. *Phytochemistry.* 49 (5): 1183–1190.

Sipes, B.S., Arakaki, A.S., Schmitt, D.P., and Hamasaki, R.T. 1999. Root-knot nematode management in tropical cropping systems with organic products. *J. Sustainable Agric.* 15: 69–76.

Stevenson, P.C., Anderson, J.C., Blaney, W.M., and Simmonds, M.S.J. 1993. Developmental inhibition of *Spodoptera litura* (Fab.) larvae by a novel caffeoylquinic acid from the wild ground, *Arachis paraguariensis* (Chhod et Hassl). *J. Chem. Ecol.* 19: 2917–2933

Summers, G., and Felton, G.W. 1994. Prooxidation effects of phenolic acids on the generalist herbivore *Helicoverpa zea*: Potential mode of action of phenolic compounds on plant anti-herbivory chemistry. *Ins. Biochem. Mol. Biol.* 24: 943–953.

Swathi, P., Rao, K.T., and Rao, P.A. 1998. Studies on control of root-knot nematode *Meloidogyne incognita* in tobacco miniseries. *Tobacco Research* 1: 26–30.

Vinken, R., Schaeffer, A., and Ji, R. 2005. Abiotic association of soil-borne monomeric phenols with humic acids. *Org. Geochem.* 36: 583–593.

Yardim, E.N., Arancon, N.Q., Edwards, C.A., Oliver, T.J., and Byrne, R.J. 2006. Suppression by vermicomposts of tomato hornworm (*Manduca quinquemaculata*) and cucumber beetle (*Acalymma vittatum*) populations and damage. *Pedobiologia* 50: 23–29.

Yardim, E.N., and Edwards, C.A. 2003. Effects of organic and synthetic fertilizer sources on pest and predatory insects associated with tomatoes. *Phytoparasitica* 31: 324–329.

CHAPTER 15

The Use and Effects of Aqueous Extracts from Vermicomposts or Teas on Plant Growth and Yields

Clive A. Edwards, Ahmed M. Askar, Marcus A. Vasko-Bennett, and Norman Arancon

CONTENTS

I Introduction 235
II Characteristics of Aqueous Extracts (Teas) from Food Waste Vermicomposts 237
III Effects of Aerated Aqueous Extracts (Teas) from Food Waste Vermicomposts on The Growth and Yields of Greenhouse Tomatoes (*Lycopersicon esculentum*) 240
IV Effects of Aerated Aqueous Solutions (Teas) from Vermicomposts on Growth and Yields of Greenhouse Cucumbers (*Cucumis sativa*) 243
V Conclusions on Effects of Vermicompost Teas on Growth and Yields 246
References 247

I INTRODUCTION

A considerable body of scientific literature describes the promotion of plant germination, growth, flowering, and yields by vermicomposts (see Chapters 9 and 10), independent of the availability of nutrients. Aqueous extracts from thermophilic composts (teas) are used widely in the United States and Europe to improve plant health and suppress plant pathogens. However, there are very few references in the literature to the use of vermicompost aqueous extracts (teas) to improve plant germination, growth, flowering, and yields. Although the chemistry and microbiology of aqueous vermicompost extracts are complex, it seems likely that the soluble mineral plant nutrients, plant growth hormones (PGHs) and regulators, microorganisms, and free enzymes that they contain may all have favorable effects on plant growth and yields.

These factors were considered by Pant et al. (2009) in studies on the effects of vermicompost aqueous extracts on the growth of pak choi (*Brassica rapa* cv. Bonsai). They concluded that all types of vermicompost teas increased the total N content of the plants consistently. They reported that populations of active microorganisms were significantly high in all types of vermicompost teas and would promote microbial activities in soils treated with them. Finally, they reported that all types of vermicompost increased the above-ground fresh weights of plants significantly compared with the controls. They associated increases in total carotenoids in plants grown with vermicompost tea treatments. Fritz et al. (2008) reported that vermicompost teas were beneficial to tomato plant growth and biomass production in the laboratory and the field, and they related these effects to increased microbiological activity. Garcia-Gomez et al. (2008), in studies of the effects of vermicompost teas on maize (*Zea mays*), concluded that the teas improved plant growth and that this was probably due more to soluble plant growth regulators (PGRs) than nutrient supply. Edwards et al. (2006) and Arancon and Edwards (2007) reported increases in growth due to vermicompost teas.

Research at the Soil Ecology Laboratory at The Ohio State University has hypothesized that these increases are at least in part due to PGHs, such as indole acetic acids, gibberellins, and kinetins, and PGRs, such as humic and fulvic acids, that are produced by the greatly increased microbial activity in organic wastes, which is due to their fragmentation during passage through the earthworm gut. PGHs are very soluble and break down rapidly in ultraviolet light so they are quite transient in soils. However, research in the Soil Ecology Laboratory has demonstrated quite clearly that in soils, PGHs can become absorbed onto humic and fulvic acids and be released gradually into soil, where they act on plants to promote plant germination, growth, and flowering and increase yields independent of nutrients (Arancon et al. 2004; Atiyeh et al. 1999). This hypothesis was confirmed by Canellas et al. 2000, who demonstrated the absorption of exchangeable auxin groups onto humic acids from cattle manure vermicompost and their gradual release to promote plant growth and crop yields. Also, Edwards and Arancon (2004); Atiyeh et al. (2000, 2002) showed that humates extracted from vermicomposts, by classic acid/alkali fractionation techniques, could yield approximately 4 g dry humic acids per kilogram of vermicompost. When these were added at a range of doses into a soilless plant growth medium such as Metro Mix 360 (MM 360), the dose/growth response pattern produced duplicated the effects of a wide range of application rates of vermicomposts, substituted into the same growth medium, which confirms that this was the probable mechanism of growth promotion (see Chapter 9).

More recent research in the Soil Ecology Laboratory at The Ohio State University has demonstrated that aqueous extracts from vermicomposts, commonly termed teas, could also influence plant growth dramatically, as well as affecting germination, flowering, and yields significantly (Edwards and Arancon 2004; Edwards et al. 2006; Arancon et al. 2007).

II CHARACTERISTICS OF AQUEOUS EXTRACTS (TEAS) FROM FOOD WASTE VERMICOMPOSTS

Vermicompost teas are usually light to dark brown in color depending on the type of vermicomposts used, the proportions of vermicompost and water, the duration of brewing, use of aeration during brewing, and the type and size of mesh by which the vermicomposts are contained during brewing (see Chapter 11). Teas prepared from food waste vermicomposts usually range from light brown in color (at 5% vermicompost concentration) to dark brown color (at 20%). During the brewing process, dark pigmentation and soluble solids resulting from decomposition of organic matter by earthworms and microorganisms constitute the final color of the vermicompost tea.

Two methods have been employed in vermicompost teas extraction: aerated and nonaerated. The primary purpose of aeration during the brewing period is to continuously supply air or dissolved oxygen needed for the microorganisms to survive and multiply. Aeration also serves as a mechanism to agitate the solid vermicompost and dissolve some of the soluble organic-matter particles and nutrient elements in the process. The length of brewing time can vary from 12 to 24 hours or longer. Research in the Soil Ecology Laboratory at The Ohio State University has shown that aerating vermicompost teas during brewing can affect some of the basic biochemical characteristics of the teas even after 24 hours of brewing. Food waste vermicompost teas brewed for 24 hours with aeration had relatively alkaline pH values (between 7.5 and 7.8) compared to nonaerated vermicompost teas, which had slightly acidic pH values of 6.6–6.8 (Figure 15.1). These values did not differ significantly between 5%, 10%, and 20% concentrations of food waste vermicompost teas. Electrical conductivity, which represents the amount of salts and mineral elements in vermicompost teas, showed that aeration during brewing produce greater electrical conductivity values, which means that the aeration is quite efficient in extraction of soluble salts and mineral nutrients from solid vermicomposts

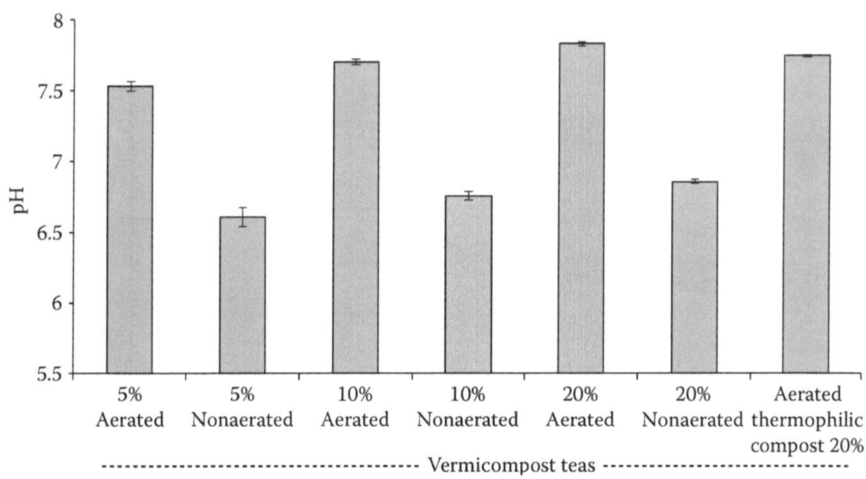

Figure 15.1 Effects of aeration on vermicompost tea pH after 24 hours.

during brewing. Electrical conductivity also increased correspondingly with the concentrations of vermicompost teas (Figure 15.2).

A major component of nutrient elements that are extracted during brewing are nitrates, and, as shown in Figure 15.3, NO_3 N values correlated positively with concentration of vermicompost tea: More nitrates were produced in aerated vermicompost teas than in those without aeration.

The amounts of microbial biomass and activity also differed quite significantly between aerated and nonaerated vermicompost teas. Microbial biomass-N, which is a measure of the total microbial content, was significantly greater in the aerated

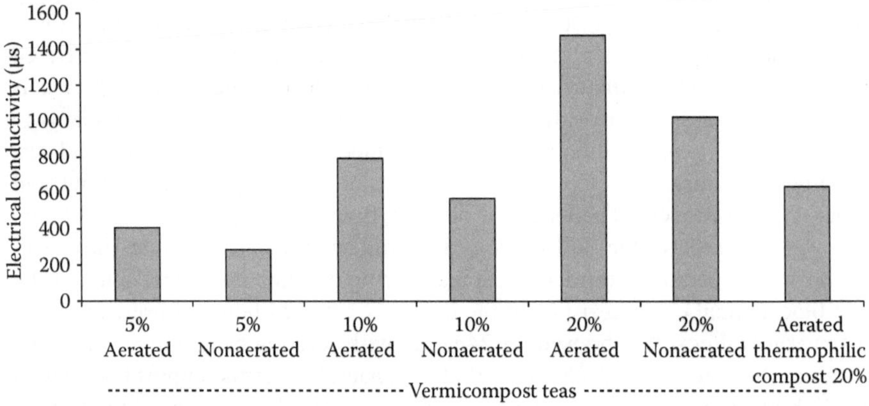

Figure 15.2 Effects of aeration on vermicompost and thermophilic tea conductivity after 24 hours.

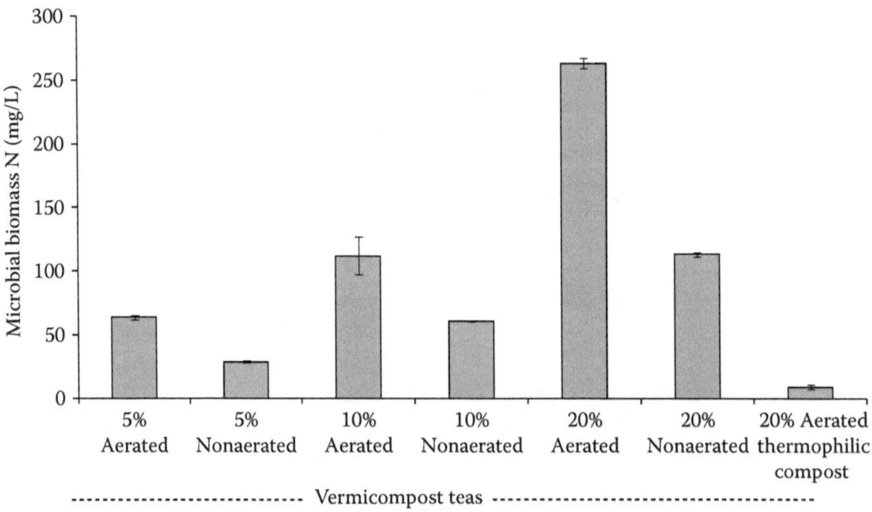

Figure 15.3 Effects of aeration on nitrate-N in vermicompost and thermophilic compost teas after 24 hours.

vermicompost teas than in those without aeration, and the values did not differ significantly between concentrations (Figure 15.4). This confirms that aeration is an efficient way of extracting microorganisms from solid vermicomposts into the teas. Dehydrogenase activity (Figure 15.5), which is a good measure of the total activity of microorganisms, also correlated positively with microbial biomass, in

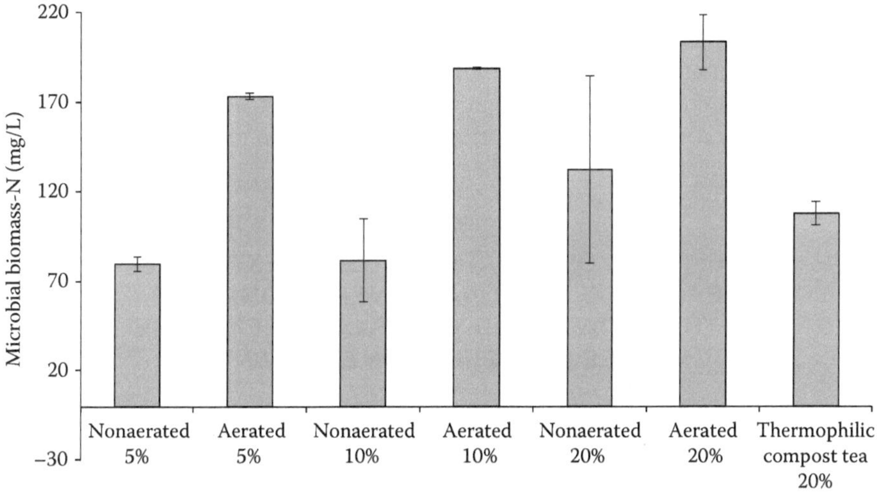

Figure 15.4 Effects of aeration on microbial biomass-N in vermicompost and thermophilic compost teas after 24 hours.

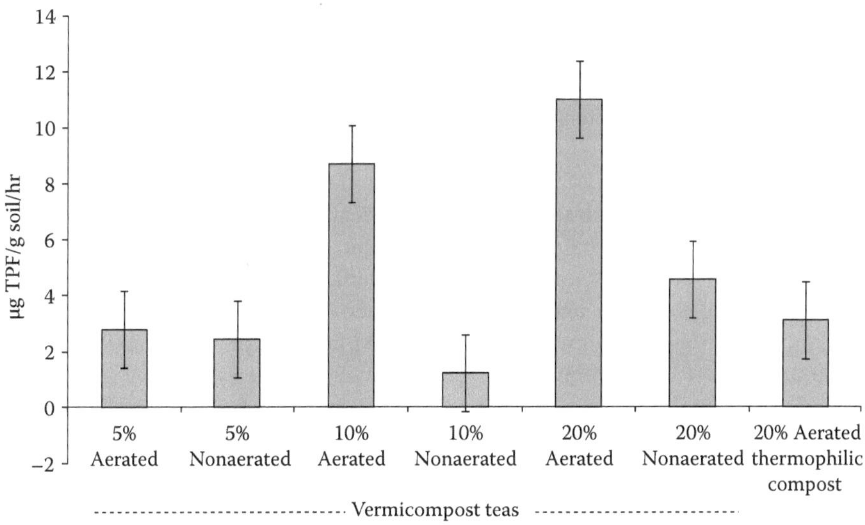

Figure 15.5 Effects of aeration on dehydrogenase activity in vermicompost and thermophilic compost teas after 24 hours.

that aeration in vermicompost tea brewing promoted greater microbial activity, except in 5% vermicompost teas. In these teas, the limited mineral elements, such as, nitrates, that could have served as the source of energy for microorganisms to support their activities could have been depleted too easily compared to 10% and 20% teas, since all of the concentrations tended to have similar amounts of microorganisms measured by microbial biomass-N. These properties are critical elements to assess the values and qualities of vermicompost teas, which could also be used to adjust production methods for extracting vermicompost teas and to predict their subsequent effects on plant growth and development. How these properties change during storage and how the optimum values are achieved and preserved as long as possible are valuable information that could add to the economic value of vermicompost teas.

III EFFECTS OF AERATED AQUEOUS EXTRACTS (TEAS) FROM FOOD WASTE VERMICOMPOSTS ON THE GROWTH AND YIELDS OF GREENHOUSE TOMATOES (*LYCOPERSICON ESCULENTUM*)

There are many methods of preparing aqueous extracts, or teas, from vermicomposts, including steeping vermicomposts in water, with or without stirring, or, in most commercial compost tea-brewing equipment, with active aeration being involved (see Chapter 11). The Ohio State University, Soil Ecology Laboratory research group used commercial tea-brewing equipment called Compost Tea System 10™, produced by Growing Solutions, Inc., Oregon, to produce aqueous tea solutions. Batches of 7.5 L (15.8 pt) of food waste vermicompost were put into 30 L (63.4 pt) of deionized water and mixed and aerated for 24 hours, and then the liquids were separated from the solids before use in the experiments. This produced a 20% concentration aqueous solution that could be diluted to 10% and 5% concentrations for use in plant growth experiments and compared with the effects of a deionized water control. All of the aqueous solutions, or teas, were brewed for 24 hours. To eliminate the effects on plant growth of soluble nutrients in the aqueous solutions, all experimental plants were provided with all needed nutrients from treatments three times a week with Peter's Professional Nutrient Solution (at 200 ppm N).

To assess the effects of soil drenches of aqueous vermicompost extracts (teas) on the growth of tomatoes and cucumbers, seeds were sown in polystyrene trays containing a soilless plant growth medium, MM 360, in four plug trays containing 50 cells (one tomato or cucumber seed per cell) per treatment. All trays were watered three times a week with Peter's Professional Nutrient Solution. Batches of four trays for each tea treatment were drenched with either (a) deionized water or (b) one of three different concentrations of vermicompost teas (5%, 10%, or 20%) after sowing and trays were kept in a mist house. Eight days later, the numbers of seedlings that had germinated were counted, the germination rates for each treatment (percentage of seeds germinating per day) were calculated, and plants were then moved into a greenhouse. Twenty days later, 10 plants were selected at random from each treatment tray. At this stage, the plant heights (distance from the soil level to the topmost node), total leaf numbers (excluding cotyledons), and leaf areas of each of the seedlings were recorded.

To extend the growth experiments to flowering and yields, two tomato or cucumber seedlings were removed at random from each tray and transplanted into small pots 10 cm (4 in) diam. filled with MM 360 potting mixture (eight replicate pots per treatment). All were watered three times a week with Peter's Professional Nutrient Solution. Each batch of plants received similar vermicompost teas or deionized water treatments as soil drenches every 2 weeks. Forty days later, the seedlings in the small pots were transplanted into MM 360 in larger pots (40 cm (15.7 in) in diameter) and watered with (a) deionized water or (b) the appropriate dose of a range of vermicompost tea treatments. Thirty days after the final transplant was made into the 40 cm pots, tomato or cucumber fruits were harvested. The numbers and weights of tomatoes or cucumbers were recorded. Tomato or cucumber fruits were harvested for 20 days after the first fruits were taken, after which the shoots of all plants were removed from the potting mixtures, oven-dried at 55°C (131°F) for 8 days, and the mean shoot dry weights and leaf areas were recorded. Leaf tissues were analyzed for nutrient content as described earlier. Roots were washed, dried at 55°C (131°F) for 5 days, and weighed.

The effects of soil drenches with 5%, 10%, or 20% vermicompost tea, compared with those of a deionized water control, on tomato plant heights are shown in Figures 15.6 and 15.7, and the effect on leaf areas in Figure 15.8. The fresh shoot weights are presented in Figure 15.9, the numbers of tomato fruits in Figure 15.10, and the mean fruit weights in Figure 15.11. All three application rates increased tomato plant heights, leaf areas, fresh shoot weights, numbers of tomato fruits, and mean weights of tomato fruits significantly ($p \leq 0.05$; Figures 15.6 through 15.11).

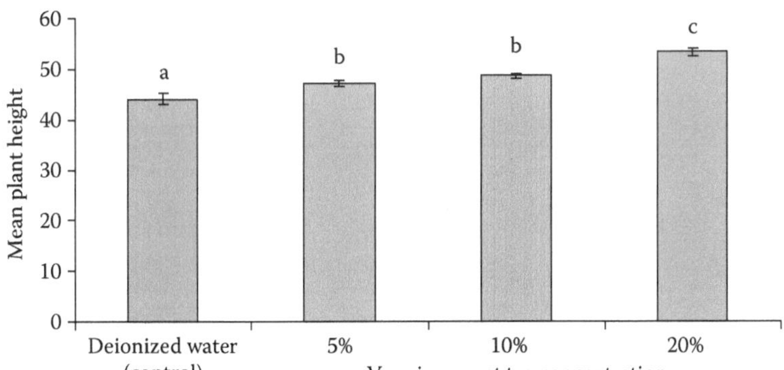

Figure 15.6 Effects of food waste vermicompost tea on plant heights of tomatoes (means ± SE). Plants were grown in MM 360 with all needed nutrients supplied. Means followed by different letters are significantly different for each treatment ($p \leq 0.05$).

Figure 15.7 Responses in growth of tomatoes to 5% vermicompost tea applications to plants growing in MM 360 in the greenhouse for 6 weeks (with all needed nutrients supplied).

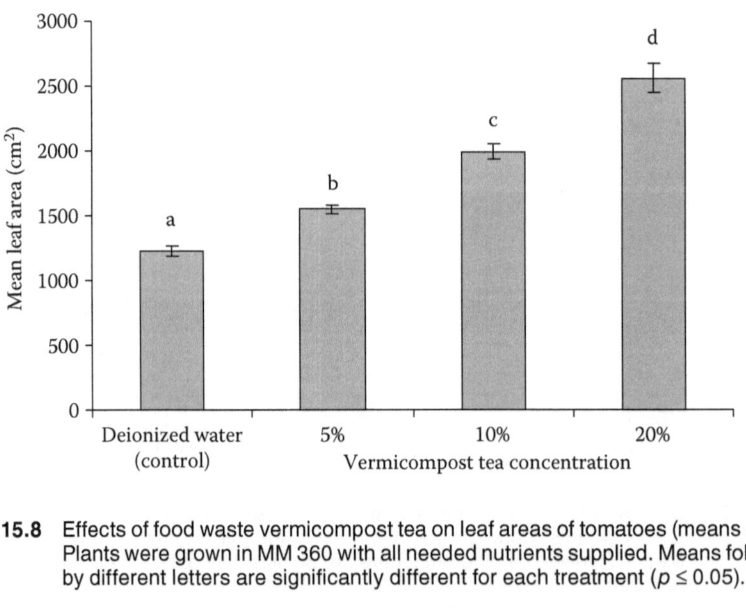

Figure 15.8 Effects of food waste vermicompost tea on leaf areas of tomatoes (means ± SE). Plants were grown in MM 360 with all needed nutrients supplied. Means followed by different letters are significantly different for each treatment ($p \leq 0.05$).

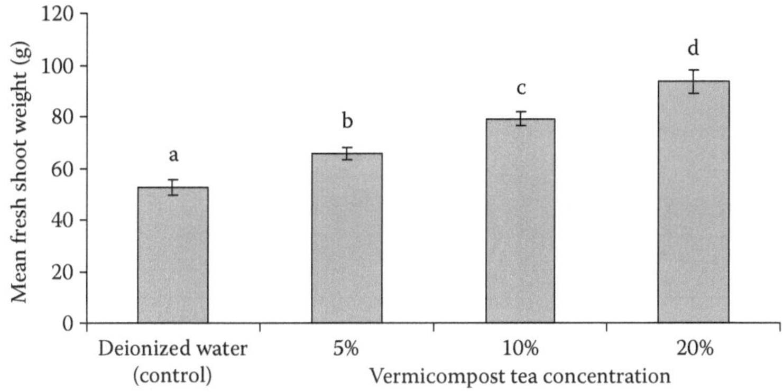

Figure 15.9 Effects of food waste vermicompost tea on fresh shoot weights of tomatoes (means ± SE). Plants were grown in MM 360 with all needed nutrients supplied. Means followed by different letters are significantly different for each treatment ($p \leq 0.05$).

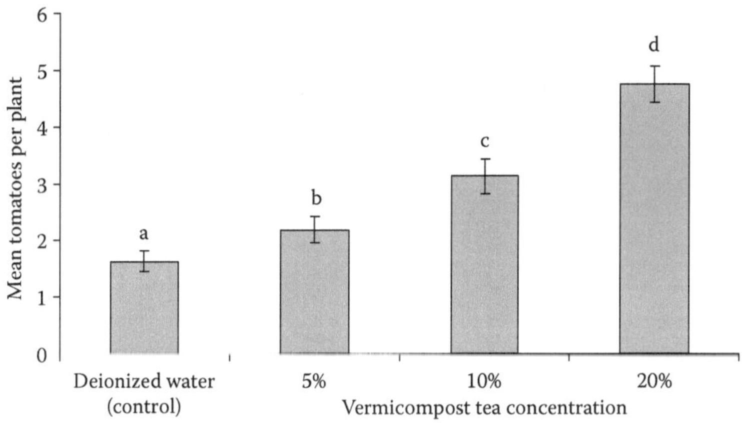

Figure 15.10 Effects of food waste vermicompost tea on numbers of tomato fruits (means ± SE). Plants were grown in MM 360 with all needed nutrients supplied. Means followed by different letters are significantly different for each treatment ($p \leq 0.05$).

IV EFFECTS OF AERATED AQUEOUS SOLUTIONS (TEAS) FROM VERMICOMPOSTS ON GROWTH AND YIELDS OF GREENHOUSE CUCUMBERS (*CUCUMIS SATIVA*)

In a similar experimental design to that used for assessing the effects of vermicompost teas on growth of tomatoes, one cucumber seed was sown into each of 10 replicate 10 cm (4 in) diam. pots filled with MM 360 soilless bedding-plant medium. All plants were given Peter's Professional Nutrient Solution three times weekly, and all had balanced nutrients. Immediately after sowing and 14 days

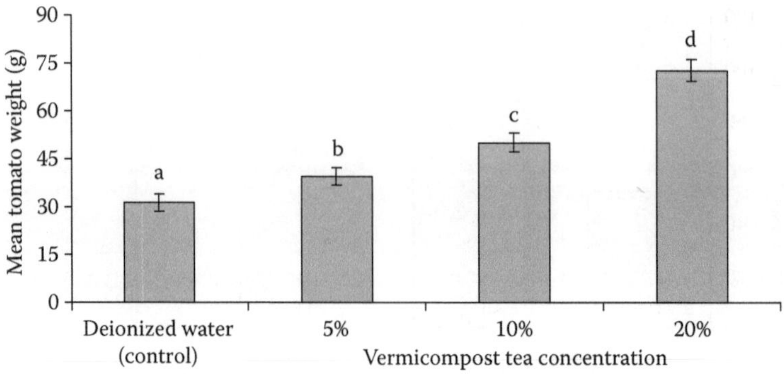

Figure 15.11 Effects of food waste vermicompost tea on mean fruit weights of tomatoes (means ± SE). Plants were grown in MM 360 with all needed nutrients supplied. Means followed by different letters are significantly different for each treatment ($p \leq 0.05$).

Figure 15.12 Early growth of cucumber seedlings treated with a range of concentrations of vermicompost teas. Plants were grown in MM 360 (all needed nutrients were supplied).

postsowing, each pot was drenched with one of a range of vermicompost teas (5%, 10%, or 20%) or the deionized water control treatment. Plant heights, leaf numbers, and leaf areas in each pot were recorded after 20 days. The plants were removed from the potting mixtures and oven-dried at 55°C (131°F) for 5 days to determine fresh and dry plant dry weights. Roots were washed, dried, and weighed.

Batches of cucumber plants were transplanted into larger pots 40 cm (15.7 in) diam. containing MM 360, with four replicate pots per treatment. The same ranges of

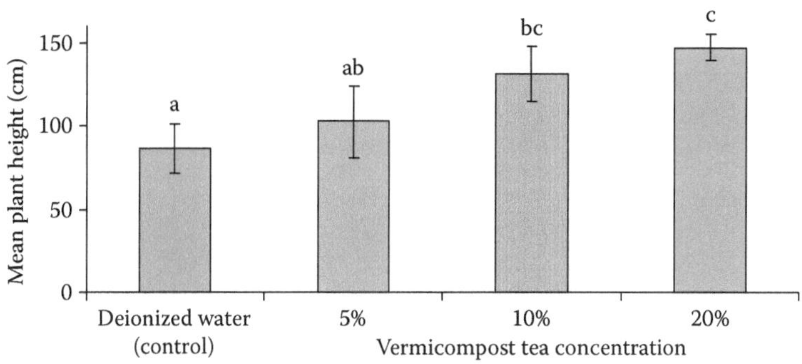

Figure 15.13 Effects of food waste vermicompost teas on plant heights of cucumbers over time (means ± SE). Plants were grown in MM 360 with all needed nutrients supplied. Means followed by different letters are significantly different for each treatment ($p \leq 0.05$).

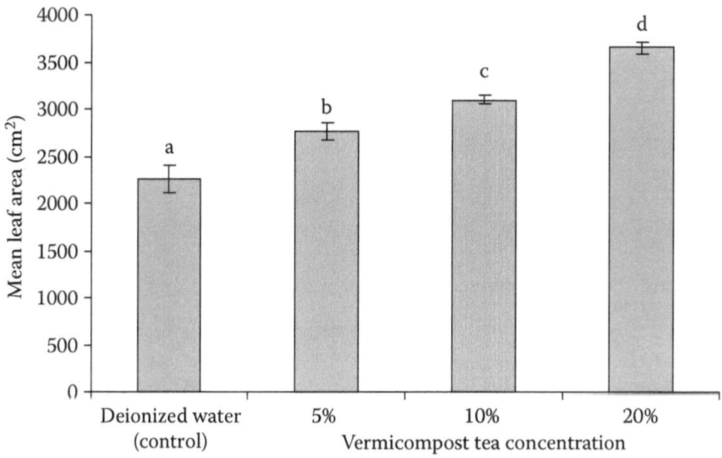

Figure 15.14 Effects of food waste vermicompost teas on mean leaf areas of cucumbers (means ± SE). Plants were grown in MM 360 with all needed nutrients supplied. Means followed by different letters are significantly different for each treatment ($p \leq 0.05$).

vermicompost tea treatments or a deionized water control were applied immediately after transplanting and thereafter every 14 days. All plants received Peter's Nutrient Solution three times weekly. Numbers and weights of all cucumbers harvested were recorded. Overall yields per treatment were assessed and analyzed statistically.

The effects of the range of food waste compost teas compared with the deionized water control on the early growth of cucumbers are illustrated in Figure 15.12. All of the food waste vermicompost teas increased the heights of the cucumber plants significantly compared with the deionized water control ($p \leq 0.05$; Figure 15.13). They also increased mean leaf areas significantly ($p \leq 0.05$; Figure 15.14). Only the 20% vermicompost tea applications increased the number of cucumbers per plant

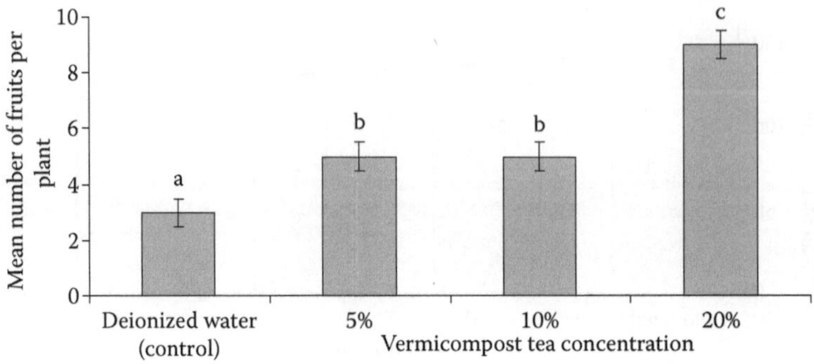

Figure 15.15 Effects of food waste vermicompost teas on numbers of cucumbers per four plants (means ± SE). Plants were grown in MM 360 with all needed nutrients supplied. Means followed by different letters are significantly different for each treatment ($p \leq 0.05$).

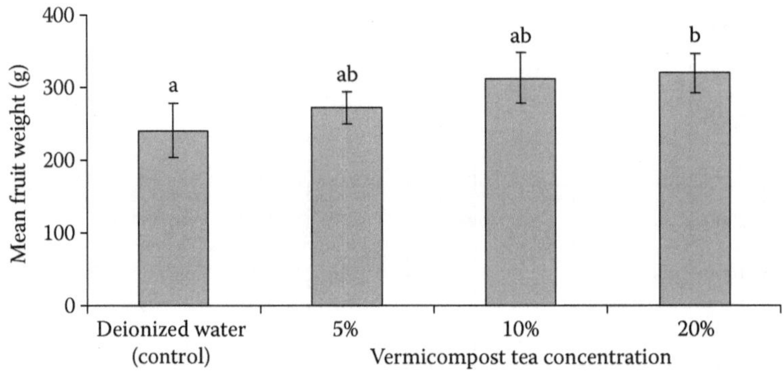

Figure 15.16 Effects of food waste vermicompost teas on fruit weights of cucumbers (means ± SE). Plants were grown in MM 360 with all needed nutrients supplied. Means followed by different letters are significantly different for each treatment ($p \leq 0.05$).

significantly ($p \leq 0.05$; Figure 15.15), whereas all three application rates increased mean cucumber weights significantly ($p \leq 0.05$; Figure 15.16).

V CONCLUSIONS ON EFFECTS OF VERMICOMPOST TEAS ON GROWTH AND YIELDS

It is clear from these research data that aqueous solutions from vermicomposts (teas) can promote the germination, growth, flowering, and yields of crops significantly and considerably. Identification of the mechanisms involved in these responses needs further research. Clearly, some of the soluble materials such as PGHs that can pass from vermicompost into the teas during the tea-brewing process and are taken up into plants must be responsible for some of the growth effects. Issues important in aqueous extracts from vermicomposts include the following:

- Aeration is very important during the aqueous extract-brewing process.
- PGHs, such as indole acetic acids, gibberellins, and cytokinins, are very soluble and can pass readily in to teas. These are produced by the extremely large populations of microorganisms in the vermicompost.
- PGRs, such as fulvic and humic acids, can pass into the teas as fine particulate matter. During vermicomposting fulvic and humic acids are produced as the organic matter stabilizes through earthworm activity.

The following factors can influence growth.

- Combinations of PGHs and PGRs can pass into the teas and persist in soil longer than the separate materials. PGHs become absorbed onto the particulate humates and fulvates and are released slowly to promote plant growth throughout the season.
- Soluble plant nutrients, particularly N (as nitrates), soluble P, K, Ca, and Mg. All of the important nutrients in vermicomposts are readily soluble and can pass easily into the teas from vermicomposts.
- Microorganisms may continue to produce PGHs and PGRs when they reach soils as soil drenches. Hence, it is important to keep the teas aerobic and microbially active so that PGHs are still produced in the teas.
- Soluble free enzymes may continue nutrient transformations. These promote conversions of nutrients into forms readily taken up by plants.

Clearly, from the evidence in this chapter these aqueous extracts, or teas, from vermicomposts can have dramatic effects on plant germination, growth, and yields. They will provide an important tool for organic growers and farmers who need to avoid the use of inorganic fertilizers in their crop production; teas can be easily applied to organic crops as soil drenches.

REFERENCES

Arancon, N.Q., and Edwards, C.A. 2007. The utilization of vermicomposts in horticulture and agriculture. In *Indo-U.S. Workshop on Vermitechnology in Human Welfare*, Kongunadu Arts and Science College, Coimbatore, India.

Arancon, N.Q., Edwards, C.A., Atiyeh, R., and Metzger, J.D. 2004. Effects of vermicomposts produced from food wastes on the growth and yields of greenhouse peppers. *Biores. Technol.* 86: 398–420.

Arancon, N.Q., Edwards, C.A., Dick, R., and Dick, L. 2007. Vermicomposts: Preparations, uses, and problems. *Biocycle* 48: 51.

Atiyeh, R.M., Lee, S., Edwards, C.A., Arancon, N.Q., and Metzger, J.D. 2002. The influence of humic acids derived from earthworms-processed organic wastes on plant growth. *Biores. Technol.* 84: 7–14.

Atiyeh, R.M., Subler, S., Edwards, C.A., Bachman, G., Metzger, J.D., and Shuster, W. 2000. Effects of vermicomposts and composts on plant growth in horticulture container media and soil. *Pedobiologia* 44: 579–590.

Atiyeh, R.M., Subler, S., Edwards, C.A., and Metzger, J. 1999. Growth of tomato plants in horticultural potting media amended with vermicompost. *Pedobiologia* 43:724–728.

Canellas, L.P., Olivares, F.L., Okorokova, A.L., and Facanha, A.R. 2000. Humic acids isolated from earthworm compost enhance root elongation, lateral root emergence, and plasma H^+ -ATPase activity in maize roots. *Plant Physiol.* 130: 1951–1957.

Edwards, C.A., and Arancon, N.Q. 2004. Interactions among organic matter, earthworms and microorganisms in promoting plant growth. In *Functions and Management of Soil Organic Matter in Agroecosystems*, ed. C.A. Edwards, F. Magdoff, and R. Weil, 327–376. CRC Press, Boca Raton, FL.

Edwards, C.A., Arancon, N.Q., and Greytak, S. 2006. Effects of vermicompost teas on plant growth and disease. *Biocycle* 47: 28–31.

Fritz, I., Haindl, S., Pruckner, M., and Braun, R. 2008. Effects of vermicompost-tea on plant growth and crop yield. *CODIS 2008:* 117–118.

Garcia-Gomez, R.C., Dendooven, L., and Gutierrez-Miceli, F.A. 2008. Vermicomposting leachate (worm tea) as liquid fertilizer for maize (*Zea mays L.*) forage production. *Asian J. Plant. Sci.* 7: 360–367.

Pant, A., Radovich, T., Hue, N., Talcott, S., and Kreneck, K. 2009. Vermicompost extracts influence growth, mineral nutrients, phytonutrients and antioxidant activity in pak choi (*Brassica rapa* cv. Bonsai, Chinensis group) grown under vermicompost and chemical fertilizer. *J. Sci. Food Agric.* 89: 2383–2392.

CHAPTER 16

Human Pathogen Reduction during Vermicomposting

Clive A. Edwards and Scott Subler

CONTENTS

I Introduction 249
II Laboratory Studies on Human Pathogen Reduction by Vermicomposting 250
III Field Studies on Human Pathogen Reductions by Vermicomposting 252
IV Definitive Study on Human Pathogen Reduction in Sewage Biosolids 255
V Possible Mechanisms of Pathogen Reduction during Vermicomposting 258
VI Conclusions 259
References 260

I INTRODUCTION

A major obstacle to the widespread acceptance of vermicomposting as a general organic-waste-management alternative is a lack of adequate scientific information about the potential for human pathogen reduction during vermicomposting of sewage biosolids or even animal manures. In contrast to conventional composting, which is often defined by the presence of a high temperature or *thermophilic* phase 50–70°C (122–158°F), vermicomposting is a *mesophilic* process. It requires that the substrate temperatures remain relatively low generally below 35°C (95°F), or else the earthworms may become inactive, flee, or die, thereby disrupting the vermicomposting process. A major benefit of higher temperatures during conventional composting is its potential reduction or elimination of many human pathogens such as *Salmonella* and *E. coli*, human viruses, and helminths. Based on a significant volume of scientific literature, U.S. Environmental Protection Agency (EPA) standards for the composting process have been developed that specify the duration and extent of raised temperatures necessary (above 55°C (161°F) for 72 hours) to eliminate the risk of human pathogen contamination in finished composts so they can be used safely as Class A materials for land use (Environmental Protection Agency [EPA] 1999). These standards have been incorporated into U.S. state and EPA regulations for the production and use of composts.

Perhaps because of this reliance on temperature management for human pathogen reduction in conventional thermophilic composting processes, questions have arisen regarding the ability of low-temperature, or mesophilic, processes, such as vermicomposting, to consistently and effectively reduce levels of human pathogens in contaminated organic wastes to safe levels. Scattered reports in the scientific literature, combined with the cumulative experience of vermiculturalists and vermicomposters, certainly suggest that effective pathogen reduction may also be achieved during vermicomposting under certain processing conditions (Eastman et al., 2001; EPA 1980). However, the available information has not been evaluated in a systematic way. In this Chapter we review the current evidence concerning the overall potential for human pathogen reduction during vermicomposting. We consider scientific reports from both the laboratory and the field, as well as data coming from actual institutional or commercial vermicomposting operations, and, finally, we discuss possible mechanisms of pathogen reduction that may be important in the vermicomposting process.

II LABORATORY STUDIES ON HUMAN PATHOGEN REDUCTION BY VERMICOMPOSTING

Initial indications of the potential of vermicomposting to reduce or eliminate human pathogens from contaminated organic wastes come from early research using earthworms to stabilize wastewater biosolids. Research into the use of earthworms to manage biosolids began at the State University of New York (SUNY), Syracuse (Hartenstein 1978), with National Science Foundation funding. In this research program, Mitchell (1978) demonstrated that aerobic biosolids that are ingested and egested as casts by the earthworms *Eisenia fetida* are decomposed and stabilized about three times as fast as noningested biosolids, apparently because of the enhancement of microbial decomposition in the casts. He found that, relative to noningested biosolids, objectionable odors in manures and biosolids disappeared much more quickly, and there was a marked reduction in populations of the pathogens *Salmonella enteritidis*, *Escherichia coli*, and other Enterobacteriaceae.

Brown and Mitchell (1981) followed these conclusions up with a series of laboratory experiments to test the effects of *E. fetida* on *Salmonella* survival. In laboratory cultures, they reported marked declines in *Salmonella* populations that were significantly greater in the presence of the earthworms than in their absence; for example, maximum rates of decrease in pathogens were twice as fast in the presence of *E. fetida* as in its absence. In trials that ranged from 4 to 28 days, *E. fetida* decreased *Salmonella* populations by 97.8% to 99.9%, compared to cultures with no earthworms. They suggested that this decrease was due to the stimulation of an endemic microflora, which when grown with *Salmonella* in liquid cultures caused nearly total elimination of the pathogen. Other work conducted in the early 1980s yielded similar results. For example, Lui (1982; cited in Aguilar 1996) reported that over 98% of human pathogens were destroyed during 17 days of vermicomposting.

Although Brown and Mitchell's (1981) early work on *Salmonella* provided a basis for optimism regarding adequate human pathogen reduction during vermicomposting, not all other laboratory results have been as clear cut. Haimi and Huhta (1987) investigated the effects of the activity of earthworms (*E. fetida*) on bacterial densities in a mixture of wastewater biosolids and bark. Stabilization of this material proceeded more quickly in the presence of earthworms than in their absence. After 7 weeks, fecal coliform densities were 40% less in earthworm-worked organic wastes than in organic wastes without earthworms. In contrast, the density of fecal streptococci tended to be greater in the earthworm-worked material. The contradictory response of these two bacterial groups during vermicomposting suggests that different mechanisms may be responsible for influencing the success or demise of different microbial species. This underscores the need to exercise caution when extrapolating limited results to cover all potential human pathogens.

Murry and Hinckley (1992) studied the effects of *E. fetida* on *Salmonella enteritidis* inoculated into culture dishes containing horse manure. Within 48 hours of inoculation, there were significant differences in the survival of *Salmonella* and other bacteria between wastes with and without earthworms. *Salmonella* numbers declined more rapidly, and populations of other bacteria also decreased, in the presence of *E. fetida*.

Other researchers have focused on the survival of bacterial pathogens during transit through the earthworm gut. For example, Finola et al. (1995) studied the survival of total and fecal coliforms, *Salmonella*, *Pseudomonas*, and other Enterobacteriaceae in the guts of earthworms (*E. fetida*) cultivated in composted poultry litter. All of these bacteria were reduced below the limits of detection of the study, prompting the authors to suggest a consistent antimicrobial response on gram-negative bacteria from the gizzard through the intestinal tract of the earthworms.

Aguilar (1996) investigated reductions in the density of fecal coliform bacteria during vermicomposting of municipal wastewater biosolids. She was interested in determining whether the vermicomposting process could reduce pathogenic bacteria in organic wastes significantly without prior treatment using thermophilic composting. In laboratory studies, she placed a range of populations of *E. fetida* (1000–8000 earthworms m^{-2} (100–800 earthworms ft^{-2})) in plastic trays containing various mixtures of biosolids and hay. Following vermicomposting periods ranging from 7 to 18 weeks, the average fecal coliform density in the vermicomposted material was 22 CFug^{-1}—over a thousand-fold reduction from the initial levels (4.7 × 104 CFug^{-1}) and well below the EPA Class A limit (1000 CFug^{-1}; 40 CFR Part 503) for land sewage disposal. A log-linear decline in fecal coliform numbers during vermicomposting is seen clearly in Figure 16.1. This work provides good evidence of the potential for effective fecal coliform reduction in organic wastes below the EPA Class A limit for land disposal after 7 to 8 weeks of vermicomposting.

In The Ohio State University Soil Ecology Laboratory, we investigated reductions of human bacterial pathogens during vermicomposting of two organic materials:

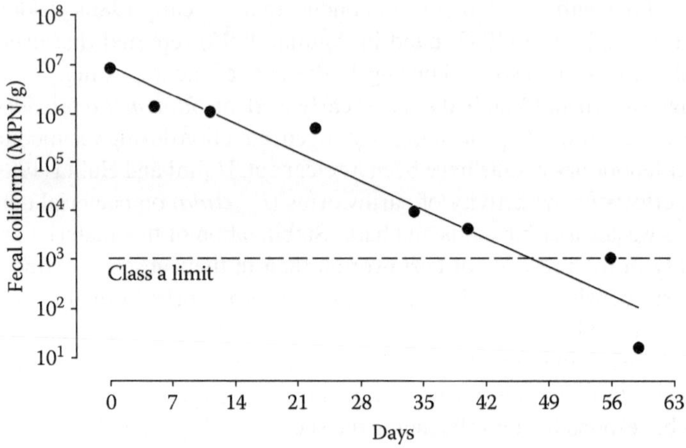

Figure 16.1 Reduction in fecal coliform bacteria during vermicomposting of wastewater biosolids in the laboratory. Earthworm (*E. fetida*) density = (1000 earthworms m^{-2} (100 earthworms ft^{-2})). (Data from Aguilar, A.L.B., *Laboratory studies of vermicomposting municipal wastewater biosolids, master's thesis*, Jackson State University, Jackson, MS, 1996.)

wastewater sewage biosolids and dairy manure. We incubated the materials in small plastic containers, either with or without earthworms (*E. fetida*). To half of the containers we added an inoculum containing *E. coli* and *Salmonella* spp. to increase the initial levels of the pathogens in the manure and sewage solids. We compared decreases in populations of the pathogens with and without earthworms with four replicates per treatment. Samples of the organic materials were taken after 1, 2, 6, and 10 weeks and analyzed for *E. coli* and *Salmonella* spp. As seen in Figures 16.2 and 16.3, populations of both *E. coli* and *Salmonella* spp. declined dramatically in the vermicomposted materials. By comparison, without earthworms, pathogen numbers declined only very slowly over the 10-week period. However, despite the early rapid reduction in human pathogen populations in the presence of earthworms, levels near to or below the Class A limits were achieved only after between 6 and 10 weeks, a time frame similar to that reported by Aguilar (1996). Although the available laboratory studies are relatively limited in numbers, taken together, they provide a fairly consistent and conclusive indication that earthworm activity can effectively reduce human pathogens in organic wastes to acceptable levels.

III FIELD STUDIES ON HUMAN PATHOGEN REDUCTIONS BY VERMICOMPOSTING

Early indications of effective human pathogen reduction during vermicomposting of sewage biosolids can be found in a U.S. EPA report describing the complete absence of *Salmonella* in biosolids-derived earthworm castings from a raw biosolids vermicomposting operation in Texas (EPA 1980). Much later, Harris et al. (1991)

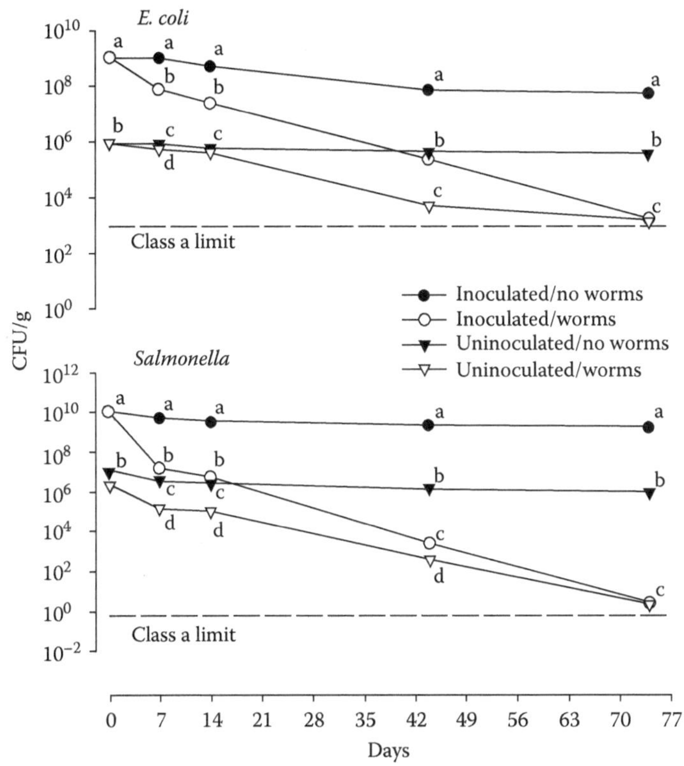

Figure 16.2 Decreases in *Salmonella* spp. and *E. coli* in sewage biosolids after laboratory vermicomposting. (Data from Soil Ecology Laboratory, The Ohio State University.)

described a full-scale vermicomposting operation to stabilize wastewater biosolids in Fallbrook, California. Following a 30-day initial thermophilic composting (static pile) phase, they used vermicomposting beds (6 months with continual feedings), and then the vermicompost was screened and cured for an additional 30 days. They reported that bacteriological analyses of the final vermicompost product showed no evidence of coliform bacteria.

Perhaps one of the most complete field studies to date was conducted by the Florida County Environmental Protection Department, in conjunction with the Florida EPA, under the direction of Bruce Eastman (Riggle 1996). In 1996, a pilot study was undertaken to assess the effectiveness of windrow vermicomposting for reducing pathogens, such as fecal coliform, *Salmonella*, enteric viruses, and viable helminth ova, in wastewater biosolids. Eight tons of dewatered sewage biosolids (about 80% moisture) were mixed with other residual biosolids and inoculated with enteric viruses to ensure sufficient levels of the pathogens in the biosolids. Two windrows 5 ft wide × 30 ft long (1.5 m × 12 m) were formed from sewage biosolids under a canopy. Twenty four kg (50 lb) of earthworms (*E. fetida*) were added to one of the rows, and the other row with no earthworms served

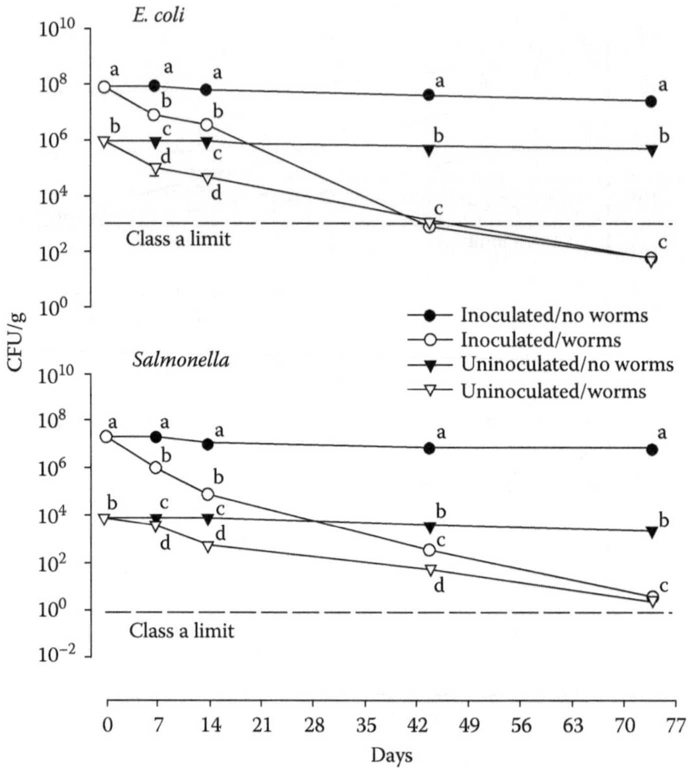

Figure 16.3 Decreases in *Salmonella* spp. and *E. coli* in cattle manure after laboratory vermicomposting. (Data from Soil Ecology Laboratory, The Ohio State University.)

as a control. Samples from the two rows were collected at the beginning of the experiment and 68 days later and analyzed for the four EPA pathogen indicators. Significant reductions in pathogen levels had occurred in both the vermicomposted windrow and the control windrow with no earthworms. Levels of *Salmonella*, enteric viruses, and viable helminth ova were below levels of detection for both windrows. Although the fecal coliforms were reduced in both windrows after 68 days, compared to the initial values, the reduction was significantly greater in the windrow with earthworms than in the control windrow, suggesting that the presence of earthworms greatly accelerated human pathogen reduction in biosolids in a relatively short period of time.

Based on these initial results, a larger-scale field study was conducted in Orange County by the same investigators. This field project has yielded hard evidence that vermicomposting can effectively reduce human pathogens in organic waste materials to Class A levels (EPA 1999), even from very high-initial levels, while also providing rapid stabilization of the biosolids in a relatively short period of time.

In addition to the published reports that have already been described, we have reviewed a number of independent laboratory analyses of vermicompost samples from commercial/institutional-scale vermicomposting facilities. Results so far have been consistent in indicating minimal levels of human pathogens in finished vermicomposts. For example, in samples of vermicomposted dairy manure (California Vermiculture, Del Mar, California) and vermicomposted pig manure (Vermicycle Organics, Charlotte, North Carolina), *Salmonella* was not detected and *E. coli* was less than 3 MPN.g^{-1}. Likewise, vermicomposted wastewater biosolids from a private treatment facility (Hume Lake Christian Camps, Hume, California) showed undetectable levels of fecal coliforms, *Salmonella*, enteric viruses, and helminth ova after less than 2 months of vermicomposting. While not carrying the same weight of evidence as reports from controlled scientific studies, these experiences from the real world are consistent with the concept that it is possible to achieve effective human pathogen reduction during vermicomposting. As with the laboratory reports, a remarkably consistent picture is emerging from the results of field studies and field operations, providing a body of accumulating evidence that weighs strongly in favor of significant human pathogen reduction occurring over a relatively short time during the vermicomposting process.

IV DEFINITIVE STUDY ON HUMAN PATHOGEN REDUCTION IN SEWAGE BIOSOLIDS

A field experiment was set up by a group of scientists to test the feasibility of vermicomposting as a method for eliminating human pathogens to obtain U.S. EPA Class A stabilization in domestic wastewater residual biosolids (Eastman et al. 2001). The experimental site was at the City of Ocoee's Wastewater Treatment Facility in Ocoee, Florida, and Class B sewage biosolids were used as the earthworm substrate. Two windrows of biosolids 6 m (20 ft) long were inoculated heavily with four human-pathogen indicators required by the U.S. EPA: fecal coliforms, *Salmonella* spp., human enteric viruses, and helminth ova. The test row was seeded with earthworms (*Eisenia fetida*) and the contral row had no earthwarms. The numbers of *E. fetida* were calculated at a 1:1.5 wet-weight earthworm biomass to biosolids ratio, and the earthworms were allowed time to consume and stabilize the biosolids.

The test indicated that all of the pathogen indicators in the test row decreased faster than in the control row within 144 hours. The test-row samples showed a 6.4-log reduction in fecal coliforms compared with the control row, which had only a 1.6-log reduction (Figure 16.4). The test-row samples showed an 8.6-log reduction in *Salmonella* spp., while the control row had a 4.9-log reduction (Figure 16.5). The test-row samples showed a 4.6-log reduction in enteric viruses while the control had only a 1.8-log reduction (Figure 16.6). The test-row samples had a 1.9-log reduction in helminth ova while the control row had only a 0.6-log reduction (Figure 16.7). The EPA indicated that a three- to four-fold reduction in indicator organisms would be sufficient to warrant serious consideration of vermicomposting as an effective biosolids-stabilization methodology. These results, in conjunction with pilot project results, indicate strong evidence that vermicomposting can be used as an alternative method for Class A biosolids stabilization.

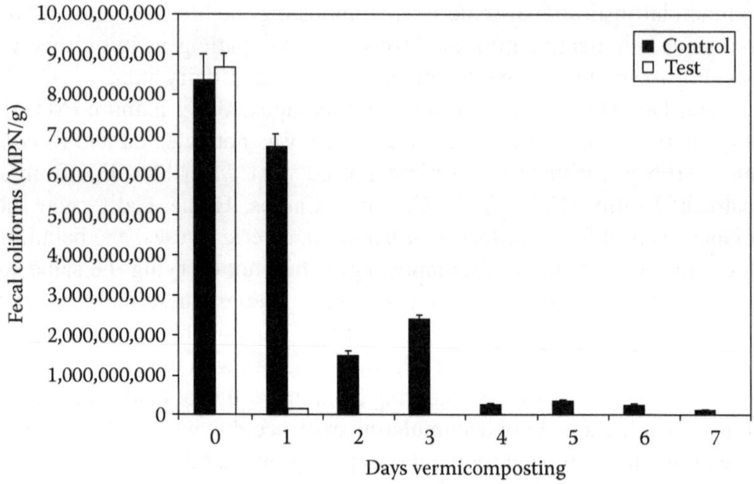

Figure 16.4 Fecal coliform average decrease (shown in linear scale) in the control and test rows during 7 days of vermicomposting (vertical T bars represent standard errors). (Adapted from Eastman, B.R., Kane, P.N., Edwards, C.A., Trytek, L., and Gunadi, B., *Compost Sci. Util.*, 9: 38–49, 2001.)

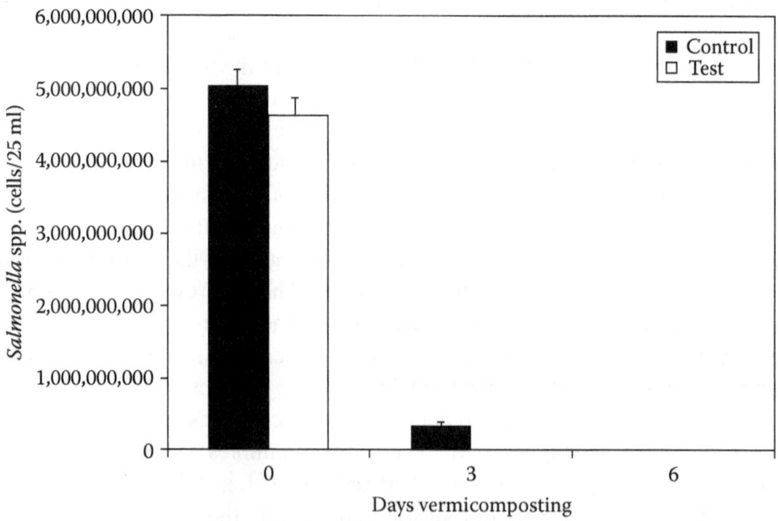

Figure 16.5 *Salmonella* spp. average decrease (shown in linear scale) in the control and test rows during 6 days of vermicomposting (vertical T bars represent standard errors). (Adapted from Eastman, B.R., Kane, P.N., Edwards, C.A., Trytek, L., and Gunadi, B., *Compost Sci. Util.*, 9: 38–49, 2001.)

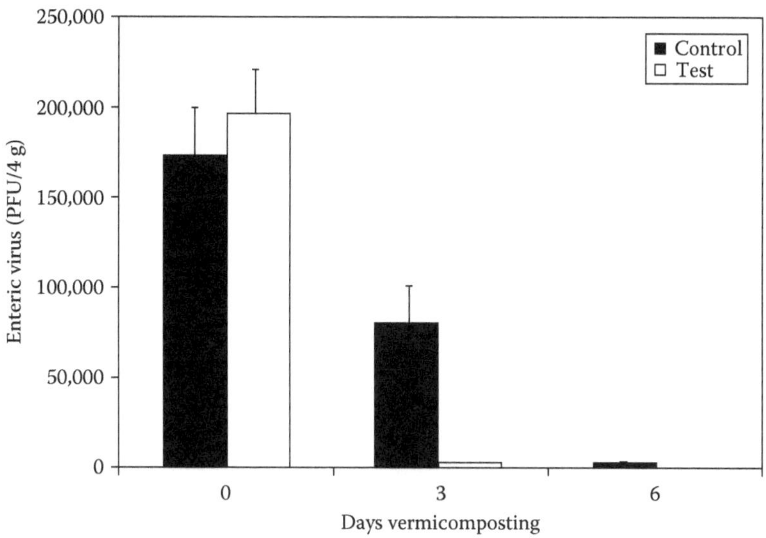

Figure 16.6 Enteric virus average decrease (shown in linear scale) in the control and test rows during 6 days of vermicomposting (vertical T bars represent standard errors). (Adapted from Eastman, B.R., Kane, P.N., Edwards, C.A., Trytek, L., and Gunadi, B., *Compost Sci. Util.*, 9: 38–49, 2001.)

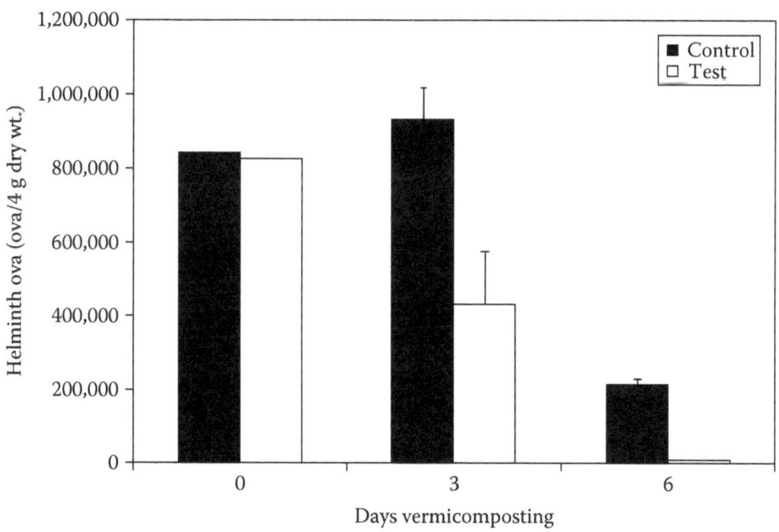

Figure 16.7 Helminth ova average decrease (shown in linear scale) in the control and test rows during 6 days of vermicomposting (vertical T bars represent standard errors).

V POSSIBLE MECHANISMS OF PATHOGEN REDUCTION DURING VERMICOMPOSTING

Hartenstein (1978) reviewed all the evidence available at the time and provided a compelling argument that bacteria, endoparasitic worms (helminths), and viruses were likely to be destroyed though the activity of earthworms on biosolids and other organic wastes. He cited Brown and Mitchell's paper (1981) as evidence of rapid destruction of bacteria in biosolids in the presence of earthworms (*E. fetida*). In their paper, Brown and Mitchell also provided solid evidence that the stimulation of endemic bacteria by earthworm activity may lead to pathogen destruction through competitive or antagonistic interactions. Hartenstein (1978) also suggested that evidence of digestive assimilation of some nematodes by earthworms (Dash et al. 1980) indicated a potential for effective destruction of helminths and their eggs (possibly even including the resistant ova of *Ascaris*). He also pointed to data suggesting increased inactivation or destruction of viruses in the presence of humates (Cliver 1983), which are produced in abundance during vermicomposting.

It has been demonstrated clearly that the composition of microbial species and their activity changes greatly on passage through the gut of earthworms (Pedersen and Hendriksen 1993; see Chapter 5). However, only a few studies have dealt specifically with human pathogens and related bacteria (Enterobacteriaceae). In most cases, numbers of Enterobacteriaceae decreased considerably during passage from the earthworm foregut to the hindgut (Day 1950; Brüsewitz 1959; Brown and Mitchell 1981; Finola et al. 1995). However, in a detailed study with *Lumbricus rubellus*, Pedersen and Hendriksen (1993) demonstrated that bacterial responses may be quite complex, with different microbial species showing widely different patterns of change during transit through the earthworm gut. In this study, numbers of *E. coli* and *Pseudomonas putida* were reduced in earthworm casts compared to the ingested material, while concentrations of *Enterobacter cloacae* and *Aeromonas hydrophila* were similar. However, concentrations of *E. cloacae* declined by four orders of magnitude in the earthworm pharynx and/or crop before increasing to the initial levels further along the gut. They suggested that this was due to the activity of antibacterial factors secreted in the foregut of earthworms. Similar antibacterial factors have been demonstrated in the coelomic fluid of *E. andrei* (Lassègues et al. 1989; Valembois et al. 1991). In addition to such humoral antibacterial responses, the immunodefense system of earthworms, such as *E. fetida*, also includes cellular defenses such as phagocytosis (Cooper 1974; Dales and Kalaç 1992).

It is clear that the influence of earthworms and the vermicomposting process on human pathogens can be quite complex. Mechanisms by which human pathogens may be reduced or eliminated include the direct effects of mechanical disruption due to ingestion and the grinding action of the gizzard, microbial inhibition by antimicrobial substances or microbial antagonists produced by the earthworms themselves, and destruction of microorganisms by enzymatic digestion and assimilation. Indirect effects may include the stimulation of endemic or other microbial species leading to competition, antagonism, phagocytic activity both within the earthworm and without, production of antimicrobial substances such as humic acids, and numerous other possible mechanisms.

VI CONCLUSIONS

For conventional thermophilic composting, standards to ensure effective human pathogen reduction that have been implemented by the EPA are based on a solid foundation of scientific research. High temperatures, 55°C–70°C (131°F–158°F) for at least 3 days, are a convenient component of the composting process that can be monitored and managed to ensure adequate pathogen reduction. Based on energetic analyses, heat production in vermicomposting may be comparable to that in composting; however, in vermicomposting organic wastes are managed to encourage heat dissipation rather than heat accumulation, and temperatures do not normally exceed 30°C–35°C (86°F–95°F) without gravely disrupting the process. Given this important constraint on the vermicomposting process, the question often arises: Can effective pathogen reduction take place during vermicomposting without high temperatures? Based on the accumulating laboratory and field evidence from studies on human pathogens within vermicomposting, the answer is clearly "yes." Even in control materials that are "cold composted" without earthworms, reduction of pathogens from initial levels has been observed consistently, although usually not to the same degree as with vermicomposted materials. Earthworm activity appears to be a critical factor leading to the rapid reduction of pathogens.

In fact, many of the mechanisms that can lead to human pathogen reduction, without the need for increased temperatures, are fairly well understood. Unfortunately, what is not so well understood is which of these mechanisms are most important in the vermicomposting process, and what process parameters are necessary to ensure their effective functioning. Research that focuses specifically on the effects of differences in methods of vermicomposting on human pathogen reduction has yet to be conducted in great detail. This type of research is essential for the identification and specification of process standards on human pathogens that can be applied to vermicomposting.

Without such research, attempts to provide operating and regulatory guidelines for human pathogen reduction in vermicomposting, possibly prescribing minimum earthworm densities or maximum material turnover rates (for example, a weight-ratio of earthworm population to organic wastes of 1:10, or a turnover rate of 1 kg of material per day per kilogram of earthworms), would currently be based mainly on speculation rather than hard scientific evidence. Until then, well-proven methods of eliminating pathogens, such as thermophilic composting for short periods before using the vermicomposting process, will be necessary. Alternatively, routine sampling and analyses of finished vermicomposts for human-pathogen indicators will be required to ensure the quality and safety of the final product. The EPA has shown willingness to grant site-specific clearances for particular operations when adequate data have been presented to them, as was done for a site in Pennsylvania.

REFERENCES

Aguilar, A.L.B. 1996. Laboratory studies of vermicomposting municipal wastewater biosolids. Master's thesis, Jackson State University, Jackson, MS.

Brown, B.A., and Mitchell, M.J. 1981. Role of the earthworm, *Eisenia foetida*, in affecting survival of *Salmonella enteritidis* ser. *typhimurium*. *Pedobiologia* 22: 434–438.

Brüsewitz, G. 1959. Untersuchungen über den Einfluss des Regenwurms auf Zahl und Leistungen von Mikroorgansimen im Boden. *Arch. Mikrobio.* 33: 52–82.

Cliver, D.O. 1983. Workshop on research needs relating to soil absorption of wastewater. CEE-8213092, National Science Foundation, Washington, DC.

Cooper, E.L. 1974. *Invertebrate Immunology*. In: E.L. Cooper (Ed.), *Continued Topics in Immunology*, Vol. 4. Plenum, New York.

Dales, R.P., and Kalaç, Y. 1992. Phagocytic defence by the earthworm *Eisenia fetida* against certain pathogenic bacteria. *Comp. Biochem. Physiol.* 101A: 487–490.

Dash, M.C., Senapati, B.K., and Mishra, C.C. 1980. Nematode feeding by a tropical earthworm. *Oikos* 34: 322–326.

Day, G.M. 1950. Influence of earthworms on soil microorganisms. *Soil Sci.* 69: 175–184.

Eastman, B.R., Kane, P.N., Edwards, C.A., Trytek, L., and Gunadi, B. 2001. The effectiveness of vermiculture in human pathogen reduction for USEPA Class A Stabilization. *Compost Sci. Util.* 9: 38–49.

Environmental Protection Agency (EPA). 1980. *Compendium on Solid Waste Management by Vermicomposting*. EPA-600/8-80-033.

Environmental Protection Agency (EPA). 1999. *Environmental Regulation and Technology: Control of Pathogens and Vector Attraction in Sewage Sludge.* E.P.A/G25?R-92/013.

Finola, M., Rodriguez, C., and Beoletto, V. 1995. Bacteriology of the gastrointestinal tract of earthworm *Eisenia fetida* (Annelida, Oligochaeta) cultivated in composted broiler litter. *Rev. Argentina Microbiol.* 27: 210–213.

Haimi, J., and Huhta, V. 1987. Comparison of composts produced from identical wastes by "vermistabilization" and conventional composting. *Pedobiologia* 30: 137–144.

Harris, G.D., Platt, W.L., and Price, B.C. 1991. Vermicomposting in a rural community. In: *BioCycle* staff (Eds.), *The Art and Science of Composting*, 143–145. JG Press, Emmaus, PA.

Hartenstein, R. 1978. The most important problem in sludge management as seen by a biologist. In: R. Hartenstein (Ed.), *Utilization of Soil Organisms in Sludge Management*, 2–8. The National Technology Information Service. PB286932, Springfield, VA.

Lassègues, M., Roch, P., and Valembois, P. 1989. Antibacterial activity of *Eisenia fetida andrei* coelomic fluid: Evidence, induction, and animal protection. *J. Invertebr. Pathol.* 53: 1–6.

Lui, T.A. 1982. Earthworms, producers of biologically active substances. *Zh. Obshch. Biol.* 24: 149–154.

Mitchell, M.J. 1978. Role of invertebrates and microorganisms in sludge decomposition. In: R. Hartenstein (Ed.), *Utilization of Soil Organisms in Sludge Management*, 35–50. The National Technology Information Service. PB286932, Springfield, VA.

Murry, A.C., Jr., and Hinckley, L.S. 1992. Effect of the earthworm *Eisenia foetida* on *Salmonella enteritidis* in horse manure. *Biores. Technol.* 41: 97–100.

Pedersen, J.C., and Hendriksen, N.B. 1993. Effect of passage through the intestinal tract of detritivore earthworms (*Lumbricus* spp.) on the number of selected Gram-negative and total bacteria. *Biol. Fertil. Soils* 16: 227–232.

Riggle, D. 1996. Worm treatment produces "Class A" biosolids. *BioCycle*, October: 67–68.

Valembois, P., Seymour, J., and Roch, P. 1991. Evidence and cellular localization of an oxidative activity in the coelomic fluid of the earthworm *Eisenia fetida andrei. J. Invertebr. Pathol.* 57: 177–183.

CHAPTER 17

Heavy Metals, Earthworms, and Vermicomposts

Andrew John Morgan

CONTENTS

I Background Context ... 263
II Introduction ... 266
III Metal Accumulation by Earthworms ... 267
IV Metal Concentration and Availability during Vermicomposting ... 270
V Effects of Metal-Contaminated Composts on Plants ... 273
VI Composts in the Remediation of Metalliferous Soils ... 274
VII Effects of Metals on Earthworms ... 275
VIII Conclusions ... 276
Acknowledgments ... 277
References ... 277

I BACKGROUND CONTEXT

The disposal and management of various organic wastes present a challenge for many countries, as human populations increase with a concomitant rise in the intensity of animal husbandry and other agricultural practices. National and international statutory and policy changes introduced during the last decade reinforce the technical and public-perception challenges. For instance, the European Union Landfill Directive (Council Directive 1999/31/EC) requires all member states to divert biodegradable municipal waste away from direct landfill disposal and into alternative forms of treatment, with targets of 25% diversion compared with 1995 levels by 2010, 50% diversion by 2013, and 65% diversion by 2020 (A. Morgan et al. 2000). Diminishing availability of suitable landfill sites coupled with escalating landfill costs are (Department for Environment food and Rural Affairs 2007) significant drivers of these initiatives.

Compost is regarded as a recycled product and, therefore, contributes toward the achievement of these ambitious targets. The directive has been further extended

Table 17.1 Sewage Sludge Production in the United States and European Union between 1992 and 2005

Parameter	Year	United States	European Union
Sludge production (dry metric ton/annum)	1998	—[a]	5,511,000
	1998	6,900,000	6,588,000
	2005	7,600,000	8,331,000
	2010[b]	8,200,000	—[a]
Sludge used for "beneficial" purposes (% of total production)	1998	60	52
	2005	66	54
	2010[b]	70	—[a]

Source: Data selected and condensed from literature compiled by Iranpour, R., Cox, H.H.J., Kearney, R.J., Clark, J.H., Pincine, A.B., and Daigger, G.J., *J. Res. Sci. Technol.,* 1: 209–222, 2004.

[a] Data not available.

[b] Predicted.

under the Waste Strategy for England 2007 (Department for Environment Food and Rural Affairs 2007) through the establishment of the following action timetable: (i) the recycling and composting of household waste (at least 40% by 2010, 45% by 2015, 50% by 2020); and (ii) the recovery of municipal waste (53% by 2010, 67% by 2015, 75% by 2020). Moreover, the high levels of sewage-sludge production in the European Union, the United States (Table 17.1), and elsewhere are of particular concern. For example, the annual production of sewage sludge in the United Kingdom is over 1 million metric tons of dry solids, 62% of which is applied to agricultural land, 19% is incinerated, 11% is used for land reclamation, and the remaining 8% is dealt with in other ways, including composting and landfill (Parliamentary Office of Science and Technology 2007). The U.K. Biomass Strategy 2007 (Department of Trade and Industry 2007) used economic and abatement criteria to bolster a move away from incineration as a means of disposing of biosolid wastes.

The judicious application of biosolids, including composted sewage sludge, can improve the nutrient status and physical characteristics of agricultural land (Epstein 2000; Petersen et al. 2003; Mantovi et al. 2005). Recycling sludge by direct land spreading, or by application after stabilization, has benefited mine spoil reclamation (Tian et al. 2006), turfgrass growth (Loschinkohl and Boehm 2001; Cheng et al. 2007), and tree plantations (Fine et al. 2006; Moffat 2006; Wang et al. 2006). However, sludges, particularly from urban sources, contain not only significant quantities of nutrients that may move into surface waters but also heavy metals and metalloids. Both essential and nonessential metals in excess can interfere with soil microbial activities and plant growth, and they may also pose direct or indirect threats to wildlife, farm animals, and human health through the consumption of tainted crops. For these reasons, and despite the many potential and realized benefits of disposing of biosolids on land, tight regulatory controls on application rates to land have been formally introduced or recommended in a number of countries to limit the exposures of the biota to metal contaminants (Table 17.2). (A recent publication indicates that organic residues, including pharmaceuticals, originating from biosolids and manure

Table 17.2 A Comparison of the Biosolids Metal (and Metalloid) Limits Applied to Agricultural Soils in China*, Canada*, Japan*, EU, and the U.S.*****

Country	China	Japan	Canada		European Union		United States (EPA)	
Metal	In Fertilizer ($mg.kg^{-1}$)	In By-Product Phosphate Fertilizers ($mg.kg^{-1}$)	Maximum Acceptable Metal Concentration in Biosolids ($mg.kg^{-1}$)	Maximum Acceptable Cumulative Additions to Soil over 45 years ($kg.ha^{-1}$)	Limit Value Ranges for Metal Concentrations in Sludge ($mg.kg^{-1}$)	Limit to Metal Amount that may be Added per Annum to Soil ($kg.ha^{-1}.yr^{-1}$)	Limit (Ceiling) Values for Metal Concentrations in Biosolids ($mg.kg^{-1}$)	Cumulative Loading Rates ($kg.ha^{-1}$)
As	50	50	75	15	—	—	75	41
Cd	8	8	20	4	20–40	0.15	85	39
Cr	500	500	—	—	1000–1500	4	—	—
Cu	—	—	—	—	1000–1750	12	4300	1500
Pb	100	100	500	100	750–1200	15	840	300
Hg	5	5	5	1	16–25	0.1	57	17
Mo	—	—	20	4	—	—	75	—
Ni	—	—	180	36	300–400	3	420	420
Se	—	—	14	2.8	—	—	100	100
Zn	—	—	1850	370	2500–4000	30	7500	2800

Source: Hogg, D., Barth, J., Favoino, E., et al. *Comparison of compost standards within the EU, North America and Australasia. The Waste and Resources Action Programme (WRAP), 2002*, Banbury, Oxon: www.wrap.org.uk.; Iranpour, R., Cox, H.H.J., Kearney, R.J., Clark, J.H., Pincine, A.B., and Daigger, G.T., *J., Res. Sci. Technol.* 1: 209–222, 2004; Evanylo, G.K., *Agricultural land application of biosolids in Virginia: Regulations, 2009*. http://pubs.ext.vt.edu/452/452-302/452-302.html. *are also useful sources of data.*

Note: Concentration values expressed on a dry mass basis; Some E.U. member states such as Belgium, Denmark, Finland, the Netherlands, and Sweden have lower metal/metalloid limits than are stipulated in E.U. Directives. For instance, the limit values for the Netherlands are As, 15 $mg.kg^{-1}$; Cd, 1.25 $mg.kg^{-1}$; Cr, 75 $mg.kg^{-1}$; Cu, 75 $mg.kg^{-1}$; Hg, 0.75 $mg.kg^{-1}$; Ni, 30 $mg.kg^{-1}$; Pb, 100 $mg.kg^{-1}$; Zn, 300 $mg.kg^{-1}$ (Iranpour, R., Cox H.H.J., Kearney, R.J., Clark, J.H., Pincine, A.B., and Daigger, G.T. *J. Res. Sci. Technol.* 1: 209–222, 2004). Composts produced to the voluntary PAS 100 standard, introduced to facilitate best practice in compost production and use in the United Kingdom, are expected to comply with the following metal concentration limits: Cd, 1.5 $mg.kg^{-1}$; Cr, 100 $mg.kg^{-1}$; Cu, 200 $mg.kg^{-1}$; Hg, 1.0 $mg.kg^{-1}$; Ni, 50 $mg.kg^{-1}$; Pb, 200 $mg.kg^{-1}$; Zn, 400 $mg.kg^{-1}$ (British Standards Institution 2005; also available at http://www.environment-agency.gov.uk/static/documents/Business/compost_proof4_1721813.pdf). Besides metals, the PAS 100 standard also contains limits for human pathogens, phytotoxins, and weed propagules. The U.K. Forestry Commission advises that composts to be used in forests meet this standard.

* http://.afpc.net/World%20Sts%2004.pdf

** European Union, 2000, *Working document on sludge, 3rd draft*. ENV.E3/LM, European Union, Brussels, April 12, 2000, http://europa.eu.int/comm/environment/sludge/report10.htm

*** U.S. Environmental Protection Agency, *Federal Register* 55: 47210–47283,1990.

can be transferred to earthworms (Kinney et al. 2008), but the implications of this important observations falls outside the remit of the present review.)

While the total metal content of sewage sludges destined for land application is important, it is increasingly realized that from the ecotoxicology and environmental-risk perspectives, the bioavailable metal fractions, often a small proportion of the total loadings, are the critical ones (Amir et al. 2004; Fuentes et al. 2006). Bioavailable metal fractions in soils or other solid substrates are products of a difficult-to-model interplay between local biotic and abiotic factors acting on the intrinsic ligand-affinities and redox chemistries of given metals that influence the metals' mobility and behavior (McBride 2003). Consequently, there is not always a direct correlation between toxicity and "total" soil metal concentrations (Leita and De Nobili 1991; Allen and Janssen 2006). Therefore, unsurprisingly, a number of ameliorants, including lime (Wong and Selvan 2006) and clay minerals (Usman et al. 2005), have been advocated for reducing metal bioavailability in co-composted sewage sludge and in soils treated with metal-contaminated sludge. Moreover, the maximum permissible loadings of, for instance, Cu and Zn in sludge-amended soils are linked to pH (Smith 1994a, 1994b, 2009), the predominant factor modulating metal bioavailability (Sauvé 2002).

II INTRODUCTION

Implicit in the background information presented in the preceding section is the notion that sewage-sludge disposal strategies must attempt to balance two opposing desirable goals. On the one hand is the pressing need to dispose of a large proportion of sewage-sludge production on land; on the other hand, the legally binding obligation to remain within national permissible soil metal limits. The cessation of marine sludge dumping at the end of 1998 has, in Europe at least, brought this dilemma into sharp focus. Stated bluntly, legislation introduced to alleviate an environmental problem in one ecosystem has transferred the problem to another. Composting, including vermicomposting, is increasingly seen as a viable means of treating biosolids and other biodegradable organic waste materials (Epstein 2003; Frederickson 2002). Vermicomposting sewage sludge containing relatively high-metal levels, however, poses a number of important and specific ecotoxicological questions, not only because earthworms are recognized as effective accumulators of several metals but also because earthworms can act as "ecosystem engineers" (Lawton 1994), exerting profound effects on the physicochemical properties of the substrates that they inhabit. Thus, this review addresses the following questions:

- Do earthworms accumulate metals in their tissues from their surroundings? If so, how readily are the metals eliminated?
- Do earthworms alter the availabilities of metals during the vermicomposting process?
- Is there any evidence to suggest that toxic metals in a vermicompost can counteract the otherwise-positive nutritional effects of the amendment on plant growth?
- Do the organic ligands in a "clean" vermicompost have the capacity to sequester and immobilize metals when added to contaminated soils? In other words, does compost application render ameliorative as well as amendment effects?

- Does the metal content of sewage sludge alter the growth and reproductive performance of earthworms during vermicomposting? Is there any evidence that the direct application of metal-contaminated sludges or composted sludges has effects on earthworm populations in the field?

The literature available to address some of these questions, particularly on relationships between earthworms and environmental metals, is relatively rich. In other areas, particularly aspects of the metal content and behavior of vermicomposts as distinct from conventional composting products, the published literature is sparse. This Chapter, therefore, provides an overview rather than an exhaustive treatment of the former, whereas in the less data-rich areas a degree of extrapolation will be given. The normal practice of mixing vermicompost with soil, or with organic materials such as coir, to produce effective plant growth media, and the burgeoning practice of thermophilic precomposting to reduce human pathogens before vermicomposting biosolids (Frederickson 2002; Frederickson et al. 2007), blurs the distinctions between composted products. A shared feature of all composts is that they contain large quantities of nutrient-rich, readily-degradable organic matter that restricts metal availability via sorption, complexation, and precipitation reactions (Speier 2008). It is, of course, eminently plausible that the organic matrices of vermicomposts and composts possess distinct cation-exchange properties independent of the feedstock and reflect aspects of the functional biology of earthworms operating in an aerobic, nonthermophilic environment. Judicious extrapolation is justified if only to serve the heuristic function of stimulating more critical research on the temporal changes in speciation of individual metals and metalloids during vermicomposting, and on the short- and long-term effects of vermicompost on the bioavailabilities and toxic potentials of essential and nonessential metals in metalliferous soils.

III METAL ACCUMULATION BY EARTHWORMS

There is ample evidence that earthworm species belonging to each of the three ecophysiological categories (epigeic, endogeic, and anecic), including epigeic vermicomposting species such as *Eisenia fetida* and *Dendrobaena veneta*, are capable of accumulating a number of essential and nonessential metals from plant growth media and soils ranging from uncontaminated "background controls" to those that are highly metalliferous due to anthropogenic activities (Morgan et al. 1993; Peijnenburg 2002; Peijnenburg and Vijver 2009). Tissue metal accumulation is a reflection of the detritivorous lifestyle of earthworms, coupled with their highly permeable body walls and an extensive tissue comprised of chloragocytes with organelles able to sequester high concentrations of certain metals in relatively insoluble states (Morgan et al. 2002). A literature survey indicated that the degree of metal accumulation by earthworms is nonlinear, with a decrease at higher soil concentrations, and thus in general can be described by log-log linear regression models (Sample et al. 1998). However, Sample et al. (1998, 1999) concluded that the accumulation of the majority of metals/metalloids (Cd, Cu, Hg, Mn, Pb, and Zn) by earthworms can be described by simple regressions (the exceptions being As, Cr, and Ni). Both soil pH and organic-matter content contribute significantly to the accumulation of certain metals, notably Pb and Cd, but not others (Peijnenburg 2002) (Table 17.3). The important general point is that bioaccumulation is both physiologically and physicochemically driven. Environmental parameters, the target species, and the properties of individual metals and metalloids combine dynamically to modulate bioaccumulation (Luoma and Rainbow 2005). Luoma and Rainbow presented a biodynamic metal-bioaccumulation model (requiring site-specific geochemical data and species-specific physiological parameters) that was found to provide good predictions of accumulated metal concentrations in several different aquatic taxa. In principle, the biodynamic model should be applicable to soil-dwelling taxa, including earthworms, but this approach appears not to have been implemented to date.

Table 17.3 Regression Models Describing the Relationships between Equilibrated Whole-Worm (*Lumbricus rubellus*) Tissue Metal Concentrations and Soil Variables

Metal	Regression Equations[a]	R^2
Cd	$\log_{10}\mathbf{y} = 1.23 + 0.553\log_{10}\mathbf{x}$	85.6
	$\log_{10}\mathbf{y} = 1.34 + 0.566\log_{10}\mathbf{x} - 0.710\log_{10}\text{pH}$	85.6
	$\log_{10}\mathbf{y} = 1.93 + 0.480\log_{10}\mathbf{x} - 0.548\log_{10}\text{OM}$	89.7
	$\log_{10}\mathbf{y} = 2.37 + 0.519\log_{10}\mathbf{x} - 0.570\log_{10}\text{pH} - 0.585\log_{10}\text{OM}$	90.1
Cu	$\log_{10}\mathbf{y} = 0.944 + 0.291\log_{10}\mathbf{x}$	73.4
	$\log_{10}\mathbf{y} = 1.480 + 0.914\log_{10}\mathbf{x} - 0.310\log_{10}\text{OM}$	82.0
Pb	$\log_{10}\mathbf{y} = 0.801 + 0.642\log_{10}\mathbf{x}$	66.5
	$\log_{10}\mathbf{y} = 1.160 + 0.916\log_{10}\mathbf{x} - 0.326\log_{10}\text{Ca}$	81.2
	$\log_{10}\mathbf{y} = 2.650 + 0.897\log_{10}\mathbf{x} - 3.560\log_{10}\text{pH}$	93.3
	$\log_{10}\mathbf{y} = 2.970 + 0.849\log_{10}\mathbf{x} - 4.460\log_{10}\text{pH} - 0.134\log_{10}\text{Ca}$	94.1
Zn	$\log_{10}\mathbf{y} = 2.200 + 0.289\log_{10}\mathbf{x}$	72.3
	$\log_{10}\mathbf{y} = 1.860 + 0.250\log_{10}\mathbf{x} - 0.643\log_{10}\text{pH}$	74.2

Source: Data selected and derived from Morgan, J.E. and Morgan, A.J., *Environ. Pollut.* 54: 123–138, 1988; Corp, N. and Morgan, A.J., *Environ. Pollut.* 74: 39–52, 1991.

Note: (Neuhauser, E.F., Cukic, Z.V., Malecki, M.R., Leohr, R.C., and Durkin, P.R., *Environ. Pollut.* 89: 293–301, 1995) presented linear regression equations describing the relationships between Cd, Cu, Pb, Ni, and Zn concentrations in earthworm tissues and a number of salient variables in (a) sludge-amended soils and (b) a variety of soils in a number of independent published studies. Comprehensive sets of multivariate regression equations describing the relationships between accumulated metal concentrations in earthworm tissues and edaphic parameters have been published by (Sample, B.E., Suter, G.W., II, Beauchamp, J.J., and Efroymson, R.A., *Environ. Toxicol. Chem.* 18: 2110–2120, 1999) and (Peijnenburg, W.J.G.M., *Bioavailability of metals in terrestrial ecosystems: Importance of partitioning for bioavailability to invertebrates, microbes, and plants*, 89–112, 2002. Pensacola, FL: Society of Environmental Toxicology and Chemistry (SETAC)). See also the public access document by (Sample, B.E., Beauchamp, J.J., Efroymson, R.A., Suter, G.W., II, and Ashwood, T.L., Development and Validation of Bioaccumulation Models in Earthworms, 1998. http://www.esd.ornl.gov/programs/ecorisk/documents/tm220.pdf) and the review by (Nahmani, J., Hodson, M.E., and Black, S., *Environ. Pollut.* 145: 402–424, 2007).

[a] **y**, whole-worm tissue metal concentrations; **x**, concentrated HNO3-extractable "total" metal concentration, mg/kg dry mass; **OM**, organic-matter content; **Ca**, "total" calcium concentration.

Concentration factors (CF_{WS}; also referred to as *bioconcentration factors*, BCFs, or *uptake factors*, UFs) are the ratio of the concentration of a metal in earthworm tissue to the "total" concentration of the metal in soil. A comparison of median CF_{WS} derived from the published earthworm literature indicates that metals display different bioaccumulative behaviors in earthworms, where $CF_{WS} < 1$ represent no biomagnifications and $CF_{WS} > 1$ represent biomagnifications: As (0.23), Cd (14.26), Cr (0.16), Cu (0.64), Hg (3.93), Mn (0.06), Ni (0.78), Pb (0.23), Zn (3.78) (Sample et al. 1998). Different bioaccumulation efficiencies are not explained simply by the metabolic essentiality or nonessentiality of a given metal; neither are they explained by ligand-seeking affinities (e.g., Cu is biologically essential, Cd is nonessential, but both are considered to be Group B metals with strong affinities for S-donating ligands such as thiols; Nieboer and Richardson 1980). In terms of using CF_{WS} for hazard assessment it is paramount that the determined value represents the intrinsic properties of the chemical under consideration and does not vary with differences in environmental conditions. Unfortunately, the CF_{WS} of a number of metals in earthworms (Neuhauser et al. 1995) and aquatic biota (McGeer et al. 2003; DeForest et al. 2007) display strong tendencies toward inverse relationships with external metal concentrations. For this and other reasons, the organization for economic co-operation and development (OECD) guidance recommends that bioaccumulation data for aquatic organisms be used with caution in hazard schemes and on a case-by-case basis (McGeer et al. 2002).

Earthworms accumulate metals from the environment, but despite positive claims (e.g., Suthar 2008), this is not a viable way of ameliorating contaminated sludges or soils because the weight of evidence indicates that progressive mineralization tends to increase the total metal concentration of metals in the substrates. In other words, it is not feasible in principle or practice to deploy earthworms in a way analogous to how metal hyperaccumulating plants phytoremediate contaminated land. Moreover, the application to land of metal-containing vermicomposts, or any other contaminated amendment, whether raw or processed, will inevitably introduce metals into terrestrial food chains via earthworms, which are significant prey organisms (e.g., Ma et al. 1991; Pankakoski et al. 1993; Roodbergen et al. 2008). The general consensus is that while earthworms can render metals available to their predators, the concentrations of metals such as Pb, Zn, Cu, and (possibly) Cd do not biomagnify during transference to higher-trophic levels. This is a general predator–prey phenomenon noted for most metals in terrestrial (Hopkin and Martin 1984, 1985; Hegstrom and West 1989; Meharg et al. 1997) and aquatic (Luoma and Rainbow 2005; DeForest et al. 2007) food chains. The metal-assimilation efficiencies of predators are relatively low due to the inert, insoluble states in which metals are accumulated by earthworms and other prey organisms. It is noteworthy that earthworms transfer metal fractions both from "internal" cellular compartments and from the ingested contents of their voluminous alimentary canal. The apparent lack of trophic biomagnification does not mean that the accumulation of metals by earthworms during vermicomposting, or during field exposure in soils receiving metal-contaminated amendments, has no potentially serious ecotoxicological impacts on consumer species (Roodbergen et al. 2008).

In toxicology and ecotoxicology the relationship between exposure and effect is a fundamental consideration. In free-living natural populations of soil-dwelling organisms "exposure" is sometimes considered synonymous with the term *bioavailability.* Estimating bioavailability, and linking it to deleterious effects on individual organisms and populations, presents a number of conceptual as well as practical challenges (Sauvé 2002; Impellitteri et al. 2003), which probably explains why the current practice is to base soil screening and risk-assessment standards on total metal concentrations rather than incorporating more site-specific bioavailability criteria (Allen 2002). While it is beyond the scope of this review to explore this important issue in detail, it is germane to note that a number of methods and models for estimating metal bioavailability and toxicity to target organisms are now available and are being refined (Luoma and Rainbow 2005; Allen and Janssen 2006; Thakali et al. 2006).

IV METAL CONCENTRATION AND AVAILABILITY DURING VERMICOMPOSTING

Leita and De Nobili (1991) showed that composting increases total metal (Cd, Cu, Pb, Zn) concentrations compared with those in the original municipal solid waste. In the cases of Pb and Zn, there was a progressive decrease to very low levels in the water-extractable fractions during the early thermophilic phase of composting; in contrast, water extracts of Cd and Cu increased markedly over the thermophilic and mesophilic composting phases, before decreasing to low levels in the finished product (Leita and De Nobili 1991). On the whole there is a high degree of consensus in the literature (Table 17.4) that composting increases metal concentrations. The losses of readily biodegradable organic-matter components, such as carbohydrates, cellulose, hemicelluloses, and lipids, leaving more recalcitrant components such as lignin, drives the metal-concentrating phenomenon. This can be estimated (Veeken and Hamelers 2002) in a straightforward manner:

$$CF = 100 \div [100 - (OM_{ini} \times (OM_{deg}/100))]$$

where CF = metal concentration factor during composting; OM_{ini} = organic-matter content (as percentage of dry mass) of the feedstock; OM_{deg} = organic-matter degradation during composting (as percentage of OM_{ini}). Moreover, the extent of metal concentration from the beginning to the end of composting (CF) is a function both of the initial organic-matter content of the feedstock and of its degradability (Veeken and Hamelers 2002).

Composting also tends to stabilize metals (Paré et al. 1999; Table 17.4), with redistribution from relatively labile to more immobilized states. But Greenway and Song (2002) demonstrated that certain metals deviate from the general pattern and express mobility behaviors reflecting to some extent the properties of the feedstock. These authors recognized three overlapping metal groupings: (i) those with low-initial availability that become progressively even less available (e.g., Cu, Pb); (ii) those that are initially readily available and whose mobility is not substantially altered by composting (e.g., As, Cd, Co, Zn); and (iii) those where mobility appears to depend on the metal level and matrix properties of the composted materials (e.g., Cr, Ni).

Table 17.4 A Selection of Publications Indicating How the Total (i.e., Concentrated Acid Digestion) and Available (i.e., Extractable) Metal Contents of Feedstocks are Altered by Composting (C) and Vermicomposting (V)

Treatment	Metal	Change in "Total" Metal Conc.	Change in Extractable Metal Conc.	Authors
C (sewage sludge; city refuse; peat residue; grape debris)	Cd, Cr, Cu, Fe, Mn, Pb, Zn	↑	↓	Garcia et al. (1990; 1995)
C (sewage sludge)	Ni &Cr	↑ (30–36%)	↓	Zheng et al. (2006)
C (sewage sludge)	Cu, Pb, Zn	↓	↓	He et al. (2009)
C (paper fiber, green waste)	Ni	–	↓	Tandy et al. (2009a)*
	Pb	–	↔	
	Cu&Zn	–	↑	
		–		
C (*in situ* stabilized sediments)	Cu, Pb, Zn	–	↓	Yu et al. (2009)
V (pig manure)	Cu&Zn	↑ (25–30%)	↓ (35 to 55%)	Dominguez et al. (1997)
V (sludges)	"metals"	↑	↑	Elvira et al. (1995)
V (sewage sludge + cow dung)	Cu, Cr, Fe, Zn	↑	–	Gupta and Garg (2008); Yadav and Garg (2009)
V (fly ash + cow dung + soil)	Cr	↓ (10–50%)	–	Jain et al. (2004)
	Cu	↓ (17–33%)	–	
	Ni	↓ (8–48%)	–	
	Pb	↓ (31–56%)	–	
	Zn	↓ (11–36%)	–	
V (fly ash + cow dung)	Cr, Cu, Ni, Pb, Zn	↓ (10–50%)	–	Gupta et al. (2005)
V (sewage sludge)	Cu	↓ (29–52%)	–	Suthar and Singh (2008)
	Fe	↓ (13–20%)	–	
	Zn	↓ (15–26%)	–	
	Pb	↓ (5–47%)	–	
V (sewage sludge)	Cu	↓ (29–58%)	–	Suthar (2008)
	Fe	↓ (24–34%)	–	
	Mn	↓ (18–46%)	–	
	Zn	↓ (21–44%)	–	

Note: ↑ = signifies an increased concentration; ↓ = signifies a decreased concentration; ↔ = signifies no significant concentration change; – = not reported; values in parentheses are the % concentration change.

* See also: Farrell, M. and Jones, D.L., *Biores. Technol.*, 100: 4423–4432, 2009.

Whether similar changes in metal concentrations and availability occur during vermicomposting (or in sludge-amended soils) has not been fully resolved. For example, Hartenstein et al. (1980) reported significant progressive increases in the total concentrations of Cd, Cu, Ni, Pb, and Zn in aerobically digested sludge over a period of 11 days, whether earthworms (*Eisenia fetida*) were present or not. These authors also observed significant and similar increases over the same period in the 0.1 N HNO_3-extractable Cd, Cu, and Zn fractions in earthworm-added and earthworm-free sludges. The unusual study by Morgan and Morgan (1992) may be instructive because when the metal concentrations in the ingesta (crop contents) and egesta (casts) of field-collected earthworms inhabiting a metalliferous site were measured, the Zn and Pb concentrations in casts were often higher than in the crop contents. This implies a concentrating effect due to the mineralization of dietary organics during gut passage analogous to the processes taking place during composting. The study also implies that vermicomposting will not necessarily reduce the metal content of the earthworm-worked product, even though earthworms are acknowledged metal accumulators, and provides a clue to enable us to rationalize the literature. The initial concentrations of metals in the feedstock, combined with the inverse relationship between metal concentration factors (i.e., CF_{WS} = ratio of earthworm-to-soil metal concentrations) and soil or compost metal concentration, probably explain some of the disparities in the vermicomposting literature, where some authors observe increases in total metal concentrations during vermicomposting while others observe decreases (Table 17.4; Shahmansouri et al. 2005). This can be illustrated by a roughly worked hypothetical example:

> Take three otherwise-similar feedstocks containing 0.00016 oz. lb^{-1} (A) (10 $mg{\cdot}kg^{-1}$ (A)), 0.0016 oz. lb^{-1} (B) (100 $mg{\cdot}kg^{-1}$ (B)), and 0.016 oz. lb^{-1} (C) (1000 $mg{\cdot}kg^{-1}$ (C)), respectively, of a given metal. If each loses 50% of its (organic) mass during composting, then, in the absence of earthworms, the concentration of the metal will be increased two-fold in the final composts. The situation is more complex during vermicomposting because earthworms can accumulate metals in their tissues. If we assume that earthworm dry mass at the end of the vermicomposting process is the same as that of the vermicompost, and that the CF_{WS} values for the metals in the three composts are 5 (for A), 1 (B), and 0.5 (C) (some actual values for a number of metals in earthworms/vermicompost are given by Suthar (2008), then it follows that the metal concentration in vermicompost A will be significantly lower than in the original feedstock, the metal concentrations in vermicompost B and feedstock will be similar, and vermicompost C will have a significantly higher metal concentration than the feedstock. The fact that the CF_{WS} values of different metals differ explains some of the other apparent disparities.

Devliegher and Verstraete (1996, 1997) made significant contributions to our appreciation of the changes that ingested materials may undergo during passage through the earthworm alimentary system. These authors showed by direct observations that "nutrient-enrichment" and "gut-associated processes" in *Lumbricus terrestris* variously increased the plant availabilities of P (P_i), Mg, Ca, Fe, Mn, Cr, Co, and Cu by 6% to 39%; smaller or negligible increases were recorded for K, Zn, Cd, and Pb. However, it is plausible that vermicomposting is a two-phase process encompassing, on the one hand, events within the gut and, on the other, external events within the casts, mediated by microbial inocula. Thus, metal mobility may be both qualitatively and quantitatively different in fresh earthworm casts compared with mature, equilibrated vermicompost. The significantly reduced extractable fractions of Cu and Zn observed by Dominguez et al. (1997) in vermicomposted swine sludge, the appreciable decrease in pH during vermicomposting (Gupta and Garg 2008; Suthar 2008; Yadav and Garg 2009) notwithstanding, makes sense in light of the physicochemical characteristics of vermicompost: high-cation-exchange capacity and chelating groups (Carrasquero Durán et al. 2006), high-surface area (porosity), coupled with a maximum adsorptive capacity (for Cd) at pH 5.0 (Pereira and Arruda 2003). However, the apparent inconsistencies in the literature concerning the modulating effects of vermicomposting on metal mobilities (Table 17.4) indicate that the metal dynamics during vermicomposting is complex, possibly metal-specific and matrix dependent, and certainly warrants further studies with arrays of metal speciation tools.

V EFFECTS OF METAL-CONTAMINATED COMPOSTS ON PLANTS

The application of metal-contaminated composts to soil obviously increases their metal contents. Chemical analyses of vertical soil profiles indicate that the downward leaching from the cultivated zone of the metal inclusions of surface-applied sewage sludge is extremely slow in all but the most acidic soils, even when measured on a time scale of several years (McGrath and Lane 1989). An important practical implication of this has been articulated by McGrath and Lane (p. 254): "the majority of the metals applied in sludge remains in the surface soil after a long period. Effectively, once the upper limits set by the DOE (UK, Department of the Environment) or CEC (Commission of the European Communities) are achieved no more metals can be added to the soil." Another issue concerns the impact of nonbiodegradable metals derived from composted feedstocks, such as sewage sludge, on plant growth and productivity. This is not a straightforward phytotoxicity issue because the nutritional input from composts may to some extent counter metal-mediated toxicity by boosting plant performance. For example, not only do earthworm casts contain more C, N, P, K, and Ca than surrounding bulk soil, but the plant availabilities of certain nutrient elements such as Mg, P, and K are significantly raised in casts at least in the short term (Kale et al. 1992; Basker et al. 1993; Sharma et al. 2005).

The putative tendency of vermicomposting to immobilize heavy metals (Table 17.4) would suggest that metal phytoavailability and consequent phytotoxicity would be relatively low in vermicompost-amended soils. The weight of published evidence based on laboratory and field observations (Gondek and Filipek-Mazur

2003; Carrasquero Durán et al. 2006; Gondek, 2008) supports this view, although, as so often in this area, there are some apparently conflicting observations (Jadia and Fulekar 2008). There are two cautionary points to be made. First, in the long term, it is possible that edaphic processes will slowly degrade even the most recalcitrant organic macromolecules in vermicompost and, thus, release into the available pools metals that were previously sequestered in unavailable forms. The significance of such temporal effects has hitherto attracted but limited attention (Jordáo et al. 2006). Second, a recent review on the impact of earthworms on soil metals, both in terms of metal mobility and availability to soil-dwelling biota (Sizmur and Hodson 2009), concluded that the balance of evidence favors the view that earthworms increase metal mobility and availability although the mechanisms are unclear. For example, earthworm activity in a metalliferous field soil enhanced Cd, Cu, and Zn (but not Pb) accumulation by snails (Coeurdassier et al. 2007). These intriguing insights create a conundrum because they attest that earthworm activities in soil promote metal mobility, while vermicomposts possess strong metal-immobilizing properties. Perhaps temporal considerations offer a resolution to the apparent contradiction: Metals in the fresh casts of soil-dwelling earthworm species may be relatively mobile and available for plant uptake, but in mature vermicomposts the metals have equilibrated with sequestration sites on the organic matrix and are less readily available. These hypothesized, relatively short-term temporal events also warrant attention.

VI COMPOSTS IN THE REMEDIATION OF METALLIFEROUS SOILS

Up to this point the review has focused on the input of metals into soils by application of contaminated composts. However, it is emerging that composts, besides fertilizing soils, may exert substantial impacts on the mobilities of metals in contaminated field soils. For example, compost incorporation enhances the stabilization of mine-associated soils, thus reducing symptoms of ecotoxicity (Gadepalle et al. 2007; Tandy et al. 2009b), although these remediative effects may be moderated or reversed in certain composts (Van Herwijnen et al. 2007). Organic constituents of vermicomposts also have high-metal affinities (Pereira and Arruda 2003; Carrasquero Durán et al. 2006), and there is an emerging interest in the use of vermicomposts for removing metals from industrial effluents (Jordão et al. 2002, 2007; Jordão, Fernandes, Ribeiro, Barros, et al. 2009; Jordão, Fernandes, Ribeiro, Nascimento, et al. 2009). In these seminal studies vermicomposts were deployed as a cation-exchange matrix. There is a pressing need to determine the efficiencies of metal adsorption by vermicomposts derived from different feedstocks and also to compare, with and without immobilizing supplements such as zeolite clays, the metal-adsorption efficiencies of vermicomposts and a range of composts. If the production costs and metal-scavenging properties of vermicomposts prove favorable, it is anticipated that the *ex situ* and *in situ* remediation of contaminated effluents and soils will become a major potential applied outlet for "clean" vermicomposts.

VII EFFECTS OF METALS ON EARTHWORMS

There is a wealth of information available describing aspects of the effects of metals on earthworms at different levels of biological organization, from molecules (Spurgeon et al. 2008) and cells (Plytycz et al. 2007) to individual performance (Spurgeon et al. 2003) and population dynamics (Bindesbøl et al. 2006; Klok et al. 2007). The connecting thread in these studies is that direct or indirect toxic effects escalate from initial molecular-genetic targets to eventual earthworm population and community perturbations. There is no justification for repeating this material here. Rather, a few concise points will be made that are pertinent to the subject of the present review.

A comparison of the sublethal data for earthworms (Table 17.5) with U.K. statutory metal-concentration limits for sludge-treated agricultural soils (Table 17.2) is revealing. Cocoon production EC_{50} (56 days) values for Cd, Cu, and Zn (Pb on the boundary; not included in this discussion) are generally lower than the biosolids metal limits for the European Union and also the U.K. metal limits for sludge-amended soils. It is, therefore, possible for earthworms to encounter Cd, Cu, and Zn levels of contamination during vermicomposting, or in soils treated with metal-contaminated sludges or composts, that do not contravene statutory limits

Table 17.5 Literature-Derived Mean Data for the Acute and Sublethal Toxicity of Metals to Earthworms (*Eisenia fetida*) Exposed under Standard Conditions to OECD Artificial Soil

Metal	LC_{50} (14 days) ($mg.kg^{-1}$)	LC_{50} (56 days) ($mg.kg^{-1}$)	Estimated NOEC (56 days) Mortality ($mg.kg^{-1}$)	EC_{50} Cocoon Production (56 days) ($mg.kg^{-1}$)	Estimated NOEC[a] Cocoon Production (56 days) ($mg.kg^{-1}$)
Cd	>300 **[740]**	>300	>300	46.3	39.2
Cu	683 **[810]**	555	210	53.3	32.0
Pb	4480 **[5200]**	3760	2190	1940	1810
Zn	1010 **[1760]**	745	289	276	199

Source: Data derived from Spurgeon, D.J., Hopkin, S.P., and Jones, D.T., *Environ. Pollut.* 84: 123–130, 1994. LC_{50} values in bold are averages calculated from data published in a number of sources for a variety of lumbricid species; the original data were expressed in a number of different ways and obtained under a range of laboratory-exposure conditions but nevertheless give an indication of the consensus acute toxicities of the four metals.

Note: Concentration values expressed on a dry mass basis. See (Kula, H. and Larink, O. *Handbook of Soil Invertebrate Toxicity Tests*, Chichester, UK: John Wiley & Sons, 95–112, 1998.) for a formal description of earthworm toxicity test protocols and for data on the acute and sublethal effects of Cu on demographic parameters.

[a] No Observed Effect Concentration.

but are capable of compromising a key life-history parameter. The reduction in reproductive output of earthworms that has been observed in metal-contaminated feedstocks (Maboeta and van Rensburg 2003) could reduce the efficiency of vermicomposting. Earthworms with multigenerational exposure histories to metal- or metalloid-contaminated soils can evolve tolerance (Langdon et al. 2009). Whether there is a case for identifying metal-resistant clones of vermicomposting species for processing heavily metal-contaminated feedstocks is an interesting theoretical consideration but perhaps of doubtful practical utility, given that the metal-enriched vermicompost will be of little intrinsic value.

Contaminated feedstocks and the composts derived from them often contain mixtures of metals that can potentially interact to modulate the availabilities and potential toxic impacts of each individual constituent (Khalil et al. 1996), yet regulatory limits are based on individual metal concentrations. To deal with this reality, it seems that either predictive models such as that presented by Jonker et al. (2005), or robust bacterial assays as described by McGrath (2002), could be appropriate to assess the quality of vermicomposts before and after soil amendment. The advantage of direct biotic assays is that they are feedstock-, compost-, and site-specific. Another future option is the deployment of genomic profiling assays (e.g., Bundy et al. 2008; Owen et al. 2008) to evaluate the health of earthworms during vermicomposting and in soil-dwelling sentinel species at intervals after compost applications. In principle these assays, besides providing insight into molecular modes of toxicity, integrate the interactive effects of contaminants with each other and with all other salient environmental variables such as pH, organic-matter quality and quantity, temperature, and so on. The continuing decline in the costs of some of these powerful "global" technologies makes the approach plausible.

Organic residuals and biosolids, if properly managed, can provide excellent alternatives to chemical fertilizers for sustainable silviculture in nutrient-poor soils (Harrison and Henry 2001). However, caution should generally be exercised when applying contaminated materials to soils even after composting since the effects of metals on biota, including earthworms, and the ecological services that they provide can be subtle and yet significant (Eijsackers et al. 2008). It is clearly important to understand fully the temporal, speciation, and toxicological consequences of metal transfer within each link of the chain from the source in feedstocks, through vermicomposting, into the environmental sinks.

VIII CONCLUSIONS

- Vermicomposting produces two products: earthworms and vermicomposts. It would appear that there is scope for the use, with sound management, of the vermicompost, even if moderately contaminated with heavy metals, as a plant growth medium or fertilizer for land improvement.
- Earthworms can promote the remediation of metalliferous soils but cannot be viewed as a means of extracting metals from contaminated feedstocks or soils in a manner equivalent to phyto(metal) extraction.

- Vermicomposts, like other types of compost, often have a higher total metal concentration than the original contaminated feedstock.
- The concentrations of heavy metals discharged into soils and sediments, by whatever means, remain elevated for a very long time after the discharge has been discontinued.
- Metals are often stabilized (i.e., reduced mobility and bioavailability) in mature or equilibrated composts. This continues to be the case in the short term when composts are applied to soils. It is possible that metals slowly and progressively become more available in amended soils with time as the organic matrices degrade.
- Vermicompost applied to metal-contaminated soils can have a stabilizing effect (i.e., reduce solubility and bioavailability) akin to ameliorants such as lime and clays.
- Putative temporal events influencing metal stability need attention: (a) short-term changes in metal mobility from the formation of fresh earthworm casts to stabilized/equilibrated mature vermicompost; and (b) long-term changes as vermicompost degrades in soil.
- Vermicompost may be useful as an ex situ and in situ medium for contaminated land remediation and for extracting metals from effluents from diverse sources.
- The realization that many species of lumbricid earthworms are exceptionally genetically diverse (King et al. 2008; Pérez-Losada et al. 2009) raises the possibility that specific genotypes or strains of composting species can be identified or bred to increase the efficiency of vermicomposting particular feedstocks, including metal-contaminated materials.
- Terrestrial risk-assessment strategies could be improved by shifting the emphasis of legislation from blanket total soil (or compost) metal concentration limits to site-specific limits based on measures of free metal ion concentrations (Allen 2002; McGrath 2002). Leachable metal concentrations are generally more environmentally relevant than total concentrations (Barker and Bryson 2002); thus predictive and high-throughput physicochemical methods for estimating (bio)available metal fractions in composts and compost-amended soils are priorities (McGrath 2002; Song and Greenway 2005). Global earthworm biomarkers based on "omic" technologies may also have risk-assessment roles.
- Conservative and pragmatic limits on heavy metals in compost may be set that encourage contaminant-reduction measures and the recycling of composted residuals on the one hand, while also protecting soil from the potentially deleterious effects of long-term heavy metal accumulation on the other (Smith 2009).

ACKNOWLEDGMENTS

I thank Drusilla Lewis and Janet Sullivan for their secretarial support. This review was written during the tenure of a research grant (F0 0407/AI) awarded by The Leverhulme Trust.

REFERENCES

Allen, H.E. 2002. Terrestrial ecosystems: An overview. In *Bioavailability of metals in terrestrial ecosystems: Importance of partitioning for bioavailability to invertebrates, microbes, and plants*, ed. H.E. Allen, 1–5. Pensacola, FL: Society of Environmental Toxicology and Chemistry (SETAC).

Allen, H.E., and Janssen, C.R. 2006. Incorporating bioavailability into criteria for metals. In *Soil and water pollution monitoring, protection and remediation*, ed. I. Twardowska, H.E. Allen, M.M. Häggblom, and S. Stefaniak, 93–105. Amsterdam: Springer Netherlands.

Amir, S., Hafidi, M., Merlina, G., and Revel, J.C. 2004. Sequential extraction of heavy metals during composting of sewage sludge. *Chemosphere* 59: 801–810.

Barker, A.V., and Bryson, G.M. 2002. Bioremediation of heavy metals and organic toxicants by composting. *Sci. World J.* 2: 407–420.

Basker, A., MacGregor, A.N., and Kirkham, J.H. 1993. Exchangeable potassium and other cations in non-ingested soil and casts of two species of pasture earthworms. *Soil Biol. Biochem.* 25: 1673–1677.

Bindesbøl, A.-M., Bayley, M., Damgaard, C., and Holmstrup, M. 2006. Life-history traits and population growth rate in the laboratory of the earthworm *Dendrobaena octaedra* cultured in copper-contaminated soil. *Appl. Soil Ecol.* 35: 46–56.

British Standards Institution. 2005. *British Standards Institution. Specification for composted materials*. London: British Standards Institution Publicly Available Specification (BSI PAS 100).

Bundy, J.G., Sidhu, J.K., Rana, F., Spurgeon, D.J., Svendsen, C., Wren, J.F., Stürzenbaum, S.R., Morgan, A.J., and Kille, P. 2008. "Systems toxicology" approach shows coordinated metabolic responses to copper in a terrestrial invertebrate. *BMC Biol.* 6: 25 (doi:10.1186/1741-7007-6-25).

Carrasquero Durán, A., Flores I., Perozo, C., and Pernalete, Z. 2006. Immobilization of lead by a vermicompost and its effect on white bean (*Vigna sinensis* var. *Apure*) uptake. *Int. J. Environ. Sci. Technol.* 3: 203–212.

Cheng, H., Xu, W., Liu, J., Zhao, Q., He, Y., and Chen, G. 2007. Application of composted sewage sludge (CSS) as a soil amendment for turfgrass growth. *Ecol. Eng.* 29: 96–104.

Coeurdassier, M., Scheifler, R., de Vaufleury, A., Crini, N., Saccomani, C., Salomon Du Mont, L., and Badot, P.-M. 2007. Earthworms influence metal transfer from soil to snails. *Appl. Soil Ecol.* 35: 302–310.

Corp, N., and Morgan, A.J. 1991. Accumulation of heavy metals from polluted soils by the earthworm *Lumbricus rubellus*: Can laboratory exposure of "control" worms reduce biomonitoring problems? *Environ. Pollut.* 74: 39–52.

DeForest, D.K., Brix, K.V., and Adams, W.J. 2007. Assessing metal bioaccumulation in aquatic environments: The inverse relationship between bioaccumulation factors, trophic transfer factors and exposure concentration. *Aquat. Toxicol.* 84: 236–246.

Department for Environment Food and Rural Affairs 2007. *Waste Strategy for England 2007*. pp.123. © Crown Copyright 2007. Norwich, England: The Stationery Office.

Department of Trade and Industry 2007. UK Biomass Strategy 2007. *Working Paper 1 – Economic analysis of biomass energy*. pp. 70. London: Energy Technologies Unit, Department of Trade and Industry, May 2007, URN 07/950. http://www.berr.gov.uk/files/file39040.pdf

Devliegher, W., and Verstraete, W. 1996. *Lumbricus terrestris* in soil core experiment: Effects of nutrient enrichment processes (NEP) and gut-associated processes (GAP) on the availability of plant nutrients and heavy metals. *Soil Biol. Biochem.* 28: 489–496.

Devliegher, W., and Verstraete, W. 1997. The effects of *Lumbricus terrestris* on soil in relation to plant growth: Effects of nutrient-enrichment processes (NEP) and gut-associated processes (GAP). *Soil Biol. Biochem.* 29: 341–346.

Dominguez, J., Edwards, C.A., and Subler, S. 1997. Comparison of vermicomposting and composting. *Biocycle* 38: 57–59.

Eijsackers, H.J.P., Groot, M., and Breure, A.M. 2008. Upgrading system-oriented ecotoxicological research. *Sci. Total Environ.* 406: 373–384.

Elvira, C., Mato, S., and Nogales, R. 1995. Changes in heavy metal extractability and organic matter fractions after vermicomposting of sludges from a paper mill industry and wastewater treatment plant. *Fresenius Environ. Bull.* 4: 503–507.

Epstein, E. 2003. *Land application of sewage sludge and biosolids.* Boca Raton, FL: CRC Press.

European Union. 2000. *Working document on sludge, 3rd draft.* ENV.E3/LM, European Union, Brussels, April 12, 2000. http://europa.eu.int/comm/environment/sludge/report10.htm

Evanylo, G.K. 2009. *Agricultural land application of biosolids in Virginia: Regulations.* http://pubs.ext.vt.edu/452/452-302/452-302.html.

Farrell, M., and Jones, D.L. 2009. Heavy metal contamination of a mixed waste compost: Metal speciation and fate. *Biores. Technol.* 100: 4423–4432.

Fine, P., Atzmon, N., Adani, F., and Hass, A. 2006. Disposal of sewage effluent and biosolids in eucalyptus plantations: A lysimeter simulation study. In *Soil and water pollution monitoring, protection and remediation*, ed. I. Twardowska, H.E. Allen, M.M. Häggblom, and S. Stefaniak, 433–453. Amsterdam: Springer Netherlands.

Frederickson, J. 2002. *Vermicomposting trial at the Worm Research Centre: Part 1—Technical evaluation; Part 2—Financial evaluation and market potential.* http://www.worms.com/worm-pdfs/worm%20research%20centre2.pdf.

Frederickson, J., Howell, G., and Hobson, A.M. 2007. Effect of pre-composting and vermicomposting on compost characteristics. *Eur. J. Soil Biol.* 43: S320–S326.

Fuentes, A., Lloréns, M., Sáez, J., Aguilar, M.I., Pérez-Marin, A.B., Ortuño, J.F., and Meseguer, V.F. 2006. Ecotoxicity, phytotoxicity and extractability of heavy metals from different stabilized sewage sludges. *Environ. Pollut.* 143: 355–360.

Gadepalle, V.P., Ouki, S.K., van Herwijnen, R., and Hutchings, T. 2007. Immobilization of heavy metals in soil using natural and waste materials for vegetation establishment on contaminated sites. *Soil Sediment Contam.* 16: 233–251.

Garcia, C., Hernández, T., and Costa, F. 1990. The influence of composting and maturation processes on the heavy-metal extractability from some organic wastes. *Biol. Wastes* 31: 291–301.

Garcia, C., Hernández, T., Costa, F., and Polo, A. 1995. Effect of composting on sewage sludges contaminated with heavy metals. *Biores. Technol.* 53: 13–19.

Gondek, K. 2008. Chromium bioaccumulation from composts and vermicomposts based on tannery sludges. *J. Cent. Eur. Agric.* 9: 129–140.

Gondek, K., and Filipek-Mazur, B. 2003. Biomass yields of shoots and roots of plants cultivated in soil amended by vermicomposts based on tannery sludge and content of heavy metals in plant tissues. *Plant Soil Environ.* 49: 402–409.

Greenway, G.M., and Song, Q.J. 2002. Heavy metal speciation in the composting process. *J. Environ. Monitor.* 4: 300–305.

Gupta, R., and Garg, V.K. 2008. Stabilization of primary sewage during vermicomposting. *J. Hazard. Mater.* 153: 1023–1030.

Gupta, S.K., Tewari, A., Srivastava, R., Murthy, R.C., and Chandra, S. 2005. Potential of *Eisenia fetida* for sustainable and efficient vermicomposting of fly ash. *Water Air Soil Pollut.* 163: 293–302.

Harrison, R.B., and Henry, C.L. 2001. Biosolids and residuals application in forests of the United States. In *Biosolids Utilization in Forestry*, ed. J.-M. Carnus and B.-M. Mohammed, 4–10. Bordeaux, France: I'institut National de la Recherche Agronomique (INRA).

Hartenstein, R., Neuhauser, E.F., and Collier, J. 1980. Accumulation of heavy metals in the earthworm *Eisenia foetida*. *J. Environ. Qual.* 9: 23–26.

He, M., Tian, G., and Liang, X. 2009. Phytotoxicity and speciation of copper, zinc and lead during aerobic composting of sewage sludge. *J. Hazard. Mat.* 163: 671–677.

Hegstrom, L.J., and West, S.D. 1989. Heavy metal accumulation in small mammals following sewage sludge application to forests. *J. Environ. Qual.* 18: 345–349.

Hogg, D., Barth, J., Favoino, E., Centemero, M., Caimi, V., Amlinger, F., Devliegher, W., Brinton, W., and Antler, S. 2002. *Comparison of compost standards within the EU, North America and Australasia. The Waste and Resources Action Programme (WRAP).* Banbury, Oxon: www.wrap.org.uk.

Hopkin, S.P., and Martin, M.H. 1984. Assimilation of zinc, cadmium, lead and copper by the centipede *Lithobius variegatus* (Chilopoda). *J. Appl. Ecol.* 21: 538–547.

Hopkin, S.P., and Martin, M.H. 1985. Assimilation of zinc, cadmium, lead, copper, and iron by the spider *Dysdera crocata*, a predator of woodlice. *Bull. Environ. Contam. Toxicol.* 34: 183–187.

Impellitterri, C.A., Saxe, J.K., Cochran, M., Janssen, G.M.C.M., and Allen, H.E. 2003. Predicting the bioavailability of copper and zinc in soils: Modeling the partitioning of potentially available copper and zinc from soil solid to soil solution. *Environ. Toxicol. Chem.* 22: 1380–1386.

Iranpour, R., Cox, H.H.J., Kearney, R.J., Clark, J.H., Pincine, A.B., and Daigger, G.T. 2004. Regulations for biosolids land application in U.S. and European Union. *J. Res. Sci. Technol.* 1: 209–222.

Jadia, C.D., and Fulekar, M.H. 2008. Phytoremediation: The application of vermicompost to remove zinc, cadmium, copper, nickel and lead by sunflower plant. *Environ. Eng. Manage. J.* 7: 547–558.

Jain, K., Singh, J., Chauhan, L.K.S., Murthy, R.C., and Gupta, S.K. 2004. Modulation of flyash-induced genotoxicity in *Vicia faba* by vermicomposting. *Ecotoxicol. Environ. Safety* 59: 89–94.

Jonker, M.J., Svendsen, C., Bedaux, J.J.M., Bongers, M., and Kammenga, J.E. 2005. Significance testing synergistic/antagonistic, dose level-dependent, or dose ratio-dependent effects in mixture dose–response analysis. *Environ. Toxicol. Chem.* 24: 2701–2713.

Jordão, C.P., Fernandes, R.B.A., Ribeiro, K.L., Barros, P.M., Fontes, M.P.F., Souza, F.M.P. 2009. A study on Al(III) and Fe(II) ions sorption by cattle manure vermicompost. *Water, Air, & Soil Poll* (doi: 10.1007/s11270-009-0223-5).

Jordão, C.P., Fernandes, R.B.A., Ribeiro, K.L., Nascimento, B.S., and Barros, P.M. 2009. Zn(II) adsorption from synthetic solution and kaolin wastewater onto vermicompost. *J. Hazard. Mater.* 162: 804–811.

Jordão, C.P., Fialho, L.L., Neves, J.C.L., Cecon, P.R., Mendonça, E.S., and Fontes, R.L.F. 2007. Reduction of heavy metal contents in liquid effluents by vermicomposts and the use of the metal-enriched vermicomposts in lettuce cultivation. *Biores. Technol.* 98: 2800–2813.

Jordáo, C.P., Nascentes, C.C., Cecon, P.R., Fontes, R.L.F., and Pereira, J.L. 2006. Heavy metal availability in soil amended with composted urban solid wastes. *Environ. Monit. Assess.* 112: 309–326.

Jordão, C.P., Pereira, D.G., Einloft, R., Santana, M.B., Bellato, C.R., and Mello, J.W.V. 2002. Removal of Cu, Cr, Ni, Zn, and Cd from electroplating wastes and synthetic solutions by vermicompost of cattle manure. *J. Environ. Sci. Health*, Part A 37: 875–892.

Kale, R.D., Mallesh, B.C., Kubra, B., and Bagyaraj, D.J. 1992. Influence of vermicompost application on the available micronutrients and selected microbial populations in a paddy field. *Soil Biol. Biochem.* 24: 1317–1320.

Khalil, M.A., Abdel-Lateif, H.M., Bayoumi, B.M., Van Straalen, N.M., and Van Gestel, C.A.M. 1996. Effects of metals and metal mixtures on survival and cocoon production of the earthworm *Aporrectodea caliginosa*. *Pedobiologia* 40: 548–556.

King, R.A., Tibble, A.L., and Symondson, W.O.C. 2008. Opening a can of worms: Unprecedented sympatric cryptic diversity within British lumbricid earthworms. *Mol. Ecol.* 17: 4684–4698.

Kinney, C.A., Furlong, E.T., Kolpin, D.W., Burkhardt, M.R., Zaugg, S.D., Werner, S.L., Bossio, J.P., and Benotti, M.J. 2008. Bioaccumulation of pharmaceuticals and other anthropogenic waste indicators in earthworms from agricultural land amended with biosolid or swine manure. *Environ. Sci. Technol.* 42: 1863–1870.

Klok, C., Goedhart, P.W., and Vandecasteele, B. 2007. Field effects of pollutants in dynamic environments. A case study on earthworm populations in river floodplains contaminated with heavy metals. *Environ. Pollut.* 147: 26–31.

Kula, H., and Larink, O. 1998. Tests on the earthworms *Eisenia fetida* and *Aporrectodea caliginosa*. In *Handbook of Soil Invertebrate Toxicity Tests*, ed. H. Løkke and C.A.M. Van Gestel, 95–112. Chichester, UK: John Wiley & Sons.

Langdon, C.J., Morgan, A.J., Charnock, J.M., Semple, K.T., and Lowe, C.N. 2009. As-resistance in laboratory-reared F1, F2 and F3 generation offspring of the earthworm *Lumbricus rubellus* inhabiting an As-contaminated mine soil. *Environ. Pollut.* 157: 3114–3119.

Lawton, J.H. 1994. What do species do in ecosystems? *Oikos* 71: 367–374.

Leita, L., and De Nobili, M. 1991. Water-soluble fractions of heavy metals during composting of municipal solid waste. *J. Environ. Qual.* 20: 73–78.

Loschinkohl, C., and Boehm, M.J. 2001. Composted biosolids incorporation improves turfgrass establishment on disturbed urban soil and reduces leaf rust severity. *Hortic. Sci.* 36: 790–794.

Luoma, S.N., and Rainbow, P.S. 2005. Why is metal bioaccumulation so variable? Biodynamics as a unifying concept. *Environ. Sci. Technol.* 39: 1921–1931.

Ma, W.-C., Denneman, W., and Faber, J. 1991. Hazardous exposure of ground-living small mammals to cadmium and lead in contaminated terrestrial ecosystems. *Arch. Environ. Contam. Toxicol.* 20: 266–270.

Maboeta, M.S., and van Rensburg, L. 2003. Vermicomposting of industrially produced woodchips and sewage sludge utilizing *Eisenia fetida*. *Ecotoxicol. Environ. Safety* 56: 265–270.

Mantovi, P., Baldoni, G., and Toderi, G. 2005. Reuse of liquid, dewatered, and composted sewage sludge on agricultural land: Effect of long-term application on soil and crop. *Water Res.* 39: 289–296.

McBride, M.B. 2003. Toxic metals in sewage-amended soils: Has promotion of beneficial use discounted the risks? *Adv. Environ. Res.* 8: 5–19.

McGeer, J.C., Brix, K.V., Skeaff, J.M., DeForest, D.K., Brigham, S.I., Adams, W.J., and Green, A. 2003. Inverse relationship between bioconcentration factor and exposure concentration for metals: Implications for hazard assessment of metals in the aquatic environment. *Environ. Toxicol. Chem.* 22: 1017–1037.

McGeer, J., Hennengsen, G., Lanno, R., Fisher, N., Sappington, K., and Drexler, J. 2002. *Issue paper on the bioavailability and bioaccumulation of metals*. http://www.epa.gov/raf/publications/pdfs/BIOFINAL81904.pdf.

McGrath, S.P. 2002. Bioavailability of metals to soil microbes. In *Soil and water pollution monitoring, protection and remediation*, ed. I. Twardowska, H.E. Allen, M.M. Häggblom, and S. Stefaniak, 69–87. Amsterdam: Springer Netherlands.

McGrath, S.P., and Lane, P.W. 1989. An explanation for the apparent losses of metals in a long-term field experiment with sewage sludge. *Environ. Pollut.* 60: 235–256.

Meharg, A.A., Shore, R.F., Weeks, J.M., Osborn, D., Freestone, P., and French, M.C. 1997. Food-chain transfers of cadmium released into a woodland ecosystem by a chemical accident. *Toxicol. Environ. Chem.* 16: 928–938.

Moffat, A. 2006. *Use of sewage sludges and composts in forestry. Information note.* 79. Edinburgh: Forestry Commission. http://www.forestresearch.gov.uk.

Morgan A.J., Morgan, J.E., Turner, M., Winters, C., and Yarwood, A. 1993. Heavy metal relationships of earthworms. In *Ecotoxicology of Metals in Invertebrates*, ed. R. Dallinger and P.S. Rainbow, 333–358. Boca Raton, FL: Lewis.

Morgan, A.J., Turner, M.P., and Morgan, J.E. 2002. Morphological plasticity in metal-sequestering earthworm chloragocytes: Morphometric electron microscopy provides a biomarker of exposure in field populations. *Environ. Toxicol. Chem.* 21: 610–618.

Morgan, A.J., Williams, K.P., and Wright, L. 2000. A review of the role of composting in existing and future U.K. waste management practices, including data from trials designed to offer engineering solutions to vermicomposting challenges. In *Innovation in waste management. Proceedings of the IV European Waste Forum, Milan (Italy)*, 65–80. Milan: Centro di Ingegneria per la Protezione dell' Ambiente (C.I.P.A.).

Morgan, J.E., and Morgan, A.J. 1988. Earthworms as biological monitors of cadmium, copper, lead and zinc in metalliferous soils. *Environ. Pollut.* 54: 123–138.

Morgan, J.E., and Morgan, A.J. 1992. Heavy metal concentrations in the tissues, ingesta and faeces of ecophysiologically different earthworm species. *Soil Biol. Biochem.* 24: 1691–1697.

Nahmani, J., Hodson, M.E., and Black, S. 2007. A review of studies performed to assess metal uptake by earthworms. *Environ. Pollut.* 145: 402–424.

Neuhauser, E.F., Cukic, Z.V., Malecki, M.R., Leohr, R.C., and Durkin, P.R. 1995. Bioconcentration and biokinetics of heavy metals in earthworms. *Environ. Pollut.* 89: 293–301.

Nieboer, E., and Richardson, D.H.S. 1980. The replacement of the nondescript term "heavy metal" by the biologically and chemically significant classification of metal ions. *Environ. Pollut. (Series B)* 1: 3–26.

Owen, J., Hedley, B.A., Svendsen, C., Wren, J., Jonker, M.J., Hankard, P.K., Lister, L.J., Stürzenbaum, S.R., Morgan, A.J., Spurgeon, D.J., Blaxter, M.L., and Kille, P. 2008. Transcriptome profiling of developmental and xenobiotic responses in a keystone soil animal, the oligochaete annelid *Lumbricus rubellus*. *BMC Genomics* 9: 266 (doi:10.1186/1471-2164-9-266).

Pankakoski, E., Hyvärinen, H., Jalkanen, M., and Koivisto, I. 1993. Accumulation of heavy metals in the mole in Finland. *Environ. Pollut.* 80: 9–16.

Paré, T., Dinel, H., and Schitzer, M. 1999. Extractability of trace metals during co-composting of biosolids and municipal solid wastes. *Biol. Fertil. Soils* 29: 31–37.

Parliamentary Office of Science and Technology. 2007. *Energy and Sewage* 282: 1–4. London: The Parliamentary Office of Science and Technology. http://www.parliament.uk/parliamentary_office/post/pubs2007.cfm.

Peijnenburg, W.J.G.M. 2002. Bioavailability of metals to soil invertebrates. In *Bioavailability of metals in terrestrial ecosystems: Importance of partitioning for bioavailability to invertebrates, microbes, and plants*, ed. H.E. Allen, 89–112. Pensacola, FL: Society of Environmental Toxicology and Chemistry (SETAC).

Peijnenburg, W.J.G.M., and Vijver, M.G. 2009. Earthworms and their use in eco(toxico) logical modelling. In *Emerging topics in ecotoxicology: Principles, Approaches and Perspectives*, Vol. 2, ed. J. Devillers, 177–204. New York: Springer US.

Pereira, M.G., and Arruda, M.A.Z. 2003. Vermicompost as a natural absorbent material: Characterization and potentialities for cadmium adsorption. *J. Braz. Chem. Soc.* 14: 39–47.

Pérez-Losada, M., Ricoy, M., Marshall, J.C., and Dominguez, J. 2009. Phylogenetic assessment of the earthworm *Aporrectodea caliginosa* species complex (Oligochaeta: Lumbricidae) based on mitochondrial and nuclear DNA sequences. *Mol. Phylogenet. Evol* 52: 293–302.

Petersen, S.O., Hendriksen, K., Mortensen, G.K., Krogh, P.H., Brandt, K.K., Sørensen, J., Madsen, T., Petersen, J., and Grøn, C. 2003. Recycling of sewage sludge and household compost to arable land: Fate and effects of organic contaminants, and impact on soil fertility. *Soil Tillage Res.* 72: 139–152.

Plytycz, B., Homa, J., Aziz, A., Molnár, L., Kille, P., and Morgan, A.J. 2007. Earthworms for monitoring metal contamination: From cells to molecules. In *Environmental Pollution: New Research*, ed. R.H. Plattenberg, 47–98. New York: Nova Science.

Roodbergen, M., Klok, C., and van der Hout, A. 2008. Transfer of heavy metals in the food chain earthworm Black-tailed godwit (*Limosa limosa*): Comparison of a polluted and a reference site in the Netherlands. *Sci. Total Environ.* 406: 407–412.

Sample, B.E., Beauchamp, J.J., Efroymson, R.A., Suter, G.W., II, and Ashwood, T.L. 1998. Development and Validation of Bioaccumulation Models in Earthworms. http://www.esd.ornl.gov/programs/ecorisk/documents/tm220.pdf.

Sample, B.E., Suter, G.W., II, Beauchamp, J.J., and Efroymson, R.A. 1999. Literature-derived bioaccumulation models for earthworms: Development and validation. *Environ. Toxicol. Chem.* 18: 2110–2120.

Sauvé, S. 2002. Speciation of metals in soils. In *Bioavailability of metals in terrestrial ecosystems: Importance of partitioning for bioavailability to invertebrates, microbes, and plants*, ed. H.E. Allen, 7–37. Pensacola, FL: Society of Environmental Toxicology and Chemistry (SETAC).

Shahmansouri, M.R., Pourmoghadas, H., Parvaresh, A.R., and Alidai, H. 2005. Heavy metals bioaccumulation by Iranian and Australian earthworms (*Eisenia fetida*) in the sewage sludge vermicomposting. *Iran. J. Environ. Health Sci. Eng.* 2: 28–32.

Sharma, S., Pradhan, K., and Vasudevan, P. 2005. Potentiality of earthworms for waste management and in other uses—a review. *J. Am. Sci.* 1: 4–16.

Sizmur, T., and Hodson, M.E. 2009. Do earthworms impact metal mobility and availability in soil?—A review. *Environ. Pollut.* 157: 1981–1989.

Smith, S.R. 1994a. Effect of soil pH on availability to crops of metals in sewage sludge-treated soils. I. Nickel, copper and zinc uptake and toxicity to ryegrass. *Environ. Pollut.* 85: 321–327.

Smith, S.R. 1994b. Effect of soil pH on availability to crops of metals in sewage sludge-treated soils. II. Cadmium uptake by crops and implications for human dietary intake. *Environ. Pollut.* 86: 5–15.

Smith, S.R. 2009. A critical review of the bioavailability and impacts of heavy metals in municipal solid waste composts compared with sewage sludge. *Environ. Int.* 35: 142–156.

Song, Q.J., and Greenway, G.M. 2005. Trace element availability in compost. *Spectrosc. Eur.* 17: 10–15.

Speier, T.W. 2008. Relationship between biochemical activity and metal concentration in soil amended with sewage sludge. In *Developments in Soil Science, 32*, ed. R. Naidu, 261–280. Oxford: Elsevier.

Spurgeon, D.J., Hopkin, S.P., and Jones, D.T. 1994. Effects of cadmium, copper, lead and zinc on growth, reproduction and survival of the earthworm *Eisenia fetida* (Savigny): Assessing the environmental impact of point-source metal contamination in terrestrial ecosystems. *Environ. Pollut.* 84: 123–130.

Spurgeon, D.J., Morgan, A.J., and Kille, P. 2008. Current research in soil invertebrate ecotoxicogenomics. In *Comparative toxicogenomics*, ed. C. Hogstrand and P. Kille, 133–163. Amsterdam: Elsevier.

Spurgeon, D.J., Weeks, J.M., and Van Gestel, C.A.M. 2003. A summary of eleven years progress in earthworm ecotoxicology. *Pedobiol.* 47: 588–606.

Suthar, S. 2008. Metal remediation from partially composted distillery sludge using composting earthworm *Eisenia fetida*. *J. Environ. Monit.* 10: 1099–1106.

Suthar, S., and Singh, S. 2008. Bioconcentrations of metals (Fe, Cu, Zn, Pb) in earthworms (*Eisenia fetida*) inoculated in municipal sewage sludge: Do earthworms pose a possible risk of terrestrial food chains contamination? *Environ. Toxicol.* 24: 25–32.

Tandy, S., Healey, J.R., Nason, M.A., Williamson, J.C., and Jones, D.L. 2009a. Heavy metal fractionation during the co-composting of biosolids, deinking paper fibre and green waste. *Biores. Technol.* 100: 4220–4226.

Tandy, S., Healey, J.R., Nason, M.A., Williamson, J.C., and Jones, D.L. 2009b. Remediation of metal polluted mine soil with compost: Co-composting versus incorporation. *Environ. Pollut.* 157: 690–697.

Thakali, S., Allen, H.E., Di Toro, D.M., Ponizovsky, A.A., Rooney, C.P., Zhao, F.J., McGrath, S.P. et al. 2006. Terrestrial Biotic Ligand Model. 2. Application to Ni and Cu toxicities to plants, invertebrates, and microbes in soil. *Environ. Sci. Technol.* 40: 7094–7100.

Tian, G., Granato, T.C., Pietz, R.I., Carlson, C.R., and Abedin, Z. 2006. Effects of long-term applications of biosolids for land reclamation on surface water chemistry. *J. Environ. Qual.* 35: 101–113.

U.S. Environmental Protection Agency. 1990. 40 CFR Part 503: National Sewage Sludge Survey: Availability of information and data, and anticipated impacts on proposed regulations; proposed rule. *Federal Register* 55: 47210–47283.

Usman, A., Kuzyakov, Y., and Stahr, K. 2005. Effect of clay minerals on immobilization of heavy metals and microbial activity in a sewage sludge-contaminated soil. *J. Soils Sediments* 5: 245–252.

Van Herwijnen, R., Laverye, T., Poole, J., Hodson, M.K., and Hutchings, T.R. 2007. The effects of organic materials on the mobility and toxicity of metals in contaminated soils. *Appl. Geochem.* 22: 2422–2434.

Veeken, A., and Hamelers, B. 2002. Sources of Cd, Cu, Pb and Zn in biowaste. *Sci. Total Environ.* 300: 87–98.

Wang, H., Kimberley, M.O., Magesan, G.N., McKinley, R.B., Lee, J.R., Lavery, J.M., Hodgkiss, P.D.F., et al. 2006. Midrotation effects of biosolids application on tree growth and wood properties in a *Pinus radiata* plantation. *Can. J. For. Res.* 36: 1921–1930.

Wong, J.W.C., and Selvan, A. 2006. Speciation of heavy metals during co-composting of sewage sludge with lime. *Chemosphere* 63: 980–986.

Yadav, A., and Garg, V.K. 2009. Feasibility of nutrient recovery from industrial sludge by vermicomposting technology. *J. Hazard. Mater.* 162: 430–439.

Yu, G., Lei, H., Bai, T., Li, Z., Yu, Q., and Song, X. 2009. *In situ* stabilisation followed by *ex situ* composting for treatment and disposal of heavy metals polluted sediments. *J. Environ. Sci.* 21: 877–883.

Zheng, G.-D., Gao, D., Chen, T.-B., and Luo, W. 2006. Stabilization of nickel and chromium in sewage sludge during aerobic composting. *J. Hazard Mater.* 142: 216–221.

CHAPTER **18**

Quality Criteria for Vermicomposts

Clive A. Edwards, Scott Subler, and Norman Arancon

CONTENTS

I Introduction 288
II Definitions of Composts 289
III Definitions of Vermicomposts 289
IV Quality Criteria for Vermicomposts 292
A Physical Characteristics of Vermicomposts 292
1 Solids Composition (Organic Matter, Ash, and Inert Materials) 292
2 Bulk Density, Total Porosity, Particle Size Distribution, and Aggregate Stability 293
3 Moisture Content 294
4 Water-Retention Characteristics 295
B Chemical Characteristics of Vermicomposts 295
1 Acidity 295
2 Cation-Exchange Capacity (CEC) 295
3 Soluble Salt Concentrations 296
C Plant Nutrient Content 296
1 Total Carbon 296
2 Total Nitrogen 297
3 Carbon:Nitrogen Ratio 297
4 Total Phosphorus and Potassium 297
5 Calcium, Magnesium, Sulfur, and Boron 297
6 Ammonium and Nitrate Nitrogen 297
7 Micronutrients (Bo, Cu, Fe, Mn, Zn) 298
8 Nutrient-Release Characteristics 298
9 Heavy Metal Contamination 298
D Biological Characteristics 298
1 Human/Animal Pathogens 298
2 Enzymatic Activity 299

3 Microbial and Faunal Community Composition in Vermicomposts....300
V Conclusions....300
References....301

I INTRODUCTION

The need for the development of quality standards for organic materials to be used as soil amendments, soil conditioners, organic fertilizers, and plant growth media has been recognized widely. Standards for thermophilically composted materials are being developed, or are already in effect, in many states in the United States, in European countries and Australia, and elsewhere around the world. Until the last 5 years, vermicomposts and earthworm castings, which are produced through the activity of epigeic species of earthworms (primarily the species *Eisenia fetida*) on organic materials, have been viewed for the most part as a very interesting, but relatively unimportant, class of products, comprising only a small fraction of the total of thermophilically composted materials that are produced and marketed.

However, more recently, increasing attention has focused on the potential of innovative vermicomposting technologies that can be utilized commercially and at institutional scales, particularly with certain types of organic materials that have produced problems for conventional thermophilic composting (for example, those with excessive moisture or odor problems). Successful examples of large-scale vermicomposting operations with pig manure in North Carolina, food wastes in Oregon, cattle manure in New York, green wastes in California, and organic wastes in other countries such as Mexico, Hong Kong, China, India, the Philippines, and Australia demonstrate the viability of these new biotechnologies and provide a positive and expanding view of the future role of vermicomposting in the organics recycling industry (see Chapters 25, 29, 31, 32, and 33).

Scientific research is also establishing a strong foundation for understanding the many pronounced benefits that vermicomposts can provide for plant growth, especially when utilized as components of container media for greenhouse crops or as organic fertilizer amendments for high-value horticultural crops. It appears that vermicomposts have many soil-fertility benefits that extend beyond total plant nutrient content, such as promoting the activity of beneficial microorganisms in the soil, increasing the overall nutrient-supplying power of the soil, suppressing certain plant diseases and pests, producing plant growth regulators, and improving overall plant health. There also appear to be interesting differences between vermicomposts and thermophilic composts produced from the same organic feedstocks, perhaps due to the distinctly different natures of the vermicomposting and composting processes. Most important is the vastly increased microbiological activity of vermicomposts compared with thermophilic composts.

The aims of this chapter are to discuss various aspects of vermicompost quality that might be important to (a) producers of vermicomposts, (b) users of vermicomposts, and (c) state and federal regulators concerned with developing quality standards for composts and soil amendments for use in marketplace.

Good plant growth media serve the major functions of supplying air, water, and nutrients to plants, as well providing an essential means of physical support for their roots. Furthermore, there is increasing recognition of the role that plant growth media play as substrates for biological activity, which can provide additional important functions such as slow nutrient release, reduced nutrient losses, enhanced micronutrient availability, suppression of and resistance to various plant diseases and pests, and production of plant growth regulating compounds (PGRs).

II DEFINITIONS OF COMPOSTS

The dictionary definition of *compost* is "Plant material that has undergone decay from the effects of bacteria, with or without chemicals, used for putting on the earth for helping the growth of plants" (Graham 1965). To the housewife or gardener, compost is any form of organic waste such as grass clippings or fallen leaves that has been decomposed in aerated containers or heaps. To regulatory agencies such as the USDA National Organic Standards Board, *compost* is defined as "Organic matter of plant and/or animal origin managed to promote aerobic decomposition and an increase in temperature, to enhance its physical and nutritious properties as a soil amendment while minimizing pathogenic organisms. Compost must achieve a minimum temperature of at least 55°C (131°F) and remain there or a minimum of 3 days." In practice thermophilic composting of organic wastes begins in the mesophilic range 10–40.5°C (50–105°F), and compost must be turned or managed (e.g., in containers) so that all of the feedstock heats to the minimum temperature required.

Chemically, thermophilic composts are extremely complex and depend on the nature of the original feedstock and the conditions under which it was decomposed. Carbon is the most abundant element in composts and may constitute as much as 50% of the total mass of a compost; by comparison, nitrogen constitutes only 1–2%. The C:N ratio is a common index used to assess the state of feedstocks and the maturity and stability of any given compost. C:N ratios in finished composts range from 12:1 to 20:1 but ideally are between 14:1 and 18:1. Microbial stability is a prerequisite for compost maturity. During curing, composting particles become smaller and the total volume of the original feedstock is reduced, usually by 30–50%. Composting helps to disinfect the compost by killing pathogens, sterilizing weed seeds, decomposing many toxic substances, and stabilizing the nutrients in the compost.

Many countries have been working to define compost quality criteria. Work on these definitions is quite well advanced in Europe but less advanced in the United States (although a number of states have developed quality standards), and it is well evolved in Australia.

III DEFINITIONS OF VERMICOMPOSTS

Vermicomposting (Latin *vermes* = worms) is a similar process to composting, featuring the addition of certain epigeic species of earthworms that are used to enhance the process of aerobic waste conversion and produce a better end product. Vermicomposting differs from composting in several ways. Mainly, vermicomposting is a mesophilic process utilizing microorganisms and earthworms that are active in a temperature range of 10–32.2°C (50–90°F). The process is considered to be faster than thermophilic composting, and because the material passes through the earthworm gut at least once, a significant but not fully understood transformation takes place, whereby the resulting earthworm castings are extremely abundant in microbial activity and plant growth regulators and fortified with pest-repellency attributes as well. In many ways it seems that the earthworms are acting as a kind of ecological engineer that consumes the organic wastes, breaks them down into very small particles, decreases the C:N ratio, and stabilizes them rapidly.

Many producers and marketers of vermicomposts prefer to distinguish their products as being different from conventional composts by using the term *vermicasts* or simply *earthworm castings* when describing and marketing their products. Although certainly more specific, this term implies that the material in question is made up entirely of organic matter that has passed through the earthworm digestive tract or gut and been excreted by earthworms (as castings). In fact, in some finished vermicomposts it may be difficult or impossible to distinguish actual earthworm castings from associated organic materials that may not have passed through the gut of an earthworm but nonetheless have undergone physical and biochemical transformations during the vermicomposting process characteristic of advanced stabilization and humification. This is a quality issue, the finer castings being classed as a much more valuable material.

We recommend the use of the term *vermicomposting* to apply specifically to those situations where the predominant process leading to the stabilization and humification of organic materials is ingestion, fragmentation, and mixing by earthworms and microorganisms. Following this definition, we suggest that the terms *vermicompost* and *earthworm castings* (or *vermicastings*) may be used interchangeably to describe the resulting products of the vermicomposting process. This would provide stricter, process-based definitions for those products that will be utilized or marketed on a commercial scale.

These definitions may become particularly important in situations where regulations and permits may apply quite differently to the two quite different processes of composting and vermicomposting. For example, in some states, such as California and Oregon, vermiculture and associated activities, such as vermicomposting, are considered to be agricultural activities and therefore not subject to the same kinds of regulations as other organic-waste-handling processes, such as thermophilic composting, even though the actual amounts and much of the handling of similar waste materials might be nearly identical. In these situations, clarifying the distinctions between the two processes can be very important. In other situations, it is necessary to evaluate the validity of claims that products were derived from vermicomposting and could be marketed as such. We need to emphasize the importance of being able to classify a process as a vermicomposting process, or to categorize a product as vermicompost (or earthworm castings), based on certain defined process standards, in the same way that attempts to develop compost standards are used to define mature thermophilic composts with regard to stability and pathogen and weed seed sanitation.

In vermicomposting, the unifying standard could be some defined measure of the degree of earthworm activity relative to the quantity of material being processed in a given period of time. The critical, unifying characteristic of institutional and commercial-scale vermicomposting is that the predominant process is the fragmentation, mixing, ingestion, and changes in particle size of the bulk of the organic material by earthworms, leading to accelerated stabilization and humification at mesophilic temperatures and under aerobic conditions. In practice, describing the degree to which a process fulfills this definition needs to take into account the mass of earthworms per area or system, the quantity of material processed per day or week, the method of processing and the length of time (i.e., rate of material throughput per unit mass of earthworms). Existing values for maximal material throughput as determined in laboratory studies vary widely. However, a general rule of thumb that appears to hold up well when compared to the available scientific literature is that 0.5 kg (1lb) of earthworms (live weight) can consume (process.) about 0.5 kg (1lb) of organic material (at 75–85% moisture) per day. A satisfactory population of earthworms has been defined as at least 9–18 kg of earthworms m^{-2} (2–4 lb ft^{-2}) (Edwards and Arancon 2004).

The process definitions given in the preceding suggest immediately a number of ways that a particular vermiculture operation could be assessed to provide useful information on its status as a vermicomposting process. The first would simply be to monitor temperatures within the processing material—more than an occasional occurrence of thermophilic temperatures above 45°C (113°F) would indicate that the predominant process was probably conventional thermophilic composting rather than vermicomposting. A second method would be to evaluate the aerobicity of the material being processed, anaerobic conditions being one indication of fermentation processes rather than aerobic vermicomposting. In practice, this may be somewhat difficult to assess, since the relatively high moisture requirements for vermicomposting can often lead to sporadic anaerobic conditions or to anaerobic layers of earthworm-stabilized material developing in the bottom layers of a vermicompost pile. A third method that provides the most direct indication of whether a process is a vermicomposting process or not would be to determine the mass of earthworms present in the processing material and calculate the ratio of earthworm mass to material throughput. If the average ratio is much less than 0.5 kg (1 lb) of earthworms for each 0.5 kg (lb) of feedstock per day, then this might indicate that earthworm activity is not the predominant process in the stabilization of these materials.

To reiterate, the mere presence of earthworms in a pile, windrow, pit, or reactor vessel full of decomposing organic materials does not, in itself, define it as a vermicomposting process. It is not at all uncommon for seemingly large numbers of earthworms to thrive close to the outer edges of compost piles or containers, even though the bulk of the material may be undergoing thermophilic microbial decomposition (i.e., conventional composting). It is also not uncommon for certain epigeic species of earthworms to be active in the aerobic surface layers of otherwise-anaerobic sludge ponds, trickling filters, or lagoons, even though the predominant stabilization process in the sludge may be anaerobic fermentation. For a process to be characterized legitimately as vermicomposting, it must meet the criteria defined in the preceding: The predominant process leading to the stabilization and humification of organic materials must be fragmentation, ingestion, mixing, and breakdown by earthworms, thereby stimulating large amounts of aerobic mesophilic microbial activity to degrade the wastes.

IV QUALITY CRITERIA FOR VERMICOMPOSTS

The intended end use of a vermicomposted waste helps to determine which of its characteristics may be important in assessing its quality, so it is important to list all of the potential quality criteria for vermicomposts and how they may be assessed. When specifying details of a vermicomposting operation, or evaluating vermicompost products, information about certain aspects of the vermicomposting process may be important. The vermicomposting system used (e.g., bin, windrow, wedge, continuous-flow reactor system, etc.) should be specified, as should the primary raw materials (feedstocks). Differences in process types and raw materials have been shown to have a very strong influence on the quality of the end product and its suitability for particular applications and value (Edwards and Arancon 2004). Any preprocessing, such as leaching or precomposting, should also be described, since this may have considerable effects on qualities such as available nutrient content, contamination with viable weed seeds, or human pathogens. Any amendments with inorganic materials made during the vermicomposting process, such as application of fertilizers, lime, sulfur, and so on, should also be noted, since these may have effects on the final chemical characteristics of the vermicompost, as well as possibly affecting its certification as an "organic" material.

Other information essential for characterizing a vermicomposting operation includes an accurate assessment of the earthworm species used and the average population density in the system, the processing time (material throughput) for a particular batch, and the duration of storage of processed material or vermicomposts. All of these can affect the quality, stability, and maturity of the resulting product. As already mentioned, the rate of material throughput per total weight of earthworms may be a very important criterion for classifying and registering vermicomposting operations and their resulting vermicomposts.

A Physical Characteristics of Vermicomposts

1 Solids Composition (Organic Matter, Ash, and Inert Materials)

A fundamental measure of vermicompost's physical composition is its organic-matter content. In reality, this measure provides only a very crude indication of the overall stability of the material. However, the portion of the organic matter that is not completely stabilized (i.e., the biodegradable organic matter) does represent an important property, since it provides the energy sources for biological activity in the soil and is the source of potentially mineralizable nutrients. In a finished vermicompost, the organic-matter content should be greater than 20–25% (but probably less than 50%). The ash content is simply the nonvolatile solids of the compost, excluding inert particulate materials such as glass, metal, plastic, gravel, and large clay aggregates. In high-quality vermicomposts, contamination with inert materials should be very low, less than 0.5–1.0% by weight.

2 *Bulk Density, Total Porosity, Particle Size Distribution, and Aggregate Stability*

Bulk density is an important physical property of vermicomposts that influences other factors critical to plant growth, such as porosity, aeration, moisture-holding capacity, and so on. The final bulk density of a vermicompost depends, to some extent, on the degree of compression within a container or other mix but can be specified in standard ways that provide information for evaluating the properties of such mixes. The total extent of pore space (as a percentage of volume) within a material is also an important characteristic affecting aeration and water relations. For organic materials suitable for potting media, pore space should occupy 70–80% of the total volume. High bulk density usually has the disadvantage of increasing the transport cost of the container medium, and reducing porosity and air capacity, which should be avoided as far as possible in commercial culture media. However, very low bulk density can cause excessive aeration of the substrate and concomitantly a decline in available water. Hence, de Boodt and Verdonck (1972) proposed optimum physical properties for an ideal substrate for plant growth: as a minimum 85% total porosity, container capacity between 55 and 75%, and air space between 20 and 30%. Table 18.1 summarizes the physical characteristics of pig manure vermicomposts and their corresponding changes as they are substituted into a soilless growth medium such as MM 360.

Table 18.1 Physical Properties of a Standard Commercial Potting Medium (MM 360) Substituted with Different Concentrations of Pig Manure Vermicompost

% (by volume) of Vermicompost in MM 360	Bulk Density[a] (g/cm^3)	Particle Density (g/cm^3)	Total Porosity (%)	Container Capacity (%)	Air Space
Control[b]	0.160 e	1.931 a	91.7 a	58.3 a	33.40 a
5	0.177 d	1.915 ah	90.7 b	68.4 be	22.33 b
10	0.182 d	1.943 a	90.6 b	67.7 e	22.96 b
25	0.202 e	1.927 ab	89.5 e–	66.5 c	22.98 b
50	0272 b	1.915 ah	85.d	70.1 ab	15.73 c
100	0.359 a	1.888 b	81.0 c	71.0 a	10.04 d

Source: Adapted from Atiyeh, R.M., Edwards, C.A., Subler, S., and Metzger J., *Biores.Technol.* 78: 11–20, 2001.

[a] Means within the same column followed by the same letter are not significantly different at $P \leq 0.05$.

[b] Control represents 100% MM 360.

The percentage container capacity of vermicompost can be defined as the percentage by volume of micropores that remain filled with water after a saturated substrate has drained (Beeson 1996), The percentage air space (10.04%) of the vermicompost can be defined as the percentage by volume of air-filled macropores in a saturated substrate (Beeson 1996). The percentage total porosity is the sum of air-filled macropores and water-filled micropores in a saturated substrate. Upon substitution of MM 360 with pig manure vermicompost, the bulk densities of the potting mixtures increased with the increasing proportions of vermicompost substituted for MM 360, and this led to gradual decreases in the total porosity, changed the pore space distribution within the container medium, and resulted in decreased air space and increased water retention (Table 18.1).

Because of the high degree of particle size decrease by grinding of organic matter in the earthworm gizzards, as well as mixing and ingestion of particles by earthworms, that occurs during the vermicomposting process, high-quality vermicomposts should typically have a relatively fine maximum particle size (less than 0.2 mm (0.007 in) diameter). It is probable that during efficient vermicomposting all of the organic matter may pass several times through the earthworm guts since earthworm can clear their guts in a few hours. Furthermore, the arrangement of individual particles in a potting medium or soil mixture determines the overall bulk density of a material as well as the size and continuity of pores between the particles. For many uses of vermicomposts, it should be essential to specify their composition, in relation to effectiveness and standardization, with regard to the distribution of particles (by weight) into different size classes. Just as important as the size distribution of water-stable aggregates in dry vermicomposts is the degree of stability of these aggregates once they are remoistened. Large aggregates may break down when wet, clogging pores and affecting the quality of a mixture. Aggregates' water stability can be measured in standard ways and may need to be specified for some uses of vermicompost.

3 Moisture Content

Moisture contents of vermicomposts should be between 75 and 90% during processing, but moisture contents may vary widely in finished vermicomposts, probably without substantially affecting their overall quality or storage potential. Nevertheless, very wet materials can be expensive to transport because of excess weight and can be difficult or messy to handle. Of course, moisture contents can also have a major impact on the value of vermicomposts that are sold by weight. Hence, it has became more common to market vermicomposts by volume, such as by the cubic yard or cubic meter. At the same time, materials that are too dry may have problems with rewetting and may suffer some reduction in plant disease suppressiveness due to changes in the composition of the microbial community, and this may affect other beneficial qualities dependent on the structure and composition of the soil food web. Some specific applications, which include blending into container media or in specialized soil mixtures, may require relatively narrow ranges of moisture in vermicomposts. Generally, acceptable moisture contents range from 30 to 50%.

4 Water-Retention Characteristics

Whereas the moisture content of a vermicompost can be modified by adding water or by drying, the capacity of a material to retain moisture is more or less fixed. The water-holding capacity of a material (often called container capacity) is defined as the amount of water that is retained after allowing a saturated material to drain freely (usually for 12–24 hours; also more specifically defined as moisture at –1/3 bar tension). However, this moisture is actually comprised of two components: (a) available water that plants can use and (b) unavailable water that is tightly retained to soil particles and is released at water tensions greater than the permanent wilting point of the plant (–15 bars). Specification of both the total and available water-holding capacity can be especially important for vermicompost end uses involving container media. A useful way of expressing the water relations of a material is to determine its moisture-retention curve, that is, a plot of the percentage moisture of a material at different water tensions, ranging from below the permanent wilting point to complete saturation. Such a curve allows calculation and assessment of the water-retention characteristics of different mixtures of container media or soil containing vermicomposts and other ingredients.

B Chemical Characteristics of Vermicomposts

1 Acidity

The acidity of a potting medium or soil has a dominant influence on many aspects of fertility and plant growth. Acceptable pH values of potting media or soils for plant growth range from 5.5 to 8.0, but preferred values range from 6.0 to 7.0. However, individual plant species may have much narrower preferred pH ranges. Vermicomposts vary in pH values depending on the parent organic materials from which they are produced; that is, vermicomposts produced from sheep manure had a pH of 8.6 (Gutierrez-Miceli et al. 2007), vermicomposts from cattle manure had a pH of 6.0 (Jordao et al. 2002) and 6.7 (Alves et al. 2001), vermicomposts from pig manure had pH of 5.7 (Atiyeh et al. 2002) and pH 5.3 (Atiyeh et al. 2001), and vermicomposts from sewage sludge had pH 7.2 (Masciandaro et al. 2002). Perhaps even more important than the pH of a vermicompost is its buffering capacity—the amount of base or acid necessary to increase or decrease the pH by one unit. Knowledge of the buffering capacity is essential for formulating potting media or soil mixtures to precise specifications. In alkaline vermicomposts specification of the $CaCO_3$ (lime) content may also be appropriate.

2 Cation-Exchange Capacity (CEC)

The negative surface charge of organic materials provides sites for retention of nutrient cations. An adequate cation-exchange capacity for vermicomposts should be in the range of 50–100 meq/L.

3 *Soluble Salt Concentrations*

The total concentration of salts (mineral anions and cations) can reach levels in vermicomposts that inhibit plant growth or are toxic to plants, particularly when vermicompost is produced from animal manure feedstocks. High soluble salt concentrations are a common problem with many composts and can be a concern for vermicomposts as well. Soluble salt concentrations (measured as electrical conductivity) in saturated extracts of high-quality plant growth media should not exceed 1–2 dS/m (100–200 mS/m) for sensitive plants and seedlings and 2–3 dS/m (200–300 mS/m) for established plants. Furthermore, the concentration of Na is a particularly important factor influencing potential damage to plants and should be specified if possible. In some cases, the damaging effect of high soluble salt concentrations may be lessened if Ca is the dominant cation. Usually, vermicomposts have low salt contents because earthworm activity is inhibited at concentrations above 0.5% (Edwards and Arancon 2004) Pig manure vermicompost had an electrical conductivity of 322 mS/m (Table 18.2; Atiyeh et al. 2001). Upon the substitution of 5%, 10%, 25%, 50%, and 100% pig manure vermicompost into Metro-Mix 360, the electrical conductivity of the container media increased linearly with the increasing concentrations of vermicompost.

C Plant Nutrient Content

1 *Total Carbon*

The total carbon content of a vermicompost is directly related to the organic-matter content. Changes in totals can therefore be an important indicator of the degree of stabilization that has occurred during the vermicomposting process.

Table 18.2 Chemical Characteristics of Pig Manure Vermicompost Compared with a Soilless Plant Growth Medium (MM 360)

	MM 360	Pig Manure Vermicompost
Conductivity (mmhos/cm)	1.35	11.76
pH	5.90	5.30
Organic C (%)	31:78	27.38
Total N (%)	0.43	2.36
P (%)	0.15	4.50
K (%)	1.59	0.40
Ca (%)	1.03	8.60
Fe (%)	2.58	0.80
Mg (%)	3.52	0.50
Cu (μg/g)	34.09	378.8
Mn (μg/g)	526.04	1170.0
Zn (μg/g)	115.26	824.7
B (μg/g)	41.85	45.3

Source: Adapted from Atiyeh, R.M., Arancon, N.Q., Edwards, C.A., and Metzger, J.D., *Biores. Technol.* 81: 103–108, 2002.

2 Total Nitrogen

The total content of nitrogen in vermicomposts can range quite widely (from 0.1 to 2–4% or more) and is an important criteria for determining the overall value of the vermicomposts as a nutrient source. However, not all of this N is immediately available for plant growth. It is important to specify the concentrations of inorganic forms of nitrogen (NH_4, $N0_3$) as well as the content of organic nitrogen. Total N in vermicomposts derived from pig manure was 0.43% (Atiyeh et al. 2002).

3 Carbon:Nitrogen Ratio

The C:N ratio is one of the most widely used and potentially useful indicators of the stability of organic materials such as composts and vermicomposts. The C:N ratio of microorganisms is generally between 15 and 25, and the C:N ratio of humus is around 11–12. Material that has been sufficiently stabilized will typically have C:N ratios below 20–22. C:N ratios much higher than this may indicate the presence of bioavailable carbon and therefore material that is not completely stabilized. Changes in the C:N ratio from the raw material feedstock to finished product may be as important as the absolute final value, since nitrogen-rich products may not necessarily be completely stabilized, even though their C:N ratios may be low.

4 Total Phosphorus and Potassium

Along with total N, it may be necessary to specify the total content as well as the form of P and K in a finished vermicompost as an indication of its overall macronutrient value. In many U.S. states, it is necessary to state the minimum contents of N, P (as phosphate), and K (as potash) on the label to be able to market vermicomposts as organic fertilizers or soil amendments. Generally, P contents of more than 0.5% are desirable. However, some plants, particularly seedlings in containers, are sensitive to high P concentrations. Container media for sensitive plants should have less than 0.1% total P.

5 Calcium, Magnesium, Sulfur, and Boron

The total contents of Ca, Mg, S, and Bo should also be specified in those situations where the nutrient value of vermicompost is important. Typical values of these elements in pig manure vermicomposts are presented in Table 18.2.

6 Ammonium and Nitrate Nitrogen

In vermicomposts, it is very common for NO_3 to be high relative to NH_4. In saturated extracts of container media, total inorganic N concentrations should be greater than 100 mg^{-1}. However, concentrations of NH_4-N should not exceed 300 mg^{-1}, and concentrations of $N0_3$ are considered to be adequate between 100–200 $mg^{-1.}$ In vermicomposts, NH_4-N should not be more than 10% of the total N or 0.04% of the total dry weight. Nitrate should be a major component of the inorganic N, with NO_3-N to NH_4N ratios greater than or equal to 0.14.

7 *Micronutrients (Bo, Cu, Fe, Mn, Zn)*

In many cases, the micronutrient value of organic amendments such as vermicomposts to soil or container media may be as important as the macronutrient content. Vermicomposts typically contain adequate amounts of most micronutrients (see Table 18.2). Humic substances in vermicomposts may be important chelating agents that enhance the availability of some micronutrients. In some situations, an overabundance of particular micronutrient elements may lead to plant toxicity problems.

8 *Nutrient-Release Characteristics*

Although total and extractable nutrient concentrations can provide important information on the overall nutrient status of vermicomposts, nutrient availability to plants is in fact a dynamic process. An understanding of the overall nutrient-supplying capability of a vermicompost, based on its nutrient-release characteristics, may be as important or more important than the size of total or extractable pools of nutrients. Numerous methods can provide information on the nutrient-release characteristics of organic materials in media or soil mixes. Most of these require incubation of the materials for weeks or months and are therefore too expensive to conduct on a routine basis. However, comparative information on the release characteristics of different mixtures may help in making decisions about which material to use. However, in general, a container-media mix should contain enough nutrients for the first 1 or 2 weeks. It has also been suggested that soil amendments such as vermicomposts should be sufficient to supply complete minor nutrients for a single growing season.

9 *Heavy Metal Contamination*

Most vermicomposts produced from food waste, paper waste, animal manures, or plant wastes are unlikely to be contaminated by heavy metals. The most common metals that may occur in vermicomposts are lead (Pb), mercury (Hg), cadmium (Cd), chromium (Cr), molybdenum (Mo), and Zinc (Zn), when the feedstocks are from unusual sources, especially sewage sludges and biosolids (see Chapter 17 and Table 18.3).

D Biological Characteristics

1 *Human/Animal Pathogens*

Unlike thermophilic composting, temperatures sufficient to kill pathogenic microorganisms are not usually attained during the vermicomposting process. Nevertheless, although the precise mechanisms are not known, there is growing evidence that the vermicomposting process can lead to effective human pathogen

Table 18.3 Typical Levels of Seven Major Micronutrients in Soils and Crops

Element	Levels Usually Found in		
	Soils* (kg/ha)	Crops (mg/kg)	Soil/crop ratio
Iron (Fe)	56,000	2	1:28,000
Manganese (Mn)	2,200	0.5	1:4,400
Zinc (Zn)	110	03	1:366
Copper (Cu)	45	0.1	1:450
Boron (B)	22	0.2	1:110
Molybdenum (Mo)	5	0.02	1:250
Chlorine (Cl)	22	2.5	1:0.9

Source: Brady, N.C., and Weil, R.R., *The Nature and Properties of Soils*, Prentice Hall, New Jersey, 1999. With permission.
*In 15 cm of topsoil.

elimination and sanitization of the processed material (see Chapter 16) (Eastman et al. 2001). Clearly the type of feedstock is an important factor in assessing the risk of vermicompost contamination with human pathogens. Some "clean" materials, such as paper or food-processing wastes, that are not contaminated with human pathogens initially may be a much less serious concern than biosolids or animal manures with high contaminant loads. Vermicomposts should meet the same health standards as thermophilic composts with regard to human pathogens (see Chapter 16) In some cases, preprocessing of feedstocks using conventional thermophilic composting for up to 14 days may be needed to ensure adequate human pathogen reduction. Alternatively, posttreatment with steam sterilization, fumigation, or some other treatment could reliably eliminate pathogens. However, such posttreatments would also kill most or all of the beneficial microorganisms that can add substantial value to the final product, so this is not recommended. Until we know more about the vermicomposting process in relation to pathogen reduction, it is not possible to specify "best management practices," as can be done with thermophilic composting based on practical and scientific experience and Environmental Protection Agency (EPA) regulation. Therefore, in some cases, it may be necessary to rely on frequent pathogen testing of vermicomposts to ensure compliance with acceptable EPA health standards. The EPA requires virtual elimination of coliform bacteria, *Salmonella*, enteric viruses, and helminth ova for material classified for class A land disposal (EPA 1980, 1999).

2 Enzymatic Activity

Research has shown that some enzyme activities are correlated with overall microbial activity, soil fertility, plant growth, and plant disease resistance. Dehydrogenase enzyme activity is often used as a parameter of the microbial activity of composts (Forster et al. 1993). During vermicomposting, earthworms enhance selectively the activities of enzymes such as invertase, urease, and alkaline phosphatases, which are of microbial origin. Measurement of enzymatic activities could constitute one of the key measurements in determining the overall microbial activity of the vermicomposts.

3 *Microbial and Faunal Community Composition in Vermicomposts*

It is not uncommon for soil- and compost-testing labs to provide information on populations and activity of selected soil microorganisms in samples of composts or vermicomposts or vermicompost-amended soil. It is generally presumed that the larger the total populations of microorganisms in composts or vermicomposts, the better. In properly prepared thermophilic composts, most microorganisms and decomposer fauna are killed during the high-temperature (thermophilic) phase. Although many microorganisms may recolonize during a lower-temperature curing phase, rates of recolonization may vary, and this process is rarely intentionally managed (with the exception of a few compost products that are purposely inoculated with selected species of microorganisms known to be involved in plant disease suppression). In vermicomposts, however, moderate (mesophilic) temperatures must be maintained, or the earthworms will become inactive, flee, or die. Although populations of some sensitive organisms may be reduced drastically or eliminated during vermicomposting, the substrate maintains an active and complex community of decomposer organisms, which, in addition to earthworms, may include enchytraeids, nematodes, springtails (Collembola), mites (Acarina), protozoans, and innumerable microorganisms. Even if it is necessary to precompost some materials to sanitize them before processing with earthworms, because active earthworm beds are normally teeming with this enormous variety of microorganisms and invertebrates, they may be ideal sites for complete and efficient reinoculation of precomposted materials with entire complex communities of beneficial soil organisms. This may be especially important for container media and in soils that are intensively chemically managed and/or impoverished in biological activity.

There is good evidence that microbial activity and food webs are much higher in vermicomposts than in thermophilic composts, but in biological assays we have shown that the microbial communities in vermicomposts and thermophilic composts are extremely different. It seems probable that mixtures of 20% vermicomposts and 80% thermophilic composts, recommended by the Soil Ecology Laboratory at The Ohio State University, are much more diversely microbially active than either component, which may improve them as plant growth media. As our understanding of soil ecology advances, determining the structure of decomposer food webs in organic amendments may become an important predictive tool in evaluating their potential beneficial qualities.

V CONCLUSIONS

It is not realistic to expect that all of these quality criteria for vermicomposts will need to be measured in all cases. However, it is important for producers, marketers, and others to be aware of these potential quality criteria and to be aware of the circumstances in which their assessment may be important, both for regulatory bodies or as a basis for marketing. Further research and discussion will be necessary to build a solid foundation for understanding the many relationships between vermicomposting process parameters and vermicompost quality and setting up firm quality criteria for vermicomposts.

REFERENCES

Alves, M.R., Landgraf, M.D., and Resende, M.O.O. 2001. Absorption and desorption of herbicide alaclor on humic acid fractions obtained from two vermicompost. *J. Environ. Sci. Health* 36: 797–808.

Atiyeh, R.M., Arancon, N.Q., Edwards, C.A., and Metzger, J.D. 2002. The influence of earthworm–processed pig manure on the growth and productivity of marigolds. *Biores. Technol.* 81: 103–108.

Atiyeh, R.M., Edwards, C.A., Subler, S., and Metzger, J. 2001. Pig manure vermicompost as a component of a horticultural bedding plant medium: Effects on physicochemical properties and plant growth. *Biores. Technol.* 78: 11–20.

Beeson, R.C., Jr. 1996. Composted yard waste as a component of container substrates. *J Environ. Hortic.* 14: 115–121.

Brady, N.C., and Weil, R.R. 1996. *The Nature and Properties of Soils*. Prentice Hall, New Jersey.

de Boodt, M., and Verdonck, O. 1972. The physical properties of the substrates in horticulture. *Acta Hortic.* 26: 37–44.

Eastman, B.R., Kane, P.N., Edwards, C.A., Trytek, L., and Gunadi, B. 2001. The effectiveness of vermiculture in human pathogen reduction for USEPA Class A Stabilization. *Compost. Sci. Util.*, 9: 38–49.

Edwards, C.A., and Arancon, N.Q., 2004. Interactions among organic matter, earthworms and microorganisms in promoting plant growth. In *Functions and Management of Organic Matter in Agroecosystems* (Eds.) Magdoff, F., and Weil, R., CRC Press. Boca Raton, FL.

Environmental Protection Agency. 1980. *Compendium on Solid Waste Management by Vermicomposting*. E.P.A.-600/8-80-033.

Environmental Protection. Agency. 1999. *Environmental Regulations and Terminology: Control of Pathogens and Vector Attraction in Sewage Sludge*. EPA/625/92/013.

Forster, J.C., Zech, W., and Wrdiger, E. 1993. Comparison of chemical and microbiological methods for the characterization of the maturity of composts from contrasting sources. *Biol. Fertil. Soils* 16: 93–99.

Graham, E.C. (ed) 1965. *The Basic Dictionary of Science* Macmillan Co. New York.

Gutierrez-Miceli, F.A., Moguel-Zamudio, B., Abud-Achila, M., and Gutierrez-Oliva, V.F. 2007. Vermicompost as a soil supplement to improve growth, yield and fruit quality of tomato (*Lycopersicum esculentum*). *Biores. Technol.* 98: 2781–2786.

Jordao, C.P., Pereira, M.G., Einloft, R., Santana, M.B., Bellato, C.R., and Vargas de Mello, J.W. 2002. Removal of Cu, Cr, Ni, Zn and Cd from electroplating wastes and synthetic solutions by vermicompost of cattle manure. *J. Environ. Sci. Health* 37: 875–892.

Masciandaro, G., Ceccanti, B., Ronchi, V., Benedicto, S., and Howard, L. 2002. Humic substances to reduce salt effect on plant germination and growth. *Commun. Soil Sci. Plant Anal.* 33: 365–378.

CHAPTER 19

The Commercial Potential and Economics of Vermicomposting

Jim Jensen, Bentley Christie, and Clive A. Edwards

CONTENTS

I Introduction 304
II Commercial History Of Vermicomposting 304
III Current Vermicomposting Market Opportunities 305
A Sale of Earthworm Castings (Vermicomposts) 305
B Sale of Aqueous Extracts from Vermicompost (Teas) 307
C Sale of Earthworms 307
D Resource Recovery 308
E Vermicomposting Systems and Supplies 309
F Market Development 310
IV Models of Enterprise 310
A Classic Earthworm Farm 310
B Home Vermicomposting 311
C Vermicomposting for Resource Recovery 311
V Vermicomposts: Impacts of Technology 312
A Levels of Technology 313
B Size and Scale of Processing Facility 313
C Technological Developments 315
D Economics of Vermicomposting 316
1 Capital Costs 316
2 Operating Costs 317
VI Factors Affecting the Commercial Potential of Vermicomposting 318
A On-Site or Off-Site Processing 319
B Location and Space 319
C Climate 319
D Human Pathogens 319
E Feedstock Loading Rates 320

F Earthworm Population Growth 320
G Earthworm Population Availability 320
H Past History 320
VII Conclusions 321
References 321

I INTRODUCTION

In the simplest terms, commercial potential is measured by profitability and returns on investment. Evaluating the commercial potential of an entire industry (or of a biological method for waste conversion) depends on a generally favorable interplay of market opportunities, available technological capabilities, and the cost and availability of raw materials versus the potential value of the product added by labor and machines as well as marketing.

This chapter explores various issues affecting the commercial potential of a variety of vermicomposting and other earthworm-related enterprises. The first section briefly describes current market opportunities. We propose to examine how the wide variety of successful earthworm businesses can be categorized into different "models of enterprise." These models provide an opportunity to discuss potential different market segments and the values of diversification. Later in this chapter we examine the impact of technology on the expanding development of vermicomposting systems and facilities. This shows the need to balance a range of factors when planning and upgrading a vermicomposting operation. The section on economics offers information related to capitalization and operating costs. A review of the environmental and economic factors that can impact the development of earthworm-related enterprises follows.

II COMMERCIAL HISTORY OF VERMICOMPOSTING

People have studied and profited from the unique contributions of earthworms to soils for many decades. A 1970s history of vermiculture determined clearly that vermiculture has come a long way during the past 70-odd years. From an industry catering only to fishermen it has become one involved in solving some of the most basic problems we face as a society in the management of waste and the maintaining of soil fertility. From an occupation pursued by a few hundred "odd-balls" scattered through rural areas, vermiculture has become a vocation or avocation for thousands of daring and imaginative, hard-working families and businessmen. From a trial-and-error, old-wives talk stage of practice; vermiculture is emerging into a pragmatic and sophisticated technology where field and laboratory are combined for best results" (Carmody 1978).

The 1970s began an exciting period for the commercial development of earthworm-related commerce and research. The same decade that brought us Earth Day, the Environmental Protection Agency (EPA), energy conservation, and solar power also brought significant investments in the establishment of earthworm farms and the start of decades of research and development into the use of earthworms to transform organic waste materials, including sewage sludge (or biosolids), animal manures, and later organic refuse from municipal solid waste, into beneficial soil-improving products. By the end of the 1970s, as many as 10,000 earthworm farms were practicing vermiculture on a part-time or full-time basis, and these numbers have increased greatly in the following decades.

In the 1980s, many successful research programs began. In the United States, pioneering research at the State University of New York (SUNY) was paralleled by the commercial vermicomposting of sewage sludge in Texas, Maryland, and Utah, and in the United Kingdom, research at the Rothamsted Experimental Station led to the startup of a large-scale commercial company—British Earthworm Technology—that manufactured vermicompost from animal manures (Knight 1989). More recently, vermiculture research at the Soil Ecology Laboratory at The Ohio State University has led the United States and the world (Edwards and Arancon 2004).

A resurgence of growth in earthworm-related commerce during and through the 1990s and into the twenty-first century was sparked on one side by increased interest and opportunity in the recovery and recycling of organic solid wastes at home and commercially on the other side by major growth in the demand for high-quality soil amendments for organic gardening and farming. As a major national news magazine reported in 1997, "Vermiculture ventures, the biggest of which use 50 million earthworms to process almost 81.65 tons (90 t) of waste per week, have boomed over the past few years" (Koerner 1997), and the appearance of new vermiculture-related enterprises in the popular media and on the world-wide web is now common (see Chapter 23).

III CURRENT VERMICOMPOSTING MARKET OPPORTUNITIES

The commercial potential of earthworm-related business and vermicomposting is enhanced by six major current market opportunities: sale of earthworms, sale of earthworm castings (or vermicomposts), sale of vermicompost teas, resource recovery services, development and marketing of vermicompost bins, and vermicomposting systems and market development. Table 19.1 summarizes some of the market opportunities described in this section.

A Sale of Earthworm Castings (Vermicomposts)

A key economic benefit of vermicomposting over other forms of organic recycling is the high quality of the final product, known as *earthworm castings* or *vermicomposts*. Earthworms obtain their nutrition from the microorganisms thriving on the decaying organic wastes. As the earthworms multiply, they consume progressively more material. Their feeding also increases the surface area of the organic material being treated, thus increasing microbial growth on it. As this process continues,

Table 19.1 Summary of Earthworm-Related Markets

Revenue	Markets	Values
Earthworms, red worms, *E. fetida*	Vermicomposting projects, farms, institutions, schools, home composters, fish bait	From as low as \$5–\$8 per pound in bulk to as high as \$25–\$35 packaged per pound.
Earthworm castings, vermicomposts	Bulk: farms, landscapes, soil reclamation Bagged: nurseries, farm markets, gardeners, homeowners	\$75–\$300 per ton (bulk) and as much as 5–10 times that for bagged products. Experience suggests that approximately 2–3 cubic yards represent a ton.
Decreased disposal fees	Municipal and public facilities, utilities, resorts, campuses, landscapers, food businesses, haulers	Vary widely from one community to another and depending on material. Tip fees for yard debris will typically be lower than for other materials. Tip fees for food scraps or mixed municipal organic materials have been reported to be as high as \$60–\$100 per ton.
Vermicomposting systems and supplies	Direct sales, mail-order sales, garden centers, marketing expos	Large-scale vermicomposting systems are valued from \$30,000 to \$100,000; manufactured bins for home vermicomposting cost \$70–\$400; books and other supplies have a wide range of values.

complex changes occur to produce fully stabilized vermicompost, a finely structured material richer in valuable plant nutrients and with more beneficial soil microbial life than thermophilic composts. There is evidence that suggests strongly that they also provide natural plant growth stimulants or regulators that could have significant economic value. High-quality vermicompost has a good physical texture and color, no odors, and few contaminants or pollutants (Camp et al. 1980b; Edwards 1988; Tomati et al. 1988; Harris et al. 1990; Dominguez 1997; Edwards and Bohlen 1996; Edwards 1998b; Edwards and Arancon 2004).

Vermicomposts are suitable both as plant growth media and as soil amendments. The Ohio State University researchers report that substituting 10–20% vermicompost into the best horticultural plant growth media increased the rates of germination, growth, flowering, and fruiting of a wide range of ornamental and vegetable crops significantly (Edwards 1998a). Vermicomposts are marketed for a broad range of purposes, including gardening, landscaping, agriculture, forestry, and horticulture and pollution management. For the high-value end markets, the vermicompost products must be very uniform and consistent in quality and nutrient content (Edwards 1988, 1998a).

Markets for vermicomposts and vermicompost teas are growing, rapidly. For example, in organic agriculture, their potential markets are enormous. "Many businesses are prospering, thanks in part to the growing popularity of organic foods, which are expected to become a multi-billion-a-year business by 2010," reported a major national news magazine. "With the U.S. Department of Agriculture estimating that 25% of Americans purchase organically grown foods at least once a week, organic farmers' demand for earthworm castings and vermicompost teas far outstrips supply."

In relation to value, the structural, nutritive, plant growth–promoting and soil-building benefits of vermicomposts translate into higher dollars when good-quality products are sold. Large producers of vermicompost in the United States have reported receiving from $75.00 to as much as $300.00 per ton (or its volume equivalent) when selling to bulk buyers. When processed and packaged attractively for retail sale, reported values of vermicomposts can be up to 5–10 times higher. This compares to $12.00–50.00 per ton received for yard debris and thermophilic composts in similar markets. The reason that many producers of vermicompost do not receive the upper range of prices for their products may be related to a general tendency on the part of many buyers to lump vermicomposts as a group together with thermophilic composts, top soils, steer manure, and even sawdust as general soil-improvement products. As more data and information about the unique properties and effects of vermicomposts become available, the average market value should rise considerably.

B Sale of Aqueous Extracts from Vermicompost (Teas)

During the last 3–4 years the use of aqueous extracts, or teas, from thermophilic composts and vermicomposts has become increasingly popular. A number of commercial tea-brewing equipment systems with a range of capacities have been marketed widely in the United States and Europe. The attraction of such systems is that tea avoids the transport and incorporation into soil of bulky vermicompost. Aqueous extracts can be applied to crops as soil drenches or foliar sprays (see Chapter 11). They have been used to promote crop growth, accelerate seed germination, and promote flowering and fruiting of a wide range of crops. They are particularly attractive to organic farmers and growers who cannot use pesticides or inorganic nutrient sources. Such products are much easier to transport and use.

Unfortunately, the rapidly increasing use of these aqueous extracts by growers has not been preceded by extensive research into their application rates and effects other than the work done in the Soil Ecology Laboratory at The Ohio State University. These workers have developed and tested aqueous-extract-preparation methods (Chapter 11), methods of application, and optimum rates of use and have demonstrated their successful use in affecting plant growth (Chapter 15), suppressing arthropod pests and nematodes (Chapter 14), and suppressing plant diseases (Chapter 13).

C Sale of Earthworms

Vermiculture refers to the science and technology of breeding earthworms. *Vermicomposting* typically refers to the use of earthworms—especially the red earthworms *Eisenia fetida, Eisenia andrei, Lumbricus rubellus, and Dendrobaena veneta* (*Eisenia hortensis*) under temperate conditions and *Eudrulus eugeniae* and *Perionyx excavatus* (the African night crawler) under higher temperatures—to process organic wastes into valuable soil amendments. Typically, in either case, earthworms are both the foundation and an end product of the operation. Markets for earthworms include: as fishing bait, for vermicomposting, and as a protein source (Chapter 20). The markets for earthworms as fishing bait are large and well established. These markets are served by many traditional, relatively small earthworm farms and involve several species of earthworms.

The focus of this chapter is on red or tiger earthworms *Eisenia fetida*, *Eisenia andrei*, and related species, which are valued highly for vermicomposting. Households and schools practicing vermicomposting represent a large and growing market for red earthworms nationwide. An increasing number of commercial vermicomposting operations use red earthworm vermicomposts to recover organic waste materials and produce earthworm castings or vermicomposts, which are valuable for soil remediation and improvement. In agriculture, many types of earthworms have been linked with increasing microbial activity and enhancing soil and plant health and will likely be the focus of greater interest in future years (Chapters 9 and 10).

The value of red earthworms for vermicomposting ranges from a low of around \$5.00–12.00 per pound half kg(1lb) for bulk sales to as much as \$25.00–35.00 per pound for earthworms packaged and sold by the pound. Mail-order sales with delivery by the U.S. Postal Service or other commercial delivery service are the predominant distribution method, but increasingly earthworms are sold on Web sites.

Production of earthworms as a protein source, for fish and animal feed, is still a potential market with much promise. British researchers in the 1980s developed extensive positive assessments of the potential for marketing earthworm meal as a source of food for aquaculture and chicken and pig feeds. Comparison of earthworm protein with other animal feedstuffs has shown it to be a highly effective substitute for meat meal or fish meal. Feeding trials with laboratory animals, pigs, poultry, and fish have shown that growth rates are similar to those of conventional feeds. Additionally, earthworm tissues contain a wide variety of long-chain fatty acids, many of which are of medical importance and are at present extracted from fish oils (Edwards 1988; Edwards and Arancon 2004) (Chapter 22). Earthworms as a source of protein feed suitable for animals or fish have not yet been found to be economical as a separate enterprise, largely because of the labor-intensive costs of separating earthworms from vermicomposts. Also, the markets for animal protein are quite competitive, and dried earthworm meal cannot compete in price with dried soybean meal or fish meal (Tomlin 1983; Edwards and Bohlen 1996; see Chapter 20). In sum, research on several fronts suggests that earthworms for protein may provide a good long-term market for vermiculture businesses, but it will require technological advances in the harvesting and cleaning of earthworms to become a profitable economic enterprise.

D Resource Recovery

Resource recovery (i.e., recycling) represents an excellent market opportunity to realize the economic benefits that business and organizations can achieve by diverting organic materials from other expensive disposal options such as landfills. The concept of avoiding disposal costs is a form of savings that is used to justify investment in vermicomposting systems. Similarly, commercial operations can invest in plants and equipment if they receive revenue in exchange for accepting or collecting organic waste materials for recycling. Waste-disposal costs around the nation have increased dramatically in the wake of increased regulation of landfill design and operation. Nationally, the average landfill-disposal tip fee is close to $32.00 per ton, with high averages over $49.00 per ton reported in the Mid-Atlantic and New England regions and $70.00–$80.00 per ton on the West Coast. When collection and transportation costs are added, the total cost of organic-waste disposal can be more that $100.00 per ton in many areas.

Support for greater recycling of wastes has been legislated in many states through the adoption of recycling goals, imposition of recycling mandates, enforcement of material bans, and investment in market development. A few states have targeted organics by banning organic residuals, such as paper or yard debris, for landfill disposal. The province of Nova Scotia, Canada, has targeted all organic materials for diversion from landfills, with a goal of 50% waste diversion by 2015. As a result, the potential for major investments in vermicomposting throughout that province was estimated by one equipment manufacturer to be as high as $12 million (Bogdanov 1997a) and is probably much greater now.

E Vermicomposting Systems and Supplies

Development and demonstration of several large-scale patented vermicomposting systems marketed by a number of U.S. and Canadian companies have occurred throughout the last 20 years. These large systems go by names such as Vermitech, OSCR, Worm Wigwam, and Worm Gin (see Chapters 6 and 24). Many incorporate top-feeding vermicomposting methods, but systems differ in the amount of air, heat, and moisture control and containerization. Their operation covers the range from considerable hand labor to full mechanization.

Similarly, the interest in home vermicomposting has promoted the manufacture and marketing of many types of earthworm bins and boxes, often made of recycled plastic or wood. Whether made by small cottage industries or major plastics manufacturers, these home-scale earthworm systems have been marketed from $50.00 to $400.00. Related products such as books, newsletters, videos, conditioners, tools, and soil screens are often sold in conjunction with earthworm bins or systems. When matched with appropriate education materials, earthworm bins and systems have been adopted by all kinds of classrooms and schools. Taken together, the home and school sectors represent a very important part of the earthworm-vermicomposting industry (see Chapter 6).

Thermophilic composting has traditionally been looked on as an alternative to vermicomposting. However, in recent years, they have been increasingly considered as complementary processes. A number of the larger commercial vermicomposting systems incorporate a 7–14 day thermophilic composting period prior to vermicomposting with the aim of killing human pathogens and weed seeds and promoting microbial diversity and activity (see Chapter 8).

F Market Development

Market-development advances for vermicomposts continue to increase considerably as their benefits become more known. Research projects around the world are showing that the benefits of vermicomposts go far beyond the usual provision of organic matter and plant nutrients, to providing plant growth stimulants and other plant- and soil-improving properties. Future advances in vermicompost market development in 2010–2020 may include, but are not limited to, the following:

- Additional research about the quantities and uses of vermicompost
- Research into the use of aqueous vermicompost extracts (teas)
- Research into the use of vermicomposts and vermicompost teas in pollutant bioremediation
- Demonstration and education directed to target markets
- Customized end products and mixtures
- Potential of inoculation of vermicomposts with disease-suppressive organisms
- Development of vermicompost standards to ensure consistent product quality
- Better ways of packaging and selling vermicomposts

IV MODELS OF ENTERPRISE

The flexibility of vermiculture and vermicomposting methods has allowed their successful development in a wide range of scenarios. Comparing a large number of earthworm-related businesses and their responses to market opportunities, it is possible to categorize them according to four basic models: the classic earthworm farm, the home vermicomposting business, the large-scale commercial operation, and the vermicomposting operation for resource recovery. Each varies to the degree that it emphasizes production and sales of earthworms, supplies, or castings Lee (1985).

A Classic Earthworm Farm

The classic earthworm farm is concerned principally with vermiculture, that is, the growing and raising of earthworms. Earthworm farms are often small-scale, part-time hobby farms or supplemental income sources for the farmer. They are often developed near consistent sources of animal manure, the primary feedstock. Earthworm farms may be located on farms as a form of manure management. Located separately, earthworm farms may pick up manure from one or more farms or have it delivered at some minimal cost. To this day, many earthworm farms use relatively low-technology methods dating back to the 1960s and 1970s. These farms typically grow earthworms in multiple bins or in outdoor windrows. The methods used to feed and harvest the earthworms tend to be labor-intensive and often involve a significant amount of labor (Bogdanov 1996; White et al. 1996).

The classic earthworm farm (as described in Chapter 7) remains a thriving section of rural economies, especially in the American West and Southeast. Although the potential profits do not often attract big companies, dozens of large-scale earthworm farms, such as Environmental Recycling Systems, WeCare Organics, in Alpine, California; Rainbow Worm Farm in Davis, California; Sonoma Valley Worm Farm, Sonoma, California; the Yelm Earthworm and Casting Farm, Yelm, Washington; Sansai International Vermitechnology, R.T. Solutions Unlimited, LLC., many having operated successfully for years, grow and ship earthworms to customers worldwide (White et al. 1996; Jensen 1988; see Chapter 23).

B Home Vermicomposting

Popularized for nearly 30 years by Mary Appelhof, author of *Worms Eat My Garbage* (1997), home vermicomposting methods attracted steadily more interest with the 1990s boom in support for recycling. Home vermicomposting is currently practiced in homes throughout North America. Home vermicomposting methods and experience have also been spread widely by thousands of volunteer "master composters," trained and supported by local governments and gardening organizations. Transmission of vermiculture information through the Web has increased dramatically. For instance, Google may provide more than 200,000 hits. As described in detail in other parts of this book, earthworm bins and systems in homes, schools, and offices now divert thousands of tons of food scraps and other organic debris from landfills and incinerators and are an important tool for waste diversion supported by solid waste managers throughout the United States and Canada and in many European countries (see Chapter 6).

This interest in home vermicomposting has given rise to another category of earthworm enterprise whose focus is on marketing earthworms, earthworm bins, kits, tools, and educational materials for small-scale vermicomposting. Often small, part-time enterprises, these businesses mainly serve local communities. Others have developed national marketing programs through garden magazines and catalogues and over the Internet, which now hosts dozens of such earthworm operations. *Casting Call* was distributed nationally for many years, and *Worm Digest* had many subscribers. The latest development is toward community-scale distribution of earthworm bins and earthworm-composting education through programs subsidized by local, state, and provincial governments.

C Vermicomposting for Resource Recovery

Utilization of animal manures has long been a standard practice in vermiculture operations. Red earthworms grow quite well on many kinds of animal manure (Edwards 1998a; Edwards and Arancon 2004). Through the 1980s, 1990s, and 2000s, greater concentrations of animals on large livestock farms and new regulation of livestock-manure-management practices has increased costs of manure disposal and encouraged research into and development of vermicomposting programs related to animal manure (Edwards 1998b; Edwards and Arancon, 2004)

In the early 1980s, increased regulations and treatment costs at sewage wastewater facilities led to research about the use of earthworms to process biosolids—a process known as *vermistabilization*. Camp, Dresser, and McKee published two reports in 1980 for the EPA to assess the technical and economic feasibility of large-scale vermicomposting of wastewater treatment sludge. The reports documented large operations in Texas, Maryland, and Utah and offered qualified support for the commercial potential of vermicomposting these waste resources. The authors looked for greater assurance about operating parameters and market information related to earthworms and castings (Appelhof 1981). However, in spite of many attempts to commercialize such biosolids facilities, only one in Pennsylvania has been successful but at a high economic cost. The main problem with an earthworm-based biosolid waste-management system is the risk of toxic chemicals being flushed into the system and killing the earthworms.

Finally, passage of local, state, and provincial recycling goals throughout the United States and Canada has increased the importance and support for recovery of organic waste materials from the municipal solid waste stream. Yard debris is one of the largest single categories of waste material. In parts of California, large-scale vermicomposting of yard debris has been attempted with varying degrees of success. Collection of tipping fees for accepting yard debris has been a main source of revenue for these operations (Jensen 1994; Bogdanov 1996).

Food waste materials represent another large, potentially recoverable resource. Several medium-scale demonstrations of nonresidential (institutional) food waste vermicomposting, such as the Food Lifeline project in King County, Washington (low-tech), and Vermitech, an originally Canadian project in Ontario, Canada (high-tech), have shown the technical and practical feasibility of using earthworms to recover food scraps on-site for beneficial reuse (Jensen 1994; Riggle 1996a).

Large-scale commercial vermicomposting of food waste in the United States has been demonstrated most successfully to date by the Oregon Soil Corporation, Oregon City, Oregon, since 1992. Using a continuous-flow vermicomposting reactor system, developed under the leadership of Dr. Clive Edwards by scientists and agricultural engineers in the United Kingdom, the Oregon Soil facility produces dark, rich-looking earthworm castings with a high market value (Bogdanov 1997b). The operation was supported by a supermarket chain that paid a fee for collection of all food wastes, and the chain also marketed the vermicompost for home and garden uses. This last type of earthworm enterprise—resource recovery and castings production—is where many near-term opportunities for commercial market expansion exist. Further development will be influenced greatly by the availability and deployment of improved technology that will minimize labor and space needs and improve processing rates and conditions. These issues are discussed in detail in the next section and in other chapters.

V VERMICOMPOSTS: IMPACTS OF TECHNOLOGY

As described in this chapter and throughout this book, vermicomposting processes offer key advantages over conventional thermophilic composting methods.

- Flexibility: Whether low- or high-technology, earthworm-based methods can adapt readily to small-, medium-, and large-scale production systems.

- Faster processing: It has been shown that earthworms and microorganisms working in combination at lower temperatures stabilize decaying material faster, which reduces operating costs and conserves nutrients and microbial activity in the end product.
- Higher value added: The vermicompost produced offers greater consistency, finer texture, more available plant nutrients, and growth-enhancing properties.
- Reduced environmental impacts: The natural decomposition of organic wastes by earthworms and their use in various vermicomposting systems can significantly reduce environmental impacts, especially odors.

However, systems that work well to produce earthworms may not be as efficient at producing earthworm castings or vermicomposts. Like composting operations, vermicomposting tends to find profitability through higher-technology production methods and economies of scale. Efficient handling of bulk volumes of materials, through all stages of collection, processing, and marketing, is a key to success.

Critical to efficiency is achieving the best combination of technical, process, and management parameters. On the technical or scientific level, consideration must be given to providing optimal food and moisture for earthworm multiplication and activity, while maintaining appropriate temperatures and other environmental conditions. For processing, operators can make choices about using generalized equipment (less costly but perhaps less effective) versus specialized equipment (perfect for the situation but much more costly). Operators must also strike a balance between using labor-intensive methods with lower start-up costs versus more technologically advanced, capital-intensive systems that provide better economic returns.

In the marketing area, critical decisions involve the assessing the relative focus on recovering organic waste materials versus producing high-quality end products. Accepting tip fees for lower-quality municipal solid waste materials generates revenue but also impacts on the cost of processing and refinement and affects the quality of the resulting vermicompost considerably.

A Levels of Technology

Table 19.2 shows major differences between low-, medium-, and high-technology vermicomposting. Any of these levels of technology can be the basis for economically viable and environmentally sound projects. Which technology is most appropriate depends on the individual circumstances of each operation and the availability of land, labor, and capital. The flexibility that vermicomposting offers gives opportunities to everyone who struggles to find an appropriate scale for his or her business. It also suggests that large-scale, high-technology vermicomposting systems are available for engineers and entrepreneurs prepared to make significant investments.

B Size and Scale of Processing Facility

Compared to thermophilic composting facilities, many of which have been constructed to handle hundreds of tons of organic residuals each day, there are only a few

Table 19.2 Summary of Vermicomposting Technologies Available

Methods Used at Different Levels of Technology	Earthworm Bins or Windrows	Enclosed Earthworm Bins or Systems with Mechanization	Fully Automated, Computer- Controlled Systems
Scale/size	Small to large scale	Small to large scale	Medium to large scale
Space per volume processed	Requires lots of surface area; typically large spaces	More efficient; reduces space requirements	Most space efficient; maximizes worm density
Aeration	Passive	Passive or active	Passive or active
Water makeup	Manual	Manual or auto	Auto
Temperature control	Passive	Passive or active	Active
Cover	Mostly outside; uncovered, with lids on bins, or with soft coverings	Under cover (e.g., greenhouse tunnel or roof) or inside enclosure or building	Enclosed system or inside building
Feedstocks	Manure, yard debris	Manure, yard debris, food wastes, and biosolids	Manure, yard debris, food wastes, and biosolids
Monitoring/controls	Manual monitoring and control	May include simple electronic monitoring and controls	Electronic monitoring and process control
Process time	6–12 months	2–4 months	1–2 months
Product quality	Fair, variable	Fairly good, consistent	Good, consistent
Capital costs	Low	Medium	High
Operating costs	Low to medium	Medium	Low
Labor requirements	Medium to high labor	Medium labor	Low labor
Equipment used (examples)	Hand tools, loader, harvester, grinder, irrigation or watering system	Grinder, mixer, loader, harvester/ screen, conveyors, blowers, temperature and moisture systems, sterilizer, or other specialized equipment	Grinder, mixer, loader, harvester/ screen, conveyors, blowers, temperature and moisture systems, sterilizer, automated systems, electronic controls, or other specialized equipment

large-scale vermicomposting facilities. One highly successful, large-scale vermicomposting operation was located in La Voulte, in southern France. This operation, known as Sovadec, was designed to sort and process the mixed household waste from an entire town. The operation handled more than 18.4 tons (20 t) of mixed solid waste each day. Recyclable materials, such as cardboard, paper, glass, metal, and plastics, are separated by various means from the mixed waste. The organic fraction is subjected to a two-phase "biosanitization" process: (1) aerobic thermophilic composting to destroy pathogens and weeds and (2) vermistabilization in large "Lombricubateurs," or "earthworm tanks." The process converts 27% of the mixed waste into a valuable vermicompost, which meets most stringent European compost standards (Bouché 1992; World Resource Foundation 1995).

The relative lack of large-scale vermicomposting operations in the United States and Canada is changing as the various advantages of vermicomposting are demonstrated and the technology for large-scale vermicomposting develops further and becomes established. According to researchers at The Ohio State University, "The conventional thermophilic composting process is suitable for the rapid treatment of large amounts of wastes, in order to eliminate contamination problems more quickly than the traditional vermicomposting systems and is currently widely utilized in this way. However, the newer, vermicomposting continuous-flow reactor systems seem to be equally applicable to large-scale, organic-waste processing. Traditional batch and bed vermicomposting may be an alternative way to manage organic wastes, produced on a smaller scale, in order to eliminate those problems and at the same time obtain a valuable organic fertilizer" (Table 19.2) (Dominguez et al. 1997; Edwards and Arancon 2004). As we learn more about the parameters for optimum efficiency in production of earthworms and vermicomposts, engineers and operators will be able to develop increasingly more efficient technology to address large waste-management and vermicompost production.

C Technological Developments

To meet the challenges of large-scale resource recovery, vermicomposting technology is undergoing significant improvements. The future vermicomposting technologies will certainly contain several of the following elements:

- Automated feeding and harvesting
- Continuous flow for consistent production
- More efficient use of vertical space to reduce the system footprint
- Controlled aeration and greater process control
- Automated moisture control
- Modularity for flexibility and expansion

The continuous-flow automated vermicomposting system already incorporates many of these elements. The earthworm bed in a continuous-flow system consists of a raised trough, 128 ft (40 m) long, 8 ft (2.4 m) wide, and about 3 ft (1 m) deep. Feedstock mixtures are added daily in thin layers by a gantry that runs lengthwise over the earthworm bed. The system is modular, so it can handle a range of volumes depending on the surface area available. The earthworm bed has a 2 × 4 in (5 × 10 cm)-aperture mesh floor, with a mechanical breaker bar above the mesh. Connected to a winch and cable, the bar automatically scrapes vermicompost from the bottom of the pile onto the concrete floor below. Once the product reaches the floor, hydraulically operated breaker bars push the vermicompost to the far end of the trough for easy collection (see Chapter 8).

The raised-bed-and-mesh-floor design maintains moisture and improves aeration. It also takes advantage of layered feeding. According to Dan Holcombe of the Oregon Soil Corporation, "The design eliminates all requirements for separation of earthworms from the waste, increases production rates and enhances the quality of the vermicomposting together with lower labor requirements and easier management" (Bogdanov 1997b).

D Economics of Vermicomposting

The economics of vermicomposting have been studied for as many years as the biology and ecology of earthworms. In the 1980s, the Environmental Protection Agency (EPA) and National Science Foundation (NSF) funded significant engineering and economic feasibility studies into vermicomposting of municipal wastewater sludge. These studies concluded that it was not economically feasible to vermicompost these wastes using the technology then in use at the operations studied (Camp et al. 1980a). However, critics pointed to cost-benefit projections that assumed an estimate of $0–15 per ton for sale of castings, because of the lack of marketing experience. They argued that plugging in values from existing markets would tip the scales toward economic viability for the 100-ton-per-day operation on which they based their figures (Appelhof 1981). Even, at that early date, the study concluded that vermicomposting of municipal wastewater sludge was not only feasible but also appeared to be competitive with the costs of other methods for small wastewater technology (Appelhof 1981).

This section of this Chapter looks at current data related to the major cost factors involved in vermicomposting. Capital costs for medium- and large-scale vermiculture projects vary significantly. Investments in land, plant, and equipment; utilization of higher levels of technology; and costs of meeting local environmental regulations are all significant factors. However, operating costs vary most significantly according to the amount of labor required to process and handle materials. The following sections look at these different costs in more detail.

1 Capital Costs

Capital costs include many items, including, but not limited to, the following:

- Land
- Permit fees (if applicable)

- Facility (engineering and construction)
- Utilities development
- General equipment, e.g., grinder, mixer, conveyors, screen
- Specialized vermicomposting systems
- Rolling stock, i.e., trucks and loaders for collecting and delivering material

Capital costs relative to processing capacity differ widely. For example, comparing several projects operating during the same time period, each having processing capacities of around 0.25 tons (227 kg) per day, we found capital costs ranging from $1,500 for a low-tech system, to $15,000 for a medium-tech system, to $50,000 for a high-tech system (Table 19.3).

2 Operating Costs

The following list provides examples of just some of the operating and maintenance expenses that vermicomposting projects will involve

- Labor: wages and benefits
- Equipment maintenance
- Fuel and utilities
- Disposal fees for unusable materials
- Insurance and taxes
- Monitoring and testing
- Marketing and sales
- Bagging (if applicable)
- Accounting and administration

Operation costs will vary widely depending on the costs of each item in different localities and regions. As described previously, faster processing times will result in lower overall processing costs than other methods for managing the same types and amounts of waste materials.

However, because many small- and medium-scale vermicomposting projects still tend to use low-technology, labor-intensive methods, the cost of labor represents one of the largest sources of operating costs for such operations. As a result, many of the successful small-scale projects to date have made use of school, volunteer, student, or prison labor (see Chapter 24). Based on experience at several facilities, projected labor costs at low- to medium-technology operations can range from $28 (student labor at $8 per hour) to a high of $80 (wage labor at $12 per hour) per ton of material processed (Jensen 1994). Economics of scale and more efficient technologies in larger-scale projects reduce the cost of labor per ton of organic wastes processed significantly. U.S. military researchers investigating substitutes for landfilling have indicated that operating costs for a continuous-flow vermicomposting reactor system handling 9.1 tons (10 t) per day (including labor at $45 per hour, benefits, maintenance, fuel, administration, etc.) may run $40–$60 per ton (Eaks 1996).

Table 19.3 Approximate Capital Costs for Indoor Continuous-Flow Reactor Facility Vermicomposting 100 Tons of Organic Waste per Day

Equipment	
35 continuous-flow vermicomposting reactors at $50,000 each	$1,750,000
Mobile gantry (multiple reactor)	$12,000
Concrete bases (12 pads, 500 m each, at $27 per m) (not needed in building)	$162,000
12 insulated polythene greenhouse-type tunnels—6000m (500 m each) (not needed in building)	$80,000
Chopping/grinding/mixing machine	$20,000
Front loader machine	$15,000
Liquid waste separator to produce solids	$35,000
Earthworm waste separator	$5000
Moving belts	$30,000
Storage bays	$10,000
Truck	$40,000
Total Cost:	**$2,159,000**
Without base and tunnels:	**$1,917,000**
Annual Returns (estimates per annum)	
Using conservative figures of $30/ton landfill recovery fees for waste disposal and the conservative and lowest sale price for vermicomposts (including a loss of 10% during processing) of $35/ton*	
Landfill fee savings	$1,050,000
Sales of Vermicompost	$1,225,000
Total:	**$2,275,000**
Less Running Costs (estimates per annum)	
Labor costs (four workers)	$140,000
Transport costs	$50,000
Energy costs	$10,000
Repair and maintenance	$20,000
Total:	**$220,000**
Potential Annual Profit:	**$2,055,000**

*Vermicomposts are usually sold at prices between $200 and $1000 per ton, depending on quality, unit size, and packaging.

VI FACTORS AFFECTING THE COMMERCIAL POTENTIAL OF VERMICOMPOSTING

Many economic and environmental factors can limit the applicability of vermicomposting methods or systems in different circumstances. Here we examine factors that affect the development of individual projects or operations and the industry as a whole.

A On-Site or Off-Site Processing

The cost of collection and transportation of raw organic waste materials is a major factor in the total cost of current off-site disposal or recovery options. This provides greater economic incentives to projects or operations that can process wastes on-site, nearer to the waste source.

B Location and Space

Many vermicomposting projects are developing in urban, suburban, and rural locations, and despite the superior ability of earthworms to control odors, such operations must be sensitive to neighboring land uses. Like low-technology thermophilic composting, vermicomposting in boxes or windrows tends to be very space and area intensive. This limits its usefulness in many urban settings where space is limited and costly.

Projects in urban locations, characterized by lack of space and proximity to neighbors, will usually accept small-scale waste diversion in low-technology earthworm boxes or invest in high-technology methods like the Canadian Vermitech system or continuous-flow reactor systems. However, many large-scale vermicomposting projects, processing hundreds or thousands of tons of material per year, have located on large sites of many acres in rural areas.

C Climate

The earthworms used in vermicomposting are readily adaptable to wide variations in temperature and climate, but they do not operate at best efficiency in extremes of climate. Like other farming enterprises, some areas of North America have proved to be more suitable for year-round vermicomposting than others. While earthworm farms and vermicomposting operations have been established in almost every state and province in the United States and Canada, the earthworm-related industry has found its greatest development on the eastern and southeastern coasts of the United States. The effective range of vermicomposting can be expanded by using enclosed high-technology systems, which can compensate for temperature swings and extremes, but at greater cost.

D Human Pathogens

Past research involving the use of earthworms to stabilize sewage sludge indicates significant and meaningful control of pathogens, including *Salmonella* bacteria, in waste (Camp et al. 1980a; Harris et al. 1990). More recent studies provide considerable scientific evidence that human pathogens do not survive the vermicomposting process and that vermicomposting can satisfy federal EPA criteria for pathogen reduction (Edwards and Bohlen 1996; Riggle 1996b; Eastman et al. 2001).

However, uncertainty still exists among some regulators about the ability of earthworm-based processing methods to provide effective control of human pathogens.

The EPA, which will approve site-specific vermicomposting operations, still insists on precomposting for many operations (Riggle and Holmes 1994).

E Feedstock Loading Rates

Feedstock refers to the manure, food wastes, or other combined waste materials prepared for feeding to earthworms. The feedstock loading rate for any vermicomposting system is governed primarily by the quantity of food that the population of earthworms can consume. Researchers have claimed that redworms can consume from one-fourth to two times their body weight per day (Haimi and Huhta 1986). However, variation in consumption rates is a source of uncertainty in design and operational planning and is attributable to variations in the nutrient value and quality of different feedstocks, temperature, moisture, pH, and other environmental conditions. As a result, greater control over these variables is important to maximizing potential loading and processing rates.

F Earthworm Population Growth

Epigeic species of earthworms that process organic wastes have a great capacity for rapid population growth. In laboratory studies, redworms have demonstrated the potential to produce 19 young earthworms per earthworm per week (Edwards and Bohlen 1996). In optimum conditions there may be many generations per year since a redworm takes 35–45 days to reach sexual maturity.

However, variability in feedstock loading rates is matched by variability in the rates of growth of earthworm populations. In field conditions, space availability, differences in food and moisture, and variation in other environmental conditions affects earthworm population growth greatly. The population growth rate is an important factor throughout a commercial operation but especially during a facility's start-up.

G Earthworm Population Availability

One factor that limits the rapid development of large-scale vermicomposting is the initial requirement to grow the large population of earthworms required to process the many tons of daily waste materials used as feedstock. Not only would purchasing tons of earthworms be cost-prohibitive, but growing a population large enough to process 50–100 tons per day would be time-consuming and has been an obstacle to the development of large operations such as Sansai International in Cleveland, Ohio.

H Past History

Finally, one of the biggest limitations to the successful commercial potential of earthworm-related enterprises may be the vermiculture industry's tarnished history. Vermiculture-based pyramid schemes in the United States since the late 1970s have tarnished the reputation of the earthworm-related industry (Machalaba 1978; Milbank 1994). Probably uncertainty concerning commercial development of vermicomposting opportunities has been affected by these past scandals. However, the recent successes of newer commercial vermicomposting companies in the United States, Canada, India, Hong Kong, and Australia have gone a long way to overcome such doubts.

VII CONCLUSIONS

Current markets for the products and services of vermiculture and vermicomposting operations are expanding rapidly. Earthworm castings (vermicomposts) have many well-demonstrated advântages over related thermophilic compost products. These advantages translate into greater value in the marketplace. When the increasing values of vermicomposts are combined with the revenues that can be generated from processing organic residuals from the solid waste stream, profitably improves and commercial ventures become more attractive.

Commercially viable applications of vermicomposting technology were first demonstrated in the processing of animal manures and biosolids. More recent applications of vermicomposting to process organic solid wastes, such as paper, cardboard, yard and garden debris, and food wastes, have been very successful. Experience shows that the potential of commercial vermicomposting is strong for both small- and large-scale projects.

The potential to improve the technologies for vermicomposting is perhaps the most exciting area for researchers and entrepreneurs working in this industry. Various research programs around the world are developing new technologies and systems for faster, more efficient processing and better production of consistent, high-quality vermicompost. Other research efforts are focused on potential end uses for vermicompost and are developing important data for market-development programs.

REFERENCES

Appelhof, Mary. 1981. "Current progress in earthworm research." *BioCycle*, November/December.

Appelhof, Mary. 1997. *Worms Eat My Garbage*. Kalamazoo, MI: Flower Press.

Bogdanov, Peter. 1996. *Commercial Vermiculture: How to Build a Thriving Business in Redworms*. Merlin, OR: Vermico.

Bogdanov, Peter. 1997a. "Oregon soil corporation contracts with Portland Metro to vermiprocess food residuals." *Casting Call*, October.

Bogdanov, Peter. 1997b. "Casting Call interview with Al Eggen." *Casting Call*, October.

Bouché, Marcel B. 1992. *Industrial Vermicomposting of Waste*, Montelimar, France: Sovadec Technologies.

Camp, Dresser, and McKee, Inc. 1980a. *Compendium on Solid Waste Management by Vermicomposting*. Cincinnati, OH: Municipal Environmental Research Lab, EPA.

Camp, Dresser, and McKee, Inc. 1980b. *Engineering Assessment of Vermicomposting Municipal Wastewater Sludges*. Cincinnati, OH: Municipal Environmental Research Lab, EPA.

Carmody, Frank. 1978. The History of Vermiculture. *Vermiculture Journal*, September. (Reprinted in *Worm Digest*, no. 14, 1996.)

Dominguez, Jorge. 1997. Testing the Impact of Vermicomposting. *BioCycle*, April, 58.

Dominguez, Jorge, Edwards, Clive, and Subler, Scott. 1997. A comparison of vermicomposting and composting. *BioCycle*, April, 57–59.

Eaks, Wallace. 1996. *Vermicomposting*. Tri-Services' Pollution Prevention Technical Library, Naval Facilities Engineering Service Center, September.

Eastman, B.R., Kane, P.N., Edwards, C.A., Trytek, L., and Gunadi, B. 2001. The effectiveness of vermiculture in human pathogen reduction for USEPA Class A Stabilization. *Compost. Sci. Util.*, 9: 38–49.

Edwards, Clive A, ed. 1988. Breakdown of animal, vegetable and industrial organic wastes by earthworms. In *Earthworms in Waste and Environmental Management*. 3–4. The Hague, the Netherlands: SPB Academic Publishing.

Edwards, Clive A., ed. 1998a. *Earthworm Ecology*. New York: CRC Press.

Edwards, Clive A. 1998b. Vermicomposting's potential: Small- to large-scale operations. *Resource Recycling*, December.

Edwards, C.A., and Bohlen, P.J. 1996. *Biology and Ecology of Earthworms*. 3rd ed. New York: Chapman & Hall.

Edwards, C.A., and Arancon, N.Q., 2004. Interactions among organic matter, earthworms and microorganisms in promoting plant growth. In *Functions and Management of Soil Organic Matter in Agroecosystems* (Eds.) Magdoff, F., and Weil, R., CRC Press. Boca Raton, FL.

Haimi, J., and Huhta, V. 1986. Capacity of various organic residues to support adequate earthworm biomass for vermicomposting. *Biol. Fertil. Soils*, Spring–Summer.

Harris, George, et al. 1990. Vermicomposting in a rural community. *BioCycle*, January.

Jensen, Jim. 1988. Worm farm takes on new challenges. *BioCycle*, January.

Jensen, Jim. 1994. The worms crawl in, the worms crawl out. *Resource Recycling*, May.

Knight, David. 1989. Nice work for a worm. *New Scientist*, July 8.

Koerner, Brendan I. 1997. It's all business for worms. *U.S. News & World Report*, September 22.

Lee, K.E. 1985. *Earthworms: Their Ecology and Relationships with Soils and Land Use*. Sydney, Australia: Academic Press.

Machalaba, Daniel. 1978. Many states worry about using worms to lure investors. *Wall Street Journal*, June 5. (Reprinted in *Worm Digest*, no. 14, 1996.)

Milbank, Dana. 1994. As the worm turns: Hungarians crushed by fertilizer scheme. *Wall Street Journal*, December 5. (Reprinted in *Worm Digest*, no. 14, 1996.)

Riggle, David. 1996a. Scaling up for commercial vermiculture. *BioCycle*, Feburary.

Riggle, David. 1996b. Worm treatment produces 'class A' biosolids. *BioCycle*, October.

Riggle, David, and Holmes, Hannah. 1994. New horizons for commercial vermiculture. *BioCycle*, October.

Tomati, U., Grappelli A., and Galli, E. 1988. The hormone-like effect of earthworm casts on plant growth. *Biol. Fertil. Soils* 5: 288–294.

Tomlin, A.D. 1983. Fact sheet: Raising earthworms. Communications Branch, Agriculture Canada.

White, Stephen, et al. 1996. Worm growers and sellers (various articles). *Worm Digest*, no. 14.

World Resource Foundation. 1995. Composting with worms—information sheet. Warmer Bulletin. Tonbridge, UK.

CHAPTER 20

The Production of Earthworm Protein for Animal Feed from Organic Wastes

Clive A. Edwards and Alan Niederer

CONTENTS

II Introduction 323
II Potential Protein Food Value of Earthworms 324
III Growth of Earthworms in Animal Wastes 324
IV Methods of Production of Earthworms 326
V Harvesting of Earthworms from Organic Wastes 327
VI Processing of Earthworms for Animal Feed 328
VII Processing of Earthworm Pastes 328
A Blanching 328
B Formic Acid Ensilation 329
C Dry Earthworm Meals 329
VIII Animal-Feeding Trials with Earthworm Protein 330
A Fish-Feeding Trials 330
B Chicken-Feeding Trials 330
C Pig-Feeding Trials 332
IX Economics of Earthworm-Protein Production 332
X Conclusions 333
References 333

I INTRODUCTION

Research in the United States (Hartenstein et al. 1979; Edwards and Neuhauser 1988) and the United Kingdom (Edwards 2004) has shown that some epigeic species of earthworms such as *Eisenia fetida*, *Perionyx excavatus*, *Eudrilus eugeniae*, and *Dendrobaena veneta* can be used to break down a range of organic wastes, such as sewage biosolids, animal manures, food wastes, and organic industrial wastes into vermicomposts and can also be used as sources of animal feed protein (Edwards

1983). The first scientists to suggest that earthworms contained sufficient protein to be considered as animal food or sources of feed protein were Lawrence and Millar (1945), and in recent years, full analyses of the body tissues of earthworms that have become available fully support this conclusion. The first successful animal-feeding trials were organized on chicken and suckling pigs (Sabine 1978), and there have been many later similar trials.

II POTENTIAL PROTEIN FOOD VALUE OF EARTHWORMS

Since the work by McInroy (1971), there have been numerous analyses of the constituents of the tissues of different species of earthworms (Schulz and Graff 1977; Sabine 1978; Yoshida and Hoshii 1978; Mekada et al. 1979; Taboga 1980; Graff 1982; Edwards 1985a, 1985b; Edwards and Niederer 1988). The overall composition of earthworm tissues does not differ greatly from that of many vertebrate tissues (Table 20.1). The essential amino acid spectrum of earthworm tissues compares well with that from other currently used sources (Table 20.2). Clearly, the mean amounts of essential amino acids recorded are very adequate for a good animal or fish feed as recommended by Food and Agriculture organization of United Nations (FAO) and World Health Organization (WHO), particularly in terms of lysine and the combinations of methionine and cysteine, phenylalanine and tyrosine, all of which are important components of animal feeds. In addition, earthworm tissues contain a preponderance of long-chain fatty acids, many of which nonruminant animals cannot synthesize, and an adequate mineral content (Figure 20.1). They have an excellent range of vitamins and are rich in niacin, which is a valuable component of good animal feeds (Table 20.3).

III GROWTH OF EARTHWORMS IN ANIMAL WASTES

The life cycles of four species of earthworms that grow and reproduce well in organic wastes have been investigated thoroughly by the lead author of this chapter. These are *Eisenia fetida* (Savigny), *Eudrilus eugeniae* (Kinberg), *Perionyx excavatus* (Gates), and *Dendrobaena veneta* (Rosa) (see Chapter 3). The growth patterns of individual earthworms or whole earthworm populations follow sigmoid growth curves, and the maximum rates of protein production have been achieved by inoculating large volumes of animal wastes or other organic wastes with relatively small

Table 20.1 Composition of Earthworm Tissues

Water	78–88%
Protein	60–70% (dry matter)
Fat	6–11% (dry matter)
Carbohydrate	5–21% (dry matter)
Minerals	2–3% (dry matter)
Gross energy	16–24 kJ g^{-1} (dry matter)

Table 20.2 Reported Essential Amino Acid Content (Grams per 100 g Protein) of Earthworm Tissues

Essential Amino Acid	WHO-FAO Requirements of Essential Amino Acids in Animal Feed		McInroy (1971)	Schulze and Graff (1977)	Sabine (1978)	Yoshida and Hoshii (1978)	Mekada et al. (1979)	Taboga (1980)	Graff (1982)	Rothamsted U.K. (1982)	Mean of All Analyses
Arginine	—	—	6.1	6.1	4.2	6.9	4.5	7.3	6.1	6.5	6.0
Cysteine	2.0	[a]	1.8	1.4	2.3	0.8	—	1.8	1.4	0.7	1.5
Histidine	—	—	2.2	2.3	1.6	4.3	1.6	3.8	2.3	3.0	2.6
Isoleucine	4.2	4.0	4.6	4.7	2.6	4.7	4.3	5.3	4.7	4.1	4.3
Leucine	4.8	7.0	8.1	8.2	4.8	8.7	5.0	6.2	8.2	8.3	7.2
Lysine	4.2	6.5	6.6	7.5	4.3	8.7	5.9	7.3	7.5	6.5	6.8
Methionine	2.2	[a]	1.5	1.8	2.2	1.6	1.9	2.0	1.8	2.8	2.0
Phenylalanine	2.8	[b]	4.0	3.5	2.3	4.4	3.4	5.1	3.5	4.0	3.8
Threonine	2.8	4.0	5.3	4.7	3.0	5.2	6.9	6.0	4.7	5.6	5.2
Tryptophan	1.4	1.0	—	—	—	1.3	—	2.1	—	0.7	1.4
Tyrosine	2.8	[b]	—	3.0	1.4	4.4	2.7	4.6	3.0	3.5	3.2
Valine	4.2	5.1	5.1	5.2	3.0	5.1	5.1	4.4	5.2	4.7	4.7

[a] 3.5 Total for methionine + cysteine
[b] 6.0 Total for phenylaniline + tyrosine

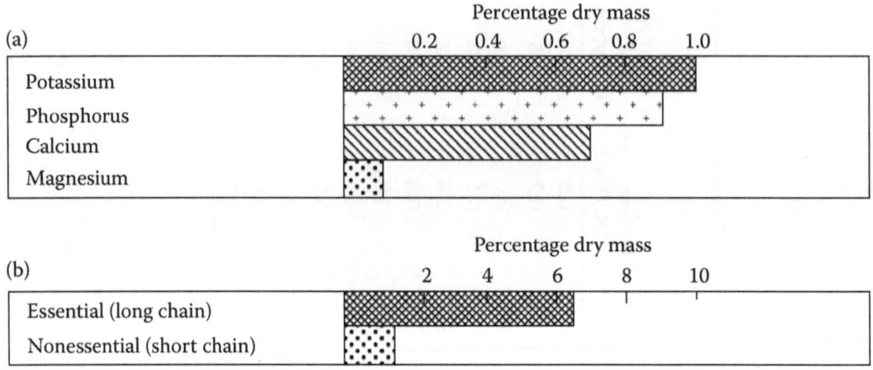

Figure 20.1 (a) Minerals in earthworms; (b) fatty acids in earthworms.

Table 20.3 Vitamin Composition (Milligrams per Kilogram) of Earthworm Tissues

Niacin	358
Riboflavin (B_2)	147
Pantothenic acid (B complex)	16
Thiamin (B_1)	15
Pyridoxine (B_6)	2
Vitamin B_{12}	4
Folic acid	0.5
Biotin (B complex)	0.35

numbers of young earthworms (Figure 20.2). Dry-matter conversion ratios of organic waste into earthworm tissues as high as 10% for cattle and pig wastes and 5% for poultry wastes have been achieved in the laboratory. These species of earthworms can withstand many adverse environmental factors, and on the basis of our research we have defined their environmental requirements within relatively broad limits.

IV METHODS OF PRODUCTION OF EARTHWORMS

The natural habitat of earthworm species that break down organic matter is heaps of decaying organic matter and compost heaps as well as trickling sewage filters. The organic matter in these habitats follows a natural microbial decomposition process, with a succession of microorganisms and changes in temperature, with temperatures of up to 70°C (158°F) being attained quite commonly in thermophilic composting systems. In such habitats, earthworm colonization is limited by the high temperatures, populations are localized, and earthworm-biomass productivity is low. To produce earthworms as a commercial source of protein, such systems must be controlled to optimum environmental conditions, at less than 35°C (95°F) for the earthworms, with organic waste added periodically in thin horizontal layers to the

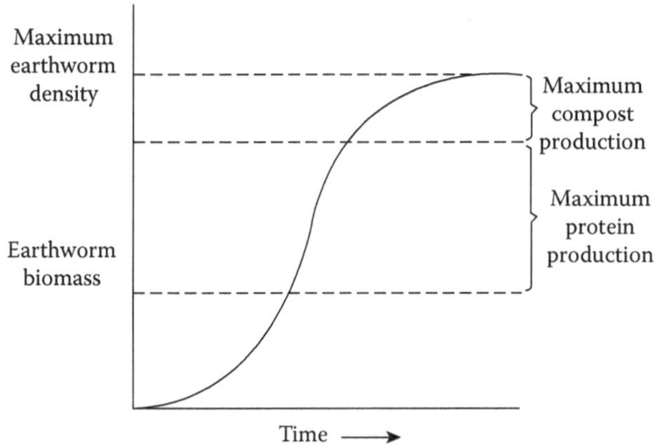

Figure 20.2 Part of population curve of classic earthworm growth for maximum protein production.

upper surface of the waste, to enable earthworms to colonize all of the organic matter and utilize the microorganisms in the organic matter on which they feed to achieve maximum productivity.

Animal wastes range from almost-liquid slurries to straw-based mixtures to relatively dry and finely dispersed mixtures such as those produced by laying chickens over deep litter systems. The more liquid animal manures can be turned into separated solids by a range of commercial press separating systems. For efficient vermicomposting and earthworm growth, the wastes must be brought to a suitable moisture content (80–90%) and optimal temperatures 15–25°C (59–77°F), and the ammonia and salt contents must be reduced to acceptable levels by leaching, composting, or some other method, before the earthworms can be grown successfully (see Chapter 3).

A range of vermicomposting systems, from low- to high-technology, using epigeic species of earthworms to process organic wastes are described in more detail in Chapters 7 and 8 of this book. They all depend on addition of thin layers of organic wastes to vermicomposting systems to avoid the high temperatures produced by thermophilic composting action. The high-technology vermicomposting systems described in Chapter 8 are extremely efficient and need a minimum of labor to keep them operating. In adapting these systems for earthworm-protein production it is important to use species of large earthworms that gain weight rapidly, such as *D. veneta* in temperate climates and *E. eugeniae* in the tropics. There is some possibility that the production of vermicomposts and earthworm protein for animal feed may be complementary and produced in the same systems.

V HARVESTING OF EARTHWORMS FROM ORGANIC WASTES

As the individual earthworms and earthworm populations grow, they fragment the animal wastes into finer and finer particles as these materials pass repeatedly through the earthworms, producing, eventually, a finely divided, stabilized material with a low C:N ratio, which has considerable potential as a plant growth medium in horticulture. A major problem is that earthworms grow best at relatively high moisture levels (80–90% moisture content), and it is not easy to separate earthworms mechanically from the finely divided organic matter at such high moisture contents, so some drying of the vermicompost may be necessary.

Specialized machinery for separating earthworms from vermicomposts has been developed at Rothamsted and the National Institute for Agricultural Engineering, Silsoe (Phillips 1985; see Chapters 7 and 8). The efficiency of this machinery in terms of percentage recovery of earthworms from vermicomposts is very high. The machine that was developed at Silsoe will separate earthworms from about one ton of vermicompost per hour, and this machinery can be automated and scaled up in size to increase this throughput. However, the production of protein for animal feed from earthworms may remain an uneconomical process due to labor costs until more efficient separation machinery is available, although the quality of the feed protein for fish, chickens, or young pigs is excellent.

VI PROCESSING OF EARTHWORMS FOR ANIMAL FEED

The earthworms collected from the separating machinery may still have small particles of vermicomposts on their bodies and are likely to contain residual waste in their guts. Hence, the first stage of all the methods of processing is to wash the earthworms thoroughly and leave them standing in water for a minimum of 8 hours, to completely evacuate the residual wastes from their guts, then blanche them in boiling water.

Various methods of processing the earthworms for animal feed have been developed and tested by the author and his colleagues. Two of the methods tested produced a paste product and the other four a dry earthworm meal; all of these materials were found to be acceptable for different animal feed uses. The ultimate choice of a method of processing must depend on (i) the type of animal feed required, (ii) the cost of production of the protein, (iii) minimal loss of dry matter, and (iv) minimal loss of nutrient value.

VII PROCESSING OF EARTHWORM PASTES

A Blanching

The first method tested was blanching the earthworms in boiling water for one minute, then incorporating them in 30% molasses together with 0.3% potassium sorbate to produce a paste. The molasses lowered the water activity to about A_W 0.90, and the addition of potassium sorbate reduced this further to approximately A_W 0.65. At this moisture level, the growth of yeasts and molds is inhibited, permitting indefinite storage of the product at room temperature.

B Formic Acid Ensilation

Another wet method of producing earthworm protein tested was to incorporate 3% formic acid with the earthworms with thorough homogenization, then allow the mixture to ensile and eventually produce a very stable paste or liquid product.

C Dry Earthworm Meals

1. In one method a dry earthworm-protein meal was produced by blanching earthworms in boiling water for 1 minute, then air-drying and grinding them into a powder.
2. Another type of dry earthworm meal was produced by freezing earthworms quickly, then freeze-drying them and grinding them into a powder.
3. Yet another type of dry meal was produced by first killing earthworms by immersing them in acetone for 1 hour or then air- and oven-drying them at 95°C (203°F) before grinding them.
4. A fourth type of dry meal was produced by killing earthworms in boiling water, drying them in an oven at 95°C (203°F), and grinding them up into a fine powder.

All of the methods tested gave a good protein product that could be used as animal feed, but there were some minor variations in the dry-matter yield (grams of dry product per 100 g fresh earthworms). Killing earthworms in boiling water and then drying them in an oven resulted in the lowest yield of dry matter (11.6%). Freeze-drying produced a meal with 13.5% dry matter. Killing earthworms in acetone and drying them in air produced a dry matter of 14.5%, but after subsequent oven-drying this fell to 12.8%. Killing and drying the earthworms in a hot-air oven at 95°C (203°F) gave the greatest dry-matter yield of 15.2%.

The effects of the different processing methods on the amounts of essential amino acids in the products are summarized in Table 20.4. Most of the methods

Table 20.4 Amounts of Essential Amino Acids (Grams per 100 g Protein) in Earthworm Tissues after Different Processing Methods

	Processing Method					
Amino Acid	**Molasses Ensiling**	**Formic Acid Ensiling**	**Freeze-Drying**	**Acetone-Drying**	**Heat-Drying**	**Blanching and Heat-Drying**
Arginine	6.5	6.7	6.4	7.0	3.8	4.1
Cysteine	0.3	0.5	0.5	0.3	0.5	0.4
Histidine	2.8	2.7	2.7	2.5	2.3	2.3
Isoleucine	4.2	4.3	4.0	4.3	4.4	4.3
Leucine	7.3	7.1	6.9	7.5	8.3	8.5
Lysine	5.4	5.9	5.7	6.1	6.2	6.5
Methionine	1.6	1.1	1.2	1.2	1.1	0.9
Phenylalanine	3.0	2.8	3.1	3.2	3.1	3.6
Threonine	4.8	5.1	8.5	5.1	5.5	5.6
Tyrosine	2.0	2.4	4.0	2.8	2.8	3.0
Valine	5.1	4.9	4.7	5.0	5.5	4.6

had relatively little effect on the overall amounts of amino acids in the final product, although the lysine content was decreased slightly by ensiling with molasses, by the use of formic acid, and by freeze-drying, compared with other methods. Clearly, a stable animal feed protein can be produced by any of the methods tested, and the choice of method must depend on the type of fish or animals to which the protein feeds are to be presented.

VIII ANIMAL-FEEDING TRIALS WITH EARTHWORM PROTEIN

The main outlets suggested for utilization of earthworm feed protein have been in fish farming and as protein supplements in feeds for poultry and pigs.

A Fish-Feeding Trials

The first trials involving the feeding of earthworms to fish were by Tacon et al. (1983), who used earthworms to feed trout. The growth of trout fed only on *E. fetida*, *Allolobophora longa* (Uge), and *Lumbricus terrestris* L. was compared with that of fish fed on a commercial protein ration. Fish fed with frozen *A. longa* and *L. terrestris* grew as well or better than fish fed on commercial trout pellets. However, trout did not grow as well on a whole diet of freeze-dried *E. fetida*, although they grew much better on *E. fetida* that had been blanched in boiling water before freezing; this probably removed a mucus that this species produces to discourage predation by birds and mammals (Stafford and Tacon 1985). However, dried earthworm meal derived from *E. fetida* that had not been blanched could still replace the fish meal component of formulated trout feed pellets at levels of inclusion between 5 and 30% without affecting the rates of growth of the trout. The conclusion reached was that earthworms have potential both as complete fish feed or a protein supplement for trout or other fish. Hilton (1983) reported that trout did not grow well on another earthworm species, *E. eugeniae*, but there are considerable doubts about his experimental techniques, because other forms of protein supplement currently used commercially, such as blood meal, would have also been unsuccessful as feed if used in the same way that he used earthworm protein. Guerrero (1983) reported that *Tilapia* fish grew better on diets containing earthworm-protein supplements from *P. excavatus* (Perrier) or *E. eugeniae* (see Chapter 29) than on those with fish meal supplements.

B Chicken-Feeding Trials

The first trials that assessed the rates of growth of chickens on earthworm protein were reported by Harwood (1976) and Sabine (1978). They compared the use of earthworm meal with meat meal as a protein supplement in chicken feed and found no significant difference in growth on the diets containing the two protein sources (Figure 20.3). Similar results were reported by Mekada et al. (1979) and Taboga (1980), who also reported that when earthworms were fed to older chickens, egg production was maintained. Jin-you and his colleagues (1982a) reported that chickens fed on earthworms put on weight faster than those given other diets (including fish meal), had more breast muscle, and consumed less food. These results have been confirmed by research by our colleague, Fisher (1985). In his experiments, chickens grew well and had a good mass gain per unit of food and an excellent nitrogen retention when fed on diets with levels of earthworm meal from 72–215 kg^{-1} (159–474 lb) (Figure 20.4).

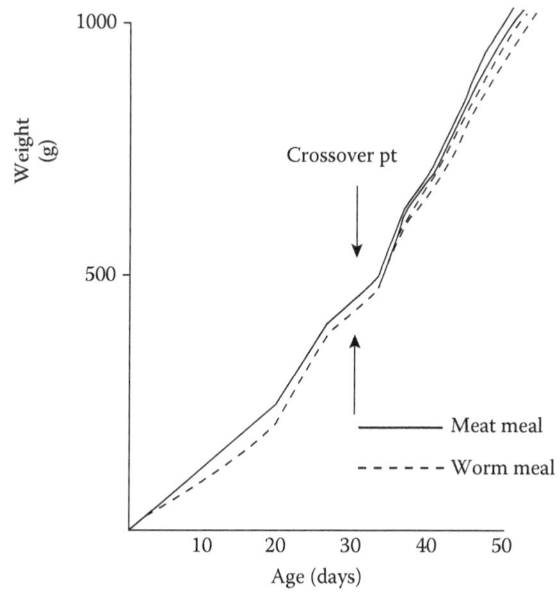

Figure 20.3 Growth of chickens on meat meal and earthworm meal. (From Sabine, J., *Proceedings of Conference on Utilization of Soil Organisms in Sludge Management*, Syracuse, N.Y.: Kalamazoo, MI., 122–130, 1978. With permission.)

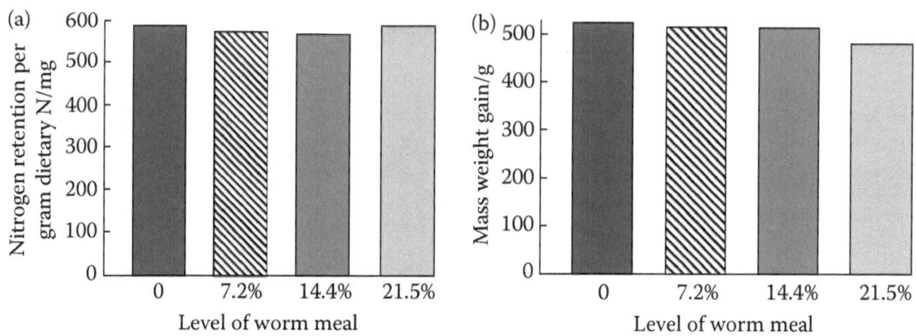

Figure 20.4 Growth of chickens on an earthworm-protein diet. (a) Nitrogen retention; (b) mass gain.

C Pig-Feeding Trials

Only two trials have reported on the growth of pigs on earthworm-protein supplements. Harwood (1976) and Sabine (1978) showed that in feeding trials with both starter and grower pigs, animals fed on an earthworm-protein supplement grew equally well (Table 20.5) and had similar feed-conversion ratios to those grown on commercial rations. Jin-you et al. (1982b) reported that piglets grew better on feeds with earthworm-protein supplements than on others and that older pigs had accelerated weaning, earlier estrus in sows, increased disease resistance, and a decreased incidence of white diarrhea.

IX ECONOMICS OF EARTHWORM-PROTEIN PRODUCTION

There have been several studies of the economics of the production of earthworm protein (e.g., Fieldson 1985). Their general conclusions were that earthworm production had the best prospects of good profits when done by larger farmers with considerable amounts of unused organic wastes available. The most important criterion is that earthworm meal for feed protein must be produced at an economical price, although the value of the vermicompost produced in the vermicomposting system can also be taken into account. Currently, the only labor-intensive part of earthworm-protein production is the harvesting process, and this remains the main economic barrier to successful commercial earthworm-protein production for animal feed, but this should be able to be resolved by developing improved technology for separating earthworms from vermicomposts or combining earthworm and vermicompost products.

Table 20.5 Growth of Young Pigs on Earthworm Protein Supplements

	Ratios between Components		
	Worm Meal Supplement	Meat Meal Supplement	Commercial Protein Supplement
Starter period (days 36–50)			
A. Mean weight gain (kg)	4.36	3.65	4.26
B. Mean % gain	48.9	41.2	48.9
C. Feed consumption (kg/pig/14 days)	10.05	10.05	10.05
D. Feed conversion	2.31	2.78	2.37
E. Mean rate of growth (kg/day)	0.31	0.26	0.30
Grower period (days 84–98)			
A. Mean weight gain	6.54	6.80	6.60
B. Mean % gain	28.1	28.2	27.0
C. Feed consumption (kg/pig/14 days)	16.78	16.78	16.78
D. Feed conversion	2.47	2.48	0.46

Table 20.6 Value of Earthworm Protein for Different Animal Feeds

Animal	Maximum Value (U.K.£ t^{-1})
Cows	130
Ducks	183–218
Broiler chickens	220–249
Turkey (starter)	1011
Turkey (finisher)	338
Trout	404
Eels	2000

In a computer analysis of the economic value of earthworm meal (Table 20.6), based on its amino acid, fatty acid, mineral, and vitamin content, it was shown that it is extremely valuable as feed for specialized animals such as eels and young turkeys and has about the same value for fish, pig, and poultry feed as fish meal or meat meal.

X CONCLUSIONS

The production of earthworm protein for feeding fish, crustaceans, chickens, and pigs is marginally uneconomical in developing countries because of the high cost of labor involved in separating earthworms from vermicomposts. However, in countries such as India (Chapter 28) and the Philippines (Chapter 29), the economics of earthworm feed protein are much more favorable due to lower labor costs, and many more commercial projects are underway.

REFERENCES

Edwards, C.A. 1983. Earthworms, organic wastes and food. *Span* 26, 106–108.

Edwards, C.A. 1985a. Production of earthworm protein for animal feed from potato waste. *Libro* 15 (3): 153–155.

Edwards, C.A. 1985b. Production of feed protein from animal waste by earthworms. In *Technology in the 1990's: Agriculture and Food*, ed. Sir Kenneth Blaxter and Sir Leslie Fowden, *Proceedings of the Royal Society*, London, U.K. 153–163.

Edwards, C.A. 2004. *Earthworm Ecology* (2nd Ed.) CRC Press, Boca Raton, FL.

Edwards, C.A., and Niederer, A. 1988. The production and processing of earthworm protein. In *Earthworms in Waste and Environmental Management*, ed. C.A. Edwards and E.F. Neuhauser, 211–220. The Hague, the Netherlands: SPB Academic Publishing.

Edwards, C.A. and Neuhauser, E.F. (Eds), 1988. Earthworms in waste and Environmental Management The Hague, The Netherlands: SPB Academic Publishing.

Fieldson, R.S. 1985. The economic viability of earthworm culture on animal wastes. In *Earthworms in Waste and Environmental Management*, ed. C.A. Edwards and E.F. Neuhauser, 145–156. SPB Academic Publishing, the Hague, the Netherlands.

Fisher, C. 1985. The nutritional value of earthworm meal for poultry. In *Earthworms in Waste and Environmental Management*, ed. C.A. Edwards and E.F. Neuhauser, 181–192. SPB Academic Publishing, the Hague, the Netherlands.

Graff, O. 1982. Vergleich der Regenwurmarten *Eisenia foetida* und *Eudrilus eugeniae* hinsichtlich ihrer Eignung zur Proteingewinnung aus Abfallstoffen. *Pedobiologia* 23, 277–282.

Guerrero, R.D. 1983. The culture and use of *Perionyx excavatus* as a protein resource in the Phillippines. In *Earthworm Ecology*, ed. J.E. Satchell, 309–313. Chapman & Hall, London.

Hartenstein, R., Neuhauser, E.F., and Kaplan, D.L. 1979. A progress report on the potential use of earthworms in sludge management. *Proc. 8th Nat. Conf. on Sludge Composting, Florida*, March 1979, pp. 238–241. *Information Transfer Inc.* FL.

Harwood, M. 1976. Recovery of protein from poultry waste by earthworm. *Proc. Austr. Poultry Stockfeed Conf.*, Sydney, pp. 138–143.

Hilton, J.W. 1983. Potential of freeze-dried worm meal as a replacement for fish meal in trout diet formulations. *Aquaculture* 32, 277–283.

Jin-you, X., Xian-Kuan, Z., Zhi-ren, P., Zhen-yong, J., Yan-hua, G., Hong-bo, I., Xue-yan, H., and Quio-ping, X. 1982a. Experimental research on the substitution of earthworm for fish meal in feeding broilers. *J. So. China Normal Coll.* 1, 88–94.

Jin-you, X., Xi-cong, H., and Wen-xi, L. 1982b. An observation on the results of using earthworms as a supplementary food for suckling pigs. *J. So. China Normal Coll.* 1, 1–8.

Lawrence, R.D., and Millar, R.H. 1945. Protein content of earthworms. *Nature, Lond.* 3939, 517.

McInroy, D.M. 1971. Evaluation of the earthworm *Eisenia foetida* as food for man and domestic animals. *Feedstuffs* 43, 46–47.

Mekada, H., Hayashi, N., Yokota, H., and Okomura, J. 1979. Performance of growing and laying chickens fed diets containing earthworms. *Jpn. Poult. Sci.* 16, 293–297.

Phillips, V.R. 1985. Engineering problems in the breakdown of animal wastes by earthworms. In *Earthworms in Waste and Environmental Management*, ed. C.A. Edwards and E.F. Neuhauser, 111–118. SPB Academic Publishing, the Hague, the Netherlands.

Sabine, J. 1978. The nutritive value of earthworm meal. In *Proceedings of Conference on Utilization of Soil Organisms in Sludge Management*, Syracuse, N.Y., ed. R. Hartenstein, Kalamazoo, MI. 122–130.

Schulz, E., and Graff, O. 1977. Zur Bewertung von Regenwurmmehl aus *Eisenia foetida* (Savigny) als Eiweissfuttermittel. *Landbauforsch. Völk.* 27, 216–218.

Stafford, E.A., and Tacon, A.G.J. 1985. The use of earthworms as food for rainbow trout *Salmo gairdneri.* In *Earthworms in Waste and Environmental Management*, ed. C.A. Edwards and E.F. Neuhauser, 193–208. SPB Academic Publishing, the Hague, the Netherlands.

Taboga, L. 1980. The nutritional value of earthworms for chickens. *Brit. Poult. Sci.* 21, 405–410.

Tacon, A.G.J., Stafford, E.A., and Edwards, C.A. 1983. A preliminary investigation of the nutritive value of three terrestrial lumbricid worms for rainbow trout. *Aquaculture* 35, 187–199.

Yoshida, M., and Hoshii, H. 1978. Nutritional value of earthworms for poultry feed. *Jpn. Poult. Sci.* 15, 308–311.

CHAPTER 21

The Use of Vermiculture for Land Improvement

Kevin R. Butt

CONTENTS

I Vermiculture 335
II Land Improvement as Related to Vermiculture 336
III Effects of Earthworms on Soils 337
IV Successful Examples of Earthworm Introduction 337
V Earthworm Ecological Relationships 338
VI Breeding Soil-Inhabiting Earthworms 339
VII Earthworms for Land Improvement 340
  A Collection Techniques 340
  B Breeding Techniques 341
VIII An Alternative Vermiculture 342
IX The Future of Earthworms for Land Improvement 344
  A Earthworm Inoculation en Masse 344
  B Intensive Cultivation 345
X Conclusions 346
References 346

I VERMICULTURE

The process of vermiculture may take a number of forms (see other chapters) but is defined simply as the growth and maintenance of earthworm cultures over a period of time. These earthworms must be self-sustaining, such that if some are removed for use elsewhere, then those remaining will continue to multiply and thus offer further opportunities for stock removal. For this process to be successful, the culture medium must be acceptable to the earthworms, and a type of food (which can form the bulk of the culture medium itself) needs to be supplied. The rate of food application will depend on a host of factors, such as its nutritional value, the species of earthworm under culture, the ambient temperature, and earthworm population density. Anyone who has attempted to culture earthworms will have found that the process, while relatively simple, needs fine-tuning in order to obtain the most from it. Scientific research over the past 25 years (see, for example, Lofs-Holmin 1985; Edwards and Neuhauser 1988) has shown that vermiculture, which has been utilized in some form for centuries by man, works best if the process is carefully monitored. To date, most vermiculture (vermicomposting) has been concerned with the processing of organic wastes in order to produce earthworm-worked material or vermicompost for plant growth (see Chapter 9) or the production of earthworm protein (see Chapter 20) or even both, but research has shown that protein production from earthworms as a sole aim is not economically viable (Fieldson 1988).

It has been suggested that earthworms could play an important role in the restoration of degraded soils, a topic originally reviewed by Curry (1988) and reexamined by Baker et al. (2006). To this end, low-technology processes using vermiculture for land improvement have been used and are still undergoing development in the light of more recent findings (Butt 1999, 2008; Snyder and Hendrix 2008).

II LAND IMPROVEMENT AS RELATED TO VERMICULTURE

In this context, land improvement will be considered as positive changes brought about in the soil and the vegetation that it supports. Thus, it excludes large-scale physical changes that may alter, for example topography, but includes small-scale physical, chemical, and biological changes that might influence, for example, drainage and soil aeration.

Many types of land that have been subjected to human activities could benefit from soil improvement, including many postindustrial landscapes (brownfield sites). In England these include ex-colliery workings, abandoned heavy industrial sites, and old mineral workings, covering a combined area of some 620 km^2 (239 $miles^2$) which is 0.4% of the total land surface area (Communities and Local Government [CLG] 2008). In many instances topsoil was removed from such sites during development, and on abandonment they may be left littered with the debris of former use. Coupled with this, any soil that does remain may be contaminated, compacted, or simply devoid of a typical fauna and flora. Amelioration of soil contamination is beyond the scope of this chapter, although it can be improved by microbial activity (e.g., Matthews 1994), but many other potential problems can be addressed simply by the addition of earthworms. Even where sites are restored to something approaching a good soil status, natural earthworm colonization may be extremely slow. Judd and Mason (1995) reported that earthworms had moved only 5 m (16 ft) into a restored landfill site after 4 years, and Butt et al. (1999) found no evidence of natural earthworm colonization in the first 3 years at another landfill site. Unlike other beneficial soil invertebrates, earthworms often require human intervention to reach restored sites since these may be surrounded, for example, by drainage ditches, which present physical barriers to natural colonization (Halle and Fattorini 2004).

III EFFECTS OF EARTHWORMS ON SOILS

The positive activities and benefits of earthworms have been appreciated since the times of Darwin (1881). By burrowing into the soil, earthworms provide channels that allow air to circulate more freely and equally permit rainwater to percolate in, rather than potentially causing erosion through surface runoff. The intimate mixing of soil layers brought about by earthworms' ingestion of soil and their castings, either within the soil profile or on the soil surface, causes mineral components and organic fragments to become closely associated. The nature of the castings produced by earthworms is unique, and the crumb structure, and the aggregates formed by the various inorganic and organic components held together by mucus, is an ideal substrate for plant growth. Earthworm casts are rich in beneficial microorganisms and nutrients compared with the contents of the surrounding soil (Orazova et al. 2003). Therefore, passage of soil through the gut of an earthworm adds to the microbial status of soil (Edwards and Bohlen 1996). In many studies earthworms have been shown to have positive effects on the growth of vegetation. When earthworms were introduced to pastures where they were previously absent, initial production of grass increased by 70% in the first year (Stockdill and Cossens 1966). Root production by fruit trees was found to be greater where earthworms had been added to Dutch orchards on polder soils (Rhee 1971), and barley's growth rate and yield were significantly increased by the presence of earthworms in direct drilled (no till) cultivation (Edwards and Lofty 1980). Senapati et al. (1994) also demonstrated that earthworm inoculation into commercial Indian tea plantations promoted fine root biomass and green leaf production.

IV SUCCESSFUL EXAMPLES OF EARTHWORM INTRODUCTION

Should it prove possible, it appears that introducing earthworms into soils where they are absent can have profound consequences. Studies have been undertaken within a number of land types including reclaimed cut-over peat, colliery spoil, upland pasture, and landfill sites. Previous accounts of such research (e.g., Curry 1988) have considered each examples in relation to particular land uses/habitats that are available for recolonization. Alternatively, it is equally applicable to examine some of the results in relation to earthworms and soil conditions and identify common factors, such as organic-matter content, pH, moisture, tillage, and human influences, that have proved helpful to or have hindered the intended earthworm colonization processes (Butt 1999, 2008).

All systems may be limiting, and the presence of suitable food for organisms within them can be one of the most fundamental problems, which will in part determine the earthworm carrying capacity (number supported per unit area) of a field. Where earthworms are to be introduced, sufficient organic matter must be present. If the site is a recently reclaimed area, this should form part of the restoration plan. Scullion and Mohammed (1991) discovered that a topdressing of poultry manure encouraged colonization by earthworms of an opencast coal site in the United Kingdom. Additions of deciduous leaf litter to poor mineral soils in Finland also aided earthworm establishment by providing a food source (Huhta 1979). Huhta also added lime to the site to raise soil pH, thought to be another limiting factor. Organic amendments that are applied at the soil surface can, however, serve a greater function than simply providing earthworm food. Dunger (1969) suggested that organic matter led to stabilization of both temperature and moisture conditions at the soil surface, both factors vital to earthworm survival. Practices of subsoiling of compacted sites, such as those present after opencast coal extraction, have also been shown to assist earthworm population development by leading to improved site drainage (Scullion and Mohammed 1991).

It should be noted that the factors identified are often interrelated and should be considered together rather than in isolation. An "earthworm's view" of the situation must be taken with the objective of making the soil conditions as hospitable to earthworms as possible. Any factors that impinge on earthworm survival and well-being need to be addressed.

The conclusions from the preceding discussion support the contention that earthworms can improve the quality of soil when introduced to areas where they are not indigenous. However, in all the examples given, the earthworms utilized were not bred specifically for the task but were collected from elsewhere and then introduced into the land that was in need of improvement. This is a practice that has occurred worldwide and also includes examples from the tropics (e.g., Lavelle et al. 2004). Only in a relatively small number of instances have earthworms been bred specifically for land improvement. Why has this been the case? What aspects of their ecology may be important in successful colonization?

V EARTHWORM ECOLOGICAL RELATIONSHIPS

All of the earthworms described to date (some 3000 or so species) can be placed into one of three major groupings: epigeic, endogeic, and anecic species (Bouché 1977). The first of these groups, the epigeic species, encompasses all of the earthworm species that inhabit the upper organic layers of the soil and may also occur within compost heaps (often referred to as "compost earthworms"). These include common species such as *Eisenia fetida* (the brandling or tiger worm) and *Dendrobaena veneta* (another earthworm) in Europe, *Eudrilus eugeniae* (the African night crawler) used in the United States, and *Perionyx excavatus* in Asia. This group of species is familiar to vermicomposters, and species in it are used extensively to break down various types of waste organic matter into vermicomposts. However, excluding *Dendrobaena veneta* and *Lumbricus rubellus*, most of these earthworms do not thrive well in mineral soils and consequently are of limited use in land-improvement schemes (even though they may be sold for this purpose by some dealers).

Earthworms that fall within the other two major ecological groups do not require such a high level of organic matter and are very productive in mineral soils. Endogeic species may be of small size and live within temporary horizontal branching burrows close to the soil surface (e.g., *Allolobophora chlorotica*, *Aporrectodea caliginosa*). They are geophagous (soil-feeding) in nature—literally eating their way through the soil. Equally, some anecic species may be larger and inhabit more vertical and permanent deep burrows (e.g., *Aporrectodea longa*, *Lumbricus terrestris*). The latter two species are the earthworms that can prove most useful in the process of land restoration. However, although their life cycles have the same three developmental stages (… adult → cocoon → hatchling → adult …), they are less prolific in population buildup than the vermicompost earthworms. They do not have either the rapid growth or the rapid reproduction shown by vermicompost earthworms and consequently take longer to produce sustainable populations (i.e., they tend to be K-selected). Also, due to their lifestyle, culturing these earthworms proves more difficult since the substrate in which they must be maintained requires not only an organic food content but also a mineral soil base. Thus a greater volume of material is needed, which may be difficult to accommodate. This is the major reason why earthworms have usually been collected for use from the field, rather than bred specifically in most research projects. However, if feasible, there may be advantages to breeding earthworms for field inoculation.

VI BREEDING SOIL-INHABITING EARTHWORMS

If this process is to prove useful to managers of degraded land, then much basic biological and ecological research is still required prior to large-scale field earthworm inoculation trials and projects. Even today, basic information relating to the life cycles of common species of earthworms is lacking (Edwards and Bohlen 1996), although this is being addressed systematically (e.g., Boström 1988; Butt 1991, 1993; Hamoui 1991; Edwards et al. 1998; Lowe and Butt 2005; Uvarov 2009). Maintenance of earthworms in culture means that controls can be achieved so that, for example, the age of earthworms and their reproductive potential are known. (After reaching sexual maturity, reproductive capacity decreases as earthworms age.)

From starter cultures of recently matured deep-burrowing earthworms such as *L. terrestris* (the European night crawler), cocoons can be produced at a rate of three per earthworm per month, and 99% of these will contain a single hatchling, which will take at least 10 weeks to emerge from the cocoon and may take 12 weeks to mature (Butt 1991). These earthworms are attractive due to their size and worth investigating further due to their important beneficial activities within the soil, particularly with no till or direct drilling. However, it must be borne in mind that these rates of reproduction were under relatively ideal conditions. To prove useful in land-improvement schemes, the introduced earthworms must maintain a viable rate of reproduction and growth in the field soils.

Two major issues face production of earthworms for land improvement once the field inoculation stage is reached:

a. Initial survival of inoculum in the field
b. Sustainable population development

If either of these does not occur, then the whole process is useless.

VII EARTHWORMS FOR LAND IMPROVEMENT

A Collection Techniques

Traditionally, earthworms have been collected in large numbers from pasture or golf greens (see Tomlin 1983); if not destined for fishing bait, they may have been used for soil improvement. For example, Judas et al. (1997) used *L. terrestris* for inoculation into a limed spruce forest in Germany. They obtained 10,000 earthworms from a commercial trader but were uncertain about their origin and homogeneity. In general, the age of the earthworms, their physiological state, and their relative survival capability at sites of reintroduction into the field are not considered by traders, and transport across, and even between, continents is not unknown. The evolutionary, environmental, and ecological adaptations to local conditions of "ecotypes" at one site within a species may prove less beneficial elsewhere under different environmental conditions. To this end, long-distance translocation of earthworms is not recommended, and it is suggested that earthworms for use in land improvement should be obtained from sites as close to the proposed site of reintroduction as feasibly possible.

Earthworm inoculations usually involve adult and juvenile earthworms being broadcast onto the soil surface, often with no thought for the establishment of a viable population or community. Should soil conditions prove to be inhospitable immediately after the time of introduction, then it is quite possible that the majority of earthworms may perish (bird predation can also be a significant factor at this stage). Spring and autumn are seen as the most appropriate times of year for earthworm introductions into field sites in temperate regions due to the prevailing soil moisture and temperature levels. Introduction of earthworms in cut pieces of soil turf is another method of earthworm inoculation, one that has been used extensively in New Zealand (e.g., Stockdill and Cossens 1966) to introduce European species and, more recently, in the United Kingdom (Morton 1993; Butt 1999). Turf transfer has the advantage over introducing earthworms alone, because cocoons may also be introduced, and these can prove to be resistant to adverse soil conditions and therefore offer, at worst, a second chance of colonization if the introduced earthworms die or, at best, an early boost to populations when good conditions for cocoon hatching prevail. However, only shallow-working and mainly juvenile deep-burrowing species will be transferred. The relative merits of large-scale collection and broadcast versus turf-cutting methods are summarized in Table 21.1.

The study of Morton (1993) is of interest as it involved an improvement to basic turf cutting and transfer. A two-stage process was used to increase the earthworm numbers in the turves. Before turf cutting, additional earthworms were added to the grassed area and allowed to colonize the potential cut turves. It was suggested that these could be specifically chosen depending on the proposed site of introduction, but in practice various field-collected species were placed at high population densities and allowed to settle under relatively natural conditions. "Plugs" of turf 6 in (15 cm) in diameter and 2–3.3 in (5–8 cm) deep were then cut and translocated to the inoculation site and dug into the soil surface. This method involves some

Table 21.1 Relative Merits of Earthworm Inoculation Techniques

Technique	Advantages	Disadvantages
Turf cutting and relaying	Protective microenvironment Cocoons transferred	Densities usually low Little control over species/numbers Mainly shallow-working worms Cutting machines/labor required Damage to collection site
Chemical/physical extraction with broadcasting	High densities possible Species selection possible	Protective microenvironment absent No cocoon transfer Mainly deep-burrowing worms Worms may be injured during extraction Laborious and expensive Damage to collection site
Earthworm Inoculation Unit (EIU) method	Protective microenvironment Species selection possible Worms of known origin Cocoons transferred High densities possible	Laborious and potentially expensive (compared with above methods)

Source: Adapted from Butt, K. R., Frederickson, J., and Morris, R.M., *Soil Biol. Biochem.* 29:251–257, 1997. With permission.

manipulation and requires not only a suitable source of turf but also a good source of earthworms. Results from such work are at best patchy. Morton (1993) found no earthworms 6 months after plugs were placed into a site in northeastern England. The timing of introduction (November) may have been the critical factor since temperatures were low 0°C (32°F), but plug shrinkage and failure to "gel" with the parent material at the inoculation site did not assist colonization by earthworms. An investigation of a site where this technique had been employed in Scotland revealed similar problems (Butt 1999). Under conditions where the soil is clay rich and likely to dry rapidly, this process is not particularly appropriate.

B Breeding Techniques

Controlled breeding of soil-dwelling earthworms means that optimal conditions of temperature, moisture, density, and feed can be set up, and with practice, best results can be achieved. Generally, deep-burrowing species such as *L. terrestris* and *Aporrectodea longa* require a minimum depth of 4–6 in (10–15 cm) of soil but can be surface-fed with rotted cattle or horse manure. If very fresh manure is used, the ammonia content can be harmful to these earthworms. Application of such feed to the soil surface should occur every 2–3 weeks or as appropriate (assessed by visual inspection of the soil surface). For rapid earthworm growth and reproduction, soil temperatures can be kept artificially higher than ambient (e.g., by the use of subunit soil-heating cables), but temperatures should not exceed 15°C (59°F) since some earthworm species, such as *A. longa*, will then enter a resting stage (diapause) and become inactive. Smaller surface-dwelling worms such as *A. chlorotica* or *A.*

caliginosa need less soil depth but are more easily maintained when soil and feed are mixed intimately. This can be achieved on setting up the system, but subsequent additions of organic matter as feed need to be at the surface to reduce disturbance to the system, which is not ideal for this group of earthworms. An obvious way to counteract this problem is to initiate a production system containing both deep-burrowing and shallow-working earthworms inoculated together. They exist in such associations under natural conditions, and it is thought that beneficial effects may result from mutualistic interactions such as the downward transport of feed materials (e.g., Butt 1998; Lowe and Butt 1999; Uvarov 2009).

Production systems such as those described in the preceding are perfectly sound in theory but may present some difficulties in practice. For a single species, for example, culturing *L. terrestris*, the optimum population density might be only 2–4 adults per liter of soil, which with a depth of 15 cm (6 in) amounts to only 150–300 earthworms m^{-2} (180 earthworms yd^{-2}). This is relatively small number compared with that of vermicomposting earthworm species and, given their slow rate of production, is potentially costly. A very large bed area would be required for a viable system of this nature, but such large bed systems have a number of associated problems. The major one may be the space requirement (hundreds of square meters), although this could be offset by use of a stacking box system of subunits. This, however, may give rise to a requirement for heavy equipment for lowering and lifting of boxes at feeding and harvesting (depending on box size). The latter can also be a time-consuming process if earthworms are to be mechanically sorted from the culture medium. Another problem is that of pathogen or even predator attacks on earthworms. Earthworms are susceptible to a number of pathogens, and reports from U.K. vermicomposting-earthworm breeders have recorded "blister disease" (see Heimpel, 1966). There are also concerns in the United Kingdom that the presence of predatory flatworms, such as *Artioposthia triangulata*, accidentally introduced from New Zealand (e.g., Blackshaw 1995) could colonize large systems with catastrophic effects. Large-scale earthworm waste-processing systems in the United States, such as the Lufkin project (Green and Penton 1981), have also been known to fail, possibly through accidental contamination by chemically treated liquids.

VIII AN ALTERNATIVE VERMICULTURE

When earthworms are to be used for land-improvement schemes, a carefully considered approach is necessary. Increased costs may result, but, if the outcome is successful, earthworm establishment and ultimately thriving populations make these acceptable. Trials over the past decade have resulted in the development of what has been termed the Earthworm Inoculation Unit (EIU) technique (Butt et al. 1997). This process combines careful earthworm rearing under controlled environmental conditions, in small-volume units (typically 5 L (1.3 gal) or less), with an effective method of inoculation into the field. Starter cultures of four to eight mature earthworms (ideally bred for the purpose but possibly field-collected) are kept within polyethylene bags filled with sterilized soil and a suitable food material. The number of earthworms used at the outset is a function of optimum population density (e.g., Butt et al. 1994) and EIU size. The EIUs are then sealed, provided with air holes, and maintained for a period of approximately 3 months at optimal conditions of temperature in an insulated, darkened polyethylene greenhouse or similar building. The time period is determined by the reproductive rate of the earthworm species and any need to avoid opening the units for refeeding. After this cultivation stage, the intact units are transported to the proposed inoculation site. Here they are deposited into holes of corresponding size dug into the soil (Figure 21.1). Prior to deposition the polyethylene covering is removed, and the contents are inoculated as a unit with as little disturbance to the soil matrix as possible. This ensures a protective microenvironment for the earthworms and also ensures that cocoons deposited in the EIU remain at the same level in relation to the soil surface (Table 21.1).

The distance between EIUs is determined by cost and required rates of spread. At the time of inoculation all three life stages—adults, cocoons, and hatchling earthworms—will be present within a EIU, maximizing the chances of successful population establishment. Results with *A. longa* have shown that this technique leads to greater survival compared with the simple broadcasting of adult earthworms over the surface of compacted landfill cover. On the same site, colonization from the point of inoculation with EIUs occurred at approximately 1 m (1.1 yd) per year (Butt et al. 1997, 1999) and gave rise to typical sustainable earthworm populations (Butt et al. 2004). Details of cultivation results from 2 L (0.53 gal) EIUs using *A. longa* and *A. chlorotica*, as single-species (mono)cultures and in paired culture, are shown in Table 21.2. These results demonstrate that the cultivation of such species is viable within these units.

Figure 21.1 Earthworm Inoculation Units (2 litre) post cultivation phase and ready for inoculation into site for restoration at Calvert, Buckinghamshire in 2003. (Butt, K. R. and Grigoropoulou, N., *Applied and Environmental Soil Science*, 2010, 562816, 12, doi: 10.1155/2010/562816, http://www.hindawi.com/journals/aess/2010/562816.html, 2010.)

Table 21.2 Mean Earthworm Numbers (Including Cocoons) in 2-Liter Earthworm Inoculation Units (EIUs) ($n = 10$ per Treatment) after 12 Weeks at 18 ± 2°C (64 ± 3°F)

Earthworm Species	No. Adults in Starter Culture	Mean Mass (g)	Mean Reproductive Output		SE	Total Earthworms per EIU at Inoculation
			Cocoons	Hatchlings		
Aporrectodea longa	4	2.50	24.6	4.4	± 4.0	33.0
Allolobophora chlorotica	6	0.33	37.8	37.8	± 8.0	81.6
A. longa	4	2.10	12.6	4.4	± 2.5	21.0
+	+ = 10					+ = 69.7
A. chlorotica	6	0.30	28.8	13.9	± 2.4	48.7

Source: Adapted from Butt, K. R., Frederickson, J., and Morris, R.M., *Soil Biol. Biochem.* 29:251–257, 1997. With permission.

IX THE FUTURE OF EARTHWORMS FOR LAND IMPROVEMENT

There is no doubt that earthworms can play a significant role in the rehabilitation of soils. Sites that are earthworm free are likely to have been damaged by industrial activity and therefore require more in the way of soil improvement than simple earthworm introductions. It is quite reasonable to introduce earthworms to fertile soils, such as those reclaimed from the sea as polders in Holland, and find rapid earthworm population development and expansion. However, it is unrealistic to expect earthworms to transform a disused industrial site without major modifications such as addition of organic matter and lime to the soil in advance of inoculation. Timing of earthworm introductions is all important, as are the earthworm species chosen. At a disused steelworks site in Scotland, a company chose to introduce *L. terrestris* into colliery spoil amended using the EIU technique. Monitoring of the site after 1 and 2 years showed that no earthworms had survived, because soil conditions were not suitable for this species at that time (Bain et al. 1999). However, with appropriate amendments, for example, lime and organic matter, positive results could have been obtained, but earthworms should still not be viewed as a universal panacea in such situations. Land improvement through vermitechnology has a future based on two potential routes, both of which require the site of inoculation to be made as hospitable as feasibly possible for the chosen earthworm species.

A Earthworm Inoculation En Masse

This should involve large-scale introductions of suitable earthworm species collected from pastureland close to the proposed inoculation site. Collection could be achieved by way of "picking" (Tomlin 1983) or by large-scale formalin (or equivalent) extraction from grassland, brought about by application of a vermifuge from a mobile tanker followed by a large number of collectors with fresh water to wash the earthworms brought to the soil surface. The latter was used successfully by researchers from Rothamsted at a landfill site in southeastern England, where 34,000 earthworms were translocated from an adjacent field (Edwards and Bater 1992; Butt et al. 1993). However, to make such translocations as successful as possible, it is necessary to collect from earthworm-rich areas, usually grass fields. To this end, the site of collection should be treated with organic matter, such as cattle dung or sewage sludge, over a lengthy period prior to collection, in order to maximize numbers of deep-burrowing earthworms present, as suggested by Curry (1988). Such a treatment could also be used to increase the numbers of shallow-working earthworm species, to permit turf cutting and transfer if these species were considered the most appropriate for inoculation.

This type of vermiculture is currently being used in both the United Kingdom and Australia. At a number of colliery sites in County Durham, United Kingdom, the Agricultural Development and Advisory Service (ADAS) encouraged the growth of earthworm populations within soil banks set aside during the process of land restoration by addition of well-rotted farmyard manure. Populations of up to 4000 earthworms m^{-2} (3344 earthworms yd^{-2}) were recorded under such conditions prior to inoculation (Collins 1998). In Australia a team led by Geoff Baker has been seeking to increase populations of *A. longa* under similar field conditions by applying organic matter to a pasture near Canberra, where this species has been introduced. The long-term goal is to use this "seed bed" as a source for further inoculations of this species in appropriate areas in the south of Australia (see Curry and Baker 1998). In southeastern England, a recent (2008) trial utilized earthworms collected by following a plow to repopulate a purposely constructed training "gallop" for racehorses. Broadcast inoculation was employed so as not to damage the carefully created soil and turf. Monitoring is still ongoing.

B Intensive Cultivation

This involves rearing earthworms using the EIU technique, but the earthworms forming starter cultures should be collected close to the point of proposed inoculation. The advantages of this method have already been stated and, coupled with a stock of earthworms of local origin, should lead to enhanced chances of survival and population development. It should be stressed, though, that subsequent monitoring of populations is vital to quantify the success of the operation.

Best practice, as viewed today, will involve using either of the preceding techniques depending on various factors. These include, cost, time, and expertise. Finance, for example, often drives commercial land-improvement operations, and those that choose to include earthworm inoculation may have only a small budget available. This could determine the level and type of work conducted. It is suggested that greater attention should be paid to soil ecology during improvement schemes and that soil health is vital to subsequent land use. Production of earthworms (vermiculture) for land improvement could help to address this using the most appropriate inoculation method as determined by site conditions.

X CONCLUSIONS

a. Earthworms can play an important part in land improvement, provided that soil types and conditions are appropriate for their inoculation.
b. Large-scale breeding of earthworms for land improvement is not recommended.
c. Large-scale inoculations should involve collection of appropriate species in an appropriate manner from a donor location close to the receptor site.
d. Prior to collection for inoculation, increases in existing earthworm populations may be achieved by addition of organic matter or other amendments to the soil.
e. Intensive cultivation of selected earthworm species in small numbers can give rise to increased chances of successful population establishment and growth under adverse soil conditions.
f. Introduction of a suite of earthworm species at appropriate times, using a variety of techniques, may be most beneficial but should be determined by site conditions and budgets.
g. Long-term monitoring of all earthworm inoculations is essential to fully gauge the success of inoculation. Knowledge gained needs to be disseminated to fellow researchers and land managers.

REFERENCES

Bain, S. O., K. R. Butt, and R. M. Morris. 1999. Survival and reproduction of *Lumbricus terrestris* L. in colliery spoil and sewage sludge. *Pedobiologia* 43:729–734.

Baker, G. H., G. Brown, K. Butt, J. P. Curry, and J. Scullion. 2006. Introduced earthworms in agricultural and reclaimed land, their ecology and influences on soil properties, plant production and other soil biota. *Biol. Invasions* 8:1301–1316.

Blackshaw, R. P. 1995. Changes in populations of the predatory flatworm *Artioposthia triangulata* and its earthworm prey in grassland. *Acta Zool. Fenn.* 196:107–110.

Boström, U. 1988. Growth and cocoon production by the earthworm *Aporrectodea caliginosa* in soil mixed with various plant materials. *Pedobiologia* 32:77–80.

Bouché, M. B. 1977. Stratégies lombriciennes. In organisms as components of Ecosystems. Eds. U. Lohm and T. Persson, Ecol. Bull., NFR Stockholm. 122–132.

Butt, K. R. 1991. The effects of temperature on the intensive production of *Lumbricus terrestris* L. (Oligochaeta: Lumbricidae). *Pedobiologia* 35:257–264.

Butt, K. R. 1993. Reproduction and growth of three deep burrowing earthworms (Lumbricidae) in laboratory culture in order to assess production for soil restoration. *Biol. Fertil. Soils* 16:135–138.

Butt, K. R. 1998. Interactions between selected earthworm species: A preliminary, laboratory-based study. *Appl. Soil Ecol.* 9:79–83.

Butt, K. R. 1999. Introduction of earthworms into reclaimed soils: The UK experience. *Land Degrad. Dev.* 10:565–575.

Butt, K. R. 2008. Earthworms in soil restoration: Lessons learnt from UK case studies of land reclamation. *Restor. Ecol.* 16:637–641.

Butt, K. R., J. Frederickson, and C. N. Lowe. 1999. Colonisation, survival and spread of earthworms on a partially restored landfill site. *Pedobiologia* 43:684–690.

Butt, K. R., J. Frederickson, and R. M. Morris. 1993. Investigations of an earthworm inoculation experiment, London Borough of Hillingdon. *Waste Plan.* 7:9–12.

Butt, K. R., J. Frederickson, and R. M. Morris. 1994. The life cycle of the earthworm *Lumbricus terrestris* L. (Oligochaeta: Lumbricidae) in culture. *Eur. J. Soil Biol.* 30:49–54.

Butt, K. R., J. Frederickson, and R. M. Morris. 1997. The Earthworm Inoculation Unit (EIU) technique, an integrated system for cultivation and soil-inoculation of earthworms. *Soil Biol. Biochem.* 29:251–257.

Butt, K. R., C. N. Lowe, J. Frederickson, and A. J. Moffat. 2004. The development of sustainable earthworm populations at Calvert Landfill Site UK. *Land Degrad. Dev.* 15:27–36.

Collins, D. J. 1998. Discussion held during the International Conference on Land Reclamation for Sustainable Benefits (Nottingham, UK, 1998).

Communities and Local Government (CLG). 2008. Previously-developed land that may be available for Development: England 2007. http://www.communities.gov.uk/publications/corporate/statistics/previouslydevelopedlan d2007 (accessed August 19, 2009).

Curry, J. P. 1988. The ecology of earthworms in reclaimed soils and their influence on soil fertility. In *Earthworms in Waste and Environmental Management*, eds. C. A. Edwards and E. F. Neuhauser, 251–261. The Hague, the Netherlands: SPB Academic Publishing.

Curry, J. P., and G. H. Baker. 1998. Cast production and soil turnover by earthworms in soil cores from South Australian pastures. *Pedobiologia* 42:283–287.

Darwin, C. 1881. *The Formation of Vegetable Mould through the Action of Worms with Observations on Their Habits.* London: John Murray.

Dunger, W. 1969. Fragen der natürlichen und experimentellen Besiedlung kulturfeindlichen Boden durch Lumbriciden. *Pedobiologia* 9:146–151.

Edwards, C. A., and J. E. Bater. 1992. The use of earthworms in environmental management. *Soil Biol. Biochem.* 24:1683–1689.

Edwards, C. A., and P. J. Bohlen. 1996. *Biology and Ecology of Earthworms.* 3rd ed. London: Chapman and Hall.

Edwards, C. A., J. Dominguez, and E. F. Neuhauser. 1998. Growth and reproduction of *Perionyx excavatus* (Perr.) (Megascolecidae) as factors in organic waste management. *Biol. Fertil. Soils* 27:155–161.

Edwards, C. A., and J. R. Lofty. 1980. Effects of earthworm inoculation upon the root growth of direct drilled cereals. *J. Appl. Ecol.* 17:533–543.

Edwards, C. A., and E. F. Neuhauser (eds.). 1988. *Earthworms in Waste and Environmental Management.* The Hague, the Netherlands: SPB Academic Publishing.

Fieldson, R. S. 1988. The economic viability of earthworm culture on animal wastes. In *Earthworms in Waste and Environmental Management*, eds. C. A. Edwards and E. F. Neuhauser, 145–156. The Hague, the Netherlands: SPB Academic Publishing.

Green, E., and S. Penton. 1981. Full scale vermicomposting at the Lufkin Water Pollution Control Plant. In *Workshop on the Role of Earthworms in the Stabilization of Organic Residues*, Vol. 1, *Proceedings*, ed. M. Appelhof, 229–231. Kalamazoo, MI: Beech Leaf Press.

Halle, S., and M. Fattorini. 2004. Advances in restoration ecology: Insights from aquatic and terrestrial ecosystems. In *Assembly Rules and Restoration Ecology*, eds. V. M. Temperton, R. J. Hobbs, T. Nuttle, and S. Halle, 10–33. Washington, DC: Island Press.

Hamoui, V. 1991. Life cycle and growth of *Pontoscolex corethrurus* in the laboratory. *Rev. Ecol. Biol. Sol.* 28:469–478.

Heimpel, A. M. 1966. A crystalline bacterium associated with a "blister disease" in the earthworm *Eisenia foetida* (Savigny). *J. Invert. Pathol.* 8:295–298.

Huhta, V. 1979. Effects of liming and deciduous litter on earthworm (Lumbricidae) populations of a spruce forest, with an inoculation experiment on *Allolobophora caliginosa*. *Pedobiologia* 19:340–350.

Judas, M., J. Schauermann, and K. Meiwes. 1997. The inoculation of *Lumbricus terrestris* L. in an acidic spruce forest after liming and its influence on soil properties. *Soil Biol. Biochem.* 29:677–679.

Judd, K. W., and C. F. Mason. 1995. Earthworm populations of a restored landfill site. *Pedobiologia* 39:107–115.

Lavelle, P., B. Pashanasi, F. Charpentier, C. Gilot, J.-P. Rossi, L. Derouard, J. Andre, J.-F. Ponge, and N. Bernier. 2004. Effects of earthworms on soil organic matter and nutrient dynamics at a landscape scale over decades. In *Earthworm Ecology*, ed. C. A. Edwards, 145–160. 2nd ed. Boca Raton, FL: CRC Press.

Lofs-Holmin, Astrid. 1985. *Vermiculture: Present Knowledge of the Art of Earthworm Farming; A Summary of Recent Literature.* Uppsala: Swedish University of Agricultural Sciences.

Lowe, C. N., and K. R. Butt. 1999. Interspecific Interactions between Earthworms: A laboratory-based investigation. *Pedobiologia* 43:808–817.

Lowe, C. N., and K. R. Butt. 2005. Culture techniques for soil dwelling earthworms: A review. *Pedobiologia* 49:401–413.

Matthews, John E. 1994. *Handbook of Bioremediation.* Boca Raton, FL: Lewis.

Morton, I. 1993. Earthworm inoculation techniques for opencast coal mining restoration. Master's thesis, Cranfield Institute of Technology.

Orazova, M. K., T. A. Semenova, and A. V. Tiunov. 2003. The microfungal community of *Lumbricus terrestris* middens in a linden (*Tilia cordata*) forest. *Pedobiologia* 47:27–32.

Rhee, J. A. van. 1971. Some aspects of the productivity of orchards in relation to earthworm activities. *Ann. Zool. Ecol. Anim. Spec. Publ.* 4:99–108.

Scullion, J., and A. R. A. Mohammed. 1991. Effects of subsoiling and associated incorporation of fertilizer on soil rehabilitation after opencast coal mining for coal. *J. Agric. Sci.* 116:256–274.

Senapati, B. K., P. K. Panograhi, S. Giri, A. Patnaik, and P. Lavelle. 1994. Restoration of degraded soil in intensive tea plantation (India). In *Conservation of Soil Fertility in Low Input Agricultural Systems of the Humid Tropics by Manipulating Earthworm Communities*, ed. P. Lavelle, 39–51. (Macrofauna Project II) Report No. 3, CEE Project, Paris.

Snyder, B. A., and P. F. Hendrix. 2008. Current and potential roles of soil macroinvertebrates (earthworms, millipedes, and isopods) in ecological restoration. *Restor. Ecol.* 16:629–636.

Stockdill, S. M. J., and G. G. Cossens. 1966. The role of earthworms in pasture production and moisture conservation. *Proc. N. Z. Grassland Assoc.* 28:168–183.

Tomlin, A. D. 1983. The earthworm bait market in North America. In *Earthworm Ecology; From Darwin to Vermiculture*, ed. J. E. Satchell, 331–338. London: Chapman & Hall.

Uvarov, A. V. 2009. Inter and intraspecific interactions in lumbricid earthworms: Their role for earthworm performance and ecosystem functioning. *Pedobiologia* 53: 1–27.

CHAPTER 22

The Potential of Earthworms Produced from Organic Wastes in Production of Pharmaceuticals

Mariappan Balamurugan, Kasi Parthasarathi, Lalpet S. Ranganathan, and Edwin L. Cooper

CONTENTS

I Introduction ... 350
II Therapeutic Properties of Earthworms ... 350
  A Antioxidative Activity of Earthworms ... 350
  B Antiulcer Properties of Earthworms ... 351
  C Anti-Inflammatory Effects of Earthworm Paste or Extracts ... 351
  D Anticarcinogenic Activity of Earthworm Extracts ... 352
  E Antibacterial Activity of the Coelomic Fluid of Earthworms ... 353
  F Anticoagulative Activity of Earthworm Extracts ... 353
III Therapeutic Molecules Isolated from Earthworms ... 354
IV Why Therapeutic Research on Earthworms? ... 358
V Therapeutic Properties of *Lampito mauritii* ... 358
  A Antioxidant Properties of *L. mauritii* ... 359
  B Anti-Inflammatory and Antipyretic Properties of *L. mauritii* ... 359
  C Hepatoprotective Properties of *L. mauritii* ... 360
  D Antimicrobial Properties of *L. mauritii* ... 360
VI Lumbrokinase ... 361
VII Summary and Conclusions ... 361
Acknowledgment ... 362
References ... 362

I INTRODUCTION

Age-old, rich civilizations have existed in many parts of Asia—Mesopotamia, Egypt, India, and China—and the diversity of earthworms in this region is far richer than in other places. This has raised interest in the possible medicinal properties of earthworms. Ancient folklore, scientific documentation, and traditional practices of village doctors are rich sources of information relating to the uses of earthworms as medicine by humans. Our prehistoric ancestors sought natural compounds to cure their ills and improve and enrich their lives. Most of these compounds were derived from plants and animals. The science of treating various ailments by using therapeutic materials obtained from animals is known as zootherapy. Animal-based medicines have been elaborated from various parts of the animal body, from their products (secretions and excrements), or from materials they produce, such as nests and cocoons. Research on traditional medicines has been vital in the search for organic pharmaceutical compounds. A total of 252 essential compounds have been selected by the World Health Organization, 11.1% derived from plants and 8.7% from animals, for use as medicines (Costa-Neto 2005).

We discuss in this Chapter how earthworms have been used in the past by various civilizations and review more recent studies seeking cures for a variety of diseases by means of earthworm products in different forms—for example, live earthworms, dried earthworms, powders, and liquid extracts (see also Chapters 29 and 34). For almost 4000 years, China has a history of research into the medicinal uses of earthworms. Earthworm preparations have been formulated by allergy research groups and used as suitable drugs for gastrointestinal disorders, particularly inflammation in the gastrointestinal tract. Earthworm tonics have also been used to balance the sympathetic and parasympathetic functions of the human central nervous system. Some earthworms have a strong taste that may result in nausea and even vomiting in some sensitive individuals. Taking earthworm powders with citrus fruits or other herbs can counter this nausea and vomiting. Various studies on earthworms have shown them to exhibit antipyretic, antispasmodic, detoxifying, diuretic, antihypertensive, antiallergic, antiasthmatic, spermatocidal, antioxidative, antimicrobial, anticancer, antiulcer, and anti-inflammatory activities (Cooper et al. 2004; Cooper and Roch 1984, 2004; Hori et al. 1974).

II THERAPEUTIC PROPERTIES OF EARTHWORMS

A Antioxidative Activity of Earthworms

An antioxidant is a substance that protects other chemicals of the body from being damaged by oxidation reactions. It does this by reacting with free radicals and other reactive oxygen forms within the body, hence hindering the process of oxidation. However, antioxidant supplies are limited, as one antioxidant molecule can react with only a single free radical. Therefore, there is a constant need to replenish antioxidant resources, whether endogenously or through supplementation.

Earthworm pastes and extracts can prevent oxidative damage because the earthworm tissues have significant amounts of antioxidants such as Glutathione (GSH), Glutathione Peroxidase (GPx), Catalase (CAT), and Glutathione Peroxidase (SOD) (Balamurugan et al. 2008). The effect of tissue homogenate glycolipoprotein (G-90) extract of *Eisenia fetida* (Savigny) as an antioxidant was investigated in cultured human fibroblasts and epithelial cells. It was found to have tissue-repairing capacities due to an antioxidative effect (Grdisa et al. 2001). Likewise, paste from *Lampito mauritii* (Kinberg) was found to enhance liver antioxidants such as GSH, GPx, CAT, and SOD and decrease lipid peroxidation in albino rats (Balamurugan et al. 2007).

B Antiulcer Properties of Earthworms

Organic and functional dyspepsia are common clinical syndromes characterized by stomach pain. Organic dyspepsia can be caused by peptic ulcer disease, a common gastrointestinal disorder. The increases in acid secretions and pepsin secretions and a decrease in mucosal resistance appear to be the basic causes of peptic ulceration. The mucosal resistance is lowered by *Helicobacter pylori*, gram-negative bacilli that colonize the stomach and duodenum. Over the last 25 years, a remarkable revolution in the pathophysiology and treatment of gastric and duodenal ulcers has occurred. Several therapies that have been successful in healing peptic ulcers include neutralizing gastric acid, inhibiting acid secretion, and protecting the gastroduodenal mucosa.

Prakash et al. (2007) reported that the administration of pastes prepared from *L. mauritii* restored gastrointestinal damage by reducing the gastric acid secretions and acidity and enhancing the pH. It also increased antioxidative enzymes such as GSH, GPx, SOD, and CAT, thereby preventing oxidative damage in the mucous membrane in the stomach of albino rats.

C Anti-Inflammatory Effects of Earthworm Paste or Extracts

Inflammation is a localized protective response elicited by injury or destruction of tissues that serves to destroy, dilute, or wall off both the injurious agent and the injured tissue. Characteristics of inflammation are redness, swelling, heat, and pain. Inflammation can have some beneficial effects, such as destruction of invading microorganisms and walling off of an abscess cavity, thus preventing spread of inflammation. It involves a complex series of events, including dilation of arterioles, capillaries, and venules with increased permeability and blood flow, exudation of fluids, and migration of leukocytes into the inflamed site.

The use of earthworm pastes and extracts against inflammation (in both the acute and chronic phase) has been demonstrated by several authors: Yegnanarayan et al. (1987, 1998), Ismail et al. (1992), and Balamurugan et al. (2007). Yegnanarayan et al. (1987) found that ethanol and petroleum extracts from the earthworm *L. mauritii* exhibited anti-inflammatory activity in albino rats, where edema was induced in the paw and granuloma by cotton pellets. Ismail et al. (1992) isolated a petroleum ether fraction of total earthworm protein from *L. mauritii* and found it had anti-inflammatory properties on albino rats where edema had been induced with carrageenan. Balamurugan et al. (2007) reported the inhibition of exudates in carrageenan-induced paw edema in rats by earthworm pastes, which also inhibited the granulomatous proliferation in chronic-phase granulomas in rats.

D Anticarcinogenic Activity of Earthworm Extracts

Cancer is an abnormal growth of cells that divide uncontrollably and may spread to other parts of the body. There are many kinds of benign and malignant cancer that can involve just about any part of the body. Nagasawa et al. (1991) found that a skin extract "lombricine" from *Lumbricus terrestris* inhibited the growth of mammary tumors in SHN mice. Hrzenjak et al. (1992) extracted a biologically active glycolipoprotein (G-90) from whole-earthworm tissue (*Eisenia fetida (Savigny), Lumbricus rubellus* (Linneus)) homogenates and found that it slowed tumor growth in mouse cells in vivo. Hrzenjak et al. (1998) reported the mitogenicity of the P I and P II fraction of G-90 on the cultures of human cell lines W138 (normal fibroblast), HeLa 9 (cervical carcinoma), and MiaPaca2 (pancreatic carcinoma) in vivo.

According to Xing et al. (1997), an extract from the earthworm *E. fetida*, isolated by acetone sedimentation and gel filtration, was found to be cytotoxic to several tumor cell lines. This earthworm extract, which is rich in trace elements, is composed mainly of protein [60.43 ± 2.36% ($n = 5$)], and bioactive molecules were identified as glycoproteins or glycopeptides. By fibrin plate assay, earthworm extract (EE) was found to be fibrinolytic and plasminogen-activating. The antitumor activity of the extract was identified by MTT assay and SRB assay in vitro. The concentrations of the extract required for 50% growth inhibition of human tumor cell strains (HCT-116, SY5Y, MGc803, and Hela) were between 60 and 120 $mg.L^{-1}$ (0.001 and 0.002 $oz.pint^{-1}$). In vivo, this earthworm extract was found to significantly prolong the life span of ascites tumor (S180)–bearing mice by 135.3 and 123.5% when the concentrations intraperitoneal (i.p.) were 28 and 36 $mg.kg^{-1}$ (0.0004 and 0.0006 $oz.lb^{-1}$), respectively.

The anticarcinogenic mechanism of earthworm extracts in vitro was investigated by several methods, including phase-contrast microscopy, fluorescence microscopy, transmission electron microscopy, ordinary agarose gel electrophoresis, and flow cytometry. When the effective concentration was 80 $mg.L^{-1}$ (0.0013 $oz.pint^{-1}$) or less (taking HCT-116 as the main tested cells), apoptosis was the main mechanism of cell death induced by the extract. With different microscopes, obvious apoptotic changes in the cell morphology, distribution and content of nucleic acids constituents, and ultrastructure of tested cells were observed in most affected cells. The DNA ladder was assayed by agarose gel electrophoresis. When the effective concentrations were 4 $mg.L^{-1}$ (0.00067 $oz.pint^{-1}$) and 80 $mg.L^{-1}$ (0.0013 $oz.pint^{-1}$), the percentage of apoptotic HCT-116 cells, which was assayed by flow cytometry, was 18.9 and 46.1%, respectively. However, cytolysis was also observed in most of the affected cells that were treated with earthworm extract (≥120 mg.L^{-1} (0.0002 oz.pint^{-1})).

E Antibacterial Activity of the Coelomic Fluid of Earthworms

A substance that kills or inhibits bacteria at a low concentration is called an antibiotic. Earthworm coelomic fluid and earthworm tissue extracts exhibit a strong antimicrobial activity. Various extracts from earthworms possess potency against some pathogenic and nonpathogenic bacteria such as *Escherichia coli*, *Streptococcus pyogenes*, *Pseudomonas aeruginosa*, *Staphylococcus aureus*, and *Salmonella enteritidis*.

Roch et al. (1984) reported an antibacterial activity of the coelomic fluid fraction (A, B, C, H, I, and J) of *Eisenia andrei* against *Bacillus megateriums*. Vaillier et al. (1985) isolated bacteriostatic molecules (molecular weight of 20,000, 40,000, and 45,000) from cell-free coelomic fluid from *E. andrei* against *B. megateriums*. Popovic et al. (2005) reported antibacterial activity of G-90 from *Eisenia* against *Streptococcus pyogenes*, *Pseudomonas aeruginosa*, *Staphylococcus aureus*, and *Salmonella enteritidis*.

F Anticoagulative Activity of Earthworm Extracts

An annelid group, the Hirudunaria, includes the leeches, which are closely related to earthworms and are well known for anticoagulant properties, which enable them to suck liquid blood from prey. Quite recently it has been discovered that extracts from earthworms are also able to prevent clotting of blood. Earthworm extract (G-90) has been shown to have anticoagulative and fibrinolytic properties in various blood samples (Cooper et al. 2004).

Hrzenjak et al. (1998) isolated two serine proteases (P I and P II) from G-90 extracted from tissue homogenates of the earthworm *E. fetida* and found they had effects on lysis of fibrin clots originating from the blood of patients with malignant tumors. Popovic et al. (2001) isolated proteolytic enzymes (P I and P II) from G-90 of *E. fetida* and found it affected the lysis of clots from venous blood of dogs with cardiopathies and malignant tumors.

Kim et al. (1998) investigated fibrinolytic and antithrombotic effects by orally administering to rats a freeze-dried powder from *L. rubellus*. The fibrinolytic activity of the plasma was determined by measuring the plasmin activity of the euglobulin fraction. It was increased two-fold, compared to the controls, at a dose of 0.5 g.kg^{-1}.day^{-1}(0.008 oz.lb^{-1}.day^{-1}) and five times with 1 g.kg^{-1}.day^{-1} (0.016 oz.lb^{-1}·day^{-1}) after a 4-day administration. The antithrombotic effect was studied in an arteriovenous shunt rat model. The thrombus weight decreased significantly from 43.2 to 32.4 mg (0.0015 to 0.001 oz) at a dose of 0.5 g.kg^{-1}.day^{-1} (0.008 oz.lb^{-1}. day^{-1}) after 8 days of treatment. The level of fibrinogen/fibrin degradation product (FDP) in serum was elevated in a dose-dependent manner during the treatment period. On the 8th day after administration, the FDP value was increased to 7.7 μg^{-1}.ml^{-1} (0.00013 oz. pint^{-1})

compared with the control value of 3.3 $\mu g^{-1}.ml^{-1}$ (5.51×10^{-5}oz.pint^{-1}) These results suggest that earthworm powders have value for preventing and/or treating thrombotic conditions.

III THERAPEUTIC MOLECULES ISOLATED FROM EARTHWORMS

In both *in vitro* and *in vivo* studies, Cooper (2002) reported the presence of phagocytic activity in the coelomic corpuscles of earthworms against foreign particles and bacteria. He reported that the efficacy of their phagocytic process, formed from phagocytic cells that reach the exterior through the earthworm dorsal pores, and also reported that immune recognition is apparent in the cellular and humoral defense functions of earthworms. Bacteriostatic molecules like A, B, C, αBC, βBC, and γBC were isolated from *E. andrei.* These molecules exhibit bactericidal activity against human pathogens (Vaillier et al. 1985).

Fetidins isolated from *E. fetida* are characterized as heat-labile, polymorphic, multifunctional proteins. They also produce cytolysis, antibacterial reactions, and clotting (Valembois et al. 1984). A hemolytic factor is one of the five proteic components located by immunoelectrophoresis in the coelomic fluid of *E. fetida.* The lipoproteic nature of the hemolytic factor suggests that the releasing cells might be the chloragogenous cells.

Roch et al. (1991) proposed the name hemolysin (or) hemolytic factor as the *Eisenia fetida andrei* factor (EFAF). From this, four molecules of different molecular weight were identified, which were capable of agglutination. These proteins are able to bind the antigen used in stimulation. Rejneck et al. (1991) demonstrated the interaction of protein from the coelomic fluid of earthworms (*L. terrestris* and *E. fetida*) with the staphylococcal binding protein (SPA), and he suggested that the coelomic protein was not homologous with vertebrate immunoglobulin (IgE).

A biologically active glycolipoprotein was isolated from whole-earthworm tissue homogenates and named G-90 by Hrzenjak et al. (1992). It shows anticoagulative and fibrinolytic activities. Eiseniapore (38 K Da) isolated from *E. fetida* was found to have lytic activity (Lange et al. 1999). Several lipid vesicles were used to determine the receptors of the Eiseniapore (lytic protein). The lysenin from *E. fetida* was found to show hemolysis through the specific binding of sphingomylein in cellular membranes (Kobayashi et al. 2004). Bilej et al. (1995) isolated a coelomic cytolytic factor from *E. fetida*. Hrzenjak et al. (1998) isolated two serine proteases (P I and P II) from G-90 extracted from tissue homogenates from the earthworms. *E. fetida* was found to lyse fibrin clots originating from the blood of patients with malignant tumors. Popovic et al. (2001) isolated proteolytic enzymes (P I and P II) from G-90 of *E. fetida* and found it lysed clots from the venous blood of dogs with cardiopathies and malignant tumors. Nagasawa et al. (1991) found a skin extract termed "lombricine" from *Lumbricus terrestris* that could inhibit the growth of mammary tumors in SHN mice. Details of these studies, the earthworm species, and the therapeutic values are summarized in Table 22.1.

Table 22.1 Various Therapeutic Properties of Earthworms

Author	Year	Earthworm	Active fraction	Activity
Balamurugan et al.	2007 2008 2009	*L. mauritii*	Earthworm extract (EE) Earthworm paste (EP)	Antipyretic, anti-inflammatory, antioxidative, antimicrobial, and hepatoprotective
Bilej et al.	1995	*E. fetida*	Coelomic cytolytic factor (CCF-1)	Antitumor
Borgeraas et al.	1996	*E. andrei* and *E. veneta*	Glutathione transferase (GST)	Ethacrynic acid (ETHA) activity
Bruhn et al.	2006	*E. fetida*	Lysenin	Cytolytic and antibacterial
Cho et al.	1998	*L. rubellus*	Lumbricin I	Antimicrobial
Cho et al.	2004	*L. rubellus*	Fibrinolytic serine proteases	Fibrinolytic
Chung and Lee	1989	*Lamnodrilus gotai hatai*	Extract	Hepatoprotective
Cooper et al.	1978–2009	Many earthworm species	Many formulations	Proved most of the pharmacological activities
Engelmann et al.	2004	*E. fetida*	Lysosomal enzymes	Antibacterial
Eue et al.	1998	*E. fetida*	Hemagglutinin 1, Hemagglutinin 2, and Hemagglutinin 3 (ECF)	Hemolytic and agglutinative
Fontt et al.	2002	*E. fetida fetida*	Coelomic cytolytic factor-1 (CCF-1)	Cytolytic
Grdisa et al.	2001	*E. fetida*	Glycoprotein complex (G-90)	Antioxidative
	2004			Skin wound healing
Hirabayashi et al.	1998	*Allolobophora japonica* and *L. terrestris*	Lectins (galactose-binding proteins)	Hemagglutinating
Hirigoyenberry et al.	1990	*E. fetida andrei*	Coelomic fluid	Antibacterial
Hong et al.	2007	*E. fetida*	Fibrinolytic enzyme	Antitumor activity on human hepatoma cells in vitro and in vivo
Honsi and Stenersen	2000	*E. fetida* and *D. veneta*	Lysosomal marker enzymes	Used as biomarkers for xenobiotic-induced lysosomal membrane damage in worms
Hrzenjak et al.	1992	*E. fetida*	Glycoprotein complex (G-90)	Mitogenic

(*Continued*)

Table 22.1 Various Therapeutic Properties of Earthworms (Continued)

Author	Year	Earthworm	Active fraction	Activity
Hrzenjak et al.	1998	*E. fetida*	Glycoprotein complex (G-90)	Fibrinolytic and anticoagulative
				Fibrinolytic activity in the blood of patients with malignant tumors
Hu and Fu	1997	*Pheretima*	Fibrinolytic enzyme	Fibrinolytic
Kauschke and Mohrig	1989	*E. fetida* and *L. terrestris*	Coelomic fluid	Hemolytic and hemagglutinating
Kauschke et al.	2001	*E. fetida*	Small coelomocytes	Perforin-like
Kim et al.	1998	*L. rubellus*	Freeze-dried powder	Antithrombotic and fibrinolytic
Kobayashi et al.	2004	*E. fetida*	Lysenin	Hemolytic, cytotoxic, and vasodepressive
Kohlerova et al.	2004	*E. fetida*	Coelomic cytolytic factor (CCF) and hemolytic factor fetidin	Antimicrobial
Lange et al.	1999	*E. fetida*	Eiseniapore	Cytolytic
Lassegues et al.	1997	*E. fetida*	Fetidin	Antibacterial
	2003	*E. fetida*	Glycosylated fibrinolytic enzyme	Fibrinolytic
Liu and Ge	2002	*E. fetida*	Fibrinolytic enzymes extracted from earthworms reared in different substracts	Fibrinolytic
Mihara et al.	1991	*L. rubellus*	Fibrinolytic enzymes	Fibrinolytic
Milochau et al.	1997	*E. fetida andrei*	Fetidins	Antibacterial, hemolytic, and hemagglutinating
Mohrig et al.	1996	*E. fetida, L. terrestris*, and *Aporrectodea caliginosa*	Hemolytic and hemagglutinating proteins in the coelomic fluid	Hemolytic and hemagglutinating
Nagasawa et al.	1991	*L. terrestris*	Lombricine	Antitumor
Nakajima et al.	1996	*L. rubellus*	Fibrinolytic protease (F-III-2)	Fibrinolytic
Nakajima et al.	2000	*L. rubellus*	Earthworm autolysate	Antioxidant
Oumi et al.	1995	*E. fetida* and *Pheretima vittata*	GGNG peptides	Myoactive
Pan et al.	2003	*E. fetida andrei*	Antibacterial compound	Antibacterial

Table 22.1 Various Therapeutic Properties of Earthworms (Continued)

Author	Year	Earthworm	Active fraction	Activity
Popovic et al.	1998	*E. fetida*	G-90 (Immunoglobulin-like structures)	Cell proliferation
Popovic et al.	2005	*E. fetida*	Glycoprotein complex (G-90)	Antibacterial
Prakash et al.	2007	*L. mauritii*	Eiseniapore (EP)	Antiulcer
Quaglino et al.	1996	*E. fetida*	Coelomocytes	Antitumor activity against cell line K562
Rejnek et al.	1991	*E. fetida andrei* and *L. rubellus*	Staphylococcal protein A-binding protein	Antibacterial
Roch et al.	1991	*E. fetida andrei*	*Eisenia fetida andrei* factor (EFAF)	Hemolytic
Roch et al.	1984	*E. fetida andrei*	Lysins	Agglutinating, hemolytic, and bacteriostatic
Roch et al.	1989	*E. fetida andrei*	Hemolysins	Hemolytic
Rosenberg and Ennor	1959	*Allolobophora caliginosa* and *Octolasium cyaneum*	Lombricine	Biological precursor molecule
Saint-Denis et al.	1998	*E. fetida andrei*	Enzyme extract	Antioxidant
Stein et al.	1982	*L. terrestris*	Coelomic fluid	Hemagglutinating
Sugimoto and Nakajima	2001	*L. rubellus*	Serine protease	Fibrinolytic
Sugiura et al.	1995	*E. fetida*	Ether-containing phospholipids; platelet-activating factor (PAF)	Platelet-activating
Vaillier et al.	1985	*E. fetida andrei*	Bacteriostatic proteins from cell-free coelomic fluid	Antibacterial, hemolytic, and hemagglutinating
Wu et al.	2002	*L. rubellus*	Fibrinolytic enzyme III-1	α_2-macroglobulin inhibition
Xing et al.	1997	*E. fetida*	Fibrinolytic enzyme (eFE-D)	Fibrinolytic
Yamaji et al.	1998	*E. fetida*	Lysenin	Sphingomyelin binding
Yang and Ru	1997	*E. fetida*	Sodium dodecyl sulfate (SDS)-activated fibrinolytic enzyme	Fibrinolytic

(*Continued*)

Table 22.1 Various Therapeutic Properties of Earthworms (Continued)

Author	Year	Earthworm	Active fraction	Activity
Zhang et al.	2006	*Pheretima*	Sp. Dilong extract	Promotive of wound healing
Zhao et al.	2003	*E. fetida*	Fibrinolytic enzyme II (EFE-II)	Hydrolysis of fibrinogen and plasminogen
Zhao et al.	2006	*E. fetida*	Earthworm protease-II and III	Fibrinolytic

IV WHY THERAPEUTIC RESEARCH ON EARTHWORMS?

Invertebrates may lack mixed leukocyte reactivity, which is a functional marker of the invertebrates. This lack of supporting data has led to some counterhypotheses. Some argue that major histocompatibility complex (MHC) the evolved in vertebrates from heat-shock proteins. But now the discovery of B_2-microglobulin–type molecules in earthworms, crustaceans, and insects does support the idea that MHC precursors may have arisen in the invertebrates. Although B_2-microglobulin in vertebrates is encoded by a gene not linked to the MHC, it associates with class I MHC molecules that may be the descendants of a single domain molecule like B_2-microglobulin that has been expanded by gene rearrangement, gene duplication, and natural selection.

Their high protein content makes earthworms a desired food material for a wide range of animals such as fish, amphibians, reptiles, birds, and mammals. The presence of plant and animal growth-promoting amino acids and other substances may have made humans, knowingly and unknowingly, first consume earthworms in various forms and later observe their therapeutic effects (Edwards and Bohlen 1996). This gave further impetus to look for molecules of curative value. So far we have been studying the dependence of earthworms on microbes, and of microbes on earthworms, and their combined activity playing vital roles in keeping and raising soil's fertility and productivity.

V THERAPEUTIC PROPERTIES OF *LAMPITO MAURITII*

Disease is a major factor contributing to human abnormalities and the result of, oxidative stress–mediated tissue damage observed in various conditions. Currently available drugs and therapies for inflammation, pyresis, oxidation, hepatic damage, and microbial attacks are always associated with side effects. Hence, we search for drugs from natural resources with low costs, more activity, and no side effects. In scientific investigations in recent years, attention has been drawn to the health-promoting activity of animal products and their active components. *L. mauritii* is a well-known earthworm species associated with various traditional medicinal values linked to the presence of active phytochemicals, biopeptides, and multiple biological activities. In our study, we evaluated the antioxidant, anti-inflammatory, antipyretic, hepatoprotective, and antimicrobial effects of earthworm extracts of *L. mauritii* by in vivo investigation of histamine- and turpentine-induced inflammation, yeast-induced pyrexia, and acetaminophen-induced hepatic damage conditions in Wistar rats and in vitro analysis of antioxidant and antimicrobial properties.

A Antioxidant Properties of *L. mauritii*

Antioxidant activity (determined by peroxide value), free radical–scavenging activity (determined by the 2,2-diphenyl-1-picrythydrazyl (DPPH) scavenging method), antioxidant capacity (determined by the reducing power), and total phenol content (determined by the Folin-Ciocalteau method) in earthworm coelomic fluid, earthworm pastes, and earthworm extracts in vitro showed earthworm extracts to have more overall antioxidant potency. Our studies (Table 22.2) suggested that earthworm extracts have potency to donate electrons to reactive free radicals, converting them into more stable, nonreactive forms and terminating the free radical chain reaction. The prevention of the chain reaction step by scavenging of free radical species seems to be the antioxidant mode of action of the earthworm extract.

B Anti-Inflammatory and Antipyretic Properties of *L. mauritii*

Research was conducted to investigate the therapeutic properties, such as anti-inflammatory and antipyretic activity, of biologically active extracts isolated from whole earthworms (*L. mauritii*). These include anti-inflammatory effects in comparison with a standard anti-inflammatory drug (indomethacin) and an antipyretic drug (acetaminophen) on Wistar albino rats (*Rattus norvegicus*). Our results showed that the administration of indomethacin 10 $mg.kg^{-1}$ (0.00016 $oz.16^{-1}$), acetaminophen 150 $mg.kg^{-1}$ (0.0024 $oz.16^{-1}$), and/or different doses of earthworm extract 50, 100, and 200 $mg.kg^{-1}$ (0.0008, 0.0016, 0.0032 $oz.lb^{-1}$) reduced and restored to normal conditions, in a dose-dependent manner, histamine- and turpentine-induced inflammations, in both the acute and chronic phase, and brewer's yeast–induced pyrexia

Table 22.2 ***In Vitro*** **Assessment of Antioxidant Properties of** ***L. mauritii***

Sample	Antioxidant Activity (%)	DPPH Scavenging Activity EC50 ($\mu g.mL^{-1}$)	Reducing Power ($\mu g.mL^{-1}$)	Total Phenolic Content ($\mu g. mL^{-1}$)
Coelomic fluid	16.4 ± 2.3^a	121 ± 2.6^a	1.52 ± 1.2^a	11.1 ± 2.4^a
Earthworm paste	32.3 ± 1.2^b	91 ± 1.4^b	2.12 ± 2.1^b	42.2 ± 2.8^b
Earthworm extract	52.4 ± 2.4^c	74 ± 0.9^c	3.71 ± 0.1^c	61.1 ± 1.3^c

Note: Each value is the mean ± SD of triplicate measurements. Values within a column with different letters differ significantly Duncan's MRT (DMRT); $p < 0.05$; DPPH=2,2 diphenyl1-picrylhydrazyl; EcSo = Median effective concentration required to induce a 50% effect.

in rats. We found that the more significant inhibition of paw edema and granuloma, normalized antioxidant activities (glutathione peroxidase, reduced glutathione, catalase, superoxide dismutase, and thio-barbituric acid reactive substances in the liver), and normalized the hematological picture (red erythrocyte corpuscles, white leukocyte corpuscles, differential levels of neutrophils and hemoglobin content, and serum biochemical contents—protein, albumin, cholesterol, and alkaline phosphatase) in inflamed rats. It also produced significant reductions in hyperpyrexia in rats when treated with standard drugs as well as different doses of earthworm extracts.

These results show the anti-inflammatory, antioxidant, and antipyretic potential of earthworm extracts, which could be due to the presence of phenolic substances in the earthworm tissues, which might scavenge the free radicals, stimulate the activities of antioxidative enzymes, and normalize the hematological and serum biochemical characteristics.

C Hepatoprotective Properties of *L. mauritii*

The hepatoprotective potential of earthworm extracts (*L. mauritii*) was evaluated with respect to acetaminophen-induced liver injury in Wistar albino rats, in comparison with a standard hepatoprotective drug (silymarin). It was observed that liver injury in the acetaminophen-treated rats $2g^{-1}.kg^{-1}$ ($.032\ oz^{-1}lb^{-1}$) resulted in a reduction in liver antioxidants like GSH, SOD, GPx, CAT, and serum total protein and an increase in serum alkaline phosphatase (ALP), serum glutamate oxalate transaminase (AST), serum glutamate pyruvate transaminase (ALT), bilirubin, and liver lipid peroxidation (TBARS). In contrast, increases in the activities of liver GSH, SOD, GPx, and CAT and in the serum total protein level and decreases in the serum ALP, AST, ALT, bilirubin, and liver TBARS were observed in rats treated with different doses of earthworm extract 100, 200, and 300 $mg.kg^{-1}$ (0.0016, 0.0032, and 0.0048 $oz.lb^{-1}$). The results were similar to those caused by the hepatoprotective drug silymarin 150 $mg^{-1}.kg^{-1}$ (0.0024 $oz.lb^{-1}$). The mode of action of earthworm extracts, as evidenced by the preceding parameters, seems to suggest, on the one hand, that it prevents the formation of reactive oxygen groups or scavenges these groups (thereby preventing damage to the hepatic cells) and, on the other hand, that they can modulate the genes responsible for the synthesis of antioxidant enzymes like GPx, CAT, and SOD in liver tissue and decrease alkaline phosphatase and aminotransferase enzymatic activities in the serum. Histopathological observations of liver tissues corroborated these findings (Balamurugan 2009).

D Antimicrobial Properties of *L. mauritii*

Studies were done to determine the antibacterial and antifungal activities of *L. mauritii* tissue extracts against eight human pathogenic bacteria and eight fungi by the disc diffusion method (Balamurugan 2009). Earthworm extracts had antibacterial activity against four Gram-negative bacteria—*E. coli*, *P. aeruginosa*, *K. pneumoniae*, and *Vibrio cholerae*—and four Gram-positive bacteria—*Staphylococcus aureus*, *Streptococcus pneumoniae*, *Staphylococcus epidermidis*, and *Bacillus subtilis*. Earthworm extracts exhibited antifungal activity against four yeast-type fungi (*Candida albicans*, *Candida tropicalis*, *Candida krusei*, and *Candida parapsilosis*), one dermatophytic fungus (*Tricophyton mentagrophytes*), and three mold-type fungi (*Aspergillus niger*, *Aspergillus flavus*, and *Aspergillus fumigatus*). These results suggest a broad spectrum of antimicrobial activity for earthworm extracts from *L. mauritii*. The extract from *L. mauritii* seems to be composed of low-molecular-weight proteins (29–66 kD), among other active principles such as phenolic substances and humic substances that have already been reported to be present in the tissues of earthworms and are known to have antibacterial and antifungal activities (Ranganathan 2006).

VI LUMBROKINASE

Lumbrokinase (LK) is a group of proteolytic enzymes, including a plasminogen activator and plasmin, separated biochemically from earthworms. The plasminogen activator (e-PA) in LK is similar to the plasminogen activator (t-PA) from other tissues. This makes it possible to show thrombolytic activity, particularly in the presence of fibrin. Thus, LK has the advantage of not causing hemorrhages due to hyperfibrinolysis during medication, as compared with streptokinase and urokinase. In addition, long-term animal toxicological experiments show no damage to hepatic or renal functions. There has been no negative influence on embryonic development, nor have teratogenic or mutagenic effects been observed in embryonic rats. In clinical experiments there have been no undesired effects on blood levels of glucose and lipids (Cooper et al. 2004).

Lipid vesicles of various compositions were used to determine whether specific lipids might serve as receptors to eiseniapore. The lysins bind to and disturb the lipid bilayer only when distinct sphingolipids consisting of a hydrophilic head group (as phosphorylcholine or galactosyl) and a ceramide backbone (e.g., sphingomyelin) are present. Cholesterol enhances eiseniapore lytic activity against sphingomyelin-containing vesicles, probably due to interaction with sphingomyelin. Leakage of vesicles was most efficient when the lipid composition resembled that of the outer leaflet of human erythrocytes.

VII SUMMARY AND CONCLUSIONS

Earthworms are well-known invertebrate animals used extensively in Indian and Chinese medicine and now becoming recognized in the West. In early periods, the use of earthworms was recommended against various ailments for which there was no effective cure. Besides this, folklore medicine claims their successful use in the treatment of cancer, ulcers, inflammation, diarrhea, fever, dysentery, toothache, and so on. Research carried out using modern pharmacological techniques of biological evaluation supports many of these claims. Presently, there is increasing interest worldwide in naturally produced medicines, accompanied by increased

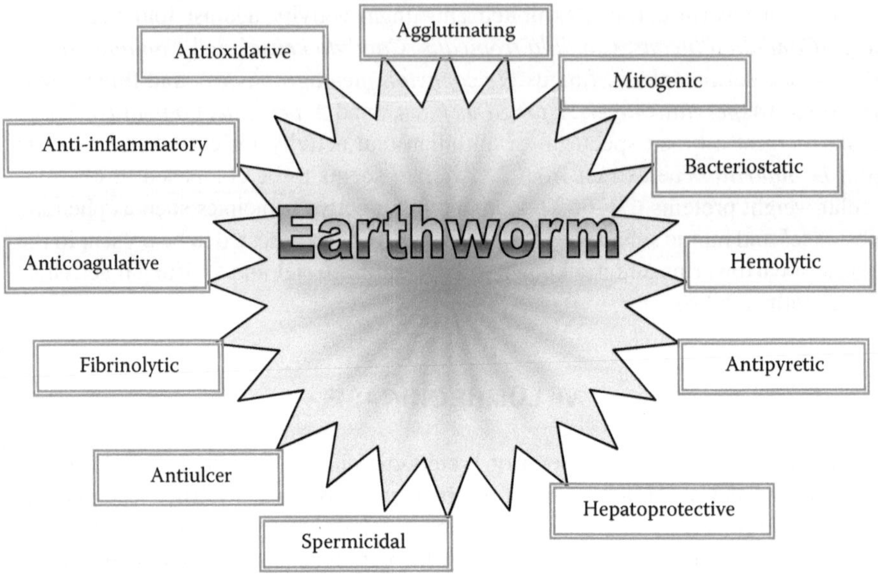

Figure 22.1 Prospective use of earthworms in human health care.

laboratory investigations into the pharmacological properties of bioactive ingredients and their ability to treat various diseases. Numerous natural drugs have entered the international pharmaceutical market through discoveries in traditional medicine (Cooper 2005, 2006). Even though scientific studies have been done on a large number of products, significantly fewer viable drugs have become effective therapeutics (Figure 22.1).

ACKNOWLEDGMENTS

We express appreciation to Mr. Jesus Heredia and Mr. David Yao for assistance in preparing the manuscript.

REFERENCES

Balamurugan, M. 2009. Studies on selective therapeutic properties of earthworm *Lampito mauritii* (kinberg). PhD diss., Annamalai University, Tamilnadu, India.

Balamurugan, M., Parthasarathi, K., Cooper E.L., and Ranganathan, L.S. 2007. Earthworm paste (*Lampito mauritii*, Kinberg) alters inflammatory, oxidative, haematological and serum biochemical indices of inflamed rat. *Eur. Rev. Med. Pharmacol. Sci.* 11 (1): 77–90.

Balamurugan, M., Parthasarathi, K., Cooper, E.L., and Ranganathan, L.S. 2009. Antiinflammatory and antipyretic activities of earthworm extract—*Lampito mauritii* (Kinberg). *J. Ethnopharmacol.* 121: 330–332.

Balamurugan, M., Parthasarathi, K., Ranganathan, L.S., and Cooper, E.L. 2008. Hypothetical mode of action of earthworm extract with hepatoprotective and antioxidant properties. *J. Zhejiang Univ. Sci.* B 9 (2): 141–147.

Bilej, M., Brys, L., Beschin, A., Lucas, R., Vercauteren, E., Hanusova, R., and Baetselier, P.D. 1995. Identification of a cytolytic protein in the coelomic fluid of *Eisenia fetida* earthworm. *Immunol. Lett.* 45: 123–128.

Borgeraas, J., Nilsen, K., and Stenersen, J. 1996. Methods for purification of glutathione transferase in the earthworm genus *Eisenia*, and their characterization. *Comp. Biochem. Physiol.* 114C (2): 129–140.

Bruhn, H., Winkelmann, J., Andersen, C., Andra, J., and Leippe, M. 2006. Dissection of the mechanisms of cytolytic and antibacterial activity of lysenin, a defence protein of the annelid *Eisenia fetida*. *Dev. Comp. Immunol.* 30: 597–606.

Cho, H., Choi, E.S., Lim, H.G., and Lee, H.H. 2004. Purification and characterization of six fibrinolytic serine-proteases from earthworm *Lumbricus rubellus*. *J. Biochem. Mol. Biol.* 37 (2): 199–205.

Cho, J.H., Park, C.B., Yoon, Y.G., Kim, S.C., and Lumbricin, I. 1998. A novel proline-rich antimicrobial peptide from the earthworm: Purification, cDNA cloning and molecular characterization. *Biochim. Biophys. Acta* 22 (1): 67–76.

Chung, Y., and Lee, H.M. 1989. Effects of earthworm (*Lamnodrilus gotai hatai*) extract on hepatotoxicity of carbon tetrachloride in rats. *Chin. Med. J.* 33: 101–110.

Cooper, E.L. 1978. Comparative and developmental immunology. In *Handbook of Cancer Immunology*. Vol. 1, ed. H. Waters, 37–62, New York, Garland STPM Press.

Cooper, E.L. 2002. Comparative immunology. *Curr. Pharm. Des.* 8: 99–110.

Cooper, E.L. 2005. CAM, eCAM, bioprospecting: The 21st century pyramid. *Evid. Based Complement Altern. Med.* 2: 125–128.

Cooper, E.L. 2006. Stem cells and CAM. *Evid. Based Complement Altern. Med.* 3 (2): 167–169.

Cooper, E.L. 2009. A Closer Look at Clinical Analyses. *Evid based Complement Alternate Med. eCAM*. 6(3): 279–81.

Cooper, E.L., Cossarizza, A., Suzuki, M.M., Salvioli, S., Capri, M., Quaglino, D., and Franceschi, C. 1995. Autogeneic but not allogeneic earthworm effector coelomocytes kill the mammalian tumor target K562. *Cell. Immunol.* 166, 113–122.

Cooper, E.L., Hrzenjak, T.M., and Grdisa, T. 2004. Alternative sources of fibrinolytic, anticoagulative, antimicrobial and anticancer molecules. *Int. J. Immunopathol. Pharmacol.* 17: 237–244.

Cooper, E.L., Kvell, K., Engelman, N., and Nemeth, P. 2006. Still waiting for the toll? *Immunol Lett.* 104 (1–2): 18–28.

Cooper, E.L., and Roch, P. 1984. Earthworm leukocyte interactions during early stages of graft rejection. *J. Exp. Zool.* 232: 67–72.

Cooper, E.L., and Roch, P. 2004. Earthworm immunity: A model of immune competence. *Pedobiol.* 47: 676–688.

Costa-Neto, E.M. 2005. Animal-based medicines: Biological prospection and the sustainable use of zootherapeutic resources. *An. Acad. Bras. Cienc.* 77 (1): 33–43.

Edwards, C.A., and Bohlen, P.J. 1996. *Biology and Ecology of Earthworms*. 3rd ed. Chapman and Hall, London.

Engelmann, P., Molnar, L., Palinkas, L., Cooper, E.L., and Nemeth, P. 2004. Earthworm leukocyte populations specifically harbor lysosomal enzymes that may respond to bacterial challenge. *Cell Tissue Res.* 316: 391–401.

Eue, I., Kauschke, E., Mohrig, W., and Cooper, E.L. 1998. Isolation and characterization of earthworm hemolysins and agglutinins. *Dev. Comp. Immunol.* 22 (1): 13–25.

Fontt, E.O., Beschin, A., Dijck, E.V., Vincent, V., Bilej, M., Lucas, R., Baetselier, P.D., and Vray, B. 2002. *Trypanosoma cruzi* is lysed by coelmic cytolytic factor-1, an invertebrate analogue of tumor necrosis factor and induces phenoloxidase activity in the coelomic fluid of *Eisenia fetida. Dev. Comp. Immunol.* 26 (1): 27–34.

Grdisa, M., Popovic, M., and Hrzenjak, T. 2001. Glycolipoprotein extract (G-90) from earthworm *Eisenia fetida* exerts some antioxidative activity. *Comp. Biochem. Physiol.* 128A: 821–825.

Grdisa, M., Popovic, M., and Hrzenjak, T. 2004 Stimulation of growth factor synthesis in skin wounds using tissue extract (G-90) from the earthworm *Eisenia fetida. Cell Biochem. Function* 22 (6): 373–378.

Hirigoyenberry, F., Lassalle, F., and Lassegues, M. 1990. Antibacterial activity of *Eisenia fetida andrei* coelomic fluid: Transcription and translation regulation of lysozyme and proteins evidenced after bacterial infestation. *Comp. Biochem. Physiol. B* 95 (1): 71–75.

Hong, C., Takahashi, S., Michio, I., Okutani, E., Zhi-guo, Z., Chayama, K., and Bao-an, C. 2007. Earthworm fibrinolytic enzyme: Anti-tumor activity on human hepatoma cells *in vitro* and *in vivo. Chin. Med. J.* 120 (10): 898–904.

Honsi, T.G., and Stenersen, J. 2000. Activity and localisation of the lysosomal marker enzymes acid phosphatase ,N-acetyl-beta-D-glucosaminidase and beta-galactosidase in the earthworms *Eisenia fetida* and *E. veneta. Comp. Biochem. Physiol. B Biochem. Mol. Biol.* 125 (3): 429–437.

Hori, M., Kondon, K., Yoshida, T., Konishi, E., and Minami, S. 1974. Studies of anti-pyretic components in the Japanese earthworm. *Biochem. Pharmacol.* 23: 1582–1590.

Hrzenjak, T., Hrzenjak, M., Kasuba, V., Efenberger-Marinculic, P., and Levanat, S.A. 1992. New source of active compounds—earthworm tissue (*Eisenia fetida, Lumbricus rubellus*). *Comp. Biochem. Physiol.* 102: 441–447.

Hrzenjak, T., Popovic, M., and Tiska Rudman, L. 1998. Fibrinolytic activity of earthworms extract (G-90) on lysis of fibrin clots originated from the venous blood of patients with malignant tumors. *Pathol. Oncol. Res.* 4 (3): 201–211.

Hu, W., and Fu, T. 1997. Isolation and properties of a novel fibrinolytic enzyme from an earthworm. *Zhong Yao Cai* 20 (2): 78–81.

Ismail, S.A., Pulandiran, K., and Yegnanarayan, R. 1992. Anti-inflammatory activity of earthworm extracts. *Soil Biol. Biochem.* 24: 1253–1254.

Kauschke, E., Komiyama, K., Moro, I., Eue, I., Konig, S., and Cooper, E.L. 2001. Evidence for perforin-like activity associated with earthworm leukocytes. *Zoology* (Jena) 104 (1): 13–24.

Kauschke, E., and Mohrig, W. 1989. Comparative analysis of haemolytic and hemagglutinating activities in the coelomic fluid of *Eisenia foetida* and *Lumbricus terrestris* (Annelida, Lumbricidae). *Dev. Comp. Immunol.* 11 (2): 331–341.

Kim, Y.S., Pyo, M.K., Park, K.M., Hahn, B.S., Yang, K.Y., and Choi, H.S. 1998. Dose dependency of earthworm powder on antithrombotic and fibrinolytic effects. *Arch. Pharm. Res.* 21 (4): 374–377.

Kobayashi, H., Ohta, N., and Umeda, M. 2004. Biology of lysenin, a protein in the coelomic fluid of the earthworm *Eisenia fetida. Int. Rev. Cytol.* 236: 45–99.

Kohlerova, P., Beschin, A., Silemva M, Baetselier, P.D. and Bilej, M 2004. Effect of experimental microbial challenge on the expression of defense molecules in *Eisenia foetida* earthworm. Dev. Comp. *Immunol.* 28:701–711.

Lange, S., Kauschke, E., Mohrig, W., and Cooper, E.L. 1999. Biochemical characteristics of Eiseniapore, a pore-forming protein in the coelomic fluid of earthworms. *Eur. J. Biochem.* 262 (2): 547–556.

Lassegues, M., Milochau, A., Doignon, F., Du Pasquier, L., and Valembois, P. 1997. Sequence and expression of an *Eisenia-fetida*-derived cDNA clone that encodes the 40-kDa fetidin antibacterial protein. *Eur. J. Biochem.* 246 (3): 756–762.

Li, L., Zhao, J., and He, R.Q. 2003. Isolation and some characterizations of a glycosylated fibrinolytic enzyme of earthworm, *Eisenia fetida*. *Protein Pept. Lett.* 10 (2): 183–190.

Liu, X., and Ge, F. 2002. Factors influencing the activity of fibrinolytic ezymes from earthworm, *Eisenia fetida*. *Zhongguo Zhong Yao Za Zhi* 27 (6): 423–426.

Mihara, H., Sumi, H., Yoneta, T., Mizumoto, H., Ikeda, R., Seiki, M., and Maruyama, M. 1991. A novel fibrinolytic enzyme extracted from the earthworm, *Lumbricus rubellus*. *Jpn. J. Physiol.* 41: 461–472.

Milochau, A., Lassegues, M., and Valembois, P. 1997. Purification, characterization and activities of two haemolytic and antibacterial proteins from coelomic fluid of the annelid *Eisenia fetida andrei*. *Biochim. Biophys. Acta* 4 (1): 123–132.

Mohrig, W., Eue, I., Kauschke, E., and Hennicke, F. 1996. Crossreactivity of haemolytic and hemagglutinating proteins in the coelmic fluid of *Lumbricidae* (Annelida). *Comp. Biochem. Physiol. A Physiol.* 115 (1): 19–30.

Nagasawa, H., Sawaki, K., Fujii, Y., Kobayashi, M., Segawa, T., Suzuki, R., and Inatomi, H. 1991. Inhibition by lombricine from earthworm (*Lumbricus terrestris*) of the growth of spontaneous mammary tumours in SHN mice. *Anticancer Res.* 11: 1061–1064.

Nakajima, N., Ishihara K., Sugimoto, M., Sumi, H., Mikuni, K., Hamada, H., 1996. Chemical modification of earthworm fibrinolytic enzyme with human serum albumin fragment and characterization of the protease as a therapeutic enzyme. *Biosci, Biotechnol. Biochem.* 60(2): 293–300.

Nakajima, N., Sugimoto, M., and Ishihara, K. 2000. Stable earthworm serine proteases: Application of the protease function and usefulness of the earthworm autolysate. *J. Biosci. Bioeng.* 90 (2): 174–179.

Oumi, T., Ukena, K., Matsushima, O., Ikeda, T., Fujita, T., Minakata, H., and Nomoto, K. 1995. The GGNG peptides: Novel myoactive peptides isolated from the gut and the whole body of the earthworms. *Biochem. Biophys. Res. Commun.* 216 (3): 1072–1078.

Pan, W., Liu, X., Ge, F., and Zheng, T. 2003. Reconfirmation of antimicrobial activity in the coelomic fluid of the earthworm *Eisenia fetida andrei* by colorimetric assay. *J. Biosci.* 28: 723–731.

Popovic, M., Grdisa, M., and Hrzenjak, T.M. 2005. Glycolipoprotein (G-90) obtained from the earthworm *Eisenia fetida* exerts antibacterial activity. *Veterinarski Archiv.* 75 (2): 119–128.

Popovic, M., Hrzenjek, T., Grdisa, M., and Vukovic, S. 1998. Adhesins of immunoglobulin-like superfamily from earthworm *Eisenia fetida*. *Gen. Pharmacol.* 30 (5): 795–800.

Popovic, M., Mihaela, T.H.R., Babic, T., Kos, J., and Grdisa, M.A. 2001. Effect of earthworm (G-90) extract on formation and lysis of clots originated from venous blood of dogs with cardiopathies and with malignant tumors. *Pathol. Oncol. Res.* 7 (3): 197–202.

Prakash, M., Balamurugan, M., Parthasarathi, K., Gunasekaran, G., Cooper, E.L., and Ranganathan, L.S. 2007. Anti-ulceral and anti-oxidative properties of "earthworm paste" of *Lampito mauritii* (Kinberg) on *Rattus norvegicus*. *Eur. Rev Med. Pharmacol. Sci.* 11 (1): 9–15.

Quaglino, D., Cooper, E.L., Salvioli, S., Capri, M., Suzuki, M.M., Ronchetti, I.P., Franceschi, C., and Cossarizza, A. 1996. Earthworm coelomocytes *in vitro*: Cellular features and granuloma formation during cytotoxic activity against the mammalian tumor cell target K562. *Eur. J. Cell Biol.* 70 (3): 278–288.

Ranganathan, L.S. 2006. Vermicomposting enhances humification, mineralisation and chelation. *J. Ann. Univ. Sci.* 42: 1–14.

Rejnek, J., Tukova, L., Zikan, J., and Tomana, M. 1991. The interaction of protein from the coelomic fluid of earthworm with staphylococcal protein-A. *Dev. Comp. Immunol.* 15: 269–277.

Roch, P., Canicatti, C., and Valembois, P. 1989. Interactions between earthworm hemolysins and sheep red blood cell membranes. *Biochem. Biophys. Acta* 983 (2): 193–198.

Roch, P., Davant, N., and Lassegnes, M. 1984. Isolation of agglutinins from lysis in earthworm coelomic fluid by gel filtration than chromatofocusing. *J. Chromatogr.* 290: 231–235.

Roch, P., Lassegver, M., and Valembois, P. 1991. Antibacterial activity of *Eisenia fetida andrei* coelomic fluid: (III)—Relationship within the polymorphic hemolysins. *Dev. Comp. Immunol.* 15 (1–2): 27–32.

Rosenberg H., and Ennor, A.H., 1960. The Isolation and Characterization of D-Serine Ethanoloamine Phosphodiester from Earthworms. *Biochem J.* 75: 179–182.

Saint-Denis, M., Labrot, F., Narbonne, J.F., and Ribera, D. 1998. Glutathione, glutathione-related enzymes, and catalase activities in the earthworm *Eisenia fetida andrei. Arch. Environ. Contam. Toxicol.* 35 (4): 602–614.

Stein, E.A., Wojdani, A., and Cooper, E.L. 1982. Agglutinins in the earthworm *Lumbricus terrestris*: Naturally occurring and induced. *Dev. Comp. Immunol.* 6 (3): 407–421.

Sugimoto, M., and Nakajima, N. 2001. Molecular cloning, sequencing and expression of cDNA encoding serine protease with fibrinolytic activity from earthworm. *Biosci. Biotechnol. Biochem.* 65 (7): 1575–1580.

Sugiura, T., Yamashita, A., Fukuda, N., Miyamoto, T., Cheng, N.N., Kishimoto, S., Waku, K., Tanaka, T., and Tsukatani, H. 1995. Platelet-activating factor and its structural analogues in the earthworm *Eisenia fetida. Biochim. Biophys. Acta* 125 (1): 19–26.

Vaillier, J., Cadoret, M.A., Roch, P., and Valembois, P. 1985. Protein analysis of earthworm coelomic fluid. III. Isolation and characterization of several bacteriostatic molecules from *Eisenia fetida anderi. Dev. Comp. Immunol.* 9: 11–20.

Valembois, P., Roch, P., and Lassegues, M. Cassanda, P. 1982. Antibacterial activity of the hemolytic system from the earthworm *Eisenia fetida andrei. J. Invert Pathol.* 40: 21–27.

Wu, C., Li, L., Zhao, J., Fan, Q., Tian, W.X., and He, R.Q. 2002. Effect of alpha2M on earthworm fibrinolytic enzyme III-1 from *Lumbricus rubellus. Int. J. Biol. Macromol.* 31:71–77.

Xing, B.D., Yin, S.M., and Ru, B.G. 1997. Purification and characterization of the fibrinolytic enzyme (eFE-D) from earthworm *Eisenia foetida. Sheng Wu Hua Xue Yu Sheng Wu Wu Li Xue Bao* (Shanghai) 29 (6): 609–612.

Yamaji, A., Sekizawa, Y., Emoto, K., Sakuraba, H., Inoue, K., Kobayashi, H., and Umeda, M. 1998. Lysenin, a novel sphingomyelin-specific binding protein. *J. Biol. Chem.* 27 (9): 5300–5306.

Yang, J.S., and Ru, B.G. 1997. Purification and characterization of an SDS-activated fibrinolytic enzyme from *Eisenia fetida. Comp. Biochem. Physiol. B Biochem. Mol. Biol.* 118 (3): 623–631.

Yegnanarayan, R., Ismail, S.A., and Shorti, D.S. 1998. Anti-inflammatory activity of two earthworm portions in carrageenan pedal oedema test in rats. *Ind. J. Physiol. Pharmacol.* 32: 72–74.

Yegnanarayan, R., Sethi, P.P., Rajhans, P.K., Pulandiran, K., and Ismail, S.A. 1987. Anti-inflammatory activity of total earthworm extracts in rats. *Ind. J. Pharm.* 19: 221–224.

Zhang, M., Li, X., Liu, Y., and Qiu, F.Y.G. 2006. Effects of extract of dilong (pheretima) on the scalded skin in rats. *J. Tradit. Chin. Med.* 26 (1): 68–71.

Zhao, J., Li, L., Wu, C., and He, R.Q. 2003. Hydrolysis of fibrinogen and plaminogen by immobilized earthworm fibrinolytic enzyme II from *Eisenia fetida. Int. J. Biol. Macromol.* 32: 165–171.

Zhao, J., Xiao, R., He, J., Pan, R., Wu, C., Liu, X., and He, R.Q. 2006. *In situ* localization and substrate specificity of earthworm protease-II and protease-III-1 from *Eisenia fetida. Int. J. Biol. Macromol* 26: 16814–16856.

[illegible]

CHAPTER 23

The Status of Vermicomposting in North America: A Rapidly Developing Technology

Rhonda L. Sherman and Peter Bogdanov

CONTENTS

I Introduction ... 370
II Vermicomposting or Vermiculture ... 370
III Feedstock as Income versus Expense ... 371
IV Temperate Regions Predominate in Vermicomposting ... 371
V Factors Influencing Differences in Vermicomposting Systems ... 372
VI Earthworm Species Used in Vermicomposting ... 372
VII Commercial Vermiculture as Private Enterprise ... 373
VIII Sales of Vermicomposts and Vermicompost Teas ... 373
IX Vermicomposts or Earthworm Castings? ... 374
X West Coast Operations—Early History in California ... 375
 A North American Bait Farms ... 375
 B Vermicomposting Biosolids in Fallbrook, California ... 376
 C Canyon Recycling, San Diego, California ... 376
 D Vermicomposting Organic Residuals from Material Recovery Facilities ... 377
 E Rainbow Worm Farm, Davis, California ... 379
 F Pacific Landscape Supply (Formerly American Resource Recovery), Vernalis, California ... 379
XI Current West Coast Operations ... 380
 A Sonoma Worm Farm, Sonoma, California ... 380
 B The Worm Farm, Durham, California ... 380

C Oregon Soil Corporation, Portland, Oregon 381
D Yelm Earthworm and Castings Farm (Yelm Earth), Yelm, Washington 381
XII Vermiculture in the Southern and Eastern United States 382
A RT Solutions LLC, Avon, New York 382
B WeCare Organics, Pennsylvania 383
C Key Packing Company (KPC), Robbins, North Carolina 383
D Windswept Worm Farm, Blue Springs, Missouri 384
E Rabbit Hill Farm, Corsicana, Texas 384
F Vermicomposting in Canada: Forterra Inc., Puslinch, Ontario 385
XIII North American Interest in Vermiculture and Vermicomposting 385
XIV Conclusions 387
References 388

I INTRODUCTION

The role of earthworms in the management of organic wastes and the production of earthworms and vermicompost in North America may be characterized by the following issues:

- Emphasis or purpose of the operation
- Feedstock (e.g., sources and types of earthworm feed)
- Region
- Vermicomposting system used
- Species of earthworm used

Once these distinguishing characteristics are described, selected sites could be examined to illustrate the history and current status of vermicomposting in North America in the first decade of the twenty-first century.

II VERMICOMPOSTING OR VERMICULTURE

When earthworms are used in waste management, the stabilization of organic residuals and the production of a marketable soil amendment (vermicompost) are generally the chief goals. When the focus is on breeding earthworms for bait, as a protein source, or for stocking vermicomposting projects, there is often a declining emphasis on the value of earthworm by-products. Thus, *vermicomposting* facilities tend to utilize earthworms as processors in resource recovery and look for markets in the soil-amendment or plant-growth-media sectors. *Vermiculture* facilities, in contrast, concentrate principally on the production of earthworms for sale or use as feed protein (see Chapter 20). However, due to increasing awareness of the value of earthworm-worked material or vermicomposts, more vermiculture operations are seeking to market their earthworm castings as a supplement to their earthworm sales.

Some earthworm businesses that harvest indigenous earthworm species to sell as fish bait are not generally concerned with culture or optimizing breeding conditions. Thus, the term *vermiculture* may not accurately describe these operations, although their marketing structure puts them into the larger category of what may be called an earthworm-related industry.

III FEEDSTOCK AS INCOME VERSUS EXPENSE

The largest vermicomposting facilities typically follow the model of large organics-recovery operations that practice conventional thermophilic composting. Tipping fees collected from landfills for incoming feedstocks usually comprise the largest part of any facility's income stream and may comprise 80% or more of its total revenues. Sales of vermicomposts also add to the profit margin of a company and can be very significant. Large vermicomposting operations process organic residuals that include yard debris, preconsumer and restaurant food wastes, manure from herbivorous animals, cardboard and paper waste, the biodegradable fraction of municipal solid waste, and biosolids from wastewater treatment plants. Precomposting of manures, yard debris, and food residuals often occurs prior to vermicomposting. Some of these vermicomposting facilities also sell earthworms. For most vermiculture operations, feedstocks are purchased, or arrangements must be made to procure feedstocks from available sources, maybe with some cost to the earthworm grower, such as delivery charges. These vermiculture facilities tend to be smaller in scale, handle smaller volumes of organic materials, are less mechanized, and are more labor-intensive than vermicomposting sites.

IV TEMPERATE REGIONS PREDOMINATE IN VERMICOMPOSTING

Because interest continues to grow throughout the United States and Canada, temperate regions seem to predominate in managed earthworm activities. Large outdoor windrow systems are in common use in U.S. West Coast operations, from the southernmost border of California to Washington State. The Atlantic seaboard, extending from North Carolina to Florida and moving throughout much of the southeastern United States, forms the second major region for vermicultural activity. Covered systems are more common here, affording greater protection from the elements. Scattered throughout the remainder of the United States and Canada are a smaller number of sites with greater provision made to control for the adverse effects of extreme temperatures. Indoor and container systems are required for effective earthworm waste management in regions where the climate is seasonally cold. Earthworms that are harvested for fish bait are often obtained from cooler climatic regions.

V FACTORS INFLUENCING DIFFERENCES IN VERMICOMPOSTING SYSTEMS

Earthworms are raised in beds, pits, trenches, bins, trays, windrows, wedges, and continuous-flow systems. Some managed systems involve outdoor piles of organic matter either on the ground (windrows, pits) or in containers. Beds may be on the ground or raised to allow space for harvesting vermicompost and capturing leachates. Indoor systems may be in permanent buildings, polyethylene tunnel structures, Quonset huts, or pole barns with adjustable sides on soil, asphalt, or concrete. Various designs of indoor and container earthworm vermicomposting systems have been manufactured and vary in their ability to control temperature, moisture, aeration, feedstock application, and separation of earthworm-worked material from incoming feedstocks. Factors accounting for differences in the efficiency of systems include geography; amounts of feedstocks to be processed; availability of capital investment and the costs of land, existing buildings, and labor; availability of water or concern for its conservation; state or local regulations; and the operation's emphasis on vermicomposting or on vermiculture. Applying feedstocks, maintaining proper moisture content in the earthworm beds, and separating earthworms from vermicompost are three ongoing activities involved in any size of operation, but techniques for achieving these differ markedly from one facility to another.

VI EARTHWORM SPECIES USED IN VERMICOMPOSTING

Eisenia fetida and the related species *Eisenia andrei* (most typically called redworms, red wigglers, or tiger worms) seem to occupy the leading role among earthworm species for use in both vermicomposting and vermiculture operations throughout the United States and Canada. Although most vermicomposting facilities use and recommend *Eisenia fetida*, vermiculture operations may breed or collect more than one earthworm species, particularly if sales are made to bait dealers. *Lumbricus terrestris* L. (the common night crawler) is generally harvested by hand from orchards or golf courses at night because creating optimum conditions and media for enhancing reproduction in bins is generally considered difficult to achieve with this earthworm species (see Chapter 21).

Increasing interest has been shown in the commercial cultivation of species other than *E. fetida* for vermicomposting. Particular interest has been shown in more recent years in *Dendrobaena veneta* (or *Eisenia hortensis*). Of lesser interest are *Eudrilus eugeniae* (African night crawler), *Perionyx excavatus* (blueworm), and *Pheretima hawayana* (Alabama jumper) because they are conditioned to warmer climate regions due to their inability to withstand cold temperatures (see Chapters 3 and 4).

Over 100 vermiculture operations in the United States and Canada are listed in the biannual edition of *Earthworm Buyer's Guide* (2008–2009). Eschewing scientific nomenclature, this directory lists vendors who market several species of earthworms, which are described only by their popular names: redworms, African night crawlers, jumpers, brown nose worms, and others. It is impossible to estimate from this directory alone the magnitude of the earthworm industry in North America. Nevertheless, it provides a rudimentary overview of the main regions where vermiculture is practiced and is suggestive of some of the practices, products, and markets in the earthworm industry.

VII COMMERCIAL VERMICULTURE AS PRIVATE ENTERPRISE

Whereas composting facilities in North America may be owned either privately, by a municipality, or by a joint partnership of the two entities, it is less common to find a vermicomposting operation that is owned by a municipality. An entrepreneurial spirit has driven most efforts in promoting the value of both earthworms and earthworm by-products. While municipal-scale and municipally funded waste management is concerned with stabilization of the organic fraction of solid waste, vermicomposting entrepreneurs must focus on the profit-making potential of their endeavors. Choosing and acquiring optimum earthworm feedstocks and giving attention to product quality motivate the entrepreneur. Municipal waste managers and many compost-facility operators do not usually enjoy the luxury of choosing a preferred waste stream and may not be able to provide the same attention to quality control as the entrepreneur. If the goal of high-value product sales drives the entrepreneur, then products must be standardized, promoted, and marketed.

In some glaring instances, gross misrepresentation of potential earthworm reproductivity, coupled with phenomenal forecasts of future demand, has meant that a few entrepreneurial business promoters have been able to dupe investors. Optimistic projections of earthworm multiplication rates have tempted some entrepreneurs to promote various dubious kinds of futures markets. There have been numerous reports of "buy-back contracts" that are sold to investors who are promised exorbitant rates of return on their investment. Claims of exaggerated earthworm reproductivity (e.g., "earthworms can double their population every month") and booming worldwide markets have been made to induce would-be earthworm growers to invest their money. Reports of these questionable activities have come from California, Oklahoma, Florida, Ohio, Idaho, Nevada, New Mexico, Pennsylvania, Kentucky, and other states.

There have also been instances of failure of some large-scale vermicomposting facilities due to undercapitalization, difficulties with state and federal regulators, unstable markets, or inadequate marketing methods. However, several entrepreneurs have demonstrated that large-scale vermiconversion of organic residues into vermicomposts can be very profitable.

VIII SALES OF VERMICOMPOSTS AND VERMICOMPOST TEAS

The marketing of earthworm-worked material, described as *vermicomposts* by some and labeled as *earthworm castings* (and, more commonly, *worm castings*) on bags of finished product by others, has some way to go before it becomes as universally accepted as peat moss, dried cow manure, soilless growth media, or other commercially available horticultural products found in nurseries and garden centers. Even thermophilic composts, which must compete with other types of soil amendments, currently seem to have greater market recognition than vermicomposts (see Chapters 2 and 18).

Vermicomposters believe that they produce a superior product to traditional composts, but a huge gap in marketing exists between the approximately 3000 composting facilities in North America and the several hundred earthworm-growing and vermicompost-producing operations in the same geographic area. This gap may likely shrink as the "blend trend" in compost marketing continues, when compost becomes a basic but proportional ingredient in a product mixture specially formulated for certain applications. There are various instances of joint-venture partnerships in which compost facilities link up with vermicomposting technology to create blended products for sale. Some soil blenders have also added vermicomposts in small quantities with other soil amendments to create unique product combinations.

A rapidly growing trend is the brewing and sale of aqueous vermicompost extracts, commonly known as "castings teas," "vermicompost teas," or "worm teas." A huge spike in sales of vermicompost teas occurred in 2009, and the demand is expected to increase very significantly in 2010, especially as the use of teas becomes progressively supported by valid scientific trials and scientific publications (see Chapters 11, 12, 13, and 14).

IX VERMICOMPOSTS OR EARTHWORM CASTINGS?

The terms *vermicomposts* and *earthworm castings* have been used interchangeably to identify earthworm-worked organic wastes (see Chapter 18). *Castings* seems to be the term of choice among industry personnel throughout North America, while *vermicompost* tends to be the accepted designation within the scientific community. An informal survey of purveyors of vermicompost in the United States finds the terms *worm castings* and *earthworm castings* in most frequent use in the labeling of soil-amendment products containing vermicompost. Some bagged products have been labeled only with the words *worm castings* or *earthworm castings* and may contain anywhere from 10 to 100% vermicompost.

A number of purveyors of vermicompost, particularly in California but in other regions as well, insist that labeling their product *earthworm castings* rather than *vermicompost* has at least two distinct advantages. First, in the State of California, an agricultural exemption exists in favor of vermiculture operations, excluding them from composting regulations. As an agricultural activity, the practice of vermiculture (and vermicomposting) results in the production of not only earthworms but also earthworm by-products, that is, earthworm castings. In the sense that cows produce milk and bees produce honey, earthworms produce castings as a marketable product. To call the earthworm-worked material *vermicompost* might associate the material with compost, a product produced for sale by state-regulated facilities. Second, some vermicompost marketers believe that the term *earthworm castings* is more descriptive and better understood in the marketplace than the term *vermicompost*, which may contribute to the perception that the material somehow contains thermophilic composts.

X WEST COAST OPERATIONS—EARLY HISTORY IN CALIFORNIA

A North American Bait Farms

Interest in vermiculture on the West Coast of the United States can be dated as early as 1936 when Thomas Barrett, a physician and "Renaissance man" of many interests, established his Earthmaster Farms in El Monte, California. Here he conducted experiments in raising earthworms and recorded his recommendations in a book, *Harnessing the Earthworm* (Barrett 1947).

In 1967, Ronald E. Gaddie, Sr., started a vermiculture business after a disabling back injury. By 1972, Gaddie's North American Bait Farms in Ontario, California, was approaching $100,000 in gross annual sales and grew to over $600,000 in 1975. Gaddie coauthored *Earthworms for Ecology and Profit* (Vols. 1 and 2), along with Donald E. Douglas. Over 750,000 copies of the first volume, subtitled *Scientific Earthworm Farming* (1975), were printed, and some 250,000 copies of the second volume, subtitled *Earthworms and the Ecology* (1977), were later published. Gaddie's vermiculture and book publishing business grew wildly throughout the second half of the 1970s, and an extensive network of earthworm growers was established throughout the United States. He reported that his network exceeded 1100 in California alone. Earthworms were shipped to Italy, France, Korea, and Japan. His Bookworm Publishing Company earned tremendous profits from the sale of his writings and of earthworm books written by others. His own books were translated into Japanese, French, and Spanish. Just as the foreign markets began to surge further in sales (an order for $170,000 of earthworms to be sent to Italy was received), Gaddie was forced to close his doors in early 1980 (Bogdanov 1996b).

The closure of North American Bait Farms was the result of costly litigation Gaddie faced as the alleged "kingpin" of a pyramid selling scheme. In 1974, the Securities and Exchange Commission began informing North American Bait Farms that a price guarantee offering to buy back earthworms from potential investors could be construed as a sale of a security that would have to be registered under the Securities Act of 1933 (Gaddie and Douglas 1975). Despite several attempts to warn others against violating Securities and Exchange Commission regulations, Gaddie became caught in litigation accusing him of participating in illegal pyramid schemes. The great cost of having to defend himself and others eventually took its toll. The once $1,000,000-per-year business in earthworms alone (not counting book sales) collapsed. Along with it, perhaps tens of thousands of other earthworm growers, by the estimate of one person who was active both then and now, found themselves in an industry locked in a tailspin (Bogdanov 1996b). Nearly 20 years later, the memory of this boom and bust cycle remains in the minds of those still associated with vermiculture in California and throughout portions of the United States. Today this story's almost mythic proportions serve as a reminder of both the immense opportunities available in vermiculture and the dire consequences that may befall even the most circumspect.

B Vermicomposting Biosolids in Fallbrook, California

In 1986, after conducting a successful pilot-scale vermicomposting program, the Fallbrook Sanitary District embarked on a full-scale program to use earthworms (*Eisenia fetida* and *Lumbricus rubellus*) for stabilization of biosolids. The district produced approximately 544 kg (0.6 dry tons) of sludge per day on its 17.4 ha (43 acres) site in a community of about 16,000 people located in northern San Diego County. The two-stage process included precomposting the material to comply with the U.S. Environmental Protection Agency's (EPA) standards to reduce pathogens. After approximately 30 days in a static pile, the material was transferred to vermicomposting beds where it was applied at the rate of 10–15 cm (4–6 in) per week to the 2.4 m (8 ft) wide windrows of varying length. To maintain porosity, straw bulking material was added about once per month. In about 6 months, windrows reached a height of approximately 0.9 m (3 ft) and were ready for harvesting. The top 15–20 cm (6–8 in) of material, containing the greatest concentrations of earthworms, was removed and used to establish new windrows. The remainder, stabilized vermicompost, was screened and placed in storage where it was allowed to cure for an additional 30 days. The district sold its static pile compost for \$20 m^{-3} (15yd^{-3}) and its vermicompost for \$46 m^{-3} (35 yd^{-3}). It reported that it could not keep up with local demand (Harris et al. 1990). However, the Fallbrook vermicomposting project was forced to close as local residential development increased, and the once-rural community became a suburb of San Diego.

C Canyon Recycling, San Diego, California

Interest in vermicomposting continued in San Diego County as Resource Conversion Corporation (RCC) obtained 2268 kg (5000 lb) of earthworms from Fallbrook and brought them to Canyon Recycling in San Diego. On an 4 ha (8-acre), Canyon Recycling established twenty-two 76.2 m (250 ft) long windrows, 3.3 m (10 ft) in width. A landfill-diversion site, Canyon received tipping fees for municipal yard trimmings; manure from the San Diego Zoo, San Diego Wild Animal Park, and Del Mar Race Track; and construction and demolition (C&D) debris. In the early 1990s, Canyon concentrated on vermiculture, building up its earthworm population. Earthworm beds were fed and split continuously until the time came to shift from vermiculture to a vermicomposting operation (Bogdanov 1996a). By March 1996, John Beerman, general manager of the facility, reported that he provided his 34,020 kg (75,000 lb) of earthworms with about 13.6–18.1 tons (15–20 t) of green waste every day (Barbour 1996). 8–10 cm (3–4 in) of feedstock were applied with manure spreaders twice a week to each windrow. Water usage amounted to between

151416.5–189270.6L (40,000–50,000 gal) per day. Earthworms were sold only rarely. After growing its earthworm inventory for about five years, the harvesting of vermicompost began in earnest, and sales of Vermigro, a blend of earthworm castings with compost, were made to nurseries, landscapers, organic farmers, and the general public. The blended product was sold in bulk (46 m^3 (35 yd^3)) and in bags ($7.00 retail for 0.028 m^3 (1 ft^3)). Canyon Recycling also sold recycled wood-fiber products to particleboard manufacturers and cogeneration facilities, and produced compost and mulch for roadside application by California's Department of Transportation (CalTrans). However, the early creation of burdensome and unmanageable indebtedness pressured RCC's directors to put Canyon Recycling up for sale in 1997. Although Canyon reported it could not make enough Vermigro to satisfy the demand, other factors contributed to the need for restructuring this facility.

D Vermicomposting Organic Residuals from Material Recovery Facilities

Pacific Southwest Farms (PSF), a 21.9 ha (54 acre) vermicomposting facility in Ontario (San Bernardino County), California, began its operation in 1994 with 10 metric tons (11 t) of earthworms transported from the failed Worm Concern project in Simi Valley, California. Owner Barry Meijer steadily built his operation into what may have been the largest project of its kind up until its closure. PSF received the biodegradable fraction of municipal solid waste, or "green material" (as defined by California's compost regulations), from up to three different material recovery facilities (MRFs) for a tipping fee. Initially, PSF took in about 68 tons (75 t) per day and increased that amount to approximately 90.7 tons (100 t) per day. Earthworm-stocked windrows measuring 2.4 m (8 ft) in width and 30.48 m (100 ft) in length were fed at the rate of 3.63 tons (4 t) of material per row each week. Situated east of Los Angeles in an arid climate, PSF's water usage amounted to 454249.4 L (120,000 gal) per day. Sources for water included residential sprinkler runoff and barn water from local dairies. Although the water was abundant and free, pumps, irrigation lines, and use of electricity added significant expense. At its zenith, PSF estimated that more than 90.74 tons (100 t) of earthworms processed organic residuals in 360 windrows. Finished vermicompost was reportedly sold to agricultural users in central California. Due to the mixed quality of feedstocks, which contained a significant portion of inert material (especially glass shards), the final product had to be screened to 3 mm (0.125 in) and was not acceptable for retail sales to the public (Bogdanov 1997d).

PSF's feedstocks were nontraditional in comparison to other vermicomposting sites. From the beginning of its operation, MRFs in nearby Orange County processed the commingled material they received and sent the biodegradable fraction to PSF. This material was approximately 95% organic but contained enough pieces of plastic to cause a problem with site and product appearance. The maximum particle size of the incoming product was later reduced from 10 cm (4 in) to 3 cm (1.25 in), which proved to work better and contained less visible and unsightly plastic. PSF also received ground paper that had come into contact with food material or other green waste. The materials that were fed to earthworms were specified as "green material," defined by the California Integrated Waste Management Board (CIWMB) as "any plant material that is either separated at the point of generation, or separated at a centralized facility [a MRF] that employs methods to minimize contamination. Green material includes, but is not limited to, manure, untreated wood wastes, paper products, and natural fiber products. Green material does not include treated wood waste, mixed demolition or mixed construction debris" (California Integrated Waste Management Board, 1997b). During the time of its operation, Meijer believed PSF was the only project using municipal solid waste for vermicomposting in California.

The San Bernardino Local Enforcement Agency (LEA) effectively shut down PSF in November 1996 by issuing a notice and order requiring PSF to obtain a solid waste facilities permit as a transfer/processing station. PSF was also told it could not "process" any of its incoming feedstock. Processing could include either blending with manure or precomposting the incoming feedstock. PSF appealed this notice and order. In February 1997, the San Bernardino County Independent Hearing Panel issued a decision that specified that the earthworm bed activity was excluded from regulation by the CIWMB's compost regulations and that PSF was not required to obtain a solid waste facilities permit.

However, PSF's problems continued. San Bernardino County attempted to close down PSF because of its location in a dairy zone, saying that it needed a conditional use permit and did not possess one. PSF filed an appeal of this ruling, and, in April 1997, the Court of Appeal, State of California, Fourth Appellate District, determined that PSF could continue its vermicomposting operation. Citing California's Food and Agricultural Code, the court agreed that vermiculture is an "agricultural use" and that PSF was in operation for the purpose of producing an "animal product" (Bogdanov 1997a).

At least two victories for PSF and the practice of vermiculture in the state of California were won by these decisions. First, vermiculture continues to enjoy an agricultural exclusion from California's composting regulations by virtue of the fact that the Food and Agriculture Code identifies vermiculture and its by-products as agriculture. And, second, precomposting of feedstock prior to application to earthworm beds does not fall under the CIWMB's compost regulations. Critics have complained that these exclusions do not allow for a level playing field for composters and vermicomposters alike. Additionally, the exclusions open the door for disguising a thermophilic composting operation by allowing it to involve a small quantity of earthworms and call it vermicomposting. To discourage the possible abuse of vermiculture exclusions, CIWMB amended its regulations to clarify what it will allow. In its Initial Statement of Reasons, CIWMB wrote: "A revision of the term 'vermicomposting' is necessary to clarify that earthworm castings, not compost, are the primary product of vermicomposting activities" (California Integrated Waste Management Board, 1997a). CIWMB maintains that an enforcement agency has the flexibility to determine whether an activity is or is not a vermicomposting activity. Incidental earthworm activity, in which significant amounts of biological decomposition that is not related to earthworm activity occurs, would not constitute "vermicomposting."

Therefore, according to the CIWMB, the presence of a few earthworms in a compost pile would not qualify the operation as a "vermicomposting activity."

Meijer's PSF won only a Pyrrhic victory, however, as time-consuming litigation during the cease and desist order forced haulers to locate other sites to transport their organic waste. Without tipping fees and feedstocks to continue his operation, Meijer was forced into shutting down the facility.

E Rainbow Worm Farm, Davis, California

From 1979 to 2004, Al Cardoza's Rainbow Worm Farm saw steady growth, largely due to Cardoza's talents and persistence in single-handedly creating a full-service operation. Cardoza obtained dairy manure from Dixon, a small community located a few miles from his vermiculture facility in Davis, west of Sacramento, California; one m (4 ft) wide windrows, called "ricks," covered 1.5 ha (3 acres) of their 10 ha (20-acre) farm. Sprinkler irrigation was used to spray a fine mist on the unshaded beds where temperatures frequently hit triple digits Fahrenheit in summer. The exclusively outdoor vermiculture operation had 30 ricks approximately 61 m (200 ft) long. The rows received about 2.5 cm (1 in) of material every two weeks, amounting to about 30–38 m^3 (40–50 yd^3) per row. *Eisenia fetida* earthworms were harvested in a trommel designed and built by Cardoza. Custom-made earthworm harvesters and blueprints were available for sale. Harvested earthworms were packaged in wax-coated cardboard boxes and shipped by ground carrier and air freight all over the world. Cardoza applied wax to the interior of the boxes, perforating each one with enough holes to allow ventilation. A specially blended bedding mixture of peat moss, shredded paper, and oyster-shell flour was used in packaging earthworms for shipment (Bogdanov 1998a). Cardoza designed a heavy-duty blender for mixing vermicompost with other ingredients to create custom potting soils for nurseries. He also designed bagging and sealing machines that packaged vermicompost in 0.28 m^3 (1 ft^3) bags. Cardoza also offered two-day seminars and consultation services.

F Pacific Landscape Supply (Formerly American Resource Recovery), Vernalis, California

At one time the largest vermicomposting operation in North America, Pacific Landscape Supply (PLS) was started in 1993 by Jim Davis and Mario Travalini. Located on an old military air base, part of its 129.5 ha (320 acres) consisted of two paved airplane landing strips covering 35 ha (75 acres) that were used for the vermicomposting operation. Earthworms were grown in windrows that were 0.9 m (3 ft) wide, up to 0.9 m (3 ft) high, and up to 360 m (400 yd) in length. The primary feedstock was short paper pulp fibers generated from recycling cardboard, supplemented with varying amounts of tomato residuals, green waste, and manure. The paper waste was saturated with high-nitrogen glues that held together the cardboard corrugations. The relatively high nitrogen and small particle size of the paper sludge solids provided a good feedstock for the earthworms. The paper sludge formed a crust on the surface of the uncovered windrows that provided gas exchange yet prevented the organic materials beneath from drying out. The crust also protected the earthworms from constantly windy conditions, hot and cold temperature extremes, and direct sunlight and prevented water loss from the beds (Slocum 2005). Over 300 tons (272.2 t) of cardboard sludge were delivered seven days a week, so a total of about 68038 tons (75,000 t) of organic residues were vermiprocessed annually. PLS began processing and selling vermicompost in 1997, shipping up to 90.7 ton (100 t) per week to retail bagging distributors, nurseries, vineyards, and farms. They began selling earthworms in 1998 and shipped them throughout the United States and to other countries. The operation closed about three years ago when the management changed.

XI CURRENT WEST COAST OPERATIONS

A Sonoma Worm Farm, Sonoma, California

In 1992, a commercial airline pilot named Jack Chambers purchased a 5-acre farm in Sonoma, California, from a chicken rancher who also raised earthworms on poultry manure. Chambers expanded his Sonoma Valley Worm Farm by adding outdoor windrows to the existing covered row system, obtaining dairy manure, installing an irrigation system, and purchasing a tractor and trommel screen (Bogdanov 1997b). In 2002 he started precomposting in an aerated bin for 10 days and began researching automated raised-bed reactors based on Clive Edwards's design. Chambers now has four flow-through reactors that are 27 m (90 ft) long, 1.5 m (5 ft) wide, and 60 cm (2 ft) deep and sit 45 cm (18 in) off the ground. The earthworms work the organic material for 60 days before it is removed from the bottoms of the reactors by a manual breaker bar. Chambers is selling vermicompost to several high-end vineyards, and they are achieving positive results in grape production. Some vineyards were losing up to 20% of their new plants, but now, after using Chambers's vermicompost, their losses are less than 1%.

B The Worm Farm, Durham, California

In 1992, Mark and Arlita Purser converted their family's 100-year-old chicken farm into a vermicomposting operation. They now have 288 m (9600 ft) of windrows, both indoors and outdoors. Each windrow is 91.4 m (300 ft) long, 2.4 m (8 ft) wide, and 0.9 m (3 ft) high. The windrows are harvested once a year, and it takes two men 3 hours to harvest one windrow. The top 8–10 cm (3–4 in) of each windrow (which contains the most earthworms) is removed with a pitchfork and placed in a new row. Then a 3 m (10 ft) wide compost turner is used to turn the old windrow. To remove the vermicompost, one man uses a pitchfork to load a harvester, and the other operates the machine. A 1 cm (0.375 in) mesh screen is used to separate the earthworms from the vermicompost. The Worm Farm sells three species of earthworms, vermicompost, worm bins, coir bricks, worm chow, books, hats, and 55 types of soil amendments. The Worm Farm Learning Center offers educational seminars for homeowners and worm-farm consulting. About 95% of the earthworms they sell are to households, and 90% of their earthworms are sold from their Internet site.

C Oregon Soil Corporation, Portland, Oregon

Dan R. Holcombe founded Oregon Soil Corporation in February 1988. His automated continuous-flow vermicomposting reactor, designed and developed by Dr. Clive Edwards and his colleagues of The Ohio State University, has been used since the early 1990s (see Chapter 8). The raised vermicomposting bed measures 40 m (128 ft) in length and 2.4 m (8 ft) in width and is 0.9 m (3 ft) deep. A manually operated, 2-ton-capacity gantry feeder, riding on rails fixed to the top of the plywood sides, disburses up 2.7 tons (3 t) of blended organic materials daily. About 80% of the feedstock is preconsumer food waste picked up from Portland-area supermarkets and food processors. Composted yard trimmings and shredded paper are blended in as bulking agents along with the wet organics. A chain-driven breaker bar mechanically scrapes vermicompost from the raised mesh floor, allowing the finished material to fall to the floor under the unit. A recovery flap scraper unit then moves the vermicompost from one end of the reactor for collection at the other end. Daily applications of thin layers of 2.5–5.0 cm (1–2 in) of organics allow earthworms to work in the upper level of the reactor as earthworm-worked material descends toward the mesh floor. Vermicompost is packaged in 0.5 kg (1 lb) cardboard boxes and 0.0283 m^3 (1 ft^3) bags and labeled as "Oregon Soil Earthworm Castings."

D Yelm Earthworm and Castings Farm (Yelm Earth), Yelm, Washington

Perhaps the largest indoor vermicomposting operation on the West Coast, Yelm Earth was converted from a mushroom-producing operation to an earthworm farm in 1991 by RCC of San Diego, California. RCC used the Yelm farm for research and development experiments, with the hope of stocking other vermicomposting projects it had planned to start in addition to its Canyon Recycling project. Earthworms bred in Yelm were sold in quantities of up to 2.26 tons (5000 lb) and shipped as far as Texas. In 1997 the farm came under the ownership of Sound Resource Management, an environmental consulting firm based in Seattle, Washington (Bogdanov 1997e). Since then, it has changed hands twice more. The current owners, Hunt McLean and Kelan Maynagh, bought the farm in August 2005 and dubbed it Yelm Earth. All of the vermicomposting takes places indoors on cement floors in insulated warehouses. Eight rooms have two windrows each measuring 18 m (60 ft) long and 1.8 m–2.4 m (6–8 ft) wide. Earthworms are fed washed dairy manure solids that have been drenched with aerobically brewed vermicompost tea. When the windrows are 1.05 m (3.5 ft) high, they are harvested using a Bobcat to load a screener with a funnel that does a three-way sort for 0.6 cm (0.25 in) particle size vermicompost, oversized material, and earthworms. Their "Barefoot Soil Earthworm Castings" products are certified organic and legal for retail in Washington, Oregon, and California. These products are sold bagged or in bulk in their on-site Soil Depot and online in the Yelm Worm Store. They also sell earthworms, topsoil, mulch, tea brewers, vermicompost tea, the Worm Wigwam and Worm Factory (medium- and small-scale bins), and other organic amendments. In 2007, Yelm Earth received the Governor's Award for Pollution Prevention and Sustainable Practices.

XII VERMICULTURE IN THE SOUTHERN AND EASTERN UNITED STATES

The highest concentration of earthworm farms east of the Rockies is in states with warmer climates. In the southern United States, Alabama, Arkansas, Louisiana, and Texas are the principal vermiculture locations, with Kentucky, Missouri, and Tennessee also well represented. North Carolina, South Carolina, Georgia, and Florida predominate along the Atlantic coast, but vermiculture is also practiced in Pennsylvania, New York, and the Midwestern and New England states.

In some aspects, vermiculture in the southeastern United States differs from West Coast operations in terms of feedstock and design. Hundreds of rabbit breeders throughout the southeastern United States use earthworms to convert manure dropped from rabbit hutches into vermicomposts. Vermiculture represents a secondary industry in many of these instances. The construction of covered pits and bins, both aboveground and in ground, is fairly common. Earthworm growers speak in terms of creating "bedding" and may use peat moss and topsoil mixed with manure. While manure from herbivorous animals is a common feedstock, pulverized grain feeds are also popular. Poultry mash, alfalfa meal, and other finely ground high-protein feeds are added in thin layers or applied in trenches. Some of the larger-scale vermicomposting operations are highlighted in the following sections.

A RT Solutions LLC, Avon, New York

RT Solutions is currently based in four buildings, although it will be expanding by 800% in 2010. Dairy manure is mixed with spent silage and precomposted for 2 weeks before being fed to earthworms in flow-through raised digesters measuring 36.6 m (120 ft) long by 2.4 m (8 ft) wide by 0.9 m (3 ft) deep (see Chapter 8). About 60 days later, finished vermicompost at the bottom of the reactors is skimmed off by a breaker bar through a mesh into a conveyor trench and transported into another building. There, the vermicompost is screened and sorted into varying grades. About 3.6 ton (4 t) of vermicompost is produced each day and shipped in wholesale quantities to nurseries, landscapers, golf courses, turf fields, farmers, and vineyards. The 2008 Greenhouse Grower of the Year uses RT Solutions's Worm Power certified organic plant growth products. The RT Solutions operation has been featured on the TV History Channel in an episode of *Modern Marvels* titled "Fertilizers of the Future" and in a PBS television episode on "Sustainability in Agriculture."

B WeCare Organics, Pennsylvania

When it was announced that the local landfill was shutting down near Granville Township, Pennsylvania, the wastewater treatment plant investigated alternatives for managing their 9.1 tons (10 t) per week of biosolids and chose vermicomposting. A $500,000 Growing Greener grant was acquired, and an Australian vermicomposting business called Vermitech was selected to vermiprocess the biosolids on-site, beginning in 2005. This was the first vermicomposting operation in the United States to process biosolids commercially (previous ones were pilot projects in other states). Around the same time as Vermitech was commissioning Granville, they contracted with another sewer authority in Waste Hanover Township near Harrisburg, Pennsylvania. That facility is approximately twice as large as the Granville one and was commissioned by Vermitech in 2007. WeCare Organics acquired both vermicomposting operations in 2008.

Currently, each facility is managed differently to accommodate for varying feed and site characteristics; however, the general Vermitech process flow for these two sites was designed as follows. The liquid biosolids are dewatered and mixed with a small amount of wood chips. Vermitech used their own design to construct raised flow-through digesters that are 30.5 m (100 ft) long, 1.5 m (5 ft) wide, and 1.5 m (5 ft) high with 76.2 cm (30 in) deep beds. Once or twice a week, a predetermined amount of dewatered biosolids is fed to the earthworms with a semiautomated feeder and retained for approximately 60–90 days. One layer of old corrugated cardboard is laid on the bottom of the digester, which consists of lengthwise steel supports (not cross-meshed like most digesters). After the old corrugated cardboard has decomposed, and the vermicompost is approximately 50.8 cm (20 in) deep, the vermicompost is harvested from the bottom with a semiautomated patented harvesting system. Generally, a 30% volume reduction is achieved through vermicomposting the biosolids. The harvested vermicompost has a 35–45% moisture content, so it is solar dried on asphalt and then screened prior to distribution. The vermicompost must pass EPA Class A testing before it can be sold on a site-specific basis, primarily to golf courses and turf fields.

C Key Packing Company (KPC), Robbins, North Carolina

A 36.4 ha (90 acre) farm owned by Gilbert Key is home to 3,000 pigs and 1359 kg (3,000 lb) of earthworms. Since August 2005, KPC has been feeding separated pig manure solids to earthworms housed in a wood and metal structure that can be enclosed during the winter. The manure is flushed out of four pig sties, and a machine called a solids separator helps KPC recover about 80% of the solid waste. The manure is fed fresh to earthworms every 24–36 hours, applying about 2.5 cm (1 in) of material to the top of the earthworm beds using a manure spreader attached to a tractor (the troughs are spaced far enough apart to accommodate the tractor wheels). It takes a total of 30 minutes to feed all of the earthworms. The earthworms live in six 53.3 cm (21 in) deep troughs that are 61 m (200 ft) long and 1.2 m (4 ft) wide. Water is misted onto the earthworm beds twice daily for about 1 minute. The facility was designed by Tom Christenberry, owner of Vermicycle Organics, who has been vermicomposting pig manure for more than 35 years. Although Christenberry used to vermicompost with flow-through reactors, he prefers the troughs because the cool underground temperatures provide refuge for the earthworms if the manure heats up. However, harvesting vermicompost from the troughs is more labor-intensive than simply slicing off castings from the bottom of a raised vermicomposting reactor. Workers remove the top 10.2 cm (4 in) of material in the troughs with pitchforks and then run a rototiller through each bed to aerate the vermicompost and help dry it out. After 4 or 5 days, the vermicompost is removed from the troughs and screened to 0.6 cm (0.25 in) particle size using a trommel built by KPC. The vermicompost is sold bagged or in bulk to a variety of markets.

D Windswept Worm Farm, Blue Springs, Missouri

Started in 2001, this is perhaps the only vermicomposting operation in North America that is done in an underground limestone cave. Doug and Becky Halphin think it is the ideal environment for vermicomposting because the humidity and temperatures, 11.6°C –15.6°C (52°F–60°F), are maintained year round. They use a variety of feedstocks—four different types of animal manures, preconsumer food produce waste, recycled newspaper, and sawdust—to increase the variety of microorganisms in their vermicomposting process and finished product. The feedstocks are blended in a pit and then loaded into a 4.6 m^3 (6 yd^3) in-vessel thermophilic composting unit manufactured by Jet Compost in Texas. The Halphins are expanding their operation, so they will soon have a 9.2 m^3 (12 yd^3) in-vessel composter. This mixture is composted for 3–5 days at over 60°C (140°F) to destroy pathogens and weed seeds. Then a skid steer loader is used to load the compost into a gantry feeder on top of continuous-flow vermicomposting reactors. The four reactors, measuring 26.5 m (87 ft) long and 2.4 m (8 ft) wide, were designed and built by the Halphins. Approximately 4.5 million *Eisenia fetida* consume about 54.4 tons (60 t) of organic wastes per month. Vermicompost is harvested from the bottom of the raised beds and screened to particle size 0.3 cm (0.12 in) in a trommel screener. Vermicompost is sold in 6.8 kg (15 lb) and 2.3 kg (5 lb) bags labeled Nature's Own Worm Castings to garden centers in five states. They also sell European night crawlers *Dendrobaena veneta*, (also known as *Eisenia hortensis*) and green glow worms for fishing, as well as red wigglers (*E. fetida*) for vermicomposting.

E Rabbit Hill Farm, Corsicana, Texas

Beginning in the early 1980s, Jay and Joanne Mertz started raising rabbits and earthworms for fishing. After a while, they realized that the manure from both animals would be beneficial for soils and plants, and a business was born. They collected waste materials such as coffee hulls, oak leaves, spoiled hay, and seaweed residues and mixed them together for earthworm feedstock. The Mertzs raised earthworms in shallow wooden bins outdoors and harvested the vermicompost in a rotating trommel screener. After experimenting with a variety of organic materials, they developed two dozen specialty soil-amendment products. Their bagged mixtures contained such combinations as earthworm casts and rabbit manure, lava sand, alfalfa meal, rock phosphate, greensand, kelp, and humate. The Mertzs recently retired and sold their company to Maestro-Gro, a family-owned and -operated fertilizer and plant-food manufacturer located in Hamilton, Texas. Maestro-Gro markets Rabbit Hill Farm products throughout the United States and several countries.

F Vermicomposting in Canada: Forterra Inc., Puslinch, Ontario

Although currently there appears to be only one large-scale vermicompost producer in Canada, in all likelihood the industry will soon expand. A primary driver is the pesticide ban in Ontario, Quebec, and Halifax, Nova Scotia. A ban on pesticides is also being considered for British Columbia, Prince Edward Island, and New Brunswick. In Ontario, more than 250 products and 80 ingredients are prohibited from sale and cosmetic use. Aimed at lawn and garden use, the restrictions are creating a significant upsurge in the number of companies offering organic lawn care. In turn, that should cause increases in the numbers of vermicomposting and composting companies trying to meet the demand for organic products.

Founded in 2003 under the name Riverdale Amendments, in 2009 Forterra emerged as the largest in the vermicomposting industry in Canada. Vermicomposting takes place in a 2322.6 m^2 (25,000 ft^2) indoor facility in trays that are 2.4 m (8 ft) long, 1.2 m (4 ft) wide, and 25.4 cm (10 in) high. Currently, 1800 trays are stacked nine high in the building, and the number of trays will increase to 2200 in 2010. Proprietary bedding substrate and feedstock are prepared in custom blending equipment designed by Forterra to ensure consistency in their vermicompost products. Rather than processing waste organic materials for a tipping fee, Forterra's focus is on producing a quality end product; thus they must pay for the bedding and feedstock. As it is very labor-intensive to harvest vermicompost from trays, in 2010 the company will automate the process and expects to need only 10% of the current labor requirement. Their products are sold in 9.1 kg (20 lb) and 22.7 kg (50 lb) bags, cubic yard totes, or by the tractor-trailer load in Canada and the United States. Forterra currently sells blended vermicompost products but will also offer 100% vermicompost in 2010. Currently producing 2721.6–3628 tons (3000–4000 t) of screened vermicompost per year, Forterra markets its products for soil enrichment to lawn-care companies, golf courses, greenhouses, vineyards, organic growers, and homeowners.

XIII NORTH AMERICAN INTEREST IN VERMICULTURE AND VERMICOMPOSTING

In 1978, an article appeared in the *Wall Street Journal* concerning securities regulators in several states who were investigating earthworm-breeding pyramid schemes. The story, which many say led to the eventual demise of a then-burgeoning industry, quoted the Arkansas securities commissioner, who said, "Millions of dollars are being ripped off from the public across the country because of the flimflam in earthworm growing arrangements" (Machalaba 1978). The article sounded the death knell for earthworm growers throughout North American. The collapse of Ronald Gaddie's network of hundreds of growers in association with North American Bait Farms, Ontario, California, which has been mentioned earlier in this chapter, is indicative of the nadir the earthworm industry was later to reach.

Perhaps ironically, R. Hartenstein and his associates at about the same time in the late 1970s began researching the use of earthworms in the stabilization of sewage sludge at the State University of New York in Syracuse (Edwards and Neuhauser 1988). In 1980, the Workshop on the Role of Earthworms in the Stabilization of Organic Residues was held at Western Michigan University. Here, a balance of background and opinion was sought in a forum of 22 academic scientists, 2 public-sector representatives, and 14 entrepreneurs (Carmody 1981).

As scientific efforts grew, popular interest in earthworms and vermicomposting did not resume until the 1990s, and then perhaps due to a confluence of factors. Mary Appelhof's *Worms Eat My Garbage* (1982/1997) grew in sales to reach 15,000 copies per year, as *Newsweek* reported (Rogers and Annin 1996). While sales were moderate when the book was first published, popular interest in vermicomposting began to escalate noticeably in the 1990s. *Worm Digest*, a quarterly periodical published in Eugene, Oregon, launched its first issue in the summer of 1993. Within 3 years, it grew in size to printing 10,000 issues of its 32-page newsprint journal. In July 1994, the Fifth International Symposium on Earthworm Ecology (ISEE 5) was held in Columbus, Ohio, organized by Clive Edwards. More than 220 scientists from 38 countries gathered to hear 165 research presentations (Edwards 1998). *BioCycle: The Journal of Composting and Organics Recycling* occasionally published articles about vermicomposting in the early 1990s. But by October 1994, its cover story proclaimed, "New Horizons for Commercial Vermiculture" (Riggle and Holmes 1994). Since then, although its chief focus centers on traditional composting throughout the United States and internationally, articles on vermicomposting appear more regularly. *BioCycle* conferences also feature presentations by vermicomposting experts. In 1995, the bimonthly publication *Casting Call* was introduced by Peter Bogdanov and was eventually sent to subscribers in almost every state and over 19 countries during its 10-year production run.

In 1996, a cadre of five individuals in southern California formed a short-lived International Worm Growers Association, a nonprofit corporation that successfully held a well-attended Worm Summit and published one newsletter before its early demise. The urge to establish a Global Vermiculture Association (GVA) was passionately discussed on Internet vermicomposting forums, later to be scrapped by a steering committee. In spite of the lack of an organized, formal association, popular interest in vermicomposting can be assessed by visiting the numerous Internet Web sites and forums dedicated to this subject. *U.S. News and World Report* in a 1997 article pointed to the resurgence of interest in vermicomposting: "Vermiculture ventures, the biggest of which involve 50 million worms chowing down on almost 90 tons of waste per week, have boomed over the past few years" (Koerner 1997, p. 53). From 1999 to 2003, Bogdanov coordinated 2-day seminars on the West Coast on commercial vermicomposting. In 2000, Rhonda Sherman coordinated the first seminar/workshop on large-scale vermicomposting on the East Coast and has continued to offer it annually since then. Her 10th Annual Vermiculture Conference in 2010 attracted an international audience of 116 people and featured 10 speakers. Interest in vermicomposting has continued to grow over the past decade, and now it is commonplace to read articles about it in print and online news.

XIV CONCLUSIONS

For vermicomposting in North America to assume a place of commercial importance on a par with the traditional composting industry and its efforts in both stabilizing organic residues and providing a worthwhile soil amendment, several issues merit attention.

1. Vermicomposting facilities, to be successful in the long run, must have a waste-management focus in which vermicomposting is presented as a viable alternative in managing organic residues. If vermicomposting, as has been shown, is a low-odor process that speeds the rate of organic-residue stabilization, then it must merit attention from the appropriate levels within the solid-waste-management sector. If it is demonstrated to be a viable alternative in organic-waste management, it deserves not only funding for pilot projects but also longer-term support through tipping fees or other funding mechanisms within municipal solid waste management.
2. Vermicomposting facilities, to enjoy long-term success, should focus greater attention on marketing vermicomposting as well as vermicomposts. Here, the income potential may be the greatest, although there is much work to do. Plant growth trials conducted at The Ohio State University (Subler et al. 1998; Arancon et al. 2007) clearly demonstrate the horticultural value of vermicompost in container media and fields (see Chapters 9 and 10). Compost marketers now have a limited but useful arsenal of good scientific data to support their efforts in promoting the advantages of vermicompost.
3. Evidence that as little as 5% (by volume) vermicompost in a blend of potting media produces substantive plant growth increases suggests that vermicompost may have a future indelible effect on both the traditional composting industry and soil blenders (see Chapters 9 and 10). Where compost marketers may be looking for an edge to distinguish their product from the competition, and where soil blenders may be looking to produce a value-added product, the addition of vermicompost to their products may provide the means to secure a greater fraction of the marketplace, particularly when there is evidence from The Ohio State University that mixtures of 80% compost and 20% vermicompost cost much less and perform better than 100% vermicompost. This may increase the economic attractions of vermicomposting.
4. Composting facilities, to maintain or increase their market share, may consider on-site vermicomposting on a limited scale. Since nearly all composting facilities have some of the available resources in place to practice vermicomposting (feedstocks, equipment, labor, utilities, permitting, etc.), all that would remain would be to inoculate earthworms into a portion of the composted material and to manage key process variables in a vermicomposting system. Such a foray into limited-scale vermicomposting might be arranged through a joint partnership with a vermiculture or vermicomposting facility nearby, as has been reported (Bogdanov 1997h). In the future, existing composting facilities may be the most

likely candidates to advance the development of large-scale vermicomposting in North America.

5. Indoor vermicomposting systems for institutional and municipal-scale organic-waste management offer the greatest promise as a future growth industry. Due to the necessity of maintaining temperature and moisture requirements in vermicomposting, indoor or covered systems are a necessity in most regions of North America, considering the seasonal variations in weather. On-site vermicomposting of food and other organic residuals makes economic sense for institutions such as prisons, military bases, hospitals, schools, nursing homes, and virtually any business or institution where organic residuals are generated (see Chapter 24). Reductions in collection costs and hauling and tipping fees would help in amortizing the costs of an indoor vermicomposting system, as would the decreased costs for soil amendments otherwise purchased for on-site landscaping. The continuous-flow reactor, in use on both the East and West coasts of the United States, has been demonstrated to be effective in handling municipally generated wastes.
6. Increasing research efforts on the ability of vermicomposting to reduce human pathogens may help establish vermicomposting as a valuable and preferable means of processing biosolids and other pathogen-contaminated organic wastes.
7. Confirmation of the biological mechanisms present in vermicompost that may be responsible for the greatly increased growth and health of many plants is a challenge facing researchers in horticulture and crop science. The current hypothesis from The Ohio State University is that plant growth regulators are produced by the increased microbial activity in vermicomposts. These compounds are relatively unstable since they are very soluble and break down in daylight but become absorbed onto the humic fraction in vermicomposts and are released slowly to promote plant growth independently of nutrient availability (Atiyeh et al. 2002). Future publication of the results of university-conducted plant growth trials that investigate the use of various types of vermicompost (e.g., from pig manure, food wastes, etc.) will serve to promote wider use for this product in agriculture and horticulture.

Finally, there may be reason to hope that a concomitant decrease in dubious investment schemes will result from an increase in the dissemination of factual information about the realities of vermicomposting in organic-waste management.

REFERENCES

Appelhof, M. 1982/1997 1st and 2nd Ed. *Worms Eat My Garbage*. Flower Press, Kalamazoo, MI.

Arancon, N. Q., Edwards, C. A., Babenko, A., Cannon, J., Galvis, P., and Metzger, J. D. 2007. "Influence of Vermicomposts Produced by Earthworms and Microorganisms from Cattle Manure, Food Waste, on Flowering of Petunias in the Greenhouse." *Appl. Soil Ecol.* 39, 91–99.

Atiyeh, R. M, Lee, S., Edwards, C. A., Arancon, N. Q., and Metzger, J. D. 2002. "The Influence of Humic Acids Derived from Earthworm Processed Organic Wastes on Plant-Growth." *Biores. Technol.* 84, 7–14.

Barbour, J. 1996. "Hard-Working Worms: Turning Green Horticulture Waste into Greenbacks." *The Register* (Orange County, CA), March 1.

Barrett, T. J. 1947. *Harnessing the Earthworm*. Wedgewood Press, Boston, MA.

Bogdanov, P. 1996a. Canyon Can't Make Enough Castings. *Casting Call*, August, 2–3.

Bogdanov, P. 1996b. *Commercial Vermiculture: How to Build a Thriving Business in Redworms*. Petros, Merlin, OR.

Bogdanov, P. 1997a. California Proposes Changes to Compost Regulations. *Casting Call*, June, 1–4.

Bogdanov, P. 1997b. Casting Call Interview: Jack Chambers, Sonoma Valley Worm Farm, Sonoma, CA. *Casting Call*, April, 8–10.

Bogdanov, P. 1997d. Interview with Barry Meijer, Pacific Southwest Farms, Ontario, CA. *Casting Call*, June, 4–8.

Bogdanov, P. 1997e. Interview with Jim Jensen. *Casting Call*, August, 5–8.

Bogdanov, P. 1997h. West Coast Vermicomposting Operations: Low to High Tech Systems. *Casting Call*, April, 1–5.

Bogdanov, P. 1998a. Interview with Al Cardoza, Rainbow Worm Farm. *Casting Call*, February, 4–8.

Bogdanov, P. 1999. Interview with Mario Travalini, American Resource Recovery. *Casting Call* 3 (6).

California Integrated Waste Management Board. 1997a. Initial Statement of Reasons. Title 14, Division 7, Chapter 3.1, Article 1, Section 17852, Subdivision (II). May 13.

California Integrated Waste Management Board. 1997b. Notice of Proposed Rulemaking. Title 14, Division 7, Chapter 3.1 Composting Operations Regulatory Requirements, Article 1, Section 17852 (u).

Carmody, F. 1981. Workshop Summary. In *Workshop on the Role of Earthworms in the Stabilization of Organic Residues*, Vol. 1, ed. Mary Appelhof, xxi–xxv. Beach Leaf Press, Kalamazoo, MI.

Earthworm Buyer's Guide 1998–99, 2008–09: A Directory of Earthworm Hatcheries in the U.S.A. and Canada. Shields Publications, Eagle River, WI.

Edwards, C. A. (ed.). 1998. *Earthworm Ecology*. CRC Press, Boca Raton, FL.

Edwards, C. A., and E. F. Neuhauser (eds.). 1988. *Earthworms in Waste and Environmental Management*. SPB Academic Publishing, the Hague, the Netherlands.

Gaddie, R. E., and D. E. Douglas. 1975. *Earthworms for Ecology and Profit*. Vol. 1, *Scientific Earthworm Farming*. Bookworm, Ontario, CA.

Gaddie, R. E., and D. E. Douglas. 1977. *Earthworms for Ecology and Profit. Vol. 2, Earthworms and the Ecology*. Bookworm, Ontario, CA.

Harris, G. D., W. L. Platt, and B. C. Price. 1990. Vermicomposting in a Rural Community. *BioCycle*, January, 48–51.

Koerner, B. I. 1997. It's All Business for Worms; Yuck! Vermiculture Is Booming Once Again. *U.S. News and World Report*, September 22, p. 53.

Machalaba, D. 1978. Many States Worry about Using Worms to Lure Investors. *Wall Street Journal*, June 5.

Rogers, A., and P. Annin. 1996. Kitchen Help: Wrigglers under the Sink. *Newsweek*, February 12, p. 76.

Slocum, K. 2005. Pacific Landscape Supply (Formerly American Resource Recovery). *Worm Digest*.

Subler, S., C. A. Edwards, and J. Metzger. 1998. Comparing Vermicomposts and Composts. *BioCycle*, July, 63–66.

CHAPTER 24

Vermicomposting for Businesses and Institutions

Rhonda L. Sherman

CONTENTS

I Introduction 392
II Vermicomposting by Businesses 394
  A Metro Hall, Toronto, Canada 394
  B National Institutes of Environmental & Health Sciences (NIEHS) 394
  C Len Foote Hike Inn 395
  D Seattle Kingdome 395
  E IKEA 395
III Vermicomposting Institutions 396
IV Vermicomposting in Colleges and Universities 397
  A The Evergreen State College 397
  B Southern Illinois University Carbondale 398
  C University of Oregon 398
  D University of Massachusetts Lowell 399
V Vermicomposting in Hospitals 399
  A Medical University of South Carolina (MUSC) 399
  B Huron Hospital 400
VI Vermicomposting in Prisons 400
  A Washington State 401
  B Cedar Creek Corrections Center (CCCC) 401
  C Stafford Creek Correctional Facility (SCCC) 402
  D Brown Creek Correctional Facility 402
  E Sampson Correctional Institution (SCI) 403
  F Broad River Correctional Institution 403

VII Vermicomposting in Schools....404
A Birch Lane Elementary School (California)....404
B Waterville Elementary School (Oregon)....404
C Cinnabar School (California)....405
D Louisiana Schnell Elementary School (Placerville, California)....405
E Lathrop E. Smith Environmental Education Center....405
VIII Vermicomposting in Military Bases....406
A Wright-Patterson Air Force Base (WPARB)....406
IX Conclusions....406
References....407

I INTRODUCTION

Businesses and institutions are interested in adopting sustainability practices while at the same time cutting costs. Vermicomposting is a viable method of diverting the organic portion of their waste streams, avoiding the costs of disposal and converting it to a value-added product. Over the past decade, an increasing number of businesses and institutions have begun using vermicomposting to manage their organic wastes. Organic materials are being vermicomposted primarily on-site, but in some cases they are transported to a centralized facility. Vermicomposting helps businesses and institutions save fees for solid waste collection and disposal, water usage costs, and wastewater disposal expenses.

Organic materials continue to be the largest component of mixed solid waste produced in the United States. Paper and paperboard are 31% of the waste stream, and yard trimmings and food scraps account for 26% of the total of 250 million tons of mixed solid waste generated in 2008 (U.S. Environmental Protection Agency [U.S. EPA] 2009b). The U.S. EPA estimated in 2008 that 35–45% of the organic wastes generated in the United States were produced by schools, businesses, and institutions (U.S. EPA 2009b). The principal organic components of waste streams for businesses and institutions are food residuals, paper, and yard debris. Some agencies already have their paper and yard waste collected for recycling (by their municipality or local businesses), and they want to focus on food-waste recycling. Others desire to recycle all three waste streams on-site.

Businesses that generate food waste include restaurants, grocery stores and chains, hotels, food processors, nursing homes, wholesale food outlets and farmers markets, shopping malls, resorts, retirement communities, and offices with dining facilities. For some businesses, such as restaurants, the majority of their waste stream is organic. Institutions generating food waste include hospitals, schools, universities and colleges, prisons, military bases, long-term care facilities, and government centers.

One of the most commonly used midscale vermicomposting units on the market is the Worm Wigwam, which is manufactured by Sustainable Agricultural Technologies, Inc. (formerly EPM Inc.) of Cottage Grove, Oregon. Hundreds of these systems have been used since 1995 in schools, restaurants, universities, prisons, and office buildings throughout North America. The recycled plastic

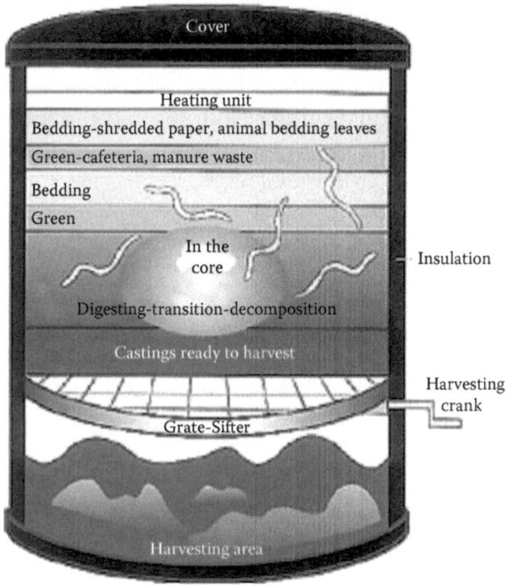

Figure 24.1 Worm Wigwam.

shell with insulated outside walls and a lid stands 0.9 m (3 ft) high and is 0.9 m (3 ft) in diameter (Figure 24.1). Inside, a galvanized steel grate mesh separates the 45.7 cm (18 in) high vermiprocessing area from the 40.6 cm (16 in) high harvest chamber. Each opening on the grate is 10 cm (4 in) by 5 cm (2 in). On top of the mesh is a breaker bar attached to a hand-operated crank on the outside of the bin. It is used to trim off about an inch of vermicompost, causing it to fall into the chamber below. The unit uses a 110-volt heater with a built-in thermostat. To begin vermicomposting in the Worm Wigwam, two sheets of newspaper are laid on the grate and 15 cm (6 in) of high-quality moist compost is put down on top of it as bedding. Earthworms are placed gently on the compost and allowed to move into the bedding on their own as they seek darkness. After a few days, food waste may be left on top of the compost and covered thoroughly with a few inches of moistened shredded paper. Up to seven tons per year of organic materials can be processed in one Worm Wigwam.

Businesses and institutions with larger organic waste streams have purchased commercial vermicomposting systems that range from 2.4 m (8 ft) to 14.4 m (48 ft) long. In the 1990s, Original Vermitech of Ontario, Canada, was the leading manufacturer of smaller flow-through reactors, but they are no longer in business. Vermitechnology Unlimited of Orange Lake, Florida, has installed insulated bins at prisons, hospitals, and businesses. Sustainable Agricultural Technologies manufactures two types of flow-through reactors that have significantly larger capacities than the Worm Wigwam. An institutional-size Model 5–6 described later in this chapter is owned by Sampson Correctional Institution. Their industrial-sized systems are 1.5 m (5 ft) wide and range in length from 1.4 m–14.4 m (8 ft–48 ft).

Some businesses and institutions have designed and built their own vermicomposting systems. Some are bins on the ground, and others are stacked in pallet racks; some are indoors, and others are outside. The following case studies describe how businesses and institutions are conserving resources and reducing their costs by vermicomposting their food residuals and other organic materials.

II VERMICOMPOSTING BY BUSINESSES

A Metro Hall, Toronto, Canada

This 28-story downtown office building with 2000 tenants houses administrative offices for the City of Toronto. In March 1996, they began vermicomposting food-preparation and postconsumer food waste from Metro's 48 eateries in addition to paper towels from restrooms. The organic materials were loaded manually into a machine that shredded and mixed the food and paper and then blew it out at a controlled rate into a vermicomposting unit. The feedstock was leveled out with a rake until a uniform layer half an inch to an inch thick covered the each earthworm bed. The vermicomposting unit was the first of its size and type in the world when it was built in March 1996. The size of a standard parking lot space, the units measured 4.2 m (14 ft) long by 2.4 m (8 ft) wide by 1.5 m (5 ft) high. The enclosed flow-through reactor was designed and built by Vermitech Systems, a Canadian company that closed its doors earlier in this decade. A thermostatically controlled ventilation system maintained room temperature in the unit. To control the temperature and aerate the unit, a fan periodically blew air into a lower sealed chamber, and the air was forced upward through the steel grate earthworm bed with spaces measuring 10 cm (4 in) by 5 cm (2 in). Approximately 100 kg (200 lbs) of earthworms in the unit were fed vegetables, fruit, bread, pastas, tea bags, coffee grounds, and paper filters. In 2004, more than 2.7 ton (3 t) of food residuals and 0.75 ton (0.68 t) of paper towels were consumed by the earthworms. A hydraulic, motor-driven breaker bar raked vermicompost on the raised bed mesh into the enclosed chamber below. The vermicompost was air dried and used on the building's flowerbeds. Metro Hall was added to the City of Toronto's Green Bin organics-collection composting program in 2004, and, shortly thereafter, the vermicomposting program was discontinued (Maher 2009).

B National Institutes of Environmental & Health Sciences (NIEHS)

In December 1996, NIEHS became the first federal agency to vermicompost its dining waste. An employee who obtained support from the director, cafeteria staff, and other key individuals at NIEHS spearheaded the program. Located in Research Triangle Park, North Carolina, they purchased two Worm Wigwams for $375 each and placed them outside under the shade of a tree (see the description of these bins earlier in this chapter). A hole was drilled in the side of the building to run an electrical line to the bins to operate their heaters. NIEHS purchased 10 kg

(20 lb) of *Eisenia fetida* earthworms and put 10,000 earthworms in each bin. Every workday, an average of 4.5–9.0 kg (10–20 lb) of cafeteria food-preparation scraps are added to the bins. A squad of volunteers dubbed Worm Wizards takes turns getting the food buckets from the cafeteria and carrying them out to the earthworm bins. After feeding the earthworms, the volunteers cover the food waste with shredded office paper, finely chopped wood waste from clean animal bedding, and leaves. The Worm Wizards harvest about 25 kg (50 lb) of vermicompost every 2–3 weeks. Vermicompost is placed in a large drum (for ground keepers to use for planters and flowers) and in smaller containers (for employees and schoolchildren; R. Sloane, e-mail to author, September 15, 1997).

C Len Foote Hike Inn

Located near Dawsonville, Georgia, this inn is a unique state park facility that is accessible only by foot over a 5-mile hiking trail. The Leadership in Energy and Environmental Design (LEED) inn was built in 1998 with sustainability in mind and features vermicomposting, solar panels, composting toilets, and rainwater harvesting. Breakfast and dinner are served to guests, and the food-preparation scraps and plate scrapings are fed to *Eisenia fetida* earthworms. In addition to food residuals, the inn vermicomposts shredded cardboard and office paper, discarded cotton and wool clothing, and even cotton mop heads. There are 10 large earthworm bins in their basement that process over 680 kg (1500 lb) of organic materials per year. Vermicompost is harvested regularly from the bins and incorporated into the inn's organic garden and soil around native plants. The inn also teaches guests how to set up their own earthworm bins (Hike Inn at Amicalola Falls.).

D Seattle Kingdome

The home of the Mariners and Seahawks embarked on a comprehensive sustainability program in the early 1990s. They decided to expand their recycling program by vermicomposting food-preparation waste, which consisted of mostly salad scraps. Twelve earthworm bins were constructed; four were made of wood and the rest were created by laying 208.2 L (55 gal) barrels sideways and cutting them in half. Leaves and shredded newspaper were used as bedding, and a total of approximately 18,000 earthworms were divided between the 12 bins. The earthworms consumed about 25 kg (50 lb) of food scraps per week (30% of the stadium's total food waste). The vermicomposting program worked smoothly, and they had no problems with odor or pests. When vermicompost was harvested every few months, it was used on the Kingdome's flowerbeds (Sherman, 1997).

E IKEA

The socially responsible Swedish furniture company IKEA requires its stores to minimize their environmental impact by continually increasing their waste-reduction rates. Every IKEA store has a restaurant for employees and customers, so food residuals are part of their waste stream. The IKEA store in Schaumburg, Illinois, decided to try vermicomposting to help reach their 90% recycling goal. They obtained a grant from the Illinois Department of Commerce and Economic Opportunity and hired a vermicomposting company called New Horizon Organics (see Chapter 23) to develop a pilot program. Limited space was a critical issue for this IKEA store, so they had to be creative about where to locate the vermicomposting bins. New Horizon designed and built four flow-through vermicomposting reactors and housed them in two trucks that could be moved in and out of the loading docks to make way for shipments. Each earthworm bin was 6.0 to 6.6 m (20–22 ft) long and 0.9 m (3 ft) wide with raised beds made of tubular steel. The beds were stocked with *Perionyx excavatus* earthworms. A winch attached to each unit pulled a breaker bar across the bottom of each reactor to loosen about an inch of vermicompost, causing it to drop into drawers on wheels under the beds. The trucks contained fans and heating and air-conditioning units. A regular house door was installed on each truck for easy access. Each truck also had a grinder/mixer to shred IKEA's cafeteria food waste and office paper. However, the pilot program experienced several problems. They struggled with inadequate ventilation and excess moisture in the reactors. Grinding capacity was limited due to their required 110-volt power supply. Adequate markets were not secured for the vermicompost, and coupled with low landfill-disposal fees, the IKEA store could not justify continuing its vermicomposting program (Saskatchewan Waste Reduction Council 2005).

III VERMICOMPOSTING INSTITUTIONS

Since the 1970s, various institutions (primarily schools and universities) have implemented vermicomposting programs. Those numbers have been growing over the past three decades, particularly in the twenty-first century with increased awareness of scarce resources. A unique program has three institutions working together with a city to convert food waste beneficial by-products. The City of Middletown, Connecticut, wanted to increase their waste-diversion rate by developing an institutional composting program. In 2002, they obtained a grant from the Connecticut Department of Environmental Protection for $36,000. The city partnered with the Alternative Incarceration Center (AIC) and Connecticut Valley Hospital (CVH) to vermicompost food waste from the kitchen and cafeteria of a homeless shelter called the Eddy Center, which is located on the grounds of CVH. The city administered the grant, helped with site preparation, and handled publicity. AIC trained workers, oversaw daily operations, and served as a liaison with CVH. Based on waste audits, they initially anticipated that 498 kg (1,095 lb) of food waste would be diverted per year. A heated and ventilated greenhouse was erected to house the vermicomposting operation. They built four wooden earthworm bins, each measuring 2.4 m (8 ft) long by 4.8 m (5.3 yd) wide by 30 cm (1 ft) high, and stacked them two high on pallet racks. About 34 kg (70 lb) of *E. fetida* (red wigglers) were stocked in the earthworm bins. Eddy Center workers put kitchen preparation wastes in 19 L (5 gal) containers with step-on lids, and AIC clients transported them to the greenhouse. The food scraps were processed through a small grinder and then spread in thin layers on the earthworm beds (American Recycler 2005).

They fed the earthworms an average of 17.8 kg (39 lb) of food wastes per week from the Eddy Center. This amount was less than expected due to poor participation by kitchen staff. Hence, they expanded the program to collect food waste from restaurants, a firehouse, and a children's home. This increased the food-waste collection by up to 45.5 kg (100 lb) per week. However, the project was plagued with transportation problems, so currently just one earthworm bin is receiving about 9.1 kg (20 lb) of food waste per week while the program is being revamped. A screener has been acquired that will enable them to sell the vermicompost produced. Two Earth Tubs (in-vessel composting units) are being hooked up so they can process more food waste. Plans are being put together to collect food waste from area cafeterias to compost in the four earthworm bins and the Earth Tubs (Connecticut Department of Environmental Protection 2005).

IV VERMICOMPOSTING IN COLLEGES AND UNIVERSITIES

There is a huge drive on campuses across the nation to "go green" and try to reduce wastes. A number of organizations have sprung up to assist their efforts, including the Campus Sustainability Planning Network, SustainableCampus.org, and the National Recycling Coalition's University and College Recycling Council. In addition, many states and regions have university sustainability initiatives that assist campuses with going green. Since the late 1980s to early 1990s, many campuses have composted their leaves and tree trimmings. Now that colleges are exploring a wider range of sustainability initiatives, more of them are choosing to compost or vermicompost their food waste.

A The Evergreen State College

Located in Olympia, Washington, Evergreen has been vermicomposting at their Organic Farm since the early 1990s. Food waste from dining services and a local restaurant used to be vermicomposted in wooden bins a total of 15.2 m (50 ft) in length. Beginning in 2008, Evergreen changed their campus organics management to two streams going to separate locations. Most campus food waste from dining halls and student housing, including compostable dinnerware, goes to Silver Springs Organics, a commercial composting facility located a short distance away in Rainier, Washington (The Evergreen State College 2000). Approximately 1817 L (480 gal) of food residuals collected each week from dining services are taken to Evergreen's Organic Farm Composting Facility. First, it is composted in an aerated, in-vessel thermophilic composting unit and then finished in a vermicomposting system. The system consists of four 2.4 m (8 ft) modular flow-through vermicomposting reactors that are connected and total 9.6 m (32 ft) long by 1.5 m (5 ft) wide. Up to 473.2 L (125 gal) of compost is added to the vermicomposting reactors weekly.

Vermicompost is released through the mesh bed by breaker bars into trays below and used on the Organic Farm and Community Gardens (Center for Ecological Living & Learning).

B Southern Illinois University Carbondale

Illinois first vermicomposting facility at a university has been operating in Carbondale since August 2006. It was funded initially by $217,000 in grants from the Illinois Department of Commerce and Economic Opportunity, the Illinois Department of Agriculture, and Jackson County Health Department (Clark 2006). The campus Plant & Service Operations and College of Agricultural Sciences research faculty administers the Vermicomposting Center. Each year, about 80,000 kg (170,000 lb) of food waste from three residence hall dining facilities is ground up and fed to approximately three million earthworms. The food is put through a pulper that grinds it up and extracts liquids from it, and then it is dumped into the earthworm bins and raked out evenly. The raised bins are flow-through reactors with mesh grates that were manufactured by Sustainable Agricultural Technologies. When the cafeterias are closed, the earthworms are fed shredded confidential paper. The center is housed in a 521 m^2 (4692 ft^2) pole barn on the campus farm that is heated with used oil from university vehicles during winter. The Department of Plant, Soil and Agricultural Systems is using the vermicompost in field tests for vegetable and turf-grass growth. In 2009, vermicompost was applied to an organic garden, and the harvested vegetables were served to the university students (Draine 2009).

C University of Oregon

For over 29 years, this university in Eugene, Oregon, has had an organic gardening program called the Urban Farm. In 1994, staff from the Urban Farm and Physical Plant Grounds Department initiated an unusual vermicomposting experiment. Instead of buying earthworms and setting up earthworm bins, they spread organic wastes on the ground to be eaten by earthworms occurring naturally in the soil. They began by clearing two areas measuring 54 m by 2.4 m (180 ft by 8 ft) of rocks and plants. On this half-acre site, they planted fruit trees and blueberry bushes in rows at 3.0 m (10 ft) intervals. About 114 kg (250 lb) of preconsumer food waste was collected from a nearby restaurant every few days and spread in a thin layer on the ground between the fruit trees. The food was applied in rotation so that each area was fed once every 14 days. The food waste was covered with a sprinkling of rock dust from a local quarry and then a layer of yard debris (grass, leaves, and shredded branches). Campus grounds crews, the city of Eugene, and private landscapers provided the yard waste. Within six months, the site was teaming with red earthworms and burrowing earthworms. The project was considered a success because in addition to diverting thousands of pounds of food waste and several tons of yard debris, there were no pests or offensive odors, and the plants on the site thrived (Sherman 1997).

D University of Massachusetts Lowell

In November 2009, University of Massachusetts Lowell announced it was launching Lowell Loam Ltd. (LLL), a university business that will vermicompost food residuals, leaves, and cardboard generated by the surrounding community and sell the vermicompost. They will also sell vermicomposting kits and earthworms for fishing bait. Beginning with food scraps generated in its own dining halls, LLL plans to also vermicompost food waste from area restaurants, prisons, hospitals, supermarkets, and schools. An enclosed educational and demonstration site, which will be built in a campus parking lot, will house the vermicomposting operation. Plans for expansion call for setting up additional vermicomposting sites on local farms. LLL participants anticipate creating six full-time jobs and producing 907.2 ton (1000 t) of vermicompost per year to sell to farmers, municipalities, and garden stores. The city of Lowell's mayor, Rita Mercier, supports the vermicomposting project, saying it will reduce trash collection expenses and help the environment (Hughes 2007).

V VERMICOMPOSTING IN HOSPITALS

There is a nationwide trend for hospitals to adopt sustainable, ecofriendly practices. Approximately 70% of the average hospital's waste stream is organic; 53% is paper and 17% is food waste and other organics (U.S. EPA 2002). Although it is common for hospitals to have recycling programs, and food waste is the second-highest component of their waste streams, very few of them have vermicomposting or composting programs (Health Care Without Harm 2001). The first hospital project that implemented vermicomposting began over 10 years ago.

A Medical University of South Carolina (MUSC)

In the late 1990s, the recycling coordinator at this Charleston, South Carolina, hospital decided to increase its waste-diversion rate by implementing a food-residuals recycling program. Constrained by limited space in its urban location, they opted for vermicomposting. In 1999, MUSC obtained funding from South Carolina's Department of Health and Environmental Control and that department's Energy Office, in addition to the Sustainable Universities Initiative. They used $25,000 to purchase a flow-through vermicomposting reactor, shredder, and earthworms from Vermitech Systems Incorporated (now defunct) of Toronto, Canada. Another $31,000 was spent to house the reactor in a 7.2 m (24 ft) by 5.4 m (18 ft) ventilated building with a sloping acrylic floor that could be easily cleaned. The vermicomposting system measured 5.4 m (18 ft) long by 2.1 m (7 ft) wide by 1.5 m (5 ft) high. Inside the system were two 2.4 m (8 ft) modular beds and a 60 cm (2 ft) section for hydraulic equipment and an air conditioner.

Preconsumer food waste was placed in a wheeled 170.3 L (45 gal) container by cafeteria kitchen staff. The container was picked up daily by a recycling team member and taken to the earthworm building for weighing and processing. The weight varied according to the types of food prepared, but on average they collected approximately 60 kg (115 lb) per day. The food waste was emptied into a Vermitech shredder/mixer, and cardboard was ground up with it to achieve the proper moisture level. The mixture was loaded into the flow-through reactor using a conveyor belt attached to the shredder. It took one hour each day to collect the food waste, weigh and shred it, and feed it to the earthworms. Another 30 minutes a week was spent inspecting the reactor environment and the vitality of the earthworms. Vermicompost was harvested from the reactor every 2–4 weeks using a hydraulically operated breaker bar to release it to the floor below. A broom and squeegee were used to collect the vermicompost, which totaled about 230 kg (500 lb) per month. MUSC's grounds department used the vermicompost as a soil conditioner and found that it reduced their need for commercial inorganic fertilizers to some extent. Currently, vermicomposting is on hold while the cafeteria space is being renovated, but it will resume when the remodeling is completed. The shredder broke and cannot be replaced, so MUSC's recycling coordinator is planning to precompost the food waste in outdoor static pile bins next to the earthworm building. The compost piles will be covered with yard waste daily.

B Huron Hospital

Part of the Cleveland Clinic, this hospital serves a poor, unhealthy community in the inner city of East Cleveland, Ohio. The hospital serves 400 meals to patients and 300 meals to customers daily. In 2005, a unique sustainability program was launched that encompasses the surrounding community. The hospital reduced waste by switching from Styrofoam disposables to compostable dinnerware and reusable china. When employees and customers finish their meals in the cafeteria, they separate compostables, recyclables, and trash. A community raised-bed garden measuring 36 m (120 ft) by 7.5 m (24 ft) grows enough vegetables to give to the community and use in the hospital's kitchen for meals and cooking classes. Some of the food waste is added to a "lasagna garden" in layers along with shredded cardboard and office paper. About 600 kg (1200 lb) to 700 kg (1400 lb) per month of preconsumer food waste is taken to Sansai Environmental (see Chapter 23) for vermicomposting at their nearby facility. Each workstation in the hospital kitchen has a 18.9 L (5 gal) bucket or clear plastic container for workers to place food-preparation waste. When full, the buckets are emptied into 166.6 L (44 gal) Rubbermaid wheeled bins located in the freezer. Sansai picks up the food waste weekly from the hospital. Postconsumer food waste used to also go to Sansai, but it contained too many contaminants, so only kitchen preparation waste goes there now.

VI VERMICOMPOSTING IN PRISONS

Correctional facilities and prisons throughout the United States are taking steps to increase sustainability and reduce their costs. In some states, the Department of

Corrections (DOC) is partnering with other state agencies (such as environmental protection) to accomplish these goals.

A Washington State

A unique collaboration is taking place between the state DOC and a local college. The State of Washington's Sustainable Prisons Project is a 2-year partnership between the DOC and Evergreen College. The project was created in July 2008 with $300,000 in funding from the DOC to help the state's prisons implement sustainable practices and train inmates for employment in "green-collar" jobs when they are released (Ulrich & Nadkarni 2008). By adopting sustainable measures, correctional facilities can save significant amounts of money by reducing costs for hauling and disposing of garbage, wastewater treatment, electricity, and water use. The project began with four prisons that represented well Washington's 15 correctional facilities—a mixture of genders, security levels, inmate populations, infrastructures, and ecosystems. The four prisons are Cedar Creek Corrections Center, McNeil Island Corrections Center, Stafford Creek Corrections Center, and Washington Corrections Center for Women. These facilities are reducing food and other types of waste, boosting composting and recycling, and decreasing water, fuel, and energy consumption. The prisons are using the vermicompost and compost they produce to raise organic vegetables in gardens and greenhouses, producing approximately 7,000 kg (15,000 lb) of vegetables per yr and saving the DOC about $17,000 annually (Lewis 2009). Many inmates vie for vermicomposting jobs at prisons because they find it meaningful and rewarding. Former Cedar Creek Correctional Facility inmate Craig Ulrich said, "It gives you a reason to go out and work and appreciate what you're doing rather than mopping the prison floor or something" (Brown 2009).

B Cedar Creek Corrections Center (CCCC)

Evergreen College's partnership with this Corrections Center led to the Sustainable Prisons Project. Among its multifaceted sustainability initiatives, Prison Administrators decided to develop a vermicomposting program that would divert food waste from disposal, improve regional groundwater quality, and produce fertilizers for the prison's vegetable garden. The 400-inmate facility began vermicomposting in 2004. They created a stacking-shelf vermicomposting system using recycled lumber. Six shelves hold three stacked earthworm bins. Each shelf is 0.92 m (3 ft) long by 0.92 m (3 ft) wide by 0.2 m (7.9 in) high. The bins are fed to a depth of 0.15 m (5.9 in). Earthworms are fed approximately 1400 kg (3000 lb) of food residuals per month from the inmate kitchen, staff food scraps, coffee grounds, and garden waste. Vermicompost is harvested every 2–3 months and used in the facility's gardens to grow vegetables that supplement prisoners' daily meals. Prisoners designed and built a hand-cranked sifting machine to separate the earthworms from the vermicompost. It costs an average of $98 per t to haul and dispose of solid waste, so vermicomposting at CCCC is saving close to $2,000 per year. By vermicomposting instead of disposing of food down the drain, CCCC reduced its per capita water consumption by about 25% in recent years.

C Stafford Creek Correctional Facility (SCCC)

This 1800-inmate correctional facility began vermicomposting to reduce waste and improve its organic gardening program. SCCC acquired a vermicomposting system from Washington's Department of Ecology during the summer of 2006. The continuous-flow reactor has electric winches on each end to pull a scraper bar across the surface of the mesh bed to harvest the vermicompost. The reactor is housed in a plastic hoop greenhouse to protect it from the weather. Vegetative food scraps are collected daily in 18.9 L (5 gal) buckets from the preparation room in the prison kitchen. When the vegetative scraps are approximately 12.7 cm (5 in) deep in the buckets, inmates use flat-nose shovels to chop the waste into pieces no larger than 0.8 cm (0.25 in) so the earthworms can consume it faster. About 48 kg (100 lb) of food waste is fed to the earthworms daily. Grass and excess vegetative waste that cannot be fed to the earthworms is composted.

D Brown Creek Correctional Facility

Located in Polkton, North Carolina, this medium-security facility with 852 inmates began an intensive waste-reduction program in October 1997. Within 3 months, the prison reduced its waste disposal by 47%, going from twice-weekly garbage pickup to just once a month and saving $1000 in monthly hauling and landfill tipping fees. The significant savings were achieved through vermicomposting, source reduction, reuse, recycling, and composting. It all started with vermicomposting. The prison superintendent at that time was Rick Jackson, and when he read about San Quentin's vermicomposting program, he decided to try it. Jackson began vermicomposting in August 1997 by asking inmates to build a 1.2 m by 0.6 m (4 ft by 2 ft) earthworm bin and adding 2.4 kg (5 lb) of earthworms. A few months later, a wooden shipping crate was divided into two 30 cm by 30 cm (2 ft by 2 ft) sections, and earthworms from the first bin were added to the new containers. The three bins are housed in a greenhouse used for a therapeutic planting program for chronically mentally ill inmates.

A little later, Jackson recovered steel I-beams that were left over from a highway bridge construction project near the prison and used them to construct an outdoor worm bin measuring 7.8 m (24 ft) long by 7.2 m (20 ft) wide. The earthworm bin was divided into three sections using salvaged concrete blocks. As more earthworms became available, earthworms from the original wooden bin were used to stock the outdoor vermicomposter by slowly adding them to small areas. One correctional officer and one inmate spend 15–30 minutes per week checking the earthworm bins and an additional 30–45 minutes twice a month adding food waste to the bins. Since the earthworm bins could not process all of the organic waste generated at the prison (including lint from their laundry operation, hair clippings from the barber shop, and unusable cotton T-shirts), the prison processed the excess in two Greendrum rotary drum in-vessel thermophilic composters. After acquiring the composting units, the prison fed select waste to the earthworms based on what they consumed most readily—coffee grounds, lettuce, banana peels, and shredded paper. The harvested vermicompost was mixed with soil and used to grow vegetables and flowers (Sherman 1999).

E Sampson Correctional Institution (SCI)

After hearing about vermicomposting at Brown Creek prison, Superintendent Steve Muller decided to try it at SCI in Clinton, which is located in southeastern North Carolina. In May 1999, they acquired two flow-through vermicomposting reactors, dubbed Model 5–6, from EPM, Inc. (now called Sustainable Agricultural Technologies). Resembling 7.3 m (8 yd) dumpsters, the reactors were 2.8 m (8 ft) long and 1.7 m (5 ft) wide, with beds 95 cm (3 ft) deep. The units have a powder-coated steel frame and pressure-treated wooden 30 cm (12 in) by 5 cm (2 in) sides. Each reactor was stocked with 48 kg (100 lb) of earthworms, and they were fed pre- and postconsumer food waste and shredded paper. When the reactors were operating in top form, they could process up to 28 kg (60 lb) of Sampson's organics daily. The prison also had a Worm Wigwam from EPM that measured 0.9 m (3 ft) in diameter and 0.9 m (3 ft) tall; it was used to test different types and amounts of feedstocks. SCI workers were surprised that it processed as much as each of the two larger bins. Initially placed outdoors, the high humidity and temperatures over 37.8°C (100°F) killed the earthworms and caused the reactor lids to disintegrate. Shade cloth covering was erected overhead, and temperatures inside the reactors were lowered 6°C (10°F) by moving fans from the top side of the bins to just above the grate. By reversing the flow of the fans, air was forced out of the units, thus drawing air through the system. Eventually the earthworm bins were housed in a climate-controlled greenhouse. By 2004, the earthworms had been replaced twice. When the summer temperatures soared and the earthworms died again, SCI decided to discontinue the vermicomposting program (Muller 2005).

F Broad River Correctional Institution

In May 1998, this prison established vermicomposting outdoors on its grounds in Columbia, South Carolina. An earthworm bin was installed that was designed and built by Vermitechnology Unlimited of Orange Lake, Florida. The insulated wooden bin had plastic panels inside the bin. It measured 10 m (34 ft) long, 2.1 m (7 ft) wide, and 45 cm (20 in) high. A divider down the middle made for 1 m (33.5 ft) of working space for easy access on either side of the bin. Because moles eat earthworms, a mesh screen was installed underneath the earthworm bin. The sun can be intense in South Carolina, so shade cloth was suspended on metal poles over the bin. Shredded food-preparation scraps were precomposted, and then approximately 400–500 kg (800–1000 lb) of partially composted materials was added to the earthworm bin every 2–3 weeks in 10–15 cm (4–6 in) layers.

Vermicompost was harvested from the earthworm bin two to three times a year. Most of it was used on the prison grounds, but for a while they also sold vermicompost in 2.3 kg (5 lb) to 6.8 kg (15 lb) bags at three local retail stores (Sherman-Huntoon 2000).

VII VERMICOMPOSTING IN SCHOOLS

Thousands of schools throughout the United States have earthworm bins in classrooms for students to learn about food webs, nutrient cycling, earthworm anatomy and behavior, and so on. More and more schools that have implemented recycling programs and set up school gardens are choosing to vermicompost their cafeteria organic waste. At many schools, the students build wooden earthworm bins with help from parent volunteers. Students are in charge of collecting food waste, weighing it, and feeding it to the earthworms. Rather than viewing it as a chore, students are excited about the opportunity to care for the earthworms and enjoy learning about them.

A Birch Lane Elementary School (California)

Before the 2000–2001 school year started, this Davis, California, school purchased four earthworm bins and set them up outdoors near the garden and science classrooms. The continuous-flow bins were the Eliminator 600 EM, and each measured 48 cm (19 in) by 64 cm (25 in) and 46 cm (18 in). The earthworm bins are portable so they can be moved to different classrooms for earthworm activity to be viewed through a Plexiglas panel on one side of each bin. Each intermediate-grade-level class divided students into eight "science teams" that took turns feeding the earthworms. Approximately 13.6–15.9 kg (30–35 lb) of postconsumer food waste was collected each lunch hour in a bucket labeled "WORM FOOD, fruits and vegetables only please." The bucket was weighed, and then a shovel was used to break up the food into small pieces. The food waste was placed into the designated earthworm bin and covered with shredded paper from the school office (Havstad & Wheeler 2002).

B Waterville Elementary School (Oregon)

Fourth-grade students at this Oregon school, with help from a parent volunteer, built five earthworm bins. The bins were set on top of concrete blocks at the edge of the school grounds next to the forest. To test the effectiveness of a variety of beddings, students put shredded newspaper in two bins, pulverized cardboard in another, old corn silage in the fourth bin, and composted horse and cow manure in the fifth bin. Students determined that the manure mixture worked the best, although newspaper also made good bedding. Cardboard got too soggy when it was moistened and became hard to handle. A local farmer provided the class with about 15 kg (30 lb) of earthworms, so students added approximately 3 kg (6 lb) of earthworms to each bin. All of the postconsumer food waste and paper napkins, straw wrappers, and portion cups are added to the earthworm bins. Most schools do not add meat and dairy products to their earthworm bins because they can cause odors and attract vermin, but Waterville decided it was acceptable because the earthworm bins are so far away from the school. Second- through fifth-grade students are assigned on a rotational basis to feed the earthworms. They scrape food waste off plates, weigh it, and haul it out to the earthworm bins. About 7–15 kg (15–30 lb) of food waste is spread each day in a thin layer on the surface of the bedding and covered with additional bedding. This maintains an aerobic environment in the earthworm bins. To harvest the vermicompost, students stop adding organic materials and leave the lid off the bin. This causes the bedding to dry out and lets light into the bin. The earthworms move downward toward the bottom of the bin where it is dark and moist, and vermicompost can be removed from the top of the bin (Sherman 1997).

C Cinnabar School (California)

This Petaluma, California, school generates approximately 15–20 kg (30–40 lb) of food waste daily 4 days a week. Students built eight wooden earthworm bins measuring 1 m (3.3 ft) by 0.9 m (3 ft) and set them outside on top of pallets. They added a variety of beddings, including horse manure, leaves, and wood shavings. 2.3 kg (5 lb) of earthworms were placed in each of the eight bins. On each of the 4 days, two bins of earthworms receive 7.5–10 kg (15–20 lb) of food per bin. One whole week passes before those two earthworm bins receive more food waste. Food waste is added to each bin in a thin layer because the students realized that thicker clumps could heat up and become anaerobic, thus giving off foul odors. Students also discovered that they could prevent pizza crust and bread from getting moldy if they soaked it before adding it to the earthworm bin (Sherman 1997).

D Louisiana Schnell Elementary School (Placerville, California)

This school's 450 students participate in the Schnell School Garden Program, which includes a large vermicomposting component. Lunch and garden scraps are fed to earthworms in the school's huge Worm Motel. School lunch staff monitor students as they separate food for the earthworms. During winter, students check out earthworms from the Worm Motel to study their anatomy. At the end of the school year, vermicompost is harvested from the Worm Motel and added to the gardens (Wessman 2004).

E Lathrop E. Smith Environmental Education Center

Local students' outdoor education trips to the Maryland Center include a tour of the Composting Discovery Garden and the Worm Garden. The Worm Garden consists of earthworm bins that receive fruit and vegetable scraps from the cafeteria. Food for the earthworms is weighed and recorded by center staff. Then students take the compostables outside behind the cafeteria to one of several large earthworm bins.

VIII VERMICOMPOSTING IN MILITARY BASES

A Wright-Patterson Air Force Base (WPARB)

In July 2002, this military base in Dayton, Ohio, began vermicomposting a daily average of 250 kg (500 lb) of vegetable food residuals and newspaper, saving approximately $25 a day on hauling and waste-disposal fees. Approximately 300,000 earthworms consume the feedstocks in a climate-controlled flow-though vermidigester similar to the one installed in Metro Hall in Toronto. Dayton acquired the digester and 250,000 earthworms at no cost from another military base that could no longer use it. Measuring 5 m (16 ft) long, 3.4 m (8 ft) wide, and 1.7 m (5 ft) tall, the digester includes air-conditioning and heating capabilities that enable it to maintain a constant temperature. The digester is housed indoors to protect it from the elements and fluctuating temperatures (Geiselman 2003). Recycling center staff discovered that the earthworms thrive when the system operates at 21°C (70°F). The labor requirement is 1 hour per day for one person to operate the system and care for the earthworms. The vermicompost is used on the base's golf course and grounds to save money on soil amendments and reduce fertilizer runoff into ground and surface waters (The Associated Press 2002).

IX CONCLUSIONS

With the current nationwide interest in going green, businesses and institutions are exploring many ways to reduce their impact on the environment. Usually their first step is to set up a recycling program for bottles, cans, and office paper. However, agencies that generate a lot of food residuals eventually realize that recycling is having a minimal effect on the amount of waste going to dumpsters, and they begin to look for ways to divert food scraps from disposal. Although conservation has become a priority for them, usually the bottom line takes higher priority, so the solution has to make economic sense for the business or institution. Unless a waste-diversion alternative is perceived as able to save money for the agency, it will choose to do something else.

If a business or institution were thinking about vermicomposting their food waste, they would need to determine how much it would cost to implement and maintain it. Then they would need to analyze the potential cost savings, which could include reduced hauling and disposal fees for waste, water usage expenses, and fertilizer or soil-amendment expenditures. As demonstrated in this chapter, many businesses and institutions have discovered that vermicomposting provides a cost-effective and environmentally beneficial solution for managing food waste, paper products, and yard debris on-site.

REFERENCES

American Recycler. 2005. Vermicomposting program continues and expands. AR Publishing. Vol. 8, Issue 12. http://www.americanrecycler.com/1205vermicomposting.shtml.

The Associated Press. 2002. Ohio base worms help in composting. U.S. Composting Council Forum, August 5. http://mailman.cloudnet.com/pipermail.compost/2002=August/009673.html.

Bogdanov, P. 1999. Vermicomposting lunchroom wastes: New manual for schools and small businesses. February, http://www.vermico.com/bookreview.html.

Brown, V. 2009. The ecologist and the prisoners. *Miller-McCune*, February 24. http://www.miller-mccune.com/science_environment/the-ecologist-and-the-prisoners-1006.

Center for Ecological Living & Learning, The Evergreen State College. Compost facility. http://www.evergreen.edu/cell/compost.htm.

Clark, E. 2006. Wormy workers: Vermicomposting turns food waste into environmental enhancer. *Daily Egyptian*, August 27. http://newshound.de.siu.edu/fall06/stories/storyReader.

Connecticut Department of Environmental Protection. 2005. Composting institutional food scraps with worms (vermicomposting). http://www.ct.gov/dep/cwp/view.asp.

Connecticut Department of Environmental Protection. 2005. As the worms churn. *P2View: Pollution Prevention View* 5 (4). http://www.ct.gov/dep/lib/dep/p2/newsletter/p2view-fall05.pdf.

Draine, C. 2009. Composting with worms helps plants stay healthy. *Rapid City Journal*, October 29. http://www.rapidcityjournal.com/lifestyles/article_a821f3ac-c41e-11de-afee-001cc4c03286.html.

Geiselman, B. 2003. Worms eat away at air force's waste. EarthWormDigest.org, February 3. www.wormdigest.org/contentview/188/2

Havstad, C., and L. Wheeler. 2002. Davis Joint Unified School District Food Waste Diversion Project FINAL REPORT. Contract #IWM-C9061E.

Health Care Without Harm. 2001. *Waste Minimization, Segregation & Recycling in Hospitals*. Pub. 4-01. Going Green: A Resource Kit for Pollution Prevention in Health Care. October 15.

Hike Inn at Amicalola Falls. http://www.hike-inn.com.

Hughes, V. 2007. Vermicomposting project goes commercial. *Lowell Sun Online*, July 5. http://www.uml.edu/Media/NewsArticles/print_1_57778.html.

Lewis, C. 2009. Sustainable prisons project begins to take shape. Washington State Department of Corrections, April 22. http://www.doc.wa.gov/news/stories/1009/042209SustainablePrisonsProject.asp.

Maher, D. 2009. Composting with worms at Metro Hall, Toronto. Fact sheet provided by author, November 9.

Saskatchewan Waste Reduction Council. 2005. IKEA takes on vermicomposting. http://www.saskwastereduction.ca/resources/Composting/IKEA-vermi.html.

Sherman, R. 1997. *Worm away your cafeteria food scraps*. Ag-551. North Carolina Cooperative Extension Service.

Sherman, R. 1999. Worms solve problems at North Carolina correctional facility. *Vermicomposting News*, no. 3 (March).

Sherman-Huntoon, R. 2000. Latest developments in mid-to-large-scale vermicomposting. *BioCycle*, November, 51.

Sherman-Huntoon, R. Medium-to-large-scale vermicomposting systems. http://www.p2pays.org/ref/11/10298.pdf.

Ulrich, C., and N. Nadkarni. 2008. Sustainability research and practices in enforced residential institutions: Collaborations of ecologists and prisoners. *Environment, Development and Sustainability*, April 29. http://www.springerlink.com/content/3074v62030633859/.

U.S. Environmental Protection Agency. 2002. *Reusable totes, blue wrap recycling and composting*. Environmental Best Practices for Health Care Facilities. November.

U.S. Environmental Protection Agency. 2009. *Municipal solid waste generation, recycling, and disposal in the United States: Facts and figures for 2008*. EPA-530-F-009-021. http://www.epa.gov/osw/nonhaz/municipal/pubs/msw2008rpt.pdf.

Wessman, K. 2004. *Garden of learning. The worm guide: A vermicomposting guide for teachers*. Publication #560-01-007. Saeramenti, CA: California Integrated Waste Management Board.

CHAPTER 25

New Developments and Insights on Vermicomposting in Spain

Jorge Dominguez and Manuel Aira

CONTENTS

I Introduction 410
II Evaluation of Organic Wastes for Vermicomposting 410
III Biology of Earthworms Suitable for Vermiculture 411
  A *Eisenia andrei* and *Eisenia fetida* are Two Different Earthworm Species 411
IV Influence of Environmental Factors on the Survival, Growth, and Reproduction of Vermicomposting Earthworms 412
V Ecology of Vermicomposting 412
  A Earthworms and Microorganisms: Disentangling the Black Box of Vermicomposting 412
  B Stimulation and Acceleration of Microbial Decomposition by Earthworms during Vermicomposting 414
VI Vermicomposting and Human Pathogen Destruction 415
VII Effects of Vermicomposts on Plant Growth 415
VIII Effects of Vermicomposts on Pesticide Degradation in Soils 415
IX Sixth International Symposium on Earthworm Ecology 416
Acknowledgments 417
References 417

I INTRODUCTION

Nowadays, the technologies of vermiculture and vermicomposting are well established, and there are many commercial enterprises in Spain. However, there are only two vermicompost research groups, the Soil Ecology Laboratory at the University of Vigo and the Department of Environmental Protection, Estación Experimental del Zaidín (EEZ), Consejo Superior de Investigaciones Cientifícas (CSIC), in Granada, both working on a range of scientific aspects of this discipline. Both of these groups have been developing a comprehensive research program in vermicomposting over the past 20 years, including many different aspects of the process and the effects of the application of vermicomposts on crop protection and pest management. This Chapter summarizes the research on vermicomposting conducted in these two laboratories in Spain. The extent of vermiculture research in Spain makes the country a leader in this new technology.

The research program at the University of Vigo that focuses on the vermicomposting of different types of organic wastes includes the following topics: biology and ecology of earthworms, reproduction and life histories of earthworms, evolutionary ecology of earthworms, ecology of vermicomposting, interactions between earthworms and other microorganisms, interactions between earthworms and soil organisms, microbial ecology of vermicomposting, influence of earthworms on nutrient cycling and nutrient dynamics during vermicomposting, fate of human pathogens during vermicomposting, effects of vermicomposts on the growth of greenhouse and field crops, production of plant growth regulators during vermicomposting, and aging and conservation of vermicomposts.

The overall goal of the research program Biotransformation of Organic Wastes, Protection of Soil and Agricultural Crops at the Department of Environmental Protection, EEZ, CSIC, in Granada, is to contribute sustainable and environmentally friendly practices to soil and crop protection through the use of low-cost bioremediation technologies and the promotion of sustainable development of agricultural systems by means of ecological alternatives. This general objective is approached from a triple perspective: (a) the development of the bioremediation processes, mainly vermicomposting, that favor the biotransformation, recycling, and recovery of polluted organic wastes, as well as the development of low-cost technologies for the protection and bioremediation of soil against organic contaminants; (b) the development of integrated management of pests in agroecosystems using vermicomposts; and (c) the development of methods including the use of vermicomposts to evaluate and maintain biodiversity in sustainable agroecosystems.

II EVALUATION OF ORGANIC WASTES FOR VERMICOMPOSTING

Since vermicomposting is a method of converting solid organic wastes into environmentally friendly and valuable resources for crop production and soil improvement, we evaluated the suitability of different types of organic wastes for the process. We found that vermicomposting works very well for processing sewage sludge and biosolids from wastewater (Elvira et al. 1997; Benítez et al. 1999a, 1999b, 2000; Domínguez et al. 2000, 2003; Plana et al. 2001), paper industry waste (Elvira, Mato, et al. 1995; Elvira, Domínguez, et al. 1995; Elvira, Goicoechea, et al. 1996; Elvira et al. 1997, 1998, 1999), urban residues, food, and animal wastes (Domínguez et al. 1996; Elvira, Domínguez, and Briones 1996; Elvira, Domínguez, and Mato 1996; Domínguez and Edwards 1997; Domínguez, Edwards, et al. 1997, 2001; Atiyeh et al. 2000; Aira et al. 2002, 2006a, 2006b, 2008, 2009; Aira, Monroy, et al. 2007a, 2007b; Aira and Domínguez 2008, 2009; Lazcano et al. 2008; Monroy et al. 2009), and food industry waste (Elvira et al. 1998, 1999; Nogales et al. 1998, 2005, 2008; Nogales, Elvira, et al. 1999; Nogales, Melgar, et al. 1999; Benítez et al. 2002, 2005; Melgar et al. 2002; Saavedra et al. 2006; Romero et al. 2007; Fernández-Bayo et al. 2008; Plaza et al. 2008; Vivas et al. 2009; Gómez-Brandón et al. 2010).

III BIOLOGY OF EARTHWORMS SUITABLE FOR VERMICULTURE

Although more than 3000 species of earthworms have been described, for the great majority of these species, we know only their names, morphologies and location and little about their biology, life cycles, or ecology. Certain epigeic earthworms, with their natural ability to colonize organic wastes and digest and assimilate organic matter, high rates of feedstock consumption, tolerance of a wide range of environmental factors, short life cycles, high reproductive rates, and endurance and tolerance of handling, show good potential for vermicomposting (see Chapters 3 and 4). Few earthworm species display all of these characteristics, and in fact only five have been used extensively in vermicomposting: *Eisenia andrei* Bouché, 1972, *Eisenia fetida* (Savigny, 1826), *Dendrobaena veneta* (Rosa, 1886), *Perionyx excavatus* Perrier, 1872 and *Eudrilus eugeniae* (Kinberg, 1867). We have studied many different aspects of the biology, ecology, and life cycles of four of these species (Elvira, Domínguez, and Briones, 1996; Domínguez and Edwards 1997, 2004; Domínguez, Briones, et al. 1997; Edwards et al. 1998; Domínguez et al. 2000, 2001, 2003, 2005; Aira et al. 2002, 2007; Monroy et al. 2003, 2005, 2006; Domínguez 2004; Pérez-Losada et al. 2005; Tato et al. 2006; Velando et al. 2006, 2008). We have also studied the biology, ecology, and life cycles of other earthworm species, such as *Lumbricus rubellus* and *Dendrodrilus rubidus* (Elvira, Domíguez, and Mato, 1996) and *Octodrilus complanatus* (Monroy et al. 2007). We are studying the ecology of a range of epigeic earthworm species and working in a laboratory screening program that is seeking other species suitable for vermicomposting (see Chapter 4).

A *Eisenia andrei* and *Eisenia fetida* are Two Different Earthworm Species

The closely related species *Eisenia fetida* (Savigny, 1826) and *Eisenia andrei* Bouché, 1972 are those used most commonly and globally for the management of organic wastes, and also in studies of ecotoxicology, physiology, and genetics. The problem of their taxonomic status remained unresolved for long time, and in much of the current literature, both species are termed indiscriminately as *E. fetida* or *E. foetida*, and it is often not clear which of the two species is being referred to. We have confirmed that they are two different biological species, reproductively isolated, and that they are also two different phylogenetic species. The reproductive isolation was determined after studying the offspring viability from inter- and intraspecific crosses of both species (Domínguez et al. 2005). Additionally, fully resolved and well-supported phylogenetic trees, based on mitochondrial (COI) and nuclear DNA sequences (28S), confirmed that they are different phylogenetic species (Pérez-Losada et al. 2005). This evidence has important considerations: For vermiculture or vermicomposting *E. andrei* is recommended most often since its growth and reproduction rates are higher than those of *E. fetida*. In ecotoxicological studies it is not possible to assume that contaminants will have the same effect on the two species, since their responses to stress factors could be different. The existence of postcopula but not precopula isolation in sympatric populations clearly affects population dynamics by reducing the fitness of the individuals. For this reason, for applied aspects of vermiculture, it is important keep the two species separated where possible, although the mixed populations often used may still function well in vermicomposting.

IV INFLUENCE OF ENVIRONMENTAL FACTORS ON THE SURVIVAL, GROWTH, AND REPRODUCTION OF VERMICOMPOSTING EARTHWORMS

The survival, reproduction, and growth of earthworms can be affected critically by environmental conditions. We have studied the influence of temperature, moisture content, ammonium content, population densities, type of food, and intra- and interspecific competition on the life histories of the four earthworm species most extensively used in vermicomposting: *E. andrei*, *E. fetida*, *P. excavatus*, and *E. eugeniae* (Elvira, Domínguez, and Briones, 1996; Elvira, Domínguez, and Mato, 1996; Domínguez and Edwards 1997, 2004; Domínguez, Briones, et al. 1997; Edwards et al. 1998; Domínguez et al. 2000, 2001; Domínguez 2004).

V ECOLOGY OF VERMICOMPOSTING

A Earthworms and Microorganisms: Disentangling the Black Box of Vermicomposting

Vermicomposting systems sustain a complex food web in organic wastes, which results in the recycling of the organic matter and release of the nutrients it contains. Biotic interactions between decomposer microorganisms (i.e., bacteria and fungi) and soil invertebrates include competition, mutualism, predation, and facilitation. The rapid changes that occur in both functional diversity and in substrate qualities are the main properties of these systems (Sampedro and Domínguez 2008). The most numerous and diverse members of this food web are microorganisms, although there are also abundant protozoa and many invertebrates of varying sizes and life-history patterns, including nematodes and microarthropods, as well as large populations of earthworms (Monroy et al. 2006; Sampedro and Domínguez 2008; Domínguez et al. 2010).

Microorganisms are largely responsible for organic-matter decomposition, but earthworms also affect rates of decomposition directly by feeding on and fragmenting the organic matter. This also affects the rates of decomposition indirectly through interactions with microorganisms, basically involving stimulation or depression of microbial biomass and activity and enzymatic activity (Domínguez 2004; Domínguez et al. 2010). We found that these processes mainly depend on earthworm population densities, with significant decreases in microbial biomass and activity related to increasing numbers of earthworms per unit area or volume of waste and time (Aira et al. 2002, 2008).

We found that the vermicomposting of animal manures with *E. fetida* comprises two separate stages, mainly associated with earthworm activities. Thus, when earthworms are present, not only is microbial biomass and activity enhanced but also rates of mineralization are increased (Aira, Monroy, et al. 2007b, 2007c); moreover, we also found significant increases in fungal populations in this stage that were associated with cellulose degradation (Aira et al. 2006b); this priming of fungal populations was observed in short-term experiments (72 hours; Aira et al. 2008). In animal waste experiments, once the earthworms moved from processed material to new batches of raw manure, the second stage begins. This part is characterized by the stabilization of the manure, with continuous decreases in microbial biomass and activity (Aira, Monroy, et al. 2007a, 2007b, 2007c; Domínguez et al. 2010). Thus, we can expect that microbial communities in the process from manure to vermicompost should change markedly, as we reported in a study on different animal manures and earthworm species (*E. andrei*, *E. eugeniae*, and *Lumbricus rubellus*). Fungal biomass increased significantly in horse manure vermicomposted by *L. rubellus* and in cow manure vermicomposted by all three earthworm species, whereas it decreased significantly in pig manure vermicomposted by *L. rubellus* and *E. eugeniae*. Furthermore, protozoa biomass, undetectable in the animal manures, increased significantly in all vermicomposts produced by the three earthworm species (Lores et al. 2006; Gómez-Brandón et al. 2010). Surprisingly, we found an effect of earthworm species, since the microbial communities in vermicomposts produced by each earthworm species were very similar, independently of the type of parent animal waste (horse, cow and pig manure) clustering together in related groups, mainly due to the above-mentioned changes together with a marked drop in bacterial biomass (Lores et al. 2006; Domínguez et al. 2010; Gómez-Brandón et al. 2010).

Results from analyses of fresh earthworm casts and their parent raw manures demonstrated increases in microbial biomass and decreases in microbial activity (Aira et al. 2006a; Aira and Domínguez 2009); these indicate that the direct effects of *E. fetida* produce changes in microbial populations that can influence the overall dynamics of organic-matter degradation. These decreases in microbial activity can be attributed to reductions in organic C and N in the wastes (Aira and Domínguez 2008). However, analyses of the gut contents of several epigeic earthworm species revealed no changes in bacterial numbers or microbial activity between species (Aira et al. 2009). We inoculated fresh manure with vermicompost to study the indirect effects of earthworms on organic-matter decomposition. We found that the inoculation of vermicomposts into animal manures modified the microbial community functions, separating clearly microbial communities depending on the type of vermicompost, size of inoculum, and time of incubation. These changes all occurred in the same directions, first an increase and then a decrease. These changes in microbial communities, and those found in our vermicomposting experiment, suggest that the indirect effects of earthworms are to alter the dynamics of animal manure decomposition (Aira and Domínguez 2010). However, the extent of these effects was not as great as those we found during vermicomposting; these observations, together with the results of our earthworm casting experiment, suggest the existence of other factors governing relationships between earthworms and microorganisms that become established during vermicomposting.

B Stimulation and Acceleration of Microbial Decomposition by Earthworms during Vermicomposting

Nutrient mineralization is governed directly by the activities of bacteria and fungi, and these activities are strongly affected by soil invertebrates that interact with the microorganisms, and also by food web interactions that determine the transfer of nutrients through the system. Although epigeic earthworms have few direct impacts on mineralization, their indirect effects on microbial biomass and microbial activity are very important in mineralization. These indirect effects include the digestion and release of readily assimilable substances, such as mucus for the microbiota, as well as the transport and dispersal of microorganisms through earthworm casting (Domínguez et al. 2010).

In studies at the University of Vigo we found that earthworms accelerated the rates of organic-matter decomposition during vermicomposting significantly (Atiyeh et al. 2000; Domínguez et al. 2003; Domínguez 2004; Aira et al. 2006b, 2008; Aira, Monroy, et al. 2007a, 2007b; Aira and Domínguez 2008, 2009). Although earthworms can assimilate C from the more labile fractions of organic wastes, their contribution to the total heterotrophic respiration is relatively low due to their relatively poor capacity for assimilation.

Nitrogen mineralization is regulated basically by the availability of dissolved organic nitrogen and ammonium, the activity of the microorganisms, and their relative requirements for C and N. In our studies we found that earthworms also have a great impact on N transformations during vermicomposting, through modifications of the environmental conditions and their interactions with microorganisms; they enhance N mineralization, thereby producing conditions in the organic wastes that favor nitrification, resulting in the rapid conversion of NH_4-N into NO_3-N (Atiyeh et al. 2000; Domínguez 2004; Aira et al. 2008; Aira and Domínguez 2008; Lazcano et al. 2008).

VI VERMICOMPOSTING AND HUMAN PATHOGEN DESTRUCTION

We found that earthworms decreased populations of total coliforms during vermicomposting greatly (see Chapter 16). Thus, passage through the gut of the earthworm species *E. andrei*, *E. fetida*, and *E. eugeniae* reduced population densities of total coliform bacteria by 98%, relative to those in fresh pig slurry (Monroy et al. 2008). We also found similar drastic reductions in the population densities of total coliforms in another experiment after 2 weeks of vermicomposting using *E. fetida* (Monroy et al. 2009).

VII EFFECTS OF VERMICOMPOSTS ON PLANT GROWTH

Earthworms have beneficial physical, biological, and chemical effects on soils, and these effects increase plant growth and crop yields in both natural and agroecosystems (Edwards and Bohlen 1996; Edwards et al. 1998;) (see Chapters 9, 10, and 15). Over the past few years, the Soil Ecology Laboratory at the University of Vigo has been developing a comprehensive research program in vermicomposting, which has included studies on the effects of vermicomposts on plant growth. The effects of vermicomposts on the growth of a variety of crops, including cereals, legumes, vegetables, ornamental and flowering plants, and trees, have been assessed in the greenhouse, and to a lesser degree in field crops (Lazcano, Arnold, et al. 2009; Lazcano, Sampedro, et al. 2009; Lazcano et al. 2010). These investigations have consistently demonstrated that vermicomposts have beneficial effects on plant growth independent of nutrient transformations and availability. Whether vermicomposts are used as soil additives, or as components of horticultural soilless bedding-plant container media, they have improved seed germination consistently, enhanced seedling growth and development, and increased plant productivity and yields, much more than would be possible from the mere conversion of mineral nutrients into more plant-available forms.

VIII EFFECTS OF VERMICOMPOSTS ON PESTICIDE DEGRADATION IN SOILS

Vermicomposting is an efficient way of utilizing organic wastes; in addition, vermicomposts can be added to soil to improve plant productivity, via nutrient additions, suppress plant pathogens, or promote of plant hormone activity, as mentioned previously (see Chapter 9). Otherwise, the research program at the Department of Environmental Protection, EEZ, CSIC, in Granada found that the application of vermicomposts to soils restored much biochemical activity and increased the bacterial diversity in a soil contaminated with trichloroethylene (Moreno et al. 2009). They also found that application of vermicomposts from the winery industries increased the retention of the insecticide imidacloprid in 10 different soils (Fernández-Bayo et al. 2007) and that the sorption capacity of vermicomposts was lower for anionic herbicides than for hydrophobic pesticides (Romero et al. 2006). In these ways, they found that the amendment of soils with vermicomposts from wastes from the winery and alcohol industries reduced the persistence of two pesticides (diuron and imidacloprid) in soils, although this effect depended on the type of pesticide and soil (Fernández-Bayo et al. 2009). Their results suggested that the application of vermicomposts reduced the availability of diuron in soils (Fernández-Bayo et al. 2008). They did not find any correlation between effects of vermicomposts and enzyme activity, which suggests that these effects were mediated more by physicochemical, and not biological, properties of vermicomposts, as repeated by Delgado-Moreno and Peña (2007). These authors studied the degradation of three sulfonylurea herbicides (chlorsulfuron, prosulfuron, bensulfuron) in soils amended with raw olive cake and vermicomposts produced from it. Their results point to chemical rather than biological degradation, probably due to interactions between the three pesticides. The same trend was found for four triazine herbicides since only the addition of vermicomposts boosted the biological degradation rate of triazines (Delgado-Moreno and Peña 2008).

IX SIXTH INTERNATIONAL SYMPOSIUM ON EARTHWORM ECOLOGY

The Sixth International Symposium on Earthworm Ecology (ISEE) was held at the University of Vigo, Spain, on August 31–September 4, 1998. The history of these major symposia on earthworm ecology began in 1981, with the first symposium held in Grange-over-Sands, England (ISEE1). The next gathering of earthworm scientists was in 1985 in Bologna, Italy (ISEE2), then 2 years later in 1987 in Hamburg, Germany (ISEE3), followed by ISEE4 in 1990 in Avignon, France. In 1994, the ISEE shifted out of Europe for the first time, to Columbus, Ohio, in the United States (ISEE5). The following conferences were held again in Europe: In 1998 ISEE6 was held in Vigo, Spain, ISEE7 in 2002 in Cardiff, United Kingdom, and ISEE8 in 2006 in Kraków, Poland. In 2010, ISEE9 will be held in the city of Xalapa in the state of Veracruz (Mexico) from September 5 to September 10, 2010.

In 1998, 200 earthworm scientists from 39 countries met in the city of Vigo, Spain, for the first time. The symposium attracted scientists working on various aspects of earthworm biology and ecology. It gave an opportunity for people researching different disciplines to get together, giving a very stimulating boost to further international research and cooperation in earthworm ecology. During the symposium, 55 oral contributions were presented, accompanied by 165 posters. The symposium was divided into the following sessions:

Session 1. Earthworm biodiversity and biogeography: 5 oral presentations and 19 posters

Session 2. Influence of earthworms on soil organic matter and nutrient dynamics: 5 oral presentations and 19 posters

Session 3. Earthworm ecotoxicology: 6 oral presentations and 25 posters

Session 4. Earthworms in agroecosystems and land use: 9 oral presentations and 22 posters
Session 5. Earthworms and waste management: 7 oral presentations and 31 posters
Session 6. Earthworms and soil physical properties and function: 7 oral presentations and 7 posters
Session 7. Earthworm biology, ecology, and physiology: 9 oral presentations and 27 posters
Session 8. Interactions of earthworms with other organisms: 6 oral presentations and 16 posters

ACKNOWLEDGMENTS

This research was partially supported by the Xunta de Galicia (07MRU023383PR) and the Spanish Ministerio de Ciencia e Innovaciòn (CTM2009-08477).

REFERENCES

Aira, M., and Domínguez, J. 2008. Optimizing vermicomposting of animal wastes: Effects of dose of manure application on carbon loss and microbial stabilization. *J. Environ. Manage.*, 88: 1525–1529.

Aira, M., and Domínguez, J. 2009. Microbial and nutrient stabilization of two animal manures after the transit through the gut of the earthworm *Eisenia fetida* (Savigny, 1826). *J. Hazard. Mater.*, 161: 1234–1238.

Aira, M., and Domínguez, J. 2010. Earthworm effect without earthworms: Inoculation of fresh organic matter with worm-worked substrates changes the dynamics of microbial biomass and activity *Bioresource Technology* (in press).

Aira, M., Domínguez, J., Monroy, J., and Velando, A. 2007. Stress promotes changes in resource allocation to growth and reproduction in a simultaneous hermaphrodite with indeterminate growth. *Biol. J. Linn. Soc.*, 91: 593–600.

Aira, M., Monroy, F., and Domínguez, J. 2006a. Changes in microbial biomass and microbial activity of pig slurry after the transit through the gut of the earthworm *Eudrilus eugeniae* (Kinberg, 1867). *Biol. Fertil. Soils*, 42: 371–376.

Aira, M., Monroy, F., and Domínguez, J. 2006b. *Eisenia fetida* (Oligochaeta, Lumbricidae) activates fungal growth, triggering cellulose decomposition during vermicomposting. *Microb. Ecol.*, 52: 738–746.

Aira, M., Monroy, F., and Domínguez, J. 2007a. Microbial biomass governs enzyme activity decay during aging of worm-worked substrates through vermicomposting. *J. Environ. Qual.*, 36: 448–452.

Aira, M., Monroy, F., and Domínguez, J. 2007b. *Eisenia fetida* (Oligochaeta: Lumbricidae) modifies the structure and physiological capabilities of microbial communities improving carbon mineralization during vermicomposting of pig manure. *Microb. Ecol.*, 54: 662–671.

Aira, M., Monroy, F., and Domínguez, J. 2007c. Earthworms strongly modify microbial biomass and activity triggering enzymatic activities during vermicomposting independently of the application rates of pig slurry. *Sci. Total Environ.*, 385: 252–261.

Aira, M., Monroy, F., and Domínguez, J. 2009. Changes in bacterial numbers and microbial activity of pig slurry during gut transit of epigeic and anecic earthworms. *J. Hazard. Mater.*, 162: 1404–1407.

Aira, M., Monroy, F., Domínguez, J., and Mato, S. 2002. How earthworm density affects microbial biomass and activity in pig manure. *Eur. J. Soil Biol.*, 38: 7–10.

Aira, M., Sampedro, L., Monroy, F., and Domínguez, J. 2008. Detritivorous earthworms directly modify the structure, thus altering the functioning of a microdecomposer food web. *Soil Biol. Biochem.*, 40: 2511–2516.

Atiyeh, R.M., Domínguez, J., Subler, S., and Edwards, C.A. 2000. Changes in biochemical properties of cow manure during processing by earthworms (*Eisenia andrei*, Bouché) and the effects on seedling growth. *Pedobiologia,* 44: 709–724.

Benítez, E., Elvira, C., Gómez, M., Gallardo-Lara, F., and Nogales, R. 1995. Leachates from a vermicomposting process: A possible new liquid fertilizer? In *Fertilizers and Environment*, ed. C. Rodriguez-Barrueco, 323–325. the Netherlands: Kluwer Academic.

Benítez, E., Nogales, R., Elvira, C., Masciandaro, G., and Ceccanti, B. 1999a. Enzyme and earthworms activities during vermicomposting of carbaryl treated sewage sludge. *J. Environ. Qual.*, 28: 1099–1104.

Benítez, E., Nogales, R. Elvira, C. Masciandaro, G., and Ceccanti, B. 1999b. Enzymes activities as indicators of the stabilization of sewage sludges composting by *Eisenia andrei. Biores. Technol.*, 67: 297–303.

Benítez, E., Nogales, R., Masciandaro, G., and Ceccanti, B. 2000. Isolation by IEF of humic – urease complexes from earthworms processed sewage sludges. *Biol. Fertil. Soils,* 31: 489–493.

Benítez, E., Sainz, H., Melgar, R., and Nogales, R. 2002. Vermicomposting of a lignocellulosic by-product from olive oil industry: a pilot scale study. *Waste Manage. Res.*, 20: 134–142.

Delgado-Moreno, L., and Peña, A. 2007. Organic amendments from olive cake as a strategy to modify the degradation of dulfonylurea herbicides in soil. *J. Agric. Food Chem.*, 55: 213–218.

Delgado-Moreno, L., and Peña, A. 2008. Compost and vermicompost of olive cake to bioremediate triazines-contaminated soil. *Sci. Total Environ.*, 407: 1489–1495.

Domínguez, J. 2004. State of the art and new perspectives on vermicomposting research. In *Earthworm Ecology*, ed. C.A. Edwards, 401–424. 2nd ed. CRC Press.

Domínguez, J., Aira, M., and Gómez-Brandón, M. 2010. Vermicomposting: Earthworms enhance the work of microbes. In *Microbes at Work*, ed. H. Insam, Franke-Whittle, I and Gobernam (eds), Boca Raton, FL. 94–115. Springer, Berlin. Insam, H., Franke-Whittle, I., and Goberna, M. (eds), Boca Raton, FL.

Domínguez, J., Briones, M., and Mato, S. 1997. Effect of the diet on growth and reproduction of *Eisenia andrei* (Oligochaeta, Lumbricidae). *Pedobiologia,* 41: 566–576.

Domínguez, J., and Edwards, C.A. 1997. Effects of stocking rate and moisture content on the growth and maturation of *Eisenia andrei* (Oligochaeta) in pig manure. *Soil Biol. Biochem.*, 29: 743–746.

Domínguez, J., and Edwards, C.A. 2004. Vermicomposting organic wastes: A review. In *Soil Zoology for Sustainable Development in the 21st century*, ed. S.H Shakir and W.Z.A. Mikhaïl, 369–396, El Cairo.

Domínguez, J., Edwards, C.A., and Ashby, J. 2001. The biology and ecology of *Eudrilus eugeniae* (Kinberg) (Oligochaeta) bred in cattle waste. *Pedobiologia,* 45: 341–353.

Domínguez, J., Edwards, C.A., and Subler, S. 1997. A comparison of vermicomposting and composting methods to process animal wastes. *BioCycle,* 5: 57–59.

Domínguez, J., Edwards, C.A., and Webster, M. 2000. Vermicomposting of sewage sludge: Effect of bulking materials on the growth and reproduction of the earthworm *Eisenia andrei. Pedobiologia,* 44: 24–32.

Domínguez, J., Elvira, C., Sampedro, L., and Mato, S. 1996. Effects of bulking agents in composting of pig slurries. In *The Science of Composting*, ed. M. Bertoldi, P. Sequi, B. Lemmes, and T. Papi, 1146–1149. Chapman & Hall, London.

Domínguez, J., Ferreiro, A., and Velando, A. 2005. Are *Eisenia fetida* (Savigny, 1826) and *Eisenia andrei* Bouché, 1972 (Oligochaeta, Lumbricidae) different biological species? *Pedobiologia,* 49: 81–87.

Domínguez, J., Parmelee, R.W., and Edwards, C.A. 2003. Interactions between *Eisenia andrei* (Oligochaeta) and nematode populations during vermicomposting. *Pedobiologia,* 47: 53–60.

Edwards, C.A., and Bohlen, P.J. 1996. Biology and Ecology of Earthworms (Third Edition). London:Chapman & Hall.

Edwards, C.A., Domínguez, J., and Neuhauser, E.F. 1998. Growth and reproduction of *Perionyx excavatus* (Perr.) (Megascolecidae) as factors in organic waste management. *Biol. Fertil. Soils,* 27: 155–161.

Elvira, C., Domínguez, J., and Briones, M.J. 1996. Growth and reproduction of *Eisenia andrei* and *E. fetida* (Oligochaeta, Lumbricidae) in different organic residues. *Pedobiologia,* 40: 377–384.

Elvira, C., Domíguez, J., and Mato, S. 1996. The growth and reproduction of *Lumbricus rubellus* and *Dendrobaena rubida* in cow manure. Mixed culture with *Eisenia andrei. Appl. Soil Ecol.,* 5: 97–103.

Elvira, C., Domínguez, J., Sampedro, L., and Mato, S. 1995. Vermicomposting for the paper pulp industry. *BioCycle,* 4: 62–63.

Elvira, C., Goicoechea, M., Sampedro, L., Mato, S., and Nogales, R. 1996. Bioconversion of solid paper pulp mill sludge by earthworms. *Biores. Technol.,* 57: 173–177.

Elvira, C., Mato, S., and Nogales, R. 1995. Changes in heavy metal extractability and organic matter fractions after vermicomposting of sludges from a paper industry and wastewater treatment plant. *Fresenius Environ. Bull.,* 4: 503–507.

Elvira, C., Sampedro, L., Benítez, E., and Nogales, R. 1998. Vermicomposting of sludges from paper mill and dairy industries with *Eisenia andrei*: A pilot scale study. *Biores. Technol.,* 63: 205–211.

Elvira, C., Sampedro, L., Domínguez, J., and Mato, S. 1997. Vermicomposting of wastewater sludge from paper-pulp industry with nitrogen rich materials. *Soil Biol. Biochem.,* 29: 759–762.

Elvira, C., Sampedro, L., and Nogales, R. 1999. Suitability of sludges from dairy and paper industries for growth and reproduction of *Eisenia andrei. Pedobiologia,* 43: 766–770.

Fernández-Bayo, J.D., Nogales, R., and Romero, E. 2007. Improvement of imidacloprid (Confidor®) soil-sorption capacity by the addition of vermicompost from spent grape marc. *Sci. Total Environ.,* 378: 95–100.

Fernández-Bayo, J.D., Nogales, R., and Romero, E. 2009. Effect of vermicomposts from wastes of the wine and alcohol industries in the persistence and distribution of imidacloprid and diuron on agricultural soils. *J. Agric. Food Chem.,* 57: 5435–5442.

Fernández-Bayo, J.D., Romero E., Schnitzler, F., and Burauel, P. 2008. Assessment of pesticide availability in soil fractions after the incorporation of winery-distillery vermicomposts. *Environ. Pollut.,* 154: 330–337.

Gómez-Brandón, M., Lazcano, C., Lores, M., and Domínguez, J. 2010. Detritivorous earthworms modify microbial community structure and accelerate plant residue decomposition. *Appl. Soil Ecol.* 44: 237–244.

Lazcano, C., Arnold, J., Tato, A., Zaller, J.G., and Domínguez, J. 2009. Compost and vermicompost as nursery pot components: Effects on tomato plant growth and morphology. *Span. J. Agric. Res.,* 7: 944–951.

Lazcano, C., Gómez-Brandón, M., and Domínguez, J. 2008. Comparison of the effectiveness of composting and vermicomposting for the biological stabilization of cattle manure. *Chemosphere,* 72: 1013–1019.

Lazcano, C., Sampedro, Zas, R., and Domínguez, J. 2009. Vermicompost enhances germination of the maritime pine (*Pinus pinaster* Ait.) *New Forests*, 39: 387–400. (in press).

Lazcano, C., Sampedro, Zas, R., and Domínguez, J. 2010. Assessment of plant growth promotion by vermicompost in different progenies of maritime pine (*Pinus pinaster* Ait.). *Compost Sci. Util.,* in press.

Lores, M., Gómez, M., Pérez, D., and Domínguez, J. 2006. Using FAME profiles for the characterization of animal wastes and vermicomposts. *Soil Biol. Biochem.,* 38: 2993–2996.

Melgar, R., Benítez, E., and Sainz, H. 2002. Vermicomposting of a lignocellulosic waste from olive oil industry: A pilot case study. *Waste Manage. Res.,* 20: 134–142.

Monroy, F., Aira, M., and Domínguez, J. 2008. Changes in density of nematodes, protozoa and total coliforms after transit through the gut of four epigeic earthworms (Oligochaeta). *Appl. Soil Ecol.,* 39: 127–132.

Monroy, F., Aira, M., and Domínguez, J. 2009. Reduction of total coliform numbers during vermicomposting is caused by short-term direct effects of earthworms on microorganisms and depends on the dose of application of pig slurry. *Sci. Total Environ.,* 407: 5411–5416.

Monroy, F., Aira, M., Domínguez, J., and Velando, A. 2006. Seasonal population dynamics of *Eisenia fetida* (Savigny, 1826) (Oligochaeta, Lumbricidae) in the field. *C.R. Biol.,* 329: 912–915.

Monroy, F., Aira, M., Gago, J.A., and Domínguez, J. 2007. Life cycle of the earthworm *Octodrilus complanatus* (Oligochaeta, Lumbricidae). *C.R. Biol.,* 330: 389–391.

Monroy, F., Aira, M., Velando, A., and Domínguez, J. 2003. Have spermatophores in *Eisenia fetida* (Oligochaeta, Lumbricidae) any reproductive role? *Pedobiologia,* 47: 526–529.

Monroy, F., Aira, M., Velando, A., and Domínguez, J. 2005. Size-assortative mating in the earthworm *Eisenia fetida* (Oligochaeta, Lumbricidae). *J. Ethol.,* 23: 69–70.

Moreno, B., Vivas, A., Nogales, R., Macci, C., Masciandaro, G., and Benítez, E. 2009. Restoring biochemical activity and bacterial diversity in a trichloroethylene-contaminated soil: The reclamation effect of vermicomposted olive wastes. *Environ. Sci. Pollut. Res.,* 16: 253–264.

Nogales, R., Cifuentes, C., and Benitez, E. 2005. Vermicomposting of winery wastes: A laboratory study. *J. Environ. Sci. Health, Part B,* 49: 659–673.

Nogales, R., Elvira, C. Benitez, E., Thompson, R., and Gómez, M. 1999. Feasibility of vermicomposting dairy biosolids using a modified system to avoid earthworm mortality. *J. Environ. Sci. Health, Part B,* 34: 151–169.

Nogales, R., Melgar, R., Guerrero, A. Lozada, G., Benítez, E., Thompson, R., and Gómez, M. 1999. Growth and reproduction of *Eisenia andrei* in dry olive cake mixed with other organic wastes. *Pedobiologia,* 43: 744–752.

Nogales, R., Saavedra, M., and Benitez, E. 2008. Recycling of wet olive cake "alperujo" through treatment with fungi and subsequent vermicomposting. *Fresenius Environ. Bull.,* 17: 1822–1827.

Nogales, R., Thompson, R., Calmet, A., Benítez, E., Gómez, M., and Elvira, C. 1998. Feasibility of vermicomposting residues from oil production obtained using two stages centrifuge. *J. Environ. Sci. Health, Part A,* 33: 1491–1501.

Pérez-Losada, M., Eiroa, J., Mato, S., and Domínguez, J. 2005. Phylogenetic species delimitation of the earthworms *Eisenia fetida* (Savigny, 1826) and *Eisenia andrei* Bouché, 1972 (Oligochaeta, Lumbricidae) based on mitochondrial and nuclear DNA sequences. *Pedobiologia,* 49: 317–324.

Plana, R., Pérez, C., Domínguez, J., Mato, S., and Aguilera, F. 2001. Development of a semi-industrial scale experimental organic wastes biodegradation reactor. In *Recycling and Reuse of Sewage Sludge*, ed. R.K. Dhir, M.C. Limbachiya, and M.J. McCarthy, 351–362. Thomas Telford, London.

Plaza, C., Nogales, R., Senesi, N., Benítez, E., and Polo, A. 2008. Organic matter humification by vermicomposting of cattle manure alone and mixed with two-phase olive pomace. *Biores. Technol.,* 99: 5085–5089.

Romero, E., Plaza, C., Senesi, N., and Nogales, R. 2007. Humic acid-like fractions in raw and vermicomposted winery and distillery wastes. *Geoderma,* 139: 397–406.

Romero, E., Salido, A., Cifuentes, C., Fernandez, J.D., and Nogales, R. 2006. Effect of vermicomposting process on pesticides sorption capability using agroindustrial wastes. *Int. J. Environ. Anal. Chem.,* 86: 289–297.

Saavedra, M., Benítez, E., Cifuentes, C., and Nogales, R. 2006. Enzyme activities and chemical changes in wet olive cake after treatment with *Pleurotus ostreatus* or *Eisenia fetida. Biodegradation,* 17: 93–102.

Sampedro, L., and Domínguez, J. 2008. Stable isotope natural abundances ($\delta^{13}C$ and $\delta^{15}N$) of the earthworm *Eisenia fetida* and other soil fauna living in two different vermicomposting environments. *Appl. Soil Ecol.,* 38: 91–99.

Tato, A., Velando, A., and Domínguez, J. 2006. Influence of size and partner preference on the female function of the earthworm *Eisenia andrei* (Oligochaeta, Lumbricidae). *Eur. J. Soil Biol.,* 42: S331–S333.

Velando, A., Domínguez, J., and Ferreiro, A. 2006. Inbreeding and outbreeding reduces cocoon production in the earthworm *Eisenia andrei. Eur. J. Soil Biol.,* 42: S354–S357.

Velando, A., Eiroa, J., and Domínguez, J. 2008. Brainless but not clueless: Earthworms boost their ejaculates when they detect fecund nonvirgin partners. *Proc. R. Soc. B,* 275: 1067–1072.

Vivas, A., Moreno, B., García-Rodríguez, S., and Benítez, E. 2009. Assessing the impact of composting and vermicomposting on bacterial community size and structure, and microbial functional diversity of an olive-mill waste. *Biores. Technol.,* 100: 1319–1326.

CHAPTER 26

Vermiculture and Vermicomposting in the United Kingdom

Kevin R. Butt and Brian Williams

CONTENTS

I Introduction ... 423
II Current Vermiculture Activities in the United Kingdom ... 425
III Vermicomposting Operations in the United Kingdom ... 428
A Case Study 1: Orm Professional Products, Wales ... 428
1 Vermiprocessing ... 429
2 Earthworms and Knowledge for Sale ... 430
3 Vermi T ... 430
B Case Study 2: BEEcycle, Lancaster ... 431
IV Current U.K. Vermiculture Research ... 432
V Future Developments in Vermicomposting in the United Kingdom ... 433
References ... 434

I INTRODUCTION

Before the publication of Darwin's (1881) seminal work on earthworms, these organisms were generally regarded as garden pests, or at best given very little consideration in nature. However, some gardeners knew better and used the activities of earthworms to assist their needs. For instance, in the preparation of leaf mulch for potting plants by the natural rotting of fallen leaves, some gardeners found that certain litter-dwelling earthworm species had a positive role in soil, and their activities were duly encouraged. This type of knowledge of organic breakdown activities remained almost something of a well-kept secret until a drive in the 1980s to utilize and indeed harness the activities of earthworms to assist in the break down and reduction of organic waste materials into vermicomposts and potentially to produce feed protein (see Chapter 20).

The work undertaken in the United Kingdom in the 1980s is well documented (Edwards 1988) and in the United States in the 1990s and 2000s (Edwards and Arancon 2004; see Chapters 7, 8, and 19). In essence, it involved the combination of three types of organizational elements (academic, industrial, and commercial) in collaboration with government. The Agriculture Research Council funded research into the potential of vermiculture and vermicomposting, with the aim of taking it from an experimental research scale to a full commercial operation. The research team led by Clive Edwards at the Rothamsted Experimental Station, Hertfordshire, investigated the biology of a range of epigeic earthworm species from temperate and tropical origins, tested a host of organic wastes, and formulated optimized regimes for the processing of organic wastes by earthworms. This work involved collaboration with the National Institute of Agricultural Engineering (NIAE) at Silsoe, Bedfordshire, which produced large-scale mechanized "reactor systems" in which the action of earthworms on organic material produced vermicomposts. NIAE also developed and manufactured innovative equipment for separating the earthworms from the vermicompost at the end of the process (see Chapter 7).

At this early stage it was thought that from an organic waste material, two products could be obtained, the earthworms (continuously reproducing and multiplying in numbers) and the vermicompost as a medium for plant growth or soil amendments. A commercial group, British Earthworm Technology (BET), was established with U.K. Government support to market and publicize this developing work. It was believed that the scientists, engineers, and commercial marketers would, as a group, develop vermiculture and transform vermicomposting into an agricultural technology, although the economics of the process showed that turning wastes into two commercially viable products was not as straightforward as it originally seemed (Fieldson 1988). Production of earthworm protein was not as inexpensive as thought and could not compete with alternative fish and animal feeds, and the production of vermicomposts appeared more promising to the economists at NIAE.

In 1984, BET, in collaboration with Dr. Clive Edwards, Dr. Ed Neuhauser of Cornell University, and their extensive team, organized a conference in Cambridge that was dedicated to vermiculture. This meeting was a major success and was attended by several hundred international delegates. The proceedings of this conference, *The Use of Earthworms in Waste and Environmental Management*, were published 4 years later (Edwards and Neuhauser 1988). However, in 1985, with the disbanding of the Rothamsted earthworm research group after the departure of key players (Edwards took an appointment at The Ohio State University in the United States), funding was restricted to only graduate scholarships in the 1980s (e.g., Knight 1987; Butt 1990). Although Edwards had obtained substantial funding from the European Economic Community to promote vermiculture in Europe, he was unable to use it, and although vermicomposting had had such a high profile for a time in the United Kingdom, it entered a period of slight depression for a decade

Figure 26.1 British Earthworm Technology (BET) earthworm beds, 128 ft × 8 ft (40 m × 3.4 m), which had been used to process turkey manure in East Anglia (no longer in operation when photographed in 1989). Note hopper for mechanized feeding of organic waste above the end of the central bed. The vermicompost was collected from below by use of a mechanized breaker bar and mesh system.

or so, with BET ceasing to trade in 1988. Some of the automated, hopper-fed earthworm beds, 128 ft × 8 ft (40 m × 2.4 m) in area (Figure 26.1; see Chapter 8), that had operated in East Anglia remained usable but were underused.

During this period, a number of vermiculture and vermicomposting companies were in operation (e.g., Original Organics; Wiggly Wigglers), offering small numbers of earthworms for vermicomposting but supplying mainly an ever-viable fishing bait market. However, some companies even offered earthworms for soil improvement (e.g., Walker Organics). In certain areas individual householders were also encouraged to set up their own home-based operations. Nevertheless, the scale of operation in the United Kingdom would probably have remained relatively small but for a more recent resurgence in environmental thinking with respect to the environmentally friendly disposal of organic waste.

II CURRENT VERMICULTURE ACTIVITIES IN THE UNITED KINGDOM

The British Government and, just as important, the European Union (EU 1999), have developed and brought into operation new legislation that closely examines current waste production and has set targets for its disposal stretching ahead to 2020. Landfill regulations in the United Kingdom implement the European Directive 1999/31/EC on the Landfill of Waste (Landfill Directive). The legislation requires that the amount of biodegradable (organic) waste sent to landfills be reduced to 75% of 1995 levels by 2010, to 50% by 2013, and to 35% by 2020 (quite demanding targets). Such organic wastes, which may also be viewed as resources, can fuel a composting stream, and this in itself can encompass an element of vermicomposting. So, although not directly driven by EU legislation, the renaissance of an interest in vermiculture and the processing of organic waste materials can in part be attributed to this.

Many local authorities in the United Kingdom now operate household waste collection schemes that oblige householders to separate the wastes at the source. In this way recyclables (metals, glass, paper, and cardboard) are collected biweekly along with "green waste." The latter usually comprises grass cuttings, hedge trimmings, and other garden-produced materials. Food waste is not normally deemed acceptable and is required to be disposed of into the standard waste stream (destined for landfill). This may seem inappropriate, given the desired legislation at the national level, but locally, authorities often seek to compost the green waste aerobically and wish to avoid materials that may attract vermin such as rats.

One way to overcome this dilemma would be to encourage householders to vermicompost their own organic wastes (green and kitchen). As with so many ideas, this is not novel. Adur District Council in West Sussex pioneered such a domestic scheme over 20 years ago, providing residents with brandling earthworms (*Eisenia fetida*; Savigny, 1826). Local accounts suggest that a number of residents still operate their own small, domestic-scale vermicomposting systems, but the pilot scheme was never developed fully. Nevertheless, a number of other local authorities across the United Kingdom have now begun to promote, support, and investigate the use of home vermicomposting. This may be through the provision of subsidized vermicomposting bins to residents or simply by promotion of home wormery use along with details of where these can be obtained. Examples of such schemes are provided in Box 26.1, which shows clearly that they are found across the United Kingdom but aimed at various levels. However, detailed information on the rates of household uptake of these local schemes is not available.

Over the past 20 years, commercial vermicomposting bins (home-scale vermicomposting units, also known as wormeries) have become more widely available in the United Kingdom. An Internet search, a horticultural mail-order catalog search, or a visit to a garden center can well reveal numerous suppliers of such systems, with a range of products to suit the desired scale (wheelie-bin size for the garage/garden to bucket size for under the kitchen sink) and price range (e.g., £20–£100, or $35–$160). Most simple types rely on the regular addition of small amounts of material to the waste surface, which the earthworms (*E. fetida* or *Dendrobaena veneta* Rosa, 1886) work up into. A tap may also be present to drain off any accumulating liquid. More sophisticated designs, including the well-known Can-O-Worms system from the United States are based on stacking containers that allow the vermicompost to be collected easily from below in removable sections, as the earthworms gradually work up through the system and away from it (see Chapter 6).

As additional such products are now available and are being promoted by local waste-disposal authorities, it is almost certain that over the next 5 years there will be a greater take-up of such systems, particularly as householders begin to think more sensitively about the environment and appreciate how they can make a difference by reducing their own organic waste production and simultaneously produce a material (vermicompost) that can enhance their gardens and potted plants.

BOX 26.1 EXAMPLES OF U.K. LOCAL AUTHORITY INVOLVEMENT WITH VERMICOMPOSTING

1 SHROPSHIRE COUNCIL: "DARWIN'S WORMS"

Darwin's Worms is an exciting environmental education project that was set up initially with funding from Shrewsbury & Atcham Borough Council. It helps to promote environmental awareness and reduce amounts of organic wastes going to landfills by increasing the use of wormeries. (Note: Charles Darwin was born in Shrewsbury—the county town of Shropshire.) In September 2008, 30 nursery schools in Shrewsbury signed up for a free wormery, education pack, and support from Shropshire's master composters and the Shropshire Wildlife Trust. The project is helping young people get in touch with their environment and learn about ecology. In the process it is helping to divert tons of organic waste away from landfills. Through this project it is believed that Darwin's legacy will be enhanced and make a positive contribution to sustainable development in the area.

2 SOUTHWARK COUNCIL

In 2006, an innovative scheme in Southwark, London, had 250 wormeries available at U.K. £5 (U.S. $8) each. This was run by the Southwark Council in partnership with CRISP (Community Recycling in Southwark Project). The wormeries, made by homeless people, were relatively simple wooden boxes that residents filled with soil and earthworms themselves. They came with a comprehensive guide, and anyone with a wormery has access to a free telephone helpline. This scheme, which was funded in part by the London Recycling Fund, allowed residents without gardens (60% in Southwark) to turn their biodegradable food waste into a nutrient-rich compost ideal for window boxes and potted plants. Councellor Richard Thomas, Southwark Council's executive member for environment and transport, said, "Around one fifth of household waste is kitchen waste. Wormeries are a really efficient way of turning this waste into a useful product and at the same time reducing the amount of rubbish we send to landfill. They make perfect environmental sense in an urban community." This same authority is now interested in supplying wormeries to schools.

3 POWYS COUNTY COUNCIL—SCHOOL SCHEME

Having a wormery in a school is not only an effective way to deal with waste, but it also provides an excellent education resource. There is a misconception that wormeries are difficult to look after, but they should be no more time-consuming than any traditional composting system. If earthworms are not given new food for some time (for example, through summer and midterm breaks), they will become dormant (stop producing cocoons). As long as they are given a good feed layer and left in an area that has a stable temperature (to retain moisture), they will survive. These are the types of instructions provided with this Welsh scheme to encourage schools to investigate waste management through vermicomposting. A full PowerPoint presentation was also made available to assist the process.

4 NORTH AYRSHIRE, SCOTLAND—"WORMS'N'WASTE"

This was a recycling project that aimed to educate residents of North Ayrshire in waste minimization and to improve the employability of unemployed young people and adults. The project, launched in 2002, initially provided 17 people with full-time jobs during its first year of operation as well as demonstrating the viability of a new approach to waste management. The project generated a great deal of interest among local residents and businesses and was expanded to provide a waste management service to the local community and several local businesses. The project also diversified into using the waste for beneficial purposes, such as using compost and earthworm castings for horticultural plant growth and using waste wood to make earthworm boxes.

III VERMICOMPOSTING OPERATIONS IN THE UNITED KINGDOM

In the United Kingdom there are currently as many as 500 small- to large-scale commercial operations utilizing earthworms to process organic waste materials. However, it is almost impossible to estimate the total turnover of these in terms of either finances or amounts of vermicompost. There may be as many as several hundred large-scale operations, with around 1000 m^2 (1200 yds^2) of outdoor earthworm beds. In addition, there are up to 10 commercial groups that are concerned with the promotion, sale, and setting up of such operations, which might be considered as earthworm farms. To illustrate the range of groups present, two examples are described that differ in scale and type of operation and the duration of work. Hopefully what should be seen is that these are not true competitors as they usually offer different services and products.

A Case Study 1: ORM Professional Products, Wales

This large-scale earthworm-related company, based in Brecon, Wales, has been in operation since 1991, employs seven staff on a permanent basis, and has an annual turnover of approximately £600,000 ($1,000,000; 2008 figures). The company began as a producer of earthworms for the fishing bait industry but has grown over the years and expanded; currently, earthworm production may account for only 25% of the business. Provision of technology to new growers, in terms of hardware and expertise, accounts for a further 20% of the business, with the final 55% related to end products of the process in solid and liquid form. The work of ORM can best be described in several subsections.

1 Vermiprocessing

ORM utilizes only one species of earthworm—*Dendrobaena veneta* (*Eisenia hortensis*). The reason for this is that it has been recognized as a particularly suitable species for processing organic wastes under temperate conditions (Edwards and Arancon 2004). ORM employees suggest that this large earthworm species tends to be more tolerant to changes in diet (e.g., kinds of organic waste offered), is pH tolerant and easy to breed, and suits the growing systems employed by them. The company utilizes a variety of organic waste materials including animal manures, processed vegetable wastes, food wastes, and sewage sludge, plus paper and cardboard, as advocated by Edwards (1988).

During vermicomposting, the main criterion is to present as consistent a feedstock as possible to the earthworms. Keeping the moisture content at about 80% appears to work well, with drier and wetter materials used as appropriate to absorb excess water or add water. All of the organic wastes fed to earthworms are macerated to a pulp to enable the earthworms to enter the waste quickly. Any feedstock that has the ability to heat up (compost thermophilically) must be added to the system in very thin layers. At the Brecon site, ORM has an annual throughput of some 200 tons (196 t) of fruit and vegetable waste and 100 tons (98t) of animal manure and straw.

In Brecon, the ORM operation is a system of 5000 m^2 (6000 yds^2) of outdoor earthworm-breeding beds that are 17 ft × 330 ft (5 m × 100 m) in addition to an indoor "earthworm-fattening system" with 300 trays that are 6 ft × 4 ft (1.8 m × 1.2 m). The latter, which utilize immature earthworms and ultimately produce high-quality active reproducing earthworms under low-population-density conditions, are maintained within an insulated heated building. It is suggested that all of the waste fed through the breeding beds is digested for at least 4 months and passes through earthworms a minimum of nine times. The outdoor vermicomposting beds (Figure 26.2) are emptied in a 4-year rotation with a harvest of 100–120 tons (98–118 t) of earthworm cast material (vermicompost) per annum per bed. This material is sold for sports turf improvement (e.g., golf courses, cricket squares), horticulture (e.g., nursery stock and bedding-plant substrates), and agriculture (e.g., fruit and vegetable growth). In addition to home-produced vermicomposts, ORM also obtains material from other subproducers (who ORM has supplied with the technology and knowledge) and uses this with specifically requested amendments to produce bespoke products for the preceding industries. However, not all of the vermicomposted material is utilized within the United Kingdom. ORM has also exported material to Ireland, France, and Holland within Europe and even as far afield as Malaysia and Ivory Coast. It is suggested that the development of this overseas market is a function of the high quality of the vermicompost produced.

Figure 26.2 Extensive outdoor vermicomposting beds at ORM Professional Products in Wales. These beds, which are 5 m × 100 m (17 ft × 330 ft), are harvested on a 4-year cycle and can produce up to 120 tons (118 t) of vermicompost on each occasion.

2 Earthworms and Knowledge for Sale

In addition to vermicomposts, some 10 tons (10t) of earthworms (wet weight) are produced by ORM on an annual basis. These are separated from the organic wastes in which they have been produced using separating equipment (similar to that developed by NIAE at Silsoe; see Chapter 7). The earthworms are sold to the fish bait market but also used as starter earthworm cultures for new vermiculture businesses, as described earlier. Here, ORM also provides appropriate hardware for any enterprising individuals or organizations. This hardware takes the form of earthworm processors (Figure 26.3), which are large containers into which the organic wastes are deposited and where the vermicomposting is done. The take-up on this knowledge is the creation of about 8–10 new businesses per year. The majority of these are farmers in rural locations seeking business diversification, and these operations tend to be based on nonclay soils (with no other obvious geographic biases within the United Kingdom).

3 Vermi T

VermiT is a liquid extraction in water derived from vermicomposts processed by earthworms or vermicompost "tea." This is applied as a feed supplement to horticultural crops but has also been shown to have beneficial attributes such as suppression of insect pest species (e.g., two-spotted spider mites and aphids). ORM produces and sells some 20,000 L (5280 gal) of this liquid per annum.

Materials, vermicompost, and VermiT are also supplied to local authorities for the purposes described earlier. ORM expects the sales of both vermicompost and

Figure 26.3 An earthworm "digester" produced to a number of standard sizes (as large as 9.8–2.4 m (32 ft × 8 ft) by ORM Professional Products, Wales. The image shows an 2.4 × 1.2 m (8 ft × 4 ft) vessel that would process 25 kg (55 lb) of waste per day. Current (2009) cost is £6500, including an associated macerator for waste processing and an earthworm starter kit.

VermiT in the horticulture market to double during 2010, and sales in the agricultural market to develop considerably.

B Case Study 2: BEEcycle, Lancaster

This relatively new company was established in 2007 by an entrepreneur with a background in biochemistry. Kenneth Cheung's Lancaster-based business seeks a niche market in the vermicomposting area. The substrate used by his "millions of workers" (epigeic earthworms) is food waste. His target markets are therefore producers of this type of waste—specifically households and schools. One product that BEEcycle sell is the OvO—a "self-composting and self-watering plantation system." This relatively simple container is marketed as a "mini ecosystem" that not only deals with food waste (through prior vermicomposting) and then allows seeds to be grown but also acts as an educational tool. This product has now been adopted by a number of local Lancashire schools, as accompanying BEEcycle-produced educational materials can be utilized by teachers. The current target for BEEcycle is households and schools, and the plan is to sell the system through outlets such as garden centers. BEEcycle currently operates two large-scale wormeries at Lancaster University, run partially by student volunteers. These vermicompost the food waste from a cafeteria and produce a vermicompost that is subsequently used to produce vegetables to be used in the same cafeteria. This, once again, demonstrates the recycling nature of the business. Even though this business may be relatively in its infancy, it demonstrates clearly that the market is not saturated. However, innovation may be required to break in successfully. BEEcycle had an annual turnover of £12,000 ($20,000) in 2008.

IV CURRENT U.K. VERMICULTURE RESEARCH

As indicated previously, the major thrust of vermicomposting research grew from the Rothamsted team; in the late 1980s, the research mantle was taken up by the Open University (OU). Here, projects continued, for example, examining the use of *E. fetida* to break down agricultural wastes (Knight 1987; Frederickson and Knight 1988). Wastes such as straw, paper pulp, and cattle dung were also fed to anecic (deep-burrowing) earthworms, including *Lumbricus terrestris* in another project at the OU (Butt 1990, 1993). This work, also considered as vermiculture, was used to investigate the potential of soil-dwelling earthworm species for land improvement (see Chapter 21). However, the vermicomposting work with epigeic earthworm species also continued and was seen as an "additional extra" to more traditional thermophilic composting (e.g., de Bertoldi 1996). Vermicomposting research sought to identify the optimum stage at which to take the partially aerobically composted material that had been acted on by microorganisms and feed this resulting material to earthworms. A small amount of such work was undertaken in Preston, funded by a local Lancashire Environmental Fund ("Landfill Tax"). This work utilized *E. fetida* but looked specifically at the characteristics of the vermicomposts produced (Butt et al. 2005). However, the OU work, led by Jim Frederickson, continued and developed over a number of years. A variety of feedstocks were used, including shredded green waste (Frederickson et al. 1997) and wastepaper sludge (Short et al. 1999). The latter worker examined the use of *D. veneta* for vermicomposting, and his findings supported use of wastepaper sludge as a suitable substrate without any nitrogen additions. The vermicomposted paper waste was also found to be a suitable plant growth medium (for radishes), but improved growth was experienced when the vermicompost was diluted with coir.

During the current millennium, workers from the OU teamed up with a vermicomposing business in Yorkshire called the Worm Research Centre. With a number of funding sources and links with industry (e.g., via Urban Mines Ltd.), the consortium set out to evaluate the technical performance of large-scale outdoor vermicomposting (Frederickson personal communication). Specific waste materials such as fish and shellfish waste were evaluated, but the general aims, for example, were to add and monitor outdoor earthworm-bed heating, assess levels of supplementary mechanization, and monitor and evaluate the environmental impact of outdoor vermicomposting systems. Part of this research was published by Frederickson and Howell (2003) at the International Symposium on Earthworm Ecology in Cardiff in 2002. One report that caused the greatest interest and consternation related to nitrogen emissions from vermicomposting facilities. It was suggested that, apart from the expected earthworm-related results, gas monitoring showed vermicomposting systems to have the capacity to emit high levels of nitrous oxide and that the earthworms appeared to be primarily responsible for this. The authors went on to claim that this major environmental impact was comparable with other agricultural sources and even that of landfills.

The latter is perhaps the most contentious vermicomposting-research issue that has arisen of late in the United Kingdom. It is a topic worthy of further consideration since it raised issues in the scientific literature. In an article entitled "Can Earthworms Harm the Planet?" Edwards (2008) took particular issue with these findings. He argued strongly that inadequate control measures were in place in the findings of Frederickson and Howell (2003)—which were later repeated in the popular media—and that the findings were not a true reflection of vermicomposting output, since comparisons drawn with other greenhouse gas–producing industries were scientifically invalid. Nevertheless, in a world that is becoming ever more conscious of our environment, further careful monitoring of potential gas production may be warranted.

Further vermicomposting research has taken place at other academic institutes in the United Kingdom and in association with one of the larger companies (ORM products—see earlier in this chapter). Linking up with the University of Wales, Bangor, ORM supported a PhD student who investigated the effects on plant growth of vermicomposts from of a variety of organic wastes (Roberts 2006). This work produced a number of publications (e.g., Roberts, Edwards, et al. 2007; Roberts, Jones, et al. 2007) that examined the growth of both horticultural fruits and also commercial flowering plants in vermicomposts. In addition, ORM also worked with the University of East Anglia and investigated the effects of *D. veneta* activities and the use of vermicomposts in the bioremediation of soils contaminated with organic pollutants (e.g., Hickman and Reid 2008).

Thus, research work on vermicomposting in the United Kingdom has continued over the past 20 years but only on a relatively small scale compared with other EU countries, such as Spain (see Chapter 25), and the United States (see Chapter 23), where significant and continuous research outputs have occurred.

V FUTURE DEVELOPMENTS IN VERMICOMPOSTING IN THE UNITED KINGDOM

Legislation in the United Kingdom (and across Europe) will force national and local governments to manage waste organic materials more sensitively, that is, reduce amounts, or ultimately prevent such materials from going into landfills. Responsibility for the processing of organic wastes will then lie with local authorities, who will no doubt devise a spectrum of schemes to suit locations, types of housing, current practices, and what may result from ongoing and future experimentation. Vermicomposting, at the household, authority-centred, or commercially operated level, will undoubtedly play a role in these schemes, alongside more traditional aerobic windrow composting. As with so many operations in the United Kingdom and beyond, the real potential of vermiculture and vermicomposting may be realized only when driven directly by legislation that outlaws outdated and nonsustainable practices. Vermicomposting may be a well-known practice or scientific technology, but it still needs to become better known and be used more by the general public and commercial organizations.

REFERENCES

BEEcycle Ltd, Lancaster. http://www.beecycle.co.uk/ (accessed August 2009).

Butt, K. R. 1990. The growth and reproduction of the earthworm *Lumbricus terrestris* L. under controlled environmental conditions. PhD diss., Open University.

Butt, K. R. 1993. Utilisation of solid paper-mill sludge and spent brewery yeast as a feed for soil-dwelling earthworms. *Biores. Technol.* 44: 105–108.

Butt, K. R., A. J. Holden, and C. N. Lowe. 2005. Compost characterization with a focus on heavy metals and a biological indicator. Proceedings Sardinia 2005, *Tenth Int. Waste Management and Landfill Symposium.* Eds R. Cossu and R. Stagman 761–762. CISA Environmental Sanitary Engineering Centre, Cagliari, Italy.

Darwin, C. 1881. *The Formation of Vegetable Mould through the Action of Worms with Observations on Their Habits*. London: John Murray.

de Bertoldi, M., P. Sequi, B. Lemmes, and T. Papi (eds.). 1996. *The Science of Composting.* London: Blackie Academic and Professional.

Edwards, C. A. 1988. Breakdown of animal, vegetable and industrial organic wastes by earthworms. In *Earthworms in Waste and Environmental Management*, ed. C. A. Edwards and E. F. Neuhauser, 21–31. The Hague, the Netherlands: SPB Academic Publishing.

Edwards, C. A. 2008. Can earthworms harm the planet? *Biocycle* 49 (12): 53–54.

Edwards, C. A., and N. Q. Arancon. 2004. The use of earthworms in the breakdown of organic wastes to produce vermicomposts and animal feed protein. In *Earthworm Ecology*, ed. C.A. Edwards, 345–379. 2nd ed. Boca Raton, FL: CRC Press.

Edwards, C. A., and E. F. Neuhauser (eds.). 1988. *Earthworms in Waste and Environmental Management.* The Hague, the Netherlands: SPB Academic Publishing.

European Union. 1999. Council Directive 1999/31/EC of 26 April 1999 on the landfill of waste.

Fieldson, R. S. 1988. The economic viability of earthworm culture on animal wastes. In *Earthworms in Waste and Environmental Management*, ed. C. A. Edwards, and E. F. Neuhauser, 145–156. The Hague, the Netherlands: SPB Academic Publishing.

Frederickson, J. n.d. Vermicomposting trial at the worm research centre: Part 1—Technical Evaluation. http://www.wormresearchcentre.co.uk/scientific/Vermicomposting%20Evaluation.pdf.

Frederickson, J., K. R. Butt, R. M. Morris, and C. Daniel. 1997. Combining vermiculture with traditional green waste composting systems. *Soil Biol. Biochem.* 29: 725–730.

Frederickson, J., and G. Howell. 2003. Large-scale vermicomposting: emission of nitrous oxide and effects of temperature on earthworm populations. *Pedobiologia* 47: 724–730.

Frederickson, J., and D. Knight. 1988. The use of anaerobically digested cattle solids for vermiculture. In *Earthworms in Waste and Environmental Management*, ed. C. A. Edwards and E. F. Neuhauser, 33–47. The Hague, the Netherlands: SPB Academic Publishing.

Hickman, Z. A., and B. J. Reid. 2008. The co-application of earthworms (*Dendrobaena veneta*) and compost to increase hydrocarbon losses from diesel contaminated soils. *Environ. Int.* 34: 1016–1022.

Knight, D. 1987. The decomposition and upgrading of agricultural wastes by the earthworm *Eisenia fetida* (Savigny 1826). PhD diss., Open University.

North Ayrshire, Scotland. http://www.sustainablescotland.com/ (accessed August 2009).

Original Organics Ltd., Devon. http://www.originalorganics.co.uk/ (accessed August 2009).

ORM Professional Products, Brecon. http://www.ormproproducts.com/ (accessed June 2009).

Powys County Council. http://www.powys.gov.uk/ (accessed August 2009).

Roberts, P. 2006. Evaluation of vermicompost from composts for agricultural and horticultural uses. PhD diss., University of Wales, Bangor.

Roberts, P., C. A. Edwards, G. Edwards-Jones, and D. L. Jones. 2007. Responses of pot-grown flower species to commercial plant media substituted with vermicompost. *Compost Sci. Util.* 15: 159–166.

Roberts, P., D. L. Jones, and G. Edwards-Jones. 2007. Yield and vitamin C content of tomatoes grown in vermicomposted wastes. *J. Sci. Food Agric.* 87: 1957–1963.

Short, J. C. P., J. Frederickson, and R. M. Morris. 1999. Evaluation of traditional windrow-composting and vermicomposting for the stabilization of waste paper sludge (WPS). *Pedobiol* 43: 735–743.

Shropshire Council. http://www.shrophire.gov.uk/ (accessed August 2009).

Southwark Council, London. http://www.southwark.gov.uk/ (accessed August 2009).

Walker Organics Ltd. http://www.walkerorganics.co.uk/ (accessed June 2009).

Wiggly Wigglers Ltd. http://www.wigglywigglers.co.uk/ (accessed June 2009).

Worm Research Centre, Yorkshire. http://www.wormresearchcentre.co.uk/ (accessed June 2009).

CHAPTER 27

Vermiculture in Australia and New Zealand: From Earthworm Production to Commercial Vermicomposting

Katie A. Webster and John C. Buckerfield

CONTENTS

I Introduction 438
II History of Vermiculture in Australia and New Zealand 438
  A Earthworms for Agriculture 438
  B Earthworms and Organic-Waste Processing 439
  C Vermiculture Industry Associations 439
III Earthworms in Australia and New Zealand 439
  A Native Earthworm Species 439
  B Agricultural Earthworm Species 439
  C Earthworm Species Used in Commercial Vermiculture 440
IV Vermiculture Products 441
  A Vermicomposting-Earthworm Sources 441
  B Earthworm Farms for Domestic Use 441
  C Sales of Vermicomposts 442
  D Vermiliquids (Aqueous Extracts from Vermicomposts) or Teas 442
  E Earthworm-Protein Production 443
V Commercial Vermicompost and Vermiliquid (Tea) Production 443
  A The Vermiculture Operation in 2009 443
  B Mid- and Large-Scale Vermicomposting Systems 444
  C Feedstocks Used in Vermicomposting 444
  D Additives and Supplements Used in Vermicompost Production 445
  E Pests and Diseases in Vermiculture 445
  F Vermiliquid (Tea) Production and Use 445

VI Vermiculture Markets and Marketing 446
A Profitability 446
B Buy-Back Schemes 447
C Role of Earthworms in Organic-Waste Management 447
D Agricultural Market-Development Trials 447
VII Standards and Testing 448
A Terminology 448
B Analyses 448
C Standards 448
VIII Future Developments 449
References 449

I INTRODUCTION

Considerable enthusiasm followed the earliest promotion of vermiculture in Australia and New Zealand in the late 1970s and 1980s. With great expectations of opportunities in waste management and substantial markets for vermicompost in agriculture, the industry hit a peak of activity in the mid- to late 1990s. The industry has since declined, perhaps brought on by the failure of earthworm "buy-back" schemes. However, there has been a recent trend with very large operations increasing in numbers and focusing their efforts on market development rather than system development. Strong sales by small domestic earthworm farms continue, but installation of midscale systems in schools, hospitals, and other institutions appears to have slowed. Despite the ups and downs in the industry over the past 20 years, the average Australian and New Zealand vermiculturist maintains a positive outlook for the future of this business and industry.

II HISTORY OF VERMICULTURE IN AUSTRALIA AND NEW ZEALAND

A Earthworms for Agriculture

Australia and New Zealand have no indigenous lumbricid earthworm species. Hence, the use of earthworms in improving agricultural land began in New Zealand, where farmers introduced exotic species such as *Aporrectodea caliginosa* to pastures during the 1930s, later developing machinery for cutting and spreading earthworm-rich turfs on new sites. This machinery was used with success to redistribute *A. caliginosa* and *Aporrectodea longa* in farming soils in Tasmania (Kingston and Temple-Smith 1989; Farquhar 1992), with economic benefits from soil improvement (Stockdill 1982; Temple-Smith 1993). Experimental introduction of *A. longa* in southeastern Australia is known to have continued over 3 years (Baker et al. 2004). However, currently, there is little market for soil improvement by earthworm species in Australia, and only limited redistribution of species in New Zealand.

B Earthworms and Organic-Waste Processing

In Australia, the promotion of the use of *Eisenia fetida* to break down organic wastes started in the late 1970s (Sabine 1978). Interest grew following the International Symposium on Earthworm Ecology in Avignon (France) in 1990. The New Zealand vermicomposting industry developed from limited sales of fish bait and soil-improving species and moved into vermicomposting for organic-waste management (Whitta 1996). European and American developments in large-scale municipal vermicomposting were influential in shaping the use of technology in large-scale vermiculture in Australia and New Zealand (Edwards and Steele 1997).

C Vermiculture Industry Associations

The Australian Worm Growers Association (AWGA) was established in 1992, and by 1997 its full membership had risen to over 700. AWGA facilitated cooperation and provided training and professional development for earthworm growers; it also sought to provide voluntary regulations, developing a "Code of Practice" (Steuart and Steele 1995). The high ambitions of AWGA were not completely fulfilled, and membership steadily fell to only a relatively small number today (www.ausworm.com).

The New Zealand Earthworm Association was established in 1995 following a seminar attended by some 200 people with wide interests in vermiculture. Membership peaked quickly, and the association's ability to encompass broad interests in earthworms for composting and waste management versus those for pasture improvement was a feature. However, as in Australia, falling membership and the retirement of key enthusiasts saw the end of the association and the loss of a valuable resource to the industry.

III EARTHWORMS IN AUSTRALIA AND NEW ZEALAND

A Native Earthworm Species

Australia has a unique earthworm fauna of native Megascolecidae; some 350 species have been described, with at least as many more in research collections awaiting description (Blakemore 2000). New Zealand lists 178 species of indigenous Megascolecidae, many with links to the native Australian earthworm fauna (Lee 1959). Indigenous earthworms have been observed in outdoor beds, but it is not clear whether these are effective as composters and can persist in organic-rich environments.

B Agricultural Earthworm Species

The earthworm fauna in agricultural (and garden) soils in Australia and New Zealand is dominated by introduced exotic species of the family Lumbricidae (Martin

1977; Blakemore 1997a). Many of these species were recorded over 100 years ago (e.g., Fletcher 1886a, 1886b). The value of some of the more common lumbricid species in agriculture was first recognized in Australia 50 years ago (Barley 1959).

In the past, there has been considerable confusion among vermiculturists, with claims that compost earthworms could improve soils. Buckerfield (1994) made the distinction between the "composters" and the "earthworkers" to clarify the separate roles of the litter-dwelling *epigeic* earthworms used in vermicomposting and the subterranean *endogeic* and surface-feeding *anecic* species (Bouché 1977).

C Earthworm Species Used in Commercial Vermiculture

Reports indicate that the common composting species *Eisenia fetida*, the tiger worm, was present in Australia as early as 1860 (Fletcher 1887, 1890), and populations have been maintained in animal manures by fishermen in Australia for at least 80 years (Abbott 1982). Earthworm growers have been encouraged to recognize the species they culture to foster better understanding of their biology and management. Despite the availability of simple identification tools such as the *Earthworm Identifier* (Baker and Barrett 1994), growers persist in using common names (Webster and Taylor 2009). A range of earthworm species are currently employed in vermiculture or hold potential for this role:

- *Eisenia fetida/Eisenia andrei*. The commercial vermiculture industry in Australia and New Zealand is based almost exclusively on *Eisenia fetida* and *Eisenia andrei* (the tiger and red tiger earthworms).
- *Eisenia hortensis* (*Dendrobaena veneta*). This species has been recorded under mulch in vineyards and orchards (Buckerfield and Webster 1996), associated with manure heaps and agricultural wastes. The mature adults are similar in appearance to *E. fetida* but are larger. This species may be more common than currently recognized in earthworm beds in Australia.
- *Lumbricus rubellus*. This is often a dominant species in agricultural soils in temperate regions, and it is also considered to be effective in processing organic wastes (Edwards and Bohlen 1996). It has often been reported from earthworm beds, but *L. rubellus* does not appear to persist in culture in Australia and New Zealand without soil contact. Confusion in the industry on the distinctions between *E. andrei* and *L. rubellus* has led to inappropriate recommendations that the red tiger is suitable for use in gardens and pastures (Murphy 1991).
- *Lumbricus castaneus*. This species lives in the soil but favors organic-rich environments; it is probably common as an itinerant in outdoor windrows, where it may be mistaken for *L. rubellus*.
- *Eudrilus eugeniae*. The reputed capacity of the African night crawler to decompose large quantities of organic wastes rapidly seems not to have been exploited much in Australia. The intolerance of this species to extended periods at temperatures below 16°C limits its use in outdoor vermiculture to the tropical and subtropical regions of the country. It is more likely to be cultured for fish bait than vermicomposting.
- *Perionyx excavatus*. This species was first identified in earthworm beds in Australia in 1990 (Blakemore 1994) and subsequently has been recorded widely by earthworm growers in New Zealand. It is noted as a minor contribution to

commercial vermiculture in Australia. where it is commonly known as the "blue" earthworm. This is essentially a tropical species from Asia, and its occurrence in temperate regions appears to be seasonal.

- *Polypheretima elongata.* In Australia, the earthworm identified as *P. elongata* (Blakemore 1997a) is active in the soil but does not favor organic-rich environments. In addition, this species appears to be restricted to tropical regions and may not survive temperate winters. Its value in vermicomposting requires further investigation.
- *Lumbricus terrestris.* This species, commonly referred to as the European night crawler, is available commercially in Europe and North America but appears not to have become established widely in Australia, although there have been confirmed occurrences in suburban gardens in Launceston, Tasmania (Blakemore 1997b). *L. terrestris* is not widespread in New Zealand, despite attempts to redistribute it. Opinions differ on the introduction of exotic species of earthworm, but in view of reports of problems related to excessive disturbance in forest soils in Canada and increased erosion through *L. terrestris* burrows in arable soils (Edwards and Shipitalo 1998), we consider that the import and redistribution of this species should be discouraged in Australia.
- *Amynthas, Dendrobaena,* and *Dendrodrilus* species. Representatives of these genera have been reported by Blakemore (1997a); their distribution in Australasia has not been documented, but it is possible that some species may already be widespread in vermiculture. Some species appear to have promise for use in Australasian vermiculture and warrant further investigation (Mitchell 1993).

IV VERMICULTURE PRODUCTS

A Vermicomposting-Earthworm Sources

The supply of earthworms for use in home vermicomposting remains strong, supported by a range of home earthworm farms and government provisions of subsidies for household vermicomposting. The supply of earthworms for larger-scale vermicomposting projects appears to have decreased, with only 34% of vermiculturists currently able to fill bulk orders of earthworms, down from 48% in 1997 (Webster and Taylor 2009).

In 2009, 46% of earthworm growers specialized in a particular earthworm species, compared to 21% in 1997 (Webster and Taylor 2009). The majority of earthworm growers still referred to species by common names, most often "reds" and "tigers," making it difficult to confirm which are the most commonly cultured species. Supply of earthworms for fishing bait continues, and the collection of the native species known as "scrubbies" still occurs on a small scale.

B Earthworm Farms for Domestic Use

In 2009, the interest in growing and supplying earthworms still remained strong, with 90% of vermiculturists identifying this as their main focus (compared to 82% in 1997). Demand for earthworms has been supported by local government initiatives, to encourage vermicomposting of food wastes by householders, in order to meet Zero Waste targets (e.g., Zero Waste South Australia 2005; Parliamentary Counsel Office New Zealand 2008). Webster and Taylor (2009) conducted a survey of 74 local government organizations and found that 8% provided subsidized earthworm farms to their residents. A range of home earthworm-farm designs are currently available:

1. ***Simple free-standing box:*** constructed of wood or using adapted objects such as plastic or styrene boxes, old bathtubs, and converter kits for old municipal garbage bins. These have a solid base, are enclosed, are free-standing, and may include ventilation and drainage.
2. ***Simple open-based box:*** essentially a basic compost bin with additional ventilation. It may incorporate a mesh base to restrict rodent access and have doors to facilitate emptying.
3. ***Vertically divided earthworm farm:*** constructed of wood, plastic, or metal mesh. As food becomes exhausted, earthworms move to one side where fresh food is being added.
4. ***Horizontally divided earthworm farm:*** usually tiered plastic boxes with mesh bases to allow earthworm movement from exhausted food material in the lower trays to fresh material being added to upper trays. These are enclosed and free-standing with adequate ventilation and a solid base tray with a tap to collect liquid. These are the more popular choice for householders and are often supplied by councils and shires. The Reln Worm Factory and Can-O-Worms are particularly popular with householders and are reported to have sales of 4000 a month. The Gardenwise Worm Farm is also widely available.
5. ***Other designs:*** The hanging Swag, constructed of canvas, is gaining in popularity (see Chapter 7). A range of other designs are available on a relatively small scale.

C Sales of Vermicomposts

1. ***Bagged quantities.*** There has been an increase in the number of vermiculturists bagging vermicompost for sale, currently 41%, up from 24% in 1997 (Webster and Taylor 2009). Growers have indicated they would be willing and able to bag more vermicompost for sale but felt there was a lack of demand due to poor understanding of the benefits of vermicompost.
2. ***Bulk quantities.*** There has been considerable growth since 1997 in the number of producers of large amounts of vermicompost (i.e., greater than 200 $m^3.yr^{-1}$ (180 yd^3 yr^{-1}). Since 1997, there has been a tendency for the industry to divide into smaller, individually run operations producing less than 50 m^3 (65 yd^3) of vermicompost yr^{-1} and much larger facilities producing from 100–1000 m^3 (130–1300 yd^3) and up.
3. ***Definitions.*** There continues to be confusion among producers and buyers over the naming of vermicompost and vermicast, despite the inclusion and definition of these terms in the relevant Australian and New Zealand standards (Standards Australia 2003; Standards New Zealand 2005).

D Vermiliquids (Aqueous Extracts from Vermicomposts) or Teas

The desire to package aqueous extracts ("teas") or vermiliquids for sale has increased since 1997, with sales in this market targeted by around 50% of growers, compared to 22% in 1997 (Webster and Taylor 2009). The interest may be largely due to ease of production and use. Vermiliquid is produced either from a solid vermicast that can be dissolved or suspended in water or as a leachate that has percolated through earthworm beds. There are currently no standards or guidelines for the production of consistent teas (vermiliquids), for application rates, or for their packaging and storage.

E Earthworm-Protein Production

Since the high protein value of earthworm meal was first promoted in Australia by Sabine (1978), there have been periodic bouts of interest in its use for domestic animal feeds. The production of protein from earthworms is relatively expensive (Tomlin 1983; Edwards 1998; see Chapter 20). At the current time, we are not aware of large-scale operations supplying in bulk to commercial animal operations, though powdered earthworm meal is available in small quantities as a high-priced pet food supplement.

V COMMERCIAL VERMICOMPOST AND VERMILIQUID (TEA) PRODUCTION

Industry surveys conducted in 1997 (AWGA 1997) and 2009 (Webster and Taylor 2009) allow us to examine changes in the industry over the last 12 years.

A The Vermiculture Operation in 2009

1. ***Regionality***. In 2009, 39% of operations were in suburban areas, which was similar to 1997, when 45% operated in a suburban area.
2. ***Size***. By 2009, although medium-sized (earthworm-bed area of 50–100 m^2) operations were still plentiful, the numbers of these operations were declining in favor of both smaller and larger operations.
3. ***Earthworm-bed design***. In 2009, 64% of earthworm beds were in the open, compared to 63% in 1997. Windrows remained the most popular method of vermicomposting for 29% of 2009 earthworm growers, compared to 35% in 1997. Other popular choices were constructed beds, boxes, and small brand-name earthworm farms.
4. ***Labor***. In 2009, 18% of vermiculture operations required more than two people to operate, compared with 13% in 1997. Around one-third of growers were spending less than 5 hours per week managing their earthworm beds, and one-third more than 20 hours.
5. ***Mechanization***. The mechanization of vermiculture processes has increased markedly, with the exception of the mechanical preparation of feedstocks (currently 31%; 36% in 1997). However, in 2009, the majority of earthworm growers still did the following tasks by hand: watering of beds (56%; 77% in 1997), working of beds (83%; 97% in 1997), separation of earthworms (74%; 95% in 1997), and screening of castings (64%; 89% in 1997).
6. ***Experience***. In 2009, the average vermiculturist is a great deal more experienced than in 1997. Only 9% have been in the industry less than 2 years compared with

61% in 1997. In 2009, 56% had been in the industry longer than 10 years, compared to just 2% in 1997.

7. ***Outlook***. The average 2009 vermiculturist is very positive about the future of their business and industry, with 71% having a positive outlook, 12% neutral, 6% unsure, 6% having difficulty with their current situation, and 5% having a negative outlook.

B Mid- and Large-Scale Vermicomposting Systems

During the 1990s the installation of midscale vermicomposting systems for on-site processing of food wastes in schools, hospitals, and restaurants expanded (Natoli 1996). A 1999 review of midscale systems (Recycled Organics Unit 2007) found a range of system including continuous-flow, tray or stacking, batching, wedge, and ecotechnology systems. The review considered the size, processing capacity, and cost of 11 models of midscale systems produced by eight manufacturers. Ten years later, the promise of these systems does not appear to have been fully realized. The likely reasons for this were foreshadowed in the 1999 review (Recycled Organics Unit 2007), with staff turnover and subsequent loss of key experience and enthusiasm a key reason.

Large-scale, raised-bed, mechanized systems were promoted in the late 1990s, with the largest installation designed to process 400 m^3 (390 yd^3) of sewage sludge weekly and produce 7000 m^3 (7780 yd^3) of vermicompost annually, installed in metropolitan Brisbane (Lotzof 1998). A similar system utilizing raised mesh beds fed from above was designed to handle 150 m^3 (167 yd^3) of piggery waste weekly at Scone (New South Wales). Other large-scale systems have included a municipal green-waste operation in Hobart, Tasmania (Williams 1994; Appelhof et al. 1996); use of vegetable-processing wastes in Newcastle, New South Wales (Appelhof et al. 1996); and processing of biosolids and green waste at Bathurst, New South Wales (Scarborough 1999).

While some of these operations no longer exist, others have persisted, and yet others have been initiated. Claiming to be Australia's largest vermiculture operation, Australian Vermiculture, based at Broken Hill (New South Wales), uses open windrows to process a variety of wastes, mainly green wastes. Worms Work Technologies at Millicent (South Australia) has taken over a continuous-flow system and concentrated its efforts in market development. The number of operations producing more than 1000 m^3 (1,110 yd^3) of vermicompost per annum has increased four-fold since 1997, now constituting 8% of all operations (Webster and Taylor 2009).

C Feedstocks Used in Vermicomposting

1. ***Composition***. The general use of horse, pig, and cow manures as feedstocks for vermicomposts has fallen since 1997, though they are still widely used. It would seem that a greater diversity of materials are being accessed, and in 2009, the "average" feedstock is composed mainly of horse and cow manure, with substantial quantities of other materials, fruit- and vegetable-processing and packing residues, and smaller amounts of pig manure, restaurant wastes, and wine-industry waste. In both 1997 and 2009, the majority of respondents (70% and 68%, respectively) did not thermophilically precompost their feedstock.

2. ***Cost and transport***. In 2009, 71% of growers sourced their feed from outside their own premises (80% in 1997), and 68% did not have to pay for materials (53% in 1997). Growers traveled on average 27 km (16.8 miles) to collect the material (22 km (13.7 miles) in 1997).
3. ***Quantity***. There has been a tendency since 1997 for operations to become either smaller or larger. In 2009, more growers were using small quantities of feed (<1 ton (1t) month^{-1}) than in 1997, and twice as many growers were using large quantities (>10 ton (10t) month^{-1}). These increases seemed to be offset by the decreased number of growers in the middle of this spectrum, those using about 5–10 ton month^{-1} (5–10t).

D Additives and Supplements Used in Vermicompost Production

1. ***pH adjustment***. Lime is relatively inexpensive and is used liberally in the maintenance of earthworm beds by almost 50% of earthworm growers (AWGA 1997). It is used primarily to correct low pH but also misguidedly to reduce populations of enchytraeids and *Drosophila*, considered as pests by some earthworm growers. There are limited options for reducing high pH, often associated with an alkaline or saline water supply. Powdered sulfur has been used, with positive responses reported by some earthworm growers.
2. ***Zeolite***. This naturally occurring mineral, mined in Australia, is used regularly on earthworm beds by more than 5% of Australian earthworm growers (AWGA 1997). With a high cation-exchange capacity, it can trap positive cations such as ammonia and has been reported as effective in deodorizing beds at recommended rates of 10–15 kg (22–33 lb) of zeolite for each metric ton of composted material.
3. ***Enzymes and microorganisms***. The benefits of vermicompost to plant growth are often attributed to the abundance of beneficial microorganisms present. Biological preparations containing microorganisms and enzymes may assist in the composting process, but their benefits in the vermiculture process have not been confirmed. The growth-promoting effects of vermicompost can be eliminated by sterilization of the material (Buckerfield et al. 1999).

E Pests and Diseases in Vermiculture

Flatworms are known to be predators of earthworms in soils (Jones and Boag 1996), and although they do not have significant effects on vermiculture, publicity about their occurrence in the 1990s prompted concern in the industry. The average vermiculturist may lack the skills to correctly identify and understand the range of other invertebrate species that occur in earthworm beds, such as enchytraieds, flatworms, Collembola, and insect larvae, and some appear to become overly concerned about the control and eradication of nonearthworm species in their systems.

F Vermiliquid (Tea) Production and Use

1. ***Current output***. For the average grower, sales of vermiliquid, or teas, have increased five-fold, making up just 2% of sales in 1997 but 10% in 2009 (Webster and Taylor 2009). Of those who were currently packaging vermiliquid for sale, the median amount was 4500 L.yr^{-1} (1189 gal), and the maximum amount was 500,000 L.yr^{-1} (132,086 gal).

2. ***Production methods***. Vermiliquid, or tea, is usually produced by one of two methods: (a) Solid vermicast is dissolved or suspended in water, or (b) leachate carried in water percolated through beds is collected. Some growers also prepare teas by "brewing" vermicompost in an aerated solution, often with other materials, to promote the rapid growth of populations of beneficial microbes.
3. ***Packaging***. Preparations for packaging may involve settling or filtration of undissolved materials. For leachates, increased concentration may be achieved with several passes of the liquid through beds. Active concentration through distillation or filtration is uncommon. There are no standards or guidelines on application rates, packaging, or shelf life of teas, though some vermiculturists have discovered that vented packaging is essential if the material is to be stored on the shelf for any length of time. Enhancement of N concentrations with a high-analysis fertilizer is sometimes seen, though this practice may render the product unfit for registration with organic certification bodies such as the Biological Farmers Association (Australia) and BioGro (New Zealand).
4. ***Product consistency***. There are no standards or guidelines to ensure minimum concentrations of nutrients, evaluate claims about plant growth-promoting substances and microbes, or set maximum concentrations of salts, ammonia, human pathogens, and toxic substances.

VI VERMICULTURE MARKETS AND MARKETING

In 1997, 72% of growers spent less than 5 hours per week in marketing and sales, compared to 46% in 2009 (Webster and Taylor, 2009). The relative proportion of earthworm growers targeting earthworm supply, vermicompost supply, and waste management has changed little from 1997 to 2009, and the focus remained on selling earthworms, with 90% of core respondents targeting earthworm supply (82% in 1997). Vermiculturists reported that on average 58% of their sales were of earthworms (68% in 1997).

Around half of 2009 earthworm growers were targeting vermicompost supply, a similar proportion to the 59% in 1997 who were targeting this market. It was noted in comments given by respondents to the 2009 survey that they would like to supply more vermicompost, but the demand was not yet sufficient, due, they felt, to customers' lack of awareness of the use and benefits of vermicompost.

A notable change since 1997 has been the rise in production of vermiliquids, or teas, with sales increasing five-fold. The desire to access this market was high, with packaging of vermiliquid and sales in this market targeted by around 50% of earthworm growers, compared to 22% in 1997. Some vermiculturists have reported that producing and packaging vermiliquids was easier than and preferable to the sale of vermicompost and that they are easier to apply.

A Profitability

In 2009, 78% of vermiculturists considered themselves as operating a commercial venture, and 71% had a positive outlook, although only 62% considered their operation to be profitable (Webster and Taylor 2009). There had been a marked change since 1997, when 84% considered themselves commercial but only 35% considered their operation profitable. In 1997, 61% of vermiculturists had been in the industry less than 2 years (compared to 9% currently), reflecting the strong push at the time to recruit new growers into earthworm-supply schemes.

B Buy-Back Schemes

In the 1990s, the industry saw the rise and subsequent collapse of earthworm-growing pyramid schemes. At the time, at least one-third of earthworm sales were committed to meeting contracts with retailers and brokers (AWGA 1997). A variety of schemes were on offer, but generally new growers purchased earthworms and bedding and contracted to sell earthworms at an agreed price for a fixed term. Other packages offered investments in earthworm beds that remained on a managed site, with investors paying an adjustment or management fee.

A common complaint of investors in these schemes was that they could not breed earthworms quickly enough to meet contractual obligations. Even where quoted breeding rates were conservative, they did not allow for reduced breeding with fluctuations in bed temperatures and mortality. Contracts requiring monthly delivery of minimum quantities of earthworms were challenged and modified by the Earthworm Association in New Zealand. Rejection or devaluing of earthworm deliveries on the basis that they did not have the right "mix" of species was also seen. The AWGA introduced business guidelines (Steuart and Steele 1995) in an effort to counter the high expectations of new growers. These earthworm-growing schemes have undoubtedly hindered the development and maturation of vermiculture technology and the industry in Australia and New Zealand (Cheal and Lewis 1997).

C Role of Earthworms in Organic-Waste Management

Although there has been much interest in the use of earthworms for organic-waste management, there has been little research on its commercial feasibility under Australasian conditions (Mitchell 1993; Recycled Organics Unit 2007). In 2009, fees collected for organic-waste management made up almost 10% of sales for vermiculture operations, and waste management was targeted by 31% (Webster and Taylor 2009).

D Agricultural Market-Development Trials

There have been several substantial trials to demonstrate the use of vermicomposts under field conditions. Positive responses have been seen in broad-scale cropping of alfalfa (Foster and Windsor 1997) and cotton (Buckerfield 1998). In horticulture, impressive yield increases and soil improvements have been seen in grapes, cherries, and pears in response to surface-applied vermicompost (e.g., Buckerfield and Webster 1998) but only when applied at relatively high rates (5–20 mm depth) and protected from drying and sunlight with a mulch, or through repeated applications (Smith et al. 1999).

VII STANDARDS AND TESTING

A Terminology

The industry has adopted the terms *vermicompost* and *vermicast*, and the properties of these have been defined in the New Zealand and Australian standards (Standards Australia 2003; Standards New Zealand 2005). However, not all vermiculturists understand the distinction or use the terms to define their products. While the term *vermiliquid* has been used throughout this chapter, it has not been officially adopted as described by the industry.

B Analyses

Around half of the respondents to a 2009 industry survey (Webster and Taylor 2009) had sent their vermicompost or vermicast to a laboratory for analysis. Of those who had not, 62% were interested in having analyses performed, compared with 78% in 1997. Analysis of the finished products is required if vermiculturists are to make claims that material meets standards (Standards Australia 2003; Standards New Zealand 2005).

C Standards

The great majority of vermiculturists (89%; Webster and Taylor 2009) do not own a copy of the *Australian Standard for Composts, Soil Conditioners and Mulches* AS4454 (Standards Australia 2003). Despite considerable effort on the part of the AWGA (1999) to have vermicompost included in the revised standard (Standards Australia 1999), ownership of the standard is the same as it was prior to the revisions (90%; AWGA 1997). The *New Zealand Standard for Composts, Soil Conditioners and Mulches* was based on the 2003 Australian standard (Standards New Zealand 2005). The standards set out general conditions for soil conditioners, fine mulch and vermicast, and additional specific parameters for vermicast:

1. **Presence of viable plant propagules (seeds, rhizomes, bulbs, roots, etc.).** The standard requires nil germination after 21 days incubation. The standard also sets out requirements for thermophilic composting that would ensure achievement of this standard, though we note that only 32% of vermiculturists claimed to thermophilically precompost their feedstock (Webster and Taylor 2009). Materials that cannot demonstrate an adequate thermophilic composting phase may present a risk of transmitting weeds to agricultural properties and home gardens.

2. **Human pathogens**. Vermicomposts and vermicasts must meet the relevant state guidelines for limits on the presence of human pathogens.
3. **Definitions**. To be classed as *vermicast*, more than 90% of the material must pass through a 1.18 mm aperture sieve. Materials with less than 90% passing through this sieve may not be labeled as *vermicast* but may still meet the standard upon meeting the requirements for a soil conditioner and be called *vermicompost*, though this would necessitate a thermophilic composting phase.

VIII FUTURE DEVELOPMENTS

Without the activity of vermiculture associations, much of the future development of the industry will be in the hands of enthusiastic individuals and large operations. There is support among the vermiculture industry for the reactivation of these associations (Webster and Taylor 2009), with vermiculturists identifying a need for functions such as research and development of product use in agriculture, dissemination of research information, provision of technical information and advice, industry representation and lobbying, earthworm-growing research, marketing advice, and access to group marketing arrangements. There is a risk that the gains made during the 1990s by industry associations will be lost; for example, representations by the AWGA led to the inclusion of vermicast in the *Australian Standard for Composts, Soil Conditioners and Mulches* AS4454 (Standards Australia 1999), which later became the basis for the New Zealand standard (Standards New Zealand 2005). The Australian standard is currently under review by a panel of industry stakeholders and experts but without representation by the vermiculture industry.

The further development of very large-scale vermiculture in Australia and New Zealand is now focused primarily on developing markets for vermicomposts and vermiliquids, and less on the systems by which these products are created. The future of large-scale vermiculture for waste management and the supply of products to agricultural markets looks promising, with a stabilization of the industry and adoption of realistic yet positive expectations.

REFERENCES

Abbott, I. 1982. The distribution of earthworms in the Perth Metropolitan Area. *Rec. West Aust. Mus.* 10:11–34.

Appelhof, M., K.A. Webster, and J.C. Buckerfield. 1996. Vermicomposting in Australia and New Zealand. *Biocycle* 37:63–66.

Australian Worm Growers Association (AWGA). 1997. Members General Questionnaire. Information Circular May 1997. *Australian Worm Growers Association*, Melbourne.

Australian Worm Growers Association (AWGA). 1999. Draft Industry Minimalist Best Practice Guidelines. Information Circular 1999. *Australian Worm Growers Association*, Melbourne.

Baker, G.H., and V.J. Barrett. 1994. *Earthworm Identifier.* Melbourne: CSIRO Publishing.

Baker, G.H., G. Brown, K. Butt, J.P. Curry, and J. Scullion. 2004. Introduced earthworms in agricultural and reclaimed land: Their ecology and influences on soil properties, plant production and other soil biota. *Biol. Inv.* 8:301–316.

Barley, K.P. 1959. The influence of earthworms on soil fertility. *Aust. J. Agric. Res.* 10:171–178.

Blakemore, R.J. 1994. Earthworms of southeast Queensland and their agnomic potential in brigalow soils. PhD diss., University of Queensland, St. Lucia.

Blakemore, R.J. 1997a. Agronomic potential of earthworms in brigalow soils of southeast Queensland. *Soil Biol. Biochem.* 29:603–608.

Blakemore, R.J. 1997b. First common earthworm found in Tasmania. *Invertebrata* 9:1, 5.

Blakemore, R.J. 2000. Native earthworms (Oligochaeta) from southeastern Australia, with the description of fifteen new species. *Rec. Aust. Mus.* 52:187–222.

Bouché, M.B. 1977. Strategies lombriciennes. In *Soil organisms as Components of Ecosystems: Proceedings of the 6th International Colloquium on Soil Zoology*, ed. U. Lohm and T. Persson, 122–123. Swedish Natural Science Research Council Ecological Bulletin No. 25. Stockholm.

Buckerfield, J.C. 1994. *Appropriate earthworms for agriculture and vermiculture*. CSIRO Australia, Division of Soils Technical Report 2. Urrbrae: CSIRO Division of Soils.

Buckerfield, J.C. 1998. Worm-worked wastes as an organic amendment for cotton soils. In *Australian Cotton Conference, 11–14th August 1998, Broadbeach, Queensland*, 767–772.

Buckerfield, J.C., T.C. Flavel, K.E. Lee, and K.A. Webster. 1999. Vermicompost in solid and liquid forms as a plant-growth promoter. *Pedobiol* 43:753–759.

Buckerfield, J.C., and K.A. Webster. 1996. Earthworms, mulch, soil moisture and grape yields. *Aust. N.Z. Wine Ind. J.* 11:47–53.

Buckerfield, J.C., and K.A. Webster. 1998. Worm-worked wastes boost grape yields—prospects for vermicompost use in vineyards. *Aust. N.Z. Wine Ind. J.* 13:73–76.

Cheal and Lewis. 1997. Vermiculture—an industry perspective. In *Compost 97: Green putrescible waste management beyond 2000. Proceedings of Compost 97 Conference Held 14–15 July 1997, Brisbane*. Brisbane: Waste Management Research Unit, Griffith University.

Edwards, C.A. 2004. The use of earthworms in the breakdown and management of organic wastes to produce vermicomposts and animal feed protein 345–438. In *Earthworm Ecology*, ed. C.A. Edwards. Boca Raton, FL: CRC Press.

Edwards, C.A., and P.J. Bohlen. 1996. *Biology and Ecology of Earthworms*. 3rd ed. London: Chapman and Hall.

Edwards, C.A., and M.J. Shipitalo. 1998. Consequences of earthworms in agricultural soils: Aggregation and porosity. In *Earthworm Ecology*, ed. C.A. Edwards. Boca Raton, FL: CRC Press.

Edwards, C.A., and J. Steele. 1997. Using earthworm systems. *Biocycle* 38:63–64.

Farquhar, B.A. 1992. *Bert's Story: The Autobiography of B.A. Farquahar.* Launceston Australia: Regal Publications.

Fletcher, J.J. 1886a. Notes on Australian earthworms part I. *Proc. Linn. Soc. New South Wales* 1:523–576.

Fletcher, J.J. 1886b. Notes on Australian earthworms part II. *Proc. Linn. Soc. New South Wales* :943–974.

Fletcher, J.J. 1887. Notes on Australian earthworms part III. *Proc. Linn. Soc. New South Wales* 2:377–402.

Fletcher, J.J. 1890. Notes on Australian earthworms part VI. *Proc. Linn. Soc. New South Wales* 4:987–1019.

Foster, G.G., and R.A. Windsor. 1997. *Report on Lucerne; Worm Castings Trial at "Illourie", Boorowa.* Canberra, Australia: ACT Waste.

Jones, H.D., and B. Boag. 1996. The distributions of New Zealand and Australian terrestrial flatworms (Platyhelminthes, Turbellaria, Tricladida, Terricola) in the British Isles. *J Nat Hist* 30:955–975.

Kingston, T.J., and M.G. Temple-Smith. 1989. Earthworm populations under Tasmanian pastureland. In *Proceedings 5th Australasian Conference on Grasslands Invertebrates and Ecology, Melbourne University, Victoria, 15–19 August 1988*, ed. P.P. Stahle. Melbourne: Melbourne University.

Lee, K.E. 1959. A Key for the Identification of New Zealand Earthworms. *New Zealand Soil Bureau Publication* 172:13–60.

Lotzof, M. 1998. Successful biosolids benefication with Vermitech's large scale commercial vermiculture facility in Redlands. *Waste Disposal and Water Management,* Australia. September/October: 3–13.

Martin, N.A. 1977. Guide to the lumbricid earthworms of New Zealand pastures. *N.Z. J. Exp. Agric.* 5:301–309.

Mitchell, A. 1993. Earthworms in the management of agricultural wastes. In *National Workshop on the Role of Earthworms in Agriculture and Land Management,* June 1993, Launceston, Tasmania, Australia.

Murphy, D.J. 1991. Vermiculture: The potential and the reality. In *Earthworms Improving Soil for Agriculture*, ed. P.J. Haines. Australian Institute of Agricultural Science. Occasional Publication 62: 35–38.

Natoli, J. 1996. Worming into the community. *Worm Digest* 13:20–23.

Parliamentary Counsel Office New Zealand. 2008. Waste Minimisation Act 2008. Public Act No. 89, Wellington, New Zealand: Parliamentary Counsel Office.

Recycled Organics Unit. 2007. *Literature Review of Worms in Waste Management.* Vol. 1. Sydney: University of New South Wales.

Sabine, J.R. 1978. The nutritive value of earthworm meal. In *Proceedings of the Conference on the Utilization of Soil Organisms in Sludge Management*, ed. R. Hartenstein, 122–130. New York: Syracuse.

Scarborough, J. 1999. Biogreen castings: Scientifically assessing the merits of vermicomposting biosolid and green waste mixes. Unpublished report for the Green Waste Challenge, Environmental Protection Authority. Bathurst: Bathurst City Council.

Smith, C.J., W.J. Bond, and W. Wang. 1999. *Waste free: "Vermicompost" to improve agricultural soils.* CSIRO Land and Water Technical Report 23/99, July 1999. Canberra: CSIRO Land and Water.

Standards Australia. 1999. *Australian Standard for Composts, Soil Conditioners and Mulches.* AS-4454. 2nd ed. Sydney: Standards Australia.

Standards Australia. 2003. *Australian Standard for Composts, Soil Conditioners and Mulches.* AS-4454. 3rd ed. Sydney: Standards Australia.

Standards New Zealand. 2005. *New Zealand Standard for Composts, Soil Conditioners and Mulches.* NZS-4454. Wellington: Standards New Zealand.

Steuart, F., and J. Steele. 1995. *Australian Worm Growers Association: Code of Practice. Information Circular 1995.* Melbourne: Australian Worm Growers Association Inc.

Stockdill, S.M.J. 1982. Effects of introduced earthworms on the productivity of New Zealand pastures. *Pedobiologia* 24:29–35.

Temple-Smith, M.G. 1993. Logistics; economics of broadscale earthworm distribution. In *National Workshop on the Role of Earthworms in Agriculture and Land Management,* June 1993. Launceston, Tasmania, Australia.

Tomlin, A.D. 1983. The earthworm bait-market in North America. In *Earthworm Ecology*, ed. J.E. Satchell. 331–338. London: Chapman & Hall.

Webster, K.A., and P.D. Taylor. 2009. *The Australian and New Zealand vermiculture industry 2009*. EcoResearch Technical Report No. 02/2009. Stirling: EcoResearch.

Whitta, R. 1996. The "worm" from New Zealand. *Worm Digest* 13:2.

Williams, T. 1994. Worm your way through waste: A financially viable, ecologically sustainable, practical alternative. In *Compost 94: Options for management of organic waste. Proceedings of Compost 94 Conference held 3–4 May 1994, Brisbane*. Brisbane, Australia: Waste Management Research Unit, Griffith University.

Zero Waste South Australia. 2005. *South Australia's Waste Strategy 2005–2010*. Adelaide: Zero Waste South Australia.

CHAPTER 28

Origins and Spread of Vermicomposting in India: Focus on Sustainable Agriculture

Radha Kale

CONTENTS

I Introduction 453
II Development of Vermiculture Technology in India 454
III Vermicompost Production 457
IV Larger-Scale Commercial Development of Vermicomposting in India 459
V Vermicompost to Support Sustainable Agriculture 460
  A Agricultural Production 460
  B Growth Induction in Plants by the Use of Vermicomposts 461
  C Pathogen and Pest Management in Response to Application of Vermicompost or Aqueous Extracts of Vermicompost 462
  D Enrichment of Vermicomposts with Microbial Inoculants 463
VI The Future of Earthworm Research in India 463
References 464

I INTRODUCTION

The importance of earthworms as biological material for research is about five decades old in the Indian subcontinent. Among papers on earthworms by Indian scientists are the works of Joshi and Kelkar (1952) and Nijhawan and Kanwar (1952), who highlighted the mobilization of N in earthworm casts and the physicochemical properties of earthworm casts that differed from surrounding soil under field conditions. Other reports are on maintaining earthworm cultures (Tambe and Dubash 1961; Bhat 1974) and microbial associations in earthworm-worked materials (Khambata and Bhat 1957; Bhat et al. 1960). Roy (1957) quantified soil turnover due to earthworm activity. In the following decade, K. P. Rao and his coworkers (Saroja and Rao 1965; Rao 1965) contributed to the understanding of the physiology of earthworms, especially in relation to changing temperature conditions. The other school in the country that has contributed to earthworm research is from Sambalpur, Orissa, India, under the leadership of Dash, who has made many contributions to ecological studies of earthworms, especially their energy relationships (Dash and Patra 1977).

In the United States, culturing of earthworms in organic wastes was popularized to meet the needs of fishing; this is a lucrative commercial venture there. The aim was to harvest the maximum biomass of earthworms rather than to use the large quantities of earthworm castings produced. When knowledge of rearing of earthworms became known in India, it was considered from a different angle, vis-à-vis an ultimate answer to problems of organic solid waste management and the use of the end product to contribute to sustainable agriculture. Emphasis was placed largely on the production of earthworm casts or vermicomposts to improve the physical conditions of overburdened agricultural lands, which had experienced excessive use of inorganic fertilizers, rather than to produce protein feed for domestic animals. Then there were certain controversies and conflicts among a few earthworm workers in different parts of the country about the employment of earthworms in organic solid waste management. Moreover, as with any other innovations, many farmers, agricultural scientists, industrialists, and farming consultants were skeptical about the potential of this line of work. Although many hurdles were cleared, there was huge competition among commercial producers and consultants to make quick profits from this activity.

II DEVELOPMENT OF VERMICULTURE TECHNOLOGY IN INDIA

India on average generates 70 million tons of organic wastes per annum, and most of this is either burned or put into landfill sites. Proper methods of waste segregation and adoption of suitable methods of composting can resolve the problems of waste accumulation and also improve the physical status of productive soil to obtain better agricultural yields.

The first attempts to utilize earthworms for waste management started in 1978 in the Department of Zoology, University of Agricultural Sciences, Gandhi Krishi Vignana Kendra (GKVK), Bangalore, India. One of three species of earthworms, *Perionyx excavatus* Perr., was found in a compost pit in a garden near Bangalore city. The compost pit was receiving the slurry from a biogas plant. These earthworms were brought to the laboratory to test their ability to survive and multiply in different organic wastes. It can be said that the use of this species was the beginning of pioneering attempts in this country to employ earthworms for the degradation of organic solid wastes.

Some of the research results were presented at the International Symposium on Earthworm Ecology held at Cumbria, United Kingdom, in 1981 to mark the centenary celebration of Darwin's work on earthworms (Kale et al. 1982). Impressed with the data presented, O. Graff of Germany, one of the participants in the symposium, donated 50 cocoons of *Eudrilus eugeniae* (Kinb) to test the possibility of their establishment in India, which has similar climatic conditions to the place of origin of this species (Nigeria). Later, it was found that the distribution of this species in India had been reported earlier (Stephensen 1923). Since 1982, *E. eugeniae* has been used for organic-waste degradation in India. *E. eugeniae* surpassed *P. excavatus* in both feeding and reproduction rates. Initially, it was thought that such vigor might be short-lived, since many organisms show a tendency to perform extremely well in a new environment until they become fully established. However, no change in the biological efficiency of this species has been observed over the last 28 years. A series of studies have shown that *E. eugeniae* is an ideal species to use for agricultural and agro-based industry waste in India (Kale 1992; Ghosh 2004; Kumar and Singh 2007; Kumar et al. 2007). However, this species's meandering nature and sensitivity to population-density pressure demands larger space for their culture. They are more sensitive to handling and adaptable to different organic substrates than *Eisenia fetida* and species of *Perionyx*. *E. fetida*, which has a wide distribution in the state of Himachal Pradesh, North-Eastern State, Sikkim, and hill stations in the south, is also an introduced exotic species. *E. fetida* along with *P. excavatus* have become the preferred species for vermicomposting activities in urban areas. *E. eugeniae* is the favored species for vermicomposting in southern peninsular India, which has a subtropical climate with minimal variations in summer and winter temperatures. *E. fetida* is preferred in the northern part of the country, where temperature extremes are reached during summer and winter months. *P. excavatus* is a widely distributed earthworm species in India.

In addition to *E. eugeniae*, *E. fetida*, and *P. excavatus*, other species of earthworms that have been used successfully for vermicomposting are *Pheretima sansibaricus* Mich., *Dichogaster curgensis* (Mich.), and *Dichogaste bolaui* (Mich.). All these species can survive in organic matter with 40–60% moisture. The organic matter has to be conditioned for 15–21 days to undergo primary decomposition, the temperature drops to around 30°C (86°F). No soil or sand is mixed into the waste or provided as bedding material at the base of the tanks or pits. These earthworms can complete their life cycle in 60–90 days.

E. eugeniae has a high reproductive rate (peak production of cocoons) from June to the end of October when the temperature is 25°C–30°C (77–86°F), and the humidity is around 80%. From November there is decline in numbers of cocoons deposited, and from December to the end of May cocoon depositions are negligible until the onset of the next monsoons. A minimum 40-fold increase in the populations of *E. eugeniae* is usually observed between June and October under the climatic conditions prevailing in Bangalore (Kale and Bano 1994). The same species in Kerala State, India, produces cocoons throughout the year except for 2 months a year (April and May) when the relative humidity is very low. Kerala State is a very humid coastal area that experiences rains nearly 9 months in a year. *E. fetida* has a peak level of cocoon production from September to the end of January, when the atmospheric temperature in Bangalore drops below 25°C (77°F). The life cycle of *P. excavatus* follows a similar pattern to that of *E. eugeniae* (Kale et al. 1982). The number of juveniles that emerge from cocoons of *E. eugeniae* and *E. fetida* ranges between three and seven, whereas *P. excavatus* produces more cocoons.

Many laboratories are pursuing research to establish the ability of *P. excavatus* to survive in different types of organic wastes, with the aim of abating organic pollution. Choudhuri et al. (2001) used different aquatic weeds along with cow dung to produce vermicompost with this species. Kaushik and Garg (2004) tested the possibility of utilizing industrial wastes such as paper sludge from the paper industry. The tolerance of earthworms to industrial wastes is very low, and earthworm composting may not be suitable for the large quantities of wastes generated in these units. Garg et al. (2006) were able to maintain *E. fetida* on manure from cows, buffaloes, horses, donkeys, sheep, goats, and camels. However, the earthworm-biomass increase was greatest in horse manure, followed by cow dung. In recent times, buffalo farms in Karnataka have been maintaining *E. eugeniae* successfully on buffalo dung and other farm wastes.

Apart from the three earthworm species already mentioned, there have been trials to use many other species of earthworms for organic-waste management. Paul and Selvi (2007) have successfully maintained *P. ceylanensis* to produce vermicompost from the biogas slurry of municipal solid waste mixed with cow dung.

Awareness of vermicompost production, and of the preferences of the various earthworm species to suit different geographic zones for the breakdown of organic matter, has been developed in India. Addition of N-fixing and P-solubilizing organisms to vermicompost, together with rock phosphate, can enhance the N and P levels in vermicompost (Kumar and Singh 2001; Kumari and Kumari 2002). The proportions in which the various farm wastes are mixed with cow dung can influence the biomass increases of *E. fetida* (Barik et al. 2002). The quality of feed in turn influences the biochemical properties of the vermicompost (Zachariah and Chhonkar 2004).

Most of the progress in the field of vermiculture has to be attributed to research activities at the University of Agricultural Sciences, Bangalore, India, whose policy is that all research results should benefit the farming community in the transfer of technology from the laboratory to the land. With this motto, details of the technology of vermicomposting were released to the press in 1984 for the benefit of farmers in particular and the public in general. However, it did not create much interest among farmers until 1990. Nevertheless, inquiries started trickling in from a small number of agriculturists and nongovernmental organizations to find out more about its merits.

From 1980 on, some social workers and commercial workers from Maharashtra State (Bombay province), India, started collecting information from the articles and books published in the United States on the commercial viability of vermicomposting technology, as well as research data from the University of Agricultural Sciences, Bangalore. Using both sources of information, they made unsupported claims about methods of vermicomposting, the nutrient status of the vermicompost, and recommendations on vermicompost use for different crops without any field trials. When these activities did not yield results, it raised criticisms from some agricultural scientists and doubts among many farmers on its viability.

There was also a view that only indigenous earthworm species should be used and not the exotic species such as *E. fetida* and *E. eugeniae*. The *Fauna of British India*—a volume on Oligochaeta (Stephensen 1923)—lists several species of earthworms as exotic or peregrine. The northern part of India has more exotic species than the southern peninsula. The genera *Dichogaster* or *Perionyx* or *Lampito*, which are recommended for vermicomposting by some workers, include "peregrine" species that originated in tropical countries other than India.

III VERMICOMPOST PRODUCTION

There has been enormous progress in research related to the use of agricultural and municipal solid organic wastes to produce vermicomposts and in studies on the use of vermicomposts on different crops. Governmental organizations and banks are funding projects related to earthworm research and extending financial support to universities, research institutes, and nongovernmental organizations as a promotional activity.

Sunshine and rains are common in many parts of India, and a lack of food sources during the dry season forces rodents and other small predatory mammals to find alternative and easily available food sources. An ant species (*Dorylus orientalis*) can have devastating effects on vermiculture. A survey conducted in Madhya Pradesh (a state in central India) revealed that nearly 95% of farmers have knowledge of vermiculture and its benefits to farmland. However, they are unable to give serious thought to vermicompost production as they are concerned about ant attacks in the culture bins, water shortages, lack of storage facilities, and, finally, marketing avenues for the product when produced in excess (Baghel et al. 2005).

In the majority of places in southern India, vermicomposting is done in strongly built shallow bins, normally 3.0 × 1.2 m × 0.9 m (3.3 × 1.3 × 1.0 yd). The materials used for the construction vary depending on availability and cost. Generally, cured mud bricks, cement, or granite slabs are used for the construction of bins or enclosures. Except for depth (not exceeding 0.9 m (1 yd)), the length, breadth, and shapes of the bins vary based on what is convenient for the user. The bins are covered with a metal mesh to keep away rats and other predators. A thatched roof or polyethylene sheet cover is erected over the bins to protect them from heavy rains and direct sun. A recent survey in some parts of Karnataka showed that with the proliferation of earthworm populations, farmers have started to release them into open pits and heaps of organic wastes in the fields. The high reproductive potential of these earthworms is helping to maintain earthworm populations irrespective of predatory pressures.

The main aim of vermiculture is the recycling of organic wastes. Organic farming and sustainable agriculture are gaining considerable momentum in India. An imbalance has been created in the nutrient status of soils after continuous applications of inorganic fertilizers without use of any organic manures. The farmers are encouraged to use all kinds of organic wastes, such as hedge trimmings, weeds, litter, hay, husks, and animal waste, available on farms for vermiculture. The available waste is heaped on the ground and mixed with a thin slurry of cattle dung. The mixture is turned every 2–3 days for about 3 weeks to provide aeration and to dissipate heat generated. After 2–3 weeks, depending on the nature of the organic wastes used, the mixture is transferred to bins, and earthworms are released onto the surface. Mixed cultures of *E. eugeniae*, *E. fetida*, *P. excavatus*, and *P. sansibaricus*, along with *Dichogaster curgensis* and *D. bolaui* are commonly introduced into culture bins. About 1500–2000 adult earthworms (or 1 kg biomass) are added to each 1 m^2 of surface area.

After the addition of the earthworms, the bins are left undisturbed for 6–8 weeks except for sprinkling of water to maintain the moisture level between 60% and 70%. Depending on the nature of the organic waste (ranging from soft green leaves and animal wastes to hard dry leaf litter and small twigs), and their proportions in the mix, 60–80% of the waste is converted into fine earthworm castings or vermicompost in 75–90 days. The bins are emptied, and the vermicompost is piled in a heap along with the earthworms. Earthworms move down and cluster at the base of the pile in a few hours. The vermicompost is cleared from the top, and earthworms are taken out carefully and added to new bins or to containers of partially decomposed organic material. In an alternative method, after the introduction of earthworms into the bins, the castings that are added to the surface by the earthworms feeding on the organic matter can be collected periodically, weekly or once every 10 days. Before collecting the castings, the watering of the bins is stopped for 2 days. When the moisture is lower in the surface materials, earthworms move down and stay at lower levels. This helps to separate only the castings that are on the surface. Collected castings and vermicomposts are piled up in the shade.

The collected vermicompost will contain earthworm cocoons and juvenile earthworms. To recover these cocoons and juveniles, small cow-dung balls are buried in the vermicompost heaps at different places with twigs as markers. After 10–15 days, the buried cow-dung balls are removed from the compost. The hatchlings and small earthworms aggregate in these balls. So when they are removed from the vermicompost, they will have almost all the earthworms in them. After the vermicompost and the earthworms are separated, the vermicompost is dried in the shade and passed through a 3–4 mm (0.12–0.16 in) aperture sieve to separate the unconsumed material from the vermicompost. The moisture content of the shade-dried vermicompost is 15–20%.

During the vermicomposting process the required moisture levels are maintained by periodic sprinkling of water on the surface. Farmers follow a rule-of-thumb method to assess the required moisture in the material in the composting bins. Frequent watering is essential during dry seasons such as summer and sometimes in winter.

The high reproductive potential of *E. eugeniae*, *E. fetida*, and *P. excavatus*, which are species used for vermicomposting, has prompted farmers to test the feasibility of using them for in situ vermicompost production in plantations and orchards. In these places the land is not disturbed much, and use of drips or sprinklers to water the plants during dry seasons is becoming very popular. Here, the plant bases are mulched with organic waste, and earthworms are released into the organic matter. Although rats and some other animals like wild boars may feed on the earthworms, the surviving earthworms add to the overall populations and work on the wastes. The mulching of land and use of earthworms for in situ production of vermicompost are used in perennial mulberry crop cultivation, banana plantations, coconut and arecanut gardens, and orchards. Coffee is another crop that has shown good responses to vermicompost applications. In cardamom plantations, when *in situ* culture of earthworms was tried, rats that entered the fields to feed on earthworms started feeding on cardamom pods and the farmer lost the crop. This is a caution to farmers: Their enthusiasm to increase earthworm populations in field conditions should not result in adverse effects on the standing crop. It is important to consider the pros and cons before beginnings in situ development of earthworms for processing of farm wastes.

IV LARGER-SCALE COMMERCIAL DEVELOPMENT OF VERMICOMPOSTING IN INDIA

M/S Terra Firma, Bangalore, was the first commercial large-scale vermicompost-production unit that was started on the outskirts of Bangalore using an innovative technology (see Chapter 8). The unit had aimed at producing 300 tons (295 t) of vermicompost per month. It has expanded its activity to two more centers, one unit near Bangalore and another in the neighboring state of Andhra Pradesh. The current managing director has linked his marketing with M/S Koramandel fertilizers, and vermicompost is still marketed as Ralli Gold. Out of the 25,000 tons (24,605 t) of annual compost output, vermicompost production is maintained at 3000 tons (2950 t) per annum. According to the managing director, management of earthworms and processing of organic waste for their acceptance are difficult, because he is handling unsegregated garbage from the city and its outskirts. However, only segregated vegetable waste from market yards and agricultural wastes are being used in this unit for vermicomposting. Nonavailability of suitable organic waste limits this unit's ability to expand its vermicompost production beyond 500 tons (492 t) per month.

In recent times, many such units, including agro-based industries throughout the country, have adopted vermicompost production as part of their commercial activities. To name a few from Karnataka State, there are the Phalada Agro Research Foundation Pvt. Ltd., Multiplex, Prajwala vermicompost, Varsha vermicompost, Jayanth vermicompost, Vermitechnology House, Basanth Vermifarm, and EID Parry from Tamil Nadu. The annual production of vermicompost ranges between 500 and 1000 tons in most of these units. As the country has no specified regulations to assess quality parameters for composts and vermicomposts produced by commercial units, it is difficult to confirm the quality of the products (see Chapter 18).

Although many commercial units have been formed in India to produce composts and vermicomposts, agricultural universities and agricultural research institutes, which come under aegis of the Indian Council of Agricultural Research (ICAR), as well as many nongovernmental organizations, have felt a need to promote vermicompost production, to help farmers become self-sufficient in nutrient production by using the resources available on their farms. Vermicomposting is a proven technology that can be adopted by rural women for their self-sustenance. This is one technology that can be practiced without much educational qualification and with minimum skills. Many farm women who are producing vermicompost are marketing it, and also earthworms, to other farmers in their neighborhood, apart from using it for their own land (Kale 2000). The list of such women beneficiaries is very long. With the decision of the government of India to promote organic farming, vermicompost production is one of the activities that have been promoted among women's groups for their self-sufficiency. Thus there will be a better scope for this activity to become an important rural industry.

V VERMICOMPOST TO SUPPORT SUSTAINABLE AGRICULTURE

A Agricultural Production

In India, where about 70% of the population lives in rural areas, the main occupation is agriculture. In recent times, rapid industrialization and the growth of cities are resulting in reductions in available fertile agricultural lands. To become self-sufficient in food production, the available land is overexploited with the use of agrochemicals and high-yielding crop varieties. This situation led to the "green revolution." But indiscriminate use of agrochemicals for long periods, without proper amendment with organic manures, has resulted in declining soil productivity. Decreasing cattle populations in rural areas, due to lack of grazing land and non-availability of farm labor, have resulted in poor availability of good organic manures. As knowledge about vermicompost production is disseminated across the country, the interest among farmers has increased since it is a means to get nutrients from organic manures and to reduce the use of agrochemicals. Scientists aim to quantify the vermicompost as well as inorganic chemicals needed to maintain a balance of nutrient requirements for different crops.

Farmers have used vermicompost, with or without fertilizer, for different cereals, pulses, oil seed crops, vegetables, spices, and varieties of fruit trees (Kale 2004). Strategic planning in terms of the integrated application of manures with inorganic fertilizers can sustain the soils and benefit the farmers (Singh et al. 2007). Farmers have had positive attitudes toward the use of vermicompost as a result of its environmental impacts and increased crop yields (Reddy et al. 2006). Nutrient uptake is better in crops when inorganic fertilizers are used with organic manure products like vermicompost compared to the application of inorganic fertilizers alone (Prakash et al. 2008).

In many experimental trials carried out by different institutions in India, vermicompost has been used in different proportions, either together with inorganic fertilizer or with other manures and biofertilizers. In the organic cultivation of tomatoes (*Lycopersicon esculentum*) vermicompost was applied with neem cake. Maximum yields and availability of nutrients in soil after harvest were achieved in a treatment where 50% of N was supplied by vermicompost and 50% by neem cake (Sable et al. 2007). Application of 50% N through vermicompost and 50% through the recommended application rates of inorganic fertilizers to sunflowers (*Helianthus annus*) and onions (*Allium cepa*) resulted in better yields compared to chemical fertilizers (Mehla et al. 2006; Guggari et al. 2007). The overall performance of crops was found to be better when the required inorganic fertilizer was reduced to 50% of the recommended level and applied together with 5–10 ton ha^{-1} (2–4 t. acre) of vermicompost for any crop (Sundaramani and Mallareddy 2007). Ghuge et al. (2007) reported that applying 2.5 $ton.ha^{-1}$ (1.01 t.acre) of vermicompost together with 50%

of the recommended fertilizers to a cabbage crop resulted in the highest yields and increased uptake of NPK. Athokpam et al. (2008) reported that the addition of micronutrients like boron 10 $kg.ha^{-1}$ (8.9 $lb.acre^{-1}$), together with vermicompost (20 quintals ha^{-1}, equivalent to 1% N), improved the yields of tomatoes and the availability of NPK and organic carbon. These reports from different zonal regions of the country indicate clearly the possibility of reducing the use of inorganic fertilizers and at the same time improving crop production.

There is a need to bring down the costs of cultivation of cut flowers to improve farmers' economic status and also stop deteriorations in soil fertility. Field trials on the use of vermicompost on different floricultural crops have shown improvements in the pH of soil, levels of available nutrients after the harvest, and the quality parameters of different crops like China asters (*Callistephus chinensis* (L) Ness)—after application of 10 ton ha^{-1} (4 $t.acre^{-1}$) vermicompost and recommended inorganic fertilizers—and tuberose (*Polianthes tuberosa* L)—after application of vermicompost at 2 $ton.ha^{-1}$ (0.8 $t.acre^{-1}$) together with *Azotobacter* and phosphate-solubilizing bacteria 2.5 $gm.m^2$ (0.0802.yd^2) and inorganic fertilizers (200:300:200 kg NPK)—in place of 10 $ton.ha^{-1}$ (4 $t.acre^{-1}$) of farmyard manure (Nethra et al. 1999; Chopde et al. 2007).

Southern states like Karnataka, Andhra, and Tamil Nadu have a climate that is well suited for silkworm rearing and silk production. Mulberry is another crop that utilizes very high levels of fertilizers because vegetative growth is of importance in this perennial crop. For monophagous silkworms, the survival of larvae, and ultimately the silk yields, depend on the caterpillar food quality, that is, mulberry leaves. Kherdekar et al. (2000) observed that application of vermicompost at 5 $ton.ha^{-1}$ (2.03 t. $acre^{-1}$) to mulberry plots reduced the use of inorganic fertilizers to 50%. The yields of leaves improved, and feeding on these leaves improved the cocoon yields of silkworms and finally the silk characteristics. Reddy et al. (2003) promoted the use of sericulture waste for vermicompost production for farmers.

B Growth Induction in Plants by the Use of Vermicomposts

Another important factor that has been demonstrated is the stimulatory effect of earthworm secretions and vermicompost extracts on plants' growth and vigor. It is possible to induce earthworms to release their body fluid by following a simple technique. Earthworms (*E. eugeniae*) weighing 500 g (17.6 oz) are added to an enamel tray containing warm water 500 mL (1.05 pt) at 40°C (104°F) for about a minute. The shock makes the earthworms release body fluids through their dorsal pores and mouth. After a minute of exposure, the earthworms are placed in another tray containing water at room temperature, and this allows the fluid sticking to their body to mix with water and helps the earthworms overcome the heat shock. The liquid contents in both the trays are termed *vermiwash*. The vermiwash thus collected was diluted further with an equal volume of water and used as a spray on the Crinkle Red variety of anthuriums (*Anthurium andreanum* Lind.). The results were compared with those from plants receiving 0.05% urea as spray. The lower concentration of vermiwash tested was the most effective in inducing vegetative growth and flowering in plants (Karuna et al. 1999). Tea cuttings (CV.UPASI-9) grown in red soil and vermicompost

(3:1, v/v) had the greatest vigor, and the duration of growth was reduced from 12 months to 8 or 9 months (Radhakrishnan and Mahendran 2007). Nielson (1965) had isolated indole acetic acid (IAA)–like substances from homogenates of the tissues of different species of earthworms. Krishnamoorthy and Vajranabhaiah (1986) adopted bioassay techniques to quantify the plant growth hormones (auxins and cytokinins) in earthworm casts. These studies showed that the biomolecules in the earthworm coelomic fluid and the activity of certain microorganisms in vermicompost contribute to plant growth, which favors its use in nursery beds and in greenhouses.

C Pathogen and Pest Management in Response to Application of Vermicompost or Aqueous Extracts of Vermicompost

The practice of monocropping has resulted in a depletion of the micronutrients essential to support proper vegetative growth of many crops and, finally, expected yields. In recent times, it has become necessary to apply micronutrients like Zn, Boron, and Mn, and sometimes secondary nutrients like S to crops like rice, tomatoes, and mulberries. As soils are becoming saline, to raise the pH, Ca has to be applied. The micronutrient requirement is increasing with the continued application of only macronutrients like NPK to soils. The imbalance created in soils with respect to nutrients for crops has resulted in problems with pests and diseases. More and more agrochemicals are being used in agriculture. Farmers' dependence on chemicals is making agriculture a more unsustainable venture. It has been found that the application of vermicomposts to fields decreases the incidence of pests and diseases (see Chapters 9 and 10).

Better nutritional quality and better balance of nutrients may be one of the reasons for resistance to pests and diseases in crops as a result of vermicomposts (Prakash and Bhadoria 2002). The enhanced production of phenols and tannins that is achieved in host plants through organic amendments can also induce resistance to pests (Rao 2003; see Chapter 14). Singh et al. (2003) observed a greater increase in levels of phenolic acids in pea plants after applying a 3–4% vermicompost extract than in the controls. Yadahalli (2007) reported responses of sugarcane to sett rot disease (*Ceratocystis paradoxa*) after application of vermicomposts; the least disease was in the plot receiving farmyard manure (see Chapter 13). Populations of sucking pests like thrips and mites could be brought under control in a chili crop (*Capsicum annuum* L.) both in a crop nursery and in fields. The nursery plots received vermicompost at 500 $gm.m^{-2}$ (14.7 $oz.yd^{-2}$) and neem cake at 100 $gm.m^{-2}$ (2.9 $oz.yd^{-2}$). The plots were treated with 2.5 $ton.ha^{-1}$ (1.0 $t.acre^{-1}$) of vermicompost and 0.5 $ton.ha^{-1}$ (1.0 $t.acre^{-1}$) of neem cake. The control of thrips and mites was better in these plots than in plots treated with normal pest-control practices (George and Giraddi 2007). The root knot nematode (*Meloidogyne incognita*) was suppressed in *Whitania somnifera*, a medicinal plant, after application of vermicompost together with *Trichoderma harzianum* (Pande and Kabra 2003). The incidence of *Fusarium* wilt in fenugreek (*Trigonella foenumgraecum*) was reduced after amending soils with vermicompost (Mathur et al. 2006). *Trichoderma*-enriched vermicompost (*vermiderma*, 40 kg (88 lb) in 400 L (105 gal) water), applied to pruned buds of tea plants, resulted in earlier bud break, 17 days in the vermicompost treatment compared to 26 days in the control, and the yield per hectare increased. Initially, blister blight was under better control in fungicide-treated plots, but in later stages the control achieved in vermicompost-treated plots was comparable (Balasubramanian et al. 2006).

Shobha (2009) studied the inhibitory effects of vermicompost extracts (both in vitro and under field conditions) on three soilborne pathogens, *Xanthomonas campestris*, *Ralstonia solanacearum*, and *Fusarium oxysporum*, that cause considerable damage to solanaceous crops. The vermicompost extract (10% aqueous extract) was found to possess both antibacterial and antifungal effects against these pathogens. The 10% vermicompost derived from using neem leaves in the feed of the earthworm *E. eugeniae* exhibited a zone of inhibition at a level of 25.3 mm (1.0 in) for *F. oxysporum*, 35.3 mm (1.4 in) for *X. campestris*, and 32.3 mm (1.67 in) for *R. solanacearum* (1 mL (0.034 oz) in the well of the 7.5 cm (3 in) Petri plate). The extract retained its antimicrobial properties after exposure to 40°C (104°F) for 30 minutes but lost this property within 10 minutes after exposure to 60°C (140°F). After increasing the moisture level of the vermicompost to 25%, the antimicrobial properties could be restored for a period of 21–40 days. In microplot studies for management of *F. oxysporum* and *R. solanacearum*, the application of fresh vermicompost at 5 $ton.ha^{-1}$ (2.0 $t.acre^{-1}$) by a broadcast method and the soaking of seeds in a vermicompost extract for one hour before sowing were effective in controlling wilt disease in tomatoes (*Lycopersicon esculentum*) and brinjal (*Solanum melongena*) crops.

D Enrichment of Vermicomposts with Microbial Inoculants

Vermicompost applications to fields enhance the activities of soil microorganisms (Kale et al. 1992). It also supports the growth of beneficial microorganisms and hence serves as a suitable carrier material for biofertilizers. When vermicompost is prepared, the materials used as feedstock influence the microbial activity in the resulting vermicompost (Bhat 1974; Kale et al. 1986; Nirmalnath et al. 2001). Vermicompost was used as a growth medium to develop a culture of *Rhizobium* (Manivannan and Daniel 2007). Saha et al. (2008) observed faster transformation of organic P on application of vermicompost, which produced a better environment for microorganisms, measured in terms of phosphatase activity.

VI THE FUTURE OF EARTHWORM RESEARCH IN INDIA

Scientific investigations are progressing all over India to overcome the adverse effects of agrochemicals at a rapid rate. Vermicomposting is an activity that can solve many problems of organic-waste disposal. It helps protect soil productivity, especially the level of organic carbon. Minimizing the use of agrochemicals decreases air, water, and soil pollution; helps farmers to be self-sufficient; and decreases the burden on the Government Exchequer.

Many researchers and the Indian government are looking forward to benefits from interactions with scientists from other countries like the United States. The

Konganadu Arts and Science College is an autonomous institution in the state of Tamil Nadu that conducted an Indo-U.S. Workshop on Vermitechnology in Human Welfare from June 4 to June 7, 2007. This program was sponsored by different funding agencies in India, including the Indo-U.S. Science and Technology Forum, New Delhi.

The participants in the workshop included representatives from all over the country who had knowledge of vermiculture and vermicomposting. Progressive farmers from Tamil Nadu also participated actively in the workshop. The scientists who addressed the participants were from the United States, India, China, and the Philippines. The two important topics covered under systematics included earthworm diversity in India (Julka) and the role of taxonomists in biotechnology and in identifying suitable earthworm species for vermicomposting (James). Three papers highlighted the nature of research carried out on vermicompost production and related extension activity in India: These included the topics of empowerment of women in rural parts of Karnataka (India) through imparting of knowledge on vermicomposting (Kale), the status of composting in India (Ismail), and a case study of strategies undertaken in a small community in rural area of Tamil Nadu (Vijayalakshmi).

The participants were exposed to information on the science and technology of vermiculture (Edwards), the utility of vermicomposts in agricultural and horticultural crops, and its influence on crop health and production along with soil productivity as practiced in the United States. It was a useful opportunity for large-scale commercial vermicompost producers in India to hear from vermicompost producers from the United States. An aspect that has not been given much serious thought in India was the use of earthworms as animal feed. The importance of earthworms in pharmaceuticals was food for thought from the presentation from China (Sun), and from the Philippines there was a report on the use of earthworms as a source of animal protein. The importance of earthworm research for the production of drugs, earthworm use as a source of animal feed, and also their contribution in land reclamation are fields that need to be explored in India. The workshop started to emphasize the importance of interdisciplinary approaches in earthworm research to meet its goals. It concluded with a greater vision to identify and focus on earthworm research in diversified fields with proper interactions for better understanding.

REFERENCES

Athokpam, H., H.S. Athokpam, N.B. Singh, N. Chongtham, and N.G. Singh. 2008. Effect of vermicompost and boron availability, yield and uptake by tomato (*Lycopersicon esculentum* Mill) and soil fertility. *Environ. Ecol.* 26: 187–90.

Baghel, S.S., S.B. Agrawal, and A. Pandey. 2005. Vermicompost production technology: Adoption, advantages and problems encountered by adopters. *JNKVV Res. J* 39: 125–26.

Balasubramanian, S., S. Parathiraj, and P. Haridas. 2006. Effect of vermicompost based *Trichoderma* (Vermiderma) on the recovery of pruned bushes and on the control of certain disease in tea (*Camellia sinensis* (L) O. Kuntze). *J. Plantation Crops* 34: 512–16.

Barik, T., K.C. Samal, R.K. Panda, S. Nanda, and S.C. Mohapatra. 2002. Study on vermicomposting of farm and kitchen waste. *J. Appl. Zool. Res.* 13: 236–37.

Bhat, J.V. 1974. Suitability of experimentation diets for earthworm culture. *Curr. Sci.* 43: 266–68.

Bhat, J.V., S. R. Khambata, G. B. Maya, C.A. Sastry, R.V. Iyer, and V. Iyer. 1960. Effect of earthworms on the microflora of the soil. *Ind. J. Agric. Sci.* 30: 106–14.

Chaudhuri, P., T.K. Pal, G. Bhattacharjee, and S.K. Day. 2001. Nutrient changes during vermicomposting by *P. excavatus* of the aquatic weed *Trapa bispinosa. Philippine J. Sci.* 130: 127–33.

Chopde, M.R., S. Pillewan, and S.A. Bhongle. 2007. Integrated nutrient management in tuberose. *Adv. Plant Sci.* 20: 443–44.

Dash, M.C. and Patra, V.C. 1977. Density, biomass and energy budget of a tropical earthworm population from a grassland site in Orissa, India. *Rev. Ecol. Biol. Sol.* 14: 461–471.

Garg, V.K., Y.K. Yadav, A. Sheoran, Subhashchand, and P. Kaushik. 2006. Live stock excreta management through vermicomposting using epigeic earthworm *E. fetida. Environmentalist* 26: 269–76.

George, S., and R.S. Giraddi. 2007. Management of chilli (*Capsicum annuum* L) thrips and mites using organics. *Kar. J. Agric. Sci.* 20: 537–40.

Ghosh, C. 2004. Integrated vermin – pisciculture an alternative option for recycling of municipal soild waste in rural India. *J. Bioresource Tech.* 93: 71–75.

Ghuge, T.D., S.B. Jadhav, and A.K. Gore. 2007. Effect of organic and inorganic sources of nutrients on uptake and nutrient availability of soil after harvesting of cabbage. *J. Soils Crops* 17: 294–98.

Guggari, A.K., D.P. Biradar, S.B. Kalaghatagi, and A.P. Biradar. 2007. Integrated nutrient management in pearl millet (*Pennisetum glaucum*)–sunflower (*Helianthus annuus*) crop rotation in northern dry zone of Karnataka. *Crop Res.* 34: 39–46.

Joshi, M.V., and B V. Kelkar. 1952. Role of earthworms in soil fertility. *Ind. J. Agric. Sci.* 22: 189–96.

Kale, R.D. 1992. Earthworms—nature's quiet and biggest gift (a popular article). *Spice India* 5: 2–4.

Kale, R.D. 2000. Women in agriculture: Tips to turn a leaf for a better tomorrow. In *National Consultation on S&T for Women. A Millennium Dialogue 20–21st December 1999 Compendium of papers*, ed. V. Sharma, S. Chatterjee, and D. Raghunandan, 156–60. New Delhi: Department of Science and Technology, Government of India.

Kale, R.D. 2004. The use of earthworms: Nature's gift for utilization of organic wastes in Asia. In *Earthworm Ecology*, ed. C.A. Edwards, 381–99. 2nd ed. Washington, DC: CRC Press.

Kale, R.D., and K. Bano. 1994. Laboratory studies on age specific survival and fecundity of *Eudrilus eugeniae. Mitt. Hamb. Zool. Mus. Inst. Band. Hamburg* 89: 139–48.

Kale, R.D., K. Bano, and R.V. Krishnamoorthy. 1982. Potential of *Perionyx excavatus* for utilization of organic wastes. *Pedobiologia* 23: 419–25.

Kale, R.D., B.C. Mallesh, K. Bano, and D.J. Bagyaraj. 1992. Influence of vermicompost application on the available macronutrients and selected microbial populations in a paddy field. *Soil Biol. Biochem.* 24: 1317–20.

Kale, R.D., K. Vinayaka, K. Bano, and D.J. Bagyaraj. 1986. Suitability of neem cake as an additive in earthworm feed and its influence on the establishment of microflora. *J. Soil Biol. Ecol.* 6: 98–103.

Karuna, K., C.R. Patil, P. Narayanaswamy, and R.D. Kale. 1999. Stimulatory effect of earthworm body fluid (vermiwash) on Crinkle Red variety of *Anthurium andreanum* Lind. *Crop Res.* 17: 253–57.

Kaushik, P., and V.K. Garg. 2004. Dynamics of biological and chemical parameters during vermicomposting of solid textile mill sludge mixed with cow dung and agricultural residues. *Biores. Technol.* 94: 203–209.

Khambata, S.R., and J.V. Bhat. 1957. A contribution to study of the intestinal microflora of Indian earthworms. *Arch. Mikrobiol.* 28: 69–80.

Kherdekar, R.C., A.T. Munde., P.S. Neharkar, and M.M. Sonkamble. 2000. Effect of fertilizers on foliage yield of mulberry, *Morus alba* L. and economic traits of silkworm, *Bombyx mori.* L. *J. Soils Crops* 10: 221–25.

Krishnamoorthy, R.V., and S.N. Vajranabhaiah. 1986. Biological activity of earthworm casts: Assessment of plant growth promoter levels in the casts. *Proc. Indian Acad. Sci.* (*Animal Sci.*) 95: 341–51.

Kumar, A., B.T. Pujari, A.S. Halepyati, and M.G. Patil. 2007. Influence of organic and inorganic sources on yield components of Bt and non Bt cotton hybrids. *Kar. J. Agric. Sci.* 20: 596–97.

Kumar, V., and A. Singh. 2007. Effect of vermicompost and VAM inoculation on vegetative growth and floral attributes in China Aster (*Callstephus chinensis* (L) Nees). *J. Ornamental Hortic.* 10: 190–92.

Kumar, V., and K.P. Singh. 2001. Enriching vermicompost by nitrogen fixing and phosphate solubilizing bacteria. *Biores. Technol.* 76: 173–75.

Kumari, M.S.S., and K.V. Kumari. 2002. Effect of vermicompost enriched with rockphosphate on growth and yield of cowpea (*Vigna unguiculata.* L Walp). *J. Ind. Soc. Soil Sci.* 50: 223–24.

Manivannan, N., and T. Daniel. 2007. Isolation and mass multiplication of *Rhizobium* sp. for nitrogen enrichment of vermicompost. *J. Pure Appl. Microbiol.* 1: 251–54

Mathur, K., R.K. Bansal, and R.B.S. Gurjar. 2006. Organic management of *Fusarium* wilt of fenugreek (*Trigonella foenumgraecum* L) a seed spice. *J. Mycol. Plant Pathol.* 36: 94–95.

Mehla, C.P., M. Ram, and I. Singh. 2006. Response of onion (*Allium cepa* L.) to various organic sources of nutrients under integrated nutrient supply system. *Haryana J. Agron.* 22: 149–50.

Nethra, N.N., K.V. Jayaprasad, and R.D. Kale. 1999. China Aster (*Callistephus chinensis* (L) Ness) cultivation using vermicompost as organic amendment. *Crop. Res.* 17: 209–15.

Nielson, R.L. 1965. Presence of plant growth substances in earthworms demonstrated by paper chromatography and Went Pea test. *Nature (Lond.)* 208: 1113–14.

Nijhawan, S.D., and J.S. Kanwar. 1952. Physico-chemical properties of earthworm castings and their effect on the productivity of the soil. *Ind. J. Agric. Sci.* 22: 357–73.

Nirmalnath, P.J., A.P. Biradar, A.B. Patil, and M.B. Patil. 2001. Vermicompost microflora as influenced by different crop residues. *Kar. J. Agric. Sci.* 14: 168–69.

Pande, R., and A. Kabra. 2003. Root knot disease of Ashwagandha (*Withania somnifera*) and its ecofriendly cost effective management. *J. Mycol. Plant Pathol.* 33: 240–45.

Paul, J. and Selvi, K. 2007. Biogas production from municipal solid waste with cow dung and vermicomposting. *Environ. Ecol.* 25: 603–605.

Prakash, Y.S., and P.B.S. Bhadoria. 2002. Influence of organic manures on crop tolerance to pathogens and pests in two varieties of rice. *Ann. Agric. Res* 23: 79–85.

Prakash, M., M. Jayakumar, and N. Karmegam. 2008. Physicochemical characteristics and fungal flora in the casts of the earthworm *P. Ceylonensis* Mich. reared in *Polyalthia longifolia* leaf litter. *J. Appl. Sci. Res.* 4: 53–57.

Radhakrishnan, B., and P. Mahendran. 2007. Effect of vermicompost on growth and development of nursery plants. *Newsletter UPASI Tea Research Foundation* 17: 5–6.

Rao, K.P. 1965. Some biochemical mechanisms of low temperature acclimation in tropical poikilotherms. In *The Cell and Environmental Temperature*, ed. A.S. Troshin, 98–112. Moscow and Oxford: USSR. Acad. Sci. Pergamon Press.

Rao, K.P. 2003. Influence of host plant nutrition on the incidence of *Spodoptera litura, Helicoverpa armigera. Ind. J. Entomol* 65: 386–92.

Reddy, B.K., D.M.R. Rao, D.L. Reddy, and N. Suryanarayana. 2003. Studies on the effect of FYM and vermicompost on quantitative and qualitative characters of mulberry (*Morus* sp.) under semi-arid conditions of Andhrapradesh. *Adv. Plant Sci.* 16: 177–82.

Reddy, B.V.C., Honniah, M.N. Venkataramana, R.D. Kale, and A.N. Balakrishna. 2006. Impact of application of vermicompost on economics of coconut cultivation in Southern Karnataka. *J. Plantation Crops* 34: 663–68.

Roy, S.K. 1957. Studies on the activities of earthworms. *Proc. Zool. Soc. Calcutta.* 10: 81–98.

Sable, C.R., T.D. Ghuge, S.B. Jadhav, and A.K. Gore. 2007. Impact of organic sources on uptake, quality and availability of nutrients after harvest of tomato. *J. Soils Crops* 17: 284–87.

Saha, S., B.L. Mina, K.A. Gopinath, S. Kundu, and H.S. Gupta. 2008. Relative changes in phosphatase activities as influenced source and application rate of organic composts in field crops. *Biores. Technol.* 99: 1750–57.

Saroja, K. and Rao, K.P. 1965. Some aspects of the mechanism of thermal acclimation in the earthworm *Lampito mauritii. Z. Vergl. Physiol.* 50: 35–54.

Shobha, S.V. 2009. Eco-friendly strategies for management of soil borne plant pathogens. PhD diss., Vinayaka Missions University India.

Singh, K.P., A. Suman, P.N. Singh, and M. Lal. 2007. Yield and soil nutrient balance of a sugarcane plant ratoon system with conventional and organic nutrient management in sub-tropical India. *Nutr. Cycling Agroecosyst.* 79: 209–19.

Singh, U.P., S. Maurya, and D.P. Singh. 2003. Antifungal activity and induced resistance in pea by aqueous extracts of vermicompost and control of powdery mildew of pea and balsam. *Ztg. Pflanzenkrank Pflanzenschutz* 110: 544–53.

Stephensen, J. 1923. *Fauna of British India—Oligochaeta.* London: Taylor and Francis.

Sundaramani, N., and K. Mallareddy. 2007. Effect of different organic manures and inorganic fertilizers on growth and yield and quality of carrot (*Daucus carota* L). *Kar. J Agric. Sci.* 20: 686–88.

Tambe, V.B., and P.J. Dubash. 1961. The Earthworms: A Review. *J. Bombay Nat. Hist. Soc.* 58: 171–201.

Yadahalli, K.B. 2007. Influence of organic amendments on sugarcane sett rot development. *J. Plant Dis. Sci.* 2: 24.

Zachariah, A.S., and P.K. Chhonkar. 2004. Biochemical properties of compost as influenced by earthworms and feed material. *J. Ind. Soc. Soil Sci.* 52: 155–59.

CHAPTER 29

Vermiculture in the Philippines

Rafael D. Guerrero III

CONTENTS

I Introduction 469
II Earthworm Species Cultured 470
III Culture of Earthworms 470
IV Production and Use of Vermicompost 471
V Production and Use of Earthworm Protein as Animal Feed and Human Food 475
VI Cost and Returns of Vermicompost and Vermimeal Production 477
VII Earthworms as a Source of Pharmaceutical Materials 477
References 478

I INTRODUCTION

The Philippines is a tropical country in Southeast Asia with an agriculture-based economy. There are more than 400 species of endemic and introduced earthworms in the country (James, personal communication). In the Philippines, vermiculture, or the farming of earthworms, started in the late 1970s with the introduction of *Pheretima asiatica* from Taiwan by the private sector (Catalan 1980). This species, however, was later confirmed to actually be *Perionyx excavatus* by Edwards (personal communication) and Blackmore (personal communication).

Interest in vermiculture in the Philippines was stimulated by the private sector with the establishment of the now-defunct Philippine Earthworm Center (PEC) in 1978. The PEC organized conventions of the Philippine earthworm growers in 1981–1983. The Philippine Council for Aquatic and Marine Research and Development convened the First Philippine Vermi Symposium-Workshop in November 2003 and the International Symposium-Workshop on Vermi Technologies for Developing Countries in November 2005. This review covers the published and unpublished reports on the culture of earthworms in the Philippines for organic-waste utilization and other uses such as vermicompost, vermimeal (earthworm protein), and vermiceuticals (earthworm pharmaceuticals).

II EARTHWORM SPECIES CULTURED

Two earthworm species, *Perionyx excavatus* ("Indian blue") and the *Eudrilus eugeniae* ("African night crawler"), have been cultured commercially by farmers in the Philippines. *P. excavatus* is a native of South Asia (James, personal communication). While no record of its introduction into the Philippines before 1978 is available, Reyes (1968) reported that *P. excavatus* was present in the country in 1968. *E. eugeniae* was introduced into the Philippines by Otto Graff of Germany in 1982.

III CULTURE OF EARTHWORMS

For the culture of *P. excavatus*, a private-sector company recommended that its clients use indoor wooden boxes (60 × 46 × 20 cm) with predecomposed bedding materials consisting of sawdust, rice hulls, and rice bran in a 1:1:1 proportion (by weight). About 2000 earthworms (2 kg) (4.41b) were stocked per box and fed with fresh animal manure and/or plant materials, such as soybean residues, with amounts equivalent to 100% of the earthworm biomass per day. The moisture content was maintained at 60%–80% and the pH at 6.5–7. The weight of the earthworms was expected to double (i.e., to 4 kg per box) (8.81b) after 45–60 days of culture (Catalan 1980).

Using 0.78 m^2 circular tanks, each with 20 kg of predecomposed Murrah buffalo waste and 200 breeding earthworms (0.45 g average weight), the average survival was 97%, the average weight of the earthworms was 0.66 g, and the average juvenile production of *P. excavatus* per tank after 4 weeks of culture was 1420. The average juvenile production per earthworm was 7.1 per month or 1.8 per week (R. Guerrero 1983). This showed an increase in both weights and numbers of earthworms with reproduction and growth in a suitable medium. An individual 0.4 g *P. excavatus* starts breeding when less than 6 weeks of age. Each breeder earthworm can produce from 17 to 31 young per month. The species can reach a mean weight of 1.5 g and have a growth rate of 0.015 g day^{-1} (R. Guerrero 1979).

Ulep (1982) noted increases in the weight of *P. excavatus* in wooden boxes (25 × 20 × 14 cm) provided with different organic substrates and stocked at 70–75 earthworms per box after 31 days of culture. The mean weights of the earthworms cultured in mixtures of sawdust + rice bran, rice straw + rice bran, and rice straw + sawdust + rice bran were significantly higher than those with the other bedding materials. This indicated that the type of bedding material can affect rates of earthworm growth considerably.

R. Guerrero et al. (1984) reported significantly greater weight gains by *E. eugeniae* stocked at 200 individuals per wooden box (0.28 m^2) with 75% pig manure and 25% sawdust (by weight) as a substrate compared to those with pig manure only

Table 29.1 Weight Gain of *E. eugeniae* Stocked at 200 g/Box with Different Substrates after 30 Days of Culture

Substrate	Weight Gain (%)
Pig manure	0.94[a]
50% Pig manure + 50% sawdust	1.57[b]
75% Pig manure + 25% sawdust	1.83[b]

Note: Figures are means of four replicates. Means with the same superscript are not significantly different at $p < 0.05$.

(Table 29.1). The initial biomass of *E. eugeniae* was also found to affect earthworm production. In 1.5 m^2 outdoor beds with 200 kg of 75% fresh grass (shredded) and 25% animal manure (dry), a stocking rate of 0.7 kg m^{-2} earthworms gave the highest net earthworm biomass production after 35 culture days compared with stocking rates at 1.3 kg m^{-2} and 2.0 kg m^{-2}.

IV PRODUCTION AND USE OF VERMICOMPOST

Vermicompost, consisting of a mixture of earthworm excrement and other organic materials produced by earthworms, has the properties of improving soil structure and fertility with its organic-matter content, plant nutrients, and diverse microbial populations (Arancon and Edwards 2006). Vermicompost has also been demonstrated to suppress plant pathogens, parasitic nematodes, and arthropod pests (Edwards et al. 2006, 2007).

In producing vermicompost, a recommended carbon to nitrogen ratio (C:N) of 30:1 was used for the substrates to increase microbial activity. The common bedding materials/substrates used for vermicomposting in the Philippines are rice straw, animal manures, grass, dried leaves of fruit trees, and leaves of leguminous trees (e.g., *Gliricidia sepium* and *Leucaena leucocephala*). The plant materials are chopped manually or shredded mechanically to reduce them to 1–2 cm particle size and increase their surface area to promote microbial action. A moisture level of 60–80% is maintained in the materials from the initial anaerobic stage of decomposition (without the earthworms) until the final aerobic decomposition stage with the earthworms.

In the anaerobic stage, the mixed and watered organic substrates are kept in a pile covered with an impervious sheet (e.g., plastic) or in a closed container to ferment for 2–3 weeks. Following this, earthworms are stocked at an appropriate density 0.5–0.7 kg.m^{-2} (1.1–1.4 lb.yd^{-2}) of a mixture of adults and juveniles or 200–300 adults m^{-2}) under aerobic conditions for a culture period of 4–6 weeks. With proper maintenance and protection from predators (e.g., centipedes, beetle grubs, and parasitic flatworms), 50–60% of the original weight of the materials can be recovered as vermicompost, with a yield of 2–3 kg.m^{-2} (4–6 lb.yd^{-2}) of earthworm biomass, which is harvested by hand or with machines (R. Guerrero 2008).

Vermicompost applications are appropriate for light soils low in organic matter. Vermicomposts have high cation-exchange capacity and provide cultured plants with available nutrients. The organic-matter and nutrient content of vermicomposts varies with the kind of substrates used to produce it (Dacayo 1981). The organic-matter and nutrient content of vermicompost produced with *P. excavatus* was higher using Murrah buffalo manure as a substrate compared to mixtures of sawdust, Murrah buffalo manure, and *L. leucocephala* leaves at a 1:1:1 proportion by weight. The response of leafy plants and root crops to vermicompost fertilization in pots also differed, with leafy vegetables had higher yields in response to higher levels of vermicompost while root crops had lower yields with higher levels of vermicompost (Table 29.2).

In field experiments, Dacayo (personal communication) reported that vermicompost can substitute for one-third of the recommended rate of N 90 $kg.ha^{-1}$ (80 $lb.acre^{-1}$) for tomato (*Lycopersicon esculentum*) production and can replace a maximum of two-thirds of the recommended rate of N for corn production. Using vermicomposts produced by *E. eugeniae* in boxes with pig manure and sawdust to fertilize *Brassica compensis* in garden plots, R. Guerrero et al. (1984) showed that the yields of plants fertilized with vermicompost at 25%, 50%, and 100% levels were significantly greater than those of plants fertilized with a complete fertilizer (14-14-14) alone. The cost of fertilization was lowest for vermicompost only. The results indicated that the use of vermicompost for the plants was more efficient and economical than the use of chemical fertilizers (Table 29.3).

The weights of cabbage heads (*Brassica oleracea*) grown in pots with 50%, 75%, and 100% vermicompost in the medium were comparable to those produced by plants fertilized with the recommended inorganic fertilizer (14-14-14) at 120 $kg.ha^{-1}$ (107 $lb.acre^{-1}$) (Table 29.4). With the same levels of vermicompost, the heights of cauliflower (*Brassica botrytis*) fertilized with 75% vermicompost were not significantly different from those of plants fertilized with the recommended chemical fertilizer (Table 29.4).

Table 29.2 Yield Response of Leafy and Root Crops to Different Levels of Vermicompost (VC) Application in Pots

	Leafy Vegetable g (0.035 oz)		Root Crop g (0.035 oz)	
Treatment	Lactuca sativa	Portulaca oleracea	Allium sepa	Arachis hypogea
100% Soil	38^{a}	56^{a}	34^{a}	23^{a}
75% Soil + 25% VC (by volume)	152^{b}	216^{b}	35^{a}	29^{a}
50% Soil + 50% VC	240^{c}	276^{c}	38^{a}	21^{a}
25% Soil + 75% VC	156^{b}	386^{d}	31^{a}	9^{b}
100% VC	120^{b}	376^{d}	3^{b}	7^{b}

Source: Adapted from Dacayo, J.B., Utilization of vermicompost/earthworm castings, paper presented at Seminar on Vermiculture, UPLB-CA, Los Banos, Laguna, Philippines, 1981.

Note: Figures are means of three replicates. Means with the same superscript are not significantly different at $p < 0.05$.

Table 29.3 Yield and Cost of Fertilization of *Brassica compensis* with Vermicompost (VC) and Complete Fertilizer (CF) in Field Plots after 30 Days of Culture

Fertilizer	Average Yield/Plot (kg)	(lb)	Cost of Fertilizer/Plot (PhP)
VC	4.11[a]	9	2.25
25% VC + 75% CF	4.62[a]	10	4.78
50% VC + 50% CF	4.66[a]	10	3.93
CF	3.67[b]	8	5.62

Note: Figures are means of three replicates. Means with the same superscript are not significantly different at $p < 0.05$. PhP, Philippine pesos.

Table 29.4 Effect of Vermicompost (VC) and Chemical Fertilizer on Weight of Cabbage Heads and Height of Cauliflower in Pots

Treatment	Cabbage Head Weight (g)	(oz)	Cauliflower Height (cm)	(ins)
14-14-14 (120 kg/ha) + 100% soil	369[a]	13.0	32[a]	12.6
25% VC + 75% soil (by volume)	145[b]	5.1	20[b]	7.8
50% VC + 50% soil	351[a]	12.3	23[b]	9.0
75% VC + 25% soil	346[a]	12.2	29[a]	11.4
100% VC	329[a]	11.6	24[c]	9.4

Source: Villegas 2003.
Note: Figures are means of four replicates. Means with the same superscript are not significantly different at $p < 0.05$.

The fruit yields of eggplants (*Solanum melongena*) fertilized with vermicompost (dried tree leaves) applied at 100 g (3.53 oz) per pot 4 ton.ha^{-1} (1.6 t.acre^{-1}) and 50% of the recommended amounts of inorganic fertilizers (14-14-14 and 46-0-0) were significantly greater than those of plants fertilized with 100% of the recommended amounts of inorganic fertilizer. The net return (based on cost of fertilization) for fruit production of the plants fertilized with vermicompost in combination with the inorganic fertilizers was 18% higher than that of the plants fertilized with only inorganic fertilizers (R. Guerrero and L. Guerrero 2006a).

Grain yields of upland rice grown in outdoor containers and fertilized with vermicompost at 5 ton.ha^{-1} (2 t.acre^{-1}) and 10 ton.ha^{-1} (4 t.acre^{-1}) in combination with 50% of the recommended rate of inorganic fertilizer were significantly greater than the yields of plants fertilized with 100% of the recommended inorganic fertilizer alone. Although the use of vermicompost at the levels tested apparently did not satisfy all the nutrient requirements of the plant, the results showed that the use of vermicompost at 5 ton.ha^{-1} (2 t.acre^{-1}) in combination with 50% of the recommended amount of inorganic fertilizer significantly increased the yields of plants while reducing the use of inorganic fertilizers (Table 29.5).

Salamanca and Aihara (2006) compared the nutrient content of vermicomposts produced with *E. eugeniae* in a medium with 25% (by weight) *G. sepium* and *L. leucocephala* leaves and twigs, 75% grass, for vegetable matter plus 10% cattle

Table 29.5 Grain Yield of Upland Rice with Vermicompost (VC) and Chemical Fertilizer (CF) in Outdoor Containers after 123 Days

Treatment	Mean Yield/Container (g)	(oz)
Soil	2.7[a]	0.09
5 t/ha VC	4.0[a]	0.14
10 t/ha VC	4.4[a]	0.15
5 t/ha VC + 50% CF	24.0[b]	0.84
10 t/ha VC + 50% CF	24.2[b]	0.85
100% CF	19.2[c]	0.67

Source: Guerrero, R.D. and Guerrero, L.A., *Asia Life Sci.*, 1, 145–149, 2008. Due acknowledgment shall be made to the Food and Agricultural Organization of the United Nations.

Note: Figures are means of five replicates. Means with the same superscript are not significantly different at $p < 0.05$.

manure and 20% garden soil, with that of a traditional compost with the same materials but without earthworms. They showed that the vermicompost had greater concentrations of macronutrients and micronutrients. The composting period with earthworms required 6 weeks compared to 12 weeks without earthworms.

The commercial applications of vermicompost in the Philippines have still to be evaluated fully, particularly for their cost-effectiveness. Data collected from private practitioners, however, can be indicative. For example, for commercial sugarcane production, Amor (personal communication) claimed that 21,000 ha (51,392 acre) of the crop had been treated with vermicompost in a "balanced fertilization scheme" in Negros Oriental, Philippines. A savings of as much as 50% on the overall cost of fertilization is said to have been gained by farmers with the use of 1 ton.ha^{-1} (0.4 t.acre^{-1}) vermicompost, in addition to 250 kg (540 lb) of 46-0-0 and 250 kg (550 lb) of 0-0-60 fertilizer ha^{-1}, compared to the usual application of chemical fertilizers only, using 250 kg (540 lb) of 18-46-0, 400 kg (880 lb) of 0-0-60, and 400 kg (880 lb) of 46-0-0. The yields of 92–108 t ha^{-1} for the sugarcane crop per planting season were comparable using both fertilization schemes.

For lowland (irrigated) rice production, the application of 1 ton (1t) of vermicompost in addition to 50 kg (110 lb) of 46-0-0 or 50 kg (110 lb) of 16-20-0 ha^{-1} (2.5 acre^{-1}) can yield a comparable harvest 5 ton.ha^{-1} (0.4 t.acre^{-1}) to that with the recommended inorganic fertilization using 100 kg (220 lb) of 16-20-0, 50 kg (110 lb) of 14-14-14, and 100 kg (220 lb) of 46-0-0 ha^{-1} (2.5 acre^{-1}) (Amor personal communication). Under commercial application conditions, Cruz (2006) reported that the cost of producing vermicompost was PhP (Philippine pesos) 3.68 (US$0.08) per kilogram. Thus, the commercial price of PhP 5.0 (US$0.12) per kilogram for the product applied to sugarcane and lowland rice as reported by Amor appears to be realistic.

Vermicompost "tea," the aqueous extract of vermicompost, contains plant growth regulators (i.e., auxins, gibberellins, and cytokinins) and many aerobic microorganisms that have arthropod pest–repelling, fungicidal, and nematocidal properties

(Edwards et al. 2007). Vermicompost tea is prepared by placing a permeable bag with 7.5 kg (16.5 lb) of vermicompost in a 30 L (8 gal) container with water. With manual stirring or bubbling with an electric aquarium aerator, the tea is ready for use after a day or two, after filtration and dilution in water at a 1:10 ratio. The tea is applied as a spray on plants and the soil at weekly or biweekly intervals depending on the infestation (Henares 2003).

In a preliminary field experiment, Barbiera (personal communication) reported the absence of rice insect pests (i.e., leafhoppers) in lowland rice fertilized only with vermicompost at 1.25 ton.ha^{-1} (0.4 t.acre^{-1}) and sprayed with vermicompost tea starting 14 days after transplanting of the seedlings and at 2-week intervals thereafter, with a yield of 9.2 ton.ha^{-1} (3.7 t.acre^{-1}). In another field that was fertilized with inorganic fertilizer (48-0-0) and sprayed with vermicompost tea and a chemical pesticide, the presence of the same rice pests were noted and the yield was 7.9 ton.ha^{-1} (2.2 t.acre^{-1}).

R. Guerrero (personal communication) tested the effect of a vermicompost extract containing 0.01% nitrogen, 0.09% phosphorus, and 0.11% potassium on the growth and insect infestation of *Brassica curcas* for 26 days in outdoor containers. The results showed that the plants treated with 20% vermicompost extract had the most leaf areas and no insect damage. R. Guerrero et al. (1999) showed that the primary root growth of water cabbages (*Ipomea aquatica*) was significantly better with 5 ml.L^{-1} (0.602 gal^{-1}) of vermicompost extract compared to the control and plants treated with lower concentrations of tea.

Villegas (2003) tested the efficacy of three kinds of vermicompost as germinating media for vegetable seeds in pots. The vermicompost produced from 75% grass and 25% *G. sepium* leaves gave the highest percentage of germination compared to the other two. In addition to its use as a fertilizer for crops and as a source of vermicompost tea, vermicompost has also been applied for fishpond fertilization in the Philippines. Corpuz (2003) reported a production of 80 fry per breeder per cycle and 80% survival of fry of Nile tilapia (*Oreochromis niloticus*) stocked at 3500 m^{-3} (4577 yd^{-3}) in freshwater ponds with the application of vermicompost at 0.75 ton.ha^{-1} (0.3 t.acre^{-1}).

V PRODUCTION AND USE OF EARTHWORM PROTEIN AS ANIMAL FEED AND HUMAN FOOD

Cultured earthworms, particularly *E. fetida*, have been shown to be a suitable source of animal feed protein for farmed poultry and swine (Sabine 1978, 1981) and fish (Tacon et al. 1983). The use of earthworms as human food has also been reported Edwards and Bohlen 1996; Sabine 1983; (see Chapter 20).

Earthworm protein has been produced in the Philippines by culturing *P. excavatus* and *E. eugeniae*. *P. excavatus* biomass produced in concrete tanks with fermented Murrah buffalo manure and dried *L. leucocephala* was washed and sun-dried or oven-dried at 50°C (122°F) for 6 hours and pulverized into powder (R. Guerrero 1983). Ulep (1982) obtained the greatest recovery of protein from *P. excavatus* meal using forced-air dehydration of the earthworm at room temperature, followed by blanching at 55°C (131°F) to kill the earthworm and sun-drying. Barcelo (1988) reported the highest mean recovery of *E. eugeniae* protein by freezing the earthworm for 8–12 hours and oven-drying them at 90°C (194°F).

The use of earthworm protein to replace fish meal in the diets of cultured aquatic species has shown potential for the aquaculture industry in the Philippines. R. Guerrero (1983) reported that the weight gain, survival, and feed conversion of cage-cultured *O. niloticus Nite Tilapia* fed with 15% *P. excavatus* meal and 10% fish meal were significantly higher than those of the fish fed with other materials (Table 29.6).

After feeding adult freshwater shrimp (*Macrobrachium idella*) in ponds for 60 days, R. Guerrero et al. (1984) reported that shrimps fed with dried *E. eugeniae* had greater total weight gain, better feed conversion, and more production of juveniles than those fed with dried *Therapon plumbeus* (a freshwater fish). The survival rate and weight gain of tiger shrimps (*Penaeus monodon*) fed a formulated diet containing 10% dried *E. eugeniae* and no fish meal were greater than those of shrimps fed a diet containing 30% fish meal (Piedad-Pascual 1985).

In a 9-week feeding trial in tanks, the growth of sex-reversed Mozambique tilapia (*O. mossambicus*) fingerlings fed with 50% earthworm protein, to replace 50% of the imported Peruvian fish meal (PFM) in the control diet, was not significantly different from that of fish fed with 100% PFM (Villesas-Papalid 2008). There were no significant differences in the survival, growth rate, and feed conversion of milkfish (*Chanos chanos*) fry fed in tanks for 60 days with diets containing 35% fish meal and those of the fish fed with diets in which 15% or 55% of the fish meal was replaced with earthworm protein. This indicates that earthworms can be used at a 20% level in the diet (Catacutan personal communication).

R. Guerrero (1983) reported that the weight gains, feed conversion rates, and net return for Japanese quail (*Coturnix coturnix*) improved with increasing levels of *P. excavatus* protein in their diet. The body weight and feed conversion efficiency of broiler chickens fed with three levels of *P. excavatus* meal (3%, 5%, and 10%) were comparable to those of the birds fed with 5% and 19% fish meal, soybean oil meal, and the commercial ration. Earthworm protein at 3% and 5% in the diet produced

Table 29.6 Weight Gain, Feed Conversion, and Survival of Caged Nile *Tilapia* Fingerlings Fed with Fish Meal (FM) and *P. excavatus* Meal (EWM) in the Diet for 60 Days

Diet	Mean Weight Gain (g)	(oz)	Feed Conversion Ratio	Survival (%)
25% FM	9.0[a]	0.32	2.12	89.0
25% EWM	11.89[a]	0.42	1.76	96.3
10% EWM + 15% FM	11.26[a]	0.40	1.96	94.6
15% EWM + 10% FM	19.57[b]	0.69	1.42	98.2

Source: Guerrero, R.D., *Earthworm Ecology*, Chapman & Hall, London, 309–314, 1983.
Note: Figures are means of three replicates. Means with the same superscript are not significantly different at $p < 0.01$.

better results than fish meal, meat and bone meal, and soybean oil meal at all of the levels tested (Ulep 1982).

Barcelo (1988) determined the growth of broilers fed with different levels of *E. eugeniae* meal (6%, 10%, and 14%) and fish meal at the same levels. The growth of the birds fed with 6% vermimeal was comparable to that of the birds fed with 6% and 10% fish meal, indicating that vermimeal can completely replace fish meal in the diet. Similarly, Oarde (personal communication) noted that broilers fed with *P. excavatus* meal to replace 25–100% fish meal in the diet had greater weights and more average weight gains than the birds fed with a diet containing fish meal only.

Although earthworms have been reported to be consumed as human food by indigenous communities in developing countries (Edwards and Bohlen 1996), there is sparse literature on their use and acceptability. M. Guerrero and Martin (1979) prepared cultured *P. excavatus* by first washing them thoroughly to remove dirt, dipping them in vinegar for 5–10 minutes, and washing them again before using them to cook meatballs. In panel tests to determine texture, flavor, and palatability, it was demonstrated that the cooked earthworms were acceptable to the consumers. Dried *P. excavatus* was found to contain appreciable amounts of nutrients, vitamins, and minerals. While earthworm protein has high nutritive value, its food safety in terms of possible pathogens and heavy metal contamination should be assured (Sabine 1983).

VI COST AND RETURNS OF VERMICOMPOST AND VERMIMEAL PRODUCTION

No in-depth studies have been conducted on the economics of vermicompost and vermimeal production in the Philippines. In a cost and return analysis of a pilot production farm under commercial conditions, Cruz (2006) reported an annual rate of return of 249%. The cost of the substrates for vermicomposting comprised 58% of the total expenses, followed by labor (29%). The income from vermicompost was 96% of the total income, while that for the vermimeal was only 4%. The cost of producing vermicompost was PhP 3.68 (US$0.08) per kilogram (2.2 lb) while that for vermimeal was PhP 7.06 (US$0.15) per kilogram (2.2 lb) (Table 29.7).

VII EARTHWORMS AS A SOURCE OF PHARMACEUTICAL MATERIALS

Earthworms contain bioactive compounds that have pharmaceutical (vermiceutical) or medicinal value (Mihara et al. 1991; Sun 2003) (see Chapter 22). In traditional Philippine medicine, Ang Lopez and Alis (2006) noted the use of earthworms by the Bukidnons of Panay Island for the relief of stomachache, labor pains, toothache, rheumatism, and arthritis. In preparing decoctions, 8–10 earthworms are washed and heated in a pot until almost burnt. They are then ground in a mortar and mixed with a liter of boiled water. When cool, the mixture is decanted and drunk by the patient.

Table 29.7 Projected Yield and Cost/Return from a 130 m² Earthworm Farm for a 2-Month Cycle

A. Expenses	Amount (PhP)
Labor	23,400.00
Sawdust (30% of substrate)	14,040.00
Cattle manure (70% of substrate)	32,760.00
Lime	499.20
Earthworm biomass	1,560.00
Equipment depreciation	833.33
Miscellaneous expenses	7,309.25
Total	80,401.79
B. Return	
Sale of vermicompost (21,840 kg @ PhP5)	109,200.00
Sale of earthworm biomass (468 kg @ PhP10)	4,680.00
Total	113,880.00
Gross profit/cycle	33,478.21
Rate of return/year	249.83%

Source: Cruz, P.S., *Vermi Technologies for Developing Countries*, Philippine Fisheries Association, Laguna, Philippines, 123–134, 2006.
Note: PhP, Philippine pesos.

R. Guerrero and L. Guerrero (2006b) reported the presence of a proteolytic enzyme in a dried extract of *E. eugeniae* that had comparable activity to that of an imported Chinese product. Ang Lopez and Alis (2006) found that a 25% crude water extract of *E. eugeniae*, added to fresh human blood, delayed the clotting time by more than one hour compared to the normal clotting time of less than five minutes. In a preliminary study, it was shown that intravenous administration of dried *E. eugeniae* suspended in normal saline solution at 20 mg kg^{-1} (0.307 $oz.lb^{-1}$) lowered the blood pressure of laboratory rats (R. Guerrero personal communication).

REFERENCES

Ang Lopez, M.J., and Alis, R. 2006. The indigenous use of earthworms as health remedies in the Philippines and notes on their chemical properties. In *Vermi Technologies for Developing Countries*, ed. R.D. Guerrero III and M.R. Guerrero-del Castillo, 125–144. Philippine Fisheries Association, Laguna, Philippines.

Arancon, N.Q., and Edwards, C.A. 2006. The science of vermiculture: The use of earthworms in organic waste management. In *Vermi Technologies for Developing Countries*, ed. R.D. Guerrero III and M.R. Guerrero-del Castillo, 32–65. Philippine Fisheries Association, Laguna, Philippines.

Barcelo, P.M. 1988. Production and utilization of earthworm feeds for broilers in the Philippines. *Tropicultura* 1, 21–24.

Catalan, G.I. 1980. *Earthworm Raising: A Gold Mine.* R.O. Tesoro & Sons, Davao City, Philippines.

Corpuz, R.S. 2003. Vermicomposting: the PBSP-CRTD experience. In *Proceedings of the First Philippine Vermi Symposium-Workshop*, comp. R.D. Guerrrero III, 30–34. PCAMRD-DOST and PhilRice-DA, Los Banos, Laguna, Philippines.

Cruz, P.S. 2006. Prospects of raising earthworms as a substitute for fish meal in aquacultural feeds. In *Vermi Technologies for Developing Countries*, ed. R.D. Guerrero III and M.R. Guerrero-del Castillo, 123–134. Philippine Fisheries Association, Laguna, Philippines.

Dacayo, J.B. 1981. Utilization of vermicompost/earthworm castings. Paper presented at Seminar on Vermiculture, UPLB-CA, Los Banos, Laguna, Philippines.

Edwards, C.A., Arancon, N.Q., Emerson, E., and Pulliam, R. 2007. Suppressing plant parasitic nematodes and arthropod pests with vermicompost "teas." *BioCycle* 48 (12), 38–39.

Edwards, C.A., Arancon, N.Q., and Greytak, S. 2006. Effects of vermicompost teas on plant growth and disease. *Biocycle* 47 (5), 28–31.

Edwards, C.A., and Bohlen, P.J. 1996. *Biology and Ecology of Earthworms.* Chapman & Hall, London.

Guerrero, M.G.S., and Martin, V.M. 1979. The nutritive value and acceptability of earthworm meal as human food. Bachelor's thesis, Centro Escolar University, Philippines.

Guerrero, R.D. 1979. Studies on vermiculture in the Philippines. Report submitted to the Asia Foundation.

Guerrero, R.D. 1983. The culture and use of *Perionyx excavatus* as a protein resource in the Philippines. In *Earthworm Ecology*, ed. J.E. Satchell, 309–314. Chapman & Hall, London.

Guerrero, R.D. 2008. Vermicomposting improves farm economics. *BioCycle* 1, 13–14.

Guerrero, R.D., and Guerrero, L.A. 2006a. Response of eggplant (*Solanum melongena*) grown in plastic containers to vermicompost and chemical fertilizer. *Asia Life Sci.* 2, 199–204.

Guerrero, R.D., and Guerrero, L.A. 2006b. Eugeton—a new natural health product from earthworms. In *Vermi Technologies for Developing Countries*, ed. R.D. Guerrero III and M.R. Guerrero-del Castillo, 145–147. Philippine Fisheries Association, Laguna, Philippines.

Guerrero, R.D., and Guerrero, L.A. 2008. Effect of vermicompost on the yield of upland rice in outdoor containers. *Asia Life Sci.* 1, 145–149.

Guerrero, R.D., Guerrero, L.A., and Cargado, A.U. 1984. Studies on the culture of the earthworm *Eudrilus eugeniae* and its use as feed for *Macrobrachium idella* and fertilizer source for *Brassica compensis. Trans. Acad. Sci. Tech.* 4, 33–42.

Guerrero, R.D., Villegas, L.G., and Guerrero, L.A. 1999. Studies on the production and utilization of vermicompost produced with the African nightcrawler (*Eudrilus eugeniae*) in the Philippines. *Philipp. Technol. J.* 24, 53–63.

Henares, P.T. 2003. Vermicomposting: The Buro-Buro experience. In *Proceedings of the First Philippine Vermi Symposium-Workshop*, comp. R.D. Guerrero III, 17–29. PCAMRD-DOST and PhilRice-DA, Los Banos, Laguna, Philippines.

Mihara, H., Sumi, H., Yoneta, T., Mizumoto, H., Ikeda, R., Seiki, M., and Maruyama, M. 1991. A novel fibrinolytic enzyme extracted from the earthworm *Lumbricus rubellus. Jpn. J. Physiol.* 41, 461–472.

Piedad-Pascual, F. 1985. An evaluation of three worms as feed ingredients in formulated diets for juvenile *Penaeus monodon* Fabricius. *Fish. Res. J. Philipp.* 10, 9–15.

Reyes, V.R. 1968. Studies on *Metastrongylus apri:* III—The earthworm fauna of some piggeries in the Philippines. *Philipp. J. Anim. Ind.* 24, 71–78.

Sabine, J.R. 1978. The nutritive value of earthworm meal. In *Proceedings of the Conference on Utilization of Soil Organisms in Sludge Management*, ed. R. Hartenstein, 122–130. State University of New York, Syracuse.

Sabine, J.R. 1981. Vermiculture as an option for resources recovery in the intensive animal industries. In *Proceedings of Workshop on the Role of Earthworms in the Stabilization of Organic Residues*, comp. M. Appelhof, 243–252. Beech Leaf Press, Kalamazoo, MI.

Sabine, J.R. 1983. Earthworms as a source of food and drugs. In *Earthworm Ecology*, ed. J.E. Satchell, 285–296. Chapman and Hall, London.

Salamanca, E.F. and Aihara, H. 2006. Vermicompost from agro-forestry farm wastes. In *Vermi Technologies for Developing Countries*, ed. R.D. Guerrero III and M.R. Guerrero-del Castillo, 89–97. Philippine Fisheries Association, Laguna, Philippines.

Sun, Z. 2003. *Vermiculture and Vermiprotein.* China Agricultural University Press, Beijing, China.

Tacon, A.G.J., Stafford, E.A., and Edwards, C.A. 1983. A preliminary investigation on the nutritive value of three terrestrial lumbricoid worms for rainbow trout. *Aquaculture* 35, 187–199.

Ulep, L.J. 1982. The production, processing and evaluation of nutritive value of the earthworm (*Perionyx excavatus*) as feed for broilers. PhD diss., UPLB, College, Laguna, Philippines.

Villegas, L.G. 2003. The BPI vermicomposting experience, in *Proceedings of the First Philippine Vermi Symposium-Workshop*, comp. R.D. Guerrero III, 14–16. PCAMRD-DOST and Phil-Rice-DA, Los Banos, Laguna, Philippines.

Villesas-Papalid, L. 2008. Replacement of fishmeal by vermimeal in the diets for sex reversal tilapia, *Oreochromis mossambicus* (Peters) fry reared in a closed recirculation system. Master's thesis, University of the Philippines Visayas.

CHAPTER 30

The Status of Vermicomposting in Indonesia

Bintoro Gunadi

CONTENTS

I Introduction ... 481
II The History of Vermiculture in Indonesia ... 483
III The Status of Vermicomposting in Indonesia ... 485
 A General ... 485
 B Case Study: Vermiculture of *E. andrei* and *E. fetida* Using Tea Leaf Waste ... 486
  1 Objectives ... 486
  2 Materials and Methods ... 487
  3 Results and Discussion ... 488
IV Growth in Earthworm Populations and Earthworm Biomass ... 490
V Competition between *E. andrei* and a Local Earthworm (*Pheretima* sp.) ... 491
VI Succession of Beneficial Decomposer Organisms during Vermicomposting ... 492
Acknowledgments ... 493
References ... 494

I INTRODUCTION

Vermiculture involves intensive earthworm culture, with domestication of the earthworm, by using earthworms in organic-waste breakdown (Neuhauser 1978). Vermicomposting is an aerobic biooxidation and stabilization process that is mesophilic and depends on earthworms to fragment, mix, and aerate the organic wastes, as well as promote microbial activity, producing earthworm casts or vermicomposts.

A large amount of organic wastes is produced from animal farming, horticulture, supermarkets and food-processing industries, breweries, tea beverage companies, fruit and potato processors, mushroom industries, etc. Most solid organic wastes are kept in land-disposal areas for eventual thermophilic composting before they are utilized on agricultural land. There, they may cause odor problems and transmit diseases and could be a potential source of groundwater pollution.

The main idea of converting green or organic wastes into a source of protein (as earthworm biomass) and into vermicomposts is a part of a bioregenerative ecological strategy. Bioregenerative life-support systems depend on living systems for food, oxygen, and water. Vermicomposting is part of self-renewing system that depends on recycling of organic matter and involves close interactions between plants, animals, microorganisms, and human beings.

Eisenia fetida (Savigny) is a surface-feeding, or epigeic, earthworm known to have a broad potential for converting organic wastes into earthworm biomass, with associated production of high-value plant growth media (termed vermicompost). The technologies that are used range from windrows to high-technology continuous-flow vermicomposting reactors. Vermicomposting uses earthworms and microorganisms as the main biological organisms, together with beneficial microarthropods (e.g., many springtails and mites) in organic-waste conversion. There are three important goals in vermiculture and vermicomposting: first, to reduce the quantities of organic wastes as a part of waste-management systems; second, to produce earthworm biomass, which has approximately 70% protein content and a low fat content of less than 10% and can be used as an alternative protein food source for prawn, fish, pig, and poultry farming and for pharmaceutical preparations; and, third, to produce earthworm casts or vermicomposts that are valuable as soil amendments promoting plant growth. In recent years, vermiculture and vermicomposting have become more popular all over the world, from the temperate regions to the tropics. Vermicomposting strategy differs when it is concentrated on organic-waste management and used to convert organic wastes into high-value plant growth media termed vermicomposts or earthworm casts or when it is used for earthworm-protein production.

The strategies used to reach these aims are summarized in Table 30.1. If the main goal is to produce earthworms, the management of the activity will be like a

Table 30.1 Different Strategies in Earthworm Production and Vermicomposting

Vermiculture Producing Earthworms	Vermicomposting Producing Vermicomposts
• Start with cocoons and juveniles of earthworms	• The age of earthworms may vary
• Additional feed other than organic wastes may be needed	• Mainly fed on organic wastes
• Usually in batch systems	• Most efficient with a continuous-flow system
• Usually in small- to medium-scale facility	• Can be a large-scale facility
• Needs special care	• Needs good management
• Environmental control is of utmost importance	• Less environmental control
• Regular harvesting of earthworms	• Low frequency of harvesting of earthworms
• Produces a high quality and quantity of earthworms	• Produces fewer earthworms
• Produces a low quantity of vermicompost but with a high quality	• Produces a high quantity of vermicompost but with less quality

farm, with special feeding, care, and regular harvesting of earthworms. If the main goal is to produce vermicomposts, it will use organic wastes as the main feedstock for the earthworms. In some countries in Europe, such as the Netherlands, a license to get permission from the government to operate the vermiculture facility is easier than building the vermicomposting facility. Recently, it was reported that Oregon Soil Corporation in the United States, which uses a mixture of thermophilic compost, food waste, and paper waste to feed the earthworms in their continuous-flow vermicomposting reactors, was able to fulfill the required organic product certification for the vermicompost produced (Holcombe 2009 personal communication).

A special feed called *wormmeel* that is used for earthworm cultures in a modern vermiculture facility, Zeeland Wormerkultur in the Netherlands (Blusse 2000 personal communication), contains 15% protein, 4% lipids, 14% carbon, and 0.5% phosphorus, as well as fiber, vitamins A, D, and E, the antioxidant BHT (butylated hydroxytoluene) or ethoxyquin, and a total moisture content of 12.5%. The quality of the feedstock is much better than common organic wastes in producing healthy earthworms. It has been reported that the earthworm body contains up to 70% protein and less than 10% lipids based on their dry weight (Sabine 1983; Edwards and Niederer 1988).

Organic wastes can contain materials that are potentially toxic to earthworms, such as ammonia or salts in animal manure or tannins and acids in green wastes (Gunadi and Edwards 2003). One way to reduce the toxic components of the organic wastes is to thermophilically precompost the wastes before vermicomposting. Precomposted wastes could be more acceptable and cause less earthworm mortality. However, organic wastes that have been precomposted for more than 2 weeks may have decreased nutrient availability for earthworm growth. This could inhibit the rates of earthworm growth and numbers of cocoons and hatchlings (Gunadi, et al. 2002).

There is concern about the possible distribution of human pathogens during vermicompost production using animal manures or agricultural wastes. There is good scientific evidence that human pathogens do not survive during the vermicomposting process (Eastman et al. 2001). Eastman et al. reported on the effectiveness of vermiculture in human pathogen reduction and its ability to reach the Class A disposal standards by the U.S. Environmental Protection Agency (EPA). They found that the vermicomposting of domestic wastewater residuals (biosolids) using *E. fetida* could reduce significantly the four key human pathogen indicators, that is, fecal coliforms, *Salmonella* spp., enteric viruses, and helminth ova. Other research shows that *E. fetida* has an antibacterial protein in its coelomic fluid called fetidin (Lassegues et al. 1997; Milochau et al. 1997).

II THE HISTORY OF VERMICULTURE IN INDONESIA

The species of earthworms in Indonesia were recorded by Dammerman (1929). *Perichaeta musica* Horst, which was discovered in the mountains of Java, reaches a length of half a meter; *Moniligaster houteni* Horst, from Sumatra, has a length of more than a meter. The genus *Pheretima* is typical of the Indo-Australian region and is common in Indonesia (Gates 1948). It is generally accepted that in the tropics there are many more species of earthworms than in temperate regions, sometimes with fewer individuals, though a few species can attain enormous populations.

Earthworms can be separated into three major groups based on their feeding and burrowing habits: aneceic, endogeic, and epigeic. The deep-burrowing earthworm species, or aneceic species, like *Perichaeta musica* Horst, *Moniligaster houteni* Horst, and *Lumbricus terrestris* (L) inhabit more or less permanent burrow systems that can extend several meters into the soil. They feed mainly on debris found at the soil surface, and their reproduction rates are low. The upper-soil earthworm species, or endogeic species, like *Aporrectodea tuberculata* (Eisen), move and live in the upper-soil strata and feed primarily on soil and associated organic matter. Most aneceic and endogeic earthworm species are not suitable for vermicomposting organic wastes because their habitat is associated mainly with soil.

The third group of earthworms, epigeic species, live in surface soil and litter; these include *Eisenia fetida* (Savigny), *Perionyx excavatus* (Perrier), *Eudrilus eugeniae* (Kinberg), and *Pheretima asiatica* (Michaelsen). These species live in or near the surface of agricultural wastes or manure, or near the surface of plant litter. They are typically about 5–8 cm (2–3 in) long and are well adapted to the very variable moisture and temperature conditions that occur in the litter layer (Gunadi et al. 2003). These earthworms common in compost piles are epigeic species that could not survive long in the low-organic-matter environment of soil.

Eisenia fetida (Savigny) and *Eisenia andrei* (Bouché) are the most common earthworm species used for vermiculture in Asia, including China, Indonesia, Japan, Korea, and Taiwan. *E. fetida* corresponds to the striped or banded form, with the area around the intersegmental groove having no pigmentation and appearing pale or yellow, hence, its common names of brandling or tiger earthworm, whereas *E. andrei*, the common red worm, is uniformly reddish. Aside from these differences in pigmentation, the two species are morphologically similar (Dominguez et al. 2005).

Epigeic earthworm species, originally from the tropics, such as *Pheretima asiatica* (Michaelsen), *Perionyx excavatus* (Perrier), and *Eudrilus eugeniae* (Kinberg), are very common species that are used extensively in vermiculture and vermicomposting in India and the Philippines. According to Gates (1972) in his monumental article on the systematics and biology of earthworms, particularly those in Southeast Asia, *E. fetida* is of European origin but was widely distributed by humans, with its dispersal partially due to the individuals' wandering propensity. The species has been carried around the world in soil and plant pots and has been distributed locally from new centers in greenhouses, arboretums, and botanical gardens for several centuries.

The first scientific article in relation to vermiculture in Asia was probably by Beddard (1883). In his article "Note on Some Earthworms from India," he stated that *E. fetida* was distributed universally. Gates (1972) called *E. fetida* a homodynamous species, without any diapause in its life history. Hence, their activities, in favorable conditions, can be continuous throughout the year. *E. fetida* has been studied extensively, in particular for its potential in vermiculture, because it is easy to culture and its life cycle is relatively short with a very fast growth rate. Reinecke and Viljoen

(1990) reported that the growth and reproductive potential of *E. fetida* was influenced not only by environmental conditions but also by the quality and availability of food. It has been reported that self-fertilization occurs in *E. fetida* in about 10% of its reproduction activities (Dominguez 2000 personal communication). Since *E. fetida* and *E. andrei* were not originally from the tropics, they may need special care in vermiculture due to the high temperatures in the tropics and attacks by possible predators like rats, frogs, lizards, ants, and birds.

The native earthworm species from tropical regions that are suitable for vermiculture, such as *Pheretima asiatica* (Michaelsen), *Perionyx excavatus* (Perrier), and *Eudrilus eugeniae* (Kinberg), have fewer predators—probably because they can move faster than *E. fetida*—but they have longer life cycles and slower growth and reproduction rates than *E. fetida*, *E. andrei*, and *Lumbricus rubellus* (Hoffmeister). Vermiculture systems function better with a single species of earthworm rather than using mixed species. Mixed cultures that are used in temperate regions do not work so well in the tropics due to the higher temperatures. Moreover, in the tropics, most vermicomposting systems become contaminated by other local or native earthworms. The population densities of earthworms used in vermicomposting in the tropics are also very high, ranging from 1 to 10 kg m^{-2} (0.2–2lb. ft^{-2}).

III THE STATUS OF VERMICOMPOSTING IN INDONESIA

A General

There was a large-scale vermicomposting facility in Central Java in the 1990s that was initiated with 2 tons (2t) of *E. fetida*, using urban wastes and designed to provide vermicomposts for humic acid extraction. About 4 kg (9 lb) of humic acid can be collected from one ton (1t) of vermicompost. The project was stopped due to an economic crisis and concern about the vermicompost containing heavy metals. Metal detoxification in earthworms involves the binding of metals to proteins such as metallothioneins or sequestration in granules. Once these detoxification mechanisms have been exceeded, metal toxicity may occur in earthworms (Lanno et al. 2002).

Humic substances, especially humic acid and fulvic acid, can be extracted from vermicomposts. Applications of small amounts of pure humic acid 3–5 kg.ha^{-1} (2.7–4.5 lb.$acre^{-1}$) can improve the soil texture and water retention significantly. This is an economical option for increasing the effects of application of vermicomposts. Humic substances act similarly to chelating agents or surfactants and can provide available C to beneficial soil microorganisms; increase soil microbial activity and populations, soil quality, and nutrient uptake; and hence stimulate plant growth. Humic substances can also be used for bioremediation of soils contaminated with hydrocarbons or salt (Ke et al. 2009) and can also help to reduce odors and enhance digestion of sludges in organic wastes contaminated with phenols and formaldehyde (Kochany and Kochany 2008).

Fulvic acid is a component of humic substances that is soluble in water under all pH conditions, has a low molecular weight, and can bind and carry more than

75 different minerals that will be bioavailable because the vermicompost is chelated. Recently, it was found that fulvic acid has some healing and regenerative influences on human and animal tissues and cells (Day and Hansen 2007).

Vermicomposting has been introduced for assessment as a sustainable sanitation system for low-income urban areas in Indonesia, especially for human waste treatment and recovery options (Malisie 2008). Since earthworms are very sensitive to urine (ammonia), ammonia must be washed out of human wastes or dispersed by thermophilic precomposting of the biosolids. Using earthworm proteins produced from *E. fetida* grown in biosolids as animal feed or for pharmaceutical purposes is still potentially safe. *E. fetida* can produce antimicrobial peptides called fetidin or eiseniapore that are not hemolytic and are safe for the vertebrate immune system (Cooper et al. 2001).

The production of earthworm biomass is a principal goal of vermiculture in Indonesia. The price of earthworms from the farmer fluctuates due to the lack of stable demand (e.g., for use as animal feed, for pharmaceutical purposes, or for large-scale vermicomposting). Recently, one buyer introduced a special food to be fed to the earthworm for 2 weeks before the farmers sell them. This is related to the pharmaceutical uses of the earthworm biomass and is intended to clean up or detoxify the bodies of the earthworms from contaminants before they are used.

Pharmaceutical materials in earthworm proteins can be extracted in the form of lumbrokinase, a fibrin-dissolving enzyme that prevents hypercoagulation of the human and animal blood and blood clots. In Jakarta, lumbrokinase has been used in many hospitals and in more than 20 provinces and cities in Indonesia, as well as in Hong Kong, Taiwan, Southeast Asia, and Europe (see Chapter 22).

Rebuilding of plant substrates such as moss by using organic amendments and by incorporating earthworms and microorganisms was reported by Aryantha et al. (2001). They developed organic substrates to promote moss growth on a reclamation site overburdened with wastes at Freeport Indonesia company. Moss, a nonvascular type of plant, is very important in initiating the succession process. The colonization of pioneer moss, followed by other plants until the vegetation would ultimately become established, was the goal of their research. It was stated that a combination of indigenous bacteria and vermicomposts produced by *Pheretima* sp. and *Lumbricus rubellus* had potential as agents to promote moss growth (*Racomitrium* sp. and *Splachnobryum* sp.) for the reclamation of a contaminated site close to copper and gold mines. A bioregenerative ecosystem strategy related to the vermicomposting of tea leaf wastes using *E. fetida* and *E. andrei*, which was conducted by Gunadi, Susanto, et al. (1998), is discussed as a case study on vermicomposting in Indonesia.

B Case Study: Vermiculture of *E. andrei* and *E. fetida* Using Tea Leaf Waste

1 Objectives

This research had three main objectives: (a) to determine the nutrient content in the main feed (tea leaf wastes) and feed supplements (soy waste and bran), the feed value of *Eisenia andrei* for domestic animals, and the quality of vermicomposts or earthworm casts produced during the process; (b) to compare the growth of *Eisenia andrei* and *Eisenia fetida* in the tea leaf wastes on a small scale in the laboratory with their growth under large-scale indoor conditions, and to make preliminary observations on possible competition between *Eisenia andrei* and a local earthworm species (*Pheretima sp.*) in a mesocosm in the laboratory; and (c) to study the preliminary succession of the beneficial decomposer organisms (actinomycetes, fungi, mites, springtails, and earthworms) during the vermicomposting of tea leaf waste.

Tea leaf wastes were the main feedstock for the earthworms. They are produced by the extraction of the dry tea leaves using boiling water at a temperature of 98°C to produce Indonesian bottled tea beverages. On an industrial scale, about 20 tons (20t) of tea leaf waste with a water content of around 80% were produced daily in Java, Bali, and Sumatra in 1997 (Gunadi, Susanto, et al. 1998). According to Gunadi et al. (1996), *E. fetida* grew well in the laboratory, fed mainly on slightly fermented tea leaf wastes. During the first 2 months, or the period of active growth, they grew from 3.9 mg to 194.5 mg per individual earthworm per day, with a daily growth rate equal to their initial body weights.

2 Materials and Methods

This work describes small-scale and large-scale culturing of adult *E. fetida* and *E. andrei* over 1 year (indicated by the possession of a clitellum and average length of 5 cm), fed mainly on slightly fermented tea leaf waste (3–7 days old). Some supplementary foods, such as soybean cake waste and bran, were added after a 9-month observation period at the level of 10% of the main food. A ratio of food to earthworm wet weight of 1:1 was used, and the earthworms were fed every week. Bedding contained 60–70 % moisture; the average temperature was between 25°C (77°F) and 30°C (86°F), and the pH between 6.5 and 7.0. Bedding was wetted at least five times a week.

Amounts of lipids, protein, and carbohydrates in the main and supplementary foods for the earthworms were measured using HCl hydrolysis, macro Kjeldahl, and Luff Schoorl methods. Tannin was measured by titration with 0.025 N $KMnO_4$. Total C and total N were measured using the method of Walkley and Black, and Kjeldahl. Macro- and micronutrients in the vermicompost (P, K, Ca, Mg, Fe, Mn, Cu, Zn, B) were measured using atomic absorption spectroscopy, and Cl was measured using an ionmeter.

Cultures of adult *E. andrei* and *E. fetida* in the laboratory were maintained in microcosms, consisting of plastic boxes that were 12.5 × 12.5 × 8.5 cm (4.9 × 4.9 × 3.3 in), starting with six *E. andrei* individuals and six *E. fetida* individuals. For larger-scale cultures an initial weight of 200 g (7 oz.) of *E. andrei* and *E. fetida* was cultured in mesocosms consisting of plastic boxes, 60 × 40 × 30 cm (24 × 16 × 12 in), with aeration holes in the sides and bottom. Large-scale indoor cultures of *E. andrei* were used in shallow beds 40 cm (1.3 ft) deep, 15 m (50 ft) long, and 2 m (6.6 ft) wide, made of concrete. The initial population density of the earthworms was 0.5 $kg.m^{-2}$ (0.1 $lb.ft^{-2}$) (1 kg of adult adult *E. andrei* is about 2500 earthworms). The beds were well drained with several holes covered with mesh screen on the bottom along both sides.

The rates of growth of the earthworms in the microcosms, mesocosms, and large indoor beds were recorded every 3 months for 1 year. Cultures of each species of earthworm in the microcosms and mesocosms were replicated five times. For the large-scale indoor treatment, the earthworm growth samples were taken by removing 0.5 × 2 m (20 × 80 in) of bedding with three replicates on every sampling date. Student's *t*-test was used to compare the growth of *E. andrei* and *E. fetida* in microcosms and mesocosms. Preliminary studies on the competition between *E. andrei* and the local earthworm *Pheretima* sp. were done only in the mesocosm and were begun with 250 adults of each species.

The succession of the dominant decomposer organisms during the vermicomposting of tea leaf waste was monitored over the first month of the observation. The dominant beneficial decomposer organisms including fungi, mites, springtails, and native earthworms, which were collected by hand sorting. The identification of the mites was done by N. Kaneko from Shimane University Japan and of the springtails by P. Greenslade from CSIRO Australia.

3 Results and Discussion

The analyses of the composition of the tea leaf wastes, feed supplements (i.e., soybean cake waste and bran), earthworm bodies, and macro- and micronutrient contents are summarized in Table 30.2. One-week-old tea waste contained larger percentages of lipids, proteins, and carbohydrates than 1-day-old tea waste. This may be due to the fermentation processes or the heterogeneity of the tea wastes. The feed supplements contained greater percentages of lipids, proteins, and carbohydrates than the tea wastes, except for the soybean cake waste, which had very low protein content since the protein was extracted for human food during the tofu (soybean cake)-manufacturing process.

E. andrei and *E. fetida* can survive in almost any moist organic wastes with the exception of acid citrus fruits (e.g., citrus skins), which are a problem because of their high acidity. Compared to relatively intractable tropical pine needles with a C:N ratio of about 40 and a water content around 50% (Gunadi, Verhoef, et al. 1998), the wet tea leaf waste after boiling at 98°C (208°F) was ready to use as feedstock for the earthworms, with a C:N ratio of about 12 and a water content around 80%. The main problem is that tea leaf waste contains low concentrations of lipids, proteins, and carbohydrates (Table 30.2), which limits the earthworms' growth rates. According to Murphy (1993), some feed supplements with protein contents in the range of 11% (e.g., wheat and barley) to 80% (e.g., bloodmeal) may be used for earthworm rearing. In this 3-month investigation, the effects of the feed supplements (i.e., soybean cake waste and bran) were not clear.

The high protein content of the *Eisenia* spp. (up to 65.4%) and the low lipid content (6.8%) indicates the potential value of these earthworms as food for domestic animals or even for human consumption. Five aspects should be taken into account when considering the use of earthworms as food for animals or humans: chemical composition, practical value in animal feeds, potential hazards, economics of production, and legal constraints (Sabine 1983).

Table 30.2 The Nutritive Value (% dry weight) of Tea Leaf Waste, Feed Supplements, *Eisenia* Tissues, and Vermicompost

	Tea Leaf Wastes		Feed Supplement			
Nutrition/ Nutrients	**One Day Old**	**One Week Old**	**Soybean Waste**	**Bran**	***Eisenia* Tissues**	**Vermicompost**
Lipid	0.99	1.42	5.53	12.46	6.77	—
Protein	3.77	5.47	1.55	11.13	65.43	—
Carbohydrate	2.74	3.28	5.83	42.63	17.60	—
Tannin	0.03	n.d.	n.d.	n.d.	0.01	—
Water content (% fresh weight)	83.97	82.69	91.45	12.08	79.86	—
Macronutrients						
N	—	—	—	—	—	5.1
P	—	—	—	—	—	3.2
K	—	—	—	—	—	2.9
Ca	—	—	—	—	—	8.1
Mg	—	—	—	—	—	1.3
Micronutrients						
Fe	—	—	—	—	—	0.3
Mn	—	—	—	—	—	0.5
Cu	—	—	—	—	—	0.2
Zn	—	—	—	—	—	1.0
B		—	—	—	—	0.2
Cl	—	—	—	—	—	0.5
C:N ratio	12.3	12.2	—	—	—	6.4

Note: n.d = not detected; — = no measurement.

On average, the tea vermicomposts contained more macronutrients than vermicomposts from other organic waste sources, that is, dairy waste sludge cake (Hatanaka et al. 1983), pig solids on straw (Edwards 1998), and cow manure (Mitchell 1997). However, the C:N ratio of the tea vermicompost (6.4) was lower than that of the other materials. The nutrient content of vermicomposts may vary depending on the parent waste and the age of the vermicompost (Gunadi, et al. 2002). However, they usually contain mineral plant nutrients in a more plant-available form than thermophilic composts. Some ornamentals, particularly chrysanthemums, salvias, and petunias, flowered much earlier in vermicompost mixtures, and this could possibly be due to hormonal effects resulting from microbial action in the earthworms' gut (Edwards 1998). Vermicompost has also been used as a "blocking" material to grow seedlings for transplanting into the field. Seedlings grown in vermicomposts produced cabbages almost twice as large at harvest than those grown in the commercial blocking material (Edwards and Burrows 1988). Vermicomposts produced by both *E. andrei* and *E. fetida* were comparable and could be returned to the tea plantation and jasmine flower plantation as an environmentally useful alternative. Application of tea vermicomposts to tea plantations, in combination with reductions in amounts of artificial inorganic fertilizers, showed promise, because the tea vermicompost produced healthier leaves (i.e., bright green color) and tended to increase the production of new tea leaves.

IV GROWTH IN EARTHWORM POPULATIONS AND EARTHWORM BIOMASS

The growth in earthworm populations and earthworm biomass in microcosms, mesocosms, and large-scale indoor systems is summarized in Table 30.3. For earthworms used for vermicomposting, the relationship between biomass production and

Table 30.3 Increases in Numbers or Weights and in Cocoons of *Eisenia andrei* and *E. fetida* in the Microcosms, Mesocosms, and Large-Scale Beds

	Mean Numbers or Weights		Total Number of Cocoons	
	E. andrei	*E. fetida*	*E. andrei*	*E. fetida*
Microcosms (Numbers)				
Initial amounts	6 ± 0	6 ± 0 NS	—	—
After 3 months	67 ± 6	81 ± 22 *	—	—
After 6 months	124 ± 10	210 ± 95 ***	150	375
After 9 months	266 ± 18	509 ± 11 ***	1900	3700
After 12 months	848 ± 11	1115 ± 54 ***	930	5895
Mesocosms (Weights)				
Initial amounts	200 ± 0 g	200 ± 0 g NS	—	—
After 3 months	410 ± 74 g	259 ± 22 g **	220	2445
After 6 months	698 ± 43 g	353 ± 95 g ***	600	1005
After 9 months	748 ± 80 g	521 ± 181 g *	—	3590
Large-scale beds (Weights)				
Initial amounts	15 ± 0 kg	—	—	—
After 3 months	20.9 ± 2.9 kg	—	526	—
After 6 months	30.6 ± 4,9 kg	—	390	—
After 9 months	49.2 ± 9.4 kg	—	—	—
After 12 months	41.6 ± 6.2 kg	—	—	—

Note: The results are ± standard deviation ($n = 5$ and 3 for large-scale beds); * $p < 0.01$, ** $p < 0.05$, *** $p < 0.001$, — = no cocoon.

rates of population increase is of utmost importance (Kale 1998). For the increases in numbers and biomass of both species of earthworms in the microcosms and mesocosms, the differences between the means were all statistically significant. *E. fetida* in the microcosms had a geometric-progression growth pattern for 6 months. In the first 3 months the population increases of both species in the microcosms were more than 1000%. In the mesocosms the growth in biomass of *E. andrei* followed an arithmetic-progression pattern every 3 months for 9 months. In the indoor beds the weight increases of *E. andrei* were less than in the mesocosms and did not follow the same arithmetic-progression pattern (Table 30.3). The increase in earthworms' weights over the first 3 months was only about 39%. The growth rates of *E. andrei* in the indoor beds were lower than in the mesocosms and much lower compared to their growth rate in the microcosms.

There are no comparable reports dealing with the use of tea leaf waste as a main feedstock for *E. andrei* and *E. fetida*. In the microcosms *E. fetida* had a significantly higher population growth rate than *E. andrei* over 1 year, and growth tended to follow a geometric-progression pattern every month for 6 months. Conversely, in the mesocosms, the increase in weights of *E. andrei* during 9 months was significantly greater than that of *E. fetida* and seemed to follow an arithmetic-progression pattern every 3 months. The percentage increases in numbers of both species of earthworms was higher in the microcosms than in the mesocosms (Table 30.3). For indoor-bed culture, the weight increases of *E. andrei* followed neither a geometric-progression nor an arithmetic-progression pattern but fluctuated. This may have been due to the presence of nocturnal predators like rats, frogs, lizards, and ants and to a more heterogeneous microclimate in the beds, because *E. andrei* were more often found in cool places. Installation of closed and open mesocosms in the field (Gunadi and Verhoef 1993) was used to test the influence of predators and microclimate preferences on the optimal growth rates of both *Eisenia* species. According to Murphy (1993), in intensive cultivation *Eisenia* can be expected to double its population in about 2 months at a suitable pH (6.5), temperature 18°C–23°C (64.4°F–73.4°F), moisture (60%), and food supply (earthworm:food ratio = 1:1). That commercial statement was supported by this research, using tea leaf waste only, at the microcosm scale. It can be concluded that culture of both *E. andrei* and *E. fetida* has good potential in the tropics, with the main challenges for large-scale production being control of temperature and the control of predators.

V COMPETITION BETWEEN *E. ANDREI* AND A LOCAL EARTHWORM (*PHERETIMA* SP.)

The results of the study of the competition between *E. andrei* and *Pheretima* sp. in the mesocosm are summarized in Table 30.4. In the well-controlled conditions of the mesocosm, with an initial population of 250 earthworms of each species, after a year the total number of *E. andrei* increased to 2524 earthworms. Conversely, the total numbers of *Pheretima* sp. after a year decreased to 45 earthworms. A local genus of *Pheretima* sp. was found to have invaded the mesocosms and the indoor

Table 30.4 Preliminary Study on the Competition between *Eisenia andrei* and *Pheretima* sp. Cultured in the Same Mesocosm

	E. andrei	*Pheretima* sp.
Initial number	250	250
After 3 months	748	348
After 6 months	825	70
After 9 months	1304	60
After 12 months	2524	45

beds. Over 1 year *Pheretima* sp. populations decreased by 82% while the *E. andrei* population increased by 910%.

Ilyas (2009) reported that during the vermicomposting of leaf litter (*Dalbergia latifolia*), using combinations of three earthworm species (*Pheretima* sp., *Eisenia fetida*, and *Lumbricus rubellus*), the highest food consumption and growth rate of the earthworms were achieved with the single species *E. fetida* and with the combination of *E. fetida* and *L. rubellus*. The cocoon incubation period for *Pheretima* sp. was shorter than for the others (i.e., 15 days vs. 19 days), and the numbers of juveniles per cocoon was lower (i.e., two juveniles per cocoon vs. four juveniles per cocoon).

A growth study of three earthworm species *Pheretima* sp., *Eisenia fetida*, and *Lumbricus rubellus* at different levels of moisture and lime applications was conducted by Brata (2006). It was reported that *Pheretima* sp. was more resistant to drought, although its growth rate was slower than that of *Eisenia fetida* or *Lumbricus rubellus*. The lime applications at rates of 0.2% and 0.4% had no effect on the growth rates of the earthworms.

According to Edwards (1998), four types of information are needed for optimization of the breakdown of sewage solids by earthworms: (a) how earthworms are affected by sludge characteristics, (b) the comparative ability of different of earthworms to grow and reproduce in sludge, (c) the need for preprocessing of the sludge to make it acceptable to earthworms, and (d) the effects of mixed species of earthworms on sewage sludge breakdown in a vermin-stabilization system. The possibility of using local species of earthworms, such as *Pheretima* sp. or mixed populations with *E. andrei*, may be needed for realistic waste-management systems (e.g., tea leaf waste) in the tropics due to the local species having greater tolerance of temperatures and predators. It has been reported that *Pheretima* sp. did not grow as rapidly and reproduced more slowly than *E. fetida* (Yulipriyanto 1993; Roslim 1994). However, further investigations are required due to the advantages of the native earthworm *Pheretima* sp., with its easy adaptation to the tropical climate.

VI SUCCESSION OF BENEFICIAL DECOMPOSER ORGANISMS DURING VERMICOMPOSTING

Many groups of soil organisms are involved during the vermicomposting of tea leaf wastes, but the dominant ones were fungi (*Mucor* sp., *Penicillium* sp.,

Aspergillus sp., and *Rhizopus* sp.), mites (*Uroboovella marginata* and *Prasitus* sp.), springtails (*Yuukianurua aphoruroides* and *Onychiurus folsomi*), and epigeic earthworms (*Eisenia fetida*, *E. andrei*, and *Pheretima* sp.). In the first succession, microorganisms such as bacteria, yeasts, actinomycetes, and fungi could grow on the tea leaf wastes. To get the most efficient early vermicomposting of tea leaf wastes, the growth of the bacteria and yeast should be controlled by reducing the moisture content of the waste to about 70% and keeping the thickness of the new layer of wastes to no more than 15 cm (6 in). These conditions, with aeration of the wastes and a neutral pH, were better for the growth of beneficial actinomycetes and fungi.

In the second succession, control of the water content and aeration in the tea wastes were very important to keep the pH near neutral. The growth of enchytraeids (white worms) and fly larvae (Diptera) should be avoided because both enchytraeids and fly larvae can compete for food and oxygen with *E. andrei* or *E. fetida*. Populations of *Eisenia* spp. will decrease in an anaerobic environment with high fermentation processes, high water content, low aeration, and high acidity. Enchytraeids reach their largest populations in mor-type soils that have a well-developed layer of undecomposed organic matter and high acidity (Dash 1990). Most fly larvae in general need moisture (Teskey 1990). In this case, an abundance of fly larvae can disturb the vermicomposting processes, and the flies will detract from the aesthetic value of the vermicompost. So, the second succession of decomposer organisms during vermicomposting of tea leaf wastes should allow beneficial springtails, oribatid mites, and epigeic earthworms to grow.

The third succession, the growth of predatory invertebrates such as centipedes, staphylinid beetles, pseudoscorpions, and predators of higher-level organisms like frogs, lizards, rats, and birds, should be avoided by controlling the environment using a closed indoor system or using equipment like ultrasonic pest control to remove the predators.

I believe that the succession of the beneficial decomposer organisms during the vermicomposting of tea leaf wastes is very important. The efficiency of the vermiculture and vermicomposting of the tea leaf waste will improve with a well-controlled system, so that large quantities of high-quality earthworm biomass and vermicompost can be harvested. It is a major challenge to develop this kind of model of the succession of beneficial decomposer organisms in the other more complicated wastes, such as food wastes, urban wastes, or animal wastes mixed with plant wastes and paper wastes.

ACKNOWLEDGMENTS

The author thanks Clive Edwards for his invitation to present this chapter and also for his support during my work in the Soil Ecology Laboratory at The Ohio State University (1998–2002); Peter Sosrodjojo for his supports of the research and trial of the vermicompost on the Sosro's tea plantation and jasmine plantation in Java (1996–1998); and Pieter Blusse, Jorge Dominguez, and Dan Holcombe for personal communications, particularly concerning the recent progress on vermiculture and vermicomposting.

REFERENCES

Aryantha, I.N.P., Adianto, Harianto, R., Humaina. 2001. Rebuilding plant substrate using organic amendment and incorporation of soil microfauna and microorganism. Final report, Department of Biology, Institute of Technology Bandung, Indonesia.

Beddard, F.E. 1883. Note on some earthworms from India. *Ann. Mag. Nat. Hist.* 5(12), 213–224.

Brata, B. 2006. Growth of three earthworm species at different watering and lime applications.[In Indonesian with abstract in English.] *J. Ilmu-Ilmu Pengetahuan Indonesia* 8(1), 69–75.

Cooper, E.L., Kauschke, E., and Cossarizza, A. 2001. Annelid humoral immunity: Cell lysis in earthworms. In: Beck, G., Sugumaran, M., and Cooper, E.L. (eds.), *Phylogenetic Perspectives on the Vertebrate Immune System. Adv. Exp. Med. Biol.* (484). Kluwer, New York: Academic/Plenum Publishers, New York.

Dammerman, K.W. 1929. *The Agricultural Zoology of the Malay Archipelago: The animals Injurious and Beneficial to Agriculture, Horticulture and Forestry in the Malay Peninsula, the Dutch East Indies and the Philippines.* J.H. de Bussy, Amsterdam.

Dash, M.C. 1990. Oligochaeta: Enchytraeidae. In: Dindal, D.L. (ed.), *Soil Biology Guide.* John Wiley & Sons, New York.

Day, K.S., and Hansen, G.A. 2007. Compositions and methods for human use containing fulvic acid. US Patent US 2007/0212434 A1. Filed: March 7, 2006. Publication: September 13, 2007.

Dominguez, J., Velando, A., and Ferreiro, A. 2005. Are *Eisenia fetida* (Savigny, 1826) and *Eisenia andrei* Bouché (1972) (Oligochaeta, Lumbricidae) different biological species. *Pedobiologia* 49(1), 81–87.

Eastman, B.R., Kane, P.N., Edwards C.A., Trytek, L., Gunadi, B., Stermer, A.L., and Mobley, J.R. 2001. The effectiveness of vermiculture in human pathogen reduction for USEPA biosolids stabilization. *Compost Sci. Util.* 9(1), 38–49.

Edwards, C.A. 1998. The use of earthworms in the breakdown and management of organic waste. In: Edwards, C.A. (ed.), *Earthworm Ecology.* St. Lucie Press, Boca Raton, FL.

Edwards, C.A., and Burrows, I. 1988. The potential of earthworm composts as plant growth media. In: Edwards, C.A., and Neuhauser, E.F. (eds.), *Earthworms in Waste and Environmental Management.* SPB Academic Publishing, the Hague, the Netherlands.

Edwards, C.A., and Niederer, A. 1988. The production and processing of earthworm protein. In: Edwards, C.A., and Neuhauser, E.F. (eds.), *Earthworms in Waste and Environmental Management.* SPB Academic Publishing, the Hague, the Netherlands.

Gates, G.E. 1948. On some earthworms from the Buitenzorg museum III: Result of the third Archold Expedition 1938–1939. Oligochaeta. *Treubia* 19(2), 136–166.

Gates, G.E. 1972. Burmese earthworms: An introduction to the systematics and biology of megadrile oligochaetes with special reference to Southeast Asia. *Trans. Am. Philos. Soc.* 62(7), 1–326.

Gunadi, B., Blount, C., and Edwards, C.A. 2002. The growth and fecundity of *Eisenia fetida* (Savigny) in cattle solids pre-composted for different periods. *Pedobiologia* 46(1), 15–23.

Gunadi, B., and Edwards, C.A. 2003. The effects of multiple applications of different wastes on the growth, fecundity and survival of *Eisenia fetida* (Savigny) (Lumbricidae). *Pedobiologia* 47(4), 321–330.

Gunadi, B., Edwards, C.A., and Arancon, N.Q. 2002. Changes in trophic structure of soil arthropods after the application of vermicomposts. *Eur. J. Soil Biol.* 38(2), 161–165.

Gunadi, B., Edwards, C.A., and Blount, C., IV. 2003. The influence of different moisture levels on the growth, fecundity and survival of *Eisenia fetida* (Savigny) in cattle and pig manure solids. *Eur. J. Soil Biol.* 39(1), 19–24.

Gunadi, B., Suhardjo, T.B.A., Sofia, D., Susanto, S.J.A., and Sosrodjojo, P.S. 1996. Using redworms (*Eisenia fetida*) to decompose the tea waste—preliminary research. In: *Proceedings of XII International Colloquium on Soil Zoology*, Dublin, July 21–26.

Gunadi, B., Susanto, S.J.A., and Sosrodjojo, P.S. 1998. Laboratory and large-scale indoor conditions for culturing two species of redworms (*Eisenia andrei* and *Eisenia fetida*) using tea leaf waste as a food. In: *Proceedings of the 6th International Symposium on Earthworm Ecology*, Vigo, Spain, August 31–September 4.

Gunadi, B., and Verhoef, H.A. 1993. The flow of nutrients in a *Pinus merkusii* forest plantation in Central Java; the contribution of soil animals. *Eur. J. Soil Biol.* 29(3–4), 133–139.

Gunadi, B., Verhoef, H.A., and Bedaux, J.J.M. 1998. Seasonal dynamics of decomposition of coniferous leaf litter in a forest plantation (*Pinus merkusii*) in Central Java, Indonesia. *Soil Biol. Biochem.* 30(7), 845–852.

Hatanaka, K., Ishioka, Y., and Feruichi, E. 1983. Cultivation of *Eisenia fetida* using dairy waste sludge cake. In: Satchell, J.E. (ed.), *Earthworm Ecology from Darwin to Vermiculture*. Chapman & Hall, New York.

Ilyas, M. 2009. Vermicomposting of Sonokeling (*Dalbergia latifolia*) leaf litter using three earthworm species (*Pheretima* sp., *Eisenia fetida,* and *Lumbricus rubellus*). [In Indonesian with abstract in English.] Master's thesis, Institute of Agriculture Bogor, Indonesia.

Kale, R.D. 1998. Earthworms: Nature's gift for utilization of organic waste. In: Edwards, C.A. (ed.), *Earthworm Ecology*. St. Lucie Press, Boca Raton, FL.

Ke, L., Bao, W., Chen, L., Wong, Y.S., and Tam, N.F.Y. 2009. Effects of humic acid on solubility and biodegradation of polycyclic aromatic hydrocarbons in liquid media and mangrove sediment slurries. *Chemosphere* 76(8), 1102–1108.

Kochany, L.E., and Kochany, J. 2008. Humic substances in bioremediation of industrial wastewater—mitigation of inhibition of activated sludge caused by phenol and formaldehyde. *J. Environ. Sci. Health A Tox. Hazard Subst. Environ. Eng.* 43(6), 619–626.

Lanno, R.P., Gunadi, B., Bradham, K.D., Basta, N.T., Vijver, M., and Peijnenburg, W. 2002. A fractionation procedure for the development of critical body residues for metals in earthworms. In: *Proceedings of the 7th International Symposium on Earthworm Ecology,* Cardiff, UK., September 1–6.

Lassegues, M., Milochau, A., Doignon, F., Du Pasquer, L., and Valembois, P. 1997. Sequence and expression of an *Eisenia fetida* derived cDNA clone that encodes the 40-kDa fetidin antibacterial protein. *Eur. J. Biochem.* 246(3), 756–762.

Malisie, A.F. 2008. Sustainability assessment on sanitation systems for low-income urban areas in Indonesia. PhD diss., Hamburg University of Technology, Germany.

Milochau, A., Lassegues, M., and Valembois, P. 1997. Purification, characterization and activities of two hemolytic and antibacterial proteins from coelomic fluid of annelid *Eisenia fetida andrei. Biochim. Biophys. Acta* 1337(1), 123–132.

Mitchell, A. 1997. Production of *Eisenia fetida* and vermicompost from feed-lot cattle manure. *Soil Biol. Biochem.* 29(3/4), 763–766.

Murphy, D. 1993. *Earthworms in Australia.* Hyland House, Victoria, Australia.

Neuhauser, E.F. 1978. The utilization of earthworms in solid waste management. In: *Proceedings of the Conference on Utilization of Soil Organisms in Sludge Management*, Syracuse, New York, June 15–17.

Reinecke, A.J., and Viljoen, S.A. 1990. The influence of feeding patterns on growth and reproduction of the vermicomposting earthworm *Eisenia fetida* (Oligochaeta). *Biol. Fertil. Soils* 10, 184–187.

Roslim, D.I. 1994. Earthworms *Pheretima sp* and *Eisenia foetida* as a casting producer. [In Indonesian (unpublished).] Paper, Institute of Agriculture Bogor, Indonesia.

Sabine, J.R. 1983. Earthworms as a source of food and drugs. In: Satchell, J.E. (ed.), *Earthworm Ecology from Darwin to Vermiculture*. Chapman and Hall, New York.

Teskey, H.J. 1990. Insecta: Diptera (larvae). In: Dindal, D.L. (ed.), *Soil Biology Guide*. John Wiley & Sons, New York.

Yulipriyanto, H. 1993. Vermicomposting of organic waste. [In Indonesian with abstract in English.] Master's thesis, Institute of Agriculture Bogor, Indonesia.

CHAPTER 31

Vermicomposting Projects in Hong Kong

David John Ellery and Tse Chi Kai

CONTENTS

I Introduction 497
II Business Model Facets 501
  A Financial 501
  B Logistics 502
  C Land Availability and Cost 502
  D Regulatory Issues 503
  E The Sunburst Hong Kong Facility 503
  F Press Releases by Sunburst Biotechnologies on Vermicomposting 503
    1 Earthworms May Hold the key to One of our Most Challenging Problems in Hong Kong: How to Dispose of Waste? 503
    2 Launch Pad 504
    3 Vermiculture 504
    4 Massive Potential 505
References 506

I INTRODUCTION

The waste-management situation in Hong Kong is no different from that in many other cities all over the world. Hong Kong produces approximately 9000 metric tons of waste per day. Of this, approximately 69% is organic. The region has no coordinated or well-organized recycling system for materials such as glass, paper, steel, or plastics. Because Hong Kong is an island, other than a small amount of mainland called the New Territories, its space for landfill is very limited. Hong Kong has four large landfills all contracted to private management but owned and monitored by the Hong Kong Government through their Environmental Protection Department

(EPD). This department has plans for a mixed solid waste incinerator and large-scale anaerobic digestion facilities, with neither being started at the time of this writing.

As in so many regions of the world, the desire to recycle waste materials is often much stronger than the resolve. This came as no surprise to the management of Sunburst Biotechnology. Governments promote the virtues of recycling activities but rarely have the regulatory infrastructure or the will to facilitate or encourage the industry operators or the technologies required.

This problem has been compounded by Hong Kong Government departments that were totally ignorant of the use of vermicomposting by earthworms in organic waste recycling, even at the household level. In most regions of the world with similar socioeconomic development, home composting and vermicomposting of organic wastes are either supported or normally promoted and encouraged. This concept has never been considered seriously in Hong Kong, hence Sunburst Biotechnology has had the additional task of having to educate the very parties that regulate its activity. This is a major task requiring considerable resources and financial depth.

Vermiculture is practiced on a small scale in low-technology systems within mainland China, as it is all over the world, but it was not practiced at all in Hong Kong. The major use for Chinese earthworm production appears to be pharmaceutical, with earthworms used as ingredients in numerous Chinese medical products but with very little focus on either organic-waste management or soil-amendment manufacture using earthworms or their by-products. Sunburst Biotechnology spent approximately 6 months gathering data on the markets and earthworm breeders prior to its decision to set up business in Hong Kong. This was followed by an additional 2 years of discussion and monitoring of the political climate for such an operation.

Prior to the Hong Kong operation, David Ellery had established an Australian company called Sunburst Nominees in 1999. This was a result of Australian Government funding designed to encourage new enterprises and to foster business clusters and networks. The Australian operation was designed to process comingled mixed solid wastes, whereby the facility operated a "dirty material recovery facility (MRF)" that targeted the organic fraction of this waste stream while extracting inorganic recyclables, which were later sold to offset the cost of processing organic wastes into suitable earthworm feed. The operating model was, however, also designed to be flexible enough to add and subtract certain activities relative to the market that it was applied to. The company spent 8 years perfecting this model prior to promoting and introducing it in Hong Kong. During this 8-year period the technology developers learned, among other things, which waste streams blended better with others, how to handle various organic-waste materials, what preparation times were required when blending mixed organic-waste streams, how to use microorganisms to enhance the preparation process, how to recycle and reuse process water, how to control odors, which equipment to use and why to use it.

The Australian Sunburst operation produced both a solid composite mix (pure vermicompost blended with premium thermophilic composts, which were also made by the company from the same organic-waste inputs), and a base earthworm liquid extract ("tea") made from solid vermicompost and put through a proprietary process developed by the company. This vermicompost liquid, or tea, was then used by farmers and growers as produced or was purchased by fertilizer manufacturers for use as a base ingredient in their own products. The company would also make special mixes specific to a customer's requirements (e.g., N-boosted, K-enhanced).

Hong Kong was chosen as the entry point for the Asian business activities of the company because it offered a stable legal structure, based on the British law system, but was also located only kilometers from the border with the People's Republic of China (PRC). The company's primary motivation was to use Hong Kong as way to demonstrate the Sunburst vermiculture methodology on a relatively small scale, in a 50 metric ton per day facility. This would be used to show the government and private parties the potential of the system as an organic-waste recycling tool and agricultural-amendment-production facility, through the processing of organic residues and their conversion to valuable plant growth stimulants or organic soil-amendments using earthworms. Unlike other commercial operators within the region, the Sunburst operation is not focused on breeding earthworms for resale, and the company has no intention to do so at this point.

The company started with extensive public relations activity to stimulate both public and commercial interest and to focus Hong Kong Government departments on the potential *and* validity of the technology and process. These can be seen below in the press releases, which were timed to coincide with the 2007 Olympic games in Hong Kong and China.

The company began its activities in Hong Kong in 2004, starting with extensive research of the market potential of both the vermicomposting system and its end products, (soil-amendments being the primary market focus). This process was facilitated and assisted extensively by the Australian Government department of AUSTRADE, which had previously assisted with the development of the technology first established in Australia. AUSTRADE facilitated meetings with and introductions to Hong Kong Government departments and other bodies through their Hong Kong and China offices and allowed Sunburst to speak with parties interested in the concept.

After numerous visits to the region by Ellery, the system developer, the company began operations on June 1, 2007. A disused piggery site was chosen and converted as a Stage 1 site that accepted and processed organic wastes and residues into a homogeneous mixed C:N earthworm feedstock. A Stage 2 site was then established in close proximity; it received this Stage 1 material for both earthworm processing and thermophilic composting.

The company utilizes a highly advanced earthworm reactor system process (as originally designed by Clive Edwards and associates in the 1980s), which was engineered and used originally in Australia, then improved further in the Hong Kong operation. The facility contains a vermicompost liquid manufacturing process as designed and used in its Australian operation. This process involves the specific aqueous extraction of both nutrients and microorganisms that are contained within the solid vermicomposts after they are harvested from the earthworm reactors via controlled agitation over a designated time frame. The extract is then enhanced with natural organic additives to assist with microbial metabolization and reproduction. Other methods of earthworm-liquid extract manufacture used by the industry use a leaching or tea bag process, but this was found to be not as effective in the experience of the Sunburst developers.

The primary objective of the Hong Kong facility was to provide a small operation that would open the door to the much larger market of mainland China. Sunburst's business plan is simple—to establish and operate a small Hong Kong facility that will attract great interest from the PRC government and private investor parties. In light of the population of China, (approximately 1.4 billion), and considering that over 70% of the country is agriculturally based but has poor soils, the Sunburst system appeared well suited to handle China's organic-waste problems and provide a viable organic soil-amendment. The company targeted 2009 as its proposed entry date into the PRC. The environmental benefits to Hong Kong, in the establishment of this initial facility, will be obvious to the reader who is familiar with the principles of vermicomposting, organic-waste recycling, and recycling in general. The amounts of organic wastes that were put into landfills decreased, which also lessened the greenhouse gas effects of these wastes. Removing organic wastes from the overall waste stream also lessens the production of leachates and their release into groundwater and the needs to treat this waste by-product in the landfill process.

The range of products produced by the Sunburst operation in Hong Kong is focused primarily on liquid vermicompost concentrates, or teas, which can be enhanced further with organic nutrient additions if required or as desired by the end market, as was practiced in Australia. A solid mix of pure vermicompost and thermophilic compost is also produced for specific clients but is not available as a general product. The range of products can be altered to suit the market they are destined for; in the case of Sunburst this is ultimately the PRC.

The waste-flow process developed by the company comes from extensive experience in waste processing (prior to the entry into the earthworm-processing arena), combined with the biological requirements of organic compounds and the use of bacteria, fungi, and other microorganisms to provide a stable, nutritious feedstock for maximum efficiency in earthworm processing, using primarily *Eisenia fetida*. The company sees earthworm waste processing as the "engine room" of its technology but has also given substantial consideration to other processes designed to enhance its operating model. Areas such as anaerobic digestion and biofuel production can also be incorporated within its operations, but these are future developments that the company will consider later as steps in its PRC development.

The major hurdle that large-scale organic waste-processing operators like Sunburst face is convincing the commercial population at large that the process is far removed from the small household earthworm unit in its operation and approach, although the basic principles and outcomes are similar: Earthworms consume organic wastes and produce vermicompost.

Because of the unique nature of vermicompost-based products and their relatively new status in the soil-amendment and plant growth stimulant market, the challenge in marketing them is considerable, considering that the PRC has experienced the same market penetration (maybe more so) of the major chemical fertilizer companies and brands, and the mentality that comes from using those types of product. Sunburst positions its products as organic (suited for certified organic growers because the company has international certifications), and also as a supplement to conventional inorganic fertilizers that enriches the soil with microorganisms that assist with the conversion of chemical nutrients into plant-available nutrients. This approach has multiple appeals to the farmer who knows his soil contains locked-up chemical nutrients after years of inorganic fertilizer use. Sunburst markets its products with a heavy emphasis on soil science, which is necessary for integration into the current agricultural markets and those emerging in the areas of organic farming, especially within the Asian region. As farmers and growers begin to understand the significance of good soil health and quality and as governments place more emphasis on efficient food production and environmentally favorable farming practices, the products of Sunburst and others in this field (provided they are professional) will continue to expand.

The vermiculture industry in China is still in its infancy. As stated previously, most Chinese earthworm practitioners are focused almost solely on the breeding and supply of earthworms to the Chinese pharmaceutical industry. Over the past 2 years there has been a noticeable shift toward the greater use of earthworms as a result of both Sunburst activity and the work of Chinese researcher Professor Sun Zhenjun located in Beijing (see Chapter 34).

In any large-scale waste-processing model, including earthworm-based facilities, there are major considerations that need to be addressed and possible solutions formulated before any progress can be made. After this, the next step is to consider the business model facets. These include, but are not limited to, the political climate (whether we like it or not), favorable commercial conditions (in the case of our business this means appropriate waste streams both in quality and quantity), and infrastructure, including logistical considerations and utility availability.

II BUSINESS MODEL FACETS

A Financial

The Sunburst business model includes income streams derived from waste-disposal fees. The logic is simple and is designed to integrate the recycling facility process into the commercial world. Aspects of the Sunburst facility activity replace a conventional landfill site. Most landfill sites charge for accepting and disposing of waste. The Sunburst facility does the same but at a reduced rate to improve its economic competitiveness. This incentive is required because although the resolve to recycle wastes may be strong, the bottom financial line still wins this battle in most regions of the world. For example, most waste producers will still opt for the cheapest method of disposal, if the recycling option is better for the environment but more expensive.

This section of the business model presented the company with a unique problem. Hong Kong has no waste-disposal fee for the disposal of mixed solid wastes (as ofs December 2009). The Hong Kong Government takes total responsibility for this cost and owns the landfill sites used for this purpose. There is no user-pays disposal system yet established in the region. Sunburst sought and negotiated with waste generators who were willing to pay a disposal fee based on the policies of their international parent companies, who know that the free disposal system in Hong Kong is not sustainable in the long term. These organizations also see the public relations benefit of being seen as corporately responsible.

B Logistics

Prior to its launch in Hong Kong, the company studied, in detail, the makeup of the Hong Kong waste streams. Information revealed that Hong Kong produced approximately the same percentages of organic wastes as most modern cities, but the key difference was that most of this was wet organic residues due to the nature of the diet of the Chinese community. Hong Kong has no agricultural sector to speak of, and therefore animal manures were not an option as a supplement to earthworm feedstock material; 98% of meat and dairy products are imported into the region. The major organic-waste streams come from large hotel and fast food chains, which are a big part of the industrial sector in the region. The wastes produced in Hong Kong are split and taken to four strategically located landfill sites in the region. These landfill sites have an estimated 3–7 years of life left. Of this total waste stream, approximately 60% is organic, which is in line with most Westernized cities.

The company then assessed how much of this waste is available to it in a "uncontaminated" form (i.e., not mixed with inorganic waste, in order to minimize waste sorting prior to pretreatment for earthworm consumption). Extensive waste auditing was done with commercial parties who have responded to the company's public relations activity, including large worldwide hotel chains with operations in Hong Kong and large tourist operators also active in Hong Kong but known worldwide.

The waste transport industry, as in most parts of the world, presented the biggest challenge. Most operators have set methods of waste collection, and most of these involve no consideration on how to collect organic waste materials in a separate system unless forced to do so by the client or regulation. For operators like Sunburst this is a major issue. The company had to decide whether to run its own waste-collection vehicles (using the same disposal fee system charged by normal transporters but taking the collected waste for recycling) or to enter into agreements with commercial well-established transporters to offer an alternative waste-disposal point (the Sunburst facility). It was decided to use a mixture of these two options.

C Land Availability and Cost

Land in Hong Kong, as in such places as Singapore and Tokyo, is at a premium. Large open tracts of land such as those available in Australia, the United States, and other regions are unknown and simply not available within the region. Less-than-ideal operating conditions are created as a result, and innovative engineering and facility design becomes critical in this case. The use of multilevel applications becomes the norm, with the emphasis on using height as opposed to length and width (area). As a consequence, the cost of either leasing or purchasing land is extremely high compared with most other regions of the world, and this cost alone has had a significant impact on Sunburst's operating profitability and is a major consideration for any similar operators contemplating entering this market.

D Regulatory Issues

Because of the limited land availability in Hong Kong, the regulatory system governing land use is complex. The process of earthworm breeding and the use of earthworms to process organic wastes and residues is commonly accepted in most countries as an agricultural activity, but this is not the case in Hong Kong, where the process has been classified as an industrial activity; therefore, the regulations governing it are far more stringent.

E The Sunburst Hong Kong Facility

The Sunburst Hong Kong facility utilizes the three main processes developed by the process inventor. These are (i) the wet preparation and blending process, (ii) the earthworm waste-processing process, and (iii) the manufacture of the plant growth stimulants (solid and liquid). Each section is distinct in its operation and position within the Sunburst cycle. Each section has its own procedures and effectively operates independently of the other processes but is linked within a Sunburst facility. This has been done deliberately to cater to situations where one integrated site cannot be set up, and a number of smaller sites are required as was the case in Hong Kong.

The major section of this process, and that which is particularly relevant to this book, is the design and operation of the Sunburst continuous-flow earthworm reactor systems. As stated earlier, the original design of the Sunburst reactors was based on the designs of Edwards and his associates. The major difference with the Sunburst system is the size and degree of precision engineering used in their construction and operation. The Sunburst earthworm reactors are automated to the degree that one operator can comfortably operate four units measuring 6 m wide by a minimum of 100 m long. Temperature, moisture, and acidity are monitored via probes and recorded on a cycle determined by the operators and technical-management team. Conventional overhead gantry feeding, which is a characteristic of the earthworm reactor, is combined with the harvesting system in a one-piece integrated piece of machinery that can be removed and redeployed on numerous reactor beds.

F Press Releases by Sunburst Biotechnologies on Vermicomposting

1 Earthworms May hold the key to one of our most challenging problems in Hong Kong: how to dispose of waste?

Waste disposal costs are skyrocketing as fast as landfill sites are filling up around the world. Sunburst Biotechnologies Ltd, however, believes it has a solution, and has chosen Hong Kong as the place to launch it.

Sunburst has used a technology known as vermiculture, using earthworms to recycle organic wastes naturally. Every day at the firm's Hong Kong plant in the New Territories, 80 million hungry earthworms gobble up tons of food scraps and other organic rubbish, turning it into powerful – and completely natural – agricultural certified organic fertilizer.

The 50 tons of waste processed each day by Sunburst is transformed into 25 tons of organic fertiliser. Nearly half of this raw material is stable waste from the Hong Kong Jockey Club, which has announced a target of recycling 100 per cent of its organic waste.

The end product, in solid and liquid form, is sold to domestic users in Hong Kong and exported to markets in the Middle East, the Chinese mainland, Japan and Europe.

The concept was developed in Australia by David Ellery, Sunburst's Managing Director. Supported by government start-up funding, he opened a small demonstration facility in rural South Australia in 1999.

2 Launch Pad

Hong Kong would be the launch pad. The firm already had a local connection in Tse Chi Kai, who had been acting as Sunburst's Hong Kong and China representative since 2001 and is now its Executive Director. In June 2007, Sunburst Biotechnologies was born.

Hong Kong's location as the gateway to China is a distinct advantage for any firm with the mainland in its sights. Sunburst had identified its market opportunity even before the Chinese government elevated agriculture and the environment to the top of its policy needs for the country.

3 Vermiculture

The environmental benefits of vermiculture, compared to conventional waste disposal methods: "Organic matter placed into landfill, as is the conventional method still practiced worldwide, decomposes in what is called an anaerobic state [no oxygen], because it is buried. When this happens, methane and carbon dioxide are produced. These gases percolate through the ground and are released into the atmosphere, producing greenhouse gases. In addition, organic matter, especially food waste, releases large volumes of water as it breaks down. This water is full of pathogens such as *Erioischio coli*, which then can make their way into the groundwater system.

"The Sunburst process stops both of these processes and produces a high-grade organic fertilizer as well for reuse."

The company's solution deals with both waste management and organic fertilizer production, an integration that has caught the attention of the public. "We have had numerous approaches made to us while operating in Hong Kong," says Mr Ellery. "These range from China, through to Malaysia and Singapore. All keep in regular contact and are at various stages of assessment for their regions."

4 Massive Potential

A key participant in the vermiculture pilot project is the Hong Kong Jockey Club, which saw it as a solution to the weighty issue of disposal its stable waste. Detritus from the daily stable cleaning and horse feeding can be put to good use through vermiculture. Waste is collected from the Sha Tin and Happy Valley stables each day and transported to the New Territories plant, where it becomes part of the earthworms' food supply chain. The resultant fertilizer is later sold to people who can use it on local organic farms and household gardens. The initiative was showcased at the 2008 Beijing Olympic Games. The Games were dubbed "the green Olympics" and Hong Kong, as host city of the equestrian events, took this opportunity to pledge its environmental commitment.

When the Hong Kong Jockey Club demonstrated its vermiculture solution to the international press in the lead-up to the Olympics, it was hailed as a world-first in the mass recycling of equestrian-related waste. The club says this is only the start and it intends to recycle 100 per cent of its organic waste.

Hong Kong's landfill problems and the success of the Hong Kong Jockey Club's experience should make vermiculture a growth industry in Hong Kong. The Australian company operating the New Territories plant certainly believes so. The technology is suitable for all kinds of organic waste, including food scraps. Some hotels and restaurant chains are recycling their food waste via the Vermiculture plant.

Aside from the Jockey Club, which began recycling its stable waste at the plant as part of its environmental commitment to the 2008 Olympic Games, fast-food chains like McDonald's and Café de Coral, and banking corporations such as HSBC are using the Sunburst system. Given the huge amounts of organic waste that the catering business generates each day around the region, the system seems a promising option for the sector which has limited landfill capacity.

The company developed the technology in Australia and brought it to Hong Kong believing this was the best platform to reach emerging markets in Asia. The end product, processed into solid and liquid forms, is sold locally at B&Q MegaBox, and exported to markets in the mainland, Japan, the Middle East and Europe. According to Tse Chi-kai, Executive Director, Sunburst Biotechnology, the organic fertilizer that vermiculture produces is rich in natural nutrients that are released by earthworms during the recycling process. Among other benefits, he says, this helps to improve the yield and quality of crops, and increase pest and disease resistance. International studies endorse the efficacy of the product. Using such technologies also helps extend the lifespan of landfills and reduces greenhouse gases. Longer term, vermiculture is expected to be incorporated into carbon credit schemes, further helping the environment. The concept of natural recycling may not be new, but only recently has it been adapted for commercial application. With governments worldwide increasingly setting their farmers organic targets, the market potential for solutions such as vermiculture seems enormous.

REFERENCES

Edwards, C.A., Arancon, N.Q., Kai, T.C., and Ellery, D. 2007. The conversion of organic wastes into vermicomposts and vermicompost 'teas' which promote plant growth and suppress pests and diseases. *Hong Kong Agriculture and Fisheries Promotion Association Annual*, pp. 33–40.

Edwards, C.A., Arancon, N.Q., Kai, T.C., and Ellery, D. 2010. The use of aqueous extracts of vermicomposts ('teas') in the growth of vegetables and management of plant diseases, plant parasitic nematodes, and arthropod pests. *Hong Kong Agriculture and Fisheries Promotion Association Annual* (in press).

Kelly, Chan L. 2008. Worms win Jockey Club waste race. *South China Morning Post*, January 3, 2008.

CHAPTER 32

Vermicomposting Research and Activities in Mexico

Eduardo Aranda-Delgado, Isabelle Barois, Martin de los Santos Bailón, and Benito Hernández-Castellanos

CONTENTS

I Introduction 508
II Historical Background 508
III The Concepts of Vermicomposting 509
IV Vermicomposting Research in Mexico 511
A Initial Studies on Vermicomposting of Coffee Pulp in the 1980s and 1990s 512
B Later Studies on Vermiculture in Mexico 512
V Application of Vermicompost Products 514
A Vermicompost 514
B Vermicompost Teas 515
VI Vermicomposting Systems 515
A General Aspects of Vermicomposting 516
B Species of Earthworms Used in Vermicomposting 516
C Substrates or Feedstocks 517
D Designs of Vermicompost Systems 518
E Human Resources for Vermicomposting 521
F Vermicomposting Operation Methods 521
G Vermicompost Production 522
H Commercial Aspects of Vermicomposts 522
I Quality Control of Vermicomposting and Its Products 523
J Promotional Aspects of Vermiculture 524
K Discussion and Conclusions from the Vermicomposting Survey 524
VII Vermicompost Quality Standards 525

VIII Vermicomposting Farms in Mexico 526
IX Vermicomposting Urban Domestic Residues 526
X Conclusions 527
Acknowledgment 527
References 527

I INTRODUCTION

With the aid of earthworms, organic waste materials can be transformed into a material rich in nutrients readily available to plants. This earthworm-worked material, *castings* or *vermicompost*, provides a beneficial microbes growing substrate, which consists of very finely structured, uniform, stable, and aggregated particles of humified organic material with excellent porosity, aeration, and water-holding capacity, rich in available nutrients, plant growth hormones, enzymes, and beneficial microbial populations. Thus, it can be added to agricultural land to improve soil structure, water retention, and fertility or used as a good-quality and marketable additive to potting soil or plant growth media (Edwards and Bohlen 1996; Aranda and Barois 2000).

Since the 1980s, vermiculture and vermicomposting have gradually been gaining recognition in Mexico. The development of this technology demonstrates indisputably that vermicomposting is an effective, efficient, and economically viable method of converting organic waste and that its application contributes considerably to the productive and economical use of organic wastes in environmentally friendly conservation programs (Aranda et al. 2009).

Many by-products can be used in a better way to produce vermicomposts, reducing their polluting potential, offering a natural recycling alternative and an excellent premium product for agricultural purposes. Coffee pulp; *cachaza*, the end by-product of sugarcane-processing mills; cattle, pig, horse, sheep, chicken, and rabbit manure; domestic food wastes; and organic municipal solid wastes are some of the most important, because of their quantities and negative environmental impact (Aranda 1995).

This chapter attempts to describe and bring together Mexican experiences in vermiculture, from vermicomposting research to practical perspectives. There is still much scientific, technological, and promotional work to be done for vermicomposting to consolidate it as an economical technological option, but here we attempt to highlight achievements and future potential throughout Mexico.

II HISTORICAL BACKGROUND

There is little scientific information or written evidence found on vermicomposting in Mexico before 1984. Perhaps as a result of the boom in the United States during the 1970s (Bouché 1987; White 1996), some information and pamphlets from northern Mexican farms can be found talking about the benefits of buying supposed "hybrid worms" to initiate earthworm farms using cow manure as a feedstock.

However, two main initial vermicomposting projects in Mexico are worthy of mention here as part of the expansion of this activity.

A private enterprise, using a vermiculture package sold by an Italian Fatoria de la Roqueta, acquired a container with hundreds of thousands of supposed *Lumbricus rubellus* (afterwards recognized as *Eisenia andrei*) and started Lombrimex in 1984 to produce vermicompost at a large dairy farm. From then on, some other associated farms grew up in Veracruz and San Luis Potosi states, to produce and sell vermicompost (Ignacio Cervantes, pers. comm.)

In the same year, in a separate project, *Eisenia fetida* (Savigny 1826) and *Perionyx excavatus* (Perrier, 1872) were found fortuitously, growing on open piles of coffee pulp in Veracruz, Mexico, where they were first studied and their potential use in the production of vermicompost was confirmed (Aranda 1988). It was then pointed out that the reduction of organic waste materials, the elimination of contamination, together with the production of vermicompost allowed for a beneficial linked application, where the main by-product of the coffee industry could be used advantageously in growing coffee seedlings and fertilizing plants, and also as a source of protein for chickens. This could be seen as a whole beneficial-sustainable coffee system.

With the confirmation that *Eisenia andrei* (Bouché 1972) was also found in Mexico, basic studies and practical experiments were initiated (Aranda 1988, 1989). The first results were presented at international coffee growers' meetings held in Panama and Colombia (Aranda 1991a, 1991b). After these, many other coffee-growing countries also initiated and confirmed the viability of this method and started production. Signs of national interest in vermicomposting in Mexico have been seen with the organization of several symposia—the Simposium Internacional—Reunión Nacional de Lombricultura y Abonos Orgánicos, celebrated from 1999 to 2004 (Martínez et al. 1999, 2002; Loza Llamas 2004), and also with promotional support given by official agricultural offices or scientific institutions, to organize workshops, research projects, and also distribute vermicompost to promotional agricultural programs.

III THE CONCEPTS OF VERMICOMPOSTING

This began with the knowledge that certain species of *epigeic* earthworms—which live in litter and feed primarily on organic matter (Lavelle 1981; Abdul and Abdul 1994)—can grow in and consume organic waste materials, converting them into an earth-like, soil-building substance that forms a beneficial growing environment for plant roots.

Thus, the practice of vermicomposting could be defined as a combination of biological processes, designs, and techniques used to systematically and intensively culture large quantities of certain species of earthworms, in order to speed up the stabilization of organic waste materials. These materials are eaten, fragmented, and digested by the earthworms in interaction with microorganisms and some anaerobic microbiota, which converts into much finer, humified, microbially active casts, where important plant nutrients are held in a form that is much more soluble and available to plants than in the parent compound (Aranda et al. 1999).

The activities of earthworms are not solely the fragmenting or digesting of the organic matter, which is the reason they have been called "litter transformers" (Lavelle 1994). The earthworms derive their nourishment from the microorganism population, which they increase within the organic material. The following benefits have been mentioned (Van Gansen 1962; Satchell 1983; Bouché 1987; Tomati et al. 1987; Edwards and Neuhauser 1988; Edwards and Bohlen 1996; Aranda et al. 1999):

- While eating, they burrow, turn over, and maintain the substrate like a sponge, in an aerobic condition, ensuring the entrance of O_2 and the release of CO_2.
- While moving through the waste, they cover the surface of the burrows with a gelatinous muco-protein substance, which enhances microbial activity and subsequently decomposition.
- They cover the surface of the beds with their casts, reducing bad odors and the presence of undesirable animals such as flies.
- They macerate the organic materials through their grinding gizzard, which strongly increases the exposed surface area and enhances the beneficial action of aerobic microorganisms. In fact, earthworms can derive all of their nourishment from the microorganisms that grow on organic materials.
- The beneficial microorganisms released from the earthworm gut continue their activity for some period outside the gut because of a favorable polysaccharide-mucoprotein medium produced, the *peritrophic membrane*, which impregnates each cast and retains many minute aggregates (Van Gansen 1962).
- Each cast, covered with the peritrophic membrane, has amphiphilic water properties, acting as a water reservoir with its outstanding water-retention capacity and at the same time as a protective surface hardener if it is dried.
- The end product (earthworm casts, castings, or turricules) preserves its own shape and aggregated structure in the soil, which ensures a slow release of nutrients, without losses, draining, or soaking.
- Different organic materials can be mixed together by earthworms, allowing an improved combination and composition of nutrients, producing a much finer, fragmented, and uniform material than by any other composting method.
- During the process, earthworms produce bioactive substances, important for the biochemical and regulating activity of the soil, such as enzymes, antibiotics, vitamins, hormones, and humic substances, of great value in plant nutrition processes (see Chapter 9).
- There is a considerable scientific evidence that human pathogens do not survive the vermicomposting process, so if materials containing pathogens are used, they are for the most part killed in passing through the earthworm gut (see Chapter 16).
- Small inorganic pieces, such as rocks, plastics, or glass, hard to collect when mixed with the organics, can be easily sieved away after vermicomposting, due to the finer size of the castings.
- Contrary to common belief, earthworms do not have many serious natural enemies, diseases, or predators.
- In addition, the process procedures give space to more earthworms to extend the vermicomposting areas and to produce high-quality protein meal, suitable for inclusion in various domestic animal feeds.

IV VERMICOMPOSTING RESEARCH IN MEXICO

Initial studies on vermicomposting in Mexico had a key role in expanding the beneficial work of earthworms, underlining the importance of recycling by-products and the ecological benefit of returning organic matter to the soil, and emphasizing the socioeconomic and environmental benefits of vermicomposting. Over the last 23 years, from 1986 to 2009, there have been a growing number of student theses, book chapters, and articles, as well as institutions that have started vermicomposting research (Figure 32.1). Thus, it is possible to recognize two historical stages in vermicomposting research in Mexico, from the 1980s to current times:

In the late 1980s to 1990s, initial studies were carried out by a small group of research institutions and provided the basis for the promotion and development of vermicomposting, with fundamental studies at the laboratory level. They attempted to set up the growth and development of earthworms as well as characterize materials and the results they produce on plant growth (Aranda et al. 1999; Aranda and Barois 2000).

Later, but broader, more extensive and dispersed results of the initial studies concluded by several universities, institutions, and research centers, in different areas of the country, which expanded knowledge and practical in different regions of the country and at organic centers (Figure 32.2). Nowadays, it is possible to find some work being done in 22 of the 32 states. Of these, five to six have produced the majority of the publications produced. The remaining 10 states in Mexico have no relevant activities as yet. Thus, we can say that vermicomposting still has a long way to go to be considered, dispersed, and applied throughout the country.

Most of the work and studies are produced at a general information level, including theses, student reports, Congress abstracts, and media outreach; only a small proportion of them have reached a depth of research and are presented in articles, chapters, and indexed papers. Every level of publication has had some beneficial effect on land implementation and has strengthened the knowledge and technological development of vermicomposting.

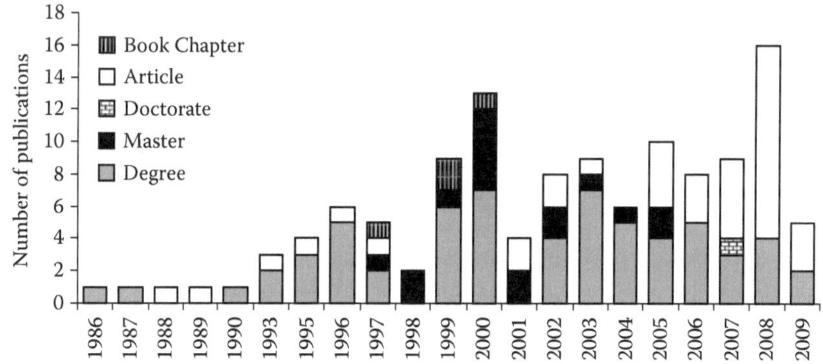

Figure 32.1 Vermicomposting studies in Mexico, 1986–2009.

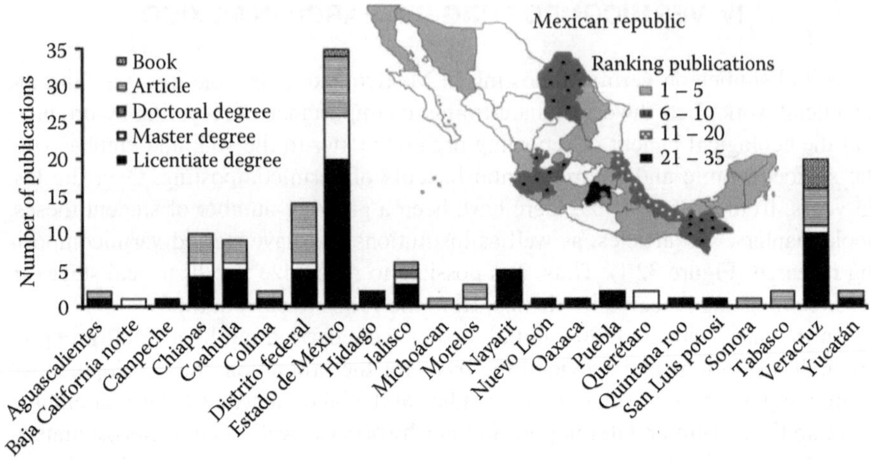

Figure 32.2 Mexican distribution of vermicomposting studies, by publication type.

A Initial Studies on Vermicomposting of Coffee Pulp in the 1980s and 1990s

One of the most influential studies was done at the Instituto de Ecología A. C. (INECOL) in Xalapa, Veracruz, providing the fundamental basis for the promotion and development of vermicomposting of coffee pulp in Mexico, which expanded across the coffee-growing areas of Latin America and the rest of the world (Barois and Aranda 1995; Aranda et al. 1999; Aranda and Barois 2000).

Together with this project, other research was done by students in other Mexican regions, who started their thesis research and then, together with their professors and organic-matter sources, started their own research programs. Although it is possible to say that coffee-pulp vermicompost is one of the most well-accepted, dependable, and consistent types of humus found on the market in Mexico, many other good products are sold, always based on the availability of organic-matter sources and the markets operating in each region.

B Later Studies on Vermiculture in Mexico

Great efforts have been made to develop broad-minded aspects of vermicomposting in Mexico. Institutions and research centers have been growing gradually in number and expanding this knowledge around the country. Hence, most of the bibliographic references come from students doing their initial thesis work, with broader and speculative projects such as the evaluation of new and different organic sources as substrates, of mixtures of materials that make them suitable for earthworms, and of biological aspects (growing, reproduction, population variables) of vermicomposting, or simply general studies of earthworm castings attempting to evaluate plant growth (Figure 32.3). Most student studies have been under the auspices of

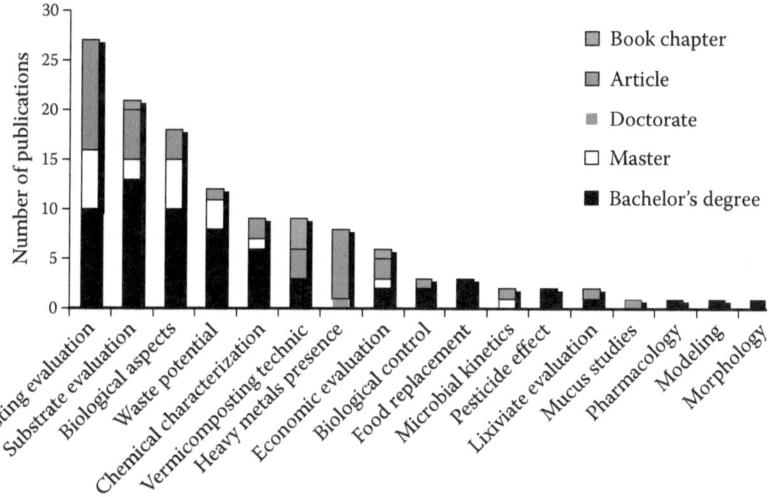

Figure 32.3 Vermicomposting studies, grouped by publication and topic.

universities or research institutions in close proximity to vermicomposting areas or specific research departments.

Some of the institutions have established research team programs and have produced students with master's and doctoral degrees, as well as chapters in books, articles, and books on very diverse subjects: the effect of earthworm casts on growth fof cultured plants, the consequences of heavy metals on the nutritional aspects or on castings, the profitability of vermicomposting, earthworm population modeling, pharmacology, and so on.

The institutions include INECOL, already mentioned; el colegia de la frontera sur (ECOSUR), a research institute in the state of Chiapas (Huerta-Lwanga 1996; León Monzón 2003; Martínez Aguilar 2004); BUAP (Benemérita Universidad Autónoma de Puebla), the state university of Puebla (Limón 1999; Klemmer González 2008); UG, the University of Guadalajara, in the state of Jalisco (Loza Llamas 2000, 2007; Aguilera Castillo 2002; Meléndez 2003); CINVESTAV-IPN, the Center of Research and Advanced Studies of the National Polytechnic Institute (Contreras-Ramos et al. 2008; Contreras-Ramos, Alvarez-Bernal, and Dendooven 2009; Contreras-Ramos, Alvarez-Bernal, Montes-Molina, et al. 2009); UACH, the Agricultural Autonomous University of Chapingo (Trejo 1995; Alviter Aguilar 2003), as well as Colegio de postgraduados (COLPOS), the postgraduados college in agriculture in Texcoco in the state of Mexico (Santamaría 2000); UAAAN, the agricultural university in the state of Coahuila (Espinoza 2003; Peña Ramos 2004; Medina 2005; Cruz Flores 2005); UV, the University of Veracruz (Irissón 1995; Salazar et al. 1995; Zamora and Martínez 1999; Hernández Sánchez 2000; Montalvo 2004; Vela 2005); and, on a minor scale, the UAT, the Tamaulipas State University; UMSNH, the University of Michoacán; UNAM, the National University of Mexico; and UADY, the University of Yucatan, among others.

V APPLICATION OF VERMICOMPOST PRODUCTS

A Vermicompost

Currently, little scientific work has been done in Mexico to test the effect of earthworm-worked materials on different crops. A successful example of the use of vermicompost is to produce and sell seedlings grown in large high-technology greenhouses. Normally, large quantities of polystyrene or plastic multicell trays are filled with peat moss and growing media for seedlings to encourage rapid germination and establishment, with eventual high yields of broccoli, carrots, asparagus, onions, garlic, tomatoes, chilies, celery, and so on.

The influence of coffee-pulp vermicompost on the growth of seedlings in greenhouses was tested under normal working conditions. Two similar tests were carried out with mixtures of about 100%, 75%, 50%, 25%, and 0% coffee-pulp vermicompost, combined with peat moss (Sunshine 3® and Fafard FPM®) to grow seedlings of broccoli (Sakata® and Peto Seed® var. Maratón®) in 392 cavity trays in two commercial greenhouses of the Bajio region (Hernandez et al. 1996). The results showed that even with the addition of the smallest proportion of coffee-pulp vermicompost, vigorous plants with dark green leaves and healthy roots were produced, and the transplanting time was reduced from 36 and 37 days for the controls to 25 and 26 days in response to 100% vermicompost.

Vermicomposts and peat moss have different but complementary properties. Castings can compensate for many of the inherent deficiencies of peat moss and be as good or better than the traditional growing-media mixtures:

- Castings easily increase nutrient contents, otherwise provided by chemical “starter nutrients” (calcium nitrate, potassium nitrate, phosphoric acid, iron + micronutrients).
- Castings also increase nutrient absorption, otherwise provided by horticultural Pearlite.
- Castings increase the water-holding capacity, otherwise provided by the “wetting agents.”
- Castings reduce the need for air drying and need less addition of superficial vermiculite.
- Castings reduce the acidity of peat moss, otherwise provided by dolomite and limestone, and regulate pH values because of their buffering capacity.
- Castings decrease the risks of plant pathogens because, although they produce substantial soil microbial activity, this is of a beneficial nature.

Mr. Menéndez, who is the manager of a large greenhouse company that has used about 300 tons of coffee-pulp-based earthworm castings per annum, mostly in a 50:50 proportion with peat moss, commented: “The vermicompost addition increases germination rates, the growth of root hairs and a better general development of the plants. The increases of nutrients in casts reduces or even eliminates the need for a soluble inorganic fertilizer, reducing transplanting time, irrigation frequency, and even the occurrence of plant pathogens. Because of the consistency and water-holding capacity of the potting mixture, the root hairs have better support and the seedlings can resist transplantation better—without "bending"—needing less time to recover their growth rate. . . . Furthermore, each time we grow the seedlings on vermicompost, we are also adding little but consistent amounts of castings to the field" (Menendez, 1998 personal communication).

When compared with other organic substrates, the success of coffee-pulp castings appears to be related to its purity, texture, and consistency; also, it is a product with a constant nutrient content. Coffee pulp has proved to be a dependable, predictable source for high-quality earthworm castings that improve potting mixes in industrial horticulture.

Another segment of the vermicompost commercial activity that is being developed includes attractively bagged retail products, which expands the market to urban areas. Because vermicomposting produces an excellent product in terms of visual appearance, touch, and smell, garden centers are increasingly interested in the product: "Once people use the product and read the bags giving an explanation of the vermicomposting method, its origin and use, homeowners understand it is a different, consistent, premium product."

B Vermicompost Teas

Similar to *compost tea*, *vermicompost tea* is an amorphous term that attempts to define an aqueous vermicompost extract produced from different feedstocks, which can contain diverse microorganisms, soluble nutrients, and suspended particles and is marketed for plant disease control, foliar nutrition, or for improvement of the recovering natural microbial composition of treated soils. Aeration, maturity, addition of nutrients, temperature, and residual fermentation are some of many variables in methods used to produce it (see Chapter 11).

It has been suggested that suppression of plant diseases or disease control, antagonistic microorganisms, microbial by-products, metabolites, antibiotics, activation of natural plant defense responses, and plant nutrients are multiple beneficial mechanisms. There are apparently many options when it comes to making, processing, and using vermicompost tea. As far as we know, vermicompost tea in Mexico doesn't have as much popularity and as high a production level as in other Latin American countries and in the United States, where the flower industry and greenhouse production are more developed.

VI VERMICOMPOSTING SYSTEMS

Up to now, vermicomposting in Mexico has been based on low-input technologies, large availability of land, low-labor costs, and a relative abundance of organic wastes. Commercial attempts at vermicomposting in Mexico have been mostly small experimental sites at agricultural school centers, universities, and research institutions but also private vermicomposting farms in different areas of Mexico. Local and governmental organizations have encouraged people by supplying earthworms, a small installation area, and some assistance.

In the next part of this chapter, we explore vermicomposting techniques in Mexico through our own experience, Internet research, and an informal survey with 76 questions aimed at collecting information on how vermiculture and vermicomposting started and developed in Mexico. The questions were divided into seven sections that included general aspects; biological aspects (which earthworm species and feedstock people use); technical aspects (use of machinery, collecting substrates, farm size, bed, and operation designs); commercial aspects (sales, price, market, distribution); quality and control of the activity and products, testing, and problems faced; promotion of vermiculture and vermicomposting; and information available on vermiculture and vermicomposting, advisors, the Internet, links, manuals, meetings, and requirements. We decided to collect the information from the Internet, where we searched for earthworm farms under the words *lombricultura* and *lombricompostaje* for the Spanish world; with this search we obtained 155 addresses, mostly from Mexico but also from other Latin American countries. The survey was from August to December 2000.

A General Aspects of Vermicomposting

The people who are involved in vermicomposting identified themselves as earthworm growers (52%), although some said they are also in other professions (25%), others are scientists, and a few are amateurs. Most of them started vermiculture because they had an environmental preoccupation (use of wastes 24%, pollution 20%, amelioration of the environment 16%), only 20% started because it could be a profitable activity to produce and sell vermicompost, and the last 20% had other reasons for being involved in this activity (some gave two reasons or more, which gave a total of 79 answers). The main ways people started vermiculture were under the guidance of an institution (34%), a friend (26%), or by reading (23%) and programs (regional and national). The majority started more than 5 years ago (from 2000 on; 46%), a few 15 years ago, and some started 2–5 years ago (31%) or less than 2 years ago (23%).

The initiation of the earthworm farm was generally with 1–10 kg (2.2–22 lbs) of earthworms (46%), but some with more than 10 kg (22 lbs) (24%) the others with fewer than 1000 individual earthworms (30%). For 60% of the people, vermiculture and composting are registered activities in their country, 50% of them said that it is regulated, 20% wrongly regulated, and 30% without any regulations. It should be noted that only 26 people answered this question (at the Latin American level), which indicates the lack of knowledge in this area.

B Species of Earthworms Used in Vermicomposting

Certainly, the most common earthworm species used at vermicomposting sites in Mexico are *Eisenia andrei* and the closely related *E. fetida*, both often confused as *E. fetida* (Jaenike 1982). The original populations of *Eisenia andrei* at INECOL were obtained from the Instituto Mexicano del Café and originally came from California in the United States. Most earthworm farmers call their earthworms California red worms (81%); the other common names they use are Californian hybrid (2%) and the African earthworm, *Eudrilus eugeniae* (9%). The rest (8%) gave other names. Although they seem to know their scientific name, the most commonly used species is *Eisenia fetida* together with *E. andrei* (used by 77% of those surveyed).

Three-quarters of the earthworm growers (74%) can recognize which species their earthworms belong to, only 4% of them do not know, and the rest were told (14%) or never did a formal identification of them (8%). It is mainly academic institutions that identify the earthworms (49%), whereas the rest of the identifications are made by scientists (31%) or friends (20%). Many of the sites also have indigenous populations of *Perionyx excavatus* (the "Oriental compost worm"), which some people in Mexico believe to be a native species although it is actually Asian. In the central area of Veracruz, *P. excavatus* is commonly the first species to invade the coffee-pulp piles in the field, because it can survive in the soil and reproduce better under the high-temperatures characteristic of the initial decomposition of the coffee pulp. This species has been used satisfactorily in open field conditions, sometimes withstanding conditions better than inoculated populations of *E. andrei.*

Although 50,000 individuals m^{-2} (41,806.4 yd^{-2}) almost 3540 gm^{-2} (104.4 oz. yd^{-2}) of *E. andrei* are commonly registered from coffee pulp under controlled conditions, for practical purposes, a good population density is assumed to be around 15,000–25,000 earthworms m^{-2} (12,542–20,903 yd^{-2}) in field conditions. This is around 2000 g m^{-2} (58.9 oz.yd^{-2}).

Based on experience, to prevent overcrowded and anaerobic conditions, the earthworms can also be used as a source of protein for animal feeds. Some good experimental tests have been made in Mexico, using earthworms to feed captive animals (Aranda and Aguilar 1995) and salmon trout (*Oncorhynchus mykiss* (Billard 1989)) as an alternative source of protein (Toscano 2000). At present, the high-coffee region of the state of Veracruz also has trout farms, which are potential consumers of the earthworms. Some attempts have been made to establish *Eudrilus eugeniae* (Kinberg 1867) populations from Cuba and Costa Rica, but although they develop fairly well, this species seems to be more attractive to ant invasions.

C Substrates or Feedstocks

A wide range of organic waste materials are available in Mexico. These come from animals (cow, pig, horse, sheep, rabbit, and chicken manure), vegetables (plants, plant residues), urban settings (municipal solid wastes, domestic residues, anaerobically digested effluents, trimmings, supermarkets), and agroindustrial processes producing residues (sugarcane-processing mill residuals, banana stems, mushroom industry residuals, horticultural and fruit-processing plant by-products, cacao residues, etc.). One of the best-known organic waste products used to produce vermicomposts in Mexico is coffee pulp, but some others are commonly used, such as cattle manure, sugarcane press mud (*cachaza*), and domestic urban organic residues.

The feedstocks used are quite diverse, although half of earthworm growers use manures (cattle 28%, horse 10%, sheep 8%, and goat 2% and others poultry, rabbit, and guinea pig). The other half of feedstocks are chiefly of vegetable origin: coffee pulp, home food wastes (15%) and market wastes (10%), sugarcane press mud (2%), and sludges (2%); 14% are other feedstocks, which include residues from the following agroindustries: corn, bean, rice, macadamias, cacao, cotton, flowers, wood (sawdust), and, finally, cardboard and garden residues. For this question we received 123 answers, showing that most of the earthworm growers use more than two feedstocks. Thus, for more than 75% of growers the vermicompost is a mixture of the different feedstocks they have, and only 24% use just one feedstock. Many of these newer substrates may have an important potential in vermicomposting on a local and regional scale although the noncontrolled mixtures can have undesired implications for the quality control and standardization of the vermicompost.

Three-quarters of the earthworm growers did not generate their own feedstocks, but only 18% of them bought them. The average volume of wastes that they use annually is between 100–1000 m^3 (130.8–1308 yd^3) (40%), while 15% use more than 1000 m^{-3} (1308 yd^{-3}) and 31% use less than 100 m^3 (130.8 yd^3). The remaining 14% evaluate their feedstock by weight: 50% of these consume 100–250 tons, 15.4% use 250–500 ton (246–492 t), and 34.6% more than 500 tons. For 75% of earthworm growers, the feedstocks are available the whole year round, 82% store them, and most of them just accumulate them (65%). Some use silage (14%), and the rest use other methods of storage (22%). Additional studies have been made to standardize the silage process of the organics (in this case coffee pulp) in order to maintain the quality of the organic matter and, therefore, the product obtained (Gaime-Perraud et al. 1991; Gaime-Perraud 1995).

D Designs of Vermicompost Systems

The first designs of vermicomposting systems practiced in Mexico were single or twin windrows constructed from building blocks, in an effort to obtain a better control of the earthworm populations. These were about 0.8 m wide by 0.6 m high, varying in length, directly on the soil or on a bed of gravel (which is not recommended). The population densities of earthworms in the majority of the farms (39%) are more than 20,000 m^{-2} (16,722.5 yd^{-2}) earthworms, 35% have 10,000–20,000 earthworms m^{-2} (8361.3–16 722.5 yd^{-2}), and 18% have fewer than 100–1000 m^2 (119.6–1195.9 yd^2) for 33% of growers, more than 1000 m^2 (1195.9 yd^2) for 22%, 11–100 m^2 (13.16–119.6 yd^2) for 25%, and less than 10 m^2 (11.96 yd^2) for 20%. Usually the width of the beds is 101–200 cm (39.8–78.7 in) (43%), although 37% use a smaller width 51–100 cm (20.1–39.4 in), and some (16%) use a width of less than a 50 cm (19.6 in). At the other extreme, there are a few (4%) that have earthworm beds more than 200 cm (78.7 in) wide.

Several variations of this system were also found using local or readily available materials like wood, wood planks, bamboo, plastic or metal meshes, and even rubber conveyor belts. The beds are placed under different kinds of shelter such as black plastic mesh, bracken ferns (*Pteridium aquilinum*), or also living creeper plants such as *chayote* (*Sechium edule*). The materials of which the beds are made are also very diverse, although 40% of the beds are made with hard materials (cement and bricks). A few (16%) do not have any wall (heaps) or are pits. Some of these beds are under cover (28%), many are in the open field, but 26% are under trees. On larger earthworm farms, the most common systems used are simple windrows on open land,

and under the shade of native or planted trees like macadamia and castor-oil plants (*Ricinus* sp.). Other farms cover the beds, making use of sugarcane tips or used jute sacks from the coffee industry. As time passes, these materials also decompose and become incorporated into the product.

Perhaps one of the simplest and most efficient examples of vermicomposting is based on earthworms working on rabbit farms. The earthworm bins are situated under the rabbit cages; the droppings, urine, water, hair, and food scraps just fall every day to nourish the earthworms. Another example developed here in Mexico in vermicomposting is using pig manure, which is associated with major problems of wastewater pollution and manure handling. Vermicomposting provides a real and beneficial integrated "pig farm waste water solution" (Curiel and Aranda 2000). Granja Porcícola Tepatlaxco is a family-owned pig farm located in a small village in a dry area, which gradually has become part of the Mexico City conurbation. Thus, in the past 10 years, its owners have become increasingly aware of the handling, treatment, and utilization of the pig manure, rather than disposal of the liquids and solids produced. The farm has around 400 sows, with a total population of about 4000 animals. The farm generates about 2600 kg (5732 lb) of fresh pig manure daily, including 100 kg (441 lb) of barley straw (bedding material) and around 13–16 m^3 (17–21 yd^3) of drainage waters (unused food plus manure and urine). Since May 1999, pig manure has been processed by vermicomposting, starting with a population of 12,000 *E. andrei* in six 20 × 2 m (21.9–2.19) open windrows. Fourteen months later the earthworm population had increased, and the initial area was expanded to 45 windrows in an effective area of around 14,000 m^{-2} (1674.4 yd^{-2}) on a corrugated concrete floor (Figure 32.4).

The pig manure is collected manually in wheelbarrows and stored in mill hoppers, with sawdust added to increase the bulk capacity and improve the C:N ratio.

Figure 32.4 Vermicomposting windrows on a corrugated concrete floor, at Granja Tepatlaxco. Courtesy of E. Aranda Delgado.

The material is then piled, turned, and aerated. Because of this, the offensive odor of ammonia and the unpleasant appearance of the feces almost disappear. After about a month, the precomposted manure is then deposited on the earthworm windrows in successive thin layers. A solids separator has been installed to extract excess moisture from the drainage waters, providing additional quantities of solid waste for the vermicomposting beds. In fact, the solids produced by this method provide the fastest and easiest way to feed the earthworms, besides leading to better general development of the earthworms. The resulting water is then sedimented and cleaned using barley straw as bedding, to reduce its excess nitrogen content, and subsequently used to moisten the earthworm beds by spraying it on with an irrigation system. The sediment of the treated waters (and barley straw) is also precomposted and added to the earthworm beds.

This solid waste treatment (omitting the water phase) has opened the possibility of installing an efficient water recycling treatment, which has been continuously and successfully operating over the last 10 years with vermicomposting and 5 years with water recycling (Figure 32.5).

Machinery is not yet commonly used on Mexican vermicomposting farms. Only the largest farms have their own trucks, tractors, or trailers to move residues and end products. The watering of windrows in the field is by irrigation or with trucks, but some still use garden hoses, and some smaller ones simply use watering cans. No sophisticated machines are used to screen or to crumble the castings, but most of them are simply adapted from generally known designs or general-purpose tools.

Figure 32.5 Wastewater and manure handling with vermicomposting, providing a real and beneficial integrated pig farm waste water solution at Granja Porcícola Tepatlaxco. Courtesy of E. Aranda Delgado.

E Human Resources for Vermicomposting

Low-input technology for vermicomposting is characterized by relative low-labor costs, together with needs for a large labor force and large areas of land. Generally speaking, for field production, it seems that one person is needed per 500 m^2 (598 yd^2) of vermicomposting beds. The number of workers commonly working full time on earthworm farms is usually two (27%), but 36% were represented farms with more than four workers (15% had four workers and 23% more than five). The workers are, in general, field workers (53%) administrators (20%), vermicompost processors (16%), and salespeople (8%). About 60% of the earthworm growers have a technical advisor for their farm, whose expertise is in biology or agronomy, and only two identified their advisor as an earthworm grower specialist.

F Vermicomposting Operation Methods

Traditional vermicomposting methods of operation in Mexico are based mostly on the addition of successive thin layers of waste to the surface of the beds. Stirring the waste together with the castings in the beds can increase the rates of colonization of earthworms and reduce the amounts of heat generated. In this basic model, earthworms always move upward to follow the new feedstock. About 3 or 4 months later, when the windrows are full, the upper layers of material, where the majority of the earthworms follow the food, can be separated to allow the deeper layers of castings to be harvested. When the castings are harvested, the upper layers of feedstock—and earthworms—can be shifted in succession to an empty row to restart vermicomposting with new food supplies.

Three-quarters of the growers feed their earthworms with thin layers of waste, and the other 25% with a large single deposit. The most common depth of the beds when the vermicompost is harvested is 21–30 cm (8.27–11.8 in), whereas those who harvest at lower depths are few (9%). The most common method for harvesting the vermicompost is by feeding the earthworms before collection (51%). 24% use a net with food, 13% sieve the material to separate the vermicompost and earthworms, and the remaining 11% employ other methods.

For vermicomposting coffee pulp, a different and simpler method has been used successfully when the sowing season of the coffee plants coincides with a single harvest of vermicompost: The coffee pulp is initially placed in lines of successive mounds as left on the ground by trucks and then shaped into windrows, with spaces in between. The earthworms are inoculated approximately 2–3 weeks after the thermophilic phase has declined. Only some periodic stirring of the upper waste layers and lateral replacement is needed to permit the earthworms to move to lower layers for food. When the whole feedstock is close to being consumed, another row of mounds from the next coffee-pulp production is then formed in the spaces left between the initial rows. The escaping behavior of *P. excavatus* on rainy nights makes them move gradually to the neighboring windrows, allowing the earthworm-worked windrows to be harvested with almost no earthworms at all.

By this method, more than 2500 trucks filled with approximately 10,000 tons of coffee pulp from a single coffee factory unit have been successfully transformed into vermicompost from each coffee crop over the last 10 years. The vermicompost produced is distributed in thousands of planting holes for new, healthy coffee plants in the coffee plantation (Aranda et al. 1999).

G Vermicompost Production

Prior to being used as a growth medium or used by the producers themselves, the earthworm-worked castings need to be harvested, dried, sieved, and bagged. Harvesting is commonly done manually, 3 or 4 days after a thin layer of feedstock is applied to induce most of the earthworms to move to the fresh materials in the upper layer. Some producers repeat the method twice or even three times to ensure that the vermicompost is almost free of earthworms. The harvested castings are still muddy and may be difficult to sieve or handle. The majority of earthworm growers express their vermicompost production in weight units (75%) and the minority in volume units. The average annual production is between 50 and 500 tons (43% of the growers surveyed); only 10% of growers produced more than 1000 tons per year. Almost all (96% of those surveyed) use electric energy during the process of producing the vermicompost.

One of the most difficult technical obstacles to solve in the production of vermicompost is drying the end product, because of the wet and rainy weather where most of the vermicomposting sites are located. On most of the farms that sell vermicompost, the material is sun-dried in patios, which is costly and slow, to reduce the initial water content of 75%–80% to about 50%–55%. Less than 45% moisture means that the casts may become dusty and gray, which diminishes the microbial activity and changes its attractive black color. It does not recover its texture after rewetting.

Using simple wooden or mechanical sieves, the casts are also screened (with a 0.5–0.6 cm (0.197–0.236 in) mesh) to remove rocks, debris, metal, wood, or even glass, in order to produce a more uniform, smooth, and high-quality product. The rejected sieved material can be mechanically ground and reincorporated into the product. Sieving is the main process to which vermicompost is submitted, followed by drying (23%); both processes are done by 22%, while only 3% grind the vermicompost. Often, some carry out more than two processes. Local sales are usually bulk sales, while bagging is for more distant regional markets. The most common way of bagging products is in new or secondhand 40 kg (88 lb) raffia (polypropylene) sacks (27%), for agricultural purposes, or 10–20 kg (22–44.1 lb) colored plastic bags (34%) for retail stores. Costs of freight and trucking are limiting factors for marketing, and a buildup of acceptance is needed to reach broader, expanding markets and to decrease shipping costs at long distances.

H Commercial Aspects of Vermicomposts

The majority of growers sell the products they generate, although 40% keep them for their own use. The sale of both vermicompost and earthworms is most common

(47%); vermicompost alone is sold by 18%, and a whole technological package by 18%. Very few growers sell only earthworms (6%). The main use of the vermicompost is for potted plants (35%), seedlings (22%), and substrate and fertilizer for cultures (25%). The main use for the earthworms is as food for animals (10%).

Only 46% of the earthworm growers have their products registered by the government; those who registered did so under a great variety of names (seven different names for vermicomposts and two for earthworms). However, the most common name was *Humus de Lombriz*, followed by *Abono Orgánico de Lombriz* ("earthworm organic amendment"; 24%), and *Lombricomposta* ("vermicompost"; 15%). The vermicompost is mainly sold pure (73%). In general terms, growers do not add any microorganisms (82%). If they do, they inoculate *Rhizobium*, mycorrhizal fungi, or enzymes. Much of the vermicompost is sold directly to the user (52%); however, some sell it to stores (24%). For the great majority (84%), vermiculture and vermicomposting are profitable, and everyone agrees that it is an activity with a great potential for profit (69%).

The prices of the products in Mexico (referred to a ton of vermicompost) vary from US$100 to US$350, with the average price about US$200. Prices of earthworms are even more variable because growers sell them by weight, individually, or with substrate. This means that for a kilogram of earthworms there is a large range of prices from US$40 to US$150. Only one grower applied a statistical sampling of earthworm populations with a supporting document, to sell a known quantity of earthworms.

I Quality Control of Vermicomposting and Its Products

Most earthworm growers say that they know the characteristics of their products in terms of quality (67%) and uses (21%). The majority (75%) perform control tests at maturity, for nutrients and the presence of pathogens in the vermicompost and on the earthworm populations; in general most growers perform one of these tests and sometimes two of them. Vermicompost is analyzed by 85% of the surveyed people, showing the main analyses performed to be chemical (30%), biochemical (18%), or microbiological (17.4%); a few analyze physical properties (15.2%), organic-matter quality (13%), and bioregulative molecules (4%). Earthworms, in contrast, were analyzed only as a food product by 28% of the earthworm growers. The analyses that they do are mainly for nutritional quality (32%) as well as biochemical (23%) and chemical (18%) analyses.

These analyses are in general done in academic institutions (universities and research institutes; 81%). Half of growers surveyed do the analysis in response to a quality regulation required by their government (35%), a national organic certifier (33%), or an international organic certifier (5.6%). However, almost all say that it is necessary to have regulation of the quality of the products.

More than 78% of the people surveyed do tests using their vermicompost to grow different plants. In contrast, fewer than 43% use earthworms as feed for animals (of these, 42% use them for chickens, 29% for fish, 11% for hogs, and 18% for other animals, such as frogs, animals in captivity, zoos, etc.). The great majority of earthworm growers (85%) said that they are linked with a research project or an academic institution.

J Promotional Aspects of Vermiculture

The great majority of earthworm growers surveyed promote their business or activity intensely (90%), which includes advising, talks, training, and courses. The public has shown a great interest (67%). When they understand more about the products of this activity, they think in general that vermicompost is excellent (48%) or good (40%). For 80% of the earthworm farms consulted, at least one new farm was being created. Most of the earthworm growers use the Internet and e-mail to obtain information (88%). Although only 45% use Internet chatting to inform themselves or to clarify their doubts, they think it is useful. In a simple Google search, it is possible to find 426,000 pages in Spanish and English from the search "*lombricultura en Mexico*" or 124,000 results for the word *lombricultura* alone, and many more with the related words *lombriz roja californiana*, *compostaje*, *humus*, *abono orgánico*, *biodigestores*, *lombricompuesto*, and so on. However, at this moment, after sifting through the Internet results (which are very repetitive), 96 earthworm farms were seen to be registered in Mexico.

Only 45% of earthworm growers belong to an association or a group, which generally is national (56%) or local (36%). Most of them attend events related to vermicomposting. These are mainly courses or conferences and may be national or regional; only 25% go to international meetings. The majority seem to have published some documents on their activity, the most common being articles in newspapers and journals or booklets (48%); only 5% are scientific articles. The impact of these is to be seen on a regional and national scale. In general, earthworm growers feel that their activity in their country is maturing (59%) or initiating (23%). The main problem they face is the commercialization of the products (40%), followed by transportation (13%), need for advice (10%), and pests (10%), particularly the *Chelyomirmex morosus* ant.

K Discussion and Conclusions from the Vermicomposting Survey

The use of the Internet for this survey led us to a group of more computer-trained earthworm growers and larger earthworm growers who are very enthusiastic, although there is a lot of doubtful information, comments, or discussion that can mislead one about the reality of the activity. However, smaller businesses were not well represented in this review, which could give a biased idea of the situation of earthworm growers, although for Mexico we have some other alternative data available. At this moment, vermiculture and vermicomposting in Mexico, and in Latin America, is growing rapidly and successfully. This success is probably linked to a better world comprehension of the importance of recycling, proper use of by-products, promotion of economic development, the build-up of environmental technology, and a wise relation to environmental protection.

Our opportunity in Mexico is perhaps the abundance and availability of organic feedstocks like coffee pulp, sugarcane press mud, municipal organic wastes, and etc. the more organic waste, the less contamination, the more uniform environmental conditions (tropical regions), the cheaper are labor costs, and less industrialization, which is in favor of developing vermicomposting. Both internationally and nationally there is a growing interest in consuming organic products, and environmental promotion of sustainable development will keep this activity growing and healthy. To conclude, it is important to strengthen the communication and collaboration among earthworm growers, scientists, and sellers of the products in order to regulate and develop their activities and by-products; in addition, the mutual interaction of earthworm growers and scientists is essential to provide the proper scientific support and quality control needed.

VII VERMICOMPOST QUALITY STANDARDS

Vermicompost produced from organic wastes depends very much on the nature of the original material that was used. It cannot be expected that a product with excellent fertilizing qualities will be obtained from an inferior-quality raw material. If vermicompost producers are to influence agriculture, they will have to establish and maintain rigid standards and consistency, support research, and listen closely to the comments of the users of their products.

Great efforts have been made in Mexico, where after two and half years of meetings held between vermicomposting producers and agricultural and standards institutions, an official standard, the Mexican quality standard for earthworm castings ("Humus de Lombriz"; NMX-FF-109-SCFI-2008) was published in June 2008, setting out the quality specifications required for earthworm castings that are produced or marketed in Mexico. This Mexican standard (of a voluntary nature) considers quality specifications for testing methods for attributes such as particle size, maturity, type of substrate, moisture content, presence of living seeds, earthworms, impurities, pathogens, and pollutants.

Standards without supervision are not standards. If vermicomposters are to gain the reputation that they deserve, monitoring systems are needed to offer support. Not only is this a job for a research institution or a state government unit, but the determining participation of a national earthworm association is also need. The association will have to assume the responsibility for deciding which standards should be maintained to ensure quality and consistency, while the recognized professional input of vermicompost producers is also indispensable.

These and many more important aspects are still waiting for a national earthworm association: to register participants; share ethics; interchange information; compare and modernize techniques, methods, and equipment; organize and prepare national surveys and reports; expand production and markets; and support research studies, tests, and technical relationships with users of their products. A close relationship between vermicomposting farms and research institutions is needed in the evolution of vermicomposting farms in order to promote and conduct worm-related studies, develop technology and scientific methods, and define standards of quality and effects on plants, as well as train people to become technicians or professionals in vermicomposting.

VIII VERMICOMPOSTING FARMS IN MEXICO

There is no official survey of the sites and earthworm farms established in Mexico, but it is possible to try to estimate the growing number of earthworm enterprises, based on sites on the Internet, general activity, and by word of mouth information. We are attempting to achieve a national survey. Without other indicative information, it can be said that there are about 10–20 large earthworm farms, around 100 medium-sized earthworm farms, and hundreds of small ones. Only the largest earthworm plants work on the order of thousands of square meters of earthworm windrows; more are in the medium rank, occupying hundreds of square meters; and most work in areas of tens of meters.

IX VERMICOMPOSTING URBAN DOMESTIC RESIDUES

The potential use of vermicomposting for domestic residues in urban areas is amazing. Although there are several successful examples, much more promotional work needs to be done in Mexico to disseminate vermicomposting methods so they can become popular in the backyard of every home, at schools and institutions, and in cities and, of course, municipalities. The separation of domestic organic residues at the source allows not only better use to be made of the organics but also better recycling of the inorganics. Because of the diversity of vegetable and fruit remains, along with yard trimmings, leaves, and garden flowers, very simple, cheap, and good home vermicomposting units can be made or are available commercially (Aranda 1997).

Although there have been some attempts to conduct vermicomposting courses in Mexico, they appear to have mostly attracted customers and sold projects. Better-supported, professional, and up-to-date information is essential. It should come from scientific and technological experts. Then, a new kind of job, new technicians with professional profiles, slowly but steadily will emerge in vermiculture. With this in mind, many scientific and promotional conferences, visits from school groups, classroom visits, television interviews, and videos have been organized by the INECOL on their own initiative or when requested by government or private entities. Similar efforts have been recently made by the agricultural center Colegio de Posgraduados in the state of Mexico, working with Claudia Martínez, who, in cooperation with a local agriculture office. has promoted workshops to improve the potential of vermiculture (Martínez 1996).

By the beginning of 1999, a Mexican book entitled *Manual de Reciclaje, Compostaje y Lombricompostaje* was printed by INECOL, is now in its third edition (Capistrán et al. 1999). It presented theoretical and practical aspects of composting and vermicomposting, using simple drawings and basic explanations to make clear the role and importance of recycling organics and inorganics. It has been useful for a wide audience but especially for students, teachers, homeowners, and organic producers.

Special attention must be paid to the municipal experience of Teocelo, Veracruz, with a population of 14,900 inhabitants, 9062 of whom live in the municipal center and the rest in 20 rural towns scattered in the vicinity over 21 miles^2 (33.8 km^2).

There, a Comprehensive Program of Separation, Handling, Education, Process and Exploitation of Solid Waste has been established (in 2000, then it was suspended and recovered again...), operating a vermicomposting unit for transforming organic wastes into fertilizers. Teocelo's experience successfully combines two elements: the development and use of low-cost ecological technologies and the participation of the community.

Approximately 10 tons of garbage is collected daily, 60% of which is source-separated organic matter, which is then precomposted and used to feed the earthworms. The product is distributed as an organic fertilizer among local producers through promotional agricultural programs (García 2003). This municipal program has generated not only a social change in culture concerning solid waste handling but also a wider interest because of its cheap implementation and technical simplicity. The National Research and Technology Council has recently implemented an interdisciplinary supporting project, with basic research and technological development components, to consolidate and replicate this experience in other small or medium-size rural municipalities (Cerdán 2008).

X CONCLUSIONS

The objectives and essential goals of modern vermicomposting are to secure and control new economic alternatives, better recycling options, and the productive utilization of organic waste, as well as to ensure significant improvements in environmental conservation and soil fertility. It is now possible to say that vermicomposting is a better-known technology, whose main product, earthworm castings, is used mainly by seedling and nursery producers, gardeners, and homeowners. In the present-day context, with global warming, carbon capture, and energy resource depletion (peak oil), vermicomposting, as well as other organics recycling methods, needs to be better recognized and heavily promoted in Mexico, to expand its potential application and benefits to the soil, to sustainable production, and of course, to wise, environmentally friendly preservation programs.

ACKNOWLEDGMENTS

We thank Clive Edwards for his stoical patience in awaiting this document and Judy Shirley for reviewing the English version of the manuscript.

REFERENCES

Abdul, R., and M. Abdul. 1994. Los gusanos de tierra y el medio ambiente. *Mundo Científico* 146(14): 408–415.

Aguilera Castillo, R.A. 2002. El potencial bactericida de la lombriz roja californiana *Eisenia andrei*. Centro Universitario de Ciencias Biológicas Agropecurias (CUCBA), División de Ciencias Biológicas y Ambientales, Universidad de Guadalajara, Mexico.

Alviter Aguilar, A. 2003. Evaluación de estiércoles de origen animal con o sin aditivo para la producción de abono orgánico. Degree thesis, Universidad Autónoma Chapingo, Estado de Mexico, Xalapa, Mexico.

Aranda, D.E. 1988. La utilización de lombrices en la transformación de la pulpa de café en abono orgánico. *Acta Zool. Mex., nueva serie* 27: 21–23.

Aranda, D.E. 1989. Perspectivas de la utilización de lombrices en la transformación de la pulpa de café en abono orgánico. *Boletin Técnico de Café, INMECAFE,* September, 1989, Xalapa, Mexico.

Aranda, D.E. 1991a. El vermicompostaje; una nueva alternativa para la transformación de la pulpa de café en abono orgánico. *Abstract in XIV Simposio de Cafeticultura Latinoamericana*, Panama.

Aranda, D.E., 1991b. El vermicompostaje, la alternativa ecológica para transformar la pulpa de café en abono orgánico. *Abstract in II Seminario Internacional de Biotecnológía en la Agroindustria Cafetalera.* Orstom-Cenicafe, Manizales. Colombia.

Aranda, D.E. 1995. El lombricompostaje, una ecotecnología para convertir los desechos organicos en abonos mejoradores de la fertilidad. In *Alternativas al manejo de laderas en Veracru,* ed E. Boege, H. García and P. Gerez, 247–252. Secretaria del Medio Ambiente, Recursos Naturales y Pesca- Fundación Friedrich Ebert Stiftung. Xalapa, Veracruz. Mexico.

Aranda, D.E. 1997. El cultivo de lombrices; una solución ecológica sencilla y eficiente para producir abono con nuestra basura orgánica doméstica. In *Reducción y Reciclaje de Residuos Sólidos Municipales*, ed F.C. Hernandez and S. Gonzalez, 257–268. Programa Universitario del Medio Ambiente, Universidad Autónoma de México, D.F. Mexico.

Aranda, D.E., and S. Aguilar. 1995. Tasa de extracción de lombrices de *Eisenia andrei* para su utilización como proteína animal. In *Utilización de lombrices en la transformación de la pulpa de café en abono orgánico*, ed. I. Barois and E. Aranda, 47–54. Final report, Conacyt-Instituto de Ecología, A.C., Xalapa, Mexico.

Aranda, D.E., and I. Barois. 2000. Coffee pulp vermicomposting treatment. In *Coffee Biotechnology and Quality*, ed. T. Sera, C.R. Soccol, A. Pandey, and S. Roussos, 489–506. Kluwer Academic, the Netherlands.

Aranda, D.E., I. Barois, P. Arellano, S. Irissón, T. Salazar, J. Rodríguez, and J.C. Patrón. 1999. Vermicomposting in the tropics. In *Earthworm Management in Tropical Agroecosystems*, ed. P. Lavelle, L. Brussard, and P. Hendrix, 253–287. CABI Publishing, UK.

Aranda, D.E., L. Durán, and E. Escamilla. 2009. Vermicomposting in coffee cultivation. In *Coffee: Growing, Processing, Sustainable Production*, ed. J. Wintgens, 328–342. 2nd ed. Wiley-VCH, Germany.

Barois, I., and E. Aranda (eds.). 1995. Utilización de lombrices en la transformación de la pulpa de café en abono orgánico. CONACYT- Instituto de Ecología A. C. (clave: 045-N9108) Final Report. Mexico.

Bouché, M. 1987. Emergence and development of vermiculture and vermicomposting: From a hobby to an industry, from marketing to a biotechnology, from irrational to credible practises. In *On Earthworms*, ed. A. M. Bonvicini Pagliai and P. Omodeo, 519–532. Selected Symposia and Monographs U.Z.I. 2. Mucchi, Modena, Italy.

Capistrán, F., E. Aranda, and J.C. Romero. 1999. *Manual de Reciclaje, Compostaje y Lombricompostaje*. Instituto de Ecología A.C., Xalapa, Veracruz, Mexico.

Cerdán, A. 2008. Tecnificación del procesamiento de los residuos sólidos municipales bajo esquemas alternativos sustentables, en una planta piloto experimental. Proyecto de Investigación y Desarrollo Tecnológico Integral, Instituto Technologico Superior de Xalapa (ITSX) Centro de Technologico Auanzanda (AC.) (CIATEQ) Instituto de Ecologia (AC.) (INECOL) ITSX-CIATEQ-INECOL. Proyecto Sectorial FORDECYT, CONACYT, Mexico.

Contreras-Ramos, S.M., D. Alvarez-Bernal, and L. Dendooven. 2008. Removal of polycyclic aromatic hydrocarbons from soil amended with biosolid or vermicompost in the presence of earthworms (*Eisenia fetida*). *Soil Biol. Biochem.* 40: 1954–1959.

Contreras-Ramos, S.M., D. Alvarez-Bernal, and L. Dendooven. 2009. Characteristics of earthworms (*Eisenia fetida*) in PAHs contaminated soil amended with sewage sludge or vermicompost. *Appl. Soil Ecol.* 41: 269–276.

Contreras-Ramos, S.M., D. Alvarez-Bernal, J.A. Montes-Molina, O. Van Cleemput, and L. Dendooven. 2009. Emission of nitrous oxide from hydrocarbon contaminated soil amended with waste water sludge and earthworms. *Appl. Soil Ecol.* 41: 69–76.

Cruz Flores, M. 2005. Reproduccion de la lombriz roja (*Eisenia fetida*) en sustratos organicos pecuarios. Master's thesis, Ingeniería-Sistemas Agrícolas, Universidad Autónoma Agraria Antonio Narro, Coahuila, Mexico.

Curiel, P., and E. Aranda. 2000. Progress in vermicomposting pig manure in México. Abstract in *Vermillennium, International Conference on Vermiculture and Vermicomposting*. Kalamazoo, MI, USA.

Edwards, C.A., and P. Bohlen. 1996. *The Biology and Ecology of Earthworms*. 3rd ed. Chapman and Hall, London.

Edwards, C.A., and E. Neuhauser (eds.). 1988. *Earthworms in Waste and Environmental Management*. Academic Publishing, the Netherlands.

Espinoza, G.L. 2003. Costos y rentabilidad de la lombricultura (*Eisenia foetida*) en el estado de Chiapas. Undergraduate thesis, Ciencias Socioeconómicas, Universidad Autónoma Agraria Antonio Narro, Coahuila, Mexico.

Gaime-Perraud, I. 1995. Cultures mixtes en milieu solide de bactéries lactiques et de champignons filamenteux pour la conservation et la décafeination de la pulpe de café. PhD diss., Université Montpellier II, Sciences et Techniques de Languedoc, France.

Gaime-Perraud, I., L. Hannibal, M. Trejo, and S. Roussos. 1991. Resultados preliminares sobre el ensilaje de la pulpa de café. Abstract in *II Seminario Internacional de Biotecnología en la Agroindustría Cafetalera (II SIBAC)*. Manizales, Colombia.

García V.G. 2003. Comprehensive handling of organic waste, Teocelo, Veracruz, México; an innovations in technology and governance case study. John F. Kennedy School of Government, Harvard University, Cambridge, MA, USA.

Hernandez, E., M.A. Niembro, and E. Aranda. 1996. Evaluación del sustrato orgánico *Terranova* en la producción de plántula en invernadero y la mezcla con sustratos comerciales Sunhine® y Fafard®. Reporte Técnico, Cotizagro-Deshidratadora La Cascada, Querétaro, Mexico.

Hernández Sánchez, M.L. 2000. *Efecto del humus de lombriz en el* desarrollo del patrón *Citrus aurantium* en etapa de semillero Xalapa. Master's thesis, Universidad Veracruzana, Mexico.

Huerta-Lwanga, E. 1996. Vermicomposteo de desechos orgánicos (Bagazo de Henequén, *Agave fourcroydes* y desechos municipales) del Municipio de Hunucma, Yucatán. Undergraduate thesis, Escuela de Biología, Universidad de Ciencias y Artes del Estado de Chiapas, Mexico.

Irissón, N.S. 1995. Calidad del abono y de la lombriz de tierra resultante del lombricompostaje de la pulpa de café. Undergraduate thesis, Facultad de Química Farmacéutica Biológica, Universidad Veracruzana, Xalapa, Veracruz, Mexico.

Jaenike, J. 1982. "*Eisenia foetida*" is two biological species. *Megadrilogica* 4(1–2): 6–7.

Klemmer González, D.M. 2008. Evaluación de dosis de sustratos vetiver-estiércol para la alimentación de la lombriz *Eisenia foetidae*. Undergraduate thesis, Facultad de Ingeniería Química, Benemérita Universidad Autónoma de Puebla, Puebla, Mexico.

Lavelle, P. 1981. Stratégies de reproduction chez les vers de terre. *Acta Oecol.* (Oecología Generalis.) 2: 117–133.

Lavelle, P. 1994. Faunal activities and soil processes: Adaptive strategies that determine ecosystems function. In *Transactions of the 15th World Congress of Soil Science*, ed. J. D. Etchevers, 189–220. Inaugural and State of the Art Conferences (ISSS) 1. Acapulco, Querrero, Mexico.

León Monzón, J.H.D. 2003. Aprovechamiento del substrato degradado por *Pleurotus ostreatus* en la alimentación de ovinos y en la producción de vermicomposta. Degree thesis, El Colegio de la Frontera Sur Chiapas, Mexico.

Limón, G.A. 1999. Obtención de biofertilizante a partir de desechos orgánicos compostaje y vermicompostaje. Degree thesis, Facultad de Ingeniería Química, Benemérita Universidad Autónoma de Puebla, Mexico.

Loza Llamas, J.M. 2000. Dinámica Poblacional de la Lombriz Roja de California *Eisenia andrei* (Bouché, 1972) en la planta Piloto de Lombricultura del Centro Universitario de Ciencias Biológicas y Agropecuarias CUCBA: Coordinacion General de Investigacion y Postgrado: Manejo en Areas de Temporal. Master's thesis, Universidad de Guadalajara, Guadalajara, Jalisco, Mexico.

Loza Llamas, J.A. 2007. La lombriz de tierra, *Eisenia andrei* (Bouche, 1972) como bioacumulador de metales pesados (Pb) durante la transformación de los desechos sólidos orgánicos de Guadalajara, Jalisco, México. PhD diss., Centro Universitario de Ciencias Biológicas y Agropecuarias, Posgrado en Ciencias Biológicas, Guadalajara, Jalisco, Universidad de Guadalajara, Mexico.

Loza Llamas, A. (Ed) 2004. Memorias de Primer Congreso International de Lombricultura y Abores Organices, Universitario de Ciencias Biologicas, Jalisco, Mexico

Martínez Aguilar, J.G. 2004. Producción de lombrices (*Eisenia fetida* S.) con diferentes sustratos orgánicos en Villaflores, Chiapas. Undergraduate thesis, Facultad de Ciencias Agronómicas, Universidad Autónoma de Chiapas, Mexico.

Martínez, C.C. 1996. *Potencial de la Lombricultura, elementos básicos para su desarrollo.* Lombricultura Técnica Mexicana, D.F., Mexico.

Martínez, C., A. Perez, L.F. Ramirez, R. Cuevas, N. Mendez, A.L. Franco, and M. Pineda (eds.). 2002. Resumenes, *II Simposium Internacional y Reunión Nacional. Lombricultura y Abonos Orgánicos.* Fac. de Ciencias Agrícolas, UAEM, Edo. de México, Mexico.

Martínez, C.C., M.R. Romero, L. Corlay, A. Trinidad, and L.F. Ramirez (eds.). 1999. Resumenes, *I Simposium Internacional y Reunión Nacional de Lombricultura y Abonos Orgánicos.* SAGARPA, Univerisdad Autonoma Chapingo Colegio Postgraduades UACH-CP, Montecillos, Edo. de México, Mexico.

Medina, A.E. 2005. Vermicomposteo de biosolidos con lombriz (*Eisenia fetida*). Undergraduate thesis, Agronomía-Procesos Ambientales, Universidad Autónoma Agraria Antonio Narro, Coahuila, Mexico.

Meléndez, G.M. 2003. Morfologia de dos especies de lombrices de tierra: *Eisenia foetida* y *Eudrilus eugeniae.* Undergraduate thesis, CUCBA, División de Ciencias Biológicas Universidad de Guadalajara, Jalisco, Mexico.

Montalvo, Q.B. 2004. Aplicaciones foliares de humus líquidos de lombriz, natural y enriquecido en papaya (*Carica papaya* L.). Undergraduate thesis, Agronomía, Universidad Veracruzana-Xalapa, Ingeniero Agrónomo, Veracruz, Mexico.

NMX-FF-109-SCFI-2008. 2008. Humus de Lombriz (Lombricomposta)—Especificaciones y métodos de prueba. Comité Técnico de Normalización Nacional de Productos Agrícolas, Pecuarios y Forestales, Secretaria de Economía, Mexico.

Peña Ramos, F.M. 2004. Un modelo de simulación para estimar la dinámica de una población ideal de *Eisenia foetida* (Lombriz de tierra, Roja California). Undergraduate thesis, Agronomía-Agrobiología, Universidad Autónoma Agraria Antonio Narro, Coahuila, Mexico.

Salazar, T., E. Aranda, and I. Barois. 1995. Estudio comparativo del lombricompostaje de la pulpa de café por *Eisenia andrei, Eisenia fetida* y *Perionyx excavatus* en condiciones de campo. In *Utilización de lombrices en la transformación de la pulpa de café en abono orgánico*, ed. I. Barois and E. Aranda, 28–41. CONACYT (clave: 045-N9108) Final Report, Mexico.

Santamaría, R.S. 2000. Escalamiento de los procesos de composteo y vermicomposteo: Aspectos biológicos y nutrimentales. Master's thesis, COLPOS Texcoco, Edo. de México, Mexico.

Satchell, J.E. (ed.). 1983. *Earthworm Ecology, from Darwin to Vermiculture*. University Press, Chapman and Hall, London, U.K.

Tomati, U., A. Grappelli, and E. Galli. 1987. The presence of growth regulators in earthworm-worked wastes. In *On Earthworms*, ed. A. M Bonvicini Pagliai and P. Omodeo, 423–435. Selected Symposia and Monographs U.Z.I. 2. Mucchi, Modena, Italy.

Toscano, S.S. 2000. Evaluación del crecimiento de la trucha arcoiris *Oncorhynchus mykiss* (Billard, 1989) utilizando lombrices *Eisenia andrei* (Bouché, 1972) como sustituto parcial de proteína en el rancho "El Roble", Coatepec, Ver. Degree thesis, Instituto Tecnológico del Mar, Boca del Río, Veracruz, Mexico.

Trejo, T.L.I. 1995. Establecimiento del cultivo de lombriz de tierra (*Eisenia foetida*) sobre estiércol bovino. Undergraduate thesis. Universidad Autónoma Chapingo, Edo México, Mexico.

Van Gansen, P. 1962. Structures et fonctions du tube digestif du Lombricien *Eisenia foetida* Savigny. *Journal de Microscopie* 1: 1–363.

Vela, D.M. 2005. Uso del humus de lombriz (*Eisenia foetida*) en el biocontrol del nematodo agallador de la raíz (*Meloidogyne incognita*). Undergraduate thesis, Facultad de Agronomía, Universidad Veracruzana, Xalapa, Mexico.

White, S. 1996. Hooked on worms! A 20 year commitment to the environment. *Worm Digest* 14: 3–9. Eugene, Oresa, USA.

Zamora, C.A., and J.M. Martínez. 1999. Uso del compost de la industria champiñonera en el establecimiento de *Eisenia fetida.* Undergraduate thesis, Facultad de Agronomía, Universidad Veraruzana, Xalapa, Veracruz, Mexico.

CHAPTER 33

The Scope of Vermiculture in Cuba

Martha Reines Alvarez and Carlos Rodriguez Aragonés

CONTENTS

I Introduction ... 533
II Start of Vermiculture in Cuba ... 534
III Vermiculture Research Program ... 534
IV Establishment of Vermiculture in Cuba ... 538
A Basic Small Units ... 538
B Basic Medium-Size Units ... 538
C Large Units ... 538
References ... 542

I INTRODUCTION

Before the 1970s, studies of the edaphic fauna in Cuba were scarce; there were only studies of a few of the most important pests, due to changes in the agricultural production systems in Cuba. By the middle of the twentieth century, only four species of native earthworms had been reported in Cuba. Zoological workers based their earthworm studies on literature, slides, and guides of foreign practices, and examples of exotic earthworm species imported from the United States to Havana University. During the second half of the twentieth century, motivated by difficulties in obtaining educational support, Dr. Aurelia Longoria noticed differences between Cuban earthworm species and the literature on morphological studies and produced a manual for the Annelida group, incorporating the knowledge into invertebrate zoology classes (González 1978). The earthworm species were not known, and they were cataloged only by numbers and descriptions. Advances in morphological, ecological, population, and systematic breeding of the Cuban soil fauna, especially the earthworms, took place in the Faculty of Biology at the Havana University, which has an earthworm collection from different regions of Cuba.

II START OF VERMICULTURE IN CUBA

Since 1970, ideas about vermiculture development began to strengthen, introduced from the Philippines through the Cuban ambassador in that country and through Dr. Martha Reinés, who visited the Second National Convention of Philippine Earthworms Growers in Manila in 1982 (Reinés and Rodríguez 1982). The beginning of vermiculture in Cuba can be divided into separate periods that were superimposed and parallel in space and time: (i) sampling and collection of earthworm species from natural ecosystems and (ii) introduction of exotic earthworm species.

III VERMICULTURE RESEARCH PROGRAM

Given the economic pressures that confronted Cuban agriculture, the Faculty of Biology at Havana University already had some earthworm experience, so they started an intensive sampling of the native earthworm species, at livestock farms and other sites, to select some native earthworm species that had the ecological characteristics for extensive breeding. From that time, studies were done of the biology of different organic sludges in the decomposition process, with the objective of using them to feed earthworms under Cuban climate conditions. In 1980, vermiculture started using the exotic species *Eudrilus eugeniae* selected from natural ecosystems. From the beginning vermiculture was conceived as an integrated biotechnology, a "recycling biosystem," accounting for site characteristics, considering all available resources, and integrating harmoniously with other practices, such as animal breeding, biogas digesters, use of nonconventional protein sources for animal feeding, and so on. In 1980 a bag of earthworm flour was produced in the Biological Educational Laboratory of the Faculty of Biology and used in experiments on diets for chickens (Camps and Reinés 1981).

E. eugeniae was the first earthworm used in vermiculture in Cuba. This earthworm was introduced from Africa during the African slave traffic and the breeding of livestock by the Spanish colonizers. By 1980, *Perionyx excavatus* was introduced to Cuba, initiated with specimens sent to Cuba in hand luggage. The breeding was established at the Biological Educational Laboratory of the Faculty of Biology of Havana University. In December 1985, a starting culture of 10 kg of the red Californian hybrid, *Eisenia andrei*, was acquired in Italy and initially classified by Dr. Carlos Rodriguez Argonés as *Eisenia fetida*. Half of the initial culture was transferred to the Biological Educational Laboratory of the Faculty of Biology at Havana University, and the other half stayed in the charge of the plan for the Institute of Poultry Investigations; this would pass later to the Soil Institute of the Ministry of Agriculture.

The scientific facilities created in the Biological Educational Laboratory of the Faculty of Biology at Havana University for ecological studies of earthworms became the first Pilot Development and Vermiculture Reproduction project involving investigation, production, and extension. The preliminary results of the investigation into the development of the earthworms *E. eugeniae* and *P. excavatus* permitted the establishment of an extensive collection of breeder species in large production areas. At the Biological Educational Laboratory, the first pilot vermiculture plant was started. It was decided later to extend this vermicomposting technology to the five pilot centers of the Farming Superior Institute of Havana and the Ministry of Agriculture before extending it to all of the country. It delivered breeding earthworms and offered training, conferences, and courses, as well as providing a manual containing all the instructions for setting up vermiculture (Reinés, et al. 1981). To the African night crawler (*E. eugeniae*) and the Taiwan redworm (*P. excavatus*) in 1989 it added the Californian redworm (*E. andrei*). These species are well established in Cuban ecosystems in the eastern region, as well as central and western parts of the country.

The first objective of vermiculture in Cuba was the production of vermicomposts or earthworm castings to substitute for inorganic fertilizers, because of the difficult economic situation in Cuba. This technology was later converted to a search for a source of proteins for domestic animal feed based on earthworms. From 1981 to 1988 new vermiculture centers were created, with the participation of various organizations from different institutions. Eighty percent of all the provinces had at least one "vermiculture unit" by the late 1980s.

In 1988, the Fertilizer and Soil National Center of the Ministry of Agriculture of Cuba started to organize vermiculture activities in response to the rapid development of the technology. In July 1987, a Program for the Progressive Development of Vermiculture in Cuba was proposed, whose objectives were to instruct people in charge of implementation of vermiculture about its production phase in the enterprises; to complete studies and investigations about the handling, reproduction, and exploitation of earthworms; to gather castings or vermicomposts for soil improvement; to produce earthworms for feeding animals and humans in Cuba; and to organize the treatment of different organic sludges using vermiculture technology. A working group at the national level was created to improve cooperation between institutions and unify projects that contributed to the development of vermiculture in Cuba.

Among the tasks assigned to this commission by the Minister of Agriculture were to develop national and international strategies for the development of vermiculture in Cuba and to produce methodologies, instructions, regulations or any relevant information, and literature for the efficient fulfillment of vermiculture (lingking directly to the provinces, enterprises, educational and private sectors, etc.). Parallel to the Vermiculture National Commission, a Vermiculture Expert National Commission was created, as advisers to the national group, to take significantly important measures to improve the development of vermiculture activities in Cuba. The National Commission administers all the plans of municipal and national vermiculture production, advised by the Expert Commission with qualified scientific-technological functions to implement the orientations given in these plans.

With three different earthworm species available for the development of vermiculture in Cuba (*P. excavatus*, *E. eugeniae*, and *E. andrei*), scientific research intensified along different themes: life cycles, behavior, biology and chemistry of the breakdown of organic matter (Prieto and Rodríguez 1989; Reinés and Rodríguez 2006), specific inter and intra-relationships, specimen population development, life table determinations, influence of different substrates on the earthworm species, biology, and so on (Rodriguez 1998), with the aim of applying information on these parameters to the large earthworm-breeding programs developed.

E. eugeniae is very widely distributed and is reproducing very well in all degradation ecosystems, cattle barn drains, and places with abundant organic wastes. In Cuba, shade is essential in earthworm-rearing beds. All species suffer in the warm climatic conditions of Cuba, *E. fetida* more than the others. When the temperatures are extreme (30–35°C), they turn inactive and move to bed bottoms until night when temperatures decrease, and they then return to surface and become active but have a cocoon fertility of less than 50% (Reines, Rodriguez, Sierra, and Vasquez 1998). *E. eugeniae* and *P excavatus* are very active and mobile. They are more productive than and survive the warmer climatic conditions better than *E. andrei.*

The species employed in vermiculture present a group effect on population development for periodical population density control. In the vermiculture systems we observed a reduction in the size of individuals with increases in the population densities. *E. eugeniae and P. excavatus* have large population densities and move to other places with lower densities when populations are up to 5 kg.m^{-2} space is limited. *E. andrei* decreases its metabolism, reduces individual size, and supports higher population densities by migration and by colonization of sites with better conditions. In experiments we probed the possibility of different species escaping from the breeding beds. The first was *P. excavatus,* the second *E. eugeniae* and the last *E. fetida*; for this reason the majority of breeders prefer *E. andrei.*

The National Vermiculture Commission proposed a research program Development of Vermiculture in Cuba, with the objective of making optimal use of all collaboration possibilities between the different organizations, as well as to use, in a rational way, all available human resources and materials, making every effort to attain perfection; develop the biological, ecological, and physiological aspects; identify natural enemies; and develop the technology of production of earthworms. All these inputs gave origin to a project that embraced five thematic areas:

- Bioecology and breeding of earthworms for vermiculture
- Studies of the earthworm as an animal and human source of protein
- Employment of earthworm castings, or vermicomposts, for improvement of soils in Cuba
- Establishment of a poultry breeding system for the development of vermiculture in Cuba
- Establishment of a watering and drainage system for vermiculture technology in Cuba

With parallel advances in all these investigations, and sometimes before publishing the results, they were used to improve the methodologies developed and instructional activities that were facilitated through the Cuban National Commissions.

More than 25 years of research by these different organizations, institutes, and universities have produced a large body of results. Some of the more important results are knowledge of the biology of commercial earthworm species; characterization of the immature stages of *E. eugeniae*, their parasitism by nematodes, the associated fauna in the earthworm-breeding beds, with lists of predators; characterization of the vermicomposts produced by different species; and investigation of their use for feeding chickens. The characterization of vermicomposts and the application of proteases and peptone, providing germicidal actions of the earthworms, and economic validation of process controls (Carrillo and Reinés 1981; Rodríguez et al. 1988; Cruz et al. 1996; Reinés 1996; Herrera Lemus 1997; Castillo et al. 2006; Reinés, et al. 2006), and so on.

Inoculated earthworm breeding must have two forms: When we know the previous population density, we can transfer the biomass with castings to new beds at a rate of 1 kg per m^{-2} (2.2 lb. yd^{-2}) or put into the bed bottom 15 cm (6 in) of degraded sludge and add 1 kg per m^{-2} (2.2 lb. yd^{-2}) uniform batches of earthworms, sprinkle them with water, and feed earthworms with sludge that previously has been aerated every 15–20 days. Every week 15 cm (6 in) of food is transferred to the bed surface depending on the need, and it is sprinkled with water. The separation of earthworms from the substrate at the end of the development cycle occurred at 90 days depending on the sludge employed in earthworm feeding. At this time the top 30 cm (12 in) contain the earthworm biomass (4–6 kg (8–12 lb)) and the vermicompost is below. We take this layer of 30 cm and split it into 4 or 6 $Kg.m^{-2}$ (3.8–5.6 yd^{-2}) and extract the earthworms from the vermicompost. A surface plastic netting is used over the treated sludge. In 24 hours the earthworms move up to colonize a new bed. To prevent the degeneration of populations it is very important to control the population density. Adverse effects typically occur in the individual populations, increments in numbers, and reduced size.

The results of all these investigations contributed to the establishment of a *Manual for the Establishment of Vermiculture in Cuba*, which was published with preliminary instructions on continuing the process (Reinés et al. 1981). Up to now five practical books (Reinés, Loza, et al. 1998; Reinés, Rodríguez, et al. 1998; Reinés et al. 2001a, 2001b; Calero and Revestí 2003) contain a compilation of details of the commercial earthworm species, characteristics and biology, influence of ecological factors on them, the technology and recommendations for the establishment of vermiculture, clarification of the values of the vermicompost, and potential for environmental decontamination by means of the ecological transformation of organic wastes by earthworms.

More than 70 presentations at national and international events were rewarded with prizes from the Academy of Sciences of Cuba and the Ministries of Higher Education and of Agriculture. The results have been applied and presented in pregrade courses and post-graduate degrees The National Work Commissions, their members, and the institutions to which they belong recognize a great contribution in the development and implementation of the Cuban biotechnology of vermiculture.

IV ESTABLISHMENT OF VERMICULTURE IN CUBA

The methods developed for vermiculture in Cuba have been extended to other countries, by means of academic exchanges with Mexico, Mozambique, Peru, Venezuela, Brazil, and Argentina. Cuban earthworm specialists are collaborating with the University of Guadalajara, where they present courses and training in vermiculture annually and carry out combined investigations and extension, in Tomatlan, Jalisco, Sinaloa, and so on. The establishment of vermiculture in new sites emphasizes the breeding of earthworms and the training of personnel in charge of earthworm activities, shops, courses, and trainings. The Faculty of Biology has incorporated graduate and postgraduate program topics on annelids and vermiculture in their syllabus. Cuban specialists carried out national workshops for producers and also participated in the organization of the Second National Congress and the First International Congress on Earthworm Culture, held in Guadalajara, Mexico.

Vermiculture sites are preferably in cattle-breeding centers or near supplies of organic wastes that can be used for earthworm food. According to their dimensions and destination, breeding units can be termed small, medium, or large, and they can be mechanically or manually managed.

A Basic Small Units

Units of small dimensions are placed in patios, gardens, work centers, small businesses, and schools. The objective is to use earthworms to decontaminate the environment, using as feed the waste from kitchens, paper wastes, or other organic wastes, and to produce a final product that can be used in the production of ornamental plants and agricultural produce. In the field, they are produced in receptacles of different sizes and designs, made from plastic, wood, asbestos, cement, masonry, stone blocks, and so on.

B Basic Medium-Size Units

Medium-scale units are larger (10,000 m^2) than the small ones. They can be the production units of a small producer, breaking down organic wastes from crops, agroindustrial or industrial organic wastes, urban solid waste, or animal manures from farm production into vermicomposts.

C Large Units

These consist of basic units (100×100 m, or 1 ha). They are designed to be fully commercialized and generally are automated. Their final aim is the full commercialization of vermicompost production. In the larger units, the earthworms are in beds on the ground. Wood, stone blocks, bricks, cement, or other materials are used to construct these beds. The breeding units are located in shady areas with good drainage, using the natural shade of trees or roofs of palm leaves, asbestos, cement, metal, or mesh shades. We can also provide a superficial shade using straw or trash.

These earthworm units can be in abandoned farm facilities with roofs of concrete. Earthworm feedstocks are obtained from the larger industrial organic-waste producers, for example, products from the sugar-processing industry. Some vermiculture units consist of more than one unit. This provides all the organic inputs for large-scale tobacco production (Reinés, et al. 1998).

Hurricanes are important and can cause difficulties in the development of vermiculture in Cuba. The mechanization used reflects the results of research by the Institute of Agricultural Mechanization. A group of equipment for the industrial development of vermiculture, and the field designs, such as ecological classifications, dimensions, food types, wind direction, drainage and watering systems. The equipment designed includes industrial feeders and animal-drawn vehicles, electric sifters for engines and tractors, and front shovels for tractors that are adapted to the activity of the vermiculture areas and the uneven land design of biogas digesters to be incorporated into integrated vermiculture. After 1990, the basic unit of vermiculture became medium-size units, close to sources of earthworm foods and easy to manipulate. Another important application of vermiculture is in urban agriculture, in areas bordering on cities.

Vermicompost can be produced from all kinds of organic biodegradable wastes suitable to feed earthworms, including animal manures (solids and liquids from cows, sheep, horses, pigs, goats, and rabbits), agroindustrial residues (trash and waste from sugar production, residues from cultivation of edible mushrooms, harvest trash, paper pulp, etc.), and urban solid residuals (residuals from human activities that undergo previous separation, reuse, refuse and separating raw material that is recyclable). The composition of the vermicomposts produced varies according to the source of food used to feed the earthworms (Reinés and Rodríguez 2006) (Table 33.1).

In the 10 years since the beginning of the vermiculture in Cuba, 58,102.2 ton of vermicompost have been produced and 138 basic vermiculture units exist in Cuba, with 116,480 m^2 of effective beds of earthworms. A total of 1361 ton of earthworm biomass was also produced for animal feed for chickens and for shrimps. The vermicomposts were applied to different crops: tobacco, coffee, vegetables, potatoes, rice, cocoa, and grasses; currently, sugarcane cane, coffee, and cultivated rice are the most important recipients of vermicomposts. Currently, the Ministry of Agriculture has a guideline about the necessity to establish vermiculture areas in all productive centers. Figure 33.1 summarizes the annual vermicompost production in Cuba up to 1989. Currently, the production is 15.0×10^6 ton per annum.

In Cuba there are two seasons, dry and rainy. In the dry season, the temperature is high, and it is necessary to sprinkle surfaces to 70–80% of the field capacity. Always is liberated some liquids, that is recollected in decanted deposit into the grown. In banana plantations application of this liquid in different concentrations (10–50%) resulted in increased size and weight of fruit. Other research showed that vermicompost extract possesses antibacterial and antifungal effects against pathogens in fertile irrigation to protect the plants against pests. Also, different intestine parts coelomatic liquid and mucus earthworm present these proprieties (Castillo et al. 2006). The bed water is used as a fungicide and stimulant to plant growth with

Table 33.1 Composition of Vermicomposts Produced with Organic Wastes

Waste Material	Organic (%)	Content (% dry weight) mg/g (parts per million)						Microbial Concentration
		N	P	K	Ca	Mg	Ph	
Cattle solids*	44.5	1.70	0.62	1.22	10.00	1.53	7.00	20×10^{11}
Cattle solids	22.8	0.24	0.10	0.39	1.25	0.12	8.08	13×10^{6}
Treated pig solids	44.0	1.89	0.50	0.34	10.80	1.46	7.00	13×10^{12}
Goat manure	37.5	1.51	0.64	0.78	4.40	1.37	7.05	ne
Cachaza (sugar industry)	68.5	2.66	2.13	0.41	4.12	0.89	7.60	ne
Coffee (pulp)	53.8	2.01	0.27	2.14	1.90	0.37	7.60	ne
Banana (stem, leaf)	65.5	2.50	0.56	3.74	2.30	1.50	7.5	15×10^{11}
Urban solids	26.5	0.90	0.44	3.60	3.60	3.10	7.00	ne

Note: ne = not evaluated.

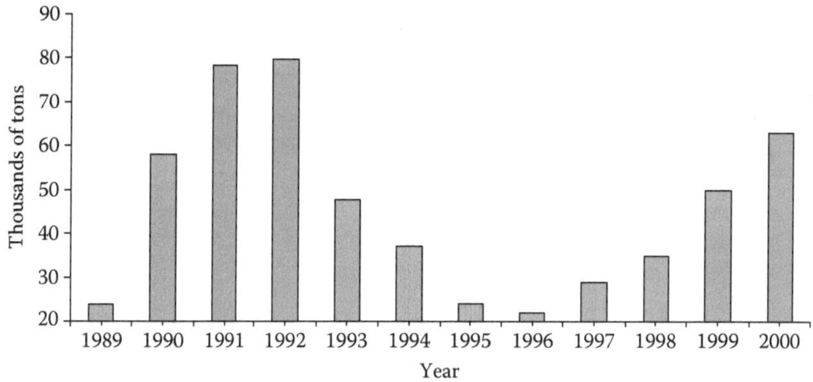

Figure 33.1 Vermicompost production in Cuba in thousands of tons, 1989–2000.

Table 33.2 Amino Acid Composition of Flours from Different Species

	Amino Acids (mg·g^{-1} (parts per thousand) protein)		
Amino Acid	***Eudrilus eugeniae***	***Eisenia fetida***	**Fish protein**
Treonine	71.06	37.60	43.60
Valine	77.10	61.40	53.60
Metionine	19.65	15.30	30.80
Isoleucine	32.61	47.30	46.30
Fenilalanine	54.64	35,40	38.70
Leucine	73.18	73.90	77.90
Lysine	36.51	12.51	78.90
Histidine	n. e.	2.51	n. e.

Note: n.e = not evaluated.

very good results on bananas, among other crops. The solid fraction contains 0.5% humics and 0.3% fulvic acids.

With respect to earthworm protein, Table 33.2 shows the amino acid composition of *E. eugeniae* biomass and *E. andrei* compared with that of fish protein. In earthworms, the amino acid lysine is deficient in comparison with that in Food and Agriculture Organization of the UN (FAO) protein feeds. The chemical composition of the tissues of two species of earthworms was good, and they can be substituted for fish in the diets of chickens, since their tissues contain enough essentials amino acids. This was tested on poultry with diets substituted with 6% fish protein and methionine amino acids to obtain high weights and good conversion rates (Table 33.3). The earthworm protein is also used in cultivation of shrimps, fish, and frogs.

Soil conservation has been a constant preoccupation, and priority is given to first erosion then, salting, and compaction of soils, as well as acidity, which are the most important problems. Over the past years there has been an exponential increase

Table 33.3 Weight, Conversion and Viability of Male and Female Chickens (Plymouth Rock) Fed with Earthworm Protein

		Inclusion of Earthworm Flour		
Parameters	**Normal Diet**	**3%**	**6%**	**8%**
Female and male weight (kg)	1.77	1.73	1.77	1.73
Female weight (kg)	1.59	1.56	1.6	1.59
Male weight (kg)	1.87	1.85	1.91	1.84
Feed intake (kg)	4.52	4.48	4.32	4.41
Conversion	2.56	2.59	2.44	2.54
Viability (%)	90.00	96.70	98.30	96.70

in vermiculture, regulation of nutrient research, production of vermicomposts, and applications to improve the soil. The biotechnology of vermiculture in Cuba contributes to agricultural and cultural development. To produce food from crops it was necessary to substitute inorganic fertilizers. Vermiculture has improved education at the university, feeding, new employment for urban people and new sources are generated for the substitution of fish protein in chickens. This biotechnology has substituted for imports. The application of vermicomposts in Cuba means a saving of US$600 per hectare, as they substitute for inorganic fertilizers and also reduce petroleum costs by decreasing the trips needed for transportation of organic matter (1 ton of vermicompost is equal to 4 ton of other organic matter). In Cuba tobacco, corn, beans, tomatoes, garlic, peppers, and onions are cultivated using organic matter. In 1996, 100,000 ton of vermicompost were obtained, saving 6358 ton per ha^{-1} of inorganic fertilizer. Tobacco production was improved between 56 and 74% to attain a superior class.

REFERENCES

Calero, B. J., and L. Revestí. (Eds.) 2003. Lombricultura, Manual Practico, Unidad de Producciones Gráficas del MINREX, pp. 99.

Camps, D. M., and M. Reinés. 1981. Utilización de la vermiharina como fuente proteica animal en pollos de engorde. *Rev. Cubana Ciencia Avícola* 13(2): 145–151.

Carrillo, O., and M. Reinés. 1979. Caracterización de las harinas de dos especies de lombrices, roja africana (*Eudrilus eugeniae*) y roja californiana (*Eisenia andrei*). Facultad de Biología, Universidad de La Habana, Cuba.

Castillo, R., A. Loza, and M. Reinés. 1979. Potencial bactericida de la lombriz de tierra roja californiana *Eisenia fetida andrei*. *Sostenibilidad* 4(1): 87–94.

Cruz, J., M. Reinés, A. Sierra, and C. Rodríguez. 1986. Valoración económica de la biotecnología, de la harina de lombriz. *Rev. Ciencias Econ.* 1(4), 12–24. Universidad de La Habana, Cuba

Del Valle, M. T., and C. Rodríguez. 1988. *Mesidionema praecomasculatis* (Rhabditida: Nematoda), un parásito de *Eudrilus eugeniae* (Oligochaeta: Eudrilidae). Prevalencia e intensidad de invasión. *Rev. Biología* 11(3): 17–23.

González, A. 1978. Anelida. Manual de Conferencias y Guías Practicas, para la Carrera de Biología, Escuela de Ciencias Biológicas, Universidad de La Habana, Cuba.

Herrera Lemus. Ma. T. 1997. Estudio de obtención de proteasa y su caracterización de tres especies d lombrices de tierra *Eudrilus eugeniae*, *Perionyx escavatus* y *Eisenia fetida*. Bachelor's thesis, Facultad de Biología, Universidad de La Habana, Cuba.

Prieto, D., and C. Rodríguez. 1989. Acarofauna asociada al sustrato de las crías de *Eudrilus eugeniae* (Oligochaeta: Eudrilidae). *Caribb. J. Sci.* 25 (1–2): 54–58.

Reinés, M. 1996. Nuevo registro de Turbelarios terrestres para Cuba. *Rev. Biología* 10: 3–8.

Reinés, M., A. Loza, and S. Contreras. 1998. *Vermicultura: Alternativa para el desarrollo sostenible*. ed. Libra, Guadalajara, Mexico.

Reinés, M., A. Loza, and S. Contreras. 2001a. Vermicultura conocer y cuidar las lombrices para obtener abono orgánico y mejorar el suelo. ed. Fundación Produce de Jalisco, México, pp. 133.

Reinés, M., A. Loza, and S. Contreras. 2001b. Vermicultura: Una biotecnología para la sustentabilidad, ISBN 970-27-0115-5, 1-54. ed. Ágata, Jalisco, México. pp. 78.

Reinés, M., and C. Rodríguez. 1982. Estudio preliminar de la cría masiva de lombrices de tierra en Cuba. Paper presented at the Second National Convention of Philippine Earthworm Growers at Manila, Philippine. (Unpublished manuscript)

Reinés, M., and C. Rodríguez. 2006. Impacto de la introducción de los resultados de investigación docencia producción en la lombricultura en Cuba al reclamo de las necesidades del país. *Revista Digitalizada Ministerio de Educación Superior, Cuba*. http:// Revistas. mes.edu; ISBN. 959-16-0458-0. pdf.

Reinés, M., C. Rodríguez, A. Loza, and S. Contreras. 2006. Nuevos avances en la biotecnología de la lombricultura, 2006. Revista digitalizada MES. Revista Digitalizada MES. http:// Revistas.mes.edu; Ministerio de Educación Superior, Cuba. pdf.

Reinés, M., C. Rodríguez, and A. Sierra. 1981. Manual de instrucciones para el desarrollo de la vermicultura en Cuba a través de la Lombriz roja africana *Eudrilus eugeniae*. Ed. Facultad de Biología, University of Havana, U.H., Cuba. pp. 48.

Reinés, M., C. Rodríguez, A. Sierra, and M. M. Vázquez. 1998. Lombrices de tierra con valor comercial, Biología y técnicas de cultivo. Universidad de La Habana, Universidad de Quintana Roo, México. pp. 59.

Rodríguez, C. 1998. Desarrollo del sistema reproductor y oviposición de *Eudrilus eugeniae*. Oligocaheta: Eudrilidae a 24oC y 30oC. *Revista biología* 5(2–3): 151–157.

Rodríguez, C. 1999. Lombrices de tierra (Oligochaeta: Haplotaxida y Moniligastrida) de Cuba. PhD thesis., Facultad de Biología, Universidad de la Habana, Cuba.

Rodríguez, C., M. E. Canetti, M. Reinés, and A. Sierra. 1986. Ciclo de vida de *Eudrilus eugeniae* a 30°C. *Poeyana* 326: 1–7.

CHAPTER 34

Commercial Applications of Vermiculture in China

Sun Zhenjun

CONTENTS

I Introduction 546
II Types of Vermiculture Used Commercially in China 547
A The Windrow Vermiculture Method for Earthworm Biomass Production in the Field 547
1 Earthworm Species or Varieties Used 547
2 Earthworm Substrates and Windrow Types 548
3 Temperature Management 548
4 Moisture Management 548
5 Earthworm Population Density and Food 548
B Pit Vermiculture Method Outdoors 549
C Multilayer-Box Vermiculture System Indoors 549
D Vermiculture in Plastic Tunnels for Low-Temperature Seasons 550
E Vermiculture with a Heat Channel System for Cold Seasons 550
1 Bed Construction 550
2 Organic-Waste Processing 551
III Case Study of Chinese High-Yield Vermiculture 1: With A Deep Bed and High-Earthworm Population Density 551
A Earthworm Species and Substrate 552
B Earthworm-Bed Construction 552
C Earthworm Substrate Processing 553
1 Pretreatment of Raw Materials 553
2 C:N Ratio of Substrate 553
3 Fermentation of the Substrate 553
4 Turning of Windrows 553
5 pH Value of the Substrate 553

D Earthworm Culture Methods.......554
1 Trial Feeding of Earthworms.......554
2 Putting the Substrate in the Earthworm Bed.......554
3 Introducing Earthworms into the Bed.......554
4 Management.......554
E Results.......555
F Technological Analyses of Earthworm Populations.......555
IV Case Study of Chinese High-Yield Vermiculture 2: Annual-Cycle Vermiproduction.......556
A Stages of Vermiculture.......556
B Results of Annual-Cycle Earthworm Production.......558
V Summary of Chinese High-Yield Vermiculture.......559
A Points Relating to Vermiculture Theory.......559
B Points in Vermiculture Techniques.......561
References.......562

I INTRODUCTION

Since the early 1970s, studies of methods of earthworm breeding under cultural conditions have been conducted in many regions and countries. In Japan during that time, the two cities of Kyushu and Hokkaido in Japan had more than 200 earthworm farms or companies, some of which were established in paper mills and livestock farms for both organic-waste disposal and production of animal protein and vermicompost (Zeng et al. 1982). Vermiculture has also developed rapidly in China, Taiwan, India, the Philippines, and other Asian countries since the 1980s.

Up to 306 species of terrestrial earthworms have been recorded in China (Huang et al. 2007). *Pheretima* sp. represents the dominant genus of the Chinese earthworm fauna, but *Allolobophora caliginosa* (Savigny 1826) and *Eisenia fetida* (Savigny 1826) are dominant in both numbers and biomass in northern China (Huang and Zhang 1980). The species *E. fetida*, *A. caliginosa* (Savigny 1826), and *Pheretima guillemi* are the most important earthworm species in culture in China. In 1979, "DAPING II" (the common Chinese name used for *E. fetida*) was introduced to China, and its culture spread very rapidly in the mid- 1980s, with 243 earthworm farms distributed over 23 provinces in China using different methods and on different scales (Sun 1993). However, most of these vermiculture methods are unsuitable for the present demands of commercial earthworm production on a large scale.

There was a burst of earthworm breeding in China in the early 1980s, but, soon thereafter, almost all earthworm farms closed due to the inefficient culture technologies that they used. The big problem in China is to improve these earthworm farms that closed down because of unsatisfactory techniques. Much of the theoretical basis of vermicultural technology, used in various earthworm farms across the world, was developed from the characteristics of earthworm populations and ecology in natural ecosystems using ecological breeding methods. The economic value of the earthworms per unit area was lower, due to a longer-growing period; there were only smaller earthworm population biomass increases; and labor-intensive methods of harvesting and processing were used. In recent years, organic-waste disposal by earthworms under intensive cultural conditions has been practiced but only on a relatively small scale in China (Brun et al. 1987; Curry 1988; Edwards 1988, 1996; Sun 1993). The key question is how great a potential for population development do earthworms have under artificial culture conditions? Although vermistabilization, as a relatively new innovative and alternative technology, has been accepted because it is environmentally sound and simple in management, there is relatively little information on the optimal technological design and practices to be used on a large commercial scale in China.

Since 1986, by studying the ecological requirements for the growth and reproduction of different developmental stages of different species of earthworms, under artificially controlled conditions, we have acquired some basic knowledge of the most important ecological characteristics for high-yield vermiculture under man-made conditions. We have established a new theory for high-yield vermiculture and a relevant technological system since 1993 (Sun 1995b, 2003).

II TYPES OF VERMICULTURE USED COMMERCIALLY IN CHINA

Owing to climatic differences between southern and northern China, many methods have been developed for commercial vermiculture, including windrows, multilayer boxes, earthworm beds, and plastic shed vermiculture. The methods adopted in different regions vary with climatic conditions and commercial objectives, whether traditional or modern in management, simple or complex in technology. The general principle for successful vermiculture is to provide earthworms with a suitable environment for their rapid growth and reproduction. On the basis of climatic variations, five types of vermiculture methods have been introduced in China:

A The Windrow Vermiculture Method for Earthworm Biomass Production in the Field

Windrow methods are used widely in China because of their low costs and labor savings.

1 Earthworm Species or Varieties Used

DAPING II earthworms, a strain of *E. fetida* originating from Japan, is the common species used for vermiculture in China. It can develop well in most climatic zones, but care should be taken in colder regions. When the temperature drops below 5°C in winter, they move toward the deeper layers in the soil or organic matter.

2 Earthworm Substrates and Windrow Types

The earthworm substrate preferred by *E. fetida* in China consists of 70% manure (preferably cattle) and 30% plant straw. The mixture is piled into windrows 1–2 m (1.2–2.2 yd) wide 0.5–0.8 m (0.55–0.87 yd) high. The length of the windrows depends on the size of the site. By this method, some indigenous earthworms can also be attracted into the substrate. However, growers recommend where possible using only cultured earthworm species and varieties in the windrows. Using this vermiculture method, the earthworms develop faster from March to October than in winter.

3 Temperature Management

Keeping the temperature in the windrows below 30°C (86°F) in summer and above 10°C (50°F) in winter is necessary. To achieve this, measures adopted in summer are to reduce the substrate thickness from 20–27 cm (7.8–10.6 in), keep out sunshine, and sprinkle with water to lower the temperatures. In winter, measures should be taken to maintain temperatures above 10°C (50°F), such as increasing the substrate thickness, adding hay to the bedding, covering with plastic film, and reducing frequency of watering.

4 Moisture Management

The relative moisture of the substrate should be maintained at 70–75%. An earthworm respires by absorbing oxygen dissolved in moisture through its skin. If the substrate moisture content is insufficient, they can die of dehydration or suffocation. Therefore, the windrow should be watered every 1–2 days during the earthworm-growing season. A simple standard for moisture content from earthworm growers is that if you squeeze a handful of vermicompost and water drips down from your fingers, there is excessive moisture in the vermicompost; if the water can be seen between your fingers, it indicates appropriate moisture.

5 Earthworm Population Density and Food

An initial stock of about 600 earthworms, with 300 $g \cdot m^{-2}$ (11 $oz \cdot yd^{-2}$) of bedding, is suitable for initiating earthworm reproduction, and a 10,000 to 15,000 bed-run of initial earthworm stock in a bed is needed for adequate earthworm weight gain to increase population biomass quickly. A suitable substrate needs to be added to the earthworm beds regularly to satisfy the earthworms' nutritional needs. When the earthworm feed appears like loose chaff and some earthworms move toward the bed wall, you should begin to add new substrate. Usually, a 7–10 cm (2.8–3.9 in) cm layer of new feedstock should be added every 7 to 10 days. The process for adding substrate is as follows: (i) The old substrate with earthworms and cocoons is laid on one half of the bed, and new substrate is added to the other half of the bed; (ii) when most earthworms move from the old organic waste into the new feed, place a thin layer of new feed on the surface of the old bedding material without mature earthworms but with cocoons; (iii) when these cocoons hatch and hatchings move into the new feedstock cover, remove the rest of the old waste, which can then be processed by sieving to a fine vermicompost in a rotating trommel, for use as organic fertilizer or as a supplement to domestic animal feed.

B Pit Vermiculture Method Outdoors

This culture method uses a pit that is partly aboveground and partly underground. The length and width of the pit depend on local conditions. The 6–10 cm (2.3–4.0 in) depth of the pit is 50 cm (20 in) underground and 16.5–33.0 cm (6.5–13.0 in) above the ground. A concrete-brick wall surrounds the pit, but the bedding contacts the soil directly at the bottom. A black plastic film covers the pit to waterproof it; maintain darkness; prevent rats, toads, and other predatory animals from entering the pit; and retain earthworms inside the pit. The management of pit vermiculture is similar to windrow vermiculture.

C Multilayer-Box Vermiculture System Indoors

Box vermiculture is used widely in China. Many kinds of wooden and plastic boxes, including packing boxes and food boxes, can be used for vermiculture. However, wood containing fragrant substances and tannic acids, which are harmful to earthworms, should not be used. The size of the box can measure 40–60 cm (16–24 in) long by 30–50 cm (12–20 in) wide by 20–30 cm (8–12 in) high, with 20 cm (8 in) of bedding. A layer of more than 30 cm (12 in) of bedding is too thick to aerate, and less than 20 cm (8 in) is too thin to retain water. There should be holes 0.7–1.5 cm (0.3–0.6 in) in diameter in the bottom and side walls of the boxes to drain excessive water and provide ventilation. Vermiculture boxes should be covered by wastepaper or paddy straw to keep the substrate moist.

To use a multilayer-box vermiculture system indoors, four to five tiers of boxes can be piled up on top of one another to form a laminated vertical structure. This makes the best use of space, is easy to manage, and increases vermicompost productivity by two to three times that of vermiculture outdoors. A vermiculture room with two aerating holes 50 cm × 25 cm (20 in × 10 in) on each side wall can keep temperatures at 18°C (64°F). Lights should be used at night to keep earthworms from migrating out of the boxes.

The earthworm population density should be 5000–10,000 earthworms m^{-2} (5980–12,000 yd^{-2}) of bedding. As the earthworms develop, the box can be divided to reduce the earthworm population density. Studies have showed that an initial stock of 2000 earthworms in a vermiculture box 60 cm × 40 cm × 20 cm (24 in × 16 in × 8 in) developed into 18,000 earthworms in 3 months under conditions of 20°C and 75–80% moisture content.

The key management measures for a multilayer-box vermiculture system are (i) a suitable earthworm–substrate composition and (ii) appropriate moisture and collection of earthworm casts on time. The substrate for box vermiculture can consist of 20 kg (44 lb) rice paddy chaff, 0.1 kg (0.2 lb) urea (carbamide), and 50 kg (110 lb) water. The urea is dissolved in the water, added to the chaff, and mixed and it then undergoes a 7-day fermentation. Only well-fermented feedstock should be used for vermiculture. When a layer of earthworm casts appears on the surface of the bedding, the casts should be taken away and a 3–5 cm (1.2–2.0 in) layer of new feedstock added.

D Vermiculture in Plastic Tunnels for Low-Temperature Seasons

Plastic tunnels used for vermiculture can be similar to those used for vegetable production. There should be two earthworm-bed windrows in each plastic tunnel; these should be no more than 6 m (6.6 yd) wide and 2 m (2.2 yd) high, a convenient width for working from either side. The length of the tunnel may be 25 m (27.3 yd), 30 m (32.8 yd), or 50 m (54.7 yd), as desired. A 30 m (32.8 yd) long tunnel is common, and usually eight tunnels are connected together to save labor and allow more convenient management. Investigations have showed that an arrangement of eight tunnels can produce 20 ton (19 t) of earthworms in one month.

Some management measures should be taken to maintain relative constant bed temperatures in the sheds: (i) When the air temperature outdoors drops to 10°C (50°F), place the plastic cover over the bed to absorb sunshine; (ii) install a windproof wall near the tunnel facing the wind and cover the tunnel with a layer of hay at night; (iii) install small arch shelters over the bed rows in the plastic tunnel; and (iv) increase the bedding depth to 40–50 cm (15.7–19.7 in). When you take these measures, even if the temperature outdoors goes as low as 14°C (68°F), the bedding temperature can be maintained at 8°C (46°F) or higher in the tunnel, which may allow a longer breeding time for the earthworms.

E Vermiculture with a Heat Channel System for Cold Seasons

1 Bed Construction

These breeding methods are widely used in the cold seasons of northern China. The key to success is to build an aerating channel system in the bottom of the bed and a heat-generating pool below the channels. The heat pool is excavated, 0.6 m long (0.7 yd) long × 2 m (2.2 yd) wide (equal to the bed width) 0.6 m (0.7 yd) deep, and is filled with green horse manure and raw straw, with a hard permeable cover (e.g., wire mesh or wood sticks) put over the channels and then a 5–8 cm (2–3 in) deep mixture of green manure with mushroom residues or sawdust laid on the surface of the bed. A brick wall painted black should surround the bed, the northern side being 80 cm (31.5 in) high, the southern one 20–25 cm (7.9–9.8 in) high. The tunnel is covered with double layers of plastic film. There are five aerating holes 120 cm (47.2 in) in diameter in the inclined walls between the northern and southern walls to aerate and adjust the temperatures (Figure 34.1).

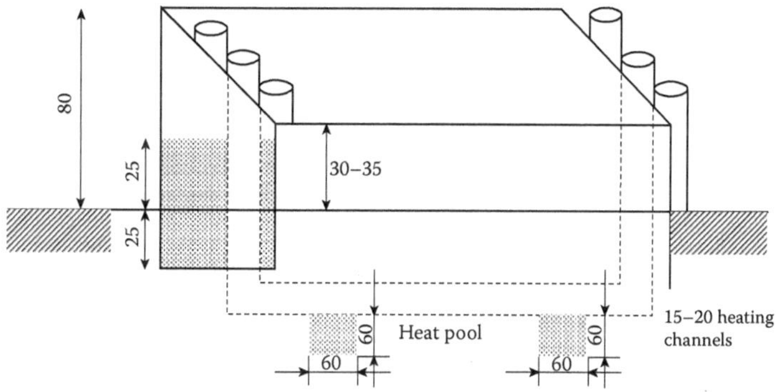

Figure 34.1 Structure of the earthworm bed with heat pool and plastic-film shed (measurements are in centimeters).

2 Organic-Waste Processing

The management for this method, except for increasing the fermentation period of the mixture to 21–28 days, is the same as described earlier. The tunnel should be covered with a straw curtain on winter nights. The drainage canal around the bed should be well excavated. The straw curtain is opened at 8 a.m. in the morning, and the bed absorbs sunshine through the plastic film. When the temperature in the bed becomes too high at noon, the aerating holes can be opened to release excess heat. The straw curtain is replaced at 3 p.m. in the afternoon. If it is raining or snowing, care is taken to maintain temperature without opening the curtain and holes.

A bed with a heat pool and plastic-film tunnel makes full use of solar energy absorbed through the plastic film to increase bed temperatures. Due to heat siphoning between the inside and outside beds, the heat from the heat-generating pool under the bed bottom passes through the heating channel system, and energy is absorbed and stored in the bedding so that the substrate temperature is increased. At the same time, air flow improves the rates of decomposition of the fermented materials in the heat pool and bed bottom. This is a new style of culture method, with optimal use of solar as well as biological energy, to ensure satisfactory temperature requirements for earthworm production during cold seasons.

III CASE STUDY OF CHINESE HIGH-YIELD VERMICULTURE 1: WITH A DEEP BED AND HIGH-EARTHWORM POPULATION DENSITY

This vermiculture system is for maximum earthworm population biomass using a deep substrate layer and a high-earthworm population density in beds with an aerating channel system, increasing the organic substrate thickness to 50 cm (19.7 in) and the earthworm population densities to 30,000–50,000 earthworms m^{-2} (28,000–46,000 yd^{-2}).

A Earthworm Species and Substrate

The populations of *E. fetida* used came from an earthworm farm. The organic-waste substrate consisted of 70% cattle manure, 20% wheat straw, and 10% mushroom residues or sawdust. The substrate materials came from an experimental station for livestock, an agricultural farm, a mushroom culture farm, and a wood mill.

B Earthworm-Bed Construction

The earthworm beds were constructed in a place sheltered from the wind and direct sunshine, with good drainage and aeration. The size of the experimental beds was 20 m^2 (24 yd^2), 20 × 1 m (22 yd × 1.0 yd) in area, and 0.55 m^2 (0.7 yd^2) in depth, with 0.3 m (0.3 yd) below ground and 0.25 m (0.3 yd) above ground level. The beds were surrounded with a brick wall, and there were five aerating channels on the bed bottoms. On the two short sides of the beds there were aerating holes. The measurements of the beds are given in Figure 34.2.

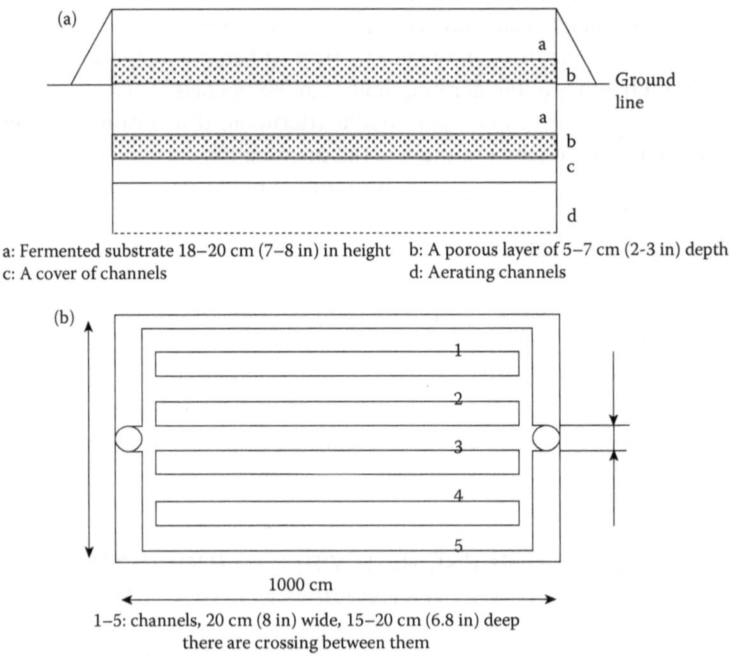

Figure 34.2 Structure of the bed with deep substrate layer and aerating channels. (a) Horizontal view. (b) Vertical view.

C Earthworm Substrate Processing

1 Pretreatment of Raw Materials

Fresh cattle dung should be spread on the ground to dry in the sunshine and to reduce the odor; wheat straw and mushroom residue should be cleaned of impurities. All materials should be macerated into 1.0 mm (0.04 in) particles, mixing cattle manure, wheat straw, and mushroom residue in a ratio of 7:2:1.

2 C:N Ratio of Substrate

Before fermentation, it is necessary to calculate and adjust the C:N ratio of materials according to a designated proportion of N and C materials, as well as the C and N content of the raw materials. The mixture was generally adjusted to a C:N ratio of 20–30 (Zeng et al. 1982).

3 Fermentation of the Substrate

A composting method was used; that is, a mixture of waste with 50% moisture was piled in a sunny place out of the wind. The windrow was 1.3 × 1.6 m (1.2 × 1.5 yd) wide, 1 m (4 in) high, and several yards (meters) long depending on the mixture and was covered with uncut wheat straw or plastic film to maintain moisture and to improve aerobic microbial decay.

4 Turning of Windrows

During the fermentation period, the temperature changes of the windrows were recorded every day. After the windrow temperature increased to 50–70°C (122°F–158°F) and then decreased, the windrows were turned, in particular by turning the surrounding parts into the center of the hill. After fermentation for 15 days and turning of the windrows two to three times, the mixture's temperature did not increase again, and the mixture appeared black brown in color, was loose textured, and had no odor.

5 pH Value of the Substrate

The composted waste should usually be pH 5–6 and may need to be adjusted to pH 6–8 using lime water. Washing the mixture has the additional function of removing harmful substances produced during fermentation and adjusting the mixture's moisture content. The fermented mixture, with 60–70% moisture to prevent drying, curdling, and mildewing, is ready for use.

D Earthworm Culture Methods

1 Trial Feeding of Earthworms

Before the substrate is placed into the earthworm bed, it is necessary to spread it and aerate it for 1–2 hours. To ensure earthworm safety, trial feeding should be conducted, putting 50 earthworms of different development stages into the trial earthworm substrate and observing the reactions of the earthworms. If all the earthworms entered into the substrate within 24 hours and survived easily, the earthworm substrate is suitable for use.

2 Putting the Substrate in the Earthworm Bed

At first, a 5–7 cm (2.0–2.8 in) layer of raw mushroom residue or sawdust, as a porous layer, was laid on the bed bottom, above the aerating channels, and, second, a 18–20 cm (7.1–7.9 in) earthworm substrate layer (fermented organic-waste mixture) was put on top of the porous layer, followed by another 5–7 cm (2.0–2.8 in) porous layer and another 18–20 cm (7.1–7.9 in) substrate layer laid on the bed (Figure 34.2a and b), so that finally the substrate height reached 50 cm.

3 Introducing Earthworms into the Bed

An earthworm mixture of cocoons, juveniles, casts, and earthworms containing, by sample calculation, 14,500 cocoons and 8000 juvenile earthworms m^{-2} (yd^{-2}) was spread on the bed below the 10 cm (4 in) feed surface. One month later the population density of the earthworms population was about 44,250 juveniles m^{-2} (yd^{-2}).

4 Management

The temperature and moisture content at a 10 cm (3.9 in) depth in the bed should be recorded daily. A moisture level of 60–70% can be maintained by spraying water on the surface once or twice per day. The pH value of the earthworm substrate should be checked 1–2 times per week. One month after putting the experimental earthworms in the bed, when the vermicompost cover on the bed surface is over 5 cm (2 in), the vermicompost on the bed surface should be cleared away and a proportional amount of fresh substrate added simultaneously to each earthworm bed. Two months later, earthworm weights and degree of development of the body clitellum should be measured. The mature earthworms should harvested at an optimum time, when earthworm clitellums have just appeared, calculated from an earthworm population curve (Sun 1995a).

The experiments lasted 90 days from May to August, with a mean substrate temperature of 24.8°C (76.6°F) and a range of 16–32°C (61–86°F), a mean substrate moisture of 64% and a range of 54%–73%, and a mean substrate pH value of 6.8.

E Results

The experiment was done in the warm season from May 3 to October 1 at an average substrate temperature of 24.8°C (76°F) and a range of 19–30°C (66–86°F), an average substrate moisture content of 64% and a range of 54–73% and an average substrate pH value of 6.8 (Table 34.1). From the results it can be seen that in the two trial beds, the mean earthworm population biomass reached 7.9 kg m^{-2} (14.5 lb. yd^{-2}), increasing the earthworm yields by 74.3% compared to conventional beds (CK1 and CK2).

F Technological Analyses of Earthworm Populations

In this study, the earthworm population density m^{-2} (yd^{-3}) in the experimental beds was twice that in the control beds. However, when calculating earthworm population density m^{-3} (yd^{-3}), due to the deep substrate layer in the experimental bed, there were relatively few differences in earthworm populations between the trial beds and control beds. Therefore, the earthworms in the trial beds had sufficient living space. It seems more reasonable to calculate the earthworm population density as the number per m^{-3} (yd^{-3}) than m^{-2} (yd^{-2}).

On average, the clitellum appeared in *E. fetida* after about 70–80 days at a mean temperature of 24.8°C temperature and 64% moisture. This is consistent with laboratory data, but in large-scale earthworm production there were differences in incubation times among cocoons, with 10–15% of earthworms still not having matured after 80 days. So, with earthworms harvested after 90 days, some earthworms may have

Table 34.1 High-Yield Vermiculture in the Warm Season in China

Items	Trial 1	Trial 2	CK1*	CK2*
Measured area (m^2)	5.4	5.4	5.4	5.4
Substrate depth (cm)	46–50	46–50	22–25	22–25
Earthworm sowing				
Time (day/month)	3/5	3/5	3/5	3/5
Number of juv. ($10^3/m^2$)	8	8	4	4
Initial wt. (kg/m^2)	1.23	1.16	0.57	0.66
Number of cocoons ($10^3/m^2$)	14.5	14.5	7.25	7.25
Harvest				
Time (day/month)	1/8	1/8	1/8	1/8
Final wt. (kg/m^2)	9.2	9.0	5.3	5.0
Weight gain (g/m^2)	7.94	7.82	4.70	4.34

* Control experiment: Two culture beds (CK1 and CK2) for control with 15 m2 and 202 were used, 30 cm deep (10 cm underground and 20 cm aboveground), without aerating channels. The bed was filled with a 25 cm earthworm substrate layer and a bed-run population density of 22,000 juveniles/m2 from a mixture of 7250 cocoons and 4000 juveniles per square meter. The ration of substrate, processing fermentation, and routine management were the same as shown in the trial beds.

exceeded the optimum harvesting time. Adequate aeration is one of the keys for high-efficiency vermiculture with a deep substrate layer and high-earthworm population densities. An aerating channel system was set up on the bed bottom, the earthworm substrate contained 10% mushroom residues, and two porous material layers were laid on the bedding. These measures were adopted to improve substrate aeration, which is essential for earthworm growth. Appropriate management measures guarantee high yields in earthworm production. The earthworm waste substrate is important in maintaining earthworm growth and reproduction. The fermented mixture of manure with plant materials such as wheat straw, mushroom residue, fruit by-products, and sawdust in appropriate ratios was characterized as soft, loose, fine, and mature and was acceptable to *E. fetida*. One month after earthworm inoculations, the bedding was turned over regularly (every 10–15 days), to make full use of the substrate, with consistent temperatures and moisture levels between the upper substrate layer and the bottom layer. Before turning over the substrate the earthworm casts should be scraped out and then that space filled with fresh substrate. However, to improve cocoon incubation, the bedding should not be moved for 1 month after earthworm incubations.

This culture method, in which earthworms are bred and grown in different earthworm beds, eradicated some shortcomings of conventional methods such as mixtures of various juveniles, cocoons, and earthworm casts in the same bed; difficulties in separating earthworms from the mixture; and lower productivity and utilization ratios. A good way of separating cocoons from bedding has not yet been developed (SIIST 1985) even though some earthworm separators have been developed (Price and Phillips 1990). In this study, we tried to keep the growing earthworms in similar life stages in each batch. There were relatively few differences in individual weights among the earthworms, because the population came from the same group of cocoons or juveniles in the earthworm-breeding bed. This made earthworm management and harvesting easier. It is suggested that through separate culture and management of breeding and growing earthworm populations, on the basis of their different ecological requirements, high yields of earthworm can be obtained. According to our results under laboratory conditions (Sun 1993), had the substrate temperature has been lowered to 18°C (64°F) and the substrate pH raised to 8–9, the efficiency in earthworm weight gains would also have been higher.

IV CASE STUDY OF CHINESE HIGH-YIELD VERMICULTURE 2: ANNUAL-CYCLE VERMIPRODUCTION

A Stages of Vermiculture

This method is used in the region of China where it is very warm in summer and very cold in winter. In the Beijing region, the annual cycle of earthworm production can be divided into four stages over two periods, with two different high-yield techniques. The earthworm beds using a heat pool and plastic-film tunnel were used in the cold period from November 1 to May 2. Two stages of earthworm culture were included in this period: The first stage was from November 1 to February 2. After construction of the earthworm beds, 15,000 juvenile earthworms and 14,000 cocoons yd^{-2} (m^{-2}), by calculation from samples, were placed into each bed on November 1; earthworms were harvested from half of the culture bedding on February 2. The second stage was from February 3 to May 2. When the mixture of adult earthworms, juveniles, casts, and earthworm substrate was sieved, some juvenile earthworms were returned to the bed. At the same time, fresh substrates were added. On May 2 all the bedding was taken out of the bed, and the earthworms and vermicompost were harvested by an illumination separation method. This method involved transferring the bedding from the bed to a plastic film to form a pile exposed to the sunshine. The result was that the earthworms in it moved to the bottom because of their negative reactions to light. The vermicompost was separated off and the plastic film replaced. Then the treatment repeated several times. After that, the vermicompost was separated, so that fresh earthworms were left in the bed (Sun 1995b).

High-yield vermiculture outdoors with a deep substrate layer and aerating channels was used in the warm period between May 3 and October 29. This period also consisted of two earthworm productive stages: The third stage was from May 3 to August 1. The contents of the earthworm beds were cleared out from the heat pool, and the channels were repaired after the aerating tubes and plastic sheets were taken away. The mixture of juveniles, cocoons, and some casts earthworms and substrate including 8200 juveniles m^{-2} (8000 juveniles yd^{-2}) and 14,500 cocoons m^{-2} (14,000 cocoons yd^{-2}) was placed into the bed. Routine management was the same as described for the first case study. In very hot summer weather, a thin straw curtain covered the tunnel to prevent the sun from shining directly on the earthworm beds at noon, and water was sprayed to reduce bed temperatures. On August 1 the earthworms and earthworm casts in the bedding were harvested. In the fourth stage, from August 2 to October 29, the bedding was replaced by fresh earthworm substrate with about 46,000 juvenile earthworms and 12,400 cocoons yd^{-2} (55,000 juvenile earthworms and 14,800 cocoons m^{-2}), by simple calculation. Adult earthworms were collected in 90 days.

The first earthworm-yield measurements were made on February 1 using a wood measuring frame 0.4 m^2 (0.35 yd^2) at three random measurement points. A mixture of earthworms, earthworm casts, and some substrate within the frame was removed onto the soil surface. The earthworms were separated by a light separating method. That is, the mixture was piled on a plastic film and exposed to the sun; due to earthworm's negative reactions to light, the earthworms were driven downward. The upper layer of earthworm casts, now with no earthworms, was scraped away, until finally only the earthworms gathered on the bottom were left and the earthworm casts were separated. Earthworms and earthworm casts were weighed, and the cocoons were sieved from the earthworm casts. When the bedding was taken out and replaced, the earthworms entered the fresh substrate rapidly. The second earthworm harvest was on May 2. Before the whole bed was harvested, earthworm-yield measurements were made with the frame at six locations with 5.4 m^2 (6.5 yd^2) total area. One culture cycle took 90 days (from earthworm inoculation to harvesting). In the first two culture stages there was only one inoculation but two earthworm harvests. The third culture stage began on May 3 and ended on August 1 with whole-bed earthworm harvesting. The fourth culture stage started on August 2 and ended on October 28. The two later culture stages took the high-yield-culture technique outdoors with a deep feed layer and aerating channels. The methods of yield measurement were basically the same for the second harvest.

B Results of Annual-Cycle Earthworm Production

From the point of view of earthworm production, it can be seen that with a one-year continual vermiculture system, harvesting by stages from November 1 to October 28 in an area of 140 m^2 (167.4 yd2), a mean yield of 22.5 kg·m^{-2} (41.4 lb·yd^{-2}) earthworms (live weight) was obtained (earthworm yield measured in 37.8 m^{-2} (45.2 yd^{-2}). The first earthworm harvest made up 9.4% of total output, the second 22.3%, the third 35%, and the fourth 33.3% (Figure 34.3). The amounts of earthworm protein and vermicomposts produced coincided approximately with the annual requirements for livestock and plant production in the region. In winter and early spring,

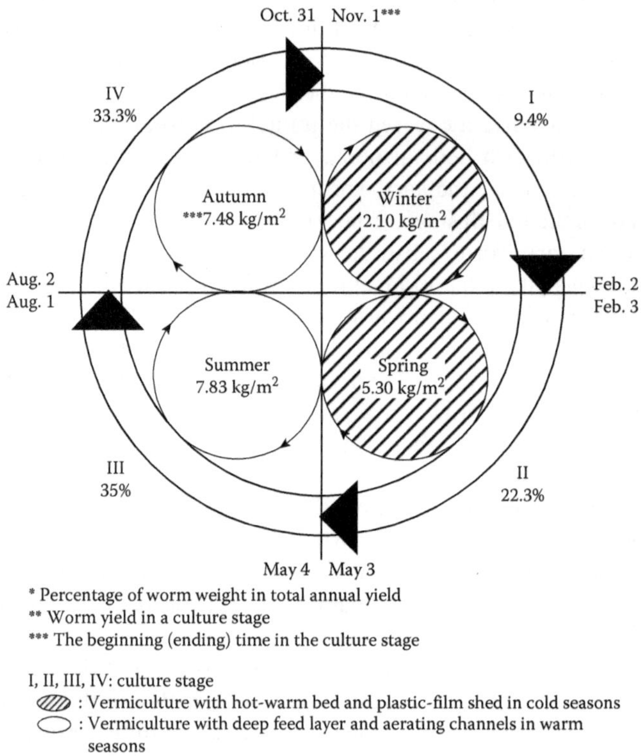

Figure 34.3 An annual cycle of vermiculture in China.

Table 34.2 Annual Vermiculture Cycle in the Trial

Culture Stage	1	2	3	4	Total
Measured area (m^2)	2.7	5.4	5.4	5.4	18.9
Earthworm inoculation					
Time (day/month)	1/11	2/2	3/5	2/8	
Number of juveniles (100 $\cdot m^{-2}$)	150	–	80	55	285
Initial wt. (kg$\cdot m^{-2}$)	2.88	–	1.23	0.79	4.8
Number of cocoons (100 $\cdot m^{-2}$)	140	–	145	148	433
Harvest					
Time (day$\cdot$month^{-1})	1/2	2/5	1/8	1/11	1/2
Harvest area (m^2)	10	20	20	20	70
Final wt. (kg$\cdot m^{-2}$)	4.99	8.19	9.18	8.26	3.06
Weight gain (kg$\cdot m^{-2}$)	2.12	5.26	7.94	7.46	22.78

vegetables and crops cannot grow well, so they do not need much vermicompost, while they will grow well in the summer and autumn seasons, and vermicompost supplies can be adequate in this period. Similarly, earthworm protein is supplied for a kind of fish feed, especially for tilapia (*Oreochromis niloticus*), and these animals also grow rapidly in the summer and autumn.

From our vermiculture model, it can be seen that outdoor vermiculture with a deep substrate layer, high-earthworm population densities, and aerating channels produced greater earthworm yields compared to the control beds (Table 34.2), with 68% of the total output produced in the third and fourth earthworm culture stages because of appropriate temperatures and the porous substrate. However, in this experiment, vermiculture using a heat pool and plastic-film tunnel did not produce the expected earthworm yields. One probable reason was a lack of cocoons in the winter breeder beds, so that incubating earthworms in the first culture stage were used for two culture stages. Thus, in the first stage, the very high-earthworm population density may have affected individual earthworm weight gains, while in the later stages low-earthworm population densities also influenced the total population biomass, after half the bedding was harvested at the end of the first culture stage. Another possible reason was the great temperature differences between day and night as well as poor aeration.

V SUMMARY OF CHINESE HIGH-YIELD VERMICULTURE

A Points Relating to Vermiculture Theory

Gaddie and Douglas (1977), Zeng et al. (1982), Wang (1983), and Brun et al. (1987) reported or reviewed various vermiculture techniques, and Edwards et al. (1988, 2004) improved vermiculture production. To investigate a new vermiculture technique with potential high yields under artificial conditions, the effects of some important factors (e.g., temperature, moisture, pH, density, and substrate composition) on earthworm growth and reproduction were studied systematically. According to Sun (1995b, 2003), a new theory for high-yield vermiculture can be established that differs from traditional methods. The main points are the following:

- Some optimum ecological factors for earthworm growth are not the best for earthworm reproduction. A temperature of 25°C (77°F), 70% moisture, and pH 6 are most suitable for *E. fetida* to produce cocoons, while a temperature of 18°C temperature, 65% substrate moisture, and pH 8–9 are most suitable to increase earthworm body weights. The earthworm *E. fetida* can live in a wide pH range from 5 to 11. The appearance of the two peaks, pH 6 and pH 9, in pH preference indicates that *E. fetida* can tolerate both acidic and alkaline environments.
- In mixtures of substrates, *E. fetida* can descend to a depth of 70 cm (0.6 yd), but 99.2% of earthworms remain within the upper 50 cm (20 in) of the substrate.
- It was suggested that earthworm substrate layers can extend to a depth of 50 cm (20 in), but *E. fetida* produces cocoons mainly within the 6–20 cm (2.5–8.0 in) substrate layer. The numbers of cocoons produced increase with deepening substrate layers within a certain range. It may be adequate for breeding earthworms to live in a 15 cm (6 in) deep substrate layer.
- A population density of 10,000 (±1,200) earthworms m^{-2} produced the most efficient reproduction by *E. fetida*.
- The developmental characteristics of *E. fetida* are as follows: (i) The clitellum of *E. fetida* appears 60–70 days after emergence from the cocoon. The maximum individual weight is reached in 90–100 days. The earthworm population density is not relative to the time when earthworms start to breed or individuals are harvested. (ii) Individual weights decrease, but overall population biomass increases with increasing population density within a certain range. Earthworm population densities for maximum weight gain can reach 55,000 earthworms m^{-2} (46,000 yd^{-2}). Generally, in normal management, 30,000–50,000 earthworms m^{-2} (25,100–41,800 yd^{-2}) was considered suitable for high-yield vermiculture. (iii) Due to the S-shape of the development curve of earthworm weight gain, there is a turning point between a fast-growing period and a slow-growing period, which can be used for a optimum harvesting time when the clitellum has just formed and individual earthworm weights are 0.38 (±0.02) g in a population density of 50,000–55,000 earthworms m^{-2} (60,000–66,000 yd^{-2}).
- With optimum earthworm-to-substrate ratios it seems that the following are true: (i) *E. fetida* prefers to live in cattle manure. It is best to mix manure with crop straw and then ferment it for 15 days to create a preferred earthworm substrate. (ii) The optimum earthworm ratio for earthworm growth and reproduction consisted of 70% cattle manure plus 30% mushroom residue or fruit by-products, probably because of its high nutrition, low odor, and the close texture of cow manure; good aeration and a more porous texture as well large microbial numbers in mushroom residues; and the sweet taste of fruit by-products, which earthworms prefer. (iii) Mixtures of manure with mushroom residues were more beneficial for earthworm reproduction, while mixtures of manure with fruit by-products are more useful for greatest earthworm weight gains.

B Points in Vermiculture Techniques

Since 1985, studies of the life cycle, growth, and reproduction of some species of earthworms have been conducted in natural ecosystems and artificial conditions in our laboratory, and we have made great progress in identifying optimal ecological factors for high-yield vermiculture (Sun 1993, 2003). On the basis of these results, gathered over 5 years with more than 10 experiments and improvements in design, a set of techniques for high-yield vermiculture was established. These techniques consist mainly of high-yield vermiculture outdoors in the warm season, and an annual cycle of production. These techniques are characterized mainly by the following:

- Vermiculture for maximum population biomass, with a deep substrate layer and high-earthworm population densities in a bed with an aerating channel system, and increasing substrate thickness from 8 in to 20 in (20 cm to 50 cm) and the population densities from 15,000–20,000 earthworms m^{-2} (12,500–16,700 yd^{-2}) to 30,000–50,000 m^{-2} (25,000–41,800 yd^{-2}).
- A novel culture model, that is, breeding mature earthworms and juvenile earthworms in different earthworm beds, providing earthworm-growing beds with cocoons or juveniles, and harvesting from earthworm-growing beds in batches.
- The optimum waste substrate for breeding *E. fetida*, consisting of cow manure, mushroom residues, and fruit by-products.
- Harvesting at the time when the earthworm clitellum appears.

These high-yield vermiculture technologies attempt to resolve the two problems of low earthworm output and intermittent earthworm reproduction, in order to provide a feasible method and model for intensive earthworm production on a commercial scale. From the result of 22.5 $kg{\cdot}m^{-2} \cdot yr^{-1}$(41.5 $lb \cdot yd^{-2} \cdot yr^{-1}$) earthworms produced under artificial conditions, it can be concluded that *E. fetida* has a great potential for population biomass development and increase.

In previous studies and practices of vermiculture, either in the laboratory or in the field, the basic principles of earthworm culture technology in China have imitated those of native earthworms in natural ecological conditions (Wang 1983). That is, most native earthworms living in soils were limited to a tillage layer of about 20 cm (7.9 in) (Zeng et al. 1982). In general, a lack of living space restricts the development of high-earthworm population densities. Conditions in our experiments satisfied the ecological requirements of *E. fetida* for growth and reproduction. Our results (Sun 1993, 2003) indicate that the main factors affecting earthworm growth and reproduction were temperature, moisture, substrate composition. and living space.

The bed with a heat pool and a plastic tunnel made earthworm culture in cold seasons possible, minimizing the limitations of vermiculture development in cold regions. Only in relatively appropriate temperatures can earthworm production be conducted all year round. From our experiments, the optimum ratio of combined cattle manure, mushroom residues, and fruit by-products was developed to suit *E. fetida* growth and reproduction on a large scale. A deep substrate layer with a high-earthworm population density was advantageous for both individual weight gain and overall population biomass increases. The relation between earthworm population density and living space was considered to be a key limiting factor to high-yield vermiculture. Key factors of the high-yield vermiculture techniques are as follows:

- Mature breeding earthworms and juvenile earthworms were cultured in different earthworm beds and managed with different measures, according to the different ecological requirements for reproduction and growth. The earthworm-breeding beds provide the earthworm-production beds with juveniles. Harvested earthworms for animal food were obtained from the earthworm-production beds in batches. From the experiment, it is found that, by application of this novel management technology, the earthworms developed with high-reproduction rates and the juveniles grew quickly with a relatively uniform body weight.
- Earthworms were harvested at the optimum time according to the S-shaped curve of weight increase of *E. fetida*, which made vermiculture with high-population densities efficient. Over a short culture period, overpopulation of earthworms in the bed was prevented, and high-economic profits were obtained.
- By renewal of breeding earthworms and periodic rotations between populations, old earthworms in beds or three generations in the same bed, which usually causes a natural decline in earthworm vitality, were prevented.

In recent years, as earthworms and vermicomposts have become widely used, not only in treatment of agricultural residues but also in the breakdown of sewage sludge and other industrial wastes, some new vermicomposting technologies have developed (Elvira et al. 1998; Dominguez et al. 2001; Edwards and Arancon 2004; Garg et al. 2006; Suthar 2007; Suthar and Singh 2008; Adi 2009; Khwairakpam 2009).

Using vermiculture, animal and vegetable wastes can be converted into animal protein for livestock feed (see Chapter 20), with an accompanied important significance in improvement of soils through addition of vermicompost. Earthworm protein meal has been proved to be a good replacement for fish meal (Edwards 1996; Sun 1995a; see Chapter 19) and a potential protein resource for human food (Sun et al. 1997). According to our results, 2.25 kg of crude protein per m^{-2} (5.95 lb. yd^{-2}) of substrate (from 10% crude protein in live weight) were produced per year, a much larger amount than from any other system. It may be concluded that these vermiculture technologies are feasible both technically and economically. Vermicomposts were reported to contain 3.43% humus, 0.184% total nitrogen, 0.248% total phosphorus, and 29.93% organic matter and have been used in China as an organic fertilizer in crops and vegetable and flower production with excellent results (Huang and Zhang 1980; Curry 1988; Sun 1993) and as an ingredient of animal feed (Sun 1997, 2003).

REFERENCES

Adi, A. J., and Z. M. Noor. 2009. Waste recycling: Utilization of coffee grounds and kitchen waste in vermicomposting. *Biores. Technol.* 100: 1027–1030.

Brun, J. J., et al. 1987. Biostimulation: perspectives et limits de L 'amelioration biologique des sols par stimulation ou introduction d'especes lombricienes. *Rev. Ecol. Biol. Sol* 24: 687–701.

Curry, J. P. 1988. The ecology of earthworms in reclaimed soils and their influence on soil fertility. In *Earthworms in Waste and Environment Management*, ed. C. A. Edwards and E. F. Neuhauser, 251–261. SPB Academic Publishing, the Hague, the Netherlands.

Dominguez, J., A. C. Edwards, and J. Ashby. 2001. The biology and population dynamics of *Eudrilus eugeniae* (Kinberg) (Oligochaeta) in cattle waste solids. *Pedobiologia* 45: 341–353.

Edwards, C. A. 1988. Breakdown of Animal, Vegetable and Industrial Organic Wastes by Earthworms. In *Earthworms in Waste and Environmental Management*, ed. C. A. Edwards and E. F. Neuhauser, 21–31. SPB Academic Publishing, the Hague, the Netherlands.

Edwards, C. A. 1996. *Soils: A New Global View*. By T. R. Paton, G. S. Humphreys, and P. B. Mitchell. Book review, *Q. Rev. Biol.* 71 (4): 596.

Edwards, C. A., and N. Q. Arancon. 2004. The use of earthworms in the breakdown of organic wastes to produce vermicomposts and animal feed protein. In *Earthworm Ecology*, ed. C. A. Edwards, 345–438. 2nd ed. CRC Press, Boca Raton, FL.

Elvira, C., L. Sampedro, E. Benítez, and R. Nogales. 1998. Vermicomposting of sludges from paper mill and dairy industries with *Eisenia andrei*: A pilot-scale study. *Biores. Technol.* 63: 205–211.

Gaddie, E. R., and E. D. Douglas. 1977. *Earthworms for Ecology and Profit*. Bookearthworm, California.

Garg, P., A. Gupta, and S. Satya. 2006. Vermicomposting of different types of waste using *Eisenia foetida:* A comparative study, *Biores. Technol.* 97: 391–395.

Huang, F.-Z., and Y.-Z. Zhang. 1980. Distribution of earthworm in China and their functions for substance cycles and nutrient accumulation in soil. *Shanxi Sci. Technol.* 18–24.

Huang, J., Q. Xu, and Z. Sun. 2007. Species abundance and zoogeographic affinities of Chinese terrestrial earthworms. *Eur. J. Soil Biol.* 43: S33–S38.

Khwairakpam, M. R. Bhargava. 2009. Vermitechnology for sewage sludge recycling. *J. Hazard. Mater.* 161: 948–954.

Price, J. S., and V. R. Phillips. 1990. An improved mechanical separator for removing live worms from worm-worked organic wastes. *Biol. Wastes* 33: 25–37.

SIIST. 1985. *Utilization and Cultivation of Earthworms*. Shanghai Academic Press.

Sun, Z.-J. 1993. Studies on a high-yielding technology of earthworm culture outdoors. *J. Laiyang Agric. College* 10: 228–231.

Sun, Z.-J. 1995a. *Earthworm production and nutritive value*. Domentijan, Novi Sad, Yugoslavia.

Sun, Z.-J. 1995b. Ecology, production and nutritive value of earthworm (*Eisenia fetida*) and use in animal feeds. PhD diss., University of Novi Sad, Yugoslavia.

Sun, Z.-J. 2003. *Vermiculture and Vermiprotein*. China Agri. Uni. Press. Beijing.

Sun, Z.-J., X.-C. Liu, L.-H. Sun, and C.-Y. Song. 1997. Earthworm as a potential protein resource. *Ecol. Food Nutr.* 36: 221–236.

Suthar, S. 2007. Vermicomposting potential of *Perionyx sansibaricus* (Perrier) in different waste materials. *Biores.. Technol.* 98: 1231–1237.

Suthar, S., and S. Singh. 2008. Feasibility of vermicomposting in biostabilization of sludge from a distillery industry. *Sci. Total Environ.* 394: 237–243.

Wang, H. 1983. A review: Vermiculture and utilization. In *A Collection of Vermiculture and Utilization Conference (Hebei, China)*, 6–14.

Zeng, Z. P., et al. 1982. *Earthworm culture.* Hubei People's Publishing House, Wuhan, China.

CHAPTER 35

Progress in Vermicomposting in Belarus, Russia, and Ukraine

Svetlana Maksimova

CONTENTS

I Introduction ... 565
II Development of Vermiculture in Russia and the Ukraine ... 566
A In Belarus ... 566
1 The Cherven Farm ... 567
2 The Brest Farm ... 567
B Gomel Region Project ... 568
C In Russia ... 572
D Vermiculture in Ukraine ... 575
References ... 576

I INTRODUCTION

Currently, the majority of the existing technologies for organic-waste recycling are not ecologically satisfactory and require large power inputs. As an alternative to existing methods, a completely new trend in the science of organic-waste bioconversion is called vermitechnology. Vermitechnology is based on the application of earthworms to organic-waste disposal. This complex method of bioconversion of environment-polluting industrial and agricultural organic wastes assumes producing two kinds of products: vermicompost and earthworm biomass. Vermiculture is a large and ecologically based business that results not only in a transition to ecological husbandry and production of high-quality food products but also helps in the development of a country's economy.

Vermicompost is the breakdown product of earthworms (synonyms *biohumus*, *vermihumus*). There are about 45 kilograms of N, P, and K in 1 ton of vermicompost, and for plant food value it greatly outperforms other organic fertilizers. Vermicompost also contains humic and fulvic acids, biologically active compounds, macro- and microelements, stimulating germination of seeds and increasing plant growth as well as crop yields and quality. Vermicomposts, enriched with nutrient elements, also contain a large and diverse community of microorganisms that play a central role in the management of soil properties and fertility.

II DEVELOPMENT OF VERMICULTURE IN RUSSIA AND THE UKRAINE

Vermitechnology has developed on the territory of three former USSR countries: Belarus, Russia and Ukraine. In 1990–1994 about 120 vermiculture farms were created in Ukraine, 136 in Russia, 10 in Belarus, and about 20 in other countries of the Commonwealth of Independent States (CIS).

A In Belarus

Vermiculture is a new direction for agricultural science in Belarus. The introduction and development of vermiculture are connected with unfavorable changes in ecology caused by humans and industry. The important role of earthworms in nature is well known. These soil animals do very important and useful work in the soil. They can convert one ton of manure into 600 kg (1323 lb) of organic fertilizer (dry substance), with 25–40% humus, 1–3% N, 1–3% P, 1–3% K, and necessary microelements contained in this fertilizer. The other 400 kg of organic substances are transformed into 100 kg (220 lb) of valuable protein in the form of the earthworm biomass. The earthworms break down the manure twice as quickly as bacteria alone.

The creation of vermiculture projects and the production of natural organic fertilizers are important links in making decisions on the problem of recycling soil organic matter. To decide this problem for Belarus soils, we must use vermicompost as an organic fertilizer to increase soil's biological activity. The vermicompost does not contain heavy metals or large amounts of nitrates. Vermicomposting of various organic wastes has become a large-scale industry, which promotes scientific studies of earthworm ecology and biology.

Belarus lacks up-to-date vermiculture production statistics. However, this type of activity is gaining popularity and offering a stable income potential to entrepreneurially minded people. There were about 100 small-scale vermiculture operations in the early 1990s, but 90% of them failed, mostly because they thought that the earthworm business was a get-rich-quick opportunity. There is much more interest and activity in vermicomposting nowadays than there was 10 years ago. Unfortunately, Belarus has only a short history in vermicompost earthworm research. Approximately five vermicompost producers are currently engaged in vermicompost production in Belarus; they produce at least 3000 tons of vermicompost per year. While some vermicompost is exported, the majority is used for domestic consumption. At this size, the vermiculture market in Belarus is still in its infancy. The vermiculture producers focus on two areas of activity—vermiculture production and management and vermiculture marketing.

1 The Cherven Farm

Vladimir Kulik, the owner, got started in the earthworm business less than 3 years ago, when he was searching for new opportunities to diversify his business. He entered into partnership with a pioneer of vermiculture technology in Belarus, who allocated him a plot at his collective farm to produce about 100 kg (220 lb) of vermicompost monthly. A failure in his first experiments in vermiculture occurred for a number of reasons, mostly from a lack of information and management within the organization. He contributed 1000 tons (100 t) of earthworms to a colleague, and some basic knowledge on how to start and run the vermiculture operations, in their partnership. Kulik invested his personal savings into purchasing the former collective farm's barn. Both of them found private individuals willing to invest $250,000 into their business. The farm currently has 15 vermicompost windrows in production and expects to harvest 100 tons of vermicompost in 2010. Facilities include 3500 m^2 (4186 yd^2) of enclosed warehouse-type space with concrete floors located throughout their 6 ha. (14.8 acres) of land. Additional facilities and land are available for expansion. Because the implementation of vermiculture technology requires careful planning and management to ensure it will be able to continually process organic materials, they have developed a business plan that calls for expanding the operation to produce 300–400 tons (300–400 t) of vermicompost monthly.

Realizing vermicompost's broad spectrum of beneficial agronomic properties and its ability to produce high crop yields, the local authorities have expressed interest in supporting the development of a vermiculture business in the region. They also, as a long-term strategy, plan to circulate the new technology among 30 farms whose employees will be trained by the Cherven producer.

2 The Brest Farm

This farm is a recently registered earthworm business. For the former director of the Brest Scientific and Technical Development Application Center, earthworms came into his plans by accident. At one of the meetings organized by the center, the owner met a Russian investor interested in developing a vermiculture business in Belarus and willing to sell earthworms to Belarusian earthworm growers. In May 2005, in what previously was a dairy barn, the Brest farm began to use 1000 earthworms to process organic materials as a conventional earthworm-growing operation: manure in, earthworms and earthworm castings or vermicompost out. In addition to some basic knowledge about the complexities of the earthworm business gained from the Russian company, the owner spent a lot of time gleaning vermicomposting information from the Internet; however, there are still some areas where he lacks knowledge. The short-term goal of the owner's new venture is to develop eight vermiculture units to use consistent sources of animal manure. It should be noted that across the country, as farms get larger and water-quality regulations stricter, the problem of what to do with animal manure worsens. Moreover, the development of vermicomposting operations has attracted considerable interest from the city Housing and Communal Services (i.e., waste authorities) and from the local government.

As for marketing, both the Cherven and Brest farms plans to focus on three primary markets for earthworms and earthworm castings or vermicomposts: local agricultural enterprises that are looking, with great interest, at the potential of earthworms to transform organic wastes into beneficial resources; home gardeners; and Russian and Belarusian marketers who use the earthworm castings as supplements with other soil amendments. The long-term goals of these two farms are to establish a firm position in the local market and to create a network of enterprises producing and supplying sufficient amounts of vermicompost at acceptable prices.

Both these agribusinesses are still using the relatively inefficient methods of ground beds or windrows to produce vermicomposts. Unfortunately, they have had limited opportunities to learn about new and innovative practices, since most of their vermiculture knowledge was obtained from relatively parochial sources that often lack the depth and objectivity to support improvement and change. Both farms plan to increase their annual production of vermicompost to 1000 tons, find reliable markets for their end products, and thus increase their total revenues from sales from 0 to 85,000,000 Br ($39,720).

B Gomel Region Project

One vermicomposting study in the territory of Gomel region (Dobrush) is very practical (Maksimova and Potylkin 2002, 2004, 2007). This region is contaminated by radionuclides, and the creation of such a technology could help to improve the health conditions of soils in the locality. Since 1998 Ltd "Polar" has processed organic wastes with the earthworms *Eisenia fetida* (Sav.). In 2000, 100 tons of vermicomposts were produced by Ltd. "Polar," and 200 tons in 2003. For the production of high-quality vermicompost they used several different kinds of organic wastes for vermiculture. For experiments, they chose the open method of vermicomposting. The base substance consisted of fermenting cattle manure (20%) with shallow straw height 15 cm (6 in). The beds were populated by *E. fetida*. They inoculated the earthworms once every 20–25 days (height of windrow 5 cm (1.6 in)). The optimal moisture content was 70–80%. In autumn, when the temperature was 7–8°C (44–47°F), the bed was covered with a thick layer of straw for winter. In spring, in April, they moved the earthworms to the new bed. After this they took away the vermicompost, and it was dried and divided into fractions on the vibration bolter. However, the effectiveness of this method was very low. They have tried to make several modifications of the base substances.

- Mix 1: Mix the base substance with wastes from the mixed fodder industry
- Mix 2: Mix the base substance with peat
- Mix 3: Mix the variant Mix 1 with peat
- Mix 4: Mix the base substance with peat and wastes from the cellulose-paper industry
- Mix 5: Mix the base substance with sawdust

The second component in the mixture composed 15–20% of the total. They put the mixtures into each bed during earthworms' active periods.

- Mix 1 was without problems, and the productivity of earthworms was higher than in the other mixes.
- Mix 2 had problems, but the problems were less than in the base substance and in Mix 5. The productivity of earthworms was lower than in Mix 1 but higher than in the other variants.
- Mix 3 had some problems, but these were less than in the base substance and in Mix 5. The productivity of earthworms was about the same as mix 2. The commercial quality of vermicompost in Mixes 1 and 2 was higher than in the other variants. The criteria were color and degree of granulation of the vermicomposts.
- Mix 4 had problems, but these were less than in the base substance and in Mix 5. The utilization of the wastes from the cellulose-paper industry improved the ecological situation in Dobrush.
- Mix 5 had problems (the same as in the base substance), and the vermicompost was of low commercial quality. Unfortunately, the time for sawdust to break down was very long. They sifted the vermicompost obtained from the base substance and harvested we had 20 % of vermicompost.

The maximal productivity of the earthworms was in Mix 1 so then propose to use such substances for future projects on growing earthworms. Then propose Mixes 2, 3, and 4 for commercial utilization. The best quality was in Mix 3, but Mix 4 was the cheapest to produce. Thus, different substances influence the quality of vermicompost and the productivity of the earthworms.

Due to ecological problems in agriculture, degradation of soils, and pollution of products, interest in alternative technologies in agriculture to create and apply safe methods of improving soil fertility and to obtain ecologically clean products for plant growth is increasing. These problems are very important in the Belarus area, which has the status of a zone of ecological disaster after the Chernobyl nuclear plant accident. The territory of Belarus belongs to the region of "risky agriculture" now.

A prospective method of all-round improvement in soil fertility and production of crops depends on a main application of vermicompost—a product obtained after the conversion of different organic wastes by earthworms. The final product, vermicompost, improves soil structure, stimulates plant growth, and facilitates ecologically clean crop production. Comparing to traditional thermophilic microbiological composting, earthworm composting is accompanied by deodorization and detoxification of the vermicompost, the organic material decomposition occurs two to five times faster, there is no need for artificial aeration, and the homogeneity of the final product increases.

The main goal of our investigations was to work out the best earthworm composting technology for the different organic wastes in our radioisotope-contaminated territory. The experiment to obtain vermicompost was done in wooden boxes on the territory of Ltd "Polar" (Dobrush, Gomel region), situated in the radioisotope-contaminated zone. To obtain vermicomposts the following organic wastes were used as the feedstocks: sewage biosolids, food wastes of vegetables, slime-lignin (wastes from a cellulose factory), cattle manures, pig manures, and chicken manures. The earthworm *E. fetida* (Savigny, 1826) was inoculated, with 1200 earthworms per square meter.

The vermicompost produced by the "Polar" process of vermicomposting (Figure 35.1) is a dark brown powder, and the content of vermicompost in it depends on the initial type of organic waste and the proportion between wastes and soil. The earthworms decreased the contents of radionuclides in soil and other toxic components by incorporating them into their body tissues. The amounts of radionuclides in the vermicomposts did not exceed the background values. The biological indices of earthworms and the indices of the germinating power of plants showed that the vermicomposts were not toxic to plants.

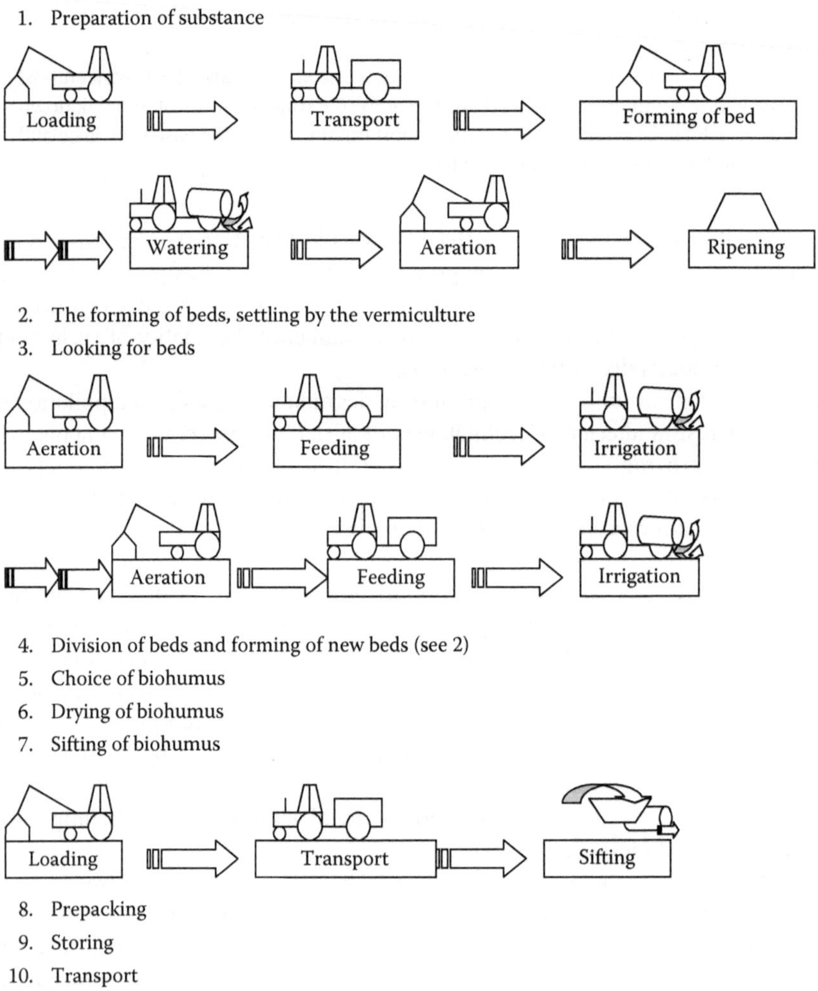

Figure 35.1 The technological process of industrial production of biohumus (Ltd "Polar").

In the Belarusian State Agricultural Academy (Tsyganov et al. 2004) the application of vermicompost to barley produced yield increases of 87 ton.ha^{-1} (35.2 t.acre^{-1}) (in comparison with the control plot), which coincides with the action of 25 ton.ha^{-1} (10 t.acre^{-1}) of bedding manure (the proportion of vermicompost and manure was 1:5). Application of vermicompost to winter rye variety Pukhovchnaka was similar in its efficiency to 30 ton.ha^{-1} (10 t.acre^{-1}) of humus (the proportion is 1:6). Yield increases were on average 70 ton.ha^{-1} (28.3 t.acre^{-1}) over 3 years. The research conducted using vermicompost on buckwheat, one of the most important grain crops, showed that its advance to the north of Belarus is quite feasible. The vegetation period was reduced, and field germination of seeds, plant survival, and the average mass of grains from 1 plant. Especially efficient was the application of vermicompost to early maturing potatoes. The yield increase of the potato variety Anosta was 627 ton. ha^{-1} (253.7 t.acre^{-1}). Vermicompost application to virus-free potato seed production improved seedling quality and the productivity of plants. Plants were noticeable for their greater height, and their increase in size and numbers of tubers (1.7–7.5 times greater). The coefficient of multiplication was observed to be 6.3–11.1% higher, and the tuber mass was 15.3–29.0% higher.

The application of vermicomposts influenced yields and the qualitative characteristics of flax fiber. Depending on the doses of the vermicompost applied, the seed yield increased by 17–30 ton.ha^{-1} (6.8–12.0 t.acre^{-1}), fiber yield by 22–30 ton.ha^{-1} (9–12 t.acre^{-1}), and the long fiber by 30–34 ton.ha^{-1} (12–14 t.acre^{-1}). The application of vermicompost for growing corn affected yields very positively. The corn yield was 130–190 dt.ha^{-1} after vermicompost applications. Similar results were obtained for vegetables: cucumbers, tomatoes, cabbages, and peppers. The plants reached their standard size 7–10 days earlier than control plants, and vegetable yields were raised by 7–10 kg.m^{-2} (12–18.4 lb.acre^{-1}) in the greenhouses. Tsyganov and Vildflush (2004) investigated the effectiveness of vermicompost applications for potatoes. Their effects were compared with the effect of manure and mineral fertilizer applications. The application of 5 ton.ha^{-1} (2 t.acre^{-1}) of vermicompost increased yields of potato tubers by 54%.ha^{-1} on average, and its effect was equal to the application of 50 ton.ha^{-1} (20 t.acre^{-1}) manure.

Zhelyazko (2004) with colleagues from the Belarusian State Agricultural Academy studied the preparation of organic fertilizers from manure slurry. At first, composting had to be started in warm periods. It was desirable that the thermal phase of organic-substance decomposition with compost heating to 55 to 60°C (131–140°F) took place at this time. A second, mesophilic phase can take place in winter. But at this time, to prevent freezing and to end the process of maturing, compost clamps should be covered by a heat insulation layer 50–80 cm (20–31.5 in) of straw, soil, turf, or sawdust.

Today, methods of vermicompost production in the CIS countries using the red Californian earthworm are limited, as far as is known, to seasonal activities, that is, the spring and summer period. Naturally, this seasonality sharply restricts capacities to develop any high-technology and industrially powerful production, let alone its mechanization and automation. With this regard, a wide-scale introduction of all-year-round industrial vermiculture complexes in Belarus is a reliable basis for the development of an integral system for future mass enrichment of croplands with organic fertilizers, especially for the remediation of desertified lands.

During the last decade, a number of Belarusian firms have accumulated sufficient experience in the seasonal (field) production of vermicomposts using the red earthworm. However, in practice, precisely this seasonality in the production of such unconventional and highly valuable organic fertilizer has prevented the development of a closed technological cycle and, consequently, has precluded any strong increases in annual output. Now we have found new varieties of manure earthworms that are adapted to local environmental conditions. They have been called the *Byelo-russian ploughman* (Mikhailova et al. 2007).

In 2001, in Belarus, a group of specialists (Kluchenovich et al. 2002, 2004) developed, and patented in "Belgospatent," an industrial method for organic-waste processing, as well as a modular system for year-round production of vermicompost using seasonal vermicomposting and vermiculturing performance data. In contrast to a well-known Russian patent, the given method allows a sharp increase (by a factor of two or three) in both vermicompost and vermiculture productivity through introduction of a processing cycle sequence and, accordingly, a sequence of technologically different manufacturing operations. As a technologically and constructively independent module, the system itself fits naturally into the closed process chain.

The industrial system consists of a line of modules, divisible by a complete processing cycle (digestion) of the organic mass by earthworms, and it is provided with not only mechanical loading/unloading systems but also with aeration and moisture and heat supply systems, automatic flow control system, and so on. The system capacity may vary over a wide range, that is, depending on the actual volume of organic-matter feedstock. According to the design data, the optimum capacity will be 20,000–40,000 megatons (MT) in terms of annual output of finished product (vermicompost). Lump-sum construction costs (excluding civil engineering works) could approximate 5–9 million units of money, but the payback period will be no more than 1.5–2 years.

C In Russia

The corporation "Green PIK" Ltd is the biggest vermicomposting corporation in Russia (Kovrov, Vladimir region). From 2000, it has produced large amounts of vermicompost. For 15 years they used the earthworms *E. fetida* of the Vladimir hybrid line—"Vladimir prospector." This hybrid was obtained by Professor A. M. Igonin. It was a result of breeding between southern (Chuiskiy) and northern (Vladimir) earthworm populations in 1982.

The composting earthworm species *E. fetida* (Savigny, 1826) is the most widely used commercial earthworm in Russia. This species seems able to adapt to a wide range of environmental changes and feedstocks. In most cases scientists use a subspecies of the compost earthworm *E. fetida* that has the commercial name "the red Californian hybrid" or "the red Californian worm." On the territory of the former USSR a number of varieties of *E. fetida* have been raised: the Vladimir, the Morevsk, the Obninsk, the Obolensk, and the Podolsk varieties (lines). However, at present, the red Californian hybrid is used widely in Russian vermitechnologies, although the ecological production importance of regional varieties, bred for the different conditions of the Russian Federation, can scarcely be exaggerated.

Vermicomposting is one of the environmentally safe ways of processing digested sewage sludges and other urban wastes (Vishnaykov et al. 2002; Pravkina et al. 2007). This method helps to solve two serious environmental problems. One of them is partial utilization of the organic wastes contained in digested sewage sludge, and the other is production of useful secondary materials. These urban have discovered that sewage sludge–based vermicomposts have no toxic effects on spring wheat seeds. Kasatikov and Nefyodova (personal communication) studied the use of manufacturing wastes from tanneries as a substrate for vermiculture. The physical characteristics of residues of manufacturing wastes generated during the period of airing at the tannery are the following: It appears on homogeneous substance of dark gray color, nonconglomerating, well accessible for clamping, with a moisture control of 72–75% and density of 1 ton.m^3. The basic substrate for vermicomposting was prepared on the base of row mixed waste (RMW) mixed with an organic filler in different proportions. Results indicated that wastes of tanneries can be used for production of vermicompost. This technology could be used worldwide at tanneries, but at first wastes should the separated into chrome-free and chrome-containing wastes. The same workers showed positive influences of sewage sludge–based vermicomposts on the agroecological properties of field agroecosystems. When vermicomposting sewage sludge, they reported organics and total N reduction as well as labile phosphorus P_2O_5, exchange potassium K_2O, and N-NO concentration increases. They also reported reductions in the total amount and distribution of mobile heavy metals: Mn, Ni, Pb, Cr, and Cd. The use of sewage sludge–based vermicompost was effective for improving the agrochemical properties of sod-podzol soil and grain productivity.

Korsunova et al. (2002) studied vermicomposts, based on a 3-month vermicomposting cycle of coniferous forestry wastes (e.g., sawdust). The first stage was microbial degradation of lignocellulose complexes with the development of active isolates of cellulose-destroying fungi, bacteria, and actinomyces. The second stage was vermicomposting. The effect of such a vermicompost on the productivity and quality of agricultural crops was studied in field experiments on potatoes and oats. Vermicompost introduction increased potato yields by 32–67%, and oats by 12–17% compared with the control. They reported that vermicomposts influenced the quality of potato tubers and green mass of oats, as well as the content of dry matter, sugar, cellulose, digestible protein, forage units, and extent of injury inflicted by pests. With vermicomposts, the rate of maturation and durability of potato tubers was accelerated; their protein content increased by 2.8–3.2% and their resistance to injuries by pests increased as well; also, the concentration of nonprotein N decreased. In experiments with oats increases in digestible proteins, carotin, Ca, and P in the green biomass were reported.

The efficiency of vermicompost usage in floriculture has been studied in open and covered crops. The outdoor crops in open ground were gladioli, with application rates of 300 g.m^{-2} (8.8 oz.yd^{-2}). Earlier flowering, larger numbers of flower buds, and more flowers on one plant have been noted. Stepanova and Stepanov (2002) carried out research on tomatoes, testing different vermicompost application rates and methods of application. The vermicompost tested was from cattle manure. The vermicompost was introduced locally and randomly, with doses from 100 g (3.5 oz) per hole up to 5 $kg.m^{-2}$ (9.2 $yd.m^{-2}$), on planted out tomato seedlings. The productivity of tomatoes depended on the amount of vermicompost applied. The use of vermicompost demonstrates its high efficiency in cultivation of tomatoes in greenhouses. When applying vermicompost at 500 g (17.6 $oz.m^2$) per hole, the productivity of tomatoes increased by 35%.

Prosyannikov et al. (2002) used vermicompost for growing cucumbers, tomatoes, and pepper seedlings. The vermicompost had the following chemical composition: moisture content 50%, organic-matter content 6.3%, pH 6.8–7.2, total N 1.4%–1.5%, P 0.71–0.8%, and K 0.50–0.54%. They used vermicompost in greenhouse soil at the rate of 30 g (1.0 oz) (1 dose), 60 g (2.0 oz) (2 doses), 90 g (3.0 oz) (3 doses), and 120 g (40.0 oz) (4 doses) in every planting hole. The use of vermicompost for planting cucumber seedlings at a rate of 30–120 g (1.0–4.2 oz) in every planting hole, and tomato seedlings at the rate of 60–120 g (2.1–4.2 oz) in every planting hole in the greenhouse soil, increased their crop yields considerably and did not decrease the quality of the crops.

Babenko et al. (2004) investigated the application of vermicomposts for growing seed potatoes. Data analyses showed that using vermicompost influenced the growth and development of potato tubers positively. On the whole, yield analyses showed that using vermicompost increased the average weight of tubers and relative numbers of large tubers. Kuzmina et al. (2004) came to the conclusion that pig manure vermicompost applied at 8 $ton.ha^{-1}$ (3.2 $t.acre^{-1}$) was the best for lettuce and radish growth on a sandy dernovo-podzolic soil.

Godokov et al. (2004) investigated the influence of different doses of vermicompost on the bioremediation of polluted soils near a petrol station. The vermicompost was applied at different doses—15, 30, and 45 $t.ha^{-1}$. To improve and to recover soil fertility and increase the productivity of perennial grasses, the optimum dose of vermicompost was 30 $t.ha^{-1}$. Research showed positive influences of sewage sludge–based vermicomposts on the agroecological properties of field crops (Kasatikov and Shabardina 2007). When vermicomposting sewage sludge, they reported organic-matter and total N reduction as well as labile phosphorus (P_2O_5), exchangeable potassium K_2O, and N-NO_3 concentration increases. They also reported reduction in total amounts and distribution of mobile heavy metals: Mn, Ni, Pb, Cr, and Cd. The use of sewage sludge-based vermicompost is effective for improving the agrochemical properties of sod-podzol soil and grain yields.

Problems related to recycling of farm and household wastes are some of the most significant issues for large cities. At present, paper and cardboard in household waste products are burned, thereby polluting the environment. The polluted paper for recycling can be processed with the help of vermiculture (Idrisova et al. 2004). As a result of processing of the polluted paper for recycling, vermiculture produces vermicomposts that can be used as fertilizer.

Boguspaev et al. (2004) investigated methods of detoxification by vermicomposts of 1,1 dimethylhydrazine (ADMH, a component of liquid rocket fuel) in polluted soils and restoration of biota in Russia and in Kazakhstan. Vermicompost was produced with standard technologies from cattle manure. In experiments with addition of vermicompost to polluted soils in the ratio 50:50 (vermicompost:soil), concentrations of ADMH were reduced to 67.5% during the first 24 hours of incubation.

D Vermiculture in Ukraine

In 1989, the first vermiculture farm was set up in Ivano-Frankovsk (Ukraine). Now the corporation Bioconversion is the biggest biotechnological organization in Ukraine. The complex methods used for organic-waste bioconversion provide an opportunity to solve conservation problems that are connected with environmental pollution in Ukraine. This method opens wide possibilities for using vermiculture in the fields of agriculture, medicine, and cattle breeding, as well as for bioremediation of polluted areas using a preparation based on BH and microorganisms, which destroys hydrocarbons xenobiotics. Not only are polluting hydrocarbons eliminated from total oil hydrocarbons but vermicomposts also rehabilitated soil-formation processes.

Gorodny et al. (1994) studied the technology of vermicompost production from pig manure. The substrate was 60% pig manure, 30% straw, and 10% soil and sand. This substrate must be composted before using for 1.5–4 months. Melnik and Karpets (1990) investigated the influence of vermicompost on the growth and development of long-fibered flax. Their results showed that seed germination was 97% on the 5th day and 100% on the 7th day with vermicompost applications (in the control, it was 17% on the 5th day and 72% on the 7th day). Investigators from Dnepropetrovsk (Kulik et al. 1994) studied vermicomposting of different substrates (mushroom substrate, plant residues with active silt and lignin). The maximum earthworm weight achieved was in the substrate with plant residues and active silt, and the minimum earthworm weight achieved was in the mushroom substrate.

Scientists from the Ukraine and Moldova (Karageorgij and Pogrebniak 1994) reported that the application of vermicompost 1.5 ton.ha^{-1} (0.6 t.acre^{-1}) increased cucumber yields by 13–24%, pepper yields by 9–19%, and onion yields by 34–36%. The localized application of vermicompost increased cucumber yields by 16–30%, pepper yields by 15–25%, and onion yields by 34–46%. Melnik and Gutsuliak (1994) reported that the optimal application rate of vermicompost for cabbage was 9 t.ha^{-1} and for peppers 6 ton.ha^{-1} (2.4 t.acre^{-1}). Such application rates increased the yields of vegetables and decreased their nitrate contents by 25–35%.

There are several reports of utilization of sewage for vermicomposting in the Ukraine (N. M. Shurova 1992; S. V. Shurova 1992; Slobodian 1992; Yutina et al. 1992). The optimal application rate of vermicompost for growing long-fibered flax was from 3–5 ton.ha^{-1} (1.2–2.0 t.acre^{-1}) in the Ukraine (Melnik and Kovalev 1992). Field results showed that vermicomposts stimulated growth and development of plants and increased plant biomass and height. A dose of 12 ton.ha^{-1} (4.8 t.acre^{-1}) of vermicompost increased the biological activity of soil by as much as a dose of 40 ton.ha^{-1} (16.2 t.acre^{-1}) of manure did (Shikula et al. 1994). The local application of vermicompost at rates of 2–4 ton.ha^{-1} (0.8–1.6 t.acre^{-1}) favored the development of potatoes (Karpenko 1994). The local application of vermicompost at a dose of 4.5 ton.ha^{-1} (1.8 t.acre^{-1})

was equal to application of 25 ton.ha^{-1} (10.1 t.acre^{-1}) of manure (Moroz and Rudenko 1994).

Cattle, pig, poultry, and horse manures; wastes from the paper industry; sewage; and urban wastes can all be used for vermicomposting. Outdoor windrow beds and bins have been used extensively in Belarus, Russia, and Ukraine. Vermicompost provides many benefits for agricultural soils, including increased ability to retain moisture, better nutrient-holding capacity, better soil structure, and higher levels of microbial activity.

REFERENCES

Babenko, A.S., Karnachouk, R.A., and Jakimov, J.E. 2004. The use of vermicompost for growing seed potato. *Proceedings of the 2nd International Conference "Earthworms and Soil Fertility*" (Russia, Vladimir), 199–200.

Boguspaev, K.K., Alimova, N.S., Batyrbekova, S.E., Baishev, K.S., and Nauryzbaev. 2004. Bioremediation of Kazkhstan soil polluted by 1,1dimethilhydrazine with use of vermicompost. *Proceedings of the 2nd International Conference "Earthworms and Soil Fertility"* (Russia, Vladimir), 241–242.

Gorodny, N.M., Serduk, A.G., and Bykin, A.V. 1994. The technological aspect of pig manure processing in vermicompost. *Bioconversion of organic wastes and preservation of surroundings* (Ukraine, Kiev), 15–18.

Idrisova, D.R., Ageev, A.Y., and Myakutiua, N. 2004. Biotechnological way of process paper recycling. *Proceedings of the 2nd International Conference "Earthworms and Soil Fertility"* (Russia, Vladimir), 121.

Karageorgij, V.V., and Pogrebniak, A.P. 1994. The vermicompost use in the link of crop rotation. *International Congress of Bioconversion of Organic Wastes* (Russia, Moscow), 15.

Karpenko, N.A. 1994. Vermicompost as factor of stability of potato production. *Bioconversion of organic wastes and preservation of surroundings* (Ukraine, Kiev), 66.

Kasatikov, V.A., and Nefyodova, E.A. 2004. Studies of using residue of manufacturing wastes of tanneries as substratum for vermiculture. *Proceedings of the 2nd International Conference "Earthworms and Soil Fertility"* (Russia, Vladimir), 112.

Kasatikov, V.A., and Shabardina, N.P. 2007. The influence of vermicomposting on sewage sludge on their agroecological properties and agrochemical effectiveness. *Vermicomposting and vermiculturing as base of ecological agriculture in the XXI century: Problems, perspectives, achievements: Collection of scientific articles*, (Belarus, Minsk), 104–106.

Kluchenovich, S.A., Spesivtsev, V.T., and Tubolets, A.A. 2002. Industrial production of vermihumus—a reliable way to high profitability. *Proceedings of the 1st International Conference "Earthworms and Soil Fertility"* (Russia, Vladimir), 227–228.

Kluchenovich, S.A., Spesivtsev, V.T., and Tubolets, A.A. 2004. Industrial vermiproduction: Intensification, process, equipment. *Proceedings of the 2nd International Conference "Earthworms and Soil Fertility"* (Russia, Vladimir), 126.

Korsunova, T.M., Zhigzhitova, I.A., Altayev, A.A., Lesnikova, L.V., and Parkhaneyeva, V.Y. 2002. Prospects of use of vermicomposts in ecologization of agriculture within the Baikal region. *Proceedings of the 1st International Conference "Earthworms and Soil Fertility"* (Russia, Vladimir), 257–259.

Kulik, A.P., Gaprmash, S.N., Buleiko, S.Y., and Chernyshova, T.P. 1994. The influence of protein content in the substrate on the vermiculture development. *International Congress of Bioconversion of Organic Wastes* (Russia, Moscow), 7–8.

Kuzmina, N.V., Nikiforova, O.V., and Verkhovtseva, N.V. 2004. The influence of vermicompost of dernovo-podzolic soil properties and agrocenosis productivity. *Proceedings of the 2nd International Conference "Earthworms and Soil Fertility"* (Russia, Vladimir), 194–195.

Maksimova, S.L., and Potylkin, V.A. 2002. Biohumus production in Belarus. *Proceedings of the 1st International Conference "Earthworms and Soil Fertility"* (Russia, Vladimir), 46–48.

Maksimova, S.L., and Potylkin, V.A. 2004. The technology of vermicomposting in Belarus. *Proceedings of the 2nd International Conference "Earthworms and Soil Fertility"* (Russia, Vladimir), 89–90.

Maksimova, S.L., and Potylkin, V.A. 2007. The using of different kinds of substratum under vermiculting. *Vermicomposting and vermiculturing as base of ecological agriculture in the XXI century: Problems, perspectives, achievements: Collection of scientific articles* (Belarus, Minsk), 101–102.

Melnik, I.A., and Gutsuliak, V.D. 1994. Biohumus and vegetable yield. *International Congress of Bioconversion of Organic Wastes* (Russia, Moscow), 15–16.

Melnik, I.A., and Karpets, I.P. 1990. Vermiculture and biohumus is her product. *Chemicalization of Agriculture* 10, 15–17.

Melnik, I.A., and Kovalev, V.V. 1992. Application of biohumus as fertilizer for long-fibred flax. *Bioconversion of organic wastes and ecological improving from a health point of view of settlement* (Ukraine, Ivano-Frankovsk), 27–29.

Mikhailova, M.E., Maksimova, S.L., Golenchenko, S.G., Volchok, N.M., and Kamysh, N.A. 2007. Estimation on the state of vermiculture gene pool in compost worm *Eisenia foetida* (Sav.) using molecular-genetic methods. *Vermicomposting and Vermiculturing as Base of Ecological Agriculture in the XXI century: Problems, Perspectives, Achievements: Collection of Scientific Articles* (Belarus, Minsk), 125–127.

Moroz, S.I., and Rudenko, G.S. 1994. Effectivity of vermicompost on potato growing in the conditions of Ukranian Polessie. *Bioconversion of Organic Wastes and Preservation of Surroundings* (Ukraine, Kiev), 67.

Pravkina, S.L., Igoshina, K.A., Levin, V.I., and Khabarova, T.V. 2007. Studying of vermicomposting influence on phytotoxicity of sewage sludge wastes. *Vermicomposting and Vermiculturing as Base of Ecological Agriculture in the XXI century: Problems, Perspectives, Achievements: Collection of Scientific Articles* (Belarus, Minsk), 136.

Prosyannikov, E.V., Sychev, S.M., Mishchenko, V.O., Orlov, A.V., and Volkov, A.V. 2002. Use of the humic preparations for the cultivation of cucumber, tomato and pepper planted in protected soil SPK "Agricultural firm "Kultura." *Proceedings of the 1st International Conference "Earthworms and Soil Fertility"* (Russia, Vladimir), 261–263.

Shikula, N.K., Fantukh, V.C., and Naumenko, V.I. 1994. The influence of vermicompost on the reproduction of the forest soils and raising of crop capacity of winter wheat. *Bioconversion of organic wastes and preservation of surroundings* (Ukraine, Kiev), 54–55.

Shurova, N.M. 1992. The perspectives of use of Oligochaeta for utilization of sewage sludge wastes. *Bioconversion of organic wastes and ecological improving from a health point of view of settlement* (Ukraine, Ivano-Frankovsk), 7.

Shurova, S.V. 1992. Use of Oligochaeta for utilization of sewage sludge wastes. *Bioconversion of organic wastes and ecological improving from a health point of view of settlement* (Ukraine, Ivano-Frankovsk), 8.

Slobodian, V.A. 1992. Spade-works on the processing of sewage sludge wastes by vermiculture. *Bioconversion of organic wastes and ecological improving from a health point of view of settlement* (Ukraine, Ivano-Frankovsk), 58–60.

Stepanova, D.I., and Stepanov, A.I. 2002. Use of vermicompost in growing of tomatoes in protected ground in conditions of the central Yakutia. *Proceedings of the 1st International Conference "Earthworms and Soil Fertility"* (Russia, Vladimir), 259–261.

Tsyganov, A.R., Ivanitsky, N.P., Vildfloosh, I.R., and Saskevich, P.A. 2004. Vermicompost production in conditions of the north-eastern part of Belarus. *Proceedings of the 2nd International Conference "Earthworms and Soil Fertility"* (Russia, Vladimir), 87.

Tsyganov, A.R., and Vildflush, I.R. 2004. The effectiveness of vermicompost application when growing grain crops and potatoes on sward-podzolic soils of the Republic of Belarus. *Proceedings of the 2nd International Conference "Earthworms and Soil Fertility"* (Russia, Vladimir), 185–186.

Vishnaykov, A.E., Popov, A.I., Gorshkov, S.I., and Nikolaenkova, N.E. 2002. Vermicomposting of sewage sludge. *Proceedings of the 1st International Conference "Earthworms and Soil Fertility"* (Russia, Vladimir), 220–225.

Yutina, A.S., Pshenichnaya, S.I., and Romatsov, A.P. 1992. Use of fertilizers on the base of sewage sludge wastes in the agriculture of Ukraine. *Bioconversion of organic wastes and ecological improving from a health point of view of settlement* (Ukraine, Ivano-Frankovsk), 5.

Zhelyazko, V.I., Mickhalchenko, N.N., and Kopytovski, V.V. 2004. Organic fertiliser preparation from manure slurry. *Proceedings of the 2nd International Conference "Earthworms and Soil Fertility"* (Russia, Vladimir), 79–80.

Author Index

A

Abbasi, S. A., 13, 117
Abbott, I., 440
Abdul, R., 509
Addabdo, T. D., 223
Aihara, H., 473
Aira, M., 13, 55, 56, 409, 410, 411, 413, 414, 415, 416, 409–421
Akhtar, M., 223
Albaladejo, J., 147
Albanell, E., 104–105, 122, 130
Albuzio, A., 122–123
Aldag, R., 121
Alfano, G., 185, 204
Alis, R., 477
Allan, D. L., 146
Allen, M. F., 103, 266, 270
Altieri, M. A., 210
Alvarez-Bernal, D., 513
Alvarez, M. R., 534, 535, 536, 537, 539, 533–543
Alves, M. R., 295
Alviter Aguilar, A., 513
Amaravadi, L., 168
Amir, S., 266
Appelhof, M., 4, 67–78, 69, 85, 312, 316, 444
Aragonés, C. R., 533–543
Arancon, N., 103, 104, 105, 106, 107, 108, 123, 103–128, 129–151, 183–204, 209–233, 235–248, 287–301
Aranda-Delgado, E., 508, 509, 510, 511, 512, 517, 519, 522, 526, 507–531
Arnold, J., 415
Arruda, M. A. Z., 273, 274
Arshad, M., 121
Aryantha, I. N. P., 486
Asciutto, K., 168, 172, 173
Ashwood, T. L., 268
Askar, A. M., 215, 183–207, 217, 209–233, 235–248
Athani, S., 133
Athokpam, H., 461
Atiyeh, R., 13, 106, 107, 109, 115, 117, 120, 121, 122, 236, 293, 295, 296, 297, 411, 414
Atkinson, D., 210
Atlavinyte, O., 121
Aviad, T., 122

B

Bååth, E., 57, 59
Bachman, G., 115, 118, 120
Baghel, S. S., 457
Bailón, M., 507–533
Baker, G. H., 336, 345, 438, 440
Balamurugan, M., 349, 350, 351, 352, 355, 360, 349–367
Balasubramanian, S., 463
Bano, K., 34, 455
Barbour, J., 376
Barcelo, P. M., 476, 477
Barik, T., 456
Barois, I., 507, 508, 511, 512–526, 507–531
Barrett, T. J., 2, 375, 440
Barros, P. M., 274
Bartlett, R. J., 122
Basker, A., 273
Bater, J. E., 345
Beauchamp, J. J., 268
Becker, J. O., 175
Beddard, F. E., 44, 484
Beerman, J., 376
Beeson, R. C., 294
Bengtsson, G., 32
Berg, G., 174
Bhadoria, P. B. S., 168, 169, 462
Bhat, J. V., 453, 454, 463
Bhonwong, A., 222
Bilej, M., 354, 355
Bilgrami, A. L., 225
Bindesbøl, A, 275
Binet, F., 166
Bisht, S. P. S., 33
Blakemore, R. J., 44, 439, 440, 441
Boehm, M. J., 173, 264
Bogdanov, P., 309, 310, 312, 316, 369–389
Boguspaev, K. K., 574
Bolton, H., 147
Bora, L. C., 175, 287, 297, 299, 462
Borneman, J., 175
Boström, U., 339
Bouché, M. B., 2–3, 19, 27, 28, 29, 41, 57, 59, 104, 315, 338, 508, 509, 510
Boulter, J, 204
Bovalo, F., 123
Bowen, G., 68
Brata, B., 492
Brinton, W. F., 154
Brun, J. J., 547, 559
Brüsewitz, G., 258
Buchanan, M. A., 130

Buckerfield, J. C., 103–128, 104, 105, 106, 123, 132, 437, 438, 439, 440, 442, 444, 448, 437–452
Bundy, J. G., 276
Businelli, M., 105
Butt, K. R., 13, 335, 345, 346, 423, 424, 425, 426, 428, 430, 432, 433, 335–348, 423–435
Bwamiki, D. P., 104

C

Calero, B. J., 537
Canellas, L. P., 123, 236
Capistrán, F., 526
Carlile, W. R., 106, 118, 146
Carlos, R. A., 534
Carmody, F., 304, 386
Carrasquero Durán, A., 273–274
Carrillo, O., 537
Carver, D., 82
Castillo, R., 5, 10, 537, 539
Catalan, G. I., 469–470
Cervantes, I., 509
Chandra, T. S., 130
Chang, C. H., 42
Chan, P. L. S., 107
Chaoui, H. I., 168, 170, 185
Chellemi, D. O., 210
Cheng, H., 264
Chen, J. H., 122, 173, 184
Chhonkar, P. K., 456
Chopde, M. R., 461
Christie, B., 82, 174, 303–322
Chrzanowski, G., 222
Clapperton, M. J., 167
Clegg, C. D., 168
Cliver, D.O., 258
Cluzeau, D., 32, 33
Coeurdassier, M., 274
Collier, J., 4
Collins, D. J., 345
Cook, B. D., 146
Cooper, E. L., 258, 349, 350, 351, 352, 353, 355, 361, 486, 349–367
Corpuz, R. S., 475
Correll, R. L., 167
Cossens, G. G., 337, 340
Crowe, M., 68
Csuzdi, C., 44
Cuevas, R., 105
Culliney, T. W., 221
Curiel, P., 519
Curl, E. A., 165
Curry, J. P., 33, 336, 337, 345, 547

D

Dacayo, J. B., 472
Dales, R. P., 258
Dammerman, K. W., 483
Darwin, C., 2, 10, 53, 80, 91, 130, 337, 423, 427
Davidson, S. K., 167
Davis, J., 379
Davoren, C. W., 167, 168
DeForest, D. K., 269
Deka, S. N., 175
Delgado-Moreno, L., 416
Dendooven, L., 513
Devi, D., 133
Devliegher, W., 272
Diaz, E., 147
Domíguez, J., 1–9, 11–25, 27–40, 16, 18, 20, 22, 34, 37, 53–66, 409–417, 484, 485
Doubé, B. M., 167, 168
Douglas, D. E., 375, 559
Dowling, D. N., 204
Drake, H. L., 62, 166
Dubash, P. J., 453
Dunger, W., 338

E

Edwards, C. A., 1–10, 11–25, 27–40, 79–90, 91–102, 103–128, 129–151, 153–163, 183–207, 209–233, 235–248, 249–261, 287–301, 312, 303–322, 323, 324
Egert, M., 166, 168
Eijsackers, H. J. P., 276
El-Akkad, S., 222
Eleftherianos, I., 222
Ellery, D. J., 497, 498, 499, 500, 501, 502, 503, 504, 505, 497–506
Elmer, W. H., 167
Elvira, C., 13, 32–33, 104–105, 122, 411, 412
Enami, Y., 167
Epstein, E., 264, 266
Erhart, E., 173
Espinoza, G. L., 513

F

Fabrizio, M. C., 168, 172
Fahy, P. C., 166, 204
Farquhar, B. A., 438
Fattorini, M., 336
Fayolle, L., 32, 33
Felicim, M., 122
Felton, G. W., 222
Fernandes, R. B. A., 274

Fernández-Bayo, J. D., 411, 415, 416
Ferrandino, F. J., 167
Ferris, H., 165
Fieldson, R. S., 332, 336, 424
Filipek-Mazur, B., 273
Finstein, M. S., 16
Flores, Cruz, M., 513
Frankenberger, W. T., 121
Franz, M. F., 3
Fraser, D. G., 147
Frederickson, J., 266, 267, 432, 433
Fritz, I., 236
Frostegård, Å., 57
Fuentes, A., 266
Fujii, K., 21
Fulekar, M. H., 274
Fu, T., 356
Fu-zhen, H., 104

G

Gaddie, R. E., 375
Gadepalle, V. P., 274
Gaime-Perraud, I., 518
Gajalakshmi, S., 13, 117
Gallaher, R. N., 223
Galli, E., 120
Garcia, C., 122, 271
Garcia-Gomez, R. C., 236
Garg, V. K., 13, 271, 273, 456
Gates, G. E., 32, 34, 324, 484
George, S., 462
Ghosh, C., 455
Ghuge, T. D., 460
Goenadi, D. H., 122
Golueke, C. G., 13
Gondek, K., 273, 274
González, A., 533
Gootas, H. B., 13
Gopal, M., 130, 168
Gorodny, N. M., 575
Graff, O., 3, 121, 324, 325
Graham, E. C., 289
Grappelli, A., 120
Grdisa, T., 351, 355
Griffiths, D. A., 107
Guastella, G. S., 168
Guerrero, R. D., 5, 10, 330, 470, 471, 472, 473, 474, 475, 476, 477, 469–480
Guggari, A. K., 460
Guimaraěs, A., 41–52
Gunadi, B., 481–496
Gupta, A., 13, 271, 273
Gutierrez-Miceli, F. A., 117, 295
Gutsuliak, V. D., 575

H

Hageman, R. H., 121
Hahn, G. E., 222
Hallatt, L., 35
Halle, S., 336
Hamelers, B., 270
Hamoui, V., 339
Han, D. Y., 204
Hansen, G. A., 486
Harrison, R. B., 276
Hartenstein, R., 3, 4, 10, 32, 166, 258, 272, 323, 386
Harwood, M., 330, 332
Hatanaka, K., 489
Haug, R. T., 13, 17
Haukioja, E., 222
Hawida, S., 222
Hawke, B. G., 168
Hayes, M. H. B., 123
Hebert, P. D., 42
Hegstrom, L. J., 269
Heimpel, A. M., 342
Henares, P. T., 475
Hendriksen, K., 258
Hendrix, P. F., 336
Henry, C. L., 276
Hernández-Castellanos, B., 507–531
Hernández Sánchez, M. L., 513
Hilton, J. W., 330
Hinckley, L. S., 251
Hirigoyenberry, F., 355
Hoitink, H. A. J., 166, 184, 204
Hopkins, W. G., 118, 269, 275
Hopp, H., 68, 78
Hori, M., 350
Horn, M. A., 62, 166
Hoshii, H., 324, 325
Howard, A., 165
Howell, G., 432, 433
Hrzenjak, T. M., 352, 353, 354, 355, 356
Huang, F. Z., 546
Huber, D. M., 165
Huelsman, M. F., 185
Huerta-Lwanga, E., 513
Huhta, V., 251, 320, 338
Hüner, N. P. A., 118

I

Idrisova, D. R., 574
Ilyas, M., 492

Irissón, S., 513
Ismail, S. A., 351, 352

J

Jack, A. L., 165–182
Jadia, C. D., 274
Jaenike, J., 43, 516
Jaffe, M., 44
James, S. W., 41–52
Janssen, C. R., 266, 270
Jenkinson, D. S., 147
Jensen, J., 303–322
Johnston, A. M., 104
Jonker, M. J., 276
Jordáo, C. P., 274, 295
Jordão, C. P., 274

K

Kai, T. C., 497–506
Kalaç, Y., 258
Kale, R. D., 5, 10, 34, 35, 146, 273, 454, 455, 459, 460, 463, 491, 453–467
Kanwar, J. S., 453
Kaplan, D. L., 3, 10
Karageorgij, V. V., 575
Karpets, I. P., 575
Karsten, G. R., 62
Karuna, K., 461
Kasatikov, V. A., 573, 574
Kaushal, B. R., 33
Kaushik, P., 13, 456
Keeney, G., 215, 217
Kelkar, B. V., 453
Kerry, B., 225
Khalil, M. A., 276
Khambata, S. R., 453
Khan, J., 185
Kherdekar, R. C., 461
Kimura, M., 166
King, R. A., 42, 43
Kingston, T. J., 438
Kinney, C. A., 266
Kloepper, J. W., 174
Klok, C., 275
Kluchenovich, S. A., 572, 575
Knuutinen, J., 222
Kobayashi, H., 354, 356
Kochany, L. E., 485
Koerner, B. I., 305, 386
Kohlerova, P., 356
Koike, T., 204
Korsunova, T. M., 573, 575
Koul, O., 222
Kovalev, V. V., 575
Krause, M. S., 204
Kretzschmar, A., 104
Kuboi, T., 21
Kulik, A. P., 567, 575
Kumaraswamy, K., 133
Kumari, M. S. S., 456
Kurowska, A., 222
Kuter, G. A., 184
Kuzmina, N. V., 574

L

Lane, P. W., 273
Langdon, C. J., 276
Lange, S., 354, 356
Lanno, R., 485
Lassègues, M., 258
Lavelle, P., 166, 338, 509, 510
Lawrence, R. D., 324
Lawton, J. H., 266
Lazcano, C., 13, 56, 411, 414, 415
Lee, K. E., 121, 122, 355, 439
León Monzón, J. H. D., 513
Limón, G. A., 513
Lindroth, R. L., 222
Liping, M. A., 184
Litterick, A. M., 166
Lofs-Holmin, A., 33, 336
Lofty, J. R., 337
Lopez, M., 168, 172
Lores, M., 14, 54, 55, 64, 66, 413
Loschinkohl, C., 264
Lotzof, M., 444
Loza, A., 537
Lui, T. A., 251
Lumsden, R. D., 173, 174
Luoma, S. N., 268, 269, 270

M

Maboeta, M. S., 276
Macdonald, D. W., 2, 10
Machalaba, D., 385
Mahaffee, W., 155, 185, 186
Mahanil, S., 222
Mahendran, P., 462
Maher, D., 394
Mahmood, I., 225
Mainoo, N., 130
Makeschin, F., 121
Maksimova, S. L., 568, 565–575
Malik, A., 223
Malisie, A. F., 486

Mallareddy, K., 460
Manivannan, N., 463
Mantovi, P., 264
Marinari, S., 146, 147
Martínez, C. C., 509, 513, 526
Martin, M. H., 269, 439, 477
Masheva, S., 223
Mason, C. F., 336
Mathur, K., 462
Ma, W. C., 269
Maynard, A., 146
Mazolla, M., 165, 204
McBride, M. B., 266
McCants, C. B., 122
McGeer, J. C., 269
McGrath, S. P., 273, 276
McKellar, M. E., 174
McSorley, R., 223
Meharg, A. A., 269
Mehla, C. P., 460
Mekada, H., 324, 325, 330
Meléndez, G. M., 513
Melgar, R., 411
Melnik, I. A., 575
Menzies, J. D., 165
Metzger, J. D., 117, 118, 120, 145, 293, 296
Mikhailova, M. E., 572
Millar, R. H., 324
Milocco, A., 2, 10
Mitchell, M. J., 3, 10, 62, 63, 250, 251, 258, 441, 447, 489
Moffat, A., 264
Mohammed, A. R. A., 337, 338
Mohrig, W., 356
Monaco, A., 21
Monroy, F., 14, 32, 55, 61, 411, 413, 414, 415
Montalvo, Q. B., 513
Morales, H., 221
Morgan, A. J., 263, 264, 266, 268, 270, 272, 274, 263–276
Morgan, J. E., 263, 267, 268, 272
Morra, L., 223
Morton, I., 340, 341
Mrinal, S., 133
Murarkar, S. R., 133
Murry, A. C., 251
Muscolo, A., 122, 123
Muyima, N. Y. O., 33
Mylonas, V. A., 122

N

Nadkarni, N., 401
Nagasawa, H., 352, 354, 356
Nahmani, J., 268
Nair, S. K., 14, 130
Nakasone, A. K., 185
Nannipieri, P., 147
Nardi, S., 122
Nascimento, B. S., 274
Natoli, J., 444
Nefyodova, E. A., 573, 575
Nelson, E. B., 172, 173, 174, 176, 177
Nenthra, N. N., 133, 146
Neuhauser, E. F., 3, 4, 9, 10, 34, 36, 268, 269, 424
Niederer, A., 323–334
Nielson, R. L., 462
Nijhawan, S. D., 453
Nirmalnath, P. J., 463
Noble, R., 204
Nogales, R., 13, 411

O

O'Gara, F., 204
Oliver, G. S., 2, 10
Orazova, M. K., 337
Orozco, S. H., 104, 105
O'Sullivan, D. J., 204
Owen, J., 276

P

Pande, R., 462
Pandey, R., 223
Pankakoski, E., 269
Pant, A., 236
Panuccio, M. R., 123
Paré, T., 270
Parthasarathi, K., 349–367
Pascual, J. A., 103, 147
Patil, M. P., 133
Patra, V. C., 454
Patriquin, D. G., 220, 221
Paul, E. A., 147, 456
Pedersen, J. C., 258
Peña, A., 416
Penton, S., 342
Pérez-Losada, M., 29, 43, 411, 412
Petersen, S. O., 264
Peterson, S. S., 222
Petrussi, F., 122
Phelan, P. L., 221
Phillips, V. R., 4, 10, 18, 84, 328, 556
Piedad-Pascual, F., 476
Plana, R., 411
Plaza, C., 411

Plytycz, B., 275
Pogrebniak, A. P., 575
Poincelot, R. P., 13
Popovic, M., 353, 354, 357
Postma, J., 174
Potylkin, V. A., 568
Pravkina, S. L., 573
Prieto, D., 536

R

Rainbow, P. S., 268, 269, 270
Ramesh, K., 220
Ranganathan, L. S., 349–367
Ravi, M., 221
Reddy, B. K., 460, 461
Reid, B. J., 433
Reinecke, A. J., 32, 34, 35, 37, 484
Rejnek, J., 357
Reneco, M., 223
Revestí, L., 537
Reyes, V. R., 470
Rey, J. R., 43
Reynolds, J. W., 28
Rhee, J. A., 337
Roberts, P., 114, 433
Rodríguez Navarro, J. A., 172, 185, 534, 536–537, 539
Rogers, A., 386
Roletto, E., 21
Roodbergen, M., 269
Roslim, D. I., 492
Rougerie, R., 42, 43
Roy, S. K., 454
Ryder, M. H., 167, 168

S

Sabine, J., 324, 325, 330, 331, 332, 439, 443, 475, 477, 483, 488
Sable, C. R., 460
Saha, S., 463
Salamanca, E. F., 473
Salazar, T., 513
Salter, C. E., 153–163
Sampedro, L., 55, 57, 166, 412, 413, 415
Sánchez-Moreno, S., 165
Santamaría, R. S., 513
Saroja, K., 454
Sauvé, S., 266, 270
Scarborough, J., 444
Scheuerell, S., 155, 171, 172, 173, 185–186
Scheu, S., 59, 66
Schmidt, O., 168
Schneider, R. W., 165
Schönholzer, F., 59, 66
Schulz, E., 324
Scott, M. A., 107, 121
Scullion, J., 337–338
Selvan, A., 266
Selvi, K., 456
Sen, B., 130
Senesi, N., 122
Sequi, P., 21, 433
Shabardina, N. P., 574
Shahmansouri, M. R., 272
Shanahan, P., 204
Sherman, R. L., 67–77, 369–387, 391–406
Shikula, N. K., 575
Shipitalo, M. J., 441
Shi-wei, Z., 104
Shobha, S. V., 463
Siddagangaiah, B. A., 107
Siddall, M. E., 42
Siddiqui, Z. A., 225
Sierra, A., 536
Singh, R. R., 137, 168–169, 171, 174, 271, 455–456, 460, 462
Sizmur, T., 274
Slobodian, V. A., 575
Slocum, K., 380
Snyder, B. A., 336
Song, Q. J., 270
Speier, T.W., 267
Spurgeon, D. J., 275
Sreenivas, C., 146
Stahl, D. A., 167
Steele, J., 439, 447
Stentiford, E., 16
Stepanova, D. I., 573, 578
Stepanov, A. I., 573
Stephensen, J., 34, 455, 456
Stephens, P. M., 167, 168
Steuart, F., 439, 447
Stockdill, S. M. J., 337, 340, 438
Stone, A. G., 173, 184, 204
Sturz, A. V., 174
Subler, S., 249–261, 287–301
Sudharama, I. M., 122
Susanto, S. J. A., 486, 487
Suthar, S., 13, 269, 271, 272, 273
Swathi, P., 223

T

Taboga, L., 324, 325, 330
Tacon, A. G. J., 330, 475

Taiz, L., 118
Tambe, V. B., 453
Tan, K. H., 122
Tantiwiramanond, D., 122
Tato, A., 411
Taylor, P. D., 440
Temple-Smith, M. G., 438
Teskey, H. J., 493
Thakali, S., 270
Thankamani, C. K., 107
Thanunathan, K., 107
Thompson, R., 14
Tian, G., 264
Tiunov, A. V., 59, 66
Tomati, U., 105, 120, 121, 306, 510
Tomlin, A. D., 308, 340, 344, 443
Toscano, S. S., 517
Tränkner, A., 185–186
Trejo, M., 513
Trigo, D., 166
Tsukamoto, J., 32
Tsyganov, A. R., 571

U

Usman, A., 266
Uvarov, A. V., 339, 342

V

Vaillier, J., 353, 354, 357
Vajrabhiah, S. N., 120
Vajranabhaiah, S. N., 462
Valdrighi, M. M., 122, 123
Valembois, P., 258, 354
Vanagas, J., 121
Van Gansen, P., 510
Van Herwijnen, R., 274
Vasko-Bennett, M. A., 183–207, 215, 217, 209–233, 235–248
Veeken, A., 270
Vela, D. M., 513
Velando, A., 411
Verdonck, O., 293
Verhoef, H. A., 488, 491
Verstraete, W., 272
Vijver, M. G., 267
Vildflush, I. R., 571
Villesas-Papalid, L., 476
Vinken, R., 222
Vishnaykov, A. E., 573

W

Wang, C., 168, 264, 559, 561
Wareing, P. F., 121
Watanabe, H., 32
Webster, K. A., 103–128, 437–452
Weltzien, H. C., 166, 184
Werner, M., 105
Wetzel, M. J., 28, 40
Wheeler, L., 404
Whipps, J. M., 165, 174
Whitta, R., 439
Williams, B., 423–435
Wilson, W. S., 106, 118, 123, 146
Windsor, R. A., 447
Windstam, S., 172
Wong, M. H., 21, 266
Worden, D. D., 4, 9
Wright, E., 168, 170, 172
Wu, C., 357

X

Xing, B. D., 352, 357

Y

Yadahalli, K. B., 462
Yadav, A., 271, 273
Yardim, F., 210, 221
Yasir, M., 130
Yegnanarayan, R., 351
Yoshida, M., 324, 325
Yulipriyanto, H., 492
Yutina, A. S., 575

Z

Zachariah, A. S., 456
Zalat, S., 222
Zaller, J. G., 117, 185
Zamora, C. A., 513
Zeiger, E., 118
Zelles, L., 57, 66
Zende, G. K., 133
Zeng, Z. P., 546, 553, 559, 561
Zhang, W., 185, 358, 546
Zhelyazko, V. I., 571
Zhenjun, S., 6, 10, 545–563
Zicsi, A., 44, 50
Zink, T. A., 103
Zucconi, F., 13, 14, 21

Subject Index

A

Aerating channel system, 550–551, 556, 561
Aeration effects of aqueous extracts of (teas) on
biomass, 239
dehydrogenase activity, 239
nitrate, 238
pH of, 237
plant growth and yields, 240–241
thermophilic tea conductivity, 238
Agricultural soils, 2–3, 12, 130, 147, 165, 440, 576
Agricultural wastes, 440, 459, 483–484
Agrochemicals, 460, 462–463
Air delivery systems, 158–159
Anecic earthworm species, 28, 338–339
Animal manures, 33, 46, 59, 64, 103, 122, 130, 140, 298, 299, 305, 310, 311, 321, 323, 384, 413, 414, 567, 568
Animal wastes, 2, 13, 324, 327, 411, 457, 458, 493
Annual-cycle earthworm production, 558
Antibacterial activity, 353, 360
Anticarcinogenic activity of earthworm extracts, 352
Anticoagulative activity of earthworm extracts, 353–354
Antioxidative activity of earthworms, 350
Antiulcer properties of earthworms, 351
Aphids, 210, 212–213, 217, 220–221, 223, 430
Aqueous extracts (teas) effects on plant growth and yields, 235, 236
aeration effects, 240, 241
biomass, 239
dehydrogenase activity, 239
nitrate, 238
pH of, 237
tea conductivity, 238
cucumbers concentration range, 244
cucumber plant heights, 245
cucumber plant mean leaf areas, 245
cucumber plant numbers, 246
cucumber plant weights, 246
effects on
tomato fresh shoot weights, 243
tomato leaf areas, 242
tomato numbers, 243
tomato plant heights, 241
tomato responses in growth, 242
tomato weights, 244
Aqueous extracts, soil-applied vermicompost, 219
Aqueous vermicompost extracts, 120, 186, 215, 217–221, 235, 240, 310, 374, 515
Arthropod pest suppression
by vermicompost aqueous solutions (teas) on, 219–223
aphid population development, 214, 217–219
citrus mealybugs effect on, 217
cucumber beetle, 213, 215
deionized water effects of, 217
fungivorous nematodes populations, 224
mealybugs, 217–219
plant parasitic, 224
soil drenches, effects of, 216
soil nematode communities, 223, 224, 226, 255
spider mites, 214, 217–219
Arthropod pests suppression by solid vermicomposts on, 210
caterpillar damage to tomato fruits, 211
mealybug development on tomato plants, 213
striped cucumber beetles on cucumber plants, 211
leaf areas of cucumber seedlings, 212
Australia, vermiculture in
commercial vermicompost and vermiliquid production
additives and supplements used in, 445
feedstocks used in, 444–445
mid-and large-scale vermicomposting systems, 444
operation in 2009, 443–444
pests and diseases in, 445
vermiliquid (tea) production and use, 445–446
earthworms used
agricultural earthworm species, 439, 440
native earthworm species, 439
species used in commercial vermiculture, 440, 441
future developments, 449
history
earthworms and organic-waste processing, 439
earthworms for agriculture, 438
industry associations, 439

markets and marketing
agricultural market-development trials, 447–448
buy-back schemes, 447
organic-waste management, earthworms role in, 447
profitability, 446–447
products
earthworm farms for domestic use, 441–442
earthworm-protein production, 443
vermicomposting-earthworm sources, 441
vermicomposts sales, 442
vermiliquids/teas, 442–443
standards and testing, 449
analysis, 448
terminology, 448
Auxins, 105, 118, 120–122, 462, 474

B

Bacteria, 15, 38, 55, 59, 64, 73, 105, 121, 130, 158, 160, 166–168, 204, 251, 258, 493
Bacteria growth rate, 59–60, 64
Barefoot soil earthworm castings products, 381
Batch systems of vermicomposting, 82, 83
Beds for vermicomposting
bedding, 68–70, 382, 385, 393, 395, 404–405, 447, 487–488, 520, 548–551, 556–557, 559
moisture, 72, 74
bottoms, 536–537, 551–552, 554, 556
earthworm
growth in, 561
production in, 562
rearing in, 536
Belarus, vermicomposting in, 566–568
Brest farm, 567–568
Cherven farm, 567
BET, *see* British Earthworm Technology (BET)
Bioavailable metal fractions, 266
Biological properties of vermicomposts, 105
Biomass earthworm productivity, 326
Black box of vermicomposting, 412
Botrytis cinerea results in
cucumber experiment, 198–199
tomato experiments, 200
Breeding earthworms, 307, 339, 370, 470, 499, 503, 536, 538, 546, 560, 562
Breeding stock, 46, 49, 51
Brewing process for teas, 80, 156, 158, 237–238, 374
British Earthworm Technology (BET), 5, 305, 424–425
Businesses vermicomposting in
IKEA, 395–396
Len Foote Hike Inn, 395
Metro hall, 394
National Institutes of Environmental & Health Sciences (NIEHS), 394–395
Seattle Kingdome, 395

C

California Integrated Waste Management Board (CIWMB), 379
California compost regulations, 378
Canyon Recycling, 376–377, 381
Casting Call, 311, 386
Cation-exchange capacity (CEC), 22, 273, 295, 445, 472
Cattle manure vermicompost, 106, 109–110, 113, 116, 123, 136, 236
Cattle wastes, 6, 18
CCF, *see* Coelomic cytolytic factor (CCF)
China, vermiculture in
annual-cycle earthworm production, 558–559
earthworm culture methods
introduction into bed, 554
management, 554
trial feeding of earthworms, 554
earthworm
bed construction, 552
populations, technological analyses, 555–556
production, 498
species and substrate, 552
earthworm substrate processing
C/N ratio of substrate, 553
fermentation of the substrate, 553
pH value of substrate, 553
pretreatment of raw materials, 553
windrow turning, 553
heat channel system for cold seasons
bed construction, 550
organic-waste processing, 551
structure, 551
high-yield vermiculture, 555, 558, 559
multilayer-box vermiculture system indoors, 549–550
pit vermiculture method outdoors, 549
plastic tunnels for low-temperature seasons, 550
structure, 552
techniques, 561–562
theory, 559–560

windrow method
earthworm population density and food, 548–549
earthworm species, 547
earthworm substrates and types, 548
moisture management, 548
temperature management, 548
Coelomic cytolytic factor (CCF), 354
Coelomic fluid of earthworms, 258, 354, 483
Coffee, studies on vermicomposting, 512
Colleges and universities, vermicomposting in
Evergreen State College, 397–398
Southern Illinois University Carbondale, 398
University of Massachusetts Lowell, 399
University of Oregon, 398
Commercial aerated compost/vermicompost tea system, 155
Commercial vermicomposting, 8, 305, 321, 386, 440–441, 443, 547
enterprise models
classic earthworm farm, 310–311
home vermicomposting, 311
resource recovery for, 311–312
factors affecting, 318
climate, 319
earthworm population growth, 320
feedstock loading rates, 320
human pathogens, 319–320
location and space, 319
on-site/off-site processing, 319
history, 304, 305, 320, 321
market opportunities
aqueous extracts sale, 307
earthworm castings sale, 305–307
earthworm-related markets, 306
earthworms sale, 307–308
market-development, 310
resource recovery, 308–309
systems and supplies, 309
technology impacts, 312
developments, 316
levels of, 313
processing facility, size and scale, 313, 315
vermicomposting technologies, 314
Common vermicompost bed organisms, 88
Complete urban waste-recycling vermicomposting systems, 99
Components, large vermicomposting systems, 405
Composition, suitable earthworm-substrate, 549
Composting processes, 14, 16, 22–23, 249, 259, 288, 445
compost and vermicompost
maturity, criteria of, 20–21
processes, 13–15, 290
composts, mature, 20–21
conventional thermophilic, 250, 315
earthworm species, 572
mass, 16–17
Compost stability, analytical methods
carbon and nitrogen ratio, 21
criteria for, 22
human pathogens, absence of, 22
humic substances, 21
plant inhibitors, absence, 21
Compost tea
additives, 162
production, 157, 162, 185
Task Force, 161
thermophilic, 184, 226, 228, 238, 239
Continuous-flow vermicomposting reactor system, 6, 18, 312
Cosmopolitan earthworms, 49
Cow manure, 456
manure vermicomposting, 60–62, 134, 140
Crops test, 107, 115, 117
Cuba, vermiculture
earthworm in
specialists, 538
species, 533
vermicompost
amino acid composition, 541
basic medium-size units, 538
basic small units, 538
composition, 540
earthworm protein, chickens feed, 542
large units, 538–542
production, 541
introduction, 534
vermiculture research program, 534–537
Cucumber beetles, 210, 215–216, 220–221
experiments, 187, 189–190, 192–193, 196, 198, 200
striped cucumber beetle, 211–212, 217
Cucumber experiments with teas on
aqueous extracts teas effects on cucumber plant growth and yields, 243
concentrations range of, 244
heights, 245
mean leaf areas, 245
numbers, 246
weights, 246
fruits, 241
increased growth, 575

Botrytis cinerea, 198
mean damage ratings, 199
mean leaf areas, 199
Pectosporium tabacinum, 196
mean damage rating, 197
mean leaf areas, 197
Phytophthora capsici, 187
mean leaf areas, 189
mean root damage, 188
Pythium ultimum
with deionized water control, 194
mean damage, 194
mean leaf areas, 194
seedling losses by soil applications, 193
vermicompost teas range effect, 194
Rhizoctonia solani, 190
mean damage, 191
mean leaf areas, 191
Sclerotinea rolfsii, 200
damage ratings of, 201
leaf areas of, 202
seedlings, 190, 193, 212, 215, 241, 244

D

Dairy manure vermicomposted, 255
Damage
ratings, 193, 201–202, 213–214, 219
suppression, 191–193
total, 187, 196, 199–202
Decomposition
biochemical, 54, 56
decomposer organisms, 300, 487–488, 492–493
organic-matter, 413–414
plant residue, 160
processes, 53, 55, 64, 534
rates, 413, 551
Dehydrogenase activity, 145, 147, 239
Deionized water control, 186–187, 189, 192, 193–194, 196, 196–197, 199–200, 202, 215, 219, 226, 230, 240–241, 245, 246
Dendrobaena veneta, 3, 29, 32–33, 36, 41, 307, 323–324, 372, 411, 429, 440
Dendrodrilus rubidus, 32, 411
Detritivorous earthworms, 64
Diseases
and parasites of earthworms, 89
plant, 166, 175
suppressive vermicompost results, 169–170
symptoms, 172–174, 196, 199–200
Domestic vermicomposting systems, 85–87
Eliminator, 85, 86, 87
Dorset Wedge vermicomposting system, 86
Drawida nepalensis, 33
Dry earthworm meals, 328–329

E

Earthworm Inoculation Unit (EIUs), 342–344
Earthworm proteins results of
animal-feeding trials with
chicken feeding, 330–331
fish-feeding, 330
pig-feeding, 332
chickens growth diet, 331
composition, 324
essential amino acid content, 325–326
essential amino acids amounts, 329
vitamin composition, 326
economics
growth, 332
value, 333
food value, 324
harvesting from organic wastes, 327–328
on meat and earthworm meal, chickens growth, 331
from organic wastes, 323–324
processing of earthworms for animal feed
blanching, 328
dry earthworm meals, 329–330
formic acid ensilation, 329
pastes, 328
production methods, 326
population curve, 327
Earthworms, 255, 548, 554, 556
activity degree of, 130, 291
adult, 32, 343, 458, 557
average population, 81, 92
beds, 4, 8, 89, 315–316, 372, 376, 383, 440, 443, 445, 520, 547–548, 551–552, 554, 556–557, 561–562
bins, 68, 70–72, 309, 311, 519, 77, 395, 406, 396, 404–405
biology, 130, 416–417, 484
biomass, 37, 327, 470–471, 482, 486, 490, 537, 539, 565–566
biosolid waste-management system, 312
breeding beds, 429, 537, 556, 562
buyer's guide, 372
by-products, 370, 373–374
castings, 8, 68, 70, 100, 105, 288, 290, 305, 307–308, 321, 371, 373–374, 377–378, 414, 535–536, 567–568
producing, 313

clitellum, 554, 561
cocoons, 76, 89, 458
coelomic fluid, 353–354, 359, 462
colonization, natural, 336
culture methods, 42, 335, 469, 483, 538, 554, 556, 561
digestion, 166–167
diseases and predators in vermicomposting systems
  bed organisms, 88
  diseases and parasites of earthworms, 89
  invertebrate earthworm enemies, 89
  vertebrate earthworm predators, 89
diversity, 350, 464
ecology, 338, 416, 432, 566
ecotoxicology, 416
Eisenia andrei, 62, 104
Eisenia fetida, 250, 568
farms, 305, 319, 381, 428, 469, 478, 508, 516, 521, 524, 526, 546, 552
feeding, 330, 458
feedstocks, 384, 499, 539
flour, 46, 534
food, 168, 338, 538–539
foregut, 258
growers, 74, 371, 375, 382, 386, 439–441, 443, 445–446, 516–518, 548
  specialist, 521
growth, 4, 35, 37, 59, 327, 341, 412, 470, 483, 488, 556, 560, 561
gut, 56, 62, 81, 104, 121, 168, 222, 236, 251, 258, 290, 510
harvesting equipment
  comb-type mechanism of separating from vermicompost, 84
  rotating screen earthworm/waste-separating system or trommel, 83–84
  Silsoe comb-type earthworm/vermicompost separator, 85
  typical rotating trommel, 83
hatching, young, 32–33
inoculations, 337, 340, 342–346, 556–557
intestinal tract, 166
introduction, 340, 344
management, 459, 556
meal, 308, 330, 331–333, 443
metals
  effects, 276
  mean data, 275
populations, 88, 95, 101, 120
  biomass, 547, 555
  densities, 14, 37, 56, 336, 413, 518, 549, 551, 555, 560–561
  sustainable, 343
powders, 350, 354
predators, 89, 445
  parasites and pathogens, 38
processing,14, 499–500
production, 332, 350, 428, 482, 546, 551, 555, 556, 558, 561
protein, 3, 308, 330, 333, 336, 424, 470, 475–477, 486, 541, 558–559
  production, 327, 332, 443, 482
reactor system process, 499, 503
rearing, 42, 342, 345, 488
reproduction, 560
  initiating, 548
  intermittent, 561
reproductivity, 373
sales, 310, 371, 447
species, 28–29, 34, 49–50, 74, 130, 324, 326, 343–344, 372, 440–441, 491–492, 516, 546
  commercial, 537
  common, 29, 339
  deep-burrowing, 484
  detritivorous, 47
  Eisenia andrei, 61
  endogeic, 45, 484
  epigeic, 2–3, 51, 63–64, 288–289, 291, 323, 411, 414, 424, 432, 484, 509
  exotic, 533–534
  harvest indigenous, 371
  holarctic, 32
  identifying suitable, 464
  less common, 46
  litter-dwelling, 423
  novel, 45
  Pheretima sp, 492
  shallow-working, 345
  soil-dwelling, 274, 432
  temperate, 33
  tropical, 35
  upper-soil, 484
  used in commercial vermiculture, 440
substrate
  layers, 554, 560
  processing, 553
survival, 338
  aeration, 37
  ammonia and salts, 37
  and growth, 35
  moisture content, 36
  pH levels, 36
  temperature, 36
systems, 74
temperature conditions, 36

tissues, 38, 269, 308, 324, 326, 329, 351, 360–361
vermicomposting species, 29–35
Polypheretima elongata, 35
temperate species, 29, 32–33
tropical species, 34–35
waste
and environmental management, 5, 424
processing, 500, 503
separator, 84
weights, 554
gain, 548, 560
windrows, 520, 526
worked material, 4, 18, 251, 336, 370, 372–373, 381, 453, 508, 514
Ecology of vermicomposting, 412
Economics of vermicomposting, 304
capital costs, 316–317
annual returns
for indoor continuous-flow reactor facility, 318
less running costs, 318
operating costs, 317
Education, 311
Effects
of aeration, 237–239
of food waste vermicompost tea on, 241–246
Botrytis cinerea damage, 198
drenches, 201–202
Fusarium oxysporum attacking tomatoes, 187
Phytophthora capsici attacking cucumbers and tomatoes, 187
Plectosporium tabacinum attacking vegetables, 196
Pythium ultimum attacking cucumbers, 193
Rhizoctonia solani attacking cucumbers, 190
Sclerotinea Rolfsii damage, 200
of vermicomposts, 202, 223, 473
of aqueous extracts, 186, 227, 236, 246
Eisenia andrei, 3, 29, 41, 58, 60–62, 74, 307–308, 353, 411, 440, 484, 486–487, 490, 509, 516, 534
Eisenia fetida, 484
earthworms, 379, 395
red earthworms, 307
tiger earthworms, 308
Enrichment of vermicomposts, 463
Enteric viruses, 253–255, 299, 483
Environmental
factors on survival and growth of earthworms, 35
protection, 396–397
Environmental Protection Agency (EPA), 68, 75, 162, 249–250, 252, 254–255, 259, 299, 304, 312, 316, 320, 376, 392, 483
Epigeic species of earthworms, 28, 32, 35, 41, 57, 63–64, 80, 411, 414, 431, 493, 509
Eudrilus eugeniae, 34
Extensive outdoor vermicomposting beds, 430

F

Fallbrook vermicomposting project, 376
Feedstocks
metal-contaminated, 276
used in vermicomposting, 444
Fees, waste-disposal, 406, 501
Fermentation, 553
Fertilizers, 2, 22, 70, 134, 137, 160, 224, 276, 292, 401, 406, 460–461, 475, 505, 523, 527
Fibrinogen/fibrin degradation product (FDP), 353
Field capacity, 186–188, 215, 539
field crops, 131, 133, 140, 148, 574
field soils, 49, 120, 131, 133, 141, 148, 223, 339
Fish meal, 308, 330–331, 333, 476–477, 562
Fish protein, 541–542
Food, 1–2, 13
crops, 156, 162
residuals, 68, 71, 73, 75, 371, 392, 394–397, 401, 406
scraps, 68, 72, 74, 76, 311, 395, 397, 399, 504, 519
wastes
applications, 132, 141
aqueous extracts, 196, 216
compost teas, 245
effects, 210, 226–227
mixtures, 213–214
postconsumer, 394, 400, 403–405
preconsumer, 381, 398–400
substitutions, 111, 213
vermicompost, 114, 140
vermicomposting, 312, 442
vermicompost tea applications, 229
vermicompost tea drenches, 201–202
vermicompost tea sprays, 189, 202
Fulvic acids, 21, 236, 485–486, 565
Fungi, 55, 73, 105, 121, 130, 158, 160, 168, 204, 225–226, 230, 360, 492–493
growth rates, 57, 59, 64, 173
Fungivorous nematodes, 224–225

G

Global Vermiculture Association (GVA), 386
Gomel region project, 568–572
  technological process, 570
Greenhouse crop production, 103
  effects of vermicomposts, 106
    Petunia seed germination, 112
    tomato, yield and sizes, 108
  factors influencing increases of
    plant growth hormones, 117–121
  humates in vermicomposts, 121–123
  nature of vermicomposts
    biological properties, 105
    characteristics of, 104
    chemical properties, 105
    physical properties, 104
Greenhouse experiments, 106, 167, 186, 215, 217, 226
Green peach aphids, 217, 219–220
Green wastes, 288, 376, 378–379, 426, 444, 483
Growth of earthworm population, 320, 345
Growth induction in plants, 461
Growth media for, 107, 113, 119
  best horticultural plant, 306
  commercial plant, 105–106
  costly peat-based plant, 106
  effective plant, 267
  good plant, 289
  high-grade plant, 96
  high-quality plant, 19, 296
  high-value plant, 482
  plants, 105, 106, 120
  producing container plants, 64
  soil-less bedding-plant, 117, 131, 212, 236, 240, 296
  suitable plant, 432
  vermicompost-substituted, 131
Growth regulators, 236, 288
Guides for vermicomposting curriculum, 77

H

Harnessing the Earthworm, 375
Harvesting earthworms, 379, 562
Harvesting vermicompost, 372, 384–385
Hatching of cocoons, 49, 51
  viability, 32–33
Heavy metals
  composts in remediation of
    metalliferous soils, 274
  concentration and availability, 270, 272–273
    publications selection, 271
  earthworm accumulation, 267–270
  metal-contaminated composts effects
    on plants, 273–274
  relationships between, 268
Home-scale earthworm systems, 309, 311
Hong Kong, vermicomposting projects in, 497
  business model facets
    financial, 501–502
    land availability and cost, 502–503
    logistics, 502
    regulatory issues, 503
Hormones for plant-growth, 160
Hospitals, vermicomposting in
  Huron hospital, 400
  Medical University of South
    Carolina (MUSC), 399–400
Household earthworm bins, 68, 69, 75
Household wastes, 264, 427, 574
Human food, 475, 477, 488, 562
Human pathogen reduction, 249
  field studies, 252–255
  laboratory studies, 250
    E. coli spp.in cattle manure, 254
    E. coli spp.in sewage biosolids, 253
    fecal coliform bacteria, 252
    Salmonella spp. in cattle manure, 254
    Salmonella spp. in sewage biosolids, 253
  mechanisms of, 258–259
  sewage biosolids, 255
    enteric virus, 257
    fecal coliform, 256
    helminth ova, 257
    Salmonella spp., 256
Human pathogens, 12, 14, 20, 22, 249–252, 255, 259, 299, 483

I

ICAR, *see* Indian Council of Agricultural Research (ICAR)
In-container vermicomposting, 19
  batch reactors, 19
  bins, 19
Indian Council of Agricultural Research (ICAR), 459
India, vermicomposting in
  development of, 454–457
  larger-scale commercial
    development, 459–460
  sustainable agriculture
    future of earthworm research, 463–464
    growth induction in plants by, 461–462
    with microbial inoculants, 463
    pathogen and pest management, 462, 463
    production, 460–461

top-harvesting vermicomposting system, 98
vermicompost production, 457–459
Indonesia, vermicomposting, 481
case study using tea leaf waste
materials and methods, 487–488
nutritive value, 489
objectives, 486–487
different strategies in, 482
growth in earthworm populations and biomass, 490–491
numbers/weights of cocoons, 490
Indoor continuous-flow reactor system vermicomposting, 318
Indoors multilayer-box vermiculture system, 549
Industrial organic wastes, 1, 3, 80, 92, 140, 538
Industry earthworm-vermicomposting, 309
Industry traditional composting, 387
Inorganic fertilizers
applications, 133, 211
recommended, 461, 472–473
rates of, 210, 223, 473
Institutions, vermicomposting, 396–397
Intensive cultivation of earthworm species, 346
Internet vermicomposting forums, 386
Invertebrate earthworm enemies, 89
In-vessel composting, 16–17

L

Laboratory vermicomposting, 253–254
Lampito mauritii therapeutic properties, 358
anti-inflammatory, 359
antimicrobial, 360–361
antioxidant, 359
antipyretic, 359
assessment of antioxidant properties, 359
hepatoprotective, 360
Large-scale commercial vermicomposting of food waste, 312
Large-scale earthworm waste-processing systems, 342, 439, 444, 449
Lumbricus rubellus, 33

M

Major micronutrients in soils and crops, 299
Manufacturing process, vermicompost liquid, 499
Markets for vermicomposts, 6, 387
Mature vermicomposts, 21, 274
stabilized/equilibrated, 277
Maximum vermicomposting efficiency, 37
Mean leaf areas, 192, 198–199, 201–203
of cucumbers, 189, 191, 194, 197, 229
of tomatoes, 203, 227
Meat meal from earthworms, 308, 330–331, 333
Medium-and high-technology vermicomposting systems, 91–92
complete urban waste-recycling, 99
earthworm toilet systems, 99
economics, 100–101
fully automated continuous-flow
breaker-bar discharge floor, 97
breaker-bar discharger and winch unit, 97
freestanding reactor, 95
gantry feed hopper, 96
reactor module, 94
reciprocating scraper, 96
Indian design
Indian top-harvesting system, 93, 94, 98
Meloidogyne hapla
mean numbers, 228
suppression on cucumber by aqueous extracts from vermicomposts, 227–228
dry shoot weights of, 229
leaf areas of plants attacked by, 229
suppression on tomatoes by aqueous extracts from vermicomposts, 226
dry shoot weights of plants infested with, 227
leaf areas of plants infested with, 227
Metal-contaminated composts
concentrations, 268–270, 272, 276
contents, 267, 272–273
extracting, 276–277
mobility, 273–274, 277
Metalliferous soils, 267, 274, 276
Metalloids, 264, 267–268
Mexico, vermicomposting research and activities
application
vermicomposts, 514–515
vermicompost teas, 515
concepts, 509–510
distribution, 512
historical background, 508–509
initial studies on coffee pulp, 512, 513
studies in, 511
vermicomposting farms, 526
vermicomposting systems, 515
aspects, 516
commercial aspects of vermicomposts, 522–523
designs, 518–520
human resources, 521

operation methods, 521–522
production, 522
promotional aspects of vermiculture, 524
quality control, 523
quality standards, 525
species of earthworms used in, 516–517
substrates or feedstocks, 517–518
survey, 524–525
urban domestic residues, 526–527
wastewater and manure handling with, 520
windrows, 519
Microbes, vermicompost-derived, 172
Microbial and faunal community composition in vermicomposts, 300
Microorganisms
disease-suppressive, 168
growth of, 156–157
range of, 89, 221
vermicompost-associated, 172
Military bases, vermicomposting in
Wright–Patterson Air Force Base, 406
Mineral plant nutrients, 489
Mites spider, 212, 214, 217, 220–221, 223
Moisture contents of vermicomposts, 294
Municipal solid waste, 270, 305, 371, 377, 456, 517
Municipal wastewater sludge, 316

N

National Organic Program (NOP), 161
National Organic Standards Board (NOSB), 156
National Organic Standards Board Compost Task Force, 163
Nematode populations, 223, 229
fungivorous, 225–226, 230
New Zealand, vermiculture in
commercial vermicompost and vermiliquid production
additives and supplements used in, 445
feedstocks used in, 444–445
mid-and large-scale vermicomposting systems, 444
pests and diseases in, 445
vermiculture operation in 2009, 443–444
vermiliquid (tea) production and use, 445–446
earthworms
agricultural earthworm species, 439–440
native earthworm species, 439
species used in commercial vermiculture, 440–441
history
earthworms and organic-waste processing, 439
earthworms for agriculture, 438
vermiculture industry associations, 439
markets and marketing
agricultural market-development trials, 447–448
buy-back schemes, 447
organic-waste management, earthworms role in, 447
profitability, 446–447
products
earthworm farms for domestic use, 441–442
earthworm-protein production, 443
vermicomposting-earthworm sources, 441
vermicompost sales, 442
vermiliquids/teas, 442–443
standards and testing, 449
analyses, 448
terminology, 448
Non-nutritional plant growth factors, 120
NOP, *see* National Organic Program (NOP)
North America vermicomposting in, 370
Bait Farms, 375–376
California early history in
biosolids in Fallbrook, 376
Canyon Recycling, 376–377
Pacific Landscape Supply, 379–380
Rainbow Worm Farm, 379
West Coast operations, 375–376
commercialization as private enterprise, 373
current West coast operations, 380–385
Oregon Soil Corporation, 381
Sonoma Worm Farm, 380
Worm Farm, 380–381
Yelm earthworm and castings farm, 381–382
differences influencing factors, 372
earthworm castings, 374–375
earthworm species used in, 372–373
Eisenia andrei, 372
Eisenia fetida, 372
feedstock income and expense, 371
material recovery facilities, 377–379
sales of, 373–374
temperate regions, 371
NOSB, *see* National organic standards board (NOSB)
Nutrients plant-available, 103, 501

O

On-site vermicomposting, 8, 387, 388
Open composting systems
forced aerated static piles
alternative ventilation, 16
bottom blowing, 16
bottom suction, 16
windrow composting, 15–16
Open vermicomposting systems, 17–18
Dorset wedge-style beds, 18
low-cost floor beds, 17
Operations, vermicompost-producing, 374
Operators, large-scale organic waste-processing, 500
Organic farm, 397–398
Organic fertilizers, 24, 130, 147, 288, 297, 315, 504–505, 527, 549, 562, 565–566, 571–572
Organic matter decomposition, 57, 64, 237
Organic-rich environments, 439–441
Organic-wastes
adding, 80, 87
agricultural, 565
breakdown, 35, 70, 80, 538
compartment, 8, 101
composting, 184
contaminated, 250
converting, 482, 508
solid, 410
disposal, 2, 309, 463, 546–547, 565
dry-matter conversion ratios, 326
exploiting, 2
fresh, 4, 20
management, 12, 23–24, 387–388, 439, 447, 456, 482, 498
materials, 3, 14, 305, 308–309, 312, 319, 424, 426, 428–429, 502, 508–509, 517
municipal, 457, 524
pathogen-contaminated, 388
polluted, 410
processing, 17, 81, 83, 88, 93, 120, 315, 336, 424, 433, 439, 499, 551, 568, 572
process factors, recommended values, 20
production, 426, 517
recycling, 457, 498
solid, 13, 35
sources, 489
suitable, 19, 96
urban, 7, 92, 98
vermicomposting, 15, 498
Organisms, biocontrol, 174–175
Outdoor vermicomposting, 70, 402–403, 432

P

Paper waste vermicomposts, 118, 134, 140, 146, 223, 298, 371, 379, 483
Pathogen-plant combination, 187, 195
Perionyx excavatus, 34–35
Peruvian fish meal, 476
Pests and diseases in vermiculture, 445
Pharmaceuticals production from earthworms
human health care perspective, 362
therapeutic properties
antibacterial activity, 353
anticarcinogenic activity, 352–353
anticoagulative activity, 353–354
anti-inflammatory effects, 351–352
antioxidative activity, 350–351
antiulcer, 351
therapeutic research, 358
molecules, 354
Pheretima asiatica, 469, 484–485
Philippines, vermiculture in
cost and returns, 477
projected yields and, 478
earthworm protein, production and use of, 475, 477
earthworm species cultured, 470
earthworm weight gain, feed conversion and survival, 476
source of pharmaceutical materials, 477–478
vermicompost, production and use of, 471, 475
effects of, 473
grain yield of upland rice with, 474
yield and cost of fertilization, 473
yield responses, 472
Phytophthora capsici in
cucumber experiments, 187
mean leaf areas, 189
mean root damage, 188
tomato experiments, 189
Pig manure
vermicompost, 107–110, 117–118, 255, 293–294, 297
chemical characteristics of, 296
Pit vermiculture method outdoors, 549
Plant growth hormones (PGHs), 235
Plant parasitic nematode populations suppression by vermicompost teas, 229
Plant pathogen suppression by
thermophilic composts, 184
thermophilic compost teas, 183, 184, 185

vermicomposts, 162, 165, 166, 168, 172–185
vermicompost teas, 185–186
vermicompost teas
mechanisms, 203–204
Plants
chicory, 122
cultured, 472, 513
dead, 53, 80, 130
diseases, 162, 203–204, 288
control, 184, 515
incidence, 165, 185
suppression, 167, 185–186, 300, 515
germination, 4, 117, 123, 235–236, 247
growth, 8, 76, 100
active, 120
amendments, 6
experiments, 240
hormones in vermicomposts, 117
horticultural, 428
inhibition, 296
media, 3, 82, 104–106, 117, 120, 131, 210, 267, 276, 288–289, 300, 306, 328, 508
stimulants, 306, 310, 499, 503, 569
trials, 387, 388
health, 161, 165, 167, 235, 288, 308
heights, 107, 122, 131, 188, 241, 244–245
inhibitors, 21
nutrients, 306, 310, 471, 515
available, 1, 22–23, 80–81, 91, 93, 235, 313, 510
ornamental, 106, 120, 162, 538
parasitic nematodes, 5, 213, 223–225, 230
pathogens, 5, 22, 168, 173, 184, 186, 204, 514
foliar, 196
suppress, 176, 204, 235, 415, 471
potted, 426–427, 523
production, 43, 558
systems, 176
productivity, 415, 571
residues, 160, 517, 575
roots, 23, 148, 173, 509
strawberry, 147
substrates, 486
tests, 106, 109, 134, 215, 218, 221
tissues, 121, 215, 220, 222
vermicompost tea, 160
Plastic
film, 548, 550–551, 553, 557
tunnels, 550, 561
Plectosporium tabacinum, 196–198
Plectosporium tabacinum
cucumber experiments, 196
mean damage rating, 197
mean leaf areas, 197
tomato experiments, 196
mean damage ratings, 198
mean leaf areas, 198
Polypheretima elongata, 35
Potato tubers, 571, 573–574
Potting medium, 172–173, 294–295
Potting mixtures, 109, 241, 244, 515
Precomposted organic wastes, 483
Prisons, vermicomposting in, 400
Broad River Correctional Institution, 403–404
Brown Creek correctional facility, 402–403
Cedar Creek Corrections Center, 401–402
Sampson Correctional Institution, 403
Stafford Creek Correctional Facility, 402
Washington State, 401
Processes
earthworm-production, 49
earthworm gut-associated, 56, 273
predominant, 290–292
processing
method, 319
time, 8–9, 101, 292
Pythium ultimum in
cucumber experiments
mean damage, 194
mean leaf areas, 194
seedling losses, 193
vermicompost teas range effects, 194
in tomato experiments
mean damage ratings of, 195
mean leaf areas of, 195

Q

Quality control of vermicomposting, 523
Quality criteria for vermicomposts, 288
biological characteristics
enzymatic activity, 299–300
faunal community composition, 300
human/animal pathogens, 298–299
microbial community composition, 300
chemical characteristics
acidity, 295
cation-exchange capacity (CEC), 295
soluble salt concentrations, 296
definitions, of composts, 289, 290, 292
physical characteristics of vermicomposts
aggregate stability, 293
bulk density, 293
moisture contents, 294
particle size distribution, 293
solids composition, 292–293

standard commercial potting medium, 293
total porosity, 293
water-retention, 295
plant nutrient content of vermicomposts
ammonium, 297–298
boron, 297
calcium, 297
carbon:nitrogen ratio, 297
heavy metal contamination, 298
magnesium, 297
micronutrients, 298
nitrate nitrogen, 297–298
nutrient-release characteristics, 298
sulfur, 297
total carbon, 296
total nitrogen, 297

R

Ratio of vermicompost to other materials, 107, 117, 154–155, 215
Reactor systems
automated continuous-flow vermicompost, 4, 5, 6, 8, 94
Reciprocating scraper vermicompost-collection system, 96
Redworms, 68–69, 74, 320, 372
Research on foliar pathogen suppression
by vermicomposts aqueous extracts (teas), 195, 196
Research on root pathogen suppression
aqueous extracts (teas), production of, 186
experiment design, 186
food waste vermicompost teas effects on
Botrytis cinerea damage, 198–200
Fusarium oxysporum attacking tomatoes, 187
Phytophthora capsici attacking cucumbers and tomatoes, 187–188
Plectosporium tabacinum attacking vegetables, 196–198
Rhizoctonia solani attacking cucumbers and tomatoes, 190–193
Sclerotinea rolfsii damage, 200–204
Fusarium oxysporum damage suppression to tomato roots, 188
Rhizoctonia solani
cucumber experiments, 190
mean damage, 191
mean leaf areas, 191
tomato experiments
mean damage, 192
mean leaf areas, 192
Russia, vermicomposting in, 572–575

S

Salmonella enteritidis, 252, 255–256, 353, 483
Schools, vermicomposting in
Birch Lane Elementary School, 404
Cinnabar School, 405
Lathrop E. Smith Environmental Education Center, 405–406
Louisiana Schnell Elementary School, 405
Waterville Elementary School, 404–405
Sclerotinea rolfsii
cucumber experiment, 200
damage ratings of, 201
leaf areas of, 202
tomato experiments, 202
mean damage ratings of, 203
mean leaf areas of, 203
Seedling
mortality, 174
survival, 172, 193
Self-fertilization by earthworms, 32, 485
Sewage biosolids, pathogen reduction during vermicomposting, 249, 252
Sewage sludge, 35–36, 103, 122, 264, 266–267, 273, 295, 298, 305, 345, 386, 429, 444, 562, 574
Shredded paper waste, 379, 381, 403–404
Shrimps, 476, 539, 541
Silsoe comb-type earthworm/vermicompost separator, 85
Sludges, 264, 279, 291, 376, 485, 492, 518, 537
vermicomposted swine, 273
vermicomposted wastepaper, 432
Small-scale systems of vermicomposting, 68, 85, 311
Soilborne plant diseases
earthworms in suppression, role of, 167–168
results in, 169–171
Soils
bioremediation of, 6, 433, 485
borne plant diseases, 167
compost-amended, 277
conditioners and mulches, 448–449
contaminated, 266
drenches, 148, 154, 187, 189, 213, 215–218, 221–222, 241, 247, 307
concentrations of vermicompost teas, 191–192

- dwelling earthworms, 51, 341
  - controlled breeding of, 341
- forest, 441
- invertebrates, 15, 53, 55, 166, 336, 412, 414
- microbial biomass, 147
- microorganisms, 146, 160, 463, 485
- polluted, 574–575
- productivity, 463–464
- sludge-amended, 266, 272, 275
- surface, 12, 28, 80, 132, 337–341, 343, 345, 484, 557
- surrounding, 121, 166, 337, 453
- vermicompost-amended, 273, 300
- vermicompost-treated, 224

Solids
- vermicomposts substitution of, 131, 220
- wastes, 373, 383, 401, 520, 527
  - facilities permit, 378

Soluble plant nutrients, 247
Solution aqueous vermicompost (teas), 186, 218
Spain, vermicomposting in
- biology of earthworms for vermiculture
  - *Eisenia andrei,* 411
  - *Eisenia fetida,* 412
- ecology of
  - earthworms and microorganisms, 412–414
  - microbial decomposition by, 414
- environmental factors on, 412
- human pathogen destruction, 415
- organic waste, evaluation, 410–411
- pesticide degradation in soils, 415–416
- plant growth, effects, 415

Specialized vermicomposting systems, 317
Species of earthworms
- cryptic, 43
- deep-burrowing, 340–341
- used in vermicomposting, 516
- *Eisenia fetida,* 288, 411
- exotic, 46, 438–439, 441, 455–456

Spider mites, 217–219
- ratings of damage, 214

Standardized vermicompost product, 3
Substitution rates, 106–107, 115, 118
Sucking pests, effects on, 220–221, 462
Suitable organic waste limits, 459
Suitable vermicomposting containers, 70
Supplementary foods, 487
Supplements to earthworm-protein, 330, 332
Suppression by vermicomposts
- of arthropod pests, 217
- pest, 220–222
- of plant pathogens, 5, 185, 204

Sustainable agriculture, 454, 457, 527
Systems
- earthworm-fattening, 429
- earthworm/waste-separating, 84

T

Targeting earthworms, 446
Tea-production systems for vermicompost
- aeration and agitation/stirring, 155
- brewing process and extraction duration, 156–157
- compost tea system, 155
- diffusion disk aeration brewer, 158
- mechanical systems, 158–159
- passive steep approach, 157
- supplemental nutrients, 156
- vermicompost sources, 154–155
- water quality and temperature, 157

Teas
- higher-application rates, 221
- leaf wastes, 486, 489
  - fermented, 487
  - using, 491
  - vermicomposting, 486, 492–493
- plantations, 490
- production systems, 157, 159
  - aerated, 158
  - commercial compost/vermicompost, 158
  - compost/vermicompost, 156
- vermicomposts, 489–490
- wastes, 488, 493

Technology for composting and vermicomposting
- best earthworm composting, 569
- bubble diffusion, 159

Temperate vermicomposting species
- *Dendrobaena veneta,* 32–33
- *Dendrodrilus rubidus,* 32
- *Drawida nepalensis,* 33
- *Eisenia andrei,* 29, 32, 411
- *Eisenia fetida,* 29, 32, 412
- *Lumbricus rubellus,* 33

Tetranychus urticae, 210, 212, 217
Thermophilic composting, 2, 12–14, 80, 93, 163, 259, 290, 298–299, 309, 448, 482, 499
- advantages, 22–23
- composts
  - traditional, 223–224
- phases, 448–449
- stages, 13, 17, 23
- systems, 99, 326

Tobacco hornworms, 210–211, 213, 215, 217, 220–221
Tomatoes
  aqueous extracts effects on plant growth and yields, 240
    on fresh shoot weights of, 243
    on leaf areas of, 242
    on numbers of, 243
    plant heights of, 241
    responses in growth of, 242
    weights of, 244
  fruits, 107–108, 210–211, 241, 243
  growth, 117, 240, 242–243
    plant, 226, 236
  numbers of plants produced, 189
  productivity of, 574
  roots, 188, 226, 228
  seedlings, 107, 117, 122–123, 574
  seeds, 107, 200, 202
Tomato experiments with
  Botrytis cinerea, 200
  Plectosporium tabacinum, 196
    mean damage ratings of, 198
    mean leaf areas of, 198
  Phytophthora capsici
    mean damage, 189
  Pythium ultimum
    mean damage ratings of, 195
    mean leaf areas of, 195
  Rhizoctonia solani
    mean damage, 192
    mean leaf areas, 192
  Sclerotinea rolfsii, 202
    mean damage ratings of, 203
    mean leaf areas of, 203
Total coliform bacteria during vermicomposting, 61
Trial feeding of earthworms, 554
Trichoderma-enriched vermicompost, 462
Trichoderma hamatum, 204
Tropical crops, 122
Tropical species
  Eudrilus eugeniae, 34
  Perionyx excavatus, 34–35
  Polypheretima elongata, 35

U

Ukraine, vermicomposting in, 575–576
United Kingdom, vermiculture and vermicomposting
  BEEcycle, Lancaster, Case Study, 431
  British Earthworm Technology (BET)
  current activities, 425–426
    local authority involvement, 427–428
  current research, 432–433
  future developments in, 433
  ORM professional products, Wales Case Study, 428
    earthworm digester, 431
    earthworms and knowledge for sale, 430
    extensive outdoor systems, 430
    vermiprocessing, 429
    VermiT, 430–431
Urban wastes, 7, 98, 493, 573, 576

V

Vegetable wastes, 3, 429, 562
Vegetative growth, 122, 220, 461–462
Vermicompost aqueous solutions (teas), 153
  application methods, 160
  compost tea, 161
    Task Force, recommendations, 161–162
  National Organic Program (NOP), 161
  tea benefits, 160
  tea-production systems
    aeration and agitation/stirring, 155
    brewer using diffusion disk aeration, 158
    brewing process and extraction duration, 156–157
    commercially available compost tea system, 155
    mechanical systems, 158–159
    passive steep approach, 157
    supplemental nutrients, 156
    vermicompost sources, 154–155
    water quality and temperature, 157
Vermicomposting process
  advantages, 23–24
  characteristics, 20
  commercial potential and economics, 313
  courses in, 526
  principles, 2–4
  processes, 13–15
  technologies
    economics, 8–9
  vermicompost
    cost of producing, 141, 474, 477
    systems designs, 518
  vermicomposts
    characteristics of, 104, 154
    chemical properties of, 105, 295
  Windrow vermicomposting systems
    characteristics, 82
Vermicompost tea direct foliar application, 160

Vermiculture, 335
- alternative to, 342–343
- commercial adoption
  - Canada, 7
  - Hong Kong, 7
  - India, 7
- earthworms
  - breeding soil-inhabiting, 339
  - ecological relationships, 338–339
  - effects on soils, 337
  - examples of, 337–338
  - inoculation units (EIUs), 343
  - numbers of, 344
- land improvement, 335–336, 344
  - breeding, 341–342
  - collection, 340–341
  - inoculation en masse, 344–345
  - intensive cultivation, 345
  - relative merits, 341
- principles, 2–4
- Russia, 7
- United Kingdom, 7
- United States, 6–7

W

Wastes
- composted, 553
- disposal, 402
- diversion, 309, 311
- earthworm-processed, 55–56
- food industry, 13, 411
- food-processing, 299
- industrial, 456, 562
- kitchen preparation, 396, 400
- layers, 87, 93
- management, 7, 304, 417, 428, 438–439, 446–447, 449, 504
  - organic, 370, 411, 454
- manufacturing, 573
- materials, 14, 92, 279, 312, 317, 432
- minimization, 428
- mixed, 315
- organic solid, 305, 321, 454
- paper, 6, 109, 493, 538
  - industry, 13, 411
- plant, 298, 493
- poultry, 326
- processing, 81, 88, 92, 385, 431, 500
- residual, 91–92, 328
- restaurant, 1, 444
- soybean cake, 487–488
- stable, 504–505
- strategy, 264, 278
- streams, 392, 396, 399, 498, 500
  - total urban, 19, 99
- surface, 32, 426
- thin layers, 521
- undigested, 14, 56
- vegetable-processing, 444
- vermicomposted, 292
  - paper, 432
- Windrow systems process, 81

Wastewater, 13, 130, 411, 520
- biosolids, 251, 253
  - municipal, 251–252
  - vermicomposted, 255
- treatment, 401

Water
- retention characteristics, 295
- tensions, 295

Wedge vermicomposting systems, 6, 8, 87–88, 96, 101

Wedge vermicomposting systems, 86
- characteristics, 88
- principles, 87

Windrow vermicomposting systems, 81–82, 254
- characteristics, 82
- earthworms
  - population density and food, 548–549
  - species, 547
  - substrates and types, 548
- management
  - moisture, 548
  - temperature, 548
- operations, 8

Worm Factory, 72–73, 86

Worms, 9, 29, 44, 74, 80, 130, 341, 561

Worm Wigwam, 309, 392–394, 403

Y

Young pigs growth on earthworm protein supplements, 332

Z

Zero waste South Australia, 442